# Undergraduate Lecture Notes in Physics

Undergraduate Lecture Notes in Physics (ULNP) publishes authoritative texts covering topics throughout pure and applied physics. Each title in the series is suitable as a basis for undergraduate instruction, typically containing practice problems, worked examples, chapter summaries, and suggestions for further reading.

ULNP titles must provide at least one of the following:

- An exceptionally clear and concise treatment of a standard undergraduate subject.
- A solid undergraduate-level introduction to a graduate, advanced, or non-standard subject.
- A novel perspective or an unusual approach to teaching a subject.

ULNP especially encourages new, original, and idiosyncratic approaches to physics teaching at the undergraduate level.

The purpose of ULNP is to provide intriguing, absorbing books that will continue to be the reader's preferred reference throughout their academic career.

More information about this series at http://www.springer.com/series/8917

Jochen Pade

# Quantum Mechanics for Pedestrians 2

## Applications and Extensions

Second Edition

 Springer

Jochen Pade
Institut für Physik
Universität Oldenburg
Oldenburg, Germany

ISSN 2192-4791 ISSN 2192-4805 (electronic)
Undergraduate Lecture Notes in Physics
ISBN 978-3-030-00466-8 ISBN 978-3-030-00467-5 (eBook)
https://doi.org/10.1007/978-3-030-00467-5

Library of Congress Control Number: 2018954852

Originally published with the title: *Quantum Mechanics for Pedestrians 2: Applications and Extensions*
1st edition: © Springer International Publishing Switzerland 2014
2nd edition: © Springer Nature Switzerland AG 2018

This work is subject to copyright. All rights are reserved by the Publisher, whether the whole or part of the material is concerned, specifically the rights of translation, reprinting, reuse of illustrations, recitation, broadcasting, reproduction on microfilms or in any other physical way, and transmission or information storage and retrieval, electronic adaptation, computer software, or by similar or dissimilar methodology now known or hereafter developed.
The use of general descriptive names, registered names, trademarks, service marks, etc. in this publication does not imply, even in the absence of a specific statement, that such names are exempt from the relevant protective laws and regulations and therefore free for general use.
The publisher, the authors and the editors are safe to assume that the advice and information in this book are believed to be true and accurate at the date of publication. Neither the publisher nor the authors or the editors give a warranty, express or implied, with respect to the material contained herein or for any errors or omissions that may have been made. The publisher remains neutral with regard to jurisdictional claims in published maps and institutional affiliations.

This Springer imprint is published by the registered company Springer Nature Switzerland AG
The registered company address is: Gewerbestrasse 11, 6330 Cham, Switzerland

# Preface to the Second Edition, Volume 2

In this second edition of Volume 2, a short introduction to the basics of quantum field theory has been added. The material is placed in the Appendix. It is not a comprehensive and complete presentation of the topic, but, in the sense of a primer, a concise account of some of the essential ideas.

Fundamentals from other areas can be found in Volume 1, i.e., outlines of special relativity, classical field theory, electrodynamics and relativistic quantum mechanics.

Oldenburg, Germany
February 2018

Jochen Pade

# Preface to the First Edition, Volume 2

In the first volume of *Quantum Mechanics for Pedestrians,* we worked out the basic structure of quantum mechanics (QM) and summarized it in the form of postulates that provided its framework.

In this second volume, we want to fill that framework with life. To this end, in eight of the 14 chapters we will discuss some key applications, what might be called the 'traditional' subjects of quantum mechanics: simple potentials, angular momentum, perturbation theory, symmetries, identical particles, and scattering.

At the same time, we want to prudently broaden the scope of our treatment, in order to be able to discuss modern developments such as entanglement and decoherence. We begin this theme in Chap. 20 with the question of whether quantum mechanics is a local-realistic theory. In Chap. 22, we introduce the density operator in order to discuss the phenomenon of decoherence and its importance for the measurement process in Chap. 24. In Chap. 27, we address again the realism debate and examine the question as to what extent quantum mechanics can be considered to be a complete theory. Modern applications in the field of quantum information can be found in Chap. 26.

Finally, we outline in Chap. 28 the most common current interpretations of quantum mechanics. Apart from one chapter, what was said in Volume I applies generally: An introduction to quantum mechanics has to take a definite stand on the interpretation question, although (or perhaps because) the question as to which one of the current interpretations (if any) is the 'correct' one it is still quite controversial. We have taken as our basis what is often called the 'standard interpretation.'

In order to formulate the postulates, we worked in the first volume with very simple models, essentially toy models. This is of course not possible for some of the 'real' systems presented in the present volume, and accordingly, these chapters are formally more complex. However, here also, we have kept the mathematical level as simple as possible. Moreover, we always choose that particular presentation which is best adapted to the question at hand, and we maintain the relaxed approach to mathematics which is usual in physics.

This volume is also accompanied by an extensive appendix. It contains some information on mathematical issues, but its principal focus is on physical topics whose consideration or detailed discussion would be beyond the scope of the main text in Chaps. 15–28.

In addition, there is for nearly every chapter a variety of exercises; solutions to most of them are given in the appendix.

# Contents

# Contents of Volume 1

# Introduction

Quantum mechanics is probably the most accurately verified physical theory existing today. To date, there has been no contradiction from any experiments; the applications of quantum mechanics have changed our world right up to aspects of our everyday life. There is no doubt that quantum mechanics 'functions'—it is indeed extremely successful. On a formal level, it is clearly unambiguous and consistent and (certainly not unimportant)—as a theory—it is both aesthetically satisfying and convincing.

The question in dispute is the 'real' meaning of quantum mechanics. What does the wavefunction stand for, and what is the role of chance? Do we actually have to throw overboard our classical and familiar conceptions of reality? Despite the nearly century-long history of quantum mechanics, fundamental questions of this kind are still unresolved and are currently being discussed in a lively and controversial manner. There are two contrasting positions (along with many intermediate views): Some see quantum mechanics simply as the precursor stage of the 'true' theory (although eminently functional); others see it as a valid, fundamental theory itself.

This book aims to introduce its readers to both sides of quantum mechanics, the established side and the side that is still under discussion. We develop here both the conceptual and formal foundations of quantum mechanics, and we discuss some of its 'problem areas.' In addition, this book includes applications—oriented fundamental topics, some 'modern' ones—for example, issues in quantum information—and 'traditional' ones such as the hydrogen and the helium atoms. We restrict ourselves to the field of nonrelativistic physics, although many of the ideas can be extended to the relativistic case.[1] Moreover, we consider only time-independent interactions.

In introductory courses on quantum mechanics, the practice of formal skills often takes priority (this is subsumed under the slogan 'shut up and calculate'). In accordance with our objectives here, we will also give appropriate space to the

---

[1] In the second edition, some essentials of relativistic quantum mechanics are added; see the Appendix.

discussion of fundamental questions. This special blend of basic discussion and modern practice is in itself very well suited to evoke interest and motivation in students. This is, in addition, enhanced by the fact that some important fundamental ideas can be discussed using very simple model systems as examples. It is not coincidental that some of the topics and phenomena addressed here are treated in various simplified forms in high-school textbooks.

In mathematical terms, there are two main approaches used in introductions to quantum mechanics. The first one relies on differential equations (i.e., analysis) and the other one on vector spaces (i.e., linear algebra); of course, the 'finished' quantum mechanics is independent of the route of access chosen. Each approach (they also may be called the Schrodinger and the Heisenberg routes) has its own advantages and disadvantages; the two are used in this book on an equal footing.

The roadmap of the book is as follows:

The foundations and structure of quantum mechanics are worked out step by step in the first part (Volume 1, Chaps. 1–14), alternatively from an analytical approach (odd chapters) and from an algebraic approach (even chapters). In this way, we avoid limiting ourselves to only one of the two formulations. In addition, the two approaches reinforce each other in the development of important concepts. The merging of the two threads starts in Chap. 12. In Chap. 14, the conclusions thus far reached are summarized in the form of quite general postulates for quantum mechanics.

Especially in the algebraic chapters, we take up current problems early on (interaction-free quantum measurements, the neutrino problem, quantum cryptography). This is possible since these topics can be treated using very simple mathematics. Thus, this type of access is also of great interest for high-school level courses. In the analytical approach, we use as elementary physical model systems the infinite potential well and free particle motion.

In the second part (Volume 2, Chaps. 15–28), applications and extensions of the formalism are considered. The discussion of the conceptual difficulties (measurement problem, locality and reality, etc.) again constitutes a central theme, as in the first volume. In addition to some more traditionally oriented topics (angular momentum, simple potentials, perturbation theory, symmetries, identical particles, scattering), we begin in Chap. 20 with the consideration of whether quantum mechanics is a local realistic theory. In Chap. 22, we introduce the density operator in order to consider in Chap. 24 the phenomenon of decoherence and its relevance to the measurement process. In Chap. 27, we continue the realism debate and explore the question as to what extent quantum mechanics can be regarded as a complete theory. Modern applications in the field of quantum information can be found in Chap. 26.

Finally, we outline in Chap. 28 the most common interpretations of quantum mechanics. Apart from this chapter, a general statement applies: While it is still a controversial issue as to which (if indeed any) of the current interpretations is the 'correct' one, an introduction to quantum mechanics must take a concrete position and has to present the material in a coherent form. In this book, we choose the version commonly known as the 'standard interpretation.'

A few words about the role of mathematics:

In describing objects that—due to their small size—are beyond our everyday experience, quantum mechanics cannot be formulated completely in terms of everyday life and must therefore remain to some extent abstract. A deeper understanding of quantum mechanics cannot be achieved on a purely linguistic level; we definitely need mathematical descriptions.[2] Of course, one can use analogies and simplified models, but that works only to a certain degree and also makes sense only if one is aware of the underlying mathematical apparatus, at least in broad terms.[3]

It is due to this interaction of the need for mathematical formulations and the lack of intuitive access that quantum mechanics is often regarded as 'difficult.' But that is only part of the truth; to be sure, there are highly formalized and demanding aspects. Many wider and interesting issues, however, are characterized by very simple principles that can be described using only a basic formalism.

Nevertheless, beginners in particular perceive the role of mathematics in quantum mechanics as discouraging. Three steps serve to counter this impression or, in the optimum case, to avoid it altogether:

First, we keep the mathematical level as simple as possible and share the usual quite nonchalant attitude of physicists toward mathematics. In particular, the first chapters go step by step, so that the initially diverse mathematics skills of the readers are gradually brought up to a common level.

In addition, we use very simple models, toy models so to speak, especially in the first part of the book, in order to treat the main physical ideas without becoming involved in complicated mathematical questions. Of course, these models are only rough descriptions of actual physical situations. But they manage with relatively simple mathematics, do not require approximation methods or numerics, and yet still permit essential insights into the fundamentals of quantum mechanics.[4] Only in Volume 2, more realistic models are applied, and this is reflected occasionally in a somewhat more demanding formal effort.

The third measure involves exercises and some support from the Appendix. At the end of almost every chapter, there is a variety of exercises, some of them dealing with advanced topics. They invite the reader to work with the material in

---

[2]This applies at least to physicists; for as Einstein remarked: 'But there is another reason for the high repute of mathematics: it is mathematics that offers the exact natural sciences a certain measure of security which, without mathematics, they could not attain.' To give a layman without mathematical training an understanding of quantum mechanics, one will (or must) rely instead on math-free approaches.

[3]Without appropriate formal considerations, it is impossible to understand, for example, how to motivate the replacement of a physical measurement variable by a Hermitian operator.

[4]We could instead also make use of the large reservoir of historically important experiments. But their mathematical formulation is in general more complex, and since in the frame of our considerations they do not lead to further-reaching conclusions than our 'toy models,' we restrict ourselves to the latter for clarity and brevity.

order to better assimilate and more clearly grasp it, as well as of course to train the necessary formal skills.[5]

The learning aids in the Appendix include chapters with some basic mathematical and physical background information; this allows the reader to refresh 'passive' knowledge without the need to refer to other sources or to become involved with new notations.

Moreover, the no doubt unusually extensive Appendix contains the solutions to many of the exercises and, in addition, some chapters in which further-reaching questions and issues are discussed; although these are very interesting in themselves, their treatment would far exceed the framework of a lecture course.

The footnotes with a more associative character can be skipped on a first reading.

A note on the term 'particle': Its meaning is rather vague in physics. On the one hand, it denotes 'something solid, not wavelike'; on the other hand 'something small', ranging from the elementary particles as structureless building blocks of matter, to objects which themselves are composed of constituent 'particles' like the $\alpha$ particle and other atomic nuclei or even macroscopic particles like sand grains. In quantum mechanics, where indeed it is often not even clear whether a particular object has mainly particle or mainly wave character, the careless use of the term may cause confusion and communication problems.

Accordingly, several terms which go beyond 'wave' or 'particle' have been suggested, such as quantal particle, wavical, wavicle, quantum object, quanton. Throughout this book, we will use the term 'quantum object,' unless there are traditionally established terms such as 'identical particles' or 'elementary particles.' The consistent use of 'quantum object' instead of 'particle' may perhaps seem somewhat pedantic, but we hope that it will help to ensure that fewer false images stick in the minds of readers; it is for this reason that this term is also found in many high-school textbooks.

Quantum mechanics is a fundamental theory of physics, which has given rise to countless applications. But it also extends deep into areas such as philosophy and epistemology and leads to thinking about 'what holds the world together at its core'; in short, it is also an intellectual adventure. The fascinating thing is that the more one becomes acquainted with quantum mechanics, the more one realizes how simple many of its central ideas really are.[6] It would be pleasing if *Quantum Mechanics for Pedestrians* could help to reveal this truth.

---

[5] 'It is a great support to studying, at least for me, to grasp everything that one reads so clearly that one can apply it oneself, or even make additions to it. One is then inclined to believe in the end that one could have invented everything himself, and that is encouraging.' Georg Christoph Lichtenberg, *Scrap Books,* Vol. J (1855).

[6] 'The less we know about something, the more complicated it is, and the more we know about it, the easier it is. This is the simple truth about all the complexities.' Egon Friedell, in *Kulturgeschichte der Neuzeit; Kulturgeschichte Agyptens und des alten Orients (Cultural history of modern times; the cultural history of Egypt and the ancient Near East).*

Let us close with a remark by Richard Feynman which holds true not only for physics in general, but even more for quantum mechanics: 'Physics is like sex: Sure, it may give some practical results, but that's not why we do it.'

# Overview of Volume 2

Now we have established the basic framework of quantum mechanics in Volume 1 in the form of postulates, we turn to two other major topics in Volume 2.

First, we want to fill the framework of quantum mechanics with life, i.e., to discuss some applications (solutions for simple potentials, angular momentum, symmetries, identical particles, scattering, quantum information). These chapters are to some extent more technical, since we cannot use toy models, as in the first volume, but instead have to consider 'real' systems.

Second, we will extend the framework of quantum mechanics prudently to address modern developments such as entanglement and decoherence. Finally, we turn again to the realism debate; the final chapter presents some of the current interpretations of quantum mechanics.

# Part II
# Applications and Extensions

# Chapter 15
# One-Dimensional Piecewise-Constant Potentials

After examining scattering by a potential step, we consider the finite potential well and the potential barrier. The physical phenomena we explore include discrete energy spectra and the tunnel effect. Finally, we show by example how to construct physically reasonable solutions by superposing (unphysical) partial solutions.

In a discussion of the stationary SEq, a major problem is that there are only very few realistic potentials for which closed solutions exist. To make analytical statements, one therefore almost always has to introduce approximations or simplifying assumptions; apart from that, one depends on numerical results.[1] This also applies to the one-dimensional case to which we restrict ourselves here. In this chapter, we simplify typical potentials by replacing them with 'steps',[2] i.e. by piecewise constant potentials; see Fig. 15.1.[3] As long as we do not assume that there are infinitely high potential walls at an arbitrary distance, we will also have to deal with continuous spectra.

Despite their schematic nature, the potentials discussed in this chapter are somewhat more realistic models of physical situations than the cases considered in Chap. 5, Vol. 1, i.e. the infinite potential well and free particle motion. We first consider a potential step, and then we will investigate in more detail bound and free states in some other potentials.

Using the example of the potential step, we discuss at the end of the chapter how to get physically reasonable solutions (also called *wave packets*) by the superposition of partial solutions (i.e. plane waves). We remind the reader that the superposition principle holds, due to the linearity of the SEq.

---

[1] Books on quantum mechanics would be significantly thinner if one could solve the SEq for arbitrary potentials in closed form.

[2] We discuss approximation techniques in Chap. 19.

[3] We obtain exact solutions in this way. In principle, one can make the subdivision finer and finer and thus approximate the 'true' potential with arbitrary accuracy, but then the computational complexity increases disproportionately.

© Springer Nature Switzerland AG 2018
J. Pade, *Quantum Mechanics for Pedestrians 2*, Undergraduate Lecture Notes in Physics, https://doi.org/10.1007/978-3-030-00467-5_15

**Fig. 15.1** Approximation of
a potential by a piecewise
constant potential. In the
region $i$, the potential is
approximated by the
constant value $V_i$

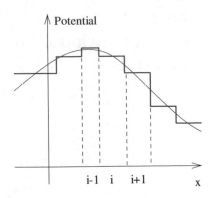

Although one-dimensional piecewise constant potentials usually are addressed in undergraduate courses on atomic physics, we will discuss them in some detail again here for the sake of completeness.

## 15.1   General Remarks

We first discuss the solutions in each region of constant potential and then consider how to put together these partial solutions in the right manner.

In the region $i$, where the potential has the constant value $V_i$, the stationary SEq reads[4]

$$E\varphi_i(x) = -\frac{\hbar^2}{2m}\varphi_i''(x) + V_i\varphi_i(x). \tag{15.1}$$

It follows that

$$\varphi_i'' = \frac{2m}{\hbar^2}(V_i - E)\varphi_i, \tag{15.2}$$

which is the well-known second-order differential equation with constant coefficients that can be solved by an exponential *ansatz*. The prefactor of $\varphi_i$ on the right-hand side is a constant for which a certain notation has become broadly established, namely $\kappa$ and $k$:

$$\kappa_i^2 = \frac{2m}{\hbar^2}(V_i - E) \ \text{ for } V_i > E$$

$$k_i^2 = -\frac{2m}{\hbar^2}(V_i - E) \ \text{ for } V_i < E, \tag{15.3}$$

where $\kappa_i, k_i > 0$ is assumed in general. Accordingly, we find two types of solutions:

---

[4]The total energy $E$ is the same everywhere, of course; the quantities which vary are the potential energy $V_i$ and the kinetic energy $E_{\text{kin}} = E - V_i$.

**Table 15.1** Scheme of the two types of solutions

| $E < V_i$ | $E > V_i$ |
| --- | --- |
| Classically forbidden region | Classically allowed region |
| $\varphi''(x) = \frac{2m}{\hbar^2}(V_i - E)\,\varphi(x) = \kappa^2\varphi(x)$ | $\varphi''(x) = -\frac{2m}{\hbar^2}(E - V_i)\,\varphi(x) = -k^2\varphi(x)$ |
| Exponential solution $e^{\pm\kappa x}$ | Oscillatory solution $e^{\pm ikx}$ |

$$\varphi_i = A_i e^{\kappa_i x} + B_i e^{-\kappa_i x} \quad \text{for} \quad V_i > E$$
$$\varphi_i = A_i e^{ik_i x} + B_i e^{-ik_i x} \quad \text{for} \quad V_i < E \tag{15.4}$$

i.e. an exponential or an oscillatory solution, depending on the sign of $V_i - E$.[5] The constants $A$ and $B$ are integration constants.

We have thus found a characteristic difference between quantum mechanics and classical mechanics. In classical mechanics, the total energy cannot be less than the potential energy, since this would imply a negative kinetic energy. The point at which is $E = V$ is called the *classical turning point*. At this point, a classical particle must turn back, i.e. it is reflected. In quantum mechanics, there is a solution for the regions with $V_i > E$, the *classically forbidden regions*. This means that the quantum object penetrates into these regions in some sense. These solutions behave exponentially, while in the *classically allowed regions* (that is $V_i < E$), they are oscillatory. The distinction between the two types of solutions is central to the present chapter. They are summarized in Table 15.1 and in Fig. 15.2.

Hence, we can specify a solution for each region $i$. But how do we put together the partial solutions from each region, or in other words, how do we determine the constants of integration? Let us assume that the potential jumps at $x_s$ (=point of discontinuity). Then we require that the different pieces of the wavefunction merge into each other 'smoothly'. This requirement is motivated by the fact that we interpret $\rho = \Psi^*\Psi$ as a probability density. In order for it to be physically reasonable, it has to be defined everywhere, e.g. it cannot have discontinuities. That is, the wavefunctions to the right and left of the discontinuity $x_s$ of the potential, $\varphi_{\text{left}}$ and $\varphi_{\text{right}}$, must be equal at $x_s$ (continuity of the wavefunction):

$$\varphi_{\text{left}}(x_s) = \varphi_{\text{right}}(x_s). \tag{15.5}$$

Likewise, we require that the probability current density

---

[5]It should again be noted that, from a physical point of view, exponential solutions $\sim e^{\pm\kappa x}$ and oscillatory solutions $\sim e^{\pm ikx}$ are worlds apart.

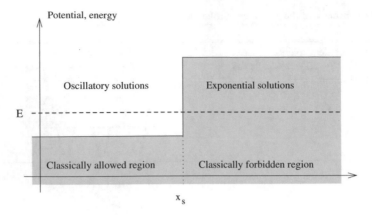

**Fig. 15.2** At the *left*: classically allowed region. At the *right*: classically forbidden region

$$j = \frac{\hbar}{2mi} \left( \varphi^* \varphi' - \varphi \varphi^{*'} \right) \tag{15.6}$$

and hence the derivative of the wavefunction be defined everywhere, which leads to:

$$\varphi'_{\text{left}} (x_s) = \varphi'_{\text{right}} (x_s) . \tag{15.7}$$

Equations (15.5) and (15.7), together with the requirement on the behavior at infinity (or the boundary conditions), allow us to determine all the integration constants— apart from one which must necessarily stay undetermined because of the linearity of the SEq. This one constant is at our disposition (we can choose it in such a way that e.g. the wavefunction is normalized).

A remark concerning infinitely high potentials: At the discontinuity of a finite to an infinite potential value, we can make a statement only about the wavefunction, not about its derivative. This means that we have only (15.5), i.e., $\varphi_{\text{left}} (x_s) = \varphi_{\text{right}} (x_s) = 0$, while (15.7) does not apply here. More on this issue in Sect. 15.5.

In principle, we have thus solved the problem. However, the calculations are quite tedious even for only roughly 'realistic' potentials. Therefore, we first address the simplest example, namely the potential step.

## 15.2  Potential Steps

A quantum object, modelled by a plane wave, is incident from the right on a potential step. The potential is given by

$$V = \begin{cases} V_0 \\ 0 \end{cases} \text{ for } \begin{array}{l} x < 0, \text{region 2} \\ x > 0, \text{region 1.} \end{array} \tag{15.8}$$

Accordingly, we have in region 1 the equation $\varphi_1'' = -k^2 \varphi_1$ with the solution

$$\varphi_1 = Ae^{ikx} + Be^{-ikx}; \quad k^2 = \frac{2mE}{\hbar^2} > 0; \quad A, B \in \mathbb{C}. \tag{15.9}$$

The term $Be^{-ikx}$ represents the incoming wave and $Ae^{ikx}$ the outgoing (scattered) wave.

We repeat the remark that the term 'wave' is actually wrong, because $e^{-ikx}$ is a time-independent spatial oscillation, not a wave. Nevertheless, this term has become prevalent since one has the factor $e^{-i\omega t}$ in mind and assumes $k > 0$ and $\omega > 0$. Otherwise, expressions like 'a plane wave, travelling to the left' for $e^{-ikx}$ would not make sense.

The solution in region 2 depends on whether $E$ is less than or greater than $V_0$. Classically, one expects the following behavior: For $E < V_0$, the particle does not have enough energy to overcome the potential step, and is simply reflected. In the case $E > V_0$, however, the particle is not reflected, but propagates across the potential step (with reduced kinetic energy or velocity). Quantum mechanically, this is different. In the first case, the quantum object is able to enter the potential step (i.e. to enter a classically forbidden region). In the second case, the quantum object can be reflected, although it actually has enough kinetic energy to overcome the potential step. These two modes of behavior are purely quantum mechanical and quite different from classical mechanics. In Fig. 15.3, the situation is sketched.

We consider first the case $E < V_0$ and then $E > V_0$.

### 15.2.1 Potential Step, $E < V_0$

In region 2 ($x < 0$), we have $\varphi_2'' = \frac{2m}{\hbar^2} (V_0 - E) \varphi_2 = \kappa^2 \varphi_2$, with the solution

$$\varphi_2 = Ce^{\kappa x} + De^{-\kappa x}; \quad \kappa^2 = \frac{2m}{\hbar^2} (V_0 - E) > 0; \quad C, D \in \mathbb{C}. \tag{15.10}$$

In this equation, we can determine one of the two coefficients by asking for physically reasonable behavior at infinity.[6] We see that in the limit $x \to -\infty$, the solutions are not bounded for $D \neq 0$. It follows that only for $D = 0$ is the solution physically acceptable.

#### 15.2.1.1 Matching at the Discontinuity

The discontinuity of the potential lies at $x = 0$; here we have $\varphi_1 = \varphi_2$ and $\varphi_1' = \varphi_2'$. This leads to

---

[6]Thus, concerning the validity of solutions, we have a criterion at hand which is not available for mathematics. This is a very nice plus for physics.

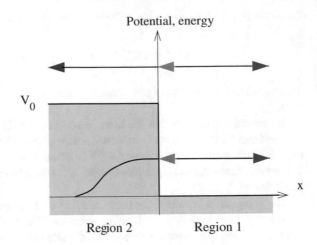

**Fig. 15.3** Situation for the potential step; *above*: $E > V_0$, *below*: $E < V_0$. The *horizontal lines* indicate an oscillation, the *curved line* an exponential decay. Incoming *green*, reflected *blue*, transmitted *red*

$$A + B = C; \quad ikA - ikB = \kappa C \qquad (15.11)$$

These are two equations with three unknowns. We solve for $A$ and $C$ as a multiple of $B$, the amplitude of the incoming wave. We obtain:

$$A = B\frac{\kappa + ik}{ik - \kappa} \quad \text{and} \quad C = \frac{2ik}{ik - \kappa}B. \qquad (15.12)$$

This leads to the result:

$$\varphi_1 = Be^{-ikx} + B\frac{\kappa + ik}{ik - \kappa}e^{ikx} \quad \text{in region 1}$$
$$\varphi_2 = \frac{2ik}{ik - \kappa}Be^{\kappa x} \quad \text{in region 2.} \qquad (15.13)$$

In region 2, the classically forbidden region, we have an exponentially decaying term. That means that there is 'something', where nothing should be according to the classical point of view. If one takes instead of the infinite potential step a potential barrier of finite width (which we will do below), this 'something' is released on the other side of the barrier—it has 'tunneled' through the barrier.

### 15.2.2 Potential Step, $E > V_0$

In region 2, we have the differential equation $\varphi_2'' = -\frac{2m}{\hbar^2}(E - V_0)\varphi_2 = -k'^2\varphi_2$, with the solution:

$$\varphi_2 = A_2e^{ik'x} + B_2e^{-ik'x}; \quad k'^2 = \frac{2m}{\hbar^2}(E - V_0) > 0; \quad A_2, B_2 \in \mathbb{C}. \qquad (15.14)$$

We see that this equation contains a wave travelling to the left ($B_2 e^{-ik'x}$) and another travelling to the right ($A_2 e^{ik'x}$). If we now let a quantum object be incident on the potential step from the right, we can exclude that a wave running from the left to the right exists in region 2, (i.e. coming from $-\infty$). Therefore, we have $A_2 = 0$ and the solution in region 2 reads:

$$\varphi_2 = B_2 e^{-ik'x}. \tag{15.15}$$

This part of the wave is called the *transmitted* wave.

### 15.2.2.1 Matching at the Discontinuity

We have again three unknown variables, namely $A$, $B$ and $B_2$. The matching conditions for $\varphi_1$ in (15.9) and $\varphi_2$ in (15.15) at $x = 0$ are $\varphi_1 = \varphi_2$ and $\varphi_1' = \varphi_2'$. It follows that

$$B_2 = A + B \quad \text{and} \quad -ik'B_2 = ikA - ikB. \tag{15.16}$$

The constants $A$ and $B_2$ are given by

$$A = B\frac{k - k'}{k + k'}; \quad B_2 = B\frac{2k}{k + k'} \tag{15.17}$$

and the solutions by:

$$\varphi_1 = Be^{-ikx} + B\frac{k - k'}{k + k'}e^{ikx}; \quad \varphi_2 = B\frac{2k}{k + k'}e^{-ik'x}. \tag{15.18}$$

### 15.2.2.2 Partial Waves: Transmission Coefficient and Reflection Coefficient

The facts can be summarized as follows: A wave coming from the right (corresponding to $e^{-ikx}$) is incident on the potential step and is transmitted, i.e. continues to travel in that direction (with a different energy or wave number). In addition, we have a reflected wave—and that is something that absolutely does not exist in classical mechanics, comparable to a truck forced off the road by a mosquito flying against its windshield.[7] Accordingly, we can identify three partial waves:

$$\varphi_{\text{in}} = Be^{-ikx}$$

$$\varphi_{\text{refl}} = B\frac{k - k'}{k + k'}e^{ikx} \tag{15.19}$$

$$\varphi_{\text{trans}} = B\frac{2k}{k + k'}e^{-ik'x}.$$

---

[7]The 'classical' historical example has the ingredients 'cannon ball' and 'snowflake'.

The one-dimensional probability current density is given by (15.6), i.e. $j = \frac{\hbar}{2mi}\left(\varphi^*\varphi' - \varphi\varphi^{*'}\right)$. Thus we obtain for the three different partial waves:

$$j_{\text{in}} = -\frac{\hbar}{m}k\,|B|^2$$

$$j_{\text{refl}} = \frac{\hbar}{m}k\left(\frac{k-k'}{k+k'}\right)^2|B|^2 \tag{15.20}$$

$$j_{\text{trans}} = -\frac{\hbar}{m}k'\left(\frac{2k}{k+k'}\right)^2|B|^2\,.$$

As a measure of the probability that a quantum object is reflected or transmitted, we define the *Transmission* and *Reflection coefficients*:

$$T = \left|\frac{j_{\text{trans}}}{j_{\text{in}}}\right| \quad \text{and} \quad R = \left|\frac{j_{\text{refl}}}{j_{\text{in}}}\right|. \tag{15.21}$$

Intuitively, these expressions indicate the relative proportions of the wavefunction which are transmitted and reflected. Their sum is always 1, since we have ruled out creation and annihilation processes.

For the present example, we obtain the expressions

$$T = \left|\frac{k'}{k}\left(\frac{2k}{k+k'}\right)^2\right| = \frac{4kk'}{(k+k')^2}$$

$$R = \left|\left(\frac{k-k'}{k+k'}\right)^2\right| = \left(\frac{k-k'}{k+k'}\right)^2. \tag{15.22}$$

Clearly, because of $4kk' + \left(k-k'\right)^2 = 4kk' + k^2 - 2kk' + k'^2 = \left(k+k'\right)^2$, we have the relation

$$T + R = 1 \tag{15.23}$$

as it indeed must be.

Finally, we investigate how $T$ and $R$ behave as functions of $E$ and $V_0$ ($E > V_0$). Using the abbreviation $z = E/V_0$, and since $1 < z$ due to $V_0 < E < \infty$, we obtain (see Fig. 15.4):

$$T = \frac{4\sqrt{\dfrac{z-1}{z}}}{\left(1+\sqrt{\dfrac{z-1}{z}}\right)^2} = 1 - \frac{1}{16z^2} - \frac{1}{16z^3} - \cdots \tag{15.24}$$

$$1 < z = \frac{E}{V_0};\quad R = 1 - T.$$

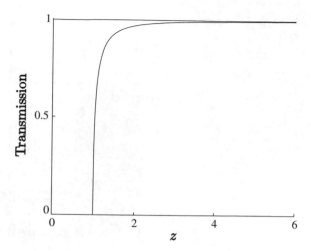

**Fig. 15.4** Potential step: transmission coefficient as a function of $z = E/V_0$

We see the following: if $E$ is very close to $V_0$, then we have $z \approx 1$, and hence the transmission coefficient is very small[8] and accordingly the reflection coefficient large. For high energies, we have $T \to 1$; in this case we have a large transmission coefficient, but always also an (albeit small) portion that is reflected.

## 15.3 Finite Potential Well

This simple example is more realistic than the infinite potential well of Chap. 5, Vol. 1: It allows for bound and free motion. We have a potential of the form (cf. Fig. 15.5):

$$V = \begin{cases} -V_0 \text{ for } -L < x < L; \quad V_0 > 0 \\ 0 \text{ otherwise} \end{cases} \qquad (15.25)$$

and thus the three SEq's:

$$\text{region 1: } x < -L \qquad E\varphi_1 = -\frac{\hbar^2}{2m}\varphi_1''$$

$$\text{region 2: } -L < x < L \quad E\varphi_2 = -\frac{\hbar^2}{2m}\varphi_2'' - V_0\varphi_2. \qquad (15.26)$$

$$\text{region 3: } x > L \qquad E\varphi_3 = -\frac{\hbar^2}{2m}\varphi_3''$$

Depending on the magnitude of $E$, we have to make a case distinction: For $E < 0$, there are only bound states, while for $E > 0$ there are only scattering states.[9] In any

---

[8]In the classical case, we would *always* have transmission 1 for $E > V_0$.

[9]This is an example of a spectrum that has both a discrete and a continuous part.

**Fig. 15.5** Finite potential
well

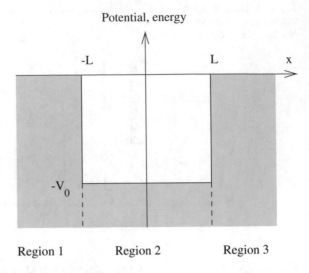

case, we can already give the solution in region 2:

$$\text{region 2: } \varphi_2(x) = Be^{ikx} + Ce^{-ikx}; \quad k^2 = \frac{2m}{\hbar^2}(V_0 + E) > 0. \tag{15.27}$$

### 15.3.1   Potential Well, $E < 0$

We first consider energies with $-V_0 < E < 0$, that is, bound motion. With

$$\kappa^2 = \frac{2m}{\hbar^2}|E| \tag{15.28}$$

the solutions are

$$\text{region 1: } \varphi_1(x) = Ae^{\kappa x} + A'e^{-\kappa x} \tag{15.29}$$
$$\text{region 3: } \varphi_3(x) = D'e^{\kappa x} + De^{-\kappa x}.$$

Since physically reasonable solutions must be bounded, we must choose $A' = 0$ and $D' = 0$. The other constants are defined by the matching conditions at the two discontinuities.

#### 15.3.1.1   Matching at the Discontinuities

At the discontinuity $x = -L$ we have the two equations

$$Ae^{-\kappa L} = Be^{-ikL} + Ce^{ikL} \quad \text{and} \quad \kappa Ae^{-\kappa L} = ikBe^{-ikL} - ikCe^{ikL} \tag{15.30}$$

and at the discontinuity $x = +L$ the two equations

$$De^{-\kappa L} = Be^{ikL} + Ce^{-ikL} \quad \text{and} \quad -\kappa De^{-\kappa L} = ikBe^{ikL} - ikCe^{-ikL}. \quad (15.31)$$

This is a homogeneous system of four equations with four unknowns. In order that the system have non-trivial solutions, the determinant of the coefficient matrix for $A$, $B$, $C$, $D$ must be equal to zero. Instead of calculating the determinant, we can also multiply the first equation in (15.30) by $\kappa$ and subtract the two equations. We obtain:

$$0 = \kappa Be^{-ikL} + \kappa Ce^{ikL} - ikBe^{-ikL} + ikCe^{ikL} = (\kappa - ik)e^{-ikL}B + (\kappa + ik)e^{ikL}C \quad (15.32)$$

and analogously from (15.31):

$$0 = \kappa Be^{ikL} + \kappa Ce^{-ikL} + ikBe^{ikL} - ikCe^{-ikL} = (\kappa + ik)e^{ikL}B + (\kappa - ik)e^{-ikL}C. \quad (15.33)$$

With (15.32) and (15.33), we have a homogeneous system for the two unknowns $B$ and $C$. It can be solved if the coefficient determinant for $B$, $C$ vanishes, i.e. for

$$(\kappa - ik)^2 e^{-2ikL} - (\kappa + ik)^2 e^{2ikL} \stackrel{!}{=} 0. \quad (15.34)$$

This equation gives the allowed energy values. We insert

$$\kappa \pm ik = \sqrt{\kappa^2 + k^2}e^{\pm i \arctan k/\kappa} \quad (15.35)$$

into (15.34) and obtain

$$\sin\left(2kL + 2\arctan\frac{k}{\kappa}\right) \stackrel{!}{=} 0. \quad (15.36)$$

The solution is evidently

$$2kL + 2\arctan\frac{k}{\kappa} \stackrel{!}{=} N\pi; \quad N = 1, 2, 3, \ldots \quad (15.37)$$

where $N$ numbers consecutively the valid solutions in such a way that the smallest energy eigenvalue has the index 1. A closer inspection of this equation is given below; here we will first determine the constants as far as possible. From (15.32) and ((15.33) would give the same information), it follows with (15.35) and because of (15.37 ) that:

$$C = -B\frac{\kappa - ik}{\kappa + ik}e^{-2ikL} = -Be^{-2i\arctan k/\kappa}e^{-2ikL}$$
$$= -Be^{-iN\pi} = B(-1)^{N+1} \quad (15.38)$$

and thus from (15.30) and (15.31)

$$A = \begin{cases} 2Be^{\kappa L}\cos kL \\ -2iBe^{\kappa L}\sin kL \end{cases} \quad \text{and} \quad D = \begin{cases} 2Be^{\kappa L}\cos kL \\ 2iBe^{\kappa L}\sin kL \end{cases} \quad \text{for } N \quad \begin{matrix} \text{odd} \\ \text{even.} \end{matrix} \qquad (15.39)$$

### 15.3.1.2  Energy Eigenvalues

Equation (15.37) is not solvable in closed form. In order to obtain solutions for specific values of $E$ and $V_0$, we must proceed numerically. Nevertheless, we can make general statements with the help of estimates. For this purpose, we rearrange (15.37) as follows:

$$2\arctan\frac{k}{\kappa} = N\pi - 2kL; \quad N = 1, 2, 3, \ldots \qquad (15.40)$$

Since $k$ and $\kappa$ are positive, we can estimate

$$0 < 2\arctan\frac{k}{\kappa} < \pi. \qquad (15.41)$$

Thus we obtain the inequality

$$0 < N\pi - 2kL < \pi \text{ or } (N-1)\pi < 2kL < N\pi. \qquad (15.42)$$

Since there are no negative terms, we can square. Substituting $k^2 = 2m(V_0 - |E|)/\hbar^2$ and subsequent rearranging yields

$$V_0 - \frac{\hbar^2}{2m}\left(\frac{N}{2L}\pi\right)^2 < |E| < V_0 - \frac{\hbar^2}{2m}\left(\frac{N-1}{2L}\pi\right)^2. \qquad (15.43)$$

This equation allows us to draw several conclusions:
  1. For $N = 1$ we have

$$V_0 - \frac{\hbar^2}{2m}\left(\frac{\pi}{2L}\right)^2 < |E| < V_0. \qquad (15.44)$$

It follows that there is always a solution. As we shall see below, it is symmetrical. This 'lowest' solution is also called the *ground state*.
  2. For $N = 2$, we have

$$V_0 - \frac{\hbar^2}{2m}\left(\frac{2\pi}{2L}\right)^2 < |E| < V_0 - \frac{\hbar^2}{2m}\left(\frac{\pi}{2L}\right)^2. \qquad (15.45)$$

It follows that there is a second state (the *first excited state*), if $V_0 > \frac{\hbar^2}{2m}\left(\frac{\pi}{2L}\right)^2$. As we shall see below, this state is antisymmetric.

3. Similarly, one sees that the $N$-th state exists if $V_0 > \frac{\hbar^2}{2m} \left( \frac{N-1}{2L} \pi \right)^2$.

4. There is an $N_0$, from which onwards the right side of the inequality is no longer satisfied. It follows that in each potential well of the kind we are considering, there is only a finite number of energy levels (see exercises).

### 15.3.1.3 Eigenfunctions

We now know that there is a finite number of solutions of the (15.36). We number them from 1 to $N_0$ (since $k$ and $\kappa$ depend on $E$, they also depend on $N$, which is indicated by a corresponding index). For each of these solutions, there is an eigenfunction; we distinguish these according to the parity of the energy quantum number $N$. With (15.38) and (15.39), it follows that

$$
\begin{aligned}
\varphi_{1,N}(x) &= 2Be^{\kappa_N L} \cos k_N L \cdot e^{\kappa_N x} \\
\varphi_{2,N}(x) &= 2B \cdot \cos k_N x \qquad\qquad \text{for } N \text{ odd} \qquad (15.46)\\
\varphi_{3,N}(x) &= 2Be^{\kappa_N L} \cos k_N L \cdot e^{-\kappa_N x}
\end{aligned}
$$

and

$$
\begin{aligned}
\varphi_{1,N}(x) &= -2i\, Be^{\kappa_N L} \sin k_N L \cdot e^{\kappa_N x} \\
\varphi_{2,N}(x) &= 2i B \cdot \sin k_N x \qquad\qquad \text{for } N \text{ even.} \qquad (15.47)\\
\varphi_{3,N}(x) &= 2i Be^{\kappa_N L} \sin k_N L \cdot e^{-\kappa_N x}
\end{aligned}
$$

The solutions in region 2, i.e. within the potential well, are evidently standing waves. The parity of the eigenfunctions alternates on climbing up the 'energy ladder', with the ground state symmetric. By the way, this parity property is a consequence of the symmetry of the problem.

## 15.3.2 Potential Well, $E > 0$

We now consider the case $E > 0$, i.e. free motion. With

$$
k'^2 = \frac{2m}{\hbar^2} E \qquad\qquad (15.48)
$$

the solutions read

$$
\begin{aligned}
&\text{region 1: } \varphi_1(x) = A'e^{ik'x} + Ae^{-ik'x} \\
&\text{region 3: } \varphi_3(x) = De^{ik'x} + Fe^{-ik'x}
\end{aligned} \qquad (15.49)
$$

and from above we have

$$
\text{region 2: } \varphi_2(x) = Be^{ikx} + Ce^{-ikx}. \qquad (15.50)
$$

If we require that the incoming wave of amplitude $F$ is incident on the potential well from the right, we have $A' = 0$, because in region 1, a wave coming from the left cannot be present. To determine the other constants, we use the matching conditions at the two discontinuities. They are

$$Ae^{ik'L} = Be^{-ikL} + Ce^{ikL}$$
$$-k'Ae^{ik'L} = kBe^{-ikL} - kCe^{ikL} \tag{15.51}$$

and

$$Be^{ikL} + Ce^{-ikL} = De^{ik'L} + Fe^{-ik'L}$$
$$kBe^{ikL} - kCe^{-ikL} = k'De^{ik'L} - k'Fe^{-ik'L}. \tag{15.52}$$

These are four equations for four unknowns (the amplitude $F$ can be chosen at will). Their solution reads

$$A = \frac{-4k'k}{N}e^{-2ik'L}F$$

$$B = \frac{2k'\left(k' - k\right)}{N}e^{ikL}e^{-ik'L}F$$

$$C = -\frac{2k'\left(k' + k\right)}{N}e^{-ikL}e^{-ik'L}F \tag{15.53}$$

$$D = \frac{2i\sin\left(2kL\right)\left(k'^2 - k^2\right)}{N}e^{-2ik'L}F$$

with

$$N = e^{2ikL}\left(k' - k\right)^2 - e^{-2ikL}\left(k' + k\right)^2$$
$$= 2i\left(k'^2 + k^2\right)\sin\left(2kL\right) - 4kk'\cos\left(2kL\right). \tag{15.54}$$

The transmission and reflection coefficients are given by

$$T = \frac{|A|^2}{|F|^2} \quad \text{and} \quad R = \frac{|D|^2}{|F|^2} \tag{15.55}$$

and this leads to

$$T = \frac{16k'^2k^2}{|N|^2} \quad \text{and} \quad R = \frac{2(k'^2 - k^2)\left(1 - \cos(4kL)\right)}{|N|^2} = 1 - T \tag{15.56}$$

with

$$|N|^2 = 2k'^4 + 12k'^2k^2 + 2k^4 - 2(k'^2 - k^2)^2\cos(4kL). \tag{15.57}$$

**Fig. 15.6** Transmission coefficient (15.60) for scattering by the potential well with $\mu = 15$ as a function of $z = \frac{E}{V_0}$

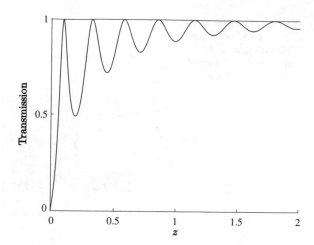

Because of

$$k'^2 = \frac{2m}{\hbar^2} E \quad \text{and} \quad k^2 = \frac{2m}{\hbar^2}(E + V_0) \tag{15.58}$$

and with

$$z = \frac{E}{V_0}; \quad \mu = \sqrt{\frac{2m}{\hbar^2} V_0 L^2}, \tag{15.59}$$

it follows for the transmission coefficient:

$$T = \frac{z(z+1)}{z(z+1) + \frac{1-\cos 4kL}{8}} = \frac{z(z+1)}{z(z+1) + \frac{1-\cos 4\mu\sqrt{z+1}}{8}}. \tag{15.60}$$

We see that $T = 1$ for $\cos 4kL = 1$, which means that $4kL = 2m\pi$. With $\lambda = \frac{2\pi}{k}$, the intuitively-clear condition $m\frac{\lambda}{2} = 2L$ follows, i.e. a kind of resonance condition, as is seen very nicely in Fig. 15.6.

## 15.4 Potential Barrier, Tunnel Effect

The potential is given by (cf. Fig. 15.7):

$$V = \begin{cases} V_0 \text{ for } -L < x < L; & V_0 > 0 \\ 0 \text{ otherwise.} \end{cases} \tag{15.61}$$

In view of the description of scattering processes, this potential is more realistic than the infinite potential step. However, the computational effort is also significantly greater.

The solutions in the different regions are

**Fig. 15.7** Potential barrier.
*Straight lines*: oscillation,
*broken line*: exponential.
*Incoming green, reflected
blue, transmitted red*

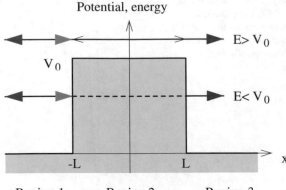

Potential, energy

$V_0$

$E > V_0$

$E < V_0$

-L                    L            x

Region 1          Region 2          Region 3

$$\begin{aligned}
x < -L: \quad & \varphi_1(x) = Ae^{ikx} + Be^{-ikx} \\
-L < x < L: \quad & \varphi_2(x) = Ce^{\gamma x} + De^{-\gamma x} \\
x > L: \quad & \varphi_3(x) = Fe^{ikx} + Ge^{-ikx}
\end{aligned} \tag{15.62}$$

with $k^2 = \frac{2m}{\hbar^2} E$. We can consider the cases $E > V_0$ and $E < V_0$ simultaneously by defining

$$\gamma = \begin{cases} \kappa \\ ik' \end{cases} \text{for} \quad \begin{matrix} E < V_0 \\ E > V_0 \end{matrix} \quad \text{with} \quad \begin{matrix} \kappa^2 = \frac{2m}{\hbar^2}(V_0 - E) \\ k'^2 = \frac{2m}{\hbar^2}(E - V_0). \end{matrix} \tag{15.63}$$

For a change, we assume this time that the incident wave comes from the left and has amplitude $A$. Then in region 3, there is no wave running from the right to the left, which means that $G = 0$. At the discontinuities $x = \pm L$, we have

$$Ae^{-ikL} + Be^{ikL} = Ce^{-\gamma L} + De^{\gamma L}$$
$$ikAe^{-ikL} - ikBe^{ikL} = \gamma Ce^{-\gamma L} - \gamma De^{\gamma L} \tag{15.64}$$

and

$$Ce^{\gamma L} + De^{-\gamma L} = Fe^{ikL}$$
$$\gamma Ce^{\gamma L} - \gamma De^{-\gamma L} = ikFe^{ikL}. \tag{15.65}$$

The calculation of the constants as multiples of $A$ is given in Appendix X, Vol. 2.

The partial waves of interest to us are

$$\varphi_{\text{in}} = Ae^{ikx}; \quad \varphi_{\text{refl}} = Be^{-ikx}; \quad \varphi_{\text{trans}} = Fe^{ikx}. \tag{15.66}$$

Transmission and reflection coefficients are given by

$$T = \frac{|F|^2}{|A|^2}; \quad R = \frac{|B|^2}{|A|^2}. \tag{15.67}$$

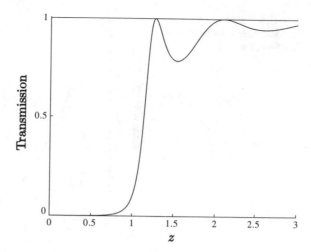

**Fig. 15.8** Potential barrier: transmission coefficient as a function of $z = E/V_0$ for $\mu = 3$

We confine the discussion to $T$. The somewhat lengthy calculation is given in Appendix X, Vol. 2. With

$$z = \frac{E}{V_0}; \quad k'L = \mu\sqrt{z-1}; \quad \kappa L = \mu\sqrt{1-z}; \quad \mu = \sqrt{\frac{2m}{\hbar^2}V_0L^2}, \quad (15.68)$$

the result reads[10]:

$$T = \begin{cases} \dfrac{8z(z-1)}{8z(z-1)+1-\cosh 4\kappa L} = \dfrac{z(z-1)}{z(z-1)+\dfrac{1-\cosh 4\mu\sqrt{1-z}}{8}} & E < V_0; \quad 0 < z < 1 \\[2em] \dfrac{8z(z-1)}{8z(z-1)+1-\cos 4k'L} = \dfrac{z(z-1)}{z(z-1)+\dfrac{1-\cos 4\mu\sqrt{z-1}}{8}} & \text{for} \quad E > V_0; \quad z > 1. \end{cases}$$

$$(15.69)$$

In Fig. 15.8, the transmission coefficient is shown as a function of $z = E/V_0$. We see that in the range $0 < z \leq 1$, it *always* holds that $T > 0$.

This means that we always have a part of the wavefunction which 'tunnels through', i.e. shows a behavior which is impossible in classical mechanics. The tunneling probability decreases of course with increasing width of the potential barrier. We illustrate this in Fig. 15.9 for the case $E = V_0/2$. This sensitive dependence of the tunneling on the potential width is responsible e.g. for the wide range of decay times observed for $\alpha$ decay; technically, it is used in tunnel diodes.

For $E = V_0$, we also find both reflected and transmitted components, because of

---

[10]Because of $\cosh iy = \cos y$ or $\cos iy = \cosh y$, one of the two expressions for $T$ is in fact sufficient for real $y$.

**Fig. 15.9** Transmission
coefficient for $E = V_0/2$ as a
function of $\mu \sim L$

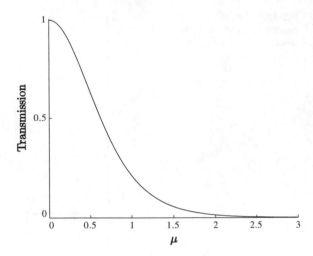

$$T(z = 1) = \frac{1}{1 + \mu^2}.\tag{15.70}$$

This is also true for all values $z > 1$, but with isolated exceptions, since for

$$z = z_m = 1 + \left(\frac{m\pi}{2\mu}\right)^2;\quad m = 1, 2, \ldots \quad \to T(z_m) = 1,\tag{15.71}$$

there are only transmitted and no reflected components.

Apart from the values $E = \left[1 + \left(\frac{m\pi}{2\mu}\right)^2\right] V_0$, for $0 < z < \infty$ we *always* obtain both reflected and transmitted portions of the wavefunction in the case of the potential barrier. For very large values of the energy, the reflection probability is indeed very small, but it is not zero, i.e. it exists in principle. In contrast, in classical mechanics we have either reflection ($E < V_0$) or transmission ($E > V_0$).

## 15.5   From the Finite to the Infinite Potential Well

In considering the infinite potential well, we assumed that the wavefunction vanishes at the potential walls. We want now to justify that assumption.

We start with a finite potential well (see Fig. 15.10):

$$V = \begin{cases} 0 & \text{for } 0 < x < L \\ V_0 > 0 & \text{otherwise.} \end{cases}\tag{15.72}$$

**Fig. 15.10** Potential well for the discussion of the limit $V_0 \rightarrow \infty$

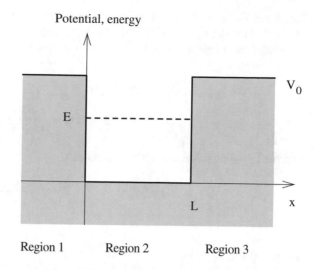

Region 1        Region 2        Region 3

We calculate the bound solutions and let then go $V_0$ to infinity. We have for the stationary SEq (region 3 is not required for the following consideration):

$$E\varphi_1 = -\frac{\hbar^2}{2m}\varphi_1'' + V_0\varphi_1 \text{ for } x < 0; \quad E\varphi_2 = -\frac{\hbar^2}{2m}\varphi_2'' \quad \text{for } 0 < x < L \quad (15.73)$$

or

$$\varphi_1'' = \frac{2m}{\hbar^2}(V_0 - E)\varphi_1 = \kappa^2\varphi_1; \quad \varphi_2'' = -\frac{2m}{\hbar^2}E\varphi_2 = -k^2\varphi_2. \quad (15.74)$$

It follows that:

$$\varphi_1(x) = Ae^{\kappa x}; \quad \varphi_2(x) = Be^{ikx} + Ce^{-ikx}$$
$$\varphi_1'(x) = \kappa Ae^{\kappa x}; \quad \varphi_2'(x) = ikBe^{ikx} - ikCe^{-ikx}. \quad (15.75)$$

At $x = 0$, we have

$$A = B + C; \quad \kappa A = ikB - ikC. \quad (15.76)$$

From these two equations, we find:

$$C = -\frac{\kappa - ik}{\kappa + ik}B \quad (15.77)$$

and therefore

$$A = B + C = \frac{2ik}{\kappa + ik}B. \quad (15.78)$$

The limit $V_0 \to \infty$ means $\kappa \to \infty$ (while $k$ remains fixed). Hence it follows that $A \to 0$. In this way, we have justified in retrospect our *ansatz* $\varphi_1(0) = 0$ at the discontinuity of the infinite potential well. For the wavefunction in region 2, we have

$$\varphi_2(0) = B + C = \frac{2ik}{\kappa + ik} B$$

$$\varphi_2'(0) = ikB - ikC = \frac{2ik\kappa}{\kappa + ik} B \qquad (15.79)$$

and for $V_0 \to \infty$ it follows, as it indeed must:

$$\varphi_2(0) \to 0; \quad \varphi_2'(0) \to 2ikB. \qquad (15.80)$$

## 15.6  Wave Packets

We have already discussed several times the fact that a plane wave cannot describe a physical object because it has the same magnitude for all positions and times. But in spite of this, it is common practice to work with this handy formulation, as we know. This is due to the linearity of the SEq and the consequent superposability of its solutions. It allows us to overlay plane waves in such a way that a physically meaningful expression results.

We want to carry this out as an example for the potential step discussed above: There is an incoming wave in the region $x > 0$, incident from the right, a transmitted wave from the right to the left in $x < 0$ and a reflected wave from the left to the right in $x > 0$. With

$$k_0 = \sqrt{\frac{2m}{\hbar^2} V_0}; \quad \gamma(k) = \begin{cases} \kappa = \sqrt{k_0^2 - k^2} \\ -ik' = -i\sqrt{k^2 - k_0^2} \end{cases} \quad \text{for } E \underset{>}{\overset{<}{\phantom{=}}} V_0 \text{ or } k \underset{>}{\overset{<}{\phantom{=}}} k_0,$$
$$(15.81)$$

we can write the solutions for fixed $k > 0$ as

$$\varphi_1 = ce^{-ikx} + c\frac{ik + \gamma}{ik - \gamma} e^{ikx}; \quad \varphi_2 = c\frac{2ik}{ik - \gamma} e^{\gamma x}. \qquad (15.82)$$

We obtain a total solution by integrating over the continuous index $k > 0$. With $\omega = \frac{\hbar k^2}{2m}$, it follows that

$$\Psi_1(x, t) = \int_0^\infty c(k) \left( e^{-ikx} + \frac{ik + \gamma}{ik - \gamma} e^{ikx} \right) e^{-i\omega t} \, dk$$

$$\Psi_2(x, t) = \int_0^\infty c(k) \frac{2ik}{ik - \gamma} e^{\gamma x} e^{-i\omega t} \, dk \qquad (15.83)$$

**Fig. 15.11** Schematic representation of the amplitude function $|c(k)|$ for comparison with the classical transmission

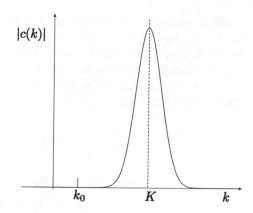

where $c(k)$ is an arbitrary function of $k$. With suitable $c(k)$, one can generate rather complicated wavefunctions. We confine ourselves to a situation that allows comparison with classical behavior (either reflection or transmission). We choose transmission.[11] From the classical perspective, this case corresponds to an object that travels from the right to the left with momentum $P$ towards the potential step, and from there continues in the same direction, but with a smaller momentum $P'$.

Since the classical particle has a definite momentum $P = \hbar K$, we choose for $c(k)$ a function that has a sharp maximum at $k = K$, has nonvanishing values only in a neighborhood of $K$, and (for the sake of simplicity) vanishes identically for $k \leq k_0$; see Fig. 15.11.[12]

Thus we can write[13]

$$\Psi_1(x, t) = \int_{k_0}^{\infty} c(k) e^{-i(kx + \omega t)} \, dk + \int_{k_0}^{\infty} c(k) \frac{k - k'}{k + k'} e^{i(kx - \omega t)} \, dk = \Psi_{\text{in}} + \Psi_{\text{refl}}$$

$$\Psi_2(x, t) = \int_{k_0}^{\infty} c(k) \frac{2k}{k + k'} e^{-i(k'x + \omega t)} \, dk = \Psi_{\text{trans}}. \tag{15.84}$$

We immediately see that we again have three types of waves[14]: incoming, reflected, and transmitted, and that—in contrast to classical mechanics—there is *always* a reflected wave.[15]

Even for very simple distributions $c(k)$, it is not possible to perform the integrations in closed form.[16] But we can make the following general observation: the magnitude of the integrals in (15.84) depends essentially on how fast the exponential

---

[11] More on wave packets can be found in Appendix D, Vol. 2.

[12] If we call $\Delta k$ the width of the function, then $\Delta k \ll K$ must apply.

[13] Due to $c(k) = 0$.

[14] We note that these are 'true' waves, functions of time and space.

[15] We recall the quantum-mechanical truck that bounces off a mosquito flying against its windshield.

[16] If $c(k)$ is given by a Gaussian curve, at least the term $\Psi_{\text{in}}$ can be calculated; see Chap. 5, Vol. 1 and Appendix D, Vol. 2.

functions oscillate in the neighborhood of $K$—the faster, the smaller the integral (or its absolute value). This is due to the fact that with a faster oscillation, the areas above the $k$ axis are better compensated by areas of opposite sign. In general, we find the biggest contribution if the exponent does not vary in the neighborhood of $K$; that is, if its derivative with respect to $k$ vanishes.[17] This means, for example for the incoming wave,

$$\frac{\mathrm{d}}{\mathrm{d}k}\,(kx+\omega t)\,\bigg/_{k=K} = x + \frac{\hbar K}{m}t = 0. \tag{15.85}$$

Thus, the incoming wave packet is particularly large for $x$ values near $x = -\frac{\hbar K}{m}t$, and this peak moves with the *group velocity* $v_g = -\frac{\hbar K}{m}$. Accordingly, for the reflected wave we have the group velocity $v_g = \frac{\hbar K}{m}$. The value of $v_g$ is obtained in both cases as $v_g = \frac{d\omega}{dk}$, which is the usual definition of group velocity, while the phase velocity $v_{ph}$ (i.e. the propagation velocity of a wave component with a well-defined oscillation frequency) is given by $v_{ph} = \frac{\omega}{k}$. We consider the transmitted wave. The stationarity of the phase

$$\frac{\mathrm{d}}{\mathrm{d}k}\,\left(k'x+\omega t\right)\,\bigg/_{k=K} = 0 \tag{15.86}$$

gives

$$x + \frac{\hbar K'}{m}t = 0;\quad K' = \sqrt{K^2 - \frac{2m}{\hbar^2}V_0}\;\;;\;\; v_g = \frac{d\omega}{dk}\bigg/_{k=K'}. \tag{15.87}$$

We have a stationary phase only in the following cases: (a) $\Psi_{\mathrm{in}}$ for $t < 0$ and $x > 0$; (b) $\Psi_{\mathrm{refl}}$ for $t > 0$ and $x > 0$; (c) $\Psi_{\mathrm{trans}}$ for $t > 0$ and $x < 0$. This means that at large negative times, only the incident wave packet provides a significant contribution; at $t \approx 0$, all three sub-packets exist with similar amplitudes; at large positive times, only the reflected and transmitted wave packets provide significant contributions. In other words: At $t \ll 0$ we have an incoming (from the right to the left) wave packet; at $t \approx 0$ there is a confusing 'wriggling', and at $t \gg 0$ we again have a clear-cut situation, namely for $x > 0$ a reflected (running to the right) and for $x < 0$ a transmitted (running to the left) wave packet.[18]

Finally, we want to address very briefly problems which may arise in explaining these relationships to laypeople, whether in schools or elsewhere. First, we have already pointed out that terms such as 'incoming wave' require bearing in mind tacitly the factor $e^{-i\omega t}$. Without this factor, the name would be misleading, because $e^{ikx}$ is not a wave travelling anywhere, but simply a time-independent spatial oscillation. In general, these facts are not considered in school classes; in addition, complex numbers

---

[17]This is why the procedure is also called the method of stationary phase.
[18]We recall that the wavefunction does not describe the object itself, but rather allows the calculation of probabilities for observing it at a particular location.

are nearly always avoided. Thus, one has to argue 'somehow' that e.g. $\cos kx$ is an incoming wave.

Another problem in this context: Teaching and learning software illustrate e.g. scattering by the potential step, but of course do this with wave packets of the form (15.83); with plane waves, one would see not too much. However, wave packets and similar formulations are usually not taught at all in school classes, so it seems difficult to establish the relationship between the mathematics and computer simulation results.

## 15.7 Exercises

1. Given the potential step

$$V(x) = \begin{cases} 0 \\ V_0 > 0 \end{cases} \text{for} \quad \begin{matrix} x > 0 \\ x \le 0. \end{matrix} \tag{15.88}$$

The incident quantum object is described as a plane wave running from the right to the left with $E > V_0$. Determine the transmission and reflection coefficients.
2. Given a finite potential well of depth $V_0$ and width $L$; estimate the number of energy levels.
3. Given a delta potential at $x = 0$; determine the spectrum (negative potential, $E < 0$) and the situation for scattering (positive potential, $E > 0$ ).
4. Given the potential barrier

$$V(x) = \begin{cases} V_0 > 0 & \text{for} \quad -L < x < L \\ 0 & \text{otherwise.} \end{cases} \tag{15.89}$$

The incident quantum object is described by a plane wave running from the left to the right. Determine the transmission and reflection coefficients.
5. Given the one-sided infinite potential well

$$V(x) = \begin{cases} 0 & L < x \\ -V_0 & \text{for} \quad 0 < x \le L \\ \infty & x \le 0 \end{cases} \tag{15.90}$$

with $V_0 > 0$. For the energy, let $-V_0 < E < 0$. Sketch the potential. Determine the stationary SEq in the different regions and deduce from them an *ansatz* for the wavefunction. Adjust the wavefunctions at the discontinuities and show that the allowed energy levels are defined by the equation $k \cot kL = -\kappa$ with $k^2 = 2m \left( V_0 + E \right) /\hbar^2$ and $\kappa^2 = -2mE/\hbar^2$. Is there always (i.e. for all $V_0$) a bound state?
6. Given the potential

**Fig. 15.12**  The potential of
(15.92)

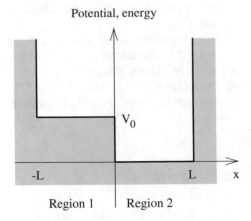

Potential, energy

$V_0$

-L                                  L        x

Region 1   |   Region 2

$$V(x) = \begin{cases} \infty & x < 0 \\ V_0 > 0 & \text{for} \quad 0 \le x \le L. \\ 0 & L < x \end{cases} \quad (15.91)$$

An object described by a plane wave passes from the right to the origin. Sketch
the potential. Calculate the wavefunction for the case $E < V_0$. Which regions
are classically allowed, which are not? Determine first the stationary SEq's in
the different regions and solve them with an appropriate *ansatz*. Are all the
mathematical solutions physically allowed? Determine the free constants using
the continuity conditions at the discontinuities of the potential.
Perform the calculations for the case $E > V_0$, also.

7. Given a potential step embedded in an infinite potential well (see Fig. 15.12):

$$V(x) = \begin{cases} 0 & 0 < x < L \\ V_0 > 0 & \text{for} \quad -L < x \le 0. \\ \infty & x \ge |L| \end{cases} \quad (15.92)$$

Calculate the spectrum for $E > V_0$.

8. (Resonances) Given a potential barrier in front of an infinite potential wall (see
Fig. 15.13):

$$V(x) = \begin{cases} \infty & x < 0 \\ V_0 > 0 & \text{for} \quad \le x \le b. \\ 0 & \text{otherwise} \end{cases} \quad (15.93)$$

The incident quantum object has the energy $E < V_0$ and comes from the right.
For which parameter values is the phase shift of the outgoing wave particularly
large/does the phase change especially fast? What is the physical explanation?

9. In this chapter, a transcendental equation of the form

**Fig. 15.13** The potential of (15.93)

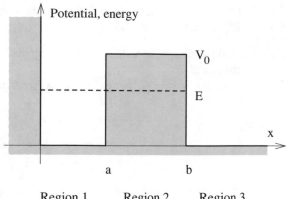

$$\tan kd = \frac{k}{\kappa} = \quad ; \kappa = \sqrt{\kappa_V^2 - k^2} \quad ; k < \kappa_V \tag{15.94}$$

occurs several times. Find an approximate solution for large $d$.

10. Given the double well potential:

$$\begin{aligned}
&\text{region 1: } -L \le x \le -a \quad V = 0\\
&\text{region 2: } -a < x < a \quad\ V = V_0 > 0.\\
&\text{region 3: } a \le x \le L. \quad\ \ V = 0
\end{aligned} \tag{15.95}$$

$V$ is infinite for $|x| > L$. We consider only energies $E$ for which $E < V_0$.

(a) Due to the symmetry of the problem ($H(x) = H(-x)$), there are symmetric and antisymmetric eigenfunctions, sS and aS (cf. Chap. 21, Vol. 2). Determine these functions and their eigenvalue equations.

(b) Show that there is no solution of the eigenvalue equations below a certain threshold value of $V_0$.

(c) Show that the ground state is symmetric.

(d) Solve the eigenvalue equations approximately for the case of a 'thick' barrier, i.e. for very large $a$.

(e) The initial state is assumed to be a linear combination of the symmetric and the antisymmetric states of the same order (for the sake of simplicity with equal amplitudes, $A_s = A_a = A$). Determine the time behavior of the wavefunction. Calculate the probabilities $P_i(t)$ of finding the object in region $i$.

(f) In the case of a thick barrier, it holds that $k_a - k_s \ll k_a + k_s$. Calculate up to and including quadratic terms in $k_a - k_s$ the quantities $R_{\text{max}}^{\text{min}} = \min(P_3)/\max(P_3)$ and $\Delta\omega$. Discuss your findings.

(g) In the ammonia molecule $NH_3$, the $N$ atom tunnels back and forth through the plane of the three $H$ atoms. This situation can be modelled by the double well potential with parameters $a = 0.2 \cdot 10^{-10}$ m, $d = 0.3 \cdot 10^{-10}$ m, $V_0 = 0.255$ eV and $m = 4 \cdot 10^{-27}$ kg (the reduced mass is $\frac{3m_H m_N}{3m_H + m_N}$).

Compute numerical values for the ground-state levels, the frequency and $R_{\text{max}}^{\text{min}}$. Discuss your findings.

11. For an illustration of the method of stationary phase, consider the (unnormalized) wavefunction

$$\psi(x, t) = \int_{-\infty}^{\infty} |A(k)| e^{i\varphi(k)} e^{i(kx - \omega t)} \, dk \qquad (15.96)$$

with

$$\omega = ck; \quad \varphi(k) = -x_0 k \qquad (15.97)$$

and

$$|A(k)| = \begin{cases} \kappa^2 - (k - K)^2 & \text{for} \quad 0 < K - \kappa \le k \le K + \kappa \\ 0 & \text{otherwise} \end{cases}. \qquad (15.98)$$

The constants $\kappa$, $K$ and $x_0$ are positive. Calculate explicitly $\psi(x, t)$ and discuss its properties. What is the physical significance of $x_0$?

# Chapter 16
# Angular Momentum

Apart from the Hamiltonian, the angular momentum operator is of one of the most important Hermitian operators in quantum mechanics. In this chapter we consider its eigenvalues and eigenfunctions in more detail.

We begin this chapter with the consideration of the orbital angular momentum. This gives rise to a general definition of angular momenta. We derive the eigenvalue spectrum of the orbital angular momentum with an algebraic method. After a brief presentation of the eigenfunctions of the orbital angular momentum in the position representation, we outline some concepts for the addition of angular momenta.

## 16.1   Orbital Angular Momentum Operator

The orbital angular momentum is given by

$$\mathbf{l} = \mathbf{r} \times \mathbf{p}. \tag{16.1}$$

As we have seen in Chap. 3, Vol. 1, it is not necessary to symmetrize for the translation into quantum mechanics (spatial representation). It follows directly that

$$\mathbf{l} = \frac{\hbar}{i} \mathbf{r} \times \nabla, \tag{16.2}$$

or, in components,

$$l_x = \frac{\hbar}{i} \left( y \frac{\partial}{\partial z} - z \frac{\partial}{\partial y} \right) \tag{16.3}$$

plus cyclic permutations ($x \rightarrow y \rightarrow z \rightarrow x \rightarrow \cdots$). All the components of $\mathbf{l}$ are observables.

© Springer Nature Switzerland AG 2018
J. Pade, *Quantum Mechanics for Pedestrians 2*, Undergraduate Lecture Notes in Physics, https://doi.org/10.1007/978-3-030-00467-5_16

We know that one can measure two variables simultaneously if the corresponding operators commute. What about the components of the orbital angular momentum? A short calculation (see the exercises) yields

$$[l_x, l_y] = i\hbar l_z; \quad [l_y, l_z] = i\hbar l_x; \quad [l_z, l_x] = i\hbar l_y, \tag{16.4}$$

or, compactly, $[l_x, l_y] = i\hbar l_z$ and cyclic permutations. This term can be written still more compactly with the Levi-Civita symbol (permutation symbol, epsilon tensor) $\varepsilon_{ijk}$ (see also Appendix F, Vol. 1):

$$\varepsilon_{ijk} = \begin{cases} 1 & ijk \text{ is an even permutation of } 123 \\ -1 & \text{if } ijk \text{ is an odd permutation of } 123 \\ 0 & \text{otherwise} \end{cases} \tag{16.5}$$

namely as

$$[l_i, l_j] = i\hbar \sum_k l_k \varepsilon_{ijk}. \tag{16.6}$$

Each component of the orbital angular momentum commutes with $\mathbf{l}^2 = l_x^2 + l_y^2 + l_z^2$ (see the exercises):

$$[l_x, \mathbf{l}^2] = [l_y, \mathbf{l}^2] = [l_z, \mathbf{l}^2] = 0. \tag{16.7}$$

## 16.2  Generalized Angular Momentum, Spectrum

We now generalize these facts by the following definition: A vector operator[1] $\mathbf{J}$ is a (generalized) angular momentum operator, if its components are observables and satisfy the commutation relation

$$[J_x, J_y] = i\hbar J_z. \tag{16.8}$$

and its cyclic permutations.[2] It follows that $\mathbf{J}^2 = J_x^2 + J_y^2 + J_z^2$ commutes with all the components:

$$[J_x, \mathbf{J}^2] = 0; \quad [J_y, \mathbf{J}^2] = 0; \quad [J_z, \mathbf{J}^2] = 0. \tag{16.9}$$

The task that we address now is the calculation of the angular momentum spectrum. We will deduce that the angular momentum can assume only half-integer and integer values. To this end, we need only (16.8) and (16.9) and the fact that the $J_i$

---

[1] Instead of $\mathbf{J}$, one often finds $\mathbf{j}$ (whereby this is of course not be confused with the probability current density). In this section, we denote the operator by $\mathbf{J}$ and the eigenvalue by $j$.

[2] The factor $\hbar$ is due to the choice of units, and would be replaced by a different constant if one were to choose different units. The essential factor is $i$.

are Hermitian operators, $J_i^\dagger = J_i$ (whereby it is actually quite amazing that one can extract so much information from such sparse initial data).

We note first that the squares of the components of the angular momentum are positive operators. Thus, we have for an arbitrary state $|\varphi\rangle$:

$$\langle\varphi|\, J_x^2\,|\varphi\rangle = \|\,J_x\,|\varphi\rangle\|^2 \geq 0. \tag{16.10}$$

Consequently, $\mathbf{J}^2$ as a sum of positive Hermitian operators is also positive, so it can have only non-negative eigenvalues. For reasons which will become clear later, these eigenvalues are written in the special form $j\,(j+1)$ with $j \geq 0$ (and not just simply $j^2$ or something similar).

Equations (16.8) and (16.9) show that $\mathbf{J}^2$ and one of its components can be measured simultaneously. Traditionally, one chooses the $z$-component $J_z$ and denotes the eigenvalue associated with $J_z$ by $m$.[3]

We are looking for eigenvectors $|j, m\rangle$ of $\mathbf{J}^2$ and $J_z$ with

$$\mathbf{J}^2\,|j, m\rangle = \hbar^2 j\,(j+1)\,|j, m\rangle\,;\; J_z\,|j, m\rangle = \hbar m\,|j, m\rangle. \tag{16.11}$$

To continue, we must now use the commutation relations (16.8). It turns out that it is convenient to use two new operators instead of $J_x$ and $J_x$, namely

$$J_\pm = J_x \pm i\,J_y. \tag{16.12}$$

For reasons that will become apparent immediately, these two operators are called *ladder operators*; $J_+$ is the *raising operator* and $J_-$ the *lowering operator*. The operators are adjoint to each other, since $J_x$ and $J_y$ are Hermitian:

$$J_\pm^\dagger = J_\mp. \tag{16.13}$$

With $J_+$ and $J_-$, the commutation relations (16.8) are written as

$$\left[J_z, J_+\right] = \hbar J_+;\;\; \left[J_z, J_-\right] = -\hbar J_-;\;\; \left[J_+, J_-\right] = 2\hbar J_z, \tag{16.14}$$

and for $\mathbf{J}^2$, we have

$$\mathbf{J}^2 = \frac{1}{2}\,(J_+ J_- + J_- J_+) + J_z^2. \tag{16.15}$$

Together with $\left[J_+, J_-\right] = 2\hbar J_z$, this equation leads to the expressions

$$J_+ J_- = \mathbf{J}^2 - J_z\,(J_z - \hbar)\,;\; J_- J_+ = \mathbf{J}^2 - J_z\,(J_z + \hbar)\,. \tag{16.16}$$

Furthermore, $\mathbf{J}^2$ commutes with $J_+$ and $J_-$ (see the exercises).

---

[3] $j$ is also called the *angular momentum quantum number* and $m$ the *magnetic* or *directional quantum number*.

Our interest in the expressions $J_+J_-$ and $J_-J_+$ is essentially due to the fact that they are positive operators. This can be easily seen since, because of (16.13), the matrix element $\langle\varphi|\,J_+J_-\,|\varphi\rangle$ is a norm and it follows that $\langle\varphi|\,J_+J_-\,|\varphi\rangle = \|J_-\,|\varphi\rangle\|^2 \geq 0$.

We apply $J_+J_-$ and $J_-J_+$ to the angular momentum states. With (16.16), it follows that

$$J_+J_-\,|j,m\rangle = \hbar^2\,[j(j+1) - m(m-1)]\,|j,m\rangle \quad \text{and}$$
$$J_-J_+\,|j,m\rangle = \hbar^2\,[j(j+1) - m(m+1)]\,|j,m\rangle. \tag{16.17}$$

Since the operators are positive, we obtain immediately the inequalities

$$j(j+1) - m(m-1) = (j-m)(j+m+1) \geq 0 \quad \text{and}$$
$$j(j+1) - m(m+1) = (j+m)(j-m+1) \geq 0 \tag{16.18}$$

which must be fulfilled simultaneously. This means that e.g. in the first inequality, the brackets $(j-m)$ and $(j+m+1)$ must both be positive or both negative. If they were negative, we would have $j \leq m$ and $m \leq -j-1$. But this is a contradiction since $j$ is positive. Hence we have $j \geq m$ and $m \geq -j-1$. If we consider in addition the second inequality, it follows that

$$-j \leq m \leq j. \tag{16.19}$$

In this way, the range of $m$ is fixed.

Now we have to determine the possible values of $j$. Let us consider the effect of $J_\pm$ on the states $|j,m\rangle$. We have

$$\mathbf{J}^2\left[J_\pm\,|j,m\rangle\right] = \hbar^2 j(j+1)\left[J_\pm\,|j,m\rangle\right] \quad \text{and}$$
$$J_z\left[J_\pm\,|j,m\rangle\right] = \hbar\,(m\pm1)\left[J_\pm\,|j,m\rangle\right]. \tag{16.20}$$

On the right-hand sides, we have the numbers $j(j+1)$ and $(m\pm1)$. This means that $J_\pm\,|j,m\rangle$ is an eigenvector of $\mathbf{J}^2$ with the eigenvalue $\hbar^2 j(j+1)$ as well as of $J_z$ with the eigenvalue $\hbar\,(m\pm1)$. It follows that

$$J_\pm\,|j,m\rangle = c^\pm_{j,m}\,|j,m\pm1\rangle. \tag{16.21}$$

The proportionality constant $c^\pm_{j,m}$ can be fixed by using (16.17) (see the exercises). The simplest choice is

$$J_\pm\,|j,m\rangle = \hbar\sqrt{j(j+1) - m(m\pm1)}\,|j,m\pm1\rangle. \tag{16.22}$$

Here, we see clearly the reason why $J_+$ and $J_-$ are called ladder operators or raising and lowering operators; for if we apply $J_+$ or $J_-$ several times to $|j,m\rangle$, the magnetic quantum number is increased or decreased by 1 each time. Hence, with the help of these operators, we can climb down or up step by step, as on a ladder.

Especially for $m = j$ or $m = -j$, we have $\|J_+ |j, j\rangle\|^2 = 0$ or $\|J_- |j, -m\rangle\|^2 = 0$. Since the norm of a vector vanishes iff it is the zero vector, it follows that

$$J_+ |j, j\rangle = 0; \quad J_- |j, -j\rangle = 0. \tag{16.23}$$

We now apply the ladder operators repeatedly and can conclude that $J_\pm^N |j, m\rangle$ is an eigenvector of $J_z$ with the eigenvalue $\hbar (m \pm N)$ (see the exercises), i.e. that it is proportional to $|j, m \pm N\rangle$ with $N \in \mathbb{N}$. In other words, if we start from any state $|j, m\rangle$ with $-j < m < j$, then after a few steps[4] we obtain states whose norm is negative (or rather would obtain them), or whose magnetic quantum number $m \pm N$ violates the inequality (16.19). This can be avoided only if the following conditions are fulfilled, 'going up' for $J_+$ and 'going down' for $J_-$:

$$m + N_1 = j \quad \text{and} \quad m - N_2 = -j; \quad N_1, N_2 \in \mathbb{N}. \tag{16.24}$$

For as we know from (16.23), $J_+ |j, j\rangle = 0$, and further applications of $J_+$ yield just zero again; and similarly for $J_-$.

The addition of the two last equations (16.24) leads to $2j = N_1 + N_2$. It follows with $j \geq 0$ that the allowed values for $j$ are given by

$$j = 0, \frac{1}{2}, 1, \frac{3}{2}, 2, \ldots \tag{16.25}$$

and for $m$ by

$$m = -j, -j + 1, -j + 2, \ldots, j - 2, j - 1, j \tag{16.26}$$

In this way, we have determined the possible eigenvalues of the general angular momentum operators $\mathbf{J}^2$ and $J_z$.

We remark that there are operators with only integer eigenvalues (e.g. the orbital angular momentum operator), or only half-integer eigenvalues (e.g. the spin-$\frac{1}{2}$ operator), but also those which have half-integer and integer eigenvalues. One of the latter operators, for example, occurs in connection with the Lenz vector; see Appendix G, Vol. 2.

As for elementary particles, nature has apparently set up two classes, which differ by their spins: those with half-integer spin are called *fermions*, those with integer spin *bosons*. General quantum objects can, however, have half-integer or integer spin, as we see in the example of helium, occurring as $^3\text{He}$ (fermion) and as $^4\text{He}$ (boson). We point out that a very general theorem (the *spin-statistics theorem*) shows the connection between the spin and quantum statistics, proving that all fermions obey Fermi-Dirac statistics and all bosons obey Bose-Einstein statistics.

---

[4]For e.g. $J_+$, $|j, m\rangle \to |j, m + 1\rangle \to |j, m + 2\rangle \to \cdots$.

We note that in the derivation of the angular-momentum eigenvalues, we have obtained little information about the eigenvectors—but we do not need it in order to derive the spectrum. It is understandable if this leaves feelings of uncertainty at first sight; but on the other hand, it is simply superb that there is such an elegant technique!

## 16.3  Matrix Representation of Angular Momentum Operators

With (16.22), it follows that

$$\langle j, m' | J_\pm | j, m \rangle = \hbar \sqrt{j(j+1) - m(m \pm 1)} \delta_{m', m \pm 1}. \tag{16.27}$$

By means of this equation, we can represent the angular momentum operators as matrices. We assume that the eigenstates are given as column vectors.

We consider the case of spin $\frac{1}{2}$, where one usually writes $s$ instead of $J$. We can represent the two possible states as[5]

$$\left| \frac{1}{2}, \frac{1}{2} \right\rangle = \begin{pmatrix} 1 \\ 0 \end{pmatrix} \quad \text{and} \quad \left| \frac{1}{2}, -\frac{1}{2} \right\rangle = \begin{pmatrix} 0 \\ 1 \end{pmatrix}. \tag{16.28}$$

Only two matrix elements do not vanish, namely

$$\left\langle \frac{1}{2}, \frac{1}{2} \right| s_+ \left| \frac{1}{2}, -\frac{1}{2} \right\rangle = \hbar \quad \text{and} \quad \left\langle \frac{1}{2}, -\frac{1}{2} \right| s_- \left| \frac{1}{2}, \frac{1}{2} \right\rangle = \hbar \tag{16.29}$$

or

$$s_+ = \hbar \begin{pmatrix} 0 & 1 \\ 0 & 0 \end{pmatrix} \quad \text{and} \quad s_- = \hbar \begin{pmatrix} 0 & 0 \\ 1 & 0 \end{pmatrix}. \tag{16.30}$$

With the definition of the ladder operators, $s_\pm = s_x \pm i s_y$ or $s_x = (s_+ + s_-)/2$ and $s_y = (s_+ - s_-)/2i$, it follows that

$$s_x = \frac{\hbar}{2} \begin{pmatrix} 0 & 1 \\ 1 & 0 \end{pmatrix} = \frac{\hbar}{2} \sigma_x \quad \text{and} \quad s_y = \frac{\hbar}{2} \begin{pmatrix} 0 & -i \\ i & 0 \end{pmatrix} = \frac{\hbar}{2} \sigma_y \tag{16.31}$$

i.e. two of the well-known spin matrices ($s$) or Pauli matrices ($\sigma$). We obtain the third one by using the equation $s_z |1/2, m\rangle = \hbar m |1/2, m\rangle$ directly as

---

[5] We omit here the distinction between $=$ and $\cong$.

$$s_z = \frac{\hbar}{2} \begin{pmatrix} 1 & 0 \\ 0 & -1 \end{pmatrix} = \frac{\hbar}{2}\sigma_z. \tag{16.32}$$

In the same way, we can obtain e.g. the matrix representation of the orbital angular momentum operator for $l = 1$; see the exercises.

We recall briefly some properties of the Pauli matrices (cf. Chap. 4, Vol. 1). With a view to a more convenient notation, one often writes $\sigma_1$, $\sigma_2$, $\sigma_3$ instead of $\sigma_x$, $\sigma_y$, $\sigma_z$. In this notation,[6]

$$[\sigma_i, \sigma_j] = 2i \sum_k \varepsilon_{ijk}\sigma_k; \; \{\sigma_i, \sigma_j\} = 2\delta_{ij}; \; \sigma_i^2 = 1; \; \sigma_i\sigma_j = i \sum_k \varepsilon_{ijk}\sigma_k \tag{16.33}$$

holds. Finally, we note that every $2 \times 2$-matrix can be represented as a linear combination of the three Pauli matrices and the unit matrix.

## 16.4 Orbital Angular Momentum: Spatial Representation of the Eigenfunctions

The eigenvalues of the orbital angular momentum are integers. For the position representation of the eigenfunctions, it is advantageous to use spherical coordinates. In these coordinates, we find that[7]:

$$\mathbf{l}^2 = -\hbar^2 \left[ \frac{1}{\sin\vartheta} \frac{\partial}{\partial\vartheta} \left( \sin\vartheta \frac{\partial}{\partial\vartheta} \right) + \frac{1}{\sin^2\vartheta} \frac{\partial^2}{\partial\varphi^2} \right]; l_z = \frac{\hbar}{i} \frac{\partial}{\partial\varphi}. \tag{16.34}$$

The eigenvalue problem (16.11) is written as

$$\left[ \frac{1}{\sin\vartheta} \frac{\partial}{\partial\vartheta} \left( \sin\vartheta \frac{\partial}{\partial\vartheta} \right) + \frac{1}{\sin^2\vartheta} \frac{\partial^2}{\partial\varphi^2} \right] Y_l^m(\vartheta, \varphi) = -l(l+1) Y_l^m(\vartheta, \varphi)$$

$$\frac{\partial}{\partial\varphi} Y_l^m(\vartheta, \varphi) = im Y_l^m(\vartheta, \varphi), \tag{16.35}$$

or, more compactly, as

$$\mathbf{l}^2 Y_l^m(\theta, \varphi) = \hbar^2 l(l+1) Y_l^m(\theta, \varphi) \text{ with } l = 0, 1, 2, 3, \ldots$$
$$l_z Y_l^m(\theta, \varphi) = \hbar m Y_l^m(\theta, \varphi) \text{ with } -l \leq m \leq l \tag{16.36}$$

---

[6]The anticommutator is defined as usual by $\{A, B\} = AB + BA$.
[7]We recall the equality

$$\nabla^2 = \frac{\partial^2}{\partial r^2} + \frac{2}{r} \frac{\partial}{\partial r} - \frac{\mathbf{l}^2}{\hbar^2 r^2}$$

See also Appendix D, Vol. 1.

where the functions $Y_l^m (\vartheta, \varphi)$ are the eigenfunctions of the orbital angular momentum in the position representation[8]; they are called *spherical functions* (or spherical harmonics). The number $l$ is the *orbital angular momentum (quantum) number*, $m$ the *magnetic* (or directional) *(quantum) number*.

With the separation *ansatz* $Y_l^m (\vartheta, \varphi) = \Theta (\vartheta) \, \Phi (\varphi)$, we obtain

$$\frac{\partial}{\partial \varphi} \Phi (\varphi) = im \Phi (\varphi)$$

$$\left[ \frac{1}{\sin \vartheta} \frac{\partial}{\partial \vartheta} \left( \sin \vartheta \frac{\partial}{\partial \vartheta} \right) - \frac{m^2}{\sin^2 \vartheta} + l \, (l+1) \right] \Theta (\vartheta) = 0. \tag{16.37}$$

The solutions of the first equation are well-known special functions (associated Legendre functions). The solutions of the second equation can immediately be written down as $\Phi (\varphi) = e^{im\varphi}$.

We will not deal with the general form of the spherical functions (for details see Appendix B, Vol. 2), but just note here some important features as well as the simplest cases. The spherical harmonics form a CONS. They are orthonormal

$$\int\limits_0^\pi d\vartheta \sin \vartheta \int\limits_0^{2\pi} d\varphi \left[ Y_l^m (\vartheta, \varphi) \right]^* Y_{l'}^{m'} (\vartheta, \varphi) = \delta_{ll'} \delta_{mm'} \tag{16.38}$$

and complete

$$\sum_{l=0}^\infty \sum_{m=-l}^l Y_l^m (\vartheta, \varphi) \left[ Y_l^m (\vartheta', \varphi') \right]^* = \frac{\delta \left( \vartheta - \vartheta' \right) \delta \left( \varphi - \varphi' \right)}{\sin \vartheta}. \tag{16.39}$$

With the notation $\Omega = (\vartheta, \varphi)$ for the solid angle and $d\Omega = \sin \vartheta d\vartheta d\varphi$ (also written as $d^2 \hat{r}$ or $d\hat{r}$, see Appendix D, Vol. 1) for its differential element, we can write the orthogonality relation as

$$\int \left[ Y_l^m (\vartheta, \varphi) \right]^* Y_{l'}^{m'} (\vartheta, \varphi) \, d\Omega = \delta_{ll'} \delta_{mm'}. \tag{16.40}$$

Because of the completeness of the spherical harmonics, we can expand any (sufficiently well-behaved) function $f (\vartheta, \varphi)$ in terms of them:

$$f (\vartheta, \varphi) = \sum_{l,m} c_{lm} Y_l^m (\vartheta, \varphi) \tag{16.41}$$

with

---

[8] Other notations: $Y_l^m (\vartheta, \varphi) = Y_l^m (\hat{r}) = \langle \hat{r} \, | l, m \rangle$. $\hat{r}$ is the unit vector and is an abbreviation for the pair $(\vartheta, \varphi)$. Moreover, the notation $Y_{lm} (\vartheta, \varphi)$ is also common.

$$c_{lm} = \int Y_l^{m*}(\vartheta, \varphi)\, f(\vartheta, \varphi)\, d\Omega. \tag{16.42}$$

Such an expansion is called a *multipole expansion*; the contribution for $l = 0$ is called the monopole (term), for $l = 1$ the dipole, for $l = 2$ the quadrupole, and generally, that for $l = n$ the $2^n$-pole.

Finally, we write down the first spherical harmonics explicitly (more spherical functions may be found in Appendix B, Vol. 2, where there are also some graphs):

$$Y_0^0 = \frac{1}{\sqrt{4\pi}}; \; Y_1^0 = \sqrt{\frac{3}{4\pi}} \cos\vartheta; \; Y_1^1 = -\sqrt{\frac{3}{8\pi}} \sin\vartheta e^{i\varphi}$$

$$Y_2^0 = \sqrt{\frac{5}{16\pi}} \left(3\cos^2\vartheta - 1\right); \; Y_2^1 = -\sqrt{\frac{15}{8\pi}} \sin\vartheta \cos\vartheta e^{i\varphi}; \; Y_2^2 = \sqrt{\frac{15}{32\pi}} \sin^2\vartheta e^{2i\varphi}. \tag{16.43}$$

For negative $m$, the functions are given by

$$Y_l^{-m}(\vartheta, \varphi) = (-1)^m\, Y_l^{m*}(\vartheta, \varphi). \tag{16.44}$$

## 16.5  Addition of Angular Momenta

This section is intended to give a brief overview of the topic. Therefore, only some results are given, without derivation. In this section, we denote the angular-momentum operators by lower-case letters **j**.

The addition theorem for angular momentum states that: If one adds two angular momenta $\mathbf{j}_1$ and $\mathbf{j}_2$, then $j$, the quantum number of the total angular momentum $\mathbf{j} = \mathbf{j}_1 + \mathbf{j}_2$ can assume only one of the values[9]:

$$j_1 + j_2, \; j_1 + j_2 - 1, \; j_1 + j_2 - 2, \ldots, \; |j_1 - j_2| \tag{16.45}$$

while the projections onto the $z$-axis are added directly:

$$m = m_1 + m_2. \tag{16.46}$$

Total angular momentum states $|jm; j_1 j_2\rangle$ can be obtained from the individual states $|j_1, m_1\rangle$ and $|j_2, m_2\rangle$ through the relation:

$$|jm; j_1 j_2\rangle = \sum_{m_1+m_2=m} \langle j_1 j_2 m_1 m_2 | jm\rangle \, |j_1, m_1\rangle \, |j_2, m_2\rangle \tag{16.47}$$

---

[9]Intuitively-clear reason: If the two angular momentum vectors are parallel to each other, then their angular-momentum quantum numbers are added to give $j = j_1 + j_2$. For other arrangements, $j$ is smaller. The smallest value is $|j_1 - j_2|$, because of $j \geq 0$.

where of course $|j_1 - j_2| \leq j \leq j_1 + j_2$ has to be satisfied. The numbers $\langle j_1 j_2 m_1 m_2 \,|\, jm \rangle$ are called *Clebsch–Gordan coefficients*[10]; they are real and are tabulated in relevant works on the angular momentum in quantum mechanics.[11] The inverse of the last equation is

$$|j_1, m_1\rangle \,|j_2, m_2\rangle = \sum_{j=|j_1-j_2|}^{j_1+j_2} \langle j_1 j_2 m_1 m_2 \,|\, jm \rangle \,|jm; j_1 j_2\rangle \,. \qquad (16.48)$$

As an application, we consider the spin-orbit coupling, i.e. the coupling of the orbital angular momentum **l** of electrons with their spins **s**. The total angular momentum of the electron is

$$\mathbf{j} = \mathbf{l} + \mathbf{s} \qquad (16.49)$$

and its possible values are $j = l + \frac{1}{2}$ and $j = l - \frac{1}{2}$ (except for $l = 0$ in which case only $j = \frac{1}{2}$ occurs). Intuitively, this means that orbital angular momentum and spin are either parallel ($j = l + \frac{1}{2}$) or antiparallel ($j = l - \frac{1}{2}$).

The spin is a relativistic phenomenon and can, if one argues from the nonrelativistic SEq, at best be introduced heuristically. A clear-cut approach starts e.g. from the relativistically correct *Dirac equation* $i\hbar \frac{\partial}{\partial t} \psi = \left( c\boldsymbol{\alpha} \mathbf{p} + \beta mc^2 + e\phi \right) \psi = H_{Dirac} \psi$. Here, **p** is the three-dimensional momentum operator $\mathbf{p} = \frac{\hbar}{i} \nabla$, while $\beta$ and the three components of the vector $\boldsymbol{\alpha}$ are certain $4 \times 4$ matrices, the *Dirac matrices*.

---

[10]There are several different notations for these coefficients, e.g. $C^{j_1 j_2}_{m_1 m_2; jm}$. The so-called *3j-symbols* are related coefficients:

$$\begin{pmatrix} j_1 & j_2 & j \\ m_1 & m_2 & m \end{pmatrix} = (-1)^{j_1-j_2-m} \frac{1}{\sqrt{2j+1}} \langle j_1 j_2 m_1 m_2 \,|\, j-m \rangle \,.$$

The 3j-symbols are invariant against cyclic permutation:

$$\begin{pmatrix} j_1 & j_2 & j \\ m_1 & m_2 & m \end{pmatrix} = \begin{pmatrix} j & j_1 & j_2 \\ m & m_1 & m_2 \end{pmatrix}, \text{ etc.}$$

[11]The two conditions $|j_1 - j_2| \leq j \leq j_1 + j_2$ and $m = m_1 + m_2$ must be met in order that a Clebsch-Gordan coefficient is nonzero. The CGC satisfy the orthogonality relations

$$\sum_{m_1, m_2} \langle j_1 j_2 m_1 m_2 \,|\, jm \rangle \langle j_1 j_2 m_1 m_2 \,|\, j'm' \rangle = \delta_{jj'} \delta_{mm'}$$

$$\sum_{j,m} \langle j_1 j_2 m_1 m_2 \,|\, jm \rangle \langle j_1 j_2 m'_1 m'_2 \,|\, jm \rangle = \delta_{m_1 m'_1} \delta_{m_2 m'_2} \,.$$

A particular CGC can in principle be calculated using the ladder operators $j_{1\pm} + j_{2\pm}$, whereby one starts e.g. from $\langle j_1 j_2 j_1 j_2 \,|\, j_1 + j_2 j_1 + j_2 \rangle = 1$. In this way, one obtains for example

$$\begin{pmatrix} A & B & A+B \\ a & b & c \end{pmatrix} = (-1)^{A-B-c} \left[ \frac{(2A)!\,(2B)!\,(A+B+c)!\,(A+B-c)!}{(2A+2B+1)!\,(A+a)!\,(A-a)!\,(B+b)!\,(B-b)!} \right]^{\frac{1}{2}}.$$

Essentially, one performs an expansion of this equation in powers of $\left(\frac{v}{c}\right)^2$ and retains only the lowest term for the nonrelativistic description. Then it turns out that in the SEq, a spin-orbit term $F(r)\,\mathbf{l}\cdot\mathbf{s}$ appears (see also Chap. 19), with $F(r)$ a radially-symmetric function. Because of

$$(\mathbf{l}+\mathbf{s})^2 = \mathbf{j}^2 = \mathbf{l}^2 + 2\mathbf{l}\mathbf{s} + \mathbf{s}^2, \tag{16.50}$$

$\mathbf{l}\cdot\mathbf{s}$ has the values

$$\mathbf{l}\cdot\mathbf{s} = \frac{1}{2}\hbar^2\left[j\,(j+1) - l\,(l+1) - s\,(s+1)\right]. \tag{16.51}$$

It follows for the spin-orbit term in the SEq:

$$F(r)\,\mathbf{l}\cdot\mathbf{s} = \begin{cases} \frac{1}{2}\hbar^2 l\,F(r) & \quad j = l + \frac{1}{2} \\[2mm] -\frac{1}{2}\hbar^2\,(l+1)\,F(r) & \quad j = l - \frac{1}{2}. \end{cases} \quad \text{for} \tag{16.52}$$

Next, we briefly discuss the corresponding eigenfunctions and write down their explicit form. We start from the states for the spin $|sm_s\rangle$ and the orbital angular momentum $|Q; lm_l\rangle$, where $Q$ stands for possible additional quantum numbers (e.g. the principal quantum number of the hydrogen atom). The total angular momentum state is then described by

$$|Q; j, m_{j;}l\rangle; \quad j = l \pm \frac{1}{2} \text{ for } l \geq 1 \,; \, j = \frac{1}{2} \text{ for } l = 0 \tag{16.53}$$

and is composed of spin and orbital angular momentum states:

$$|Q; j, m_{j;}l\rangle = \sum_{m_l + m_s = m_j} \langle lsm_l m_s \,|\, jm_j\rangle |Q; lm_l\rangle\,|sm_s\rangle \tag{16.54}$$

Calculating the Clebsch-Gordan coefficients yields

$$|Q; j = l + 1/2, m_{j;}l\rangle$$
$$= \sqrt{\tfrac{l+m_j+1/2}{2l+1}}|Q; l, m_j - 1/2\rangle|s, 1/2\rangle + \sqrt{\tfrac{l-m_j+1/2}{2l+1}}|Q; l, m_j + 1/2\rangle|s, -1/2\rangle$$
$$|Q; j = l - 1/2, m_{j;}l\rangle$$
$$= \sqrt{\tfrac{l-m_j+1/2}{2l+1}}|Q; l, m_j - 1/2\rangle|s, 1/2\rangle + \sqrt{\tfrac{l+m_j+1/2}{2l+1}}|Q; l, m_j + 1/2\rangle|s, -1/2\rangle \tag{16.55}$$

with $m_j = l + 1/2, \ldots, -(l+1/2)$ in the upper line and $m_j = l - 1/2, \ldots, -(l-1/2)$ in the lower line. With the notation as column vectors, we obtain using

$$|s, 1/2\rangle \equiv |1/2, 1/2\rangle = \begin{pmatrix} 1 \\ 0 \end{pmatrix} \quad \text{and} \quad |s, -1/2\rangle \equiv |1/2, -1/2\rangle = \begin{pmatrix} 0 \\ 1 \end{pmatrix} \tag{16.56}$$

the explicit formulations

$$|Q; j = l + 1/2, m_j; l\rangle = \begin{pmatrix} \sqrt{\frac{l+m_j+1/2}{2l+1}} \,|Q; l, m_j - 1/2\rangle \\ \sqrt{\frac{l-m_j+1/2}{2l+1}} \,|Q; l, m_j + 1/2\rangle \end{pmatrix}$$

$$\text{with } m_j = l + 1/2, \dots, -(l + 1/2) \qquad (16.57)$$

and

$$|Q; j = l - 1/2, m_j; l\rangle = \begin{pmatrix} -\sqrt{\frac{l-m_j+1/2}{2l+1}} \,|Q; l, m_j - 1/2\rangle \\ \sqrt{\frac{l+m_j+1/2}{2l+1}} \,|Q; l, m_j + 1/2\rangle \end{pmatrix}$$

$$\text{with } m_j = l - 1/2, \dots, -(l - 1/2) . \qquad (16.58)$$

## 16.6   Exercises

1. For which $K, N, M$ are the spherical harmonics (in spherical coordinates)

$$f(\vartheta, \varphi) = \cos^K \vartheta \cdot \sin^M \vartheta \cdot e^{iN\varphi} \qquad (16.59)$$

   eigenfunctions of $\mathbf{l}^2$?
2. Write out the spherical harmonics for $l = 1$ using Cartesian coordinates, $x, y, z$.
3. Show that:

$$\mathbf{l} \cdot \hat{\mathbf{r}} = \hat{\mathbf{r}} \cdot \mathbf{l} = 0 \qquad (16.60)$$

4. Show that the components of $\mathbf{l}$ are Hermitian.
5. Show that for the orbital angular momentum, it holds that

$$[l_x, l_y] = i\hbar l_z; \quad [l_y, l_z] = i\hbar l_x; \quad [l_z, l_x] = i\hbar l_y. \qquad (16.61)$$

6. Show that $[A, BC] = B[A, C] + [A, B]C$ holds. Using this identity and the commutators $[l_x, l_y] = i\hbar l_z$ plus cyclic permutations, prove that $[l_x, \mathbf{l}^2] = 0$.
7. Show that:

$$[\mathbf{J}^2, J_\pm] = 0. \qquad (16.62)$$

8. We have seen in the text that

$$J_\pm |j, m\rangle = c_{j,m}^\pm |j, m \pm 1\rangle . \qquad (16.63)$$

Using

$$J_+J_-|j,m\rangle = \hbar^2 [j(j+1) - m(m-1)]|j,m\rangle$$
$$J_-J_+|j,m\rangle = \hbar^2 [j(j+1) - m(m+1)]|j,m\rangle, \qquad (16.64)$$

show that for the coefficients $c_{j,m}^\pm$,

$$c_{j,m}^\pm = \hbar\sqrt{j(j+1) - m(m\pm1)} \qquad (16.65)$$

holds.

9. Given the Pauli matrices $\sigma_k$,

   (a) Show (once more) that

   $$[\sigma_i, \sigma_j] = 2i\varepsilon_{ijk}\sigma_k; \quad \{\sigma_i, \sigma_j\} = 2\delta_{ij}; \quad \sigma_i^2 = 1; \quad \sigma_i\sigma_j = i\varepsilon_{ijk}\sigma_k; \qquad (16.66)$$

   (b) Prove that

   $$(\sigma A)(\sigma B) = AB + i\sigma(A \times B) \qquad (16.67)$$

   where $\sigma$ is the vector $\sigma = (\sigma_1, \sigma_2, \sigma_3)$ and $A, B$ are three-dimensional vectors;

   (c) Show that every 2x2 matrix can be expressed as a linear combination of the three Pauli matrices and the unit matrix.

10. Given the orbital angular momentum operator $l$ and the spin operator $s$, show that $[l_z, s \cdot l] \neq 0$; $[s_z, s \cdot l] \neq 0$; $[l_z + s_z, s \cdot l] = 0$.

11. The ladder operators for a generalized angular momentum are given as $J_\pm = J_x \pm iJ_y$.

   (a) Show that $[J_z, J_+] = \hbar J_+$, $[J_z, J_-] = -\hbar J_-$, $[J_+, J_-] = 2\hbar J_z$, as well as $J^2 = \frac{1}{2}(J_+J_- + J_-J_+) + J_z^2$.

   (b) Show that it follows from the last equation that:

   $$J_+J_- = J^2 - J_z(J_z - \hbar); \quad J_-J_+ = J^2 - J_z(J_z + \hbar) \qquad (16.68)$$

   and hence

   $$J_+J_-|j,m\rangle = \hbar^2 [j(j+1) - m(m-1)]|j,m\rangle$$
   $$J_-J_+|j,m\rangle = \hbar^2 [j(j+1) - m(m+1)]|j,m\rangle. \qquad (16.69)$$

   (c) Show that from the last two equations, it follows that:

   $$j(j+1) - m(m-1) = (j-m)(j+m+1) \geq 0$$
   $$j(j+1) - m(m+1) = (j+m)(j-m+1) \geq 0 \qquad (16.70)$$

**Fig. 16.1** Rotation about an
axis $\hat{a}$

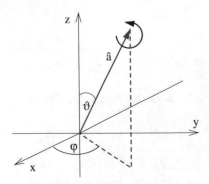

and hence

$$- j \le m \le j. \tag{16.71}$$

12. What is the matrix representation of the orbital angular momentum for $l = 1$?
13. Consider the orbital angular momentum $l = 1$. Express the operator $e^{-i\alpha L_z/\hbar}$ as
    sum over dyadic products (representation-free). Specify this for the bases

$$|1, 1\rangle \cong \begin{pmatrix} 1 \\ 0 \\ 0 \end{pmatrix}; \ |1, 0\rangle \cong \begin{pmatrix} 0 \\ 1 \\ 0 \end{pmatrix}; \ |1, -1\rangle \cong \begin{pmatrix} 0 \\ 0 \\ 1 \end{pmatrix} \tag{16.72}$$

and

$$|1, 1\rangle \cong \pm\frac{1}{\sqrt{2}} \begin{pmatrix} 1 \\ i \\ 0 \end{pmatrix}; \ |1, 0\rangle \cong \begin{pmatrix} 0 \\ 0 \\ 1 \end{pmatrix}; \ |1, -1\rangle \cong \mp\frac{1}{\sqrt{2}} \begin{pmatrix} 1 \\ -i \\ 0 \end{pmatrix}. \tag{16.73}$$

14. Calculate the term

$$e^{-i\frac{\gamma\hat{a}L}{\hbar}} = e^{-i\gamma\hat{a}l} \tag{16.74}$$

for the orbital angular momentum $l = 1$ and the basis (16.73).[12] $\hat{\mathbf{a}}$ is the rotation axis (a unit vector), $\gamma$ the rotation angle. For reasons of economy, use the
'simplified' angular momentum $\mathbf{l} = \mathbf{L}/\hbar$, i.e. the theoretical units system.

(a) Express the rotations around the $x$-, $y$- and $z$-axis as matrices.
(b) Express the rotations about an axis $\hat{a}$ with rotation angle $\gamma$ as matrices (the
    angles in spherical coordinates are $\vartheta$ and $\varphi$; see Fig. 16.1).

---

[12]Of course, all the calculations may also be performed representation-free.

# Chapter 17
# The Hydrogen Atom

Up to now, we have discussed *one* quantum object moving in a potential. We now begin to consider more than one quantum object. In this chapter, we address the simplest case, namely two quantum objects whose interaction depends only on their distance. It turns out that such systems are equivalent to a one-body problem.

The hydrogen atom is a concrete example of the general case of a two-body problem, in which two interacting bodies are considered without any external forces. The total energy of this system is composed of the kinetic energies of the two bodies and their potential energy, i.e. the energy of interaction $V$ between them[1]:

$$E = E_{\text{kin}} + E_{\text{pot}} = \frac{\mathbf{p}_1^2}{2m_1} + \frac{\mathbf{p}_2^2}{2m_2} + V. \tag{17.1}$$

We assume in the following that the potential depends only on the *relative coordinate* $\mathbf{r}_1 - \mathbf{r}_2$, i.e. that we have $V = V(\mathbf{r}_1 - \mathbf{r}_2)$.[2] Under this assumption, one can reduce the problem to an equivalent one-body problem. If we further specialize to the Coulomb interaction of point charges, this leads quantum mechanically to the familiar form of the hydrogen spectrum.

To simplify the problem, we introduce new coordinates, namely the *center-of-mass coordinate* $\mathbf{R}$ as well as *the relative coordinate* $\mathbf{r}$:

$$\mathbf{R} = \frac{m_1\mathbf{r}_1 + m_2\mathbf{r}_2}{m_1 + m_2} \text{ and } \mathbf{r} = \mathbf{r}_1 - \mathbf{r}_2$$
$$\mathbf{R} = (X, Y, Z) \text{ and } \mathbf{r} = (x, y, z). \tag{17.2}$$

---

[1] The relativistic case is treated in Appendix F, Vol. 2.

[2] Hence, the interaction of the two bodies does not depend on their absolute position in space, but only on their relative positions w.r.t. each other.

© Springer Nature Switzerland AG 2018
J. Pade, *Quantum Mechanics for Pedestrians 2*, Undergraduate Lecture
Notes in Physics, https://doi.org/10.1007/978-3-030-00467-5_17

Furthermore, we define the *total mass* $M$ and the *reduced mass* $\mu$:

$$M = m_1 + m_2 \quad \text{and} \quad \mu = \frac{m_1 m_2}{m_1 + m_2}. \tag{17.3}$$

One can now first perform the transformation of (17.1) to $\mathbf{r}$ and $\mathbf{R}$ classically, and then go to quantum mechanics; or first go to quantum mechanics and then perform the transformation. The results are of course identical. The calculation can be found in Appendix E, Vol. 2. One obtains:

$$E = \frac{\mathbf{P}^2}{2M} + \frac{\mathbf{p}^2}{2\mu} + V(\mathbf{r}), \tag{17.4}$$

where $\mathbf{P} = M\dot{\mathbf{R}}$ is the center-of-mass momentum and $\mathbf{p} = \mu\dot{\mathbf{r}}$ the relative momentum. We now go into the *center-of-mass system*. Since here $\mathbf{P} = 0$, we have

$$E = \frac{\mathbf{p}^2}{2\mu} + V(\mathbf{r}) \text{ in the center-of-mass system} \tag{17.5}$$

or

$$E\psi(\mathbf{r}) = -\frac{\hbar^2}{2\mu}\nabla^2\psi(\mathbf{r}) + V(\mathbf{r})\,\psi(\mathbf{r})\,. \tag{17.6}$$

Equation (17.6) is the *equivalent one-body problem*. It differs from the problem of a body of mass $m$ in a potential $V$ only insofar as here not the mass $m$, but rather the reduced mass $\mu$ enters.[3] By the way, a similar efficient simplification does not exist for three or more bodies.

## 17.1   Central Potential

The general treatment of the equivalent one-body problem is quite complex. Therefore we now concentrate on the important special case of a *central potential*, in which the potential is radially symmetric, $V(\mathbf{r}) = V(r)$. In view of this symmetry, we choose spherical coordinates $(r, \vartheta, \varphi)$ instead of Cartesian coordinates. Then the Laplacian is given by (see also Appendix D, Vol. 1)[4]:

---

[3] If one of the bodies has a much greater mass than the other ($m_1 \gg m_2$, e.g. hydrogen atom), we have $\mu \approx m_2$; if the two masses are equal ($m_1 = m_2$, e.g. positronium), we have $\mu = m_1/2$.

[4] An even more compact notation (which we however will not need in the following) can be formulated using the *radial momentum*

$$p_r = \frac{\hbar}{i}\frac{1}{r}\frac{\partial}{\partial r}r = \frac{\hbar}{i}\left(\frac{\partial}{\partial r} + \frac{1}{r}\right),$$

namely

$$\nabla^2 = \frac{\partial^2}{\partial r^2} + \frac{2}{r}\frac{\partial}{\partial r} - \frac{l^2}{\hbar^2 r^2} \tag{17.7}$$

with the angular-momentum operator (see Chap. 16)

$$l^2 = -\hbar^2 \left[ \frac{1}{\sin\vartheta}\frac{\partial}{\partial\vartheta}\left(\sin\vartheta\frac{\partial}{\partial\vartheta}\right) + \frac{1}{\sin^2\vartheta}\frac{\partial^2}{\partial\varphi^2} \right]. \tag{17.8}$$

Thus, for the Hamiltonian or the SEq of the central potential, it follows that:

$$H = -\frac{\hbar^2}{2\mu}\left(\frac{\partial^2}{\partial r^2} + \frac{2}{r}\frac{\partial}{\partial r}\right) + \frac{l^2}{2\mu r^2} + V(r) \quad \text{or} \quad E\psi(\mathbf{r}) = H\psi(\mathbf{r}). \tag{17.9}$$

As we have seen in the exercises for Chap. 9, Vol. 1, we have $[H, l] = 0$ for a radially-symmetric potential $V(r)$. Because of $[l^2, l_z] = 0$, the three Hermitian operators $H$, $l^2$ and $l_z$ have common eigenfunctions; more about this issue below.

Due to $V = V(r)$, the angles $\vartheta$ and $\varphi$ occur only in the angular-momentum operator. This suggests again a separation *ansatz* for $\psi(\mathbf{r})$. Because of the occurrence of $l^2$, we choose an expansion in terms of spherical harmonics (multipole expansion, see Chap. 16). For a given energy $E$, we obtain:

$$\psi(\mathbf{r}) = \sum_{l,m} c_{lm} R_{E;lm}(r) Y_l^m(\vartheta, \varphi) \tag{17.10}$$

where we characterize the dependence on the energy via the index[5] $E$. Inserting this expression into (17.9) gives, with $l^2 Y_l^m(\vartheta, \varphi) = \hbar^2 l(l+1) Y_l^m(\vartheta, \varphi)$, the equation

$$\sum_{l,m} c_{lm} E R_{E;lm}(r) Y_l^m(\vartheta, \varphi)$$

$$= \sum_{l,m} c_{lm} \left[ -\frac{\hbar^2}{2\mu}\left(\frac{\partial^2}{\partial r^2} + \frac{2}{r}\frac{\partial}{\partial r}\right) + \frac{\hbar^2 l(l+1)}{2\mu r^2} + V(r) \right] R_{E;lm}(r) Y_l^m(\vartheta, \varphi). \tag{17.11}$$

Because of the orthonormality of the spherical harmonics,[6] we obtain the *radial equation*:

---

$$\mathbf{p}^2 = p_r^2 + \frac{l^2}{r^2} \quad \text{or} \quad \nabla_\mathbf{r}^2 = -\frac{p_r^2}{\hbar^2} - \frac{l^2}{\hbar^2 r^2}.$$

It holds that $[r, p_r] = i\hbar$.

[5] The dependence on $\mu$ and on the details of the potential is not noted in general.

[6] We have $\int Y_l^{m*}(\vartheta, \varphi) Y_L^M(\vartheta, \varphi)\, d\Omega = \delta_{Ll}\delta_{Mm}$.

$$-\frac{\hbar^2}{2\mu}\left(\frac{d^2}{dr^2}+\frac{2}{r}\frac{d}{dr}\right)R_{E;lm}(r)+\left(\frac{\hbar^2 l(l+1)}{2\mu r^2}+V(r)\right)R_{E;lm}(r)=ER_{E;lm}(r).$$
$$(17.12)$$

We see that the magnetic quantum number $m$ does not appear in this equation. Thus we can omit the index $m$, i.e. we write $R_{E;l}(r)$ instead of $R_{E;lm}(r)$. We also see that from the start, a degeneracy exists with respect to $m$; due to $-l \leq m \leq l$, the degree of degeneracy is $g_l = 2l + 1$.

In order to simplify, we define using $\frac{\partial^2}{\partial r^2}+\frac{2}{r}\frac{\partial}{\partial r}=\frac{1}{r}\frac{\partial^2}{\partial r^2}r$:

$$R_{E;l}(r)=\frac{u_{E;l}(r)}{r},\qquad(17.13)$$

and obtain for the *radial wavefunction* $u_{E;l}(r)$ the equation

$$-\frac{\hbar^2}{2\mu}\frac{d^2 u_{E;l}(r)}{dr^2}+\left(\frac{\hbar^2 l(l+1)}{2\mu r^2}+V(r)\right)u_{E;l}(r)=Eu_{E;l}(r).\qquad(17.14)$$

The *centrifugal term* (also called *centrifugal barrier*) caused by the angular momentum is usually combined with the potential $V(r)$ to give the *effective potential*:

$$V_{\text{eff}}=V(r)+V_{\text{centrifugal}}=V(r)+\frac{\hbar^2 l(l+1)}{2\mu r^2}.\qquad(17.15)$$

We thus have reduced the original problem (two-body problem) with its 6 independent coordinates $\mathbf{r}_1$, $\mathbf{r}_2$ to an ordinary differential equation with *one* independent variable $r$.[7]

Some remarks:

1. This is a one-dimensional problem, which is formally similar to the previously discussed one-dimensional problems. But there is one important difference: In the cases treated previously (square-well potential, etc.), we had in principle $-\infty \leq x \leq \infty$, while for the independent variable $r$, we have $0 \leq r \leq \infty$.
2. In order to be square integrable, the wavefunction must obey certain conditions at $r = \infty$ and $r = 0$. As can be shown (see exercises), it must obey

$$u_{E;l}(r)\underset{r\to\infty}{\sim}r^{\alpha}\text{ with }\alpha<-\frac{1}{2}.\qquad(17.16)$$

In other words, $u_{E;l}$ must vanish at infinity faster than $1/\sqrt{r}$. For $r = 0$, we have

$$u_{E;l}(r)\underset{r\to 0}{\sim}r^{l+1}.\qquad(17.17)$$

---

[7]Reductions of this kind make life considerably easier, both in theoretical and in computational terms.

The radial wavefunction thus vanishes at the origin.[8] To make this perhaps more plausible, one can argue that nothing can change in (17.14) for $r \geq 0$ if one inserts an infinitely high potential at $r = 0$. It follows, as we have previously shown, that the wavefunction must vanish at the origin.

3. We have seen that the radial equation (17.12) is independent of $m$. This means that also the energy depends (if at all) only on $l$, but not on $m$. Hence, we have for a given $l$ a degree of degeneracy of the energy, $g_l = 2l + 1$. This degeneracy is called *essential degeneracy* and exists for all potentials $V(r)$. Depending on the potential, additional degeneracies may occur which are called *accidental degeneracies* .

4. To denote the wavefunctions for a given angular momentum, the following terms are common for historical reasons: $l = 0 : s$ wave, $l = 1 : p$ wave, $l = 2 : d$ wave, $l = 3 : f$ wave (s as in 'sharp', p as in 'principal', d as in 'diffuse', and f as in 'fundamental'). Thereafter, the terms continue in alphabetical order: $l = 4 : g$ wave, etc.

## 17.2 The Hydrogen Atom

There are no closed-form solutions of the radial equation (17.14) for arbitrary $V(r)$. To continue, we need to specify $V(r)$; we choose the Coulomb interaction.[9] The physical system consists therefore of two pointlike objects with *opposite* charges, $Z_1 e$ and $-Z_2 e$ (and the masses $m_i$):

$$V(r) = -\frac{1}{4\pi\varepsilon_0}\frac{Z_1 Z_2 e^2}{r} = -\frac{1}{4\pi\varepsilon_0}\frac{Z e^2}{r} = -\frac{\gamma}{r}; \gamma > 0. \qquad (17.18)$$

One speaks usually of *hydrogenic atoms* to signify that not only hydrogen itself (one electron plus one proton, $Z = 1$), but also other systems (a positron plus an electron, multiply-charged nuclei, etc.) can be described by this approach. The effective potential reads:

$$V_{\text{eff}} = V(r) + V_{\text{centrifugal}} = -\frac{\gamma}{r} + \frac{\hbar^2 l(l+1)}{2\mu r^2}. \qquad (17.19)$$

For $l = 0$ (zero angular momentum), $V_{\text{eff}}$ has a negative pole at the origin and then rises monotonically with increasing $r$ towards zero. For $l > 0$, the effective potential has a positive pole at the origin, a zero at $r_0 = \hbar^2 l(l+1)/(2\mu\gamma)$, and a minimum $V_{\text{eff min}} = -\mu\gamma^2/\left\{l(l+1)\hbar^2\right\}$ at $2r_0$; from there on, the potential increases

---

[8] The irregular solution behaves at the origin as $\sim r^{-l}$.

[9] We repeat the remark of Chap. 1, Vol. 1, that the term $V$ in the SEq, although in fact the potential energy, is usually called just 'potential'. In electrodynamics, $V(r) = \frac{1}{4\pi\varepsilon_0}\frac{q}{r}$ is the potential of a point charge $q$. Since in quantum mechanics $V(r)$ denotes the potential energy, there must be *two* charges.

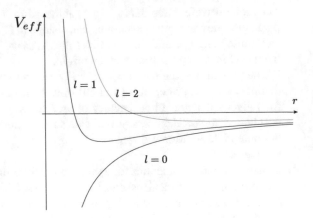

**Fig. 17.1** Effective potential (hydrogen atom, schematic representation)

for increasing $r$ monotonically to zero. In other words, the effective potential is (provided $l > 0$) repelling for $r < r_0$, and for $r > r_0$, it forms a 'well' which becomes flatter with increasing angular momentum (i.e. the bonding of the electron becomes weaker); see Fig. 17.1.

A parenthesis: If one substitutes in $r_0$ the values for the hydrogen atom, this quantity for $l = 1$ is (almost) identical with the *Bohr radius*,[10] commonly denoted as $a_0$:

$$a_0 = 4\pi\varepsilon_0 \frac{\hbar^2}{m_{\text{el.}}q_{\text{el.}}^2} = 0.529177249 \times 10^{-10}\,\text{m} \approx 0.5\,\text{Å}.$$

Almost, because in the equations of the equivalent one-body problem, the reduced mass and not the electron mass occurs. The deviation is of the order of $m_{\text{el.}}/m_{\text{proton}}$, i.e. about $1/2000$. However, some books ignore this distinction.

Since in the following, we are interested only in the bound states, we consider only negative energies:

$$E = -|E|. \tag{17.20}$$

Then (17.14) can be written as

$$\frac{\mathrm{d}^2 u_{E;l}(r)}{\mathrm{d}r^2} - \left( \frac{l(l+1)}{r^2} - \frac{2\mu\gamma}{\hbar^2 r} \right) u_{E;l}(r) = \frac{2\mu}{\hbar^2}\,|E|\,u_{E;l}(r). \tag{17.21}$$

A detailed solution of this differential equation (step by step) is given in Appendix F, Vol. 2.[11] For this method, one uses essentially a power series *ansatz* for the solution.

---

[10]Somewhat misleadingly sometimes also called the *radius of the hydrogen atom* (actually the hydrogen atom does not have a well-defined radius).

[11]Another possibility to find a solution would be e.g. to look up a book on 'special functions', such as Abramowitz, and to convince oneself that the solution can be formulated in terms of special functions, here the Laguerre polynomials; see below. An entirely different option is the algebraic approach, keyword Lenz vector (see Appendix G, Vol. 2).

One can then show that $u_{E;l}(r)$ remains bounded in the case that the power series terminates, i.e. it is not an infinite series, but a finite polynomial in $r$. This cut-off condition has the consequence that the energies of bound states can have only the following form:

$$E_n = -\frac{\mu\gamma^2}{2\hbar^2}\frac{1}{n^2} = -\frac{\mu}{2\hbar^2}\left(\frac{Ze^2}{4\pi\varepsilon_0}\right)^2\frac{1}{n^2}; n = 1, 2, \ldots \qquad (17.22)$$

with

$$0 \leq n - l - 1 \qquad (17.23)$$

The expression (17.22) is also called the *Balmer formula*, and the number $n = 1, 2, 3, \ldots$ is the *principal quantum number*. Particularly in spectroscopy, the levels with $n = 1$ are also referred to as the K shell; with $n = 2$, as the L shell; $n = 3$ as the M shell, and so on.

Because of the dependence of the energy (17.22) on the principal quantum number, one commonly uses the index $n$ (instead of $E$ or $R_{E;l} \rightarrow R_{nl}$) from the start. Thus, one writes the total wavefunction for a certain energy $E_n$ as

$$\psi(\mathbf{r}) = \sum_{l,m} c_{lm} R_{nl}(r) Y_l^m(\vartheta, \varphi) \qquad (17.24)$$

and the radial equations as

$$-\frac{\hbar^2}{2\mu}\left(\frac{d^2}{dr^2} + \frac{2}{r}\frac{d}{dr}\right)R_{nl}(r) + \left(\frac{\hbar^2 l(l+1)}{2\mu r^2} + V(r)\right)R_{nl}(r) = E_n R_{nl}(r)$$

$$-\frac{\hbar^2}{2\mu}\frac{d^2 u_{nl}(r)}{dr^2} + \left(\frac{\hbar^2 l(l+1)}{2\mu r^2} + V(r)\right)u_{nl}(r) = E_n u_{nl}(r); u_{nl}(r) = r R_{nl}(r).$$

$$(17.25)$$

In order to write (17.22) more compactly, there are various abbreviations. One is the *Rydberg constant* $R_M$[12]

$$R_M = \frac{e^4\mu}{2\hbar^2(4\pi\varepsilon_0)^2} = \frac{e^4 m_e}{2\hbar^2(4\pi\varepsilon_0)^2}\frac{1}{1 + \frac{m_e}{M}} \qquad (17.26)$$

where $m_e$ is the mass of the electron and $M$ the mass of the positive partner (the nucleus or, in the case of positronium, the positron, etc.). For the hydrogen atom,[13] it follows that:

---

[12] The convention that the same symbol $R$ is used for the radial function and for the Rydberg constant may not be very clever didactically, but it is well established.

[13] Because of the dependence on the reduced mass, the Rydberg constant of positronium has only half the value of that for hydrogen. Also, for 'normal' and heavy hydrogen, the Rydberg constants differ due to the dependence on the masses. This allows one to determine spectroscopically the proportions of the two isotopes. Incidentally, the notation

$$E_n = -\frac{R_H}{n^2} \text{ with } R_H \approx 13.6\text{eV}. \tag{17.27}$$

The ionization energy of the hydrogen atom thus equals the Rydberg constant and has the value $\approx 13.6$ eV.

Another important constant that occurs in this context is the (Sommerfeld) *fine structure constant* $\alpha$ :

$$\alpha = \frac{e^2}{4\pi\varepsilon_0 \cdot \hbar c} \approx \frac{1}{137.036}. \tag{17.28}$$

It is related to the Rydberg constant by

$$R_M = \frac{\mu c^2}{2}\alpha^2. \tag{17.29}$$

Let us look more closely at the energy spectrum. Evidently, the energy levels depend only on the principal quantum number $n$, but not on $l$ or $m$. In addition to the essential degeneracy which occurs for every central potential ($g_l = 2l + 1$), there is an accidental degeneracy which is a 'speciality' of the $1/r$ potential.[14] The total degree of degeneracy is given by

$$g_n = \sum_{l=0}^{n-1}(2l + 1) = n^2. \tag{17.30}$$

The term diagram[15] looks roughly as in Fig. 17.2.

Finally, we want to look a little closer at the radial functions. They depend on the quantum numbers $n$ and $l$ and are given by

$$u_{nl}(r) = r \cdot R_{nl} = r \cdot \sqrt{\frac{(n - l - 1)! \, (2\kappa)^3}{2n\,((n + l)!)^3}} \, (2\kappa r)^l \, e^{-\kappa r} L_{n+l}^{2l+1}\,(2\kappa r) \tag{17.31}$$

with

$$\kappa = \frac{\mu\gamma}{\hbar^2 n} = \frac{Z}{na_0}. \tag{17.32}$$

The functions $L_{n+l}^{2l+1}\,(y)$ are the *associated Laguerre polynomials*; they can be calculated from

---

$$R_\infty = \frac{m_e e^4}{2\hbar^2\,(4\pi\varepsilon_0)^2}$$

is also common; it refers to an infinite nuclear mass.

[14]The reason is that a further conserved quantity exists: the Lenz vector (see Appendix G, Vol. 2).
[15]If one describes the interaction more realistically, the term diagram changes in subtle ways (cf. Chap. 19).

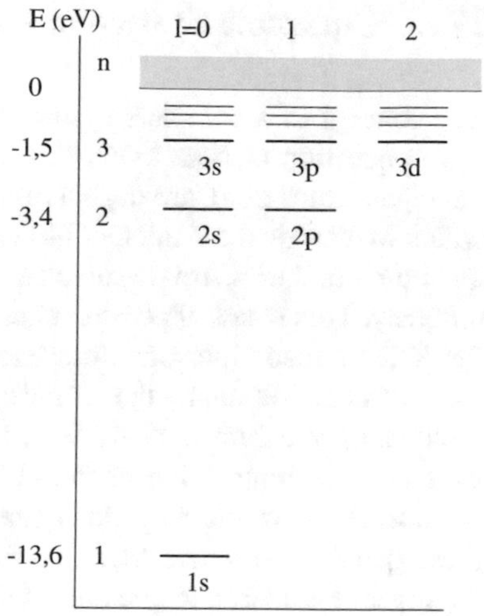

**Fig. 17.2** Spectrum of the Hydrogen atom (schematically). The yellow part indicates the continuous spectrum. Scales are not preserved

$$L_r^s (y) = \left(-\frac{d}{dy}\right)^s e^y \left(\frac{d}{dy}\right)^r e^{-y} y^r. \tag{17.33}$$

We give the first radial functions explicitly; further properties and graphic representations can be found in Appendix B, Vol. 2:

$$\text{K shell, s orbital: } R_{10}(r) = 2 \left(\frac{Z}{a_0}\right)^{\frac{3}{2}} e^{-\frac{Zr}{a_0}}$$

$$\text{L shell, s orbital: } R_{20}(r) = 2 \left(\frac{Z}{2a_0}\right)^{\frac{3}{2}} \left(1 - \frac{Zr}{2a_0}\right) e^{-\frac{Zr}{2a_0}}$$

$$\text{L shell, p orbital: } R_{21}(r) = \frac{1}{\sqrt{3}} \left(\frac{Z}{2a_0}\right)^{\frac{3}{2}} \frac{Zr}{a_0} e^{-\frac{Zr}{2a_0}}. \tag{17.34}$$

As mentioned above, the total wavefunction for a given energy eigenvalue reads

$$\psi(\mathbf{r}) = \sum_{l,m} c_{lm} R_{nl}(r) Y_l^m(\vartheta, \varphi). \tag{17.35}$$

The probability density for the quantum numbers $(n, l, m)$ is given by $\rho = |R_{nl}|^2 |Y_l^m|^2$. The probability $w(r_1, r_2)$ in a spherical shell with inner and outer radius $r_1$ and $r_2$ is given by

$$w(r_1, r_2) = \int_{r_1}^{r_2} |R_{nl}|^2 r^2 dr \int |Y_l^m|^2 d\Omega = \int_{r_1}^{r_2} |R_{nl}|^2 r^2 dr. \tag{17.36}$$

Therefore, one refers to the term $r^2 |R_{nl}|^2 = |u_{nl}|^2$ also as the *radial probability density*.

## 17.3    Complete System of Commuting Observables

The concept of a complete system of commuting observables (CSCO) is related
to the question of how to classify states by means of quantum numbers, even if
the eigenvalues of $H$ are degenerate.[16] As an example we take the hydrogen atom
which we have just considered: Its energy levels, which are classified by the principal
quantum number $n$, are degenerate. Thus, fixing $n$ is not sufficient to define a state
uniquely. This means that we cannot consider only the observable $H$ to decompose
the Hilbert space into one-dimensional subspaces. We can take a step further, if
we consider additionally the orbital angular momentum operator; the orbital angular
momentum quantum number $l = 0, 1, \ldots, n - 1$ further specifies the states and gives
us a finer decomposition of the Hilbert space. Finally, we can take into account
the operator $l_z$ or the magnetic quantum number $m$.[17] With the specification of all
these quantum numbers $(n, l, m)$, the state is uniquely described. In other words,
the eigenvectors of a single one of these observables do not form a complete basis
set of the Hilbert space—one has to consider all three operators simultaneously.
We point out again that commuting observables have common eigenfunctions, see
Chap. 27, Vol. 1. From this fact, the name is derived: 'complete' means that the
common eigenvectors of the observables constitute a basis for the complete Hilbert
space. In other words, by means of the CSCO we can decompose the Hilbert space
into one-dimensional subspaces.

A general formulation is as follows: A set of observables forms a CSCO if
(1) all observables commute *pairwise*[18] and (2) the specification of the eigenvalues of all these operators is sufficient to determine a common eigenvector (up to a
factor), whereby these common eigenvectors are not degenerate. One refers implic-
itly to a 'minimum set' of observables, i.e. those sets that are no longer a CSCO if
one removes one observable. For a given physical system, there are in general several
different CSCO's.

From the hydrogen atom, we know that in order to characterize a state the specification of the quantum numbers (=eigenvalues) is sufficient. Thus for observables
$A, B, C, \ldots$, one often denotes the ket as $|a_p, b_q, c_r, \ldots\rangle$ or, if the position of
the quantum numbers is clear, simply $|p, q, r, \ldots\rangle$, i.e. for the hydrogen atom for
example $|n, l, m\rangle$. In the spatial representation, these are of course the functions
$R_{nl}(r) Y_l^m (\vartheta, \varphi)$.

---

[16] If they are not degenerate, it is sufficient to specify the energy eigenvalue.

[17] And the spin quantum number $s_z$, if the appropriate operators are incorporated into the Hamiltonian, see Chap. 19.

[18] If two operators $A$ and $B$ commute with a third operator $C$, that does not necessarily mean
$[A, B] = 0$. An example of this behavior, called *contextuality*, is provided by the components of the
angular-momentum operator $[l_x, \mathbf{l}^2] = 0$, $[l_y, \mathbf{l}^2] = 0$, $[l_x, l_y] \neq 0$. Thus, for a CSCO, observables
are required which commute pairwise and with $H$.

Contextuality means that the result of a measurement depends on other measurements performed
at the same time; see also Chap. 27.

## 17.4 On Modelling

With the example of the hydrogen atom we can nicely illustrate the typical practice of physics in treating complex problems, i.e. the 'hierarchy of models' mentioned at the end of Chap. 14, Vol. 1.

We consider an isolated hydrogen atom, i.e. one that is separated from all interactions with the rest of the world. The idea that it makes sense to consider parts of the environment 'on their own' is of course a strong abstraction which may be challenged. On the other hand, the method has shown its usefulness; the success of modern science is based on it.

In this context, the simplest physical model describes the H atom as two structureless point masses that interact only via the Coulomb potential, which depends only on the relative distance, $\sim 1/r$. We obtain the corresponding Hamiltonian e.g. by replacing the classical variables $(\mathbf{r}, \mathbf{p})$ which appear in the energy by $\left(\mathbf{r}, \frac{\hbar}{i}\nabla\right)$ (correspondence principle). Since this model is too simple and describes only the essentials (and that incompletely), the calculated spectrum is only approximately consistent with experimentally determined data.

We can improve the model by considering the spin of the electron (which as a relativistic phenomenon must be 'grafted' onto the nonrelativistic Hamiltonian), leading to the *fine structure correction* of the spectrum. We take a further step in the modelling by taking into account that the nucleus has its own structure (spin, finite size), which leads to the *hyperfine structure corrections*.

If one wants to tackle the problem relativistically from the start, one has to turn to the Dirac equation or the Hamiltonian $H_{Dirac} = \frac{\hbar}{i}c\alpha\nabla + \beta mc^2 + e\phi$. Here, the rest mass $m$ of the electron enters, and not the reduced mass $\mu$, since there is no equivalent one-body problem for the Dirac equation. If for certain cases one wants to consider the vector potential, the replacement $\nabla \to \nabla - \frac{i}{\hbar}e\mathbf{A}$ can be performed both in the relativistic and in the nonrelativistic cases.[19]

But the Dirac equation is also not yet the endpoint. Only quantum electrodynamics (QED) provides a more comprehensive, relativistically correct quantum theory. QED in turn is again only an approximation to a higher-level theory (electroweak theory), which likewise may derive from a parent theory (grand unified theory or GUT). And this finally might stem from a further higher-level theory (theory of everything, TOE, quantum gravity).

In any case—in this chapter, we have addressed the beginning of the 'hierarchy model'; we take a further concrete step in Chap. 19 with the fine- and hyperfine-structure corrections and in the appendix with the relativistic topics.

---

[19] See also the remarks on the Galilean transformation in Appendix L, Vol. 2 and the discussion of relativistic topics in Appendix U, Vol. 1, and Appendix W, Vol. 2.

## 17.5   Exercises

1. Derive (17.14) from (17.11).
2. Show that

$$u_{E;l}(r) \underset{r \to \infty}{\sim} r^{\alpha} \text{ with } \alpha < -\frac{1}{2} \tag{17.37}$$

   must hold.
3. Hydrogen atom: the probability density of the electron in a volume element $d^3r = r^2 dr d\Omega$ around the point $(r, \vartheta, \varphi)$ is given by

$$d^3 w(r, \vartheta, \varphi) = |R_{nl}(r)|^2 |Y_l^m(\vartheta, \varphi)|^2 r^2 dr d\Omega = |u_{nl}(r)|^2 |Y_l^m(\vartheta, \varphi)|^2 dr d\Omega. \tag{17.38}$$

   Find graphical representations, as illustrative as possible, of the probability densities for the various orbitals with $n = 1$ and $n = 2$.

# Chapter 18
# The Harmonic Oscillator

The algebraic approach leads immediately to the spectrum of the harmonic oscillator. The eigenfunctions are derived in the position representation.

The harmonic oscillator is one of the most important systems of physics. It occurs almost everywhere where vibration is found—from the ideal pendulum to quantum field theory. Among other things, the reason is that the parabolic oscillator potential is a good approximation of a general potential $V(x)$, if we consider small oscillations around a stable equilibrium position $x_0$. Thus, in this case we can approximate $V(x)$ by the first terms of the Taylor series:

$$V(x) = V(x_0) + (x - x_0) V'(x_0) + \frac{1}{2}(x - x_0)^2 V''(x_0) + \cdots \tag{18.1}$$

Because $x_0$ represents a minimum, we have $V'(x_0) = 0$ and $V''(x_0) > 0$. By a redefinition of the energy scale we can set $V(x_0) = 0$. If in addition we select $x_0$ as the new coordinate origin, we see that the potential of the harmonic oscillator $\sim ax^2$ with $a > 0$ is the first approximation to a general potential $V(x)$. An instructive illustration is given in Fig. 18.1.

To arrive at the quantum-mechanical formulation, we start from the classical one-dimensional harmonic oscillator. The position variable is $q = x - x_0$ (the notation $q$ for the position variable is traditional in this context). With the mass $m$ and spring constant $D$, we obtain the familiar equation of motion $m\ddot{q} + Dq = 0$, and from that with $D = m\omega^2$:

$$\ddot{q} + \omega^2 q = 0. \tag{18.2}$$

We cannot translate this classical equation directly into quantum mechanics, because it is an equation relating forces ($\sim \ddot{q}$), while the SEq equates energies ($\sim \dot{q}^2$). We solve this by multiplying (18.2) by $m\dot{q}$ and integrating (see the exercises). With $\dot{q} = p/m$, this leads to the well-known formulation

© Springer Nature Switzerland AG 2018
J. Pade, *Quantum Mechanics for Pedestrians 2*, Undergraduate Lecture Notes in Physics, https://doi.org/10.1007/978-3-030-00467-5_18

**Fig. 18.1** Approximation of a potential *(blue)* near the equilibrium point $x_0$ by a harmonic potential *(red)*; here, as an example for $V(x) = -1/x + 1/x^2$. With $x_0 = 2$, it follows that $V_{\text{harm}}(x) = x(x-4)/16$, and with $q = x - 2$ we have $V_{\text{harm}}(q) + 1/4 = q^2$

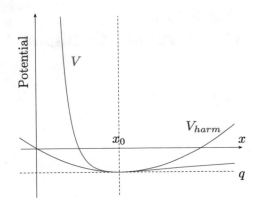

$$E = \frac{p^2}{2m} + \frac{1}{2}m\omega^2 q^2. \tag{18.3}$$

With $p = \frac{\hbar}{i}\frac{d}{dq}$, we obtain the Hamiltonian for the quantum-mechanical harmonic oscillator:

$$H = -\frac{\hbar^2}{2m}\frac{d^2}{dq^2} + \frac{1}{2}m\omega^2 q^2. \tag{18.4}$$

In the following, we first determine the energy spectrum in an algebraic manner. Even though this method works for only a few systems (we have already used it in the case of angular momentum), it is of interest in itself. Moreover, it provides a starting point for further formulations of quantum mechanics, going beyond the scope of this book. The analytical approach, that is the determination of the position functions as solutions of the SEq, can be found at the end of this chapter.

## 18.1  Algebraic Approach

### 18.1.1  Creation and Annihilation Operators

We now transform (18.4). The basic idea is to split the left side as in $b^2 + c^2 = (b + ic)(b - ic)$, although here we have the 'slight complication' that $p$ and $q$ do not commute, $[q, p] = i\hbar$. We therefore define the operator

$$a := \frac{1}{\sqrt{2\hbar}}\left\{\sqrt{m\omega}q + i\frac{p}{\sqrt{m\omega}}\right\}. \tag{18.5}$$

The particular choice of constants will become clear later, likewise the name of this operator. It is in fact called the *annihilation operator* or *lowering operator*. We recall that $q$ and $p$ are Hermitian operators, i.e. $q^\dagger = q$ and $p^\dagger = p$. Hence we have

$$a^\dagger = \frac{1}{\sqrt{2\hbar}} \left\{ \sqrt{m\omega} q - i \frac{p}{\sqrt{m\omega}} \right\}. \tag{18.6}$$

The operator $a^\dagger$ is called the *creation operator* or *raising operator*, and the operators $a$ and $a^\dagger$ in general *ladder operators*. As in the case of angular momentum, the reason for this notation is due to their effect on states, as we shall see shortly.[1]

Since $q$ and $p$ do not commute, probably also $a$ and $a^\dagger$ will not either. We calculate the commutator $[a, a^\dagger]$ starting with $aa^\dagger$ and $a^\dagger a$. We first find

$$aa^\dagger = \frac{1}{2\hbar} \left\{ m\omega q^2 + ipq - iqp + \frac{p^2}{m\omega} \right\} \tag{18.7}$$

and hence

$$aa^\dagger = \frac{1}{\hbar\omega} \left\{ \frac{p^2}{2m} + \frac{1}{2}m\omega^2 q^2 + \frac{1}{2}\hbar\omega \right\} \tag{18.8}$$

and analogously

$$a^\dagger a = \frac{1}{\hbar\omega} \left\{ \frac{p^2}{2m} + \frac{1}{2}m\omega^2 q^2 - \frac{1}{2}\hbar\omega \right\}. \tag{18.9}$$

This yields for the desired commutator

$$[a, a^\dagger] = 1. \tag{18.10}$$

We can draw two conclusions at this point. Firstly, one sees that $a^\dagger a$ and $aa^\dagger$ are not only *Hermitian* but also *positive* operators. Secondly, we have $a^\dagger a = \{H - \hbar\omega/2\} / (\hbar\omega)$ or

$$H = \hbar\omega \left\{ a^\dagger a + \frac{1}{2} \right\}. \tag{18.11}$$

Of course, a similar formulation holds for $aa^\dagger$. However, it plays no role, as we will see shortly.

---

[1]These names arise less from the simple harmonic oscillator as we treat it here, but rather from quantum field theory. see Appendix W, Vol. 2. There, one uses the ladder operators to describe the creation and annihilation of photons, phonons, and so on.

Generalized ladder operators may also be defined in general one-dimensional potentials. This leads to supersymmetric quantum mechanics (see, e.g. Schwabl, p. 351ff; Hecht, p. 130, and other relevant literature).

## 18.1.2   *Properties of the Occupation-Number Operator*

We now consider the *occupation-number operator* $\hat{n}$ (or particle number operator)

$$\hat{n} = a^\dagger a. \tag{18.12}$$

In the following, we will be concerned with the eigenvalue spectrum and the eigenfunctions of this Hermitian and positive operator (and because of (18.11), also those of the Hamiltonian). We write the eigenvalue problem as

$$\hat{n}\,|\nu\rangle = \nu\,|\nu\rangle \tag{18.13}$$

and assume without loss of generality that $|\nu\rangle$ is normalized. Since $\hat{n}$ is Hermitian, the eigenvalues are real, $\nu \in \mathbb{R}$. Since $\hat{n}$ is positive (not negative), the eigenvalues are *non-negative*. We show that again. We have:

$$\langle\nu|\,\hat{n}\,|\nu\rangle = \nu\,\langle\nu\,|\nu\rangle = \nu$$
$$\langle\nu|\,\hat{n}\,|\nu\rangle = \langle\nu|\,a^\dagger a\,|\nu\rangle = \|a\,|\nu\rangle\|^2 \geq 0 \tag{18.14}$$

and therefore $\nu \geq 0$.

For what follows it is convenient to know the commutator of $\hat{n}$ with $a^\dagger$ and $a$. It can easily be shown that:

$$[\hat{n}, a] = -a; \quad [\hat{n}, a^\dagger] = a^\dagger \tag{18.15}$$

or

$$\hat{n}a = a\,(\hat{n} - 1)\,; \quad \hat{n}a^\dagger = a^\dagger\,(\hat{n} + 1) \tag{18.16}$$

or, somewhat more generally,

$$\hat{n}a^l = a^l\,(\hat{n} - l)\,; \quad \hat{n}a^{\dagger l} = a^{\dagger l}\,(\hat{n} + l)\,; \quad l = 0, 1, 2, \dots \tag{18.17}$$

These equations can be proven by mathematical induction over $l$ (see the exercises).

## 18.1.3   *Derivation of the Spectrum*

Quite similarly to the case of angular momentum, ladder operators here also provide the desired information on the spectrum. We consider how the occupation-number operator acts on the vector $a\,|\nu\rangle$. Due to (18.16), we have

$$\hat{n}\,(a\,|\nu\rangle) = (\nu - 1)\,(a\,|\nu\rangle)\,. \tag{18.18}$$

We have put $a\,|\nu\rangle$ into superfluous parentheses, in order to clearly indicate that $a\,|\nu\rangle$ is an eigenvector of $\hat{n}$ for the eigenvalue $\nu - 1$. For the norm of this eigenvector, we have

$$\|a\,|\nu\rangle\|^2 = \langle\nu|\,a^\dagger a\,|\nu\rangle = \nu. \tag{18.19}$$

That is, there is an eigenvalue $\nu - 1$, if $a\,|\nu\rangle$ is not the zero vector and if $\nu \geq 1$ (because the eigenvalues cannot be negative, see above).

Analogously, we have:

$$\hat{n}\left(a^\dagger\,|\nu\rangle\right) = (\nu + 1)\left(a^\dagger\,|\nu\rangle\right) \tag{18.20}$$

where $a^\dagger\,|\nu\rangle$ is an eigenvector of $\hat{n}$ for the eigenvalue $\nu + 1$. For the norm, we find:

$$\left\|a^\dagger\,|\nu\rangle\right\|^2 = \langle\nu|\,aa^\dagger\,|\nu\rangle = \nu + 1. \tag{18.21}$$

**Plan for the Proof**

Now, our plan for the proof is similar to the case of the angular momentum. We start with a state $|\nu\rangle$ and apply the annihilation operator $a$ to it. This results in a state corresponding to the eigenvalue $\nu - 1$, to which we will again apply $a$, to obtain a state corresponding to the eigenvalue $\nu - 2$, then $\nu - 3$, $\nu - 4$ and so on. If we do this often enough, we will eventually find a negative eigenvalue, which however does not exist, as we have shown above. Therefore, there must be an eigenvector $|\nu_{\min}\rangle$ for the smallest eigenvalue $\nu_{\min}$, for which $a\,|\nu_{\min}\rangle = 0$. To this 'minimal' eigenvector we now apply the creation operator $a^\dagger$ repeatedly to obtain one by one the eigenstates with eigenvalues $1, 2, 3, \ldots$ Clearly, these are the natural numbers. It is important for the method that the spectrum not be degenerate, which we assume for the time being and will prove later using the analytic approach.

A remark on notation: In general one does not write $|\nu_{\min}\rangle$, but rather $|0\rangle$. This state, called the *vacuum state*, is *not* the zero vector (which is usually written simply 0), but rather the energetic ground state; it obeys $\langle 0\,|0\rangle = 1$.

**Detailed Description of Some Steps in the Proof**

**Ladder operators**: Since the spectrum is not degenerate, we must conclude that the eigenvectors $|\nu + 1\rangle$ and $a^\dagger\,|\nu\rangle$ must be proportional to each other, because they belong to the same eigenvalue. The same is true for $|\nu - 1\rangle$ and $a\,|\nu\rangle$. In other words, we have

$$a^\dagger\,|\nu\rangle = b_{\nu+1}\,|\nu + 1\rangle \quad \text{and} \quad a\,|\nu\rangle = c_{\nu-1}\,|\nu - 1\rangle. \tag{18.22}$$

To determine the unknown constants $b_{\nu+1}$ and $c_{\nu-1}$, we write the adjoint equations:

$$\langle \nu | \, a = b_{\nu+1}^* \, \langle \nu + 1 | \quad \text{and} \quad \langle \nu | \, a^\dagger = c_{\nu-1}^* \, \langle \nu - 1 | \qquad (18.23)$$

and thus obtain

$$\langle \nu | \, aa^\dagger \, | \nu \rangle = |b_{\nu+1}|^2 \quad \text{and} \quad \langle \nu | \, a^\dagger a \, | \nu \rangle = |c_{\nu-1}|^2 . \qquad (18.24)$$

From the last equation, it follows immediately with $\hat{n} = a^\dagger a$ that $|c_{\nu-1}|^2 = \nu$. Rearranging the other equation with the help of the commutator (18.10), we obtain $|b_{\nu+1}|^2 = \nu + 1$. We make the simplest choice for the coefficients, namely $b_{\nu+1} = \sqrt{\nu + 1}$ and $c_{\nu-1} = \sqrt{\nu}$, and obtain finally

$$a^\dagger \, | \nu \rangle = \sqrt{\nu + 1} \, | \nu + 1 \rangle \quad \text{and} \quad a \, | \nu \rangle = \sqrt{\nu} \, | \nu - 1 \rangle . \qquad (18.25)$$

Now at least one can see very clearly the reason for the name *ladder operators*—if they are applied to a state, they allow us to step up or down by one rung on the ladder of eigenvalues. However, at this point is still not clear what the possible values of $\nu$ are. This will be determined next.

**The smallest eigenvalue of $\hat{n}$ is zero**: We know that $\nu$ is non-negative and that we can generate eigenvectors belonging to the eigenvalues $\nu - 1, \nu - 2$ etc. by repeated application of the annihilation operator $a$ to the eigenvector $| \nu \rangle$:

$$a^l \, | \nu \rangle = \sqrt{\nu \, (\nu - 1) \ldots (\nu - l + 1)} \, | \nu - l \rangle . \qquad (18.26)$$

Choosing $l$ sufficiently large, we eventually enter the range of negative eigenvalues $\nu - l$. This can be avoided only if (a) the smallest eigenvalue is zero, or (b) if there is a state $| \nu_{min} \rangle$ with $a \, | \nu_{min} \rangle = 0$. In the case (c), we have $\langle \nu_{min} | \, a^\dagger a \, | \nu_{min} \rangle = \langle \nu_{min} | \, \hat{n} \, | \nu_{min} \rangle = \nu_{min} = 0$. So we see that in any case, the smallest eigenvalue $\nu_{min}$ is zero. Therefore, we can now write $|0\rangle$ instead of $| \nu_{min} \rangle$. In a later section of this chapter (position representation), we will show that there is only *one* minimal eigenvector.

**The spectrum is unlimited upwards and has integer values**: By repeated application of the creation operator $a^\dagger$ to the ground state, we can derive all the other states with eigenvalues $1, 2, 3, \ldots$ From (18.25), it follows that:

$$|n\rangle = \frac{1}{\sqrt{n!}} \left( a^\dagger \right)^n |0\rangle . \qquad (18.27)$$

This last equation implies that the spectrum contains only integers and is unlimited upwards:

$$\hat{n} \, |n\rangle = n \, |n\rangle ; \quad n = 0, 1, 2, \ldots \qquad (18.28)$$

Another proof of these two properties of the spectrum is found in the exercises.

### 18.1.4 Spectrum of the Harmonic Oscillator

We now summarize: The Hamiltonian for the harmonic oscillator is

$$H = \hbar\omega\left(\hat{n} + \frac{1}{2}\right). \tag{18.29}$$

The eigenvalues are discrete and have the form

$$E_n = \hbar\omega\left(n + \frac{1}{2}\right); \quad n = 0, 1, 2, \ldots \tag{18.30}$$

The energy $E_0 = \frac{\hbar\omega}{2}$ is called the *zero-point energy*[2]; it is always present in the system. In other words, one cannot cool the harmonic oscillator to absolute zero. That is quite reasonable, since a system at absolute zero has no kinetic energy and could therefore be accurately localized. This would contradict the uncertainty principle.

Note: At this point, we know almost nothing about the eigenfunctions - but that is simply unnecessary for the determination of the spectrum.[3]

## 18.2 Analytic Approach (Position Representation)

To obtain the eigenvectors in the position representation, we go back to the definition (18.5) of the annihilation operator:

$$a = \frac{1}{\sqrt{2\hbar}}\left\{\sqrt{m\omega}q + i\frac{p}{\sqrt{m\omega}}\right\}. \tag{18.31}$$

We apply it to the vacuum state $|0\rangle$, which we write in the position representation as $\varphi_0(q)$. With $a|0\rangle = 0$, we find $\left\{\sqrt{m\omega}q + i\frac{p}{\sqrt{m\omega}}\right\}\varphi_0(q) = 0$; and, using $p = \frac{\hbar}{i}\frac{d}{dq}$:

$$\frac{d\varphi_0}{dq} + \frac{m\omega}{\hbar}q\varphi_0 = 0. \tag{18.32}$$

With the *oscillator length*[4]

$$L = \sqrt{\frac{\hbar}{m\omega}} \tag{18.33}$$

we can write the normalized solution:

---

[2] In three dimensions, it is $\hbar\omega\left(n + \frac{3}{2}\right)$.

[3] This is quite similar to the case of the angular momentum.

[4] The oscillator length $L$ essentially specifies the positions of the classical turning points; see the exercises.

$$\varphi_0\,(q) = \left(\frac{m\omega}{\hbar\pi}\right)^{1/4} \cdot e^{-\frac{q^2}{2L^2}}. \tag{18.34}$$

There are no other solutions which are linearly independent of this solution. Therefore, the ground state is not degenerate. Because the other states can be generated by applying the creation operator to the ground state, we can conclude that the entire spectrum is nondegenerate.

The problem becomes more transparent (to write down, at least) on rescaling: $q \rightarrow x$ with $x = \frac{q}{L}$ and $\varphi_0\,(q) = \psi_0\,(x)$. Note that here $x$ as given by $\frac{q}{L}$ is a *dimensionless* variable. Then the ladder operators read

$$a = \frac{1}{\sqrt{2}}\left(x + \frac{\mathrm{d}}{\mathrm{d}x}\right); \quad a^\dagger = \frac{1}{\sqrt{2}}\left(x - \frac{\mathrm{d}}{\mathrm{d}x}\right) \tag{18.35}$$

and the vacuum state is determined via

$$\left(x + \frac{\mathrm{d}}{\mathrm{d}x}\right)\psi_0(x) = 0 \tag{18.36}$$

with the solution

$$\psi_0\,(x) = \left(\frac{m\omega}{\hbar\pi}\right)^{1/4} \cdot e^{-\frac{x^2}{2}}. \tag{18.37}$$

The other states are obtained by (repeated) application of the creation operator as in (18.27); the result is

$$\psi_n\,(x) = \frac{\left(a^\dagger\right)^n}{\sqrt{n!}}\psi_0\,(x) = \frac{1}{\sqrt{n!\cdot 2^n}}\left(x - \frac{\mathrm{d}}{\mathrm{d}x}\right)^n \psi_0\,(x). \tag{18.38}$$

This may be written as

$$\psi_n\,(x) = \left(\frac{m\omega}{\hbar\pi}\right)^{1/4}\frac{1}{\sqrt{n!\cdot 2^n}} \cdot e^{-\frac{x^2}{2}} \cdot H_n(x) \tag{18.39}$$

where the *Hermite polynomials* $H_n\,(x)$ are defined as

$$H_n(x) = e^{\frac{x^2}{2}}\left(x - \frac{\mathrm{d}}{\mathrm{d}x}\right)^n e^{-\frac{x^2}{2}}. \tag{18.40}$$

The first Hermite polynomials are (see the figures in Appendix B, Vol. 2):

$$H_0 = 1; \quad H_1 = 2x; \quad H_2 = 4x^2 - 2; \quad H_3 = 8x^3 - 12x. \tag{18.41}$$

Other polynomials can be calculated recursively by using

$$H_{n+1}(x) = 2x\,H_n(x) - 2n\,H_{n-1}(x). \tag{18.42}$$

The Hermite polynomials belong to the important class of *orthogonal polynomials*, and they obey the orthonormality relation

$$\int_{-\infty}^{\infty} e^{-x^2} H_n(x) H_m(x) dx = \sqrt{\pi} n! 2^n \delta_{nm}.$$  (18.43)

Their parity is given by

$$H_{2n}(-x) = H_{2n}(x); \quad H_{2n+1}(-x) = -H_{2n+1}(x).$$  (18.44)

## 18.3 Exercises

1. Show explicitly that the eigenvalues of $\hat{n}$ are positive.
2. Show that

$$a^l |\nu\rangle = \sqrt{(\nu - l)! \binom{\nu}{l}} |\nu - l\rangle \quad \text{and} \quad a^{\dagger k} |\nu\rangle = \sqrt{\nu! \binom{\nu + k}{k}} |\nu + k\rangle.$$  (18.45)

3. Determine $a^{\dagger k} a^l |\nu\rangle$ and $a^l a^{\dagger k} |\nu\rangle$.
4. Show that the oscillator length $L$ yields essentially the positions of the classical turning points.
5. Proofs by contradiction:

    (a) Show by proof of contradiction: There is no largest eigenvalue $\nu_{max}$.
    (b) Show by proof of contradiction: The eigenvalues are integers.
    (c) Show that to avoid negative eigenvalues, either (a) the smallest eigenvalue has to be zero or (b) there must be a state $|\nu_{min}\rangle$ with $a |\nu_{min}\rangle = 0$. Show that in case b), $\nu_{min} = 0$.

6. Show that $[q, p] = i\hbar$ ($q$ is the position, $p$ the momentum, $\frac{\hbar}{i} \frac{d}{dq}$).
7. Given

$$a := \frac{1}{\sqrt{2\hbar}} \left\{ \sqrt{m\omega} q + i \frac{p}{\sqrt{m\omega}} \right\};$$  (18.46)

    (a) Derive

$$a^{\dagger} = \frac{1}{\sqrt{2\hbar}} \left\{ \sqrt{m\omega} q - i \frac{p}{\sqrt{m\omega}} \right\}.$$  (18.47)

    (b) Show that

$$[a, a^{\dagger}] = 1$$  (18.48)

    and

$$H = \hbar\omega \left\{ a^\dagger a + \frac{1}{2} \right\}.$$ (18.49)

(c) Given the eigenvalue problem

$$\hat{n} \, |\nu\rangle = \nu \, |\nu\rangle \, ; \quad \hat{n} = a^\dagger a,$$ (18.50)

show that

$$\| a \, |\nu\rangle \|^2 = \nu; \quad \| a^\dagger \, |\nu\rangle \|^2 = \nu + 1.$$ (18.51)

(d) Derive

$$[\hat{n}, a] = -a; \quad [\hat{n}, a^\dagger] = a^\dagger.$$ (18.52)

(e) Show that

$$\hat{n} a^l = a^l \, (\hat{n} - l); \quad \hat{n} a^{\dagger l} = a^{\dagger l} \, (\hat{n} + l); \quad l = 0, 1, 2, \ldots$$ (18.53)

(Proof by mathematical induction.)

(f) Prove

$$\hat{n} a \, |\nu\rangle = (\nu - 1) \, a \, |\nu\rangle \, ; \quad \hat{n} a^\dagger \, |\nu\rangle = (\nu + 1) \, a^\dagger \, |\nu\rangle \, .$$ (18.54)

(g) Derive

$$a \, |\nu\rangle = \sqrt{\nu} \, |\nu - 1\rangle \, ; \quad a^\dagger \, |\nu\rangle = \sqrt{\nu + 1} \, |\nu + 1\rangle \, .$$ (18.55)

(h) Show that

$$a^l \, |\nu\rangle = \sqrt{\nu \, (\nu - 1) \ldots (\nu - l + 1)} \, |\nu - l\rangle \, .$$ (18.56)

# Chapter 19
# Perturbation Theory

Due to the lack of analytic solutions of physical problems, several perturbative methods have been developed. The time-independent perturbation theory discussed in this chapter is an important and much-used technique by which we can calculate the fine structure of the spectrum of the hydrogen atom.

Studying physics, one may get in the beginning the impression that there are closed analytical solutions for all problems. That impression is deceptive, as is well known.[1] All in all, in physics, the set of explicitly and exactly solvable problems is of measure zero; and this is particularly relevant to quantum mechanics. There are a handful of potentials for which one can specify an explicit analytic solution of the SEq, but that's about the end of it. If we pick at random any more or less physically reasonable model potential of an appropriate function space, the chance that we know an explicit analytic solution is practically zero. For this reason, one either depends on numerical calculations or, if one wants to have more or less analytic results, on some form of approximation. There are various methods[2]; here, we address the so-called *perturbation theory*.[3]

This method can be applied when the interaction being considered can be decomposed into a part $V$ which covers the essential physical effects, and another relatively small part $W$ (the 'perturbation') which describes more detailed structures. Of course it is an especially favorable case when there exist closed analytic solutions for $V$.

One distinguishes between time-independent (=stationary) and time-dependent perturbation theory. Since we consider only time-independent potentials in this text, we restrict ourselves in the following to stationary perturbation theory.

---

[1]Physics is regarded as an exact science. That does not mean that physical models are always exact or can be solved exactly. Physical models are inherently approximations, and only a few can indeed be solved exactly. A main characteristic of physics is that it deliberately keeps track of the inaccuracies of its approaches. To repeat a quote from Bertrand Russell: "Although this may be seen as a paradox, all exact science is dominated by the idea of approximation".

[2]In Chap. 23, we will discuss the Ritz variational principle, a different approximation procedure.

[3]Also called Rayleigh–Schrödinger perturbation theory.

© Springer Nature Switzerland AG 2018
J. Pade, *Quantum Mechanics for Pedestrians 2*, Undergraduate Lecture
Notes in Physics, https://doi.org/10.1007/978-3-030-00467-5_19

## 19.1    Stationary Perturbation Theory, Nondegenerate

We start with a Hamiltonian $H$:

$$H = -\frac{\hbar^2}{2m}\nabla^2 + V + W = H^{(0)} + W. \tag{19.1}$$

The prerequisite for the following considerations is that $W$ be sufficiently 'small', i.e. that for all states occurring in the calculation, it holds that: $|\langle\varphi| W |\psi\rangle| \ll |\langle\varphi| H^{(0)} |\psi\rangle|$. In this case, we can write

$$H = H^{(0)} + W = H^{(0)} + \varepsilon\hat{W} \tag{19.2}$$

with the smallness parameter $\varepsilon \ll 1$, where the matrix elements $\left|\langle\varphi| \hat{W} |\psi\rangle\right|$ and $\left|\langle\varphi| H^{(0)} |\psi\rangle\right|$ are of the same order of magnitude.

The superscripted zero denotes quantities in the unperturbed problem (which, at best, is itself analytically solvable). Eigenvalues and eigenvectors of $H^{(0)}$ are solutions of

$$H^{(0)} \left|\varphi_k^{(0)}\right\rangle = E_k^{(0)} \left|\varphi_k^{(0)}\right\rangle; \quad k = 1, 2, \ldots \tag{19.3}$$

We assume that the spectrum of $H^{(0)}$ is discrete and nondegenerate, and that its eigenvectors form a CONS. The initial state is $\left|\varphi_n^{(0)}\right\rangle$ and the initial energy is $E_n^{(0)}$.

If we now 'turn on' the perturbation $W$, we no longer have the unperturbed eigenvalue problem (19.3). Instead, states and eigenvalues are determined by

$$H |\varphi\rangle = E |\varphi\rangle. \tag{19.4}$$

We assume that for sufficiently weak $W$, we have approximately

$$|\varphi\rangle \approx \left|\varphi_n^{(0)}\right\rangle; \quad E \approx E_n^{(0)} \text{ for } \varepsilon \text{ sufficiently small.} \tag{19.5}$$

We formalize this by assuming that we can expand states and energies in power series of the smallness parameter $\varepsilon$. Thus, we use the *ansatz*:

$$|\varphi\rangle = \left|\varphi_n^{(0)}\right\rangle + \varepsilon\left|\varphi_n^{(1)}\right\rangle + \varepsilon^2 \left|\varphi_n^{(2)}\right\rangle + \ldots; \quad E = E_n^{(0)} + \varepsilon E_n^{(1)} + \varepsilon^2 E_n^{(2)} + \cdots \tag{19.6}$$

The basic idea of perturbation theory is now to insert these formulations into the SEq (19.4) and to sort by powers of $\varepsilon$. Terms with $\varepsilon^0$ give the unperturbed system, which is labelled by the superscript $^{(0)}$; the terms with $\varepsilon^1$ give the first-order corrections which are labelled by $^{(1)}$, and so on. The procedure is simple in principle, although, at first glance, it looks perhaps somewhat opaque due to the accumulation of indices. In addition, it should be noted that the convergence of the series (19.6) in $\varepsilon$ is difficult to establish. We assume again that quantum mechanics and the method considered here are 'well behaved', and estimate the quality of our approximation a posteriori by comparison with experimental data. Apart from that, such an approximation method in general makes sense in practice only if the essential corrections can be described with a few terms (proportional to $\varepsilon^1$, or at most to $\varepsilon^2$).

One more note before we go through perturbation theory: We can assume from the outset that the correction terms are orthogonal to the initial state

$$\langle \varphi_n^{(0)} | \varphi_n^{(j)} \rangle = 0; \quad j \neq 0. \tag{19.7}$$

For if it would turn out during the calculation that this were not the case (implying that $|\varphi_n^{(1)}\rangle = c \, |\varphi_n^{(0)}\rangle + |\psi_{\text{rest}}\rangle$ with $|\psi_{\text{rest}}\rangle$ orthogonal to $|\varphi_n^{(0)}\rangle$), we could add this part to the undisturbed state and renormalize:

$$\left| \varphi_n^{(0)} \right\rangle + \varepsilon c \left| \varphi_n^{(0)} \right\rangle + \varepsilon \left| \psi_{\text{rest}} \right\rangle \rightarrow \frac{1}{\sqrt{1 + \varepsilon^2 \, |c|^2}} \left| \varphi_n^{(0)} \right\rangle + \varepsilon \left| \psi_{\text{rest}} \right\rangle = \left| \varphi_n^{(0)} \right\rangle_{\text{new}} + \varepsilon \left| \psi_{\text{rest}} \right\rangle , \tag{19.8}$$

so that with this new initial state $|\varphi_n^{(0)}\rangle_{\text{new}}$, the orthogonality relation (19.7) is satisfied. We point out that this issue plays a role in the following argument.

We insert the power series (19.6) into (19.4) and obtain, sorted by powers of $\varepsilon$,

$$H^{(0)} \left| \varphi_n^{(0)} \right\rangle + \varepsilon \left[ H^{(0)} \left| \varphi_n^{(1)} \right\rangle + \hat{W} \left| \varphi_n^{(0)} \right\rangle \right] + \cdots$$
$$= E_n^{(0)} \left| \varphi_n^{(0)} \right\rangle + \varepsilon \left[ E_n^{(0)} \left| \varphi_n^{(1)} \right\rangle + E_n^{(1)} \left| \varphi_n^{(0)} \right\rangle \right] + \cdots \tag{19.9}$$

To make the principle clear, it will suffice to consider the terms $\sim \varepsilon^0$ and $\sim \varepsilon^1$. For terms $\sim \varepsilon^2$, see the exercises. Comparing powers of $\varepsilon$ leads to

$$H^{(0)} \left| \varphi_n^{(0)} \right\rangle = E_n^{(0)} \left| \varphi_n^{(1)} \right\rangle \tag{19.10}$$

$$H^{(0)} \left| \varphi_n^{(1)} \right\rangle + \hat{W} \left| \varphi_n^{(0)} \right\rangle = E_n^{(0)} \left| \varphi_n^{(1)} \right\rangle + E_n^{(1)} \left| \varphi_n^{(0)} \right\rangle. \tag{19.11}$$

Equation (19.10) is automatically satisfied. From (19.11), the correction terms $|\varphi_n^{(1)}\rangle$ and $E_n^{(1)}$ have to be calculated.

### 19.1.1  Calculation of the First-Order Energy Correction

In the first step, we multiply (19.11) from the left by $\langle \varphi_n^{(0)} |$ and find, with $\langle \varphi_n^{(0)} | \varphi_n^{(1)} \rangle = 0$ and $\langle \varphi_n^{(0)} | H^{(0)} | \varphi_n^{(1)} \rangle = E_n^{(0)} \langle \varphi_n^{(0)} | \varphi_n^{(1)} \rangle = 0$, immediately as the first correction term for the energy the matrix element:

$$E_n^{(1)} = \langle \varphi_n^{(0)} | \hat{W} | \varphi_n^{(0)} \rangle. \tag{19.12}$$

Therefore, the energy in first-order correction is given by

$$E = E_n^{(0)} + \varepsilon \langle \varphi_n^{(0)} | \hat{W} | \varphi_n^{(0)} \rangle = E_n^{(0)} + \langle \varphi_n^{(0)} | W | \varphi_n^{(0)} \rangle. \tag{19.13}$$

## 19.1.2  *Calculation of the First-Order State Correction*

In the second step, we determine the correction to the state vector in the lowest order. To calculate the correction term $|\varphi_n^{(1)}\rangle$, we multiply (19.11) from the left by $\langle\varphi_m^{(0)}|$, with $m \neq n$:

$$\langle\varphi_m^{(0)}|\, H^{(0)}\, |\varphi_n^{(1)}\rangle + \langle\varphi_m^{(0)}|\, \hat{W}\, |\varphi_n^{(0)}\rangle = \langle\varphi_m^{(0)}|\, E_n^{(0)}\, |\varphi_n^{(1)}\rangle + \langle\varphi_m^{(0)}|\, E_n^{(1)}\, |\varphi_n^{(0)}\rangle.$$

$$(19.14)$$

We can transform this to (note: $m \neq n$)

$$E_m^{(0)} \langle\varphi_m^{(0)}\, |\varphi_n^{(1)}\rangle + \langle\varphi_m^{(0)}|\, \hat{W}\, |\varphi_n^{(0)}\rangle = E_n^{(0)} \langle\varphi_m^{(0)}\, |\varphi_n^{(1)}\rangle, \qquad (19.15)$$

and this gives

$$\langle\varphi_m^{(0)}\, |\varphi_n^{(1)}\rangle = \frac{\langle\varphi_m^{(0)}|\, \hat{W}\, |\varphi_n^{(0)}\rangle}{E_n^{(0)} - E_m^{(0)}}; \quad m \neq n \qquad (19.16)$$

or

$$\sum_{m\neq n} |\varphi_m^{(0)}\rangle\langle\varphi_m^{(0)}\, |\varphi_n^{(1)}\rangle = \sum_{m\neq n} \frac{\langle\varphi_m^{(0)}|\, \hat{W}\, |\varphi_n^{(0)}\rangle}{E_n^{(0)} - E_m^{(0)}} |\varphi_m^{(0)}\rangle. \qquad (19.17)$$

To make use of the completeness relation of the eigenvectors $|\varphi_m^{(0)}\rangle$, we add on the left $|\varphi_n^{(0)}\rangle\langle\varphi_n^{(0)}\, |\varphi_n^{(1)}\rangle = 0$ (this follows from (19.7)) and obtain the correction term

$$|\varphi_n^{(1)}\rangle = \sum_{m\neq n} \frac{\langle\varphi_m^{(0)}|\, \hat{W}\, |\varphi_n^{(0)}\rangle}{E_n^{(0)} - E_m^{(0)}} |\varphi_m^{(0)}\rangle. \qquad (19.18)$$

Correspondingly, the state with first-order correction is given by

$$|\varphi\rangle = |\varphi_n^0\rangle + \varepsilon \sum_{m\neq n} \frac{\langle\varphi_m^{(0)}|\, \hat{W}\, |\varphi_n^{(0)}\rangle}{E_n^{(0)} - E_m^{(0)}} |\varphi_m^{(0)}\rangle = |\varphi_n^0\rangle + \sum_{m\neq n} \frac{\langle\varphi_m^{(0)}|\, W\, |\varphi_n^{(0)}\rangle}{E_n^{(0)} - E_m^{(0)}} |\varphi_m^{(0)}\rangle.$$

$$(19.19)$$

With (19.12) and (19.18), the corrections which are of first order in the smallness parameter $\varepsilon$ are known. In principle and if necessary, one can perform the calculation for higher powers of $\varepsilon$, but we will not do that here.

## 19.2   Stationary Perturbation Theory, Degenerate

We now assume that the initial spectrum is degenerate:

$$H^{(0)} \left| \varphi_{n,i}^{(0)} \right\rangle = E_n^{(0)} \left| \varphi_{n,i}^{(0)} \right\rangle; \; i = 1, \ldots, g_n \tag{19.20}$$

where $g_n$ is the degree of degeneracy of $E_n^{(0)}$. The states are pairwise orthogonal[4]:

$$\left\langle \varphi_{n,j}^{(0)} \middle| \varphi_{n,k}^{(0)} \right\rangle = \delta_{jk}. \tag{19.21}$$

The initial state of energy $E_n^{(0)}$ is then a superposition of all the degenerate states:

$$\left| \varphi_n^{(0)} \right\rangle = \sum_i c_i \left| \varphi_{n,i}^{(0)} \right\rangle; \; c_i \in \mathbb{C}. \tag{19.22}$$

For simplicity, we restrict ourselves to the calculation of the energy correction. We start from (19.11) and multiply from the left by $\left\langle \varphi_{n,k}^{(0)} \right|$. This gives

$$\left\langle \varphi_{n,k}^{(0)} \middle| H^{(0)} \middle| \varphi_n^{(1)} \right\rangle + \left\langle \varphi_{n,k}^{(0)} \middle| \hat{W} \middle| \varphi_n^{(0)} \right\rangle = \left\langle \varphi_{n,k}^{(0)} \middle| E_n^{(0)} \middle| \varphi_n^{(1)} \right\rangle + \left\langle \varphi_{n,k}^{(0)} \middle| E_n^{(1)} \middle| \varphi_n^{(0)} \right\rangle. \tag{19.23}$$

Again, we have

$$\left\langle \varphi_{n,i}^{(0)} \middle| \varphi_n^{(1)} \right\rangle = 0; \; i = 1, \ldots, g_n \tag{19.24}$$

The first terms on both sides vanish. What remains is

$$\sum_i \left\langle \varphi_{n,k}^{(0)} \middle| \hat{W} \middle| \varphi_{n,i}^{(0)} \right\rangle c_i = \sum_i \left\langle \varphi_{n,k}^{(0)} \middle| E_n^{(1)} \middle| \varphi_{n,i}^{(0)} \right\rangle c_i = E_n^{(1)} c_k. \tag{19.25}$$

On the left, we use the abbreviation $\left\langle \varphi_{n,k}^{(0)} \middle| \hat{W} \middle| \varphi_{n,i}^{(0)} \right\rangle = \hat{W}_{ki}$. The equation (dimension $g_n$) is then written as follows:

$$\sum_i \hat{W}_{ki} c_i = E_n^{(1)} c_k; \; i, k = 1, \ldots, g_n \tag{19.26}$$

and this is an eigenvalue problem with the matrix $\mathbb{W} = \left( \hat{W}_{ki} \right)$ and the column vector $\mathbf{c}$:

$$\mathbb{W} \mathbf{c} = E_n^{(1)} \mathbf{c}. \tag{19.27}$$

---

[4]We can always make this assumption. It is guaranteed by standard methods of linear algebra that one can construct such states, if necessary.

At this point, the usual machinery of linear algebra takes over. We obtain the characteristic polynomial; according to the fundamental theorem of algebra, there are $g_n$ solutions for $E_n^{(1)}$ which can partially coincide. If all eigenvalues are unequal, the degeneracy is removed completely, otherwise only partly. The simplest case is that $\mathbb{W}$ is diagonal from the outset and the $\hat{W}_{kk}$ are all unequal; then the solution values $E_n^{(1)}$ are just the $g_n$ different diagonal elements $\hat{W}_{kk}$, $k = 1, \ldots, g_n$.

## 19.3  Hydrogen: Fine Structure

In this section, we want to look at the hydrogen spectrum in more detail using perturbation theory. We first describe some correction terms $W$ and then explore their consequences.

Concerning the Hamiltonian and the spectrum of the hydrogen atom, we have thus far not considered the electron's *spin*. In fact, the spin is a purely relativistic phenomenon and can be 'patched' into the SEq only heuristically. As we have already noted in Chap. 17, the hydrogen atom is described in a relativistically correct manner by the Dirac equation (see also Appendix U, Vol. 1, and Appendix F, Vol. 2). Performing an expansion in terms of powers of $(v/c)^2$ of this equation, one finds relativistic corrections of various kinds to the nonrelativistic Hamiltonian $H^{(0)}$. This operator, which was the starting point of our consideration of the hydrogen atom in Sect. 17.2, is the unperturbed Hamiltonian for the following perturbation calculation[5]:

$$H^{(0)} = \frac{\mathbf{p}^2}{2m} - \frac{\gamma}{r}; \quad \gamma = \frac{e^2}{4\pi\varepsilon_0}; \quad E_n^{(0)} = -\frac{mc^2\alpha^2}{2n^2}. \tag{19.28}$$

There are three different correction terms:

$$H = H^{(0)} + W_{mp} + W_{ls} + W_D. \tag{19.29}$$

### 19.3.1  Relativistic Corrections to the Hamiltonian

The term $W_{mp}$ takes into account the relativistic dependence of the mass on its velocity in a first approximation. We expand $E = \sqrt{m^2c^4 + p^2c^2}$ for small momentum and find $E = mc^2 + \frac{\mathbf{p}^2}{2m} - \frac{\mathbf{p}^4}{8m^3c^2} + \cdots$, i.e.

$$W_{mp} = -\frac{\mathbf{p}^4}{8m^3c^2}. \tag{19.30}$$

---

[5] We use here the rest mass $m$ and not the reduced mass $\mu$, since an equivalent one-body problem does not exist for the Dirac equation. For nuclear charge (proton number) $Z \neq 1$, we have $\gamma = \frac{Ze^2}{4\pi\varepsilon_0}$.

The term $W_{ls}$ (called the spin-orbit coupling) describes the interaction between the electron's orbital angular momentum **l** and its spin **s**. Heuristically, one can explain this effect by the fact that a magnetic moment is associated with each of the two angular momenta; these moments interact. The result is[6]

$$W_{ls} = \frac{1}{2m^2c^2}\frac{1}{r}\frac{dV(r)}{dr}\mathbf{l}\cdot\mathbf{s} = \frac{1}{2m^2c^2}\frac{\gamma}{r^3}\mathbf{l}\cdot\mathbf{s}. \qquad (19.31)$$

The term $W_D$ (called the Darwin term[7]) also follows from the Dirac equation. It is given by

$$W_D = \frac{\hbar^2}{8m^2c^2}\nabla^2 V(r) = \frac{\pi\hbar^2\gamma}{2m^2c^2}\delta(\mathbf{r}) \qquad (19.32)$$

where we have used[8] $\nabla^2\frac{1}{r} = -4\pi\delta(\mathbf{r})$. Due to the delta function, this term affects only $s$ orbitals, since only for these is $\psi(0) \neq 0$ (because of $R_{nl} \sim r^l$ for $r \to 0$).

Taking into account the correction terms in (19.29) has immediately two consequences: First, the spin *must* occur in the wavefunction. In the simplest case, the space-dependent part, which we have hitherto considered exclusively, is multiplied by a two-component vector which describes the two possibilities of spin orientation (similar to the polarization states for light). The degeneracy of the $n$ levels then increases by a factor of 2, to $2n^2$.

Secondly, we must look for a new CSCO. For the unperturbed problem (19.28), the three commuting Hermitian operators $H^{(0)}$, $\mathbf{l}^2$ and $l_z$ form a CSCO, as we have seen in Sect. 17.3. Accordingly, we can denote the states by the three quantum numbers $n$, $l$ and $m_l$, for example in the form $|n, l, m_l\rangle$. If we consider also the spin **s** in the eigenfunctions of (19.28), we can write $|n, l, m_l, m_s\rangle$ (since the total spin value $s$ does not change, one omits the $\frac{1}{2}$ in $|n, l, m_l, \frac{1}{2}, m_s\rangle$).[9] The CSCO thus consists of $H^{(0)}$, $\mathbf{l}^2$, $l_z$ and $s_z$.

But for the problem (19.29), $H$, $\mathbf{l}^2$, $l_z$ and $s_z$ do *not form* a CSCO. This is due to the spin-orbit coupling $W_{ls}$, which prevents $l_z$ and $s_z$ from commuting with $H$, $[H, l_z] \neq 0$ and $[H, s_z] \neq 0$. The total angular momentum **j**, i.e. the sum of orbital angular momentum and spin

$$\mathbf{j} = \mathbf{l} + \mathbf{s} \qquad (19.33)$$

provides a remedy, since $\mathbf{j}^2$ and $j_z$ commute with $H$ and $\mathbf{l}^2$ (see the exercises). Hence, $H, \mathbf{j}^2, j_z$ and $\mathbf{l}^2$ form a (new) suitable CSCO and the states can be classified according

---

[6]See also Chap. 16 (angular momentum). There, we abbreviated the prefactor of $\mathbf{l}\cdot\mathbf{s}$ by $F(r)$.

[7]No, not an evolutionary term. Charles Galton Darwin (1887–1962), physicist, was a grandson of *the* Charles Darwin. Also known as the *Zitterbewegung* ('dithering motion', from the German).

[8]See Appendix F, Vol. 1.

[9]In the position representation, the states take the form

$$\psi_{n,l,m_l,m_s}(\mathbf{r}) = \langle\mathbf{r}|n, l, m_l\rangle\begin{pmatrix}m_{s\uparrow}\\m_{s\downarrow}\end{pmatrix}$$

to the four quantum numbers $n$, $j$, $m_j$ and $l$. So we can write $|n; j, m_{j;}l\rangle$. These states form a CONS,

$$\langle n; j', m'_{j;}l' \, |n; j, m_{j;}l\rangle = \delta_{jj'}\delta_{m_jm'_j}\delta_{ll'}. \tag{19.34}$$

According to the rules of angular-momentum addition, we start out from the following states:

$$|\varphi_n^{(0)}\rangle = |n; j, m_{j;}l\rangle; \quad j = l \pm \frac{1}{2} \text{ for } l \geq 1 \text{ ; } j = \frac{1}{2} \text{ for } l = 0. \tag{19.35}$$

Note that in the position representation, the radial component is given by the function $R_{nl}(r)$, introduced in Chap. 17. Although we do not need the explicit form of the states which we derived in Chap. 16 for the following, we give it here for the sake of completeness:

$$|n; j = l \pm 1/2, m_{j;}l\rangle = \begin{pmatrix} \pm\sqrt{\frac{l\pm m_j+1/2}{2l+1}}\,|n; l, m_j - 1/2\rangle \\ \sqrt{\frac{l\mp m_j+1/2}{2l+1}}\,|n; l, m_j + 1/2\rangle \end{pmatrix}$$
$$\text{with } m_j = l \pm 1/2, \ldots, -(l \pm 1/2). \tag{19.36}$$

## 19.3.2  Results of Perturbation Theory

With the functions (19.35), we now carry out the perturbation treatment for degenerate states,[10] cf. (19.26). We have seen there that the corrections $E_n^{(1)}$ can be quite easily calculated, if the matrix $\left(\hat{W}_{ki}\right)$ is diagonal. As it turns out, this is the case for all three relativistic corrections. For brevity, we use the notation $\langle A \rangle = \langle n; j', m'_j; l | A |n; j, m_j; l\rangle$. In particular, we obtain (a somewhat more detailed analysis is given in Appendix H, Vol. 2):

$$\langle W_{mp}\rangle = -\frac{1}{2mc^2}\left\{\left(E_n^{(0)}\right)^2 + 2E_n^{(0)}\gamma\left\langle\frac{1}{r}\right\rangle + \gamma^2\left\langle\frac{1}{r^2}\right\rangle\right\}\delta_{j'j}\delta_{m'_jm_j}. \tag{19.37}$$

The next term exists for $l \neq 0$ only:

$$\langle W_{ls}\rangle = \frac{\gamma\hbar^2}{2m^2c^2}\frac{1}{2}\left[j(j+1) - l(l+1) - \frac{3}{4}\right]\left\langle\frac{1}{r^3}\right\rangle\delta_{j'j}\delta_{m'_jm_j}. \tag{19.38}$$

The last term occurs only for $l = 0$; it is given by

---

[10]In the following, we treat the degeneracy of $n$ and $l$; the removal of the $m$-degeneracy is possible only by applying external fields.

$$\langle W_D \rangle = \frac{\pi \hbar^2 \gamma}{2m^2 c^2} |R_{n0}(0)|^2 \, \delta_{j'j} \delta_{m'_j m_j}. \tag{19.39}$$

We add these fine-structure corrections and obtain initially

$$E_n^{(1)} = -\frac{1}{2mc^2} \left\{ \begin{array}{c} \left(E_n^{(0)}\right)^2 + 2E_n^{(0)} \gamma \left\langle \frac{1}{r} \right\rangle + \gamma^2 \left\langle \frac{1}{r^2} \right\rangle \\ -\frac{\gamma \hbar^2}{m} \frac{1}{2} \left[ j(j+1) - l(l+1) - \frac{3}{4} \right] \left\langle \frac{1}{r^3} \right\rangle \\ -\frac{\pi \hbar^2 \gamma}{m} |R_{n0}(0)|^2 \end{array} \right\} \delta_{j'j} \delta_{m'_j m_j}. \tag{19.40}$$

The mean values within the brackets can be calculated (see Appendix B, Vol. 2). We do not need to do this explicitly here and simply adopt the results. For the proton number $Z = 1$ (i.e. $\gamma = \frac{e^2}{4\pi\varepsilon_0}$), we obtain

$$E_{n,j = l \pm \frac{1}{2}, l}^{(1)} = \frac{mc^2 \alpha^4}{2n^4} \left\{ \frac{3}{4} - \frac{n}{j + \frac{1}{2}} \right\} \tag{19.41}$$

where $\alpha$ is the fine-structure constant, $\alpha \approx 1/137$.

### 19.3.3  Comparison with the Results of the Dirac Equation

With these corrections, we obtain the following energy levels for the hydrogen atom:

$$E_{nj} = -mc^2 \frac{\alpha^2}{2n^2} \left\{ 1 - \frac{\alpha^2}{n^2} \left( \frac{3}{4} - \frac{n}{j + \frac{1}{2}} \right) \right\}. \tag{19.42}$$

For comparison: the Dirac equation gives (see Appendix F, Vol. 2)

$$E_{nj} = mc^2 \left\{ 1 + \alpha^2 \left[ n - j - \frac{1}{2} + \sqrt{\left(j + \frac{1}{2}\right)^2 - \alpha^2} \right]^{-2} \right\}^{-\frac{1}{2}} - mc^2. \tag{19.43}$$

Expanding this expression in powers of $\alpha^2$ and retaining only the terms $\sim \alpha^2$ and $\sim \alpha^4$, one obtains the approximate expression (19.42).[11] We see that the energy corrections are smaller than the initial values by a (relative) factor of $\alpha^2 \approx 5 \cdot 10^{-5}$—hence the name 'fine structure'. In addition, the energy levels evidently depend not only on the principal quantum number $n$ but also on $j$. So we have a partial lifting of the degeneracy of the hydrogen levels. These levels are denoted by the quantum

---

[11]Numerical values: $mc^2 \alpha^2 = 2 \cdot 13.6 \, \text{eV}$, $mc^2 \alpha^4 = 1.45 \cdot 10^{-3} \, \text{eV}$ (see also Appendix B, Vol. 1).

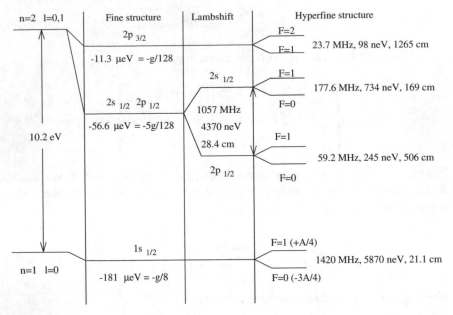

**Fig. 19.1** Fine and hyperfine structure of the hydrogen atom. Abbreviations: $g = mc^2\alpha^4 = 1.45 \cdot 10^{-3}$ eV, $A = 1,420$ MHz. The finite size of the nucleus is not taken into account. Scales are not preserved

numbers $n, l, j$ in the form $nl_j$, where for $l$ the nomenclature $s, p, d, \ldots$ is used.[12] As a result of perturbation theory, we have for $n = 1$ a slight lowering of the $1s_{1/2}$ levels and for $n = 2$ a splitting into a $2p_{3/2}$ level and a degenerate $2s_{1/2} - 2p_{1/2}$ level. Similar statements hold for higher principal quantum numbers. In the context of the assumptions made here, the $2s_{1/2} - 2p_{1/2}$ degeneracy is valid for any power of $\alpha^2$ (compare the relativistically correct expression (19.43)). The changes in the spectrum are shown schematically in Fig. 19.1.

## 19.4  Hydrogen: Lamb Shift and Hyperfine Structure

A particularly accurate (denoted by 'hyperfine') consideration of the hydrogen spectrum shows that further corrections have to be applied to the previously obtained results. They are also shown schematically in Fig. 19.1. The reasons for these additional corrections are:

---

[12]Another common notation uses $s, l, j$ and reads $^{2s+1}l_j$.

1. Quantum electrodynamic effects remove the $2s_{1/2} - 2p_{1/2}$ -degeneracy. This so-called Lamb shift is about $4 \cdot 10^{-6}$ eV.

2. Hyperfine structure: the spins of the nucleus and of the electron interact with each other. They can add up to the total spin $F = 0$ (singlet) or $F = 1$ (triplet). The interaction term is proportional to $\mathbf{s}_N \cdot \mathbf{s}_e = \frac{\mathbf{F}^2 - \mathbf{s}_N^2 - \mathbf{s}_e^2}{2} = \frac{\hbar^2}{2}\left[F(F+1) - \frac{3}{2}\right]$ and leads to a splitting of the $1s_{1/2}$ levels of the form

$$E_{F=1} = E^{(0)} + \frac{1}{4}A; \quad E_{F=0} = E^{(0)} - \frac{3}{4}A; \quad E_{F=1} - E_{F=0} = A$$

$$\text{with} \quad A = (1420405751.768 \pm 0.001) \text{ Hz.} \tag{19.44}$$

The term $A \approx 1,420$ MHz is one of the most precisely measured quantities in physics; the theory describes correctly the first six digits. The transition $E_{F=1} - E_{F=0}$ is used in the hydrogen maser, but also plays an important role in astrophysics. By detecting it, one gains information about the interstellar hydrogen clouds, which at 10–50% account for a significant proportion of the mass of galaxies. In this context, one also speaks of the 21 cm line; see the conversion table for energy units in Appendix B, Vol. 1. Because of their greater probability density at the nucleus, the hyperfine correction is most noticeable for $s$ levels; in addition, the splitting of energy levels is proportional to $n^{-3}$.

3. Equation (19.28) contains the Coulomb or point interaction $-\frac{2}{r}$, but the nucleus has finite dimensions. This effect also leads to a correction, namely to a shift of the levels, which again is most marked for the $s$ levels and is proportional to $n^{-3}$. For the lowest $s$ level, the correction is about $4 \cdot 10^{-9}$ eV. Finally, there are isotope effects in the hyperfine structure; on the one hand, isotopes lead to different reduced masses; on the other hand, different nuclear isotopes have different charge distributions (volume effect).

A practical application of the interaction of nuclear spins with their environment is *nuclear magnetic resonance spectroscopy* (NMR spectroscopy). The simplified operating principle is as follows: A sample contains hydrogen atoms whose nuclear spins (nucleus = proton) are aligned by an external homogeneous magnetic field. The sample is additionally irradiated with a single radio frequency pulse (RF pulse) or with a sequence of RF pulses. After the decay of the RF pulse, the protons in the sample exchange energy with each other and with the surrounding environment. This leads to a return to the equilibrium state (relaxation), i.e. to a measurable change in the (nuclear) magnetization. The decay time of this signal depends on the environment of the proton—in solid matter, the damping is much stronger than in liquids, for instance.

## 19.5   Exercises

1. Given
$$H = H^{(0)} + F(r)\mathbf{l} \cdot \mathbf{s} = \frac{\mathbf{p}^2}{2m} + V(r) + F(r)\mathbf{l} \cdot \mathbf{s}. \qquad (19.45)$$

   (a)  Show that:
   $$\left[H^{(0)}, l_z\right] = \left[H^{(0)}, s_z\right] = 0; \qquad (19.46)$$

   (b)  Show that:
   $$\left[H, l_z\right] \neq 0; \quad \left[H, s_z\right] \neq 0; \quad \left[H, j_z\right] = 0. \qquad (19.47)$$

   Hint: See the exercises for Chap. 16.

2. Expand the expression for the relativistic energy levels of the hydrogen atom:

$$E_{nj} = mc^2 \left\{ 1 + \alpha^2 \left[ n - j - \frac{1}{2} + \sqrt{\left(j + \frac{1}{2}\right)^2 - \alpha^2} \right]^{-2} \right\}^{-\frac{1}{2}} - mc^2$$
$$(19.48)$$

   and compare with the approximation deduced in the text.

3. Given the Hamiltonian

$$H \left| \varphi \right\rangle = \left( H^{(0)} + W \right) \left| \varphi \right\rangle = \left( H^{(0)} + \varepsilon \hat{W} \right) \left| \varphi \right\rangle = E \left| \varphi \right\rangle, \qquad (19.49)$$

   where the states and the eigenvalues of $H^{(0)} \left| \varphi_n^{(0)} \right\rangle = E_n^{(0)} \left| \varphi_n^{(0)} \right\rangle$ are known (discrete, nondegenerate). The initial state is $\left| \varphi_n^{(0)} \right\rangle$ and the corresponding energy is $E_n^{(0)}$. States and energies are expanded in terms of $\varepsilon$

$$\left| \varphi \right\rangle = \left| \varphi_n^{(0)} \right\rangle + \varepsilon \left| \varphi_n^{(1)} \right\rangle + \varepsilon^2 \left| \varphi_n^{(2)} \right\rangle + \cdots; \quad E = E_n^{(0)} + \varepsilon E_n^{(1)} + \varepsilon^2 E_n^{(2)} + \cdots \quad (19.50)$$

   We can assume from the outset that the correction terms are orthogonal to the initial state, $\left\langle \varphi_n^{(0)} \middle| \varphi_n^{(j)} \right\rangle = 0$ for $j \neq 0$. Calculate the corrections to the energy and the state to first order ($\sim \varepsilon^1$, repetition) and to second order ($\sim \varepsilon^2$).

4. We add a perturbation $\sim q^3$ to the Hamiltonian of the harmonic oscillator:

$$H = H^0 + W = -\frac{\hbar^2}{2m} \frac{d^2}{dq^2} + \frac{1}{2} m\omega^2 q^2 + \varepsilon q^3. \qquad (19.51)$$

   Calculate the correction term of the energy $E_n = \hbar\omega \left(n + \frac{1}{2}\right)$ to first order.

5. Finite nuclear size: For a hydrogen atom, we model the finite core size by the potential

$$V(r) = \begin{cases} -\frac{\gamma}{r} & \text{for } r \geq r_0 \\ \frac{\gamma}{2r_0}\left[\left(\frac{r}{r_0}\right)^2 - 3\right] & \text{for } r \leq r_0 \end{cases} \qquad (19.52)$$

(Thus, we replace the point nucleus by a homogenously-charged sphere of radius $r_0$ with the charge density $\rho_0$). Calculate the corrections to the energy in first order. Assume that the radial functions $R_{nl}(r)$ can be approximated for $r \leq r_0$ by $R_{nl}(0)$.

# Chapter 20
# Entanglement, EPR, Bell

For two or more quantum objects, there exist states that are typically quantum-mechanical and have no classical analogues. These entangled states are of central importance for the understanding of quantum mechanics and especially for modern developments such as quantum computers.

Up to now, our discussion of practical problems was confined to *one* quantum object.[1] In this chapter, we consider systems of two quantum objects that each can assume different states independently of one another (e.g. spin, polarization...). The ideas discussed can easily be generalized to several quantum objects.

## 20.1 Product Space

First, some words about the state space of a system of two quantum objects, such as two photons or an electron and a positron. The two Hilbert spaces are $\mathcal{H}_1$ and $\mathcal{H}_2$, with the dimensions $N$ and $M$. The state of the overall system is then determined by the simultaneous specification of the vectors $|\varphi\rangle \in \mathcal{H}_1$ and $|\chi\rangle \in \mathcal{H}_2$; the pair $\{|\psi\rangle, |\chi\rangle\}$ can be regarded as a vector of a vector space of dimension $N \cdot M$. This vector space is called the *product space* of the spaces $\mathcal{H}_1$ and $\mathcal{H}_2$ (or *tensor product* of the two spaces) and is denoted by $\mathcal{H}_1 \otimes \mathcal{H}_2$. If the spaces $\mathcal{H}_1$ and $\mathcal{H}_2$ have the bases $\{|n\rangle\}$ and $\{|m\rangle\}$, then the basis system of the product space is the set of all pairs $\{|n\rangle |m\rangle\}$[2]; these vectors are written as $|n \otimes m\rangle$ or $|n\rangle \otimes |m\rangle$. The vectors of the

---

[1]In Chap. 17, we treated the hydrogen atom as a system of two quantum objects (nucleus and electron), but we reduced it to *one* equivalent quantum object.

[2]In detail, $|1\rangle |1\rangle, |1\rangle |2\rangle, \ldots, |2\rangle |1\rangle, |2\rangle |2\rangle, \ldots, |3\rangle |1\rangle, \ldots$.

© Springer Nature Switzerland AG 2018
J. Pade, *Quantum Mechanics for Pedestrians 2*, Undergraduate Lecture
Notes in Physics, https://doi.org/10.1007/978-3-030-00467-5_20

individual spaces are given by $|\varphi\rangle = \sum_{n=1}^{N} c_n |n\rangle$ and $|\chi\rangle = \sum_{m=1}^{M} d_m |m\rangle$; the total state is then $|\varphi \otimes \chi\rangle = \sum_{n,m} c_n d_m |n \otimes m\rangle$.[3] A general state vector therefore takes the form

$$|\psi\rangle = \sum_{n,m} a_{nm} |n \otimes m\rangle . \tag{20.1}$$

By the way, if there is no danger of misunderstanding, one often writes simply $|nm\rangle$ or a similar formulation, instead of $|n \otimes m\rangle$.

Some remarks:

1. Up to now, if we wrote e.g. $|j, m\rangle$ for angular momentum states, we meant *two* quantum numbers for *one* quantum object. In this chapter, the notation $|nm\rangle$ always means *two* quantum objects, the first in the state $|n\rangle$, the second in the state $|m\rangle$.

2. The order of the product states is not changed on taking the adjoint: $|nm\rangle^{\dagger} = \langle nm|$.

3. Regarding the notation, one uses indices where appropriate. There are various equivalent formulations such as

$$|nm\rangle = |n\rangle |m\rangle = |n_1 m_2\rangle = |n_1\rangle |m_2\rangle = |n\rangle_1 |m\rangle_2 . \tag{20.2}$$

   In the following, we will use that notation which is best suited for the corresponding topic at hand.

4. In addition, one can use the explicit representation as a column vector in certain cases; we have e.g.[4]

$$\begin{pmatrix} a_1 \\ a_2 \end{pmatrix} \otimes \begin{pmatrix} b_1 \\ b_2 \end{pmatrix} = \begin{pmatrix} a_1 b_1 \\ a_1 b_2 \\ a_2 b_1 \\ a_2 b_2 \end{pmatrix} . \tag{20.3}$$

5. More on the basics of tensor products including some examples can be found in Appendix C, Vol. 2.

## 20.2  Entangled States

We now consider *entangled states*.[5] As we shall see, the individual quantum objects have no well-defined status in this case. It is only through measurement that they obtain definite properties.

---

[3] Strictly speaking, this applies initially only if the two systems are independent of each other. But even in the presence of interactions, we make the plausible assumption that the state space is the product space $\mathcal{H}_1 \otimes \mathcal{H}_2$.

[4] Mnemonic: 'The right index changes the fastest.'

[5] The term was coined 1935 by E. Schrödinger; it is possibly related to 'clasped hands'. In that situation, one cannot change one hand without changing the other one.

## 20.2.1 Definition

To work out the essentials, we restrict ourselves to *two* quantum objects that may exist in two states. As a concrete example, we consider the two linear polarization states $|h\rangle$ and $|v\rangle$ of two photons.[6] The basis states of the four-dimensional product space are then $|hh\rangle$, $|hv\rangle$, $|vh\rangle$, $|vv\rangle$. Accordingly, a general state of the product space reads

$$|\Psi\rangle = a_{hh} |hh\rangle + a_{hv} |hv\rangle + a_{vh} |vh\rangle + a_{vv} |vv\rangle . \qquad (20.4)$$

In such states, the question arises as to whether the single quantum object 1 or 2 is in a well-defined state, i.e. if one can say that photon 1 has a specific linear polarization state. In any case, this is not directly visible from (20.4). If each of the quantum objects is in a well-defined linear polarization state, we can write

$$|\varphi_1\rangle = a_{1h} |h\rangle + a_{1v} |v\rangle \,; \; |\varphi_2\rangle = a_{2h} |h\rangle + a_{2v} |v\rangle . \qquad (20.5)$$

It follows that

$$|\varphi_1\varphi_2\rangle = a_{1h}a_{2h} |hh\rangle + a_{1h}a_{2v} |hv\rangle + a_{1v}a_{2h} |vh\rangle + a_{1v}a_{2v} |vv\rangle . \qquad (20.6)$$

One says that this state *factorizes*, so it may be written as a product of two individual states (20.5) and is therefore also called a *product state*. However, the state (20.4) has this form only if

$$a_{hh} = a_{1h}a_{2h}; \; a_{hv} = a_{1h}a_{2v}; \; a_{vh} = a_{1v}a_{2h}; \; a_{vv} = a_{1v}a_{2v}. \qquad (20.7)$$

From this, we obtain immediately the condition

$$a_{hh} \cdot a_{vv} = a_{hv} \cdot a_{vh} \; \hat{=} \; \text{state is product state.} \qquad (20.8)$$

If $|\Psi\rangle$ in (20.4) is not of the form (20.6), i.e. it does not factorize, one speaks of an *entangled* state:

$$a_{hh} \cdot a_{vv} \neq a_{hv} \cdot a_{vh} \; \hat{=} \; \text{state is entangled.} \qquad (20.9)$$

An example is the vector

$$|\Psi\rangle = \frac{|hv\rangle - |vh\rangle}{\sqrt{2}}. \qquad (20.10)$$

Evidently, it holds that $a_{hh} \cdot a_{vv} = 0$ and $a_{hv} \cdot a_{vh} = -1$; condition (20.8) is not fulfilled and the state is not factorized, but rather entangled.[7]

---

[6]We recall that $\{|h\rangle , |v\rangle\}$ is a CONS.

[7]One cannot demonstrate this situation in an intuitive manner because it is set in a four-dimensional space.

For systems of more than two quantum objects, one can, if necessary, set up equations similar to (20.8). In general, a system of several quantum objects is called 'entangled' if the total state cannot be represented as a product of the individual states, i.e. if it is not factorized.[8]

We emphasize once again that the equations imply that the individual components of an entangled state have no well-defined properties. In today's view, this fact does not mean that our knowledge or quantum mechanics as a theory is (still) not sophisticated enough, but rather it is due to the structure of quantum mechanics itself (in the end, to the fact that the SEq is linear). Entangled states are typical of quantum mechanics—they do not exist in classical mechanics, and they are not even imaginable there.[9] In quantum mechanics, however, they not only exist, but are fundamental to some modern applications such as quantum computers and quantum teleportation.[10]

Entangled states can nowadays be produced routinely (see Appendix I, Vol. 2).[11] To this end, one generates pairs of quantum objects which move apart from each other and are entangled with respect to a certain property; an example is polarization-entangled photons.[12] There are of course other methods; for example, one can

---

[8]We take up again at this point the remark made in the Introduction that the popular-scientific presentation of quantum mechanics is possible only to a very limited extent, and this applies especially to purely quantum-mechanical phenomena such as entanglement. As an example, we illustrate this fact by a quotation from the French scientific journal 'La Recherche': "The term (i.e. entanglement) was introduced to refer to a pair of particles which are separated from each other, but have interacted at a previous time with each other, and whose state is described more completely by their common properties (called 'entangled' or 'correlated') than by their individual properties." (Anton Zeilinger and Markus Aspelmeyer, *The incredible illusion of reality*, in La Recherche, dossier 38, Feb. 2010, p. 19, translated). The authors are physicists and internationally accepted experts in the field of basic quantum-mechanics research; they also know very well how to present their field of research in a popular-scientific way. And yet their explanation is fuzzy (one might say, of course, that it has to be), and in the end not very helpful because it fits everything possible and does not grasp the essential point of entanglement. By contrast, 'factorization' is a precise definition, which of course presupposes that the reader knows at least some basic aspects of the mathematical apparatus (in this case, as the example shows, one can describe the situation without formulas). For both 'explanations', the consequences of entanglement are not clear at this point. But it is only the access via 'factorization' which leads on to further conclusions.

[9]They can, however, excite the fantasies of authors. Philippe Djian writes for example in *Vers chez les blancs*: "Edith's presence made me more human, not so petty; I had often noticed that. By the way, in physics the phenomenon of interlacing is well established. In contrast to Bohr, Einstein never believed in it. But the action-at-a-distance which connects two widely-separated particles has in the meantime been observed using photons. Edith and I were »entangled« with each other; that was the truth in its most extreme and pitiless certainty."

[10]See Chap. 26 (quantum information). Entangled photons may also be used in quantum cryptography.

[11]We note by the way that entanglement may also be generated by means of the quantum Zeno effect (cf. Appendix L, Vol. 1); see Nathan S. Williams & Andrew N. Jordan, 'Entanglement genesis under continuous parity measurement', *Phys. Rev.* A 78, 062322 (2008).

[12]Currently (2017), ten photons can be entangled; see for instance Xi-Ling Wang et al., 'Experimental Ten-Photon Entanglement', *Phys. Rev. Lett.* 117, 219502 (2016). ((The paper claims to report the first experimental demonstration of entanglement among ten spatially separated single photons.))

entangle a positron-electron pair using a double Mach–Zehnder interferometer (Hardy's experiment, see Appendix J, Vol. 2).

### 20.2.2 Single Measurements on Entangled States

Only an appropriate measurement forces the two quantum objects to assume well-defined properties. As an example, we illustrate this fact by means of the state (20.10). As usual, we describe the measurement by a projection onto the final state; each quantum object can be measured as $|h\rangle$ or $|v\rangle$. If we detect both photons, we obtain the possible amplitudes

$$\langle hh\,|\Psi\rangle = 0; \quad \langle hv\,|\Psi\rangle = \tfrac{1}{\sqrt{2}}$$
$$\langle vh\,|\Psi\rangle = -\tfrac{1}{\sqrt{2}}; \quad \langle vv\,|\Psi\rangle = 0. \tag{20.11}$$

In other words, we find two possible readings, each with probability $1/2$. Either the quantum object 1 is in the state $|h\rangle$ and quantum object 2 in $|v\rangle$, or *vice versa*. The other two options, $|hh\rangle$ or $|vv\rangle$, are eliminated.

The really interesting question however is: What happens if we measure the state of only *one* of the quantum objects? In order to treat this clearly, we denote for the moment e.g. the state in which quantum object 1 is horizontally and quantum object 2 is vertically polarized by $|h_1\rangle\,|v_2\rangle$. The state (20.10) is then written as

$$|\Psi\rangle = \frac{|h_1\rangle\,|v_2\rangle - |v_1\rangle\,|h_2\rangle}{\sqrt{2}}. \tag{20.12}$$

We now measure the state of only the first quantum object, and ask if it is horizontally polarized. As usual, we represent the measurement by the corresponding projection operator, i.e. by $|h_1\rangle\,\langle h_1|\otimes I_2$, where $I_2$ is the identity (the 1-operator) in the space 2. We find

---

With a completely different technique, namely detecting whether photons share polarizations under certain conditions, the entanglement of an even larger number of photons may be demonstrated; see for instance T. Sh. Iskhakov et al., 'Polarization-Entangled Light Pulses of 105 Photons', *Phys. Rev. Lett.* 109, 150502 (2012).

Entanglement is not restricted to photons, see e.g. F. Fröwis et al., Experimental certification of millions of genuinely entangled atoms in a solid, *Nature Communications* 8, Article number: 907 (2017) https://doi.org/10.1038/s41467-017-00898-6 (Oct 2017) where the entanglement of 16 million atoms in an one-centimeter crystal is reported. Stationary objects can also be entangled, or even stationary and propagating objects; see e.g.W.B. Gao et al., 'Observation of entanglement between a quantum dot spin and a single photon', *Nature* 491, 426–430 (2012). Moreover, a scheme was proposed to entangle the motion of two macroscopically separated objects: C. Gneiting and K. Hornberger, 'Bell test for the free motion of material particles', *Phys. Rev. Lett.* 101, 260503 (2008).

**Fig. 20.1** Two entangled
photons

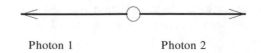

Photon 1                              Photon 2

$$|h_1\rangle \langle h_1| \otimes I_2 |\Psi\rangle = \frac{|h_1\rangle \langle h_1| h_1\rangle |v_2\rangle - |h_1\rangle \langle h_1| v_1\rangle |h_2\rangle}{\sqrt{2}} = \frac{|h_1\rangle |v_2\rangle}{\sqrt{2}}. \quad (20.13)$$

This means the following:

1. On measuring, we observe a well-defined state—quantum object 1 is horizontally, quantum object 2 is vertically polarized. Before the measurement, we could not make this statement. This means that a measurement forces the measured system from the realm of *possibility* into the realm of *actuality* or, viewed in the product Hilbert space, into the subspace of factorized states.
2. After the measurement on quantum object 1, we have two possible states, each with probability $1/2$: After the measurement per $|h_1\rangle \langle h_1|$, the state $|h_1\rangle |v_2\rangle$; and after the measurement per $|v_1\rangle \langle v_1|$, the state $|v_1\rangle |h_2\rangle$.
3. The new aspect of entangled states is this: The measurement of *one* quantum object defines the properties of *both* quantum objects—and this fact has far-reaching consequences.

We discuss this using the example of two entangled photons, for instance in the state (20.12), moving in different directions, say photon 1 to the left and photon 2 to the right; cf. Fig. 20.1.[13] If we now measure the polarization of photon 1 as in (20.13), the polarization of photon 2 is automatically determined—regardless of their mutual distance.[14] It is not hard to imagine experimental arrangements in which this 'collapse' of the state from (20.12) to (20.13) *must* be faster than the speed of light. In other words, the entanglement leads to the *non-locality*[15] of quantum mechanics—a

---

[13]This case (or the similar one using electrons instead of photons) is the illustrative 'standard scenario' for this topic.

[14]For example, in an experiment performed in 1997, entangled photons were sent in opposite directions in glass fiber cables. When they were far apart (about 10 km), they were detected. Despite the distance, a measurement on one photon impacted on the other photon as expected. Some 10 years later, a record for distance propagation in the open air was established between two of the Canary Islands. A Bell state of two photons of the form $|\Psi^-\rangle = \frac{|hv\rangle - |vh\rangle}{\sqrt{2}}$ was produced at La Palma; one photon was measured there, the other on Tenerife, 144km away. Details in R. Ursin et al., 'Entanglement-based quantum communication over 144 km', *Nature Physics* 3 (2007), 481–486. Meanwhile it is about other distances, see e.g. Juan Yin et al., 'Satellite-based entanglement distribution over 1200 kilometers', *Science* 356 (2017), 1140–114416 Jun 2017: Vol. 356, Issue 6343, pp., https://doi.org/10.1126/science.aan3211

Moreover, the 'speed' of entanglement as a quantum mechanical, nonlocal connection was experimentally determined to be at least 4 orders of magnitude greater than the speed of light; see D. Salart et al., 'Testing spooky action at a distance', *Nature* 454 (2008), 861–864.

[15]Locality refers to the following requirement: when two (sub-) systems $A$ and $B$ cannot interact with each other (for example due to a correspondingly large spatiotemporal separation), then modifications of $A$ cannot lead to changes in $B$. See also the section 'EPR', below.

fact which was not accepted by many physicists for a long time. Einstein was one of them; he regarded non-locality as a hardly-credible 'spooky action at a distance'. His objections and those of others are summarized in the EPR paradox, which we discuss below.

To sum up: With entangled states, one of the peculiarities of quantum mechanics, it does not make sense to speak of well-defined properties of one of the two partners. Only when a measurement is performed do the states of the *single* quantum objects become defined and the entanglement is broken. Thus, the measurement of *one* of the two partners also fixes the corresponding properties of the *other* partner, regardless of their mutual distance. Mathematically, the special point of entangled states manifests itself in the fact that they do not factorize, i.e. they cannot be written as products of the individual states (they cannot be 'separated', being instead 'entangled').

So we have found another 'special' concept of quantum mechanics. To those that we have already considered, such as the necessity of probability, the effect of measurements, and state reduction, we now must add entanglement. We note that, despite the difficulties that they may cause for our understanding, entangled states are mathematically just 'normal' states. They can, for example, form a basis of the state space, for instance in the form of *Bell states*:

$$\left|\Psi^+\right\rangle = \frac{|hv\rangle + |vh\rangle}{\sqrt{2}} \ ; \ \left|\Psi^-\right\rangle = \frac{|hv\rangle - |vh\rangle}{\sqrt{2}}$$
$$\left|\Phi^+\right\rangle = \frac{|hh\rangle + |vv\rangle}{\sqrt{2}} \ ; \ \left|\Phi^-\right\rangle = \frac{|hh\rangle - |vv\rangle}{\sqrt{2}}. \tag{20.14}$$

It is a basis of $\mathcal{H}_1 \otimes \mathcal{H}_2$, like any other.[16]

Finally, we want to point out that an entangled state cannot be 'disentangled' by a subtle change of the basis of the single quantum objects; the property of entanglement is preserved even in a different basis (see exercises).

### 20.2.3 Schrödinger's Cat

This is a famous example that shows what problematic consequences may result from entanglement—at least when one transfers it unthinkingly to macroscopic systems.

We first consider a physically harmless case; cf. Fig. 20.2.

A photon is incident horizontally, i.e. in the state $|H\rangle$, on a beam splitter BS. After passing through the beam splitter, it is in the state $\frac{(1+i)}{2}[|H\rangle + i\,|V\rangle]$ (see Chap. 6, Vol. 1). In shorthand notation, we write this as $|H\rangle \to \frac{(1+i)}{2}[|H\rangle + i\,|V\rangle]$. Now we incorporate—this is the new feature—the two detectors in the description also, namely by a 'detector'-ket with the following properties: If neither detector

---

[16] $\left|\Psi^-\right\rangle = \frac{|hv\rangle - |vh\rangle}{\sqrt{2}}$ is a singlet state (the global sign changes on interchange of 1 and 2), the other form is a triplet (the global sign is invariant on interchange of 1 and 2).

**Fig. 20.2** A photon is
incident on a beam splitter
BS and is detected in one of
the detectors DH or DV

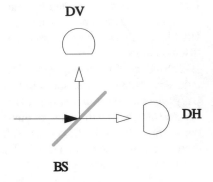

clicks, we have $|00\rangle$; if DH or DV clicks, the state reads $|10\rangle$ or $|01\rangle$. Then we can
divide the process into three stages as follows:

$$|H\rangle\,|00\rangle \;\rightarrow\; \frac{(1+i)}{2}\,[|H\rangle + i\,|V\rangle)]\,|00\rangle \;\rightarrow\; \frac{(1+i)}{2}\,[|H\rangle\,|10\rangle + i\,|V\rangle\,|01\rangle)]\,.$$

$$(20.15)$$

In words: At the beginning, the photon is in the state $|H\rangle$, no detector is activated
and the total state is factorizable. This is initially also the case when the photon has
passed through the beam splitter; eventually, with probability 50 %, it impinges on
one of the two detectors and activates it. As we can see directly, the final state in
(20.15) is entangled.

This may not seem particularly remarkable—but what happens if we choose a
cat as detector? Indeed, in the course of the debate about the EPR paradox (see
below), Erwin Schrödinger[17] published a thought experiment in 1935 pointing out
the deficiencies which in his view existed in quantum mechanics. To illustrate these,
he introduced a very special measuring apparatus:

> One can even set up quite ridiculous cases. A cat is trapped in a steel chamber, along with
> the following device (which must be secured against direct interference by the cat): adjacent
> to a Geiger counter, there is a tiny bit of a radioactive substance, so little that in the course
> of an hour, perhaps only one of its atoms decays; but, with equal probability, perhaps none
> decays. If a decay occurs, the counter tube discharges, and through a relay a hammer is
> released and shatters a small flask of hydrocyanic acid. If this entire system is left to itself
> for an hour, one would say that the cat is still alive if meanwhile no atom has decayed.
> The psi-function of the entire system would express this by containing both the live and the
> dead cat (s.v.v.)[18] mixed or smeared out in two equal parts. It is typical of such examples
> that an indeterminacy originally restricted to the atomic domain has been transferred to a
> macroscopic indeterminacy, which can then be resolved by direct observation. This prevents
> us from naively accepting a 'fuzzy model' as valid for representing reality.

We look at this in more detail, using the notation based on the example of the beam
splitter. For this purpose we describe the states of the radioactive atom by $|A\neg d\rangle$

---

[17]Erwin Schrödinger, *Naturwissenschaften* 23 (1935), p. 812.

[18]Latin '*sit venia verbo*': pardon the expression.

(not decayed)[19] and $|Ad\rangle$ (decayed) and the states of the cat by $|Ca\rangle$ (alive) and $|Cd\rangle$ (dead). We start at time $t = 0$ with the state $|A\neg d\rangle\,|Ca\rangle$, which changes to a linear combination of the form:

$$|A\neg d\rangle\,|Ca\rangle \rightarrow |A\neg d\rangle\,|Ca\rangle + |Ad\rangle\,|Cd\rangle \qquad (20.16)$$

(the kets are time dependent, and the normalization does not matter here). Obviously, we have here a normal entangled state like the one in the example of the beam splitter (20.15). The difference is just that in this state, a cat appears simultaneously to be dead and alive, which contradicts our everyday experience.

We are therefore faced with the situation that on the one hand, entanglement occurs in microscopic systems, and *must* occur—for instance, the Pauli principle for two electrons necessarily requires entanglement (see Chap. 23). On the other hand, entanglement is never observed in cats, chess boards, socks or grand pianos. In the macroscopic view, the problem is not only with macroscopic superpositions, but also with the fact that entangled states obtain objective properties only through a measurement. Entanglement suggests a holistic structure of the world. This is in conflict with our everyday experience and the (reductionist) method of the natural sciences based on it, whose success is partly due to the fact that one can examine individual subsystems of large integrated systems.[20]

A way out of this tricky situation is offered for example by the theory of decoherence, which we will examine in more detail in Chap. 24. According to it, state superpositions collapse due to interactions with the *environment*—not only through (possibly man-made) measurements. But interactions with the environment cannot be excluded in practice for a macroscopic system—it would have to be totally shielded from the outside world. This is extremely difficult even for microscopic systems, as the slow pace of development of quantum computers clearly illustrates. So decoherence effects prevent paradoxical mixed states containing both life and death, not only for cats.

### 20.2.4  A Misunderstanding

We want to warn of an obvious misunderstanding, which we illustrate with an example of suitcases and socks. Suppose we put a yellow and a blue sock into each of two suitcases, outwardly indistinguishable, so that we cannot say after closing the suitcases which one contains which sock. Then we send one suitcase to Greenland and take the other one with us to Tasmania. Here, we open the suitcase that we brought

---

[19]The symbol $\neg a$ is the (logical) negation of the property $a$. We assume that the relevant variables are dichotomous or binary, i.e. they take on one of two values, either $a$ or $\neg a$. Intermediate forms cannot occur, a third possibility does not exist (*tertium non datur*). This applies of course also to the pair $(a, d)$; states *between* 'alive' and 'dead' are excluded.

[20]To describe the motion of a simple pendulum in the lab, we do not need any information about, for example, solar flares, or the total number of penguins in the world.

**Fig. 20.3** Bertlmann's
socks. From John Stewart
Bell, 'Bertlmann's socks and
the nature of reality', in
*J. Phys. Colloq.* 42, C22
(1981) pp. C2.41–C2.62

along, see a yellow sock and know at the same moment with certainty that the blue
sock is in Greenland (classically correlated events). In this case, the measurement
(=opening of the suitcase + looking) annuls our lack of knowledge of the system—
and this is *quite a different matter* from the assignment of values by measurement in
the case of entangled quantum objects. In the example of the suitcases, entanglement
would mean that the two socks do not *have* a well-defined color[21]; instead, they
would *obtain* it (either yellow or blue) only when the suitcase is opened (quantum-
mechanically correlated events). So in the quantum-mechanical case, the problem is
not that we do not know the color of the sock in a particular suitcase; but rather the
fact that the color 'emerges' only due to the opening of the suitcase—only then, that
is at the moment of 'measurement', do the socks receive (and show) their colors.

Apropos, the use of socks as examples in this type of problem has a long tradition.
In 1981, John Bell wrote the article 'Bertlmann's socks and the nature of reality'.
Reinhold Bertlmann, then working at CERN, always wore two different-colored
socks. When one saw a pink sock, it was known with certainty that the other sock
was *not* pink. Bell contrasted this with entanglement in quantum mechanics. Compare
Fig. 20.3.

## 20.3   The EPR Paradox

The view that one cannot always assign well-defined properties to objects, or that
there are non-local processes, was the subject of serious discussions even in the
early days of quantum mechanics. It contrasted with the belief that quantum objects

---

[21]The allocation of the mixed color green (=yellow + blue) before the measurement would also
not be correct, because then the color of the socks would be a mixture of two definite states. At best,
in a universe in which there are only yellow and blue and no other colors, i.e. in which green is an
unknown color, one could argue using green socks (however, precisely this color does not exist in
that universe).

must have an autonomous reality that is independent of measurements—i.e. that an electron has spin, position and so on, even if it is not observed (boldly transferred to macroscopic conditions in Einstein's question, "Do you really think the moon isn't there if you aren't looking at it?").[22]

This dissatisfaction with the unusual perspective of quantum mechanics was focussed by the famous 'EPR paper'.[23] In 1935, Albert Einstein, Boris Podolsky and Nathan Rosen published an article intending to show that quantum mechanics does not satisfy the requirements which an acceptable physical theory generally has to fulfill. Essentially, these requirements are the following:

- *Reality*: A physical quantity whose value can be predicted with certainty is a property of a physical system $A$. In other words, the system 'has' or 'owns' this property, independently of measurements. This property is therefore an element of physical reality (Einstein reality).
- *Locality*: The result of a measurement on a system $A$ is not influenced by manipulations of other systems $B$ which are space-like separated[24] from $A$—or, in another formulation: What exists in $B$ does not depend on what is measured in $A$ (Einstein locality). Realism and locality together are also referred to as *local realism*. Of course, in order for the term locality to be meaningful, the system $A$ must be characterizable in terms of its own intrinsic properties, regardless of the properties of the other systems $B$ ( *separability*).
- *Completeness*: A physical theory must be complete within the limits of its validity. This means that there must be a corresponding theoretical counterpart (within the frame of the model) for each element of physical reality. If quantum mechanics were not complete, then there must exist in addition to the state $|\psi\rangle$ further (albeit unknown) variables $\Lambda$, the so-called *hidden variables*. Knowledge of those variables would allow a complete description, since then each observable $A$ has an *objectively existing* value $A(\psi, \Lambda)$.

The EPR paper formulated its objections on the basis of the non-commuting variables *momentum* and *position*. A simpler design of the relevant (at that time) *thought* experiment was introduced in 1952 by the U.S. physicist D. Bohm. We consider a system with a total spin of zero, which decays into two *entangled* quantum objects, each with spin $\frac{1}{2}$ (e.g. an electron—positron pair). One of them (Q1) moves off to the left, the other one (Q2) to the right; cf. Fig. (20.4).[25] When the quantum objects are so far apart that they no longer interact with each other, we measure the

---

[22]The (current, i.e. 1981) response of the U.S. physicist David Mermin: "We now know that the moon is demonstrably not there when nobody looks". This answer is correct under microscopic conditions (at least, according to the majority view of the physics community), but not under macroscopic ones (due to decoherence, see Chap. 24). In this sense it is a witty, eye-catching answer to a striking question, a *bon mot* with a certain truth content. And if the question of the moon appears too banal—is the rainbow also there when nobody looks?

[23]EPR is an acronym for the last names of the three authors.

[24]This means that the one system is not within the light cone of the other.

[25]Since this experiment is discussed generally with electrons/spin, we present it here in this way. Of course, one could instead use photons/polarization. For this, recall the polarization operators,

**Fig. 20.4** Decay of a singlet state into two oppositely-oriented spins

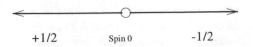

+1/2                    Spin 0                    -1/2

orientation of one spin, for instance that of Q1. Suppose we first measure $s_x$ and obtain $+\frac{\hbar}{2}$. Then we know for sure that Q2 is in the state $-\frac{\hbar}{2}$. We also know that the two quantum objects do not affect each other, because of their sufficiently large spatial separation. Then it follows as a consequence that the state $-\frac{\hbar}{2}$ of Q2 is an element of physical reality in the above sense.

The same reasoning can now be applied to the other two spin components $s_y$ and $s_z$. Thus, the *three* spin components of Q2 are elements of reality, as we can predict their values with certainty without measuring them. But this is contrary to quantum mechanics, which states that only *one* spin component can be determined, since the three spatial components of the angular momentum do not commute. So EPR concludes that quantum mechanics is not complete.[26]

Niels Bohr then responded by noting that the term 'physical reality' can refer only to situations in which the experimental setup is completely determined; but this was not the case here, because the system is perturbed by the decision of the experimenter to measure e.g. the spin component along the $x$-axis instead of the $z$-axis. Therefore, only the specific experimental setup, i.e. the context of measurement (contextuality), determines to which quantity physical reality can be attributed.

Are the observed correlations thus classical or quantum mechanical? In the classical case, the spin components of Q1 and Q2 would be equal and opposite, because the total spin is zero. But that would result from the fact that the spin vectors have well-defined values and directions *from the beginning*, and the process of measuring Q1 would not disturb Q2 in any way. In this case (and the incompleteness of quantum mechanics in the view of EPR suggests this assumption), there would therefore have to be a theory 'underlying' quantum mechanics, i.e. a theory with hidden variables.

On the other hand, one can argue quite clearly that the individual components of entangled states are not in a well-defined state from the beginning. For this, we consider two polarization-entangled photons moving apart in the state

$$|\Phi^+\rangle = \frac{|hh\rangle + |vv\rangle}{\sqrt{2}}. \qquad (20.17)$$

Each photon is incident on an analyzer. With the same, otherwise arbitrary orientation $\alpha$ of the two analyzers, either *both* photons pass through the analyzers (with probability $\cos^2 \alpha$), or they are *both* absorbed. If now the polarization direction of the two photons were determined from the start, then $\cos^2 \alpha$ would be the probability for each photon to pass its analyzer, *independently* of one another. Consequently it would be observed from time to time that e.g. only *one* photon is absorbed. Since this

---

defined in Chap. 4, Vol. 1, for linear, circular, and 45°-rotated linear polarization, i.e. $P_L = |h\rangle$
$\langle h| - |v\rangle \langle v| = \sigma_z$, $P_C = |r\rangle \langle r| - |l\rangle \langle l| = \sigma_y$ and $P_{L'} = |h'\rangle \langle h'| - |v'\rangle \langle v'| = \sigma_x$.

[26]However, whether a complete description exists at all is not discussed in the EPR paper.

is never observed, the assumption seems natural that the photons have *no* definite polarization before the measurement, but that it is determined only by the measurement.

However, as we said above, this is just an (albeit very plausible) assumption. We can directly confirm it in this experiment only if we show for *all* the photon pairs that they both pass or are both absorbed. But it is obviously impossible to verify this assumption. To proceed, we need a falsifiable statement. This is Bell's inequality, which we will discuss in the next section.

## 20.4   Bell's Inequality

How can we distinguish experimentally between classical and quantum correlations? It was John Bell who gave an answer to this question in 1964 by formulating a general inequality.[27,28] There are two remarkable points about this inequality: on the one hand, it is very simple, and on the other hand, it can be *checked experimentally*. The results of the experiments show that quantum mechanics and local realism are not compatible.

### 20.4.1   Derivation of Bell's Inequality

We consider a set whose elements can be characterized by three dichotomous properties, namely $a, b, c$. For example, $n(a, \neg b)$ means the set of elements that have the property $a$, but not $b$ (nothing is said about $c$ ); $n(a, b, \neg c)$ is the set of all elements with the properties $a$ and $b$, but not $c$. We emphasize that it is quite essential that the elements have these properties, regardless of whether we measure (or inquire into) them or not. Then we have

$$n(a, c) = n(a, \neg b, c) + n(a, b, c)$$
$$n(b, \neg c) = n(a, b, \neg c) + n(\neg a, b, \neg c). \qquad (20.18)$$

Since all $n$ are positive, it follows that

$$n(a, c) \geq n(a, b, c)$$
$$n(b, \neg c) \geq n(a, b, \neg c). \qquad (20.19)$$

The addition of the last two inequalities leads to

---

[27]In fact, there are several ways to establish such inequalities, which is why often the plural 'Bell's inequalities' is used.

[28]John Stewart Bell, 1928–1990, Northern Irish physicist.

$$n(a, c) + n(b, \neg c) \geq n(a, b, c) + n(a, b, \neg c) = n(a, b), \tag{20.20}$$

or, compactly,

$$n(a, b) \leq n(a, c) + n(b, \neg c). \tag{20.21}$$

This is *Bell's inequality* (or one possible formulation thereof).[29]

A concrete example: In a population, we distinguish the categories of female–male ($f$ and $\neg f$), eye color blue–not blue ($b$ and $\neg b$), and size short–not short ($\leq 170\,$cm and $> 170\,$cm; $s$ and $\neg s$). Then (20.21) states that the number of women with blue eyes is less than or equal to the number of short women plus the number of tall persons with blue eyes. Even if this fact perhaps is not obvious at first glance, it is still correct.

It should be emphasized once more that Bell's inequality (20.21) is based exclusively on the fact that objects *have* (in the sense of possess or own) uniquely fixed characteristics or properties. In this context, the inequality applies generally and is not restricted to the realm of quantum mechanics.

One of the simplest possibilities for testing the inequality (20.21) is given by considering the polarization of entangled photons. In this case, the characteristics $a$, $b$ and $c$ are represented by three different polarization (or analyzer) settings.

### 20.4.2   EPR Photon Pairs

To begin with, we consider two entangled photons which are emitted by a source and move apart in opposite directions, whereby each photon impinges on a polarization analyzer. The state of the system is

$$\left| \Phi^+ \right\rangle = \frac{|hh\rangle + |vv\rangle}{\sqrt{2}}. \tag{20.22}$$

We use *different* orientations of the analyzers 1 and 2, i.e. $\alpha \neq \beta$, see Fig. 20.5. To avoid a 'collusion' of the two photons, the analyzer settings are chosen only after both photons are underway so that agreement would require a superluminal exchange of information. If the analyzer settings are different, the two photons will not necessarily suffer the same fate (passing or being absorbed), but of course, probability statements are also possible for this setup.

For reasons of clarity, we assume that photon 1 arrives first at its analyzer. Since the state $\left| \Phi^+ \right\rangle$ includes any direction, so to speak, photon 1 passes its analyzer with a probability of $\frac{1}{2}$ (and it is absorbed with the same probability).[30] If it was not absorbed, then photon 1 has a well-defined polarization, namely $\alpha$. Thus, the polarization of photon 2 is also fixed—it is also $\alpha$. Consequently, the angle between the analyzer

---

[29] A derivation of this inequality based on set theory is given in Appendix K, Vol. 2.

[30] For the explicit calculation of the probabilities used in this paragraph, see the exercises.

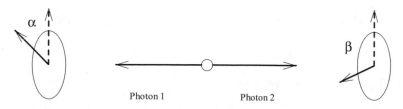

**Fig. 20.5** Two entangled photons are incident on differently-oriented analyzers

setting and the polarization direction is $\alpha - \beta$, and we can write the probability for the passage of photon 2 with the help of Malus' law as $\cos^2 (\alpha - \beta)$. The probability that both photons pass their analyzers is thus given by

$$p(\alpha, \beta) = \frac{1}{2} \cos^2(\alpha - \beta). \tag{20.23}$$

This probability depends on the difference of the angles and is symmetric w.r.t. reversal of the angles, as it indeed should be.

Analogously, we obtain the probability that one photon is absorbed. Photon 1 passes with probability $\frac{1}{2}$, photon 2 is absorbed with probability $1 - \cos^2 (\alpha - \beta) = \sin^2 (\alpha - \beta)$. With the notation $\neg\beta$, meaning absorption at the angle setting $\beta$, it follows that

$$p(\alpha, \neg\beta) = \frac{1}{2} \sin^2(\alpha - \beta). \tag{20.24}$$

To verify experimentally these two equations, one has to measure many pairs of photons (strictly speaking, an ensemble), $N \gg 1$. Then we have

$$n(\alpha, \beta) = \frac{N}{2} \cos^2(\alpha - \beta); \quad n(\alpha, \neg\beta) = \frac{N}{2} \sin^2(\alpha - \beta). \tag{20.25}$$

### 20.4.3 EPR and Bell

We can now apply the form of Bell's inequality derived above if we assume that it is an objective property of the photon (i.e. determined even before the measurement) to pass an analyzer at a given orientation (or not to pass). We need *three*[31] analyzer settings, $\alpha$, $\beta$ and $\gamma$, where the three differences $\alpha - \beta$, $\alpha - \gamma$, $\beta - \gamma$ should occur on the average with equal frequencies (so that the frequencies $n$ in (20.25) for the three different angle differences each are the same). The directions of the analyzers are chosen only when the photons are already underway.

---

[31]Note that this consideration is not based on two parameters (or one relative parameter) as in Bohm's thought experiment, but rather on three (or two relative) parameters.

For the triple of analyzer settings $(\alpha, \beta, \gamma)$, Bell's inequality reads

$$n\,(\alpha, \beta) \leq n\,(\alpha, \gamma) + n\,(\beta, \neg\gamma)\,. \tag{20.26}$$

With (20.25), this can be written as

$$\cos^2(\alpha - \beta) \leq \cos^2(\alpha - \gamma) + \sin^2(\beta - \gamma)\,. \tag{20.27}$$

In order to show more clearly that this inequality is not fulfilled for certain angles, and which angles those are, we now simplify, setting $\alpha = 0$ without loss of generality. It follows that

$$\cos^2\beta \leq \cos^2\gamma + \sin^2(\beta - \gamma)\,. \tag{20.28}$$

We can transform this to give[32]:

$$\sin(\gamma + \beta)\sin(\gamma - \beta) \leq \sin^2(\gamma - \beta)\,. \tag{20.29}$$

Rearranging this last expression yields

$$\sin(\gamma - \beta)\cos\gamma\sin\beta \leq 0\,. \tag{20.30}$$

Since the orientation is not fixed, we can set $0 < \beta < \pi$. This means $\sin\beta > 0$; therefore the inequality

$$\sin(\gamma - \beta)\cos\gamma \leq 0\,. \tag{20.31}$$

must hold if our assumption is valid that there exist hidden variables. In other words: if there are angles $\beta$ and $\gamma$ for which the inequality

$$0 < \sin(\gamma - \beta)\cos\gamma \tag{20.32}$$

is fulfilled, that assumption is *not* valid.

Now one sees directly that the last inequality is satisfied for $0 < \gamma - \beta < \pi$ and $0 < \gamma < \pi/2$. This proves that the postulated hidden variables do not exist. In fact, the function $f(\beta, \gamma) = \sin(\gamma - \beta)\cos\gamma$ always has positive ranges for $\beta \neq \pi/2$ (see exercises), as shown by an example in Fig. 20.6 for two different values of $\beta$. Hence, the view of EPR is not sustainable; the results of quantum mechanics cannot be explained by a local-realistic theory.

What do experiments say about Bell's inequalities? Their violation by quantum mechanics has been studied experimentally since the late 1960s; however, at first the results seemed not to be taken very seriously. The breakthrough came with an

---

[32]For the transformations, see the exercises.

**Fig. 20.6** The function $\sin(\gamma - \beta)\cos\gamma$ for $\beta = \frac{\pi}{24}$ (*red*) and $\beta = \frac{\pi}{3}$ (*blue*). The positive components shown here demonstrate that quantum mechanics violates local realism

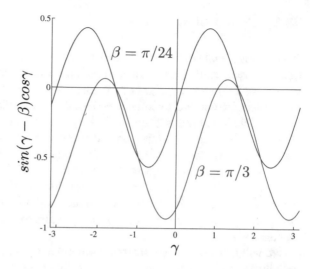

experiment conducted in 1981 in Paris.[33] It convinced (almost) all the doubters. Since then, the facts have been confirmed in a series of experiments. Among others, in an experiment performed in 1998, the polarizing filters were 400 m apart, so that a 'collusion' of the entangled photons was impossible due to the finite speed of light.[34]

An addendum: In 1989, an experiment was presented with three quantum objects, entangled in a certain way (Green, Horne, Zeilinger, GHZ). Here, a *single* measurement is sufficient to refute local realism, while for Bell's inequalities, one needs a large number of measurements to check the inequalities. The GHZ states are discussed in more detail in Chap. 27.

---

[33] Alain Aspect, Philippe Grangier and Gérard Roger, 'Experimental Realization of Einstein-Podolsky-Rosen-Bohm Gedankenexperiment: A New Violation of Bell's Inequalities', *Phys. Rev. Lett.* 49, 91–94 (1982).

[34] It is not surprising that the experimental reality is more complicated than described here. Upon closer inspection it turned out, for example, that the conditions of the relevant experiments did not fully satisfy the hypothesis underlying the Bell inequality. This means that there are so-called loopholes, i.e. possibilities of explaining the experimental results by means of local-hidden-variable theories. These loopholes are related (i) to the separation between the local measurements and (ii) to the detection efficiency. For instance, in certain experiments, not all the photons were detected. Thus, one can argue (at least in principle) that only the detected photons agree with quantum mechanics, while the entire ensemble satisfies Bell inequalities. Of course one may assume, in contrast, that each sample of pairs of photons detected is representative of all the pairs emitted (the 'fair sampling assumption'); or, in other words, that nature is not malicious, but this does not refute the argument. However, in the meantime a series of 'loophole-free' experiments have been carried out, for instance: M. Giustina et al., 'Bell violation with entangled photons, free of the fair-sampling assumption', arXiv:1212.0533 (2012); A. Cabello and F. Sciarrino, 'Loophole-Free Bell Test Based on Local Precertification of Photon's Presence', *Phys. Rev. X* 2, 021010 (2012); B. Wittman et al., 'Loophole-free Einstein-Podolsky-Rosen experiment via quantum steering', *New J. of Phys.* 14, 053030 (2012); W. Rosenfeld et al., 'Event-Ready Bell Test Using Entangled Atoms Simultaneously Closing Detection and Locality Loopholes', *Phys. Rev. Lett.* 119, 010402 (Jul 2017). The current state of discussion is found in 'The BIG Bell Test Collaboration', *Nature* 557, 212 (2018).

## 20.5   Conclusions

We note once again that our derivation of Bell's inequalities is independent of quantum mechanics and is based upon the assumptions of local realism: (1) measurement values 'really' exist, regardless of whether they are measured or not; (2) properties of a system are not directly influenced by other space-like separated systems.[35]

The violation of Bell's inequality[36] by quantum mechanics can therefore in principle have three causes: Either not all measurement values are fixed prior to the measurement; or the measurement results depend non-locally on (arbitrarily) distant random decisions; or quantum mechanics is neither realistic nor local (in the Einsteinian sense). At this point, we cannot discern what is really the cause. But in any case, it is certain that either quantum mechanics is complete, or the hidden variables have to have very exotic properties. A return to the concrete world of classical mechanics is ruled out.

We will take up these considerations again in Chap. 27. Here, we want to note in anticipation that it is now largely accepted that reality simply is not independent of measurement. All known facts seem to indicate that we must definitely abandon the idea that the properties of quantum objects exist independently of our observations or, more generally, of their environment (as we have indeed always assumed).[37]

Finally, a word about non-locality. Locality is apparently violated in the measurement of entangled states, since we then have a nonlocal (i.e. faster than light) change

---

[35] In principle, one could explain an experimental violation of the Bell inequality by models based on hidden influences propagating at a finite (and sufficiently high) speed $v > c$. But as may be shown, such models predict, for any finite speed $v$ with $c < v < \infty$, correlations that can be exploited for faster-than-light communication. Thus, assuming the impossibility of using nonlocal correlations for superluminal communication, any possible explanation of quantum correlations in terms of influences propagating at any finite speed can be excluded. See J-D. Bancal et al., 'Quantum non-locality based on finite-speed causal influences leads to superluminal signalling', *Nature Physics* 8 867–870 (2012), https://doi.org/10.1038/nphys2460.

[36] In addition to the Bell inequalities, there are other inequalities based on certain classical properties which can be proven wrong in quantum mechanics. An example are the Leggett–Garg inequalities (named for A.J. Leggett & A. Garg, 'Quantum Mechanics versus macroscopic realism: is the flux there when nobody looks?', *Phys. Rev. Lett.* 54, 857 (1985)); for details see e.g. A.J. Leggett, 'Testing the limits of quantum mechanics: motivation, state of play, prospects', *J. Phys. Condens. Matter* 14 (2002) R415–R451. An experimental realization was reported by George C. Knee et al., 'Violation of a Leggett-Garg inequality with ideal non-invasive measurements', *Nature Communications* 3, Article Number 606; https://doi.org/10.1038/nscomms1614 (2012). The results demonstrate clearly the necessity of a non-classical picture.

[37] The question 'What is real?' has been discussed in philosophy for thousands of years—in this sense, one can indeed say with some justification that this basic research into quantum mechanics is a kind of experimental philosophy.

The subject has also fascinated writers over and over. One of countless examples: "Imagine a man in a library. The books are all empty, until he pulls one out. Then the simulators—or whoever—fill the book with print. But only as long as he is leafing through it, and with a minimum of words, just in time, just as his eye turns to the page. If he returns the book and looks for a different one, the simulators make that book exist. Most of the library is a bluff, just a lot of book covers, which don't even have titles if you aren't looking too closely." Jonathan Lethem: *Chronic City* (2011), p. 281.

of state. However, locality is not violated in the sense that using entangled photons would make a superluminal *information exchange* possible. If one for example looks only at analyzer 2, half of the photons pass, half are absorbed. But this behavior does not permit conclusions about the setting of analyzer 1. Only in retrospect can the comparison (which can be carried out at most with the speed of light) tell us whether or not entanglement existed. In short, there are instantaneous correlations, but no instantaneous interactions.[38]

## 20.6 Exercises

1. Given two matrices $A$ and $B$ with

$$A = \begin{pmatrix} 1 & 3 \\ 2 & 1 \end{pmatrix}; \quad B = \begin{pmatrix} 1 & 0 \\ 2 & 1 \end{pmatrix}. \tag{20.33}$$

Determine $A \otimes B$.

2. Represent the Bell states (20.14) as column vectors. Show in this representation that the Bell states are entangled and that they form a CONS.

3. Two photons are in the state

$$|\Psi\rangle = \frac{|hv\rangle - |vh\rangle}{\sqrt{2}}. \tag{20.34}$$

(a) Show explicitly that it is an entangled state.

(b) Photon 1 passes an analyzer for right-handed circular polarization (the corresponding state reads $\frac{|h\rangle + i|v\rangle}{\sqrt{2}}$). Show that through a measurement, the state $|\Psi\rangle$ is changed into a product state.

4. Show that the Bell states can be transformed into each other by applying the Pauli matrices to a subsystem.

5. Show that the Bell states are eigenvectors of products of the same Pauli matrices.

6. Transform the inequality (20.27)

$$\cos^2(\alpha - \beta) \leq \cos^2(\alpha - \gamma) + \sin^2(\beta - \gamma) \tag{20.35}$$

for $\alpha = 0$ and $0 < \beta < \pi$ to give

$$\sin(\gamma - \beta)\cos\gamma \leq 0. \tag{20.36}$$

7. Given the function

$$f(\gamma, \beta) = \sin(\gamma - \beta)\cos\gamma; \tag{20.37}$$

---

[38] A very comprehensive and readable overview is found in A. Zeilinger, 'Light for the quantum. Entangled photons and their applications: a very personal perspective', *Physica Scripta* 92 072501 (2017).

determine the position of its zeros and the positions and values of its maxima with respect to $\gamma$.

8. A system of two photons is in one of the Bell states. The photon Q1 is incident on an analyzer for horizontal polarization, rotated by an angle $\alpha$. What is the probability that Q1 passes the analyzer?

9. Two photons in the state $|h_0\rangle$ are rotated by the angles $\alpha$ and $\beta$ to give the states $|h_\alpha\rangle$ and $|h_\beta\rangle$. How does the projection operator referring to $|h_\alpha h_\beta\rangle$ act on the Bell states?

10. Given two quantum objects Q1 and Q2, with an $N$-dimensional CONS $\{|\varphi_i\rangle\}$ for Q1 and $\{|\psi_i\rangle\}$ for Q2 (due to this notation we can omit the index for the number of the quantum object). The initial state is

$$|\chi\rangle = \sum_{ij} c_{ij} |\varphi_i\rangle |\psi_j\rangle. \tag{20.38}$$

What is the probability of measuring Q1 in some state $|\lambda\rangle$ (no matter which state)?

11. Show that entangled states such as the Bell states cannot be 'disentangled' by a reversible transformation of the single-quantum-object basis; entanglement is preserved even in a different basis.

12. Determine the behavior of the Bell states under reversible transformations. Consider the case of rotations.

# Chapter 21
# Symmetries and Conservation Laws

It is important and helpful to know all the conserved quantities for a given problem. In quantum mechanics, this means finding all the operators that commute with $H$. As in classical mechanics, these quantities are closely related to the symmetries of the problem. In the following, this relationship is examined in detail.

*Symmetry* means simply that there are different 'perspectives' from which a physical system looks the same. In other words, the system is invariant under a certain symmetry operation such as rotation or reflection, and its mathematical description does not change as a result of the transformation.[1] This leads to a remarkable coupling of the geometric and the dynamic properties of a system. In theoretical physics, symmetries are fundamental, because, in a sense, the basic laws of nature result from them, and they are regarded as the most successful principle for the unification of theories.[2]

There are *continuous* and *discrete* symmetries. Continuous symmetry transformations are characterized by a continuous parameter (or possibly several). The following four general continuous symmetries are central in physics[3]:

1. Homogeneity of time or time-translation invariance (the choice of the zero of time (starting time) does not matter).
2. Homogeneity of space or space-translation invariance (the choice of the origin of spacial coordinates (center point) does not matter).

---

[1] In general, there are two viewpoints in considering symmetry transformations: the passive viewpoint (the system remains unchanged, but the axes are changed accordingly), and the active viewpoint (the axes remain unchanged, but the system is transformed—which is of course not meant to be a dynamic rotation). Which point of view one prefers is a matter of taste. The best-known example is perhaps the passive and active rotation in two dimensions, which we already mentioned in Chap. 2, Vol. 1.

[2] P.W. Anderson, Nobel Laureate 1972: "It is only a slight exaggeration to say that physics is the study of symmetries."

[3] In special problems, there may of course be additional symmetries (e.g. the Lenz vector in the case of the hydrogen atom).

© Springer Nature Switzerland AG 2018
J. Pade, *Quantum Mechanics for Pedestrians 2*, Undergraduate Lecture
Notes in Physics, https://doi.org/10.1007/978-3-030-00467-5_21

3. Isotropy of space or space rotational invariance (the choice of the spatial direction does not matter).
4. Principle of relativity or invariance under specific Galilean transformations[4] (the choice of inertial frame does not matter).

The discrete symmetry transformations are:

1. Invariance under parity inversion, $\mathbf{r} \to -\mathbf{r}$ (the mirror images of all physical processes must be equally possible).
2. Invariance under time reversal $t \to -t$ (physical processes must occur equally both forwards and backwards in time).

One might think that all these symmetries are obvious, but that is not necessarily true. For example, it was previously believed that parity conservation applies to all interactions, but we know now that the weak interaction does not conserve parity (and is also not invariant under time reversal).[5]

Note that all transformations are *Galilean transformations*, i.e. transformations from one non-accelerated reference frame (inertial frame) to another one. Mathematically, the Galilean transformations form a group, the *Galilean group*. The *proper Galilean group*[6] contains the translations in space and time (shifts of the zero points), the rotations about constant angles, and the transformations into a non-rotating reference frame moving uniformly along a straight line with constant velocity $v$ (special Galilean transformations). The full group also includes the reflections (inversions) of space and time.

Continuous symmetries are associated with the existence of conserved quantities, as is summarized in Table 21.1.

This fact is known from classical physics through Noether's theorem, which states that 'To every continuous symmetry of a physical system, there belongs a conserved quantity, and *vice versa*.' In quantum mechanics, also, symmetries lead to conserved quantities[7]—in principle, the situation is the same as in classical mechanics, but it has to be formulated somewhat differently. For if the physics is to remain unchanged by a symmetry operation, then measurable quantities, i.e. eigenvalues and probabilities, must not change. This means that in quantum mechanics, such symmetry transformations are described by *unitary* operators in the Hilbert space (see Chap. 13, Vol. 1).

However, this is not the whole truth, since we must also consider anti-unitary transformations. We have already discussed the reason for this in Chap. 14, Vol. 1:

---

[4]Galilean and not Lorentz transformations, because we are considering here nonrelativistic quantum mechanics.

[5]Even Wolfgang Pauli took the validity of the symmetries for granted and consequently declared it as a priori absurd to search for parity-violating processes: "I cannot believe that God is a weak left-hander." He had to revise his views, as is known, after Madame Wu et al., demonstrated parity violation in the beta decay of cobalt-60 atoms in 1956.

[6]Number of parameters: $3 + 1 + 3 + 3 = 10$.

[7]For example, we have seen in an exercise for Chap. 9, Vol. 1 that the angular momentum is conserved in spherically symmetric problems.

**Table 21.1** Invariances and conserved quantities

| Invariance under | Conserved quantity |
|---|---|
| Temporal shift | Energy |
| Spatial shift | Momentum |
| Spatial rotation | Angular momentum |
| Change of inertial frame | Velocity of center of mass |

All normalized vectors $|\varphi\rangle$ with arbitrary phase are physically equivalent, e.g. the ray $e^{i\alpha}|\varphi\rangle$. Therefore, we cannot require that symmetry transformations be represented by unitary operators only, since this requirement stems from the assumption that the scalar products between states must remain unchanged, i.e. $\langle\varphi'|\psi'\rangle = \langle\varphi|\psi\rangle$. For rays, we need to weaken this assumption to $|\langle\varphi'|\psi'\rangle|^2 = |\langle\varphi|\psi\rangle|^2$. In this situation, a theorem of *Wigner* applies, stating that the last equation can be guaranteed only by *unitary* and *anti-unitary* transformations. With $|\varphi'\rangle = U|\varphi\rangle$, we have for unitary $U$

$$\langle\varphi'|\psi'\rangle = \langle U\varphi|U\psi\rangle = \langle\varphi|\psi\rangle; \tag{21.1}$$

while for an *anti-unitary* operator $U$, it holds that

$$\langle\varphi'|\psi'\rangle = \langle U\varphi|U\psi\rangle = \langle\varphi|\psi\rangle^* = \langle\psi|\varphi\rangle. \tag{21.2}$$

We see that an anti-unitary transformation leaves not the scalar products themselves invariant, but, as required, their absolute values.

We note that anti-unitarity is the exception; all transformations considered in the following are unitary, except time reversal, which is anti-unitary.

## 21.1 Continuous Symmetry Transformations

### 21.1.1 General: Symmetries and Conservation Laws

Continuous symmetry transformations $S$ are characterized by a continuous parameter $\alpha$. For $\alpha = 0$, $S$ represents the unit map. The parameter is additive, so that

$$S(\alpha_1 + \alpha_2) = S(\alpha_2)S(\alpha_1). \tag{21.3}$$

An example is a rotation about the $z$ axis by the angle $\alpha$. It is obviously the same if we rotate first by $\alpha_1$ and then by $\alpha_2$ or immediately by $\alpha_1 + \alpha_2$.

This symmetry operation is represented in quantum mechanics by a corresponding operator $U_S$, which acts in the Hilbert space $\mathcal{H}$. We have

$$U_S(\alpha_1 + \alpha_2) = U_S(\alpha_2)U_S(\alpha_1), \tag{21.4}$$

where $U_S$ is a *unitary* (i.e. not an anti-unitary) operator. This can be seen because $U_S(\alpha)$ for $\alpha = 0$ is the unit map (which is certainly unitary), from which the transformation for $\alpha \neq 0$ can be continuously deduced.

If (21.4) holds, then *Stone's theorem*[8] applies; it states that a Hermitian operator $T_S$ exists, such that[9]:

$$U_S(\alpha) = e^{-i\alpha T_S}. \tag{21.5}$$

$T_S$ is called the *infinitesimal generator* of the transformation[10] (see the discussion in Chap. 13, Vol. 1 about the propagator). We assume below that $U_S$ and thus also $T_S$ are not time dependent.

What leads to the name 'infinitesimal generator', and what is the relation of symmetry transformations and conservation laws in quantum mechanics? Let us consider the unitary transformation $|\psi'\rangle = U_S |\psi\rangle$. For the matrix element $\langle \psi| H |\psi\rangle$, it follows that

$$\langle \psi| H |\psi\rangle = \langle \psi| U_S^\dagger U_S H U_S^\dagger U_S |\psi\rangle = \langle \psi'| U_S H U_S^\dagger |\psi'\rangle := \langle \psi'| H' |\psi'\rangle. \tag{21.6}$$

Now we require that $H$ be invariant under the symmetry transformation represented by the unitary operator $U_S$. This means that $H = H' = U_S H U_S^\dagger$. Together with (21.5) and because of $U_S U_S^\dagger = 1$, it follows that

$$U_S H = H U_S \text{ or } [U_S, H] = 0 \text{ or } \left[ e^{-i\alpha T_S}, H \right] = 0. \tag{21.7}$$

This equation must be valid for all parameters $\alpha$, especially for very small or infinitesimal $\alpha$. In that case, we can expand[11]:

$$e^{-i\alpha T_S} = 1 - i\alpha T_S + O(\alpha^2). \tag{21.8}$$

This leads directly to

$$[T_S, H] = 0. \tag{21.9}$$

Since we have assumed that $T_S$ is not time dependent, $T_S$ is a constant of the motion, according to Ehrenfest's theorem.

So we have (under the above restrictions and clarifications) the following result: If $H$ is invariant under a continuous symmetry operation that is described in the Hilbert space $\mathcal{H}$ by the unitary operator $U_S$, then $T_S$ is a conserved quantity, defined

---

[8]The full version can be found in Appendix I, Vol. 1.

[9]This relationship is suggested, *inter alia*, by the fact that a function $f(x)$ with the property $f(x + y) = f(x)f(y)$ is given by a (generalized) exponential function $a^x$.

[10]The sign and any other multiplicative constants cannot be determined at this point ($e^{i\alpha T_S}, e^{-i\alpha T_S/\hbar}$, etc.).

[11]Why is one not content with $[U_S, H] = 0$? The answer is because $U$ is unitary, but $T$ is Hermitian and therefore is possibly a measurable variable.

by $U_S = e^{-i\alpha T_S}$. In this way, we have found a direct relationship between symmetries (or geometry) and conserved quantities (or dynamics) in quantum mechanics.

Of course, $T_S$ as a Hermitian operator is a good candidate for a physical observable, and we will indeed discuss below only the latter. For the fundamental transformations mentioned in the introduction, we have:

Time translation (we already know this operator from Chap. 13, Vol. 1)

$$U(t) = e^{-i\frac{tH}{\hbar}}. \tag{21.10}$$

Space translation by a distance $\mathbf{a}$ ($\mathbf{p}$ is the momentum operator):

$$U(\mathbf{a}) = e^{-i\frac{\mathbf{p}\mathbf{a}}{\hbar}}. \tag{21.11}$$

Spatial rotation about an axis $\hat{\mathbf{n}}$ by an angle $\vartheta$ ($\mathbf{j}$ is the angular-momentum operator):

$$U(\hat{\mathbf{n}}, \vartheta) = e^{-i\frac{\vartheta \mathbf{j} \cdot \hat{\mathbf{n}}}{\hbar}}. \tag{21.12}$$

Space translation with constant velocity $\mathbf{v}$ (special Galilei transformation or *boost*):

$$U(\mathbf{v}) = e^{-i\frac{\mathbf{v} \cdot \mathbf{G}}{\hbar}} \tag{21.13}$$

with $\mathbf{G} = \mathbf{p}t - m\mathbf{x}$. Due to space and time constraints, the special Galilean transformation is described in detail only in Appendix L, Vol. 2.

One can also interpret the facts in such a way that these equations *define* the operators representing the physical quantities of energy, momentum, angular momentum, and position. The denominator $\hbar$ ensures that the exponents are dimensionless. For all these transformations, we have $U(-\xi) = U^\dagger(\xi)$, where $\xi$ is the corresponding set of parameters of the transformation.

### 21.1.2 Time Translation

The time evolution operator

$$U(t) = e^{-i\frac{tH}{\hbar}}. \tag{21.14}$$

was already derived in Chap. 13, Vol. 1; the precondition was $\frac{\partial H}{\partial t} = 0$. For infinitesimal $\tau$, we have

$$U(\tau) = 1 - i\frac{\tau}{\hbar}H, \tag{21.15}$$

and we can regard $H$ as the infinitesimal generator of a time translation. Since the time in nonrelativistic quantum mechanics is not a measurable variable like the position, there is no commutator as in (21.9); instead, we have simply the result that $\frac{\partial H}{\partial t} = 0$ implies the conservation of energy.

**Fig. 21.1** Shift of the
function $f(x)$ by $a$

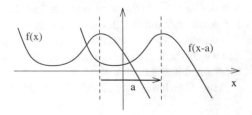

### 21.1.3  Spatial Translation

#### 21.1.3.1  Translation and Conservation of Momentum

By means of a particularly simple example, we first want to illustrate once more
how a transformation leads to a unitary operator. We start from an (infinitely often
differentiable) function $f(x)$. If we shift this function to the right by $a$, we obtain
the new function $f_a(x)$; see Fig. 21.1. This active transformation, which corresponds
to a shift of the coordinate system by $-a$, gives the new position $x \to x_a = x + a$.
Because of $f_a(x_a) = f(x)$, it follows that $f_a(x) = f(x - a)$. At the position $x - a$,
we have the Taylor expansion

$$f(x - a) = \sum_n (-1)^n \frac{a^n}{n!} \frac{d^n}{dx^n} f(x). \tag{21.16}$$

With the position representation of the momentum operator,

$$p = \frac{\hbar}{i} \frac{d}{dx}, \tag{21.17}$$

it follows that

$$f(x - a) = \sum_n \frac{a^n}{n!} \left( -\frac{ip}{\hbar} \right)^n f(x) = e^{-i \frac{pa}{\hbar}} f(x). \tag{21.18}$$

Thus we have found (for the one-dimensional case) the unitary operator that describes
the space translation. It holds that

$$f_a(x) = f(x - a) = e^{-i \frac{pa}{\hbar}} f(x). \tag{21.19}$$

In words: The application of the operator $e^{-i \frac{pa}{\hbar}}$ to a function $f(x)$ shifts it to the right
by $a$. The generalization to the three-dimensional result (21.11) and to the abstract
momentum operator is carried out analogously.

The infinitesimal generator of the spatial translation is thus **p**. From (21.9), we
have conservation of momentum if and only if

$$[\mathbf{p}, H] = 0. \tag{21.20}$$

Correspondingly, if we consider just *one* quantum object, momentum can be conserved only if there is no potential (or more precisely if $\nabla V = 0$).

For $N$ quantum objects, the transformation reads $e^{-i\frac{\mathbf{P}a}{\hbar}}$, with the total momentum $\mathbf{P} = \sum_{i=1}^{N} \mathbf{p}_i$. We consider a closed system, in which all interactions depend only on the mutual distances between the $N$ quantum objects. Hence, the Hamiltonian does not vary due to a spatial shift. According to the above, this means $[\mathbf{P}, H] = 0$, and we see explicitly that the total momentum is conserved, because the Hamiltonian of a closed system is invariant with respect to space translations.

### 21.1.3.2  Commutation Relation

Using the connection between physical transformations and the associated unitary operators, one can also derive *commutation relations*. We look more closely at the pair position/momentum (one-dimensional). To avoid confusion, we denote the abstract position and momentum operators as $X$ and $P$, while $x$ and $p$ are eigenvalues. The position operator, for example, fulfills the (idealized) eigenvalue equation $X |x\rangle = x |x\rangle$.

For the derivation of $[X, P] = i\hbar$, we need three elements:

1. We consider a quantum object in the state $|\varphi_0\rangle$, localized around the mean position $x_0$ with a certain spread $\Delta x$. Hence we have

$$\langle \varphi_0 | X | \varphi_0 \rangle = \langle X \rangle_0 = x_0. \tag{21.21}$$

2. To perform a translation by $a$, we apply the operator $U(a) = e^{-i\frac{Pa}{\hbar}}$:

$$|\varphi_0\rangle \rightarrow |\varphi_a\rangle = U(a) |\varphi_0\rangle = e^{-i\frac{Pa}{\hbar}} |\varphi_0\rangle. \tag{21.22}$$

3. We know that in this translation, the mean position changes according to

$$x_0 \rightarrow x_0 + a. \tag{21.23}$$

Thus we obtain for the mean position of the quantum object shifted by $a$:

$$\langle X \rangle_a = \begin{cases} (21.22) \ \langle \varphi_a | X | \varphi_a \rangle = \langle \varphi_0 | U^{-1}(a) X U(a) | \varphi_0 \rangle \\ (21.23) \ x_0 + a = \langle X \rangle_0 + a = \langle \varphi_0 | X + a | \varphi_0 \rangle \end{cases} \tag{21.24}$$

Since $|\varphi_0\rangle$ is arbitrary, it follows by comparison that

$$U^{-1}(a) X U(a) = X + a. \tag{21.25}$$

Expanding in powers of $a$ and taking the limit $a \to 0$ (see exercises) leads to the canonical commutation relation for the two *abstract operators* $X$ and $P$:

$$[X, P] = i\hbar. \tag{21.26}$$

We emphasize that this equation follows *directly* from general principles—without any correspondence principles or similar devices, but only from the three steps (1) to (3), and thus from (21.25) or, in the end, from translational invariance.

From (21.25), one can also deduce (see exercises) that for a function of the position operator $f(X)$, it holds that:

$$[P, f(X)] = \frac{\hbar}{i} \frac{df(X)}{dX}, \tag{21.27}$$

not by assuming a priori that $P = \frac{\hbar}{i} \frac{d}{dX}$—this equation follows rather as a *result* of the derivation, which uses only the abstract operators $X$ and $P$. Analogously, we obtain:

$$[X, f(P)] = i\hbar \frac{df(P)}{dP}. \tag{21.28}$$

Finally, as a concrete representation, we choose the position representation with $X \to x$, $P \to \frac{\hbar}{i} \frac{d}{dx}$. It follows that

$$\left[ x, \frac{\hbar}{i} \frac{d}{dx} \right] = i\hbar. \tag{21.29}$$

By the way, all representations of the canonical commutation relations, e.g. (21.26) and (21.29), may be transformed into each other by unitary transformations,[12] and are therefore equivalent.

### 21.1.4   Spatial Rotation

A spatial rotation about an axis $\hat{\mathbf{a}}$ (unit vector) by the angle $\gamma$ is described by the unitary operator

$$U_{\hat{\mathbf{a}}}(\gamma) = e^{-i \frac{\gamma \mathbf{J}\hat{\mathbf{a}}}{\hbar}}. \tag{21.30}$$

This is quite analogous to translation (and can be formally written with the usual replacements momentum → angular momentum, etc.). A simple explicit example of a constructive derivation is given in the exercises.[13]

---

[12] We repeat the remark that a unitary transformation corresponds to a change of basis in $\mathcal{H}$.

[13] Using as example the matrix representation of $e^{-i \frac{\gamma \mathbf{J}\hat{\mathbf{a}}}{\hbar}}$ for orbital angular momentum $\mathbf{l} = 1$; see Appendix X.2, Vol. 2.

**Fig. 21.2** Derivation of the commutation relations for angular momentum

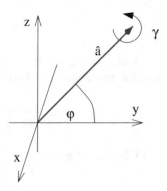

The infinitesimal generator of the spatial rotations is thus the angular momentum (operator) **j**. According to (21.9), conservation of angular momentum is assured if and only if

$$[\mathbf{j}, H] = 0. \tag{21.31}$$

Here, too, we can derive commutation relations directly. We assume that the rotation axis $\hat{\mathbf{a}}$ lies in the $y - z$ plane and makes an angle $\varphi$ with the $y$ axis; see Fig. 21.2. A rotation about $\hat{\mathbf{a}}$ by the angle $\gamma$ can then be described as follows:

First we bring the axis of rotation onto the $y$ axis by a rotation through $-\varphi$ around the $x$ axis. Then we rotate the $y$ axis by an angle $\gamma$. Finally, we bring the rotation axis back to its old position. In summary, we have for this rotation the two equivalent representations

$$e^{-i\frac{\gamma \mathbf{j} \cdot \hat{\mathbf{a}}}{\hbar}} = e^{-i\frac{\varphi j_x}{\hbar}} e^{-i\frac{\gamma j_y}{\hbar}} e^{i\frac{\varphi j_x}{\hbar}}. \tag{21.32}$$

Again we assume infinitesimal angles $\varphi$ and $\gamma$. As shown below in an exercise, the latter equation leads directly to the commutation relation

$$\left[ j_x, j_y \right] = i\hbar j_z. \tag{21.33}$$

The remaining two relations follow by cyclic permutation.

Also of interest are the commutation relations of scalar and vector operators with the angular momentum. A scalar operator $S$ is defined as an operator whose mean value remains invariant under a rotation, while for a vector operator **V**, the mean value must transform as a vector **v** (i.e. like the position vector). As shown in the exercises, a scalar operator $S$ commutes with the angular momentum

$$[\mathbf{j}, S] = 0. \tag{21.34}$$

For a vector operator **V** (such as the position or the momentum), we obtain the commutation relations[14]:

---

[14]More on vector operators is found in Appendix L, Vol 2.

$$[j_i, V_k] = i\hbar \sum_l \varepsilon_{ikl} V_l. \tag{21.35}$$

Finally, a remark concerning spin 1/2. We have the commutation relations of angular momentum (see Chap. 16)

$$[s_i, s_j] = i\hbar \sum_k \varepsilon_{ijk} s_k \tag{21.36}$$

or of the Pauli matrices

$$[\sigma_i, \sigma_j] = 2i \sum_k \varepsilon_{ijk} \sigma_k. \tag{21.37}$$

The unitary rotation transformation may be determined explicitly (see the exercises); the result reads

$$e^{-i\frac{\gamma}{\hbar}\mathbf{s}\cdot\hat{\mathbf{a}}} = e^{-i\frac{\gamma}{2}\sigma\hat{\mathbf{a}}} = \cos\frac{\gamma}{2} - i\sigma\hat{\mathbf{a}}\sin\frac{\gamma}{2}. \tag{21.38}$$

For a rotation by $\gamma = 2\pi$, it follows that

$$e^{-i\pi\sigma\hat{\mathbf{a}}} = -1. \tag{21.39}$$

Hence, a rotation through $2\pi$ does not yield the original system; this is obtained only by a rotation through $4\pi$. One might think that this is another oddity of quantum mechanics, but that is not true. Indeed, it is the case also for certain objects in our visual space, such as a Möbius strip. An ant which is running along a Möbius strip has to cover an angle of $4\pi$ to arrive back at its starting point; see Fig. (21.3). The real identity rotation for an object in relation to its environment is obviously not the rotation through $2\pi$, but rather through $4\pi$; this situation is not peculiar to quantum mechanics.

An addendum: The question may arise as to why the identity rotation of a photon, which also can assume only two (polarization) states, is not $4\pi$ as for the electron with its two (spin) states. The answer is most easily formulated using the *helicity* $h$, that is the component of the spin of a quantum object in the direction of its momentum, $h = \mathbf{s} \cdot \mathbf{p}/ |\mathbf{p}|$. Relativistic considerations show that for a quantum object with $m_0 \neq 0$ (nonzero rest mass), there are $2s + 1$ different possibilities ($s, s - 1, \ldots, -s$), while

**Fig. 21.3** Möbius strip

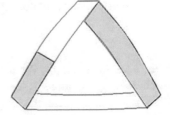

for $m_0 = 0$ (e.g. a photon), the helicity can assume only the values $h = \pm s$. The spin is responsible for the behavior under a rotation. Since it is $\frac{1}{2}$ for the electron and 1 for the photon, the identity rotation for spin 1 (thus also for a photon with two states) is given by $2\pi$. See also the exercises for Chap. 16.

### 21.1.5  Special Galilean Transformation

Due to space and time limitations, we consider this case in Appendix L, Vol. 2, and just give the result here. It states that the most general form of a Hamiltonian in three dimensions, compatible with the special Galilean transformation, is given by:

$$H = \frac{1}{2m} \left( \mathbf{p} - \mathbf{f}(\mathbf{x}) \right)^2 + V(\mathbf{x}). \tag{21.40}$$

$\mathbf{f}(\mathbf{x})$ can signify e.g. the vector potential $q\mathbf{A}(\mathbf{x})$.

## 21.2  Discrete Symmetry Transformations

In contrast to the continuous transformations, one cannot deduce the discrete transformations continuously from the unit map. Thus, there are no infinitesimal generators and it is not clear a priori whether the transformations are unitary. The two transformations $\mathcal{T}$ discussed in the following have the property $\mathcal{T}^2 = c$ with $|c| = 1$.[15]

### 21.2.1  Parity

The parity operator $\mathcal{P}$ reverses the sign of the position, $\mathbf{r} \to -\mathbf{r}$. This process can be represented as a reflection in the plane $z = 0$, followed by a rotation around the $z$ axis by $\pi$ (for typographic convenience, we use row vectors); see Fig. 21.4:

$$(x, y, z) \underset{\text{reflection}}{\to} (x, y, -z) \underset{\text{rotation by } \pi}{\to} (-x, -y, -z). \tag{21.41}$$

Because rotational invariance is valid in general, one can express parity conservation by the catchy formulation that the *mirror image* of any physical process must be physically possible.

To answer the question of whether the parity operator $\mathcal{P}$ is unitary or anti-unitary, we start with

---

[15]If $\mathcal{T}^2 = 1$, then the transformation clearly cannot come from a continuous group. For instance, we have for a rotation $(\mathcal{T}_\alpha)^2 = \mathcal{T}_{2\alpha} \neq 1$ for arbitrary angles $\alpha$.

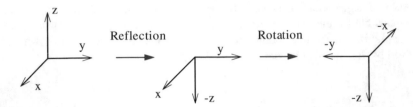

**Fig. 21.4** The parity transformation as first a reflection through the $x - y$ plane, followed by a 180° rotation around the $z$ axis

$$\mathcal{P}^2 = 1 \rightarrow \mathcal{P} = \mathcal{P}^{-1}. \tag{21.42}$$

We have for a position state $|\mathbf{r}\rangle$:

$$\mathcal{P}\,|\mathbf{r}\rangle = |-\mathbf{r}\rangle \quad \text{or} \quad \langle\mathbf{r}|\,\mathcal{P}^\dagger = \langle-\mathbf{r}|, \tag{21.43}$$

and for a general state $|\psi\rangle$:

$$\langle\mathbf{r}|\,\mathcal{P}\,|\psi\rangle = \psi\,(-\mathbf{r}) \quad \text{and} \quad \langle-\mathbf{r}|\,\psi\rangle = \psi\,(-\mathbf{r})\,. \tag{21.44}$$

It follows that
$$\langle-\mathbf{r}| = \langle\mathbf{r}|\,\mathcal{P}. \tag{21.45}$$

The comparison with (21.43) yields finally

$$\langle\mathbf{r}|\,\mathcal{P}^\dagger = \langle\mathbf{r}|\,\mathcal{P} \quad \text{or} \quad \mathcal{P}^\dagger = \mathcal{P} = \mathcal{P}^{-1}. \tag{21.46}$$

Hence, $\mathcal{P}$ is unitary. We have[16]:

$$\mathcal{P}\mathbf{r}\mathcal{P} = -\mathbf{r};\ \mathcal{P}\mathbf{p}\mathcal{P} = -\mathbf{p};\ \mathcal{P}^\dagger\mathcal{P} = 1. \tag{21.47}$$

Because of $\mathcal{P}^2 = 1$, $\mathcal{P}$ has the eigenvalues $\pm 1$. If $[H, \mathcal{P}] = 0$, one can always find common eigenfunctions with well-defined parity:

$$H\,|\varphi_\pm\rangle = E_\pm\,|\varphi_\pm\rangle\ ;\ \mathcal{P}\,|\varphi_\pm\rangle = \pm\,|\varphi_\pm\rangle \tag{21.48}$$

i.e. states with even (+) or odd (−) parity. For e.g. the one-dimensional harmonic oscillator, $H$ depends only on $x^2$ and thus commutes with $\mathcal{P}$. The eigenstates $|n\rangle$ are therefore simultaneously eigenstates of $\mathcal{P}$ and $H$; $\mathcal{P}\,|n\rangle = (-1)^n\,|n\rangle$. We have seen other examples of eigenfunctions with well-defined parity, e.g. in discussing

---

[16]Because of e.g. $\mathcal{P}\mathbf{r}f\,(\mathbf{r}) = -\mathbf{r}\mathcal{P}f\,(\mathbf{r})$; from this, it follows that $\mathcal{P}\mathbf{r} = -\mathbf{r}\mathcal{P}$, and thus $\mathcal{P}\mathbf{r}\mathcal{P} = -\mathbf{r}$.

the infinite potential well in Chap. 5, Vol. 1, or in the angular-momentum eigenstates $|l, m\rangle$ (or spherical harmonics) in Chap. 16.[17]

Finally, it remains to ask whether any interaction is parity conserving (i.e. $[H, \mathcal{P}] = 0$). According to current knowledge, this is the case for the strong and electromagnetic interactions, but not for the weak interaction, which occurs e.g. in nuclear beta decay. Essentially, this can be attributed to the fact that there are polar and axial vectors (see Appendix F, Vol. 1), which are affected differently by the parity operator. A *polar vector*, which is transformed like the position vector, is e.g. the momentum:

$$\mathbf{r} \to -\mathbf{r}; \quad \mathbf{p} \to -\mathbf{p}. \tag{21.49}$$

*Axial vectors* (= *pseudo vectors*) such as the angular momentum vector transform under $\mathbf{r} \to -\mathbf{r}$ in accordance with[18]:

$$\mathbf{l} \to \mathbf{l}. \tag{21.50}$$

Since the angular momentum $\mathbf{l}$ is an axial vector and $\mathbf{l}^2$ is a scalar, both operators commute with $\mathcal{P}$, and we have for the angular momentum eigenstates $\mathcal{P} |l, m\rangle = (-1)^l |l, m\rangle$ because of $r \to r, \vartheta \to \pi - \vartheta, \varphi \to \varphi + \pi$. Parity violation may occur in a physical process if axial and polar vectors occur simultaneously, for example $\mathbf{p}^2$ and $\mathbf{r} \cdot \mathbf{l}$. The requirement of invariance under parity reversal therefore implies a restriction of which terms may occur in $H$.

## *21.2.2   Time Reversal*

Time reversal has nothing to do with time travel into the past and the like; a better name would be *reversal of motion*. Invariance under time reversal means the following: Suppose that a system starts in a parameter space at time $T$ at point $A$ and is at point $B$ at time $T + t$. Then invariance under time reversal states that the 'reverse' process is possible, namely that a system starts at time $T + t$ at point $B$ and ends up at time $T + 2t$ at point A (for $T$, one usually chooses $-t$), cf. Fig. 21.5. That this is not self-evident is shown e.g. by the motion of an electron in a homogeneous magnetic field, which is not invariant under time reversal.[19]

---

[17]Of course, quite generally every state $\psi(\mathbf{r})$ may be divided into an even and an odd part: $\psi(\mathbf{r}) = \psi_+(\mathbf{r}) + \psi_-(\mathbf{r})$, with $\psi_\pm(\mathbf{r}) = \frac{\psi(\mathbf{r}) \pm \psi(-\mathbf{r})}{2}$.

[18]Because of $\mathbf{l} = \mathbf{r} \times \mathbf{p} \to \mathbf{l} = (-\mathbf{r}) \times (-\mathbf{p}) = \mathbf{r} \times \mathbf{p}$. Generally speaking, the product of two polar vectors is a pseudovector.

[19]When forces are conservative, the orbits are time-reversal invariant. Magnetic fields are not conservative.

**Fig. 21.5** Principle of time reversal. One usually sets $t = -T$

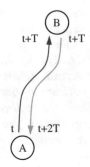

The following plausibility argument suggests that time reversal is an anti-unitary operation[20]: We assume a state with well-defined energy whose time behavior is given by $e^{-i\omega t}$. Under time reversal, this becomes $e^{i\omega t}$, which is obviously equal to the complex conjugate of $e^{-i\omega t}$, $e^{i\omega t} = \left(e^{-i\omega t}\right)^*$. But we know that the complex conjugation is an anti-linear operation, and this should thus also apply to time reversal.

We want to show this fact now in two different ways.

**First method**: We start from classical mechanics, namely

$$m\frac{d^2\mathbf{r}(t)}{dt^2} = \mathbf{F}; \quad \frac{\partial\mathbf{F}}{\partial t} = 0. \tag{21.51}$$

For each solution $\mathbf{r}(t)$, there is a time-reversed solution $\mathbf{r}'(t) = \mathbf{r}(-t)$. We have

$$\mathbf{v}'(t_0) = \left(\frac{d\mathbf{r}'(t)}{dt}\right)_{t=t_0} = -\left(\frac{d\mathbf{r}(-t)}{d(-t)}\right)_{t=t_0} = -\left(\frac{d\mathbf{r}(t')}{dt'}\right)_{t'=-t_0} = -\mathbf{v}(t_0). \tag{21.52}$$

In words: if time is reversed, the position remains the same, but the speed or the momentum changes sign, as is indeed intuitively obvious.

In quantum mechanics, we start from the SEq:

$$i\hbar\frac{\partial}{\partial t}\psi(\mathbf{r}, t) = H(\mathbf{r})\psi(\mathbf{r}, t); \quad \frac{\partial H}{\partial t} = 0; H \in \mathbb{R} \tag{21.53}$$

and obtain (note: $H$ does not depend on $t$)[21]:

$$i\hbar\frac{\partial\psi(\mathbf{r}, -t)}{\partial(-t)} = H(\mathbf{r})\psi(\mathbf{r}, -t) \to i\hbar\frac{\partial\psi^*(\mathbf{r}, -t)}{\partial t} = H(\mathbf{r})\psi^*(\mathbf{r}, -t). \tag{21.54}$$

---

[20]We repeat the remark that an anti-unitary operator $B$ is anti-linear, i.e. $BB^\dagger = B^\dagger B = 1$ and $B\alpha|\varphi\rangle = \alpha^* B|\varphi\rangle$.

[21]The considerations for time-dependent Hamiltonians are more complex and require new concepts (time-ordering operator, etc.); but time reversal is an anti-unitary operator in this case, also.

In other words, with $\psi\,(\mathbf{r},t)$, also $\mathcal{T}\psi\,(\mathbf{r},t) = \psi'\,(\mathbf{r},t) = \psi^*\,(\mathbf{r},-t)$ is a solution, where $\mathcal{T}$ is the *time reversal operator*. Since complex conjugation is anti-unitary, this holds also for the the time reversal operator. As in classical mechanics, in quantum mechanics the position remains the same and the momentum changes:

$$\mathbf{r}\,(t) \rightarrow \mathbf{r}\,(-t)\,;\ \mathbf{p}\,(t) \rightarrow -\mathbf{p}\,(-t)\,. \tag{21.55}$$

**Second method**: We write the time evolution of an arbitrary state as $e^{-iHt/\hbar}\,|\varphi\rangle$ to obtain

$$\mathcal{T}e^{-iHt/\hbar}\,|\varphi\rangle = e^{iHt/\hbar}\mathcal{T}\,|\varphi\rangle \rightarrow \mathcal{T}e^{-iHt/\hbar}\mathcal{T}^{-1} = e^{iHt/\hbar}. \tag{21.56}$$

For a Hermitian operator $B$ and a unitary or anti-unitary operator $U$ or $A$, it holds that (see exercises[22]):

$$e^{iUBU^{-1}} = Ue^{iB}U^{-1}\,;\ e^{iABA^{-1}} = Ae^{-iB}A^{-1}. \tag{21.57}$$

If we assume that $\mathcal{T}$ is unitary (i.e. we identify in (21.57) $\mathcal{T} = U$ and $B = -Ht/\hbar$), it follows that

$$\mathcal{T}e^{-iHt/\hbar}\mathcal{T}^{-1} = \begin{cases} e^{iHt/\hbar} \\ e^{-i\mathcal{T}Ht/\hbar\mathcal{T}^{-1}} \end{cases} \text{because of} \begin{array}{c} (21.56) \\ (21.57). \end{array} \tag{21.58}$$

If we assume, however, that $\mathcal{T}$ is anti-unitary (this means in (21.57) that $\mathcal{T} = A$ and $B = Ht/\hbar$), it follows that

$$\mathcal{T}e^{-iHt/\hbar}\mathcal{T}^{-1} = \begin{cases} e^{iHt/\hbar} \\ e^{i\mathcal{T}Ht/\hbar\mathcal{T}^{-1}} \end{cases} \text{because of} \begin{array}{c} (21.56) \\ (21.57). \end{array} \tag{21.59}$$

In summary, we obtain

$$\begin{array}{c} e^{iHt/\hbar} = e^{-i\mathcal{T}Ht/\hbar\mathcal{T}^{-1}} \\ e^{iHt/\hbar} = e^{i\mathcal{T}Ht/\hbar\mathcal{T}^{-1}} \end{array} \text{for} \begin{array}{c} \mathcal{T}\ \text{unitary} \\ \mathcal{T}\ \text{anti-unitary}. \end{array} \tag{21.60}$$

For infinitesimal times, we have

$$\begin{array}{c} 1 + i\frac{Ht}{\hbar} = 1 - i\mathcal{T}\frac{Ht}{\hbar}\mathcal{T}^{-1} \\ 1 + i\frac{Ht}{\hbar} = 1 + i\mathcal{T}\frac{Ht}{\hbar}\mathcal{T}^{-1} \end{array} \rightarrow \begin{array}{c} H = -\mathcal{T}H\mathcal{T}^{-1} \\ H = \mathcal{T}H\mathcal{T}^{-1} \end{array} \text{for} \begin{array}{c} \mathcal{T}\ \text{unitary} \\ \mathcal{T}\ \text{anti-unitary}. \end{array} \tag{21.61}$$

Ultimately, from (21.56) and (21.57), it follows that

---

[22]In general, every anti-unitary operator $A$ may be represented as the product of a unitary operator $U$ with the operator of complex conjugation $K$, i.e. $A = UK$. We have $A^{-1} = KU^\dagger$ and $Ki = -iK$.

$$\begin{matrix} \mathcal{T}H + H\mathcal{T} = 0 \\ \mathcal{T}H - H\mathcal{T} = 0 \end{matrix} \text{ for } \begin{matrix} \mathcal{T} \text{ unitary} \\ \mathcal{T} \text{ anti-unitary.} \end{matrix} \qquad (21.62)$$

Now one can argue that the Hamiltonian of a free quantum object contains only $\mathbf{p}^2$, so $H$ must commute with $\mathcal{T}$. Hence, $\mathcal{T}$ is anti-unitary.

The requirement of invariance under time reversal implies a limitation (similar to the case of parity conservation) on which expressions can occur in $H$. For instance, $\mathbf{p}^2$ and $\mathbf{p}$ or $\mathbf{r} \cdot \mathbf{l}$ cannot occur simultaneously. According to our current knowledge, only the Hamiltonian of the weak interaction is not invariant under time reversal, as indeed it is not invariant under the parity transformation. Another important symmetry is charge conjugation (a better term would be particle-antiparticle exchange). A fundamental theorem of quantum field theory states that any Lorentz-invariant local field theory must be invariant if we simultaneously perform parity inversion, time reversal and charge conjugation (CPT theorem).[23]

Time-reversal invariance leads for certain systems to further-reaching statements. An example is the theorem of Kramers about the degeneracy of eigenstates (Kramers doublets); we sketch it in Appendix M, Vol. 2.

## 21.3  Exercises

1. Derive the commutation relation (21.26) for position and momentum.
2. Consider the relation between symmetries and conserved quantities by means of the spatial translational invariance of an isolated system of two quantum objects whose interaction depends only on their distance $\mathbf{r}_1 - \mathbf{r}_2$.
3. Let $B$ be a Hermitian operator and $U$ and $A$ a unitary and an anti-unitary operator, resp. Show that:

$$e^{iUBU^{-1}} = Ue^{iB}U^{-1}; \quad e^{iABA^{-1}} = Ae^{-iB}A^{-1}. \qquad (21.63)$$

4. Show with the help of the propagator $U$ that eigenvalues of $A$ are conserved, if $[H, A] = 0$.
5. Consider the translation $\mathbf{r}' = \mathbf{r} + \mathbf{a}$ or $T(\mathbf{a})\,\mathbf{r} = \mathbf{r} + \mathbf{a}$. Show that it can be represented by the unitary transformation $U_{T(\mathbf{a})} = \lim_{n \to \infty} \left(1 - \frac{i}{\hbar}\frac{\mathbf{a}\mathbf{p}}{n}\right)^n = e^{-\frac{i}{\hbar}\mathbf{a}\mathbf{p}}$.

---

[23] Only recently, the violation of time-reversal symmetry was detected directly for the first time (CPT invariance, however, remains valid); see J.P. Lee et al., 'Observation of Time-Reversal Violation in the $B^0$ Meson System', *Phys. Rev. Lett.* 109, 211801 (2012). In this context, entirely new concepts, such as that of the 'time crystal', are emerging; cf. F. Wilczek, 'Quantum Time Crystal', *Phys. Rev. Let.* 109, 160401 (2012). Along with David Gross and David Politzer, Frank Wilczek was awarded the Nobel Prize for Physics in 2004 for the discovery of asymptotic freedom in the theory of strong interactions. See also J. Zhang et al., Observation of a discrete time crystal, *Nature* 543, 217–220 (Mar 2017), and S. Choi et al., Observation of discrete time-crystalline order in a disordered dipolar many-body system, *Nature* 543, 221–225 (Mar 2017).

6. Determine the commutator of $P$ with an arbitrary function of $X$, without using $P = \frac{\hbar}{i} \frac{d}{dX}$ from the outset (this is to be derived). Use

$$U^{-1}(a) X^2 U(a) = U^{-1}(a) XU(a) U^{-1}(a) XU(a) = (X+a)^2 ; \quad (21.64)$$

and analogously
$$U^{-1}(a) X^n U(a) = (X+a)^n \qquad (21.65)$$

as well as the power-series expansion of the function $f(X) = c_0 + c_1 X + c_2 X^2 + \dots$.

7. Show that a rotation through the angle $\varphi$ around the $z$ axis is represented by $e^{-i\alpha l_z}$.

8. Using (21.32),

$$e^{-i\frac{\gamma \mathbf{j} \hat{\mathbf{a}}}{\hbar}} = e^{-i\frac{\varphi j_x}{\hbar}} e^{-i\frac{\gamma j_y}{\hbar}} e^{i\frac{\varphi j_x}{\hbar}} , \qquad (21.66)$$

derive the commutation relations for the angular momentum.

9. A scalar operator is defined as an operator whose mean value is invariant under a rotation. Derive the result $[\mathbf{j}, S] = 0$.

10. A vector operator is an operator $\mathbf{V}$ whose mean value transforms like a vector $\mathbf{v}$ under a rotation through an angle $\gamma$ about an axis $\hat{\mathbf{a}}$, i.e. as

$$\mathbf{v}' = \cos\gamma \cdot \mathbf{v} + \sin\gamma \cdot (\hat{\mathbf{a}} \times \mathbf{v}) + (1 - \cos\gamma)(\hat{\mathbf{a}} \times \mathbf{v}) \cdot \hat{\mathbf{a}}. \qquad (21.67)$$

Derive $[j_i, V_k] = i\hbar \sum_l \varepsilon_{ikl} V_l$.[24]

11. Formulate explicitly the unitary operator $e^{-i\frac{\gamma}{2}\sigma\hat{\mathbf{a}}}$ for spin 1/2; $\sigma$ is the vector $\sigma = (\sigma_1, \sigma_1, \sigma_1)$ and $\hat{\mathbf{a}}$ a 3-dimensional unit vector.

---

[24]More on vector operators in Appendix G, Vol. 2.

# Chapter 22
# The Density Operator

The density operator is the most general representation of states in quantum mechanics. It allows the description of those systems which are only partially known.

Up to now, we have characterized a quantum-mechanical system by a vector $|\psi\rangle$ of the Hilbert space. In the following, we will extend the concept of state, as we already promised in Chap. 14, Vol. 1. We will introduce the density operator or density matrix, the most general representation of states in quantum mechanics. This tool allows us to describe also states for which we do not have complete information, and which therefore cannot be represented by a vector in the Hilbert space. That such a description is useful or necessary may be surprising at first, but we will see that this formulation is quite handy, especially with regard to the discussion of the measurement process in quantum mechanics.

A remark on nomenclature: If a system can be represented by a vector in a Hilbert space $\mathcal{H}$, one speaks of a *pure state*. In this case, the maximum amount of information is available (one-dimensional subspace of $\mathcal{H}$). Otherwise, one speaks of a *mixed state*.

## 22.1 Pure States

We start by noting that we can characterize a quantum-mechanical system by the normalized state[1] $|\psi\rangle$, but just as well by a dyadic product called the *density operator* $\rho$ or *statistical operator* [2]

$$\rho = |\psi\rangle \langle\psi|. \tag{22.1}$$

---

[1] A dependence on time is not mentioned explicitly in the following.

[2] Actually, it would be better to say 'state operator', but the term 'density operator' has become established—and unfortunately also the use of $\rho$. Note: Previously, $\rho$ was used exclusively for the probability density, i.e. $\rho = \langle\psi|\psi\rangle$. Which $\rho$ is meant in a particular situation should follow unambiguously from the context.

© Springer Nature Switzerland AG 2018
J. Pade, *Quantum Mechanics for Pedestrians 2*, Undergraduate Lecture Notes in Physics, https://doi.org/10.1007/978-3-030-00467-5_22

The information content is the same whether we specify the state or the density operator. There is a small difference, however: We know that (normalized) states are determined up to an arbitrary global phase[3]:

$$|\psi\rangle \rightarrow e^{i\alpha} |\psi\rangle. \tag{22.2}$$

This phase disappears when we use the density operator

$$\rho = |\psi\rangle \langle\psi| \rightarrow e^{i\alpha} |\psi\rangle \langle\psi| e^{-i\alpha} = \rho, \tag{22.3}$$

and this may be seen at this point to represent a certain advantage of this formulation.

By the way, one is in good company if one does not immediately see the need to introduce the concept of the density operator which shows the following quote: "By introducing an operator $\rho$ called the density matrix, expectation values can be written in a form that at first sight is opaque but which turns out to be very powerful".[4] In fact, it is only with the help of this operator that e.g. the extension of the term 'state' can be accomplished, as we shall see below.[5]

The density operator introduced in this way is obviously a projection operator (normalization of $|\psi\rangle$ presupposed):

$$\rho^2 = |\psi\rangle \langle\psi| \psi\rangle \langle\psi| = \rho; \ \rho^\dagger = \rho. \tag{22.4}$$

Just as with the state $|\psi\rangle$, one can formulate all those statements for a quantum-mechanical system which are accessible to measurements by using the density operator $\rho$. For example, we can calculate mean or expected values. Here, the term *trace* comes into play. We assume that there is a CONS $\{|n\rangle\}$. Then the trace $tr$ of $A$ is defined as[6]:

$$tr A = tr(A) = \sum_n \langle n| A |n\rangle. \tag{22.5}$$

With this notation, the mean value of an operator $A$ for the state $|\psi\rangle$ is given by

$$\langle A\rangle = \langle\psi| A |\psi\rangle = \sum_{n,m} \langle\psi| n\rangle \langle n| A |m\rangle \langle m| \psi\rangle = \sum_{n,m} \langle m| \psi\rangle \langle\psi| n\rangle \langle n| A |m\rangle$$

$$= \sum_{n,m} \langle m| \rho |n\rangle \langle n| A |m\rangle = \sum_m \langle m| \rho A |m\rangle = tr\,(\rho A). \tag{22.6}$$

---

[3] Indeed, states are strictly speaking not determined by vectors of $\mathcal{H}$, but instead by rays, see Chap. 14, Vol. 1.

[4] Gottfried and Yan, *Quantum Mechanics: Fundamentals*, p. 46.

[5] Another example for the quasi natural occurrence of the density operator is the theorem of Gleason, which deals with the question of how mean values (or probabilities) can be defined in quantum mechanics. It states in essence that in Hilbert spaces of dimension $\geq 3$, the mean values of projection operators $P$ can be described only by $\langle P\rangle = tr\,(\rho P)$ (see Appendix T, Vol. 2).

[6] See also Appendix F, Vol. 1.

Knowledge of the density operator thus allows us to determine averages by taking the trace.[7] As a little exercise, we now calculate the trace of the density operator itself (which of course must be 1, since $tr\,(\rho \cdot 1) = \langle 1 \rangle$). We have:

$$tr\,(\rho) = \sum_n \langle n| \rho |n \rangle = \sum_n \langle n| \psi \rangle \langle \psi |n \rangle = \sum_n \langle \psi |n \rangle \langle n| \psi \rangle = \langle \psi |\psi \rangle = 1.$$

(22.7)

We can also start from

$$|\psi \rangle = \sum_n \langle n| \psi \rangle |n \rangle = \sum_n c_n |n \rangle$$

(22.8)

with $\sum_n |c_n|^2 = 1$ due to $\langle \psi | \psi \rangle = 1$. It follows that:

$$tr\,(\rho) = \sum_n \langle n| \psi \rangle \langle \psi |n \rangle = \sum_n |c_n|^2 = 1.$$

(22.9)

A few remarks on the representation of the density operator as *density matrix*. With (22.8), we have:

$$\rho = |\psi \rangle \langle \psi | = \sum_{n,m} c_n c_m^* |n \rangle \langle m|.$$

(22.10)

Evidently we have the same information content whether we use the formulation (22.10) or the density matrix[8]:

$$\rho_{nm} = c_n c_m^* \text{ or } \rho \cong \begin{pmatrix} c_1 c_1^* & c_1 c_2^* \cdots \\ c_2 c_1^* & c_2 c_2^* \cdots \\ \cdots & \cdots \cdots \end{pmatrix}.$$

(22.11)

In fact, the two terms 'density operator' and 'density matrix' are quite often used synonymously (and one writes just $=$ instead of $\cong$; we have occasionally neglected this distinction, also), and depending on the context, $\rho$ refers to the abstract operator or the matrix (22.11). A note on nomenclature: the diagonal elements are also called *populations* or *occupation numbers*, the off-diagonal elements *coherences* or *interference terms*. In addition, we see that the trace of the density matrix is given by $\sum |c_n|^2$, and thus by exactly the same term as in (22.9), as indeed it must be. We point out that the trace of a matrix is equal to the sum of its eigenvalues.[9] A comment and two examples using two-dimensional systems follow:

---

[7] It holds that $tr\,(\rho A) = tr\,(A\rho)$; see the exercises.

[8] The density matrix is Hermitian, of course.

[9] Since the density matrix is Hermitian, its eigenvalues are real.

1. In the density matrix

$$\rho = \begin{pmatrix} c_1 c_1^* & c_1 c_2^* \\ c_2 c_1^* & c_2 c_2^* \end{pmatrix}, \tag{22.12}$$

we clearly cannot ensure that the interference terms vanish, while the two diagonal terms are nonzero. Either there are four non-zero entries or only one in this matrix; any other possibility is excluded.[10] (As always, we have discarded from the outset the trivial solution with $c_1 = c_2 = 0$). As is easily verified, we have $\rho^2 = \rho$ (see the exercises).

2. Two examples using polarization states:

   (a) Let $|\psi\rangle = |h\rangle$. Then it holds that:

$$\rho = |h\rangle \langle h| \quad \text{or} \quad \rho = \begin{pmatrix} 1 & 0 \\ 0 & 0 \end{pmatrix} \tag{22.13}$$

   (b) For $|\psi\rangle = \frac{|h\rangle - i|v\rangle}{\sqrt{2}}$, it follows that

$$\rho = \frac{|h\rangle \langle h| + i |h\rangle \langle v| - i |v\rangle \langle h| + |v\rangle \langle v|}{2} \quad \text{or} \quad \rho = \frac{1}{2}\begin{pmatrix} 1 & i \\ -i & 1 \end{pmatrix}. \tag{22.14}$$

The relation $\rho^2 = \rho$ of course holds also for these two examples.

## 22.2 Mixed States

We obtain the projection onto a pure state by means of the projection operator $|\psi\rangle \langle \psi|$. Here, it is assumed that we know the state exactly, e.g. by characterizing it using a complete system of commuting observables. But often, one has only incomplete information about a system or simply does not want to know everything—just think of the kinetic theory of gases, whose strength lies precisely in the fact that one does not have to take into account all $10^{23}$ particles explicitly. In statistical physics, one compensates ignorance of the precise state of a system by the introduction of *classical probabilities*.[11] That is what we do in quantum mechanics as well, if we do not have all the necessary information. One speaks of a *mixed state* or a *statistical mixture* as opposed to a pure state.

For a given mixture, we know only that the system is in the states $|\varphi_n\rangle$ with *classical probabilities* $p_n$ ($0 \leq p_n \leq 1$ and $\sum_n p_n = 1$). The states $|\varphi_n\rangle$ are normalized, but

---

[10]This does not change if we diagonalize the matrix (that is possible, since it is Hermitian). For the diagonal elements of the resulting diagonal matrix are the eigenvalues, and in this case they are 0 and 1 (because of $|c_1|^2 + |c_2|^2 = 1$); see the exercises.

[11]In contrast to the probabilities which are inevitable in quantum mechanics.

need not necessarily be mutually orthogonal. As the density operator for this system, we can choose the weighted superposition $\rho_{\text{stat}} = \sum_n p_n |\varphi_n\rangle \langle\varphi_n|$. For clarity, we denote this density operator temporarily by $\rho_{\text{stat}}$.

To make life easier for us, we will not discuss the general case of non-orthogonal bases,[12] but instead refer to the CONS $\{|n\rangle\}$ introduced above:

$$\rho_{\text{stat}} = \sum_n p_n |n\rangle \langle n| \tag{22.15}$$

and

$$tr\left(\rho_{\text{stat}}\right) = \sum_m \langle m| \rho_{\text{stat}} |m\rangle = \sum_{mn} p_n \langle m| n\rangle \langle n| m\rangle = \sum_n p_n = 1. \tag{22.16}$$

We note that there are no interference terms (coherences) as in (22.10); the corresponding density matrix has only diagonal entries.

So we know with the classical probability $p_n$ that the system is in state $|n\rangle$. It follows that

$$\rho_{\text{stat}}^2 = \sum_{n,m} p_n p_m |n\rangle \langle n| m\rangle \langle m| = \sum_n p_n^2 |n\rangle \langle n| \tag{22.17}$$

and

$$tr\left(\rho_{\text{stat}}^2\right) = \sum_{n,m} p_n^2 \langle m |n\rangle \langle n| m\rangle = \sum_n p_n^2. \tag{22.18}$$

The comparison of (22.15) and (22.17) shows immediately that $\rho_{\text{stat}}^2 = \rho_{\text{stat}}$ can apply only if all of the probabilities $p_n$ vanish except for one,[13] $p_n = \delta_{nN}$. For a pure state, $\rho_{\text{stat}}$ is a projection operator, but with mixtures, $\rho_{\text{stat}}^2 \neq \rho_{\text{stat}}$ applies—and this means that $\rho_{\text{stat}}$ in this case is *not* a projection operator.

Considering the trace of the density operator, we see that for a mixture (i.e. two or more $p_n \neq 0$), it holds that $tr\left(\rho_{\text{stat}}^2\right) = \sum p_n^2 < 1$. All in all, we have[14]:

---

[12]For the expansion in terms of not-necessarily-orthogonal states, see the exercises.

[13]The case that all the probabilities are zero is trivial.

[14]Note: For a change of basis to another CONS $\{|\varphi_m\rangle\}$ with

$$|n\rangle = \sum_m c_{nm} |\varphi_m\rangle,$$

off-diagonal terms can of course occur in the density matrix:

$$\rho_{\text{stat}} = \sum_{n,m,m'} p_n c_{nm} c_{n'm'}^* |\varphi_m\rangle \langle\varphi_{m'}| = \sum_{m,m'} \sum_n \left(p_n c_{nm} c_{n'm'}^*\right) |\varphi_m\rangle \langle\varphi_{m'}| = \sum_{m,m'} \alpha_{mm'} |\varphi_m\rangle \langle\varphi_{m'}|.$$

The occurrence of non-trivial off-diagonal terms is not crucial as a distinguishing feature between $\rho$ and $\rho_{\text{stat}}$, but rather the criteria $\rho_{\text{stat}}^2 \neq \rho_{\text{stat}}$ and $tr\left(\rho_{\text{stat}}^2\right) < 1$ (thus the denotation of off-diagonal terms as 'coherences' or 'interference terms' is a bit unfortunate).

$$\rho_{\text{stat}}^2 \neq \rho_{\text{stat}}; \; tr\left(\rho_{\text{stat}}^2\right) < 1 \;\; \text{for a mixed state.} \tag{22.19}$$

In addition, we have the relation between the trace and the mean value:

$$tr\left(\rho_{\text{stat}} A\right) = \sum_{n,m} p_n \langle m | n \rangle \langle n | A | m \rangle$$

$$= \sum_{n,m} p_n \delta_{nm} \langle n | A | m \rangle = \sum_m p_m \langle m | A | m \rangle = \langle A \rangle . \tag{22.20}$$

We point out that in the latter equation, there are two fundamentally different averaging processes, one being the *quantum-mechanical* mean, $\langle m | A | m \rangle$ of the operator $A$ over the states $|m\rangle$, the other one being the *classical* mean value of this quantum-mechanical average, given by $\sum_m p_m \langle m | A | m \rangle$.

For clarity, we emphasize again the differences between the two density operators:

1. If one can describe the system by a state in the Hilbert space with the basis $\{|n\rangle\}$ (pure state), e.g. by $|\psi\rangle = \sum_n c_n |n\rangle$, it is in a quantum-mechanical superposition of basis states (unless all $c_n$ are zero except one). With respect to this basis, the system therefore has no defined state; that results only from a measurement. For the density operator $\rho$, $\rho^2 = \rho$ and $tr\left(\rho^2\right) = 1$ hold.
2. In a statistical mixture, the system is in a well-defined state; however, we do not know in which one. So this is a very different situation from that described in 1), in which it is simply not reasonable to ask about the status of the system—it is not defined in relation to the selected basis. For a statistical mixture, we have $\rho_{\text{stat}}^2 \neq \rho_{\text{stat}}$ and $tr\left(\rho_{\text{stat}}^2\right) < 1$.

We can now summarize the properties of the density operator $\rho$ (we have derived some points only for the simpler case (22.15), but they hold true in general). In order to standardize the terminology, we dispense with the labelling $\rho_{\text{stat}}$ and, following the usual practice, denote all density operators just by $\rho$.[15]

1. $\rho$ is Hermitian, $\rho = \rho^\dagger$, and positive, $\langle \varphi | \rho | \varphi \rangle \geq 0$ for all $|\varphi\rangle$. The trace of $\rho$ is 1, $tr\left(\rho\right) = 1$. Generally, an operator $A$ with the three properties $A = A^\dagger$, $A$ positive and $tr(A) = 1$ is called a density operator.
2. $\rho$ is a projection operator if and only if the system is in a pure state ($p_n = \delta_{n,N}$). In this case, $\rho^2 = \rho$ and $tr(\rho^2) = 1$ applies, while a mixture is characterized by $\rho^2 \neq \rho$ and $tr(\rho^2) < 1$. This result does not depend on the choice of basis, because the trace is invariant under unitary transformations (see the exercises). The main criterion for a pure or mixed state is $tr(\rho^2) = 1$ or $tr(\rho^2) < 1$.
3. For the mean value of an observable $A$, we have $\langle A \rangle = tr\left(\rho A\right)$. By the way, from this it follows directly that $\langle \rho \rangle = tr\left(\rho^2\right)$.
4. The probability of finding the system in state $|m\rangle$ is given by $p_m = tr(\rho |m\rangle \langle m|)$ (see the exercises).

---

[15] What is precisely meant must follow from the context, if necessary.

5. The time behavior of $\rho$ is described by the *von Neumann equation*[16] (see the exercises):

$$i\hbar\partial_t\rho = [H, \rho]. \tag{22.21}$$

Due to

$$|\psi(t)\rangle = U(t)|\psi(0)\rangle, \tag{22.22}$$

the solution of this equation is

$$\rho(t) = U(t)\rho(0)U^\dagger(t). \tag{22.23}$$

6. With the density operator, we have expanded upon the state concept, as promised in Chap. 14 of Vol. 1; we can now also describe systems whose state is not known in detail and which can therefore *not* be described by a state vector of $\mathcal{H}$.

## 22.3 Reduced Density Operator

For the next topic, we start from a system of two quantum objects. We have the product space $\mathcal{H}_1 \otimes \mathcal{H}_2$; $\{|n_1\rangle\}$ and $\{|m_2\rangle\}$ are CONS in $\mathcal{H}_1$ and $\mathcal{H}_2$. Using the notation $|n_1\rangle \otimes |m_2\rangle = |n_1m_2\rangle$, it follows for an arbitrary state:

$$|\psi\rangle = \sum_{n,m} c_{nm}|n_1m_2\rangle, \tag{22.24}$$

and thus for the density of a pure state:

$$\rho = \sum_{n,m,n',m'} c_{nm}c^*_{n'm'}|n_1m_2\rangle\langle n'_1m'_2|. \tag{22.25}$$

---

[16]Compare this with the Liouville equation in classical mechanics:

$$\partial_t\rho = \{H, \rho\}_{\text{Poisson}}$$

where the Poisson bracket of two quantities $A$ and $B$ is defined by

$$\{F, G\}_{\text{Poisson}} = \sum_k \left(\frac{\partial F}{\partial q_k}\frac{\partial G}{\partial p_k} - \frac{\partial G}{\partial q_k}\frac{\partial F}{\partial p_k}\right).$$

This motivates once more the transition from classical mechanics to quantum mechanics by the substitution (keyword canonical quantization; see Appendix W, Vol. 2)

$$\{,\}_{\text{Poisson bracket}} \rightarrow \frac{1}{i\hbar}[,]_{\text{commutator}}$$

Due to

$$\sum_{n,m} |c_{nm}|^2 = 1, \tag{22.26}$$

we have $\rho^2 = \rho$. So far there is nothing new. But we know that in the case of two quantum objects, entangled states can occur which *cannot* be represented as product states. This leads to a new question, namely: What is the density operator $\rho^{(1)}$ for the quantum object 1? This question *must* lead to something new, because the individual quantum objects in an entangled state do not have well-defined properties; this fact must have some impact on the density operator.

We consider an operator $A_1$ which acts only in the space $\mathcal{H}_1$; $I_2$ denotes the identity in $\mathcal{H}_2$. The total operator is $A_1 \otimes I_2$, and its average is given by

$$\langle A_1 \otimes I_2 \rangle = tr \left( \rho \left[ A_1 \otimes I_2 \right] \right) = \sum_{n_1,m_2} \langle n_1 m_2 | \rho \left[ A_1 \otimes I_2 \right] | n_1 m_2 \rangle$$

$$= \sum_{n_1,m_2} \langle n_1 | \langle m_2 | \rho | m_2 \rangle A_1 | n_1 \rangle = \sum_{n_1} \langle n_1 | \left( \sum_{m_2} \langle m_2 | \rho | m_2 \rangle \right) A_1 | n_1 \rangle. \tag{22.27}$$

We have used a) $\langle m_2 | \rho [A_1 \otimes I_2] | m_2 \rangle = \langle m_2 | \rho I_2 | m_2 \rangle A_1$ (since the operator $A_1$ acts in $\mathcal{H}_1$only),[17] and b) $\langle m_2 | \rho I_2 | m_2 \rangle = \langle m_2 | \rho | m_2 \rangle$, because of $I_2 | m_2 \rangle = | m_2 \rangle$. The expression in parentheses means that we take the average of the density operator $\rho$ (which acts in $\mathcal{H}_1$ *and* $\mathcal{H}_2$) *only* in $\mathcal{H}_2$, not in both spaces. Such a partial averaging is called a *partial trace* $tr_k (\rho)$, where the index $k$ indicates the space in which the averaging is performed. The result of this partial averaging is an operator that acts only in $\mathcal{H}_1$, namely the *reduced density operator* $\rho^{(1)}$. We have therefore[18]:

$$\rho^{(1)} = \sum_{m_2} \langle m_2 | \rho | m_2 \rangle = tr_2 (\rho). \tag{22.28}$$

We note that $\rho^{(1)}$ is a 'proper' density operator (Hermitian, positive, trace 1). For the mean value of $A_1 \otimes I_2$, we obtain

$$\langle A_1 \otimes I_2 \rangle = \sum_{n_1} \langle n_1 | \rho^{(1)} A_1 | n_1 \rangle = tr_1 \left( \rho^{(1)} A_1 \right). \tag{22.29}$$

The salient point is that the reduced density operator $\rho^{(1)}$ can have properties which are very different from those of the density operator $\rho$. For example, in a $2 \times 2$ matrix (as in (22.12)), it is not necessarily true that either one or all entries exist; it may happen that only the off-diagonal elements are equal to zero, which is impossible for a density operator with $tr \left( \rho^2 \right) = 1$. In the reduced density operator,

---

[17]We note that $\langle m_2 | \rho | m_2 \rangle$ is an operator in $\mathcal{H}_1$, and therefore, it generally holds that $\sum_{m_2} \langle m_2 | \rho | m_2 \rangle A_1 \neq A_1 \sum_{m_2} \langle m_2 | \rho | m_2 \rangle$.

[18]Other notations are $\hat{\rho}$, $\rho_{red}$, $\rho_1$ or similar forms.

we describe in general not a pure state, but a mixed state or a statistical mixture. This manifests itself in the fact that it applies in general that $\left(\rho^{(1)}\right)^2 \neq \rho^{(1)}$.

Before we continue with our general considerations, we illustrate these findings by an example.

## 22.3.1  Example

We consider a two-state system whose single components have the basis $\{|a\rangle, |b\rangle\}$. The total state is

$$|\psi\rangle = c_{11} |a_1 a_2\rangle + c_{12} |a_1 b_2\rangle + c_{21} |b_1 a_2\rangle + c_{22} |b_2 b_2\rangle \quad \text{with} \quad \sum |c_{ij}|^2 = 1. \tag{22.30}$$

The density matrix is correspondingly a $4 \times 4$ matrix; we will not write it explicitly here (but see the exercises). The reduced density operator is constructed via

$$\rho^{(1)} = \langle a_2| \psi\rangle \langle \psi |a_2\rangle + \langle b_2| \psi\rangle \langle \psi |b_2\rangle, \tag{22.31}$$

and obviously corresponds to a $2 \times 2$ matrix (see exercises). With

$$\langle a_2 |\psi\rangle = c_{11} |a_1\rangle + c_{21} |b_1\rangle$$
$$\langle b_2 |\psi\rangle = c_{12} |a_1\rangle + c_{22} |b_1\rangle, \tag{22.32}$$

it follows that

$$\rho^{(1)} = \left[c_{11}^* c_{11} |a_1\rangle + c_{21} |b_1\rangle\right]\left[c_{11}^* \langle a_1| + c_{21}^* \langle b_1|\right]$$
$$+ \left[c_{11}^* c_{12} |a_1\rangle + c_{22} |b_1\rangle\right]\left[c_{12}^* \langle a_1| + c_{22}^* \langle b_1|\right]. \tag{22.33}$$

Multiplying and collecting similar terms leads to the density matrix

$$\rho^{(1)} = \begin{pmatrix} |c_{11}|^2 + |c_{12}|^2 & c_{11}c_{21}^* + c_{12}c_{22}^* \\ c_{21}c_{11}^* + c_{22}c_{12}^* & |c_{21}|^2 + |c_{22}|^2 \end{pmatrix}. \tag{22.34}$$

It is clear that $tr\left(\rho^{(1)}\right) = 1$. For the determination of $tr\left(\rho^{(1)2}\right)$, we notice that we can represent the reduced density matrix as[19]:

$$\rho^{(1)} = CC^\dagger; \ C = \begin{pmatrix} c_{11} & c_{12} \\ c_{21} & c_{22} \end{pmatrix}. \tag{22.35}$$

One can show (see exercises) that it then holds that

---

[19]In this representation, we can see directly that $\rho^{(1)}$ is Hermitian and positive.

$$tr\left(\rho^{(1)2}\right) = 1 - 2\left|\det C\right|^2.$$
(22.36)

We therefore have $tr\left(\rho^{(1)2}\right) = 1$, if and only if $\left|\det C\right| = 0$, i.e. for $c_{11}c_{22} = c_{12}c_{21}$. But this is just the way we defined a product state in Chap. 20. In other words, for entangled states, $(c_{11}c_{22} \neq c_{12}c_{21})$ is always $tr\left(\rho^{(1)2}\right) < 1$; thus, the reduced density matrix describes a statistical mixture in this case.

### 22.3.2  Comparison

We want to compare again the various density matrices of our standard example:

We first consider a single quantum object. For a pure state of the form $|\psi\rangle = c_1|a\rangle + c_2|b\rangle$, the density matrix is

$$\rho = \begin{pmatrix} |c_1|^2 & c_1 c_2^* \\ c_2 c_1^* & |c_2|^2 \end{pmatrix}; \quad |c_1|^2 + |c_2|^2 = 1.$$
(22.37)

As stated above, either one entry or four entries are nonzero.

For a statistical mixture, there is a basis in which it holds that:

$$\rho = \begin{pmatrix} p_1 & 0 \\ 0 & p_2 \end{pmatrix}; \quad p_1 + p_2 = 1.$$
(22.38)

Here, we have that either one entry or two entries are nonzero (in another basis, four entries could also be nonzero). If there is only one nontrivial entry, we have a pure state.

Next, we consider two quantum objects. For the reduced density matrix we start from the state $|\psi\rangle = c_{11}|a_1 a_2\rangle + c_{12}|a_1 b_2\rangle + c_{21}|b_1 a_2\rangle + c_{22}|b_2 b_2\rangle$. It follows that

$$\rho^{(1)} = \begin{pmatrix} |c_{11}|^2 + |c_{12}|^2 & c_{11}c_{21}^* + c_{12}c_{22}^* \\ c_{21}c_{11}^* + c_{22}c_{12}^* & |c_{21}|^2 + |c_{22}|^2 \end{pmatrix}; \sum|c_{ij}|^2 = 1.$$
(22.39)

In this case, there are choices for the coefficients (or unitary transformations of $\rho^{(1)}$), such that the coherences vanish, but not the two populations, which is impossible for the density matrix of a pure state. We choose, for example, $c_{22} = -c_{21}c_{11}^*/c_{12}^*$ and obtain in this special case

$$\rho^{(1)}_{\text{special}} = \left(|c_{11}|^2 + |c_{12}|^2\right) \begin{pmatrix} 1 & 0 \\ 0 & \frac{|c_{21}|^2}{|c_{12}|^2} \end{pmatrix}.$$
(22.40)

Especially for $|c_{21}|^2 = |c_{12}|^2$, we have equidistribution.

### 22.3.3  General Formulation

In this section, we consider the reduced density operator for general dimensionality. We have two quantum systems $|n_1\rangle$ and $|m_2\rangle$ in the superposition state

$$|\psi\rangle = \sum_{n,m} c_{nm} |n_1 m_2\rangle, \tag{22.41}$$

or the density operator:

$$\rho = \sum_{n,m,n',m'} c_{nm} c^*_{n'm'} |n_1 m_2\rangle \langle n'_1 m'_2| \quad \text{with} \quad \sum_{n,m} |c_{nm}|^2 = 1. \tag{22.42}$$

For the reduced density operator $\rho^{(1)}$, it follows[20] that:

$$\rho^{(1)} = \sum_M \langle M_2| \rho |M_2\rangle = \sum_M \langle M_2| \sum_{n,m,n',m'} c_{nm} c^*_{n'm'} |n_1 m_2\rangle \langle n'_1 m'_2| M_2\rangle$$

$$= \sum_{n,n'} \left( \sum_M c_{nM} \cdot c^*_{n'M} \right) |n_1\rangle \langle n'_1| = \sum_{n,n'} \rho^{(1)}_{n,n'} |n_1\rangle \langle n'_1|. \tag{22.43}$$

To keep the notation transparent, we use the matrix representation. With $\rho^{(1)}_{n,n'} = \sum_M c_{nM} \cdot c^*_{n'M}$, it follows that

$$\rho^{(1)} = CC^\dagger; C = (c_{nm}) = \begin{pmatrix} c_{11} & c_{12} & c_{13} & \cdots \\ c_{21} & c_{22} & \cdots & \cdots \\ c_{31} & \vdots & \ddots & \cdots \\ \vdots & \vdots & \vdots & \ddots \end{pmatrix}. \tag{22.44}$$

In this reduced density operator, off-diagonal terms also occur, but the fact that it nevertheless describes a statistical mixture is shown by considering $\left[\rho^{(1)}\right]^2$. We examine under which conditions it holds that

$$\left[\rho^{(1)}\right]^2 = \rho^{(1)}. \tag{22.45}$$

Since $\rho^{(1)}$ is a projection operator in this case, it can be written as

$$\rho^{(1)} = |A\rangle \langle A| = |A\rangle \langle B| B\rangle \langle A|, \tag{22.46}$$

---

[20]In the following chapters, we will consider three or more quantum objects. The reduced density operator for quantum object 1 is then given by taking the trace over all other quantum objects (colloquially, 'tracing out' these degrees of freedom).

where $|A\rangle$ and $|B\rangle$ are suitable normalized vectors,

$$|A\rangle \cong \begin{pmatrix} a_1 \\ a_2 \\ \vdots \end{pmatrix} \quad \text{and} \quad |B\rangle \cong \begin{pmatrix} b_1 \\ b_2 \\ \vdots \end{pmatrix}. \tag{22.47}$$

Thus, we can identify by comparison with (22.44) (only the structure matters here):

$$C = |A\rangle \langle B| \quad \text{or} \quad c_{nm} = a_n b_m. \tag{22.48}$$

We insert this condition into (22.41) and obtain

$$|\psi\rangle = \sum_{n,m} c_{nm} |n_1 m_2\rangle = \sum_{n,m} a_n b_m |n_1 m_2\rangle = \sum_n a_n |n_1\rangle \sum_m b_m |m_2\rangle, \tag{22.49}$$

i.e. a factorized state. This means that the reduced density operator for entangled states always describes a statistical mixture; for factorized states, however, it is a projection operator onto the state in space 1.

## 22.4  Exercises

1. Write the density operator

$$\rho = \sum_n |\varphi_n\rangle \, p_n \, \langle \varphi_n| \tag{22.50}$$

   with normalized, but not necessarily orthogonal states $|\varphi_n\rangle$ when it is transformed unitarily.
2. Show that $tr\,(AB) = tr\,(BA)$.
3. Show that the trace is cyclically invariant, i.e.

$$tr\,(ABC) = tr\,(BCA) = tr\,(CAB). \tag{22.51}$$

4. Prove that the trace is invariant under unitary transformations.
5. Show that the trace is independent of the basis. (This *must* apply, since a basis transformation is unitary.)
6. Given a CONS $\{|n\rangle\}$ and a state $|\psi\rangle = \sum_n c_n |n\rangle$ with $\sum_n |c_n|^2 = 1$, show that the probability of finding the system in the state $m$ is given by $p_m = tr(\rho\,|m\rangle \langle m|) = tr(\rho P_m)$.
7. Show that for the reduced density operator $\rho^{(1)}$, it holds in general that $tr\left(\left[\rho^{(1)}\right]^2\right) \leq 1$; hence, we have a mixture if the strict inequality applies.

8. Write the density operator in the position representation (cf. Chaps. 12 and 13, Vol. 1).

9. Show explicitly for

$$\rho = \begin{pmatrix} c_1 c_1^* & c_1 c_2^* \\ c_2 c_1^* & c_2 c_2^* \end{pmatrix} \tag{22.52}$$

that

$$\rho^2 = \rho \tag{22.53}$$

applies; using this matrix, show explicitly that $\rho^2 = \rho$. Here, it must hold that $|c_1|^2 + |c_2|^2 = 1$.

10. Show that the eigenvalues $\lambda_{1/2}$ of the matrix

$$\rho = \begin{pmatrix} c_1 c_1^* & c_1 c_2^* \\ c_2 c_1^* & c_2 c_2^* \end{pmatrix} \tag{22.54}$$

are 0 and 1.

11. Given the density matrix for a statistical mixture in the form $\rho = p_h \, |h\rangle \, \langle h| + p_v \, |v\rangle \, \langle v|$ or

$$\rho = \begin{pmatrix} p_h & 0 \\ 0 & p_v \end{pmatrix}; \tag{22.55}$$

How does this read in the circularly-polarized basis?

12. Given two quantum objects Q1 and Q2 with the respective $N$-dimensional CONS $\{|\varphi_i\rangle\}$ for Q1 and $\{|\psi_i\rangle\}$ for Q2 (by the choice of notation, we can omit the index for the number of the quantum object). The initial state is

$$|\chi\rangle = \sum_{ij} c_{ij} \, |\varphi_i\rangle \, |\psi_j\rangle. \tag{22.56}$$

Calculate the probability $w\,(\lambda)$ of measuring the quantum object 1 in a state $|\lambda\rangle$, and formulate it in terms of the reduced density operator $\rho^{(1)}$.

13. Given the density operator $\rho = \sum_n p_n \, |\varphi_n\rangle \, \langle \varphi_n|$, where it holds that $i\hbar \partial_t \, |\varphi_n\rangle = H \, |\varphi_n\rangle$. Show that the time behavior of $\rho$ is described by the von-Neumann equation:

$$i\hbar \partial_t \rho = [H, \rho]. \tag{22.57}$$

14. Using the example of a polarized photon, show explicitly that a given density matrix does not allow a unique decomposition.

   (a) First formulate the projection operators for the states $|h\rangle$, $|v\rangle$, $|r\rangle$ and $|l\rangle$.

   (b) Given the density matrix $\rho = \frac{1}{2} \begin{pmatrix} 1 & 0 \\ 0 & 1 \end{pmatrix}$; now formulate the decomposition of $\rho$ in terms of linearly- and circularly-polarized states.

15. The spin state of an electron is represented (in the basis of eigenstates of the spin matrix $s_z = \frac{\hbar}{2}\sigma_z$) by the density matrix $\rho = \begin{pmatrix} a & 0 \\ 0 & b \end{pmatrix}$, with $a + b = 1$; $a \geq 0, b \geq 0$.

(a) What is the probability of obtaining $\pm\frac{\hbar}{2}$, if one measures $s_x$?
(b) Calculate the expectation value of $s_x$ and compare it with the trace formalism.

16  Given a system of two quantum objects; the basis states are in each case $|1\rangle$ and $|2\rangle$ .[21]

(a) How is the general total state $|\psi\rangle$ formulated?
(b) Give explicitly the density matrix for this system.
(c) Starting from this matrix, calculate the reduced density matrix $\rho^{(1)}$.
(d) Show that $tr\left(\rho^{(1)}\right) = 1$ holds.
(e) Show that $\rho^{(1)} = CC^\dagger$ with $C = \begin{pmatrix} c_{11} & c_{12} \\ c_{21} & c_{22} \end{pmatrix}$ holds.
(f) Calculate $\rho^{(1)2}$.
(g) Show that $tr\left(\rho^{(1)2}\right) = 1 - 2\,|\det C|^2$ is true.

17. $\{|\varphi_i\rangle, i = 1, \ldots, N\}$ are normalized, but not necessarily orthogonal states. Show that the density matrix $\rho = \frac{1}{N}\sum_{i=1}^{N} |\varphi_i\rangle\langle\varphi_i|$ describes a pure state, iff these $N$ states are equal up to a phase.

(a) Let $|\varphi_n\rangle = e^{i\delta_n}|\varphi\rangle$. Show that $\rho^2 = \rho$.
(b) Let $\rho^2 = \rho$; show that the $N$ states $|\varphi_i\rangle$ are equal up to a phase.

---

[21]One can imagine e.g. two photons, and by $|1\rangle$ and $|2\rangle$ e.g. $|h\rangle$ and $|v\rangle$ or $|r\rangle$ and $|l\rangle$.

# Chapter 23
# Identical Particles

The fact that there are identical quantum objects has far-reaching consequences, e.g. the
Pauli principle. We also take a closer look at the spectrum of the helium atom and make the
acquaintance of another approximation method, namely Ritz's method.

We have thus far always tacitly assumed that the quantum objects we are dealing with
are distinguishable. This is familiar from classical physics, where it is always possible
to distinguish between two particles (for example by coloring them differently),
without changing their measurable properties. Quantum mechanics is different: Here,
identical quantum objects are not distinguishable *in principle*.[1] This fact leads, among
other things, to the Pauli principle and to the exchange energy in helium.

A note on nomenclature: Whenever possible, we have consistently avoided speaking of 'particles', and have instead used the term 'quantum objects'. This is intended
to emphasize that we are usually dealing not with a particle or a wave, but with something new and peculiar to quantum mechanics. In the everyday parlance of physics,
however, the term 'particle' is well established in many contexts (which in part is for
historical reasons, and partly is due to the convenience of language and physics folk-
lore). This is the case here also, where the expression 'identical particles' does not
point up the particle nature of the objects in question, but rather it parallels the term
'identical quantum objects'. However, since 'identical particles' is an established
standard term, we will continue to use it in the following sections.[2]

Finally, a remark on notation. In Chap. 20, using the notation $|nm\rangle$ or the like, we
described the situation that quantum object 1 is in the state $|n\rangle$ and quantum object
2 in the state $|m\rangle$. In this chapter, the situation is more complicated in that quantum
objects can exchange their states, which means that we must ideally number both

---

[1] Some authors express this fact explicitly as a postulate of quantum mechanics, see Appendix R,
Vol. 1.

[2] Similar nomenclature problems exist in other disciplines. For example, Egon Friedell writes about
the transcriptions of Oriental proper names in the Bible: "Almost all the other names are mangled
in a similar way, but since they were adopted by Luther in his Bible translation and are therefore
now naturalized as fixed terms, it would be pure harassment and learned affectation to try to correct
them." (*Cultural History of Egypt and the Ancient Orient*, p. 1082).

© Springer Nature Switzerland AG 2018
J. Pade, *Quantum Mechanics for Pedestrians 2*, Undergraduate Lecture
Notes in Physics, https://doi.org/10.1007/978-3-030-00467-5_23

the objects and their states independently of each other. In addition, we consider in general not just two objects, but a larger number $N$. In the literature there are various notations for this purpose. We choose in the following $|1 : \alpha_1, 2 : \alpha_2, \ldots\rangle$ for the situation that object 1 is in state $\alpha_1$ and so on. The $\alpha_i$'s signify all the quantum numbers that are necessary for the unique description of the $i$th state. An exchange of the $i$th and $j$th state (i.e. object $i$ is in state $\alpha_j$, object $j$ is in state $\alpha_i$) is then written as

$$|1 : \alpha_1, \ldots, i : \alpha_i, \ldots, j : \alpha_j, \ldots\rangle \to |1 : \alpha_1, \ldots, i : \alpha_j, \ldots, j : \alpha_i, \ldots\rangle.$$
(23.1)

## 23.1    Distinguishable Particles

For this case, we have provided the necessary ingredients in Chap. 20. We link the Hilbert spaces $\mathcal{H}_i$ of the $N$ individual quantum objects[3] to give the product space[4]:

$$\mathcal{H}_{(N)} = \mathcal{H}_{1(1)} \otimes \mathcal{H}_{2(1)} \otimes \cdots \otimes \mathcal{H}_{N(1)}.$$
(23.2)

In each Hilbert space $\mathcal{H}_{i(1)}$, there is a CONS $\{|i : a_i\rangle\}$, from which the $N$-particle product states

$$|1 : \alpha_1, 2 : \alpha_2, \ldots, N : \alpha_N\rangle = |1 : \alpha_1\rangle |2 : \alpha_2\rangle \ldots |N : \alpha_N\rangle$$
(23.3)

may be generated. The states (23.3) form a CONS in $\mathcal{H}_{(N)}$, in terms of which any of the $N$ particles can be expanded:

$$|\psi_N\rangle = \sum_{\alpha_1 \ldots \alpha_N} c_{\alpha_1 \ldots \alpha_N} |1 : \alpha_1, 2 : \alpha_2, \ldots, N : \alpha_N\rangle.$$
(23.4)

Scalar products refer to the same space:

$$\langle 1 : \beta_1, 2 : \beta_2, \ldots, N : \beta_N | 1 : \alpha_1, 2 : \alpha_2, \ldots, N : \alpha_N\rangle$$
$$= \langle 1 : \beta_1 | 1 : \alpha_1\rangle \langle 2 : \beta_2 | 2 : \alpha_2\rangle \ldots \langle N : \beta_N | N : \alpha_N\rangle.$$
(23.5)

---

[3]We denote the particle number $n$ by a subscript, the number $m$ of particles (if necessary) by a bracketed subscript and the dimension $d$ by a superscript, i.e. $\mathcal{H}^d_{n(m)}$.

[4]Or more compactly, $\mathcal{H}_{(N)} = \bigotimes\limits_{n=1}^{N} \mathcal{H}_{n(1)}$.

**Fig. 23.1** By a measurement as shown (arrow position), it is not possible to conclude unambiguously which electron was observed

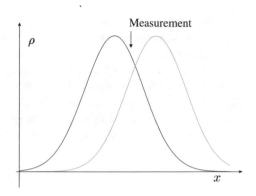

## 23.2 Identical Particles

We assume that we have two free electrons. Their probability densities are given by Gaussian curves (see Chap. 5, Vol. 1) which overlap as in Fig. 23.1. This means that if we detect an electrona somewhere, we do not know which one of the two we have observed—unless the two differ in their spin orientations, and we measure them as well. Thus, we assume that identical quantum objects agree in *all* their properties, even if we do not observe these in detail. In contrast to classical mechanics, identical quantum objects are *indistinguishable*—there is no way (no matter how sophisticated) to distinguish them.

Now we *must* perform a particle numbering in the formal description, so to speak for accounting purposes. How are we to number indistinguishable quantum objects? In any case, this must be done in such a way that experimentally-detectable quantities do *not* depend on the method of numbering.

### 23.2.1 A Simple Example

We consider the simplest example, namely a two-particle product state $|1 : \alpha_1\rangle$ $|2 : \alpha_2\rangle = |1 : \alpha_1; 2 : \alpha_2\rangle$. If we interchange the two particles, we obtain the state $|1 : \alpha_2; 2 : \alpha_1\rangle$. From these two states, we can now construct total states for which, as intended, the physics does not depend on the method of numbering. These are (this is intuitively obvious, but is treated explicitly in an exercise):

$$|\psi_\pm\rangle = \frac{|1 : \alpha_1, 2 : \alpha_2\rangle \pm |1 : \alpha_2, 2 : \alpha_1\rangle}{\sqrt{2}} \quad \text{for } \alpha_1 \neq \alpha_2 \quad (23.6)$$

and

$$|\psi\rangle = |1 : \alpha_1, 2 : \alpha_2\rangle \quad \text{for } \alpha_1 = \alpha_2. \quad (23.7)$$

Interchanging the two particles, we have $|\psi_+\rangle \rightarrow |\psi_+\rangle$ (symmetric state) and $|\psi_-\rangle \rightarrow -|\psi_-\rangle$ (antisymmetric state). We recall that a global phase (e.g. here the factor $-1$) is not observable. The two states (23.6) are clearly entangled, and consequently, the two particles do not have individual features, as we emphasized in Chap. 20.

We illustrate this by means of two electrons. For electron $i$, we denote the state 'spin up' by $|i : \frac{1}{2}\rangle$ and 'spin down' by $|i : -\frac{1}{2}\rangle$.[5] Two spins of $\frac{1}{2}$ can be added to give a total spin of $S = 1$ or $0$. The corresponding spin states $|S, m_S\rangle$ for total spin $S = 1$ form a triplet $|1, m_S\rangle$ (symmetric), with $m_S = -1, 0, 1$, i.e.

$$|1, 1\rangle = \left|1 : \frac{1}{2}, 2 : \frac{1}{2}\right\rangle$$

$$|1, 0\rangle = \frac{\left|1 : \frac{1}{2}, 2 : -\frac{1}{2}\right\rangle + \left|1 : -\frac{1}{2}, 2 : \frac{1}{2}\right\rangle}{\sqrt{2}} \qquad (23.8)$$

$$|1, -1\rangle = \left|1 : -\frac{1}{2}, 2 : -\frac{1}{2}\right\rangle;$$

and for total spin $S = 0$, a singlet $|0, 0\rangle$ (antisymmetric):

$$|0, 0\rangle = \frac{\left|1 : \frac{1}{2}, 2 : -\frac{1}{2}\right\rangle - \left|1 : -\frac{1}{2}, 2 : \frac{1}{2}\right\rangle}{\sqrt{2}}. \qquad (23.9)$$

### 23.2.2  The General Case

For systems of more than two particles, it is convenient to use *permutations*. A permutation changes the order within am $n$-tuple; for instance, $(1, 4, 3, 2)$ is a permutation of $(1, 2, 3, 4)$. Every permutation can be written as the product of *transpositions*, i.e. permutations which interchange just two positions, such as in $(1, 2, 3, 4) \rightarrow (1, 3, 2, 4)$.

The transposition operator $P_{ij}$ interchanges the $i$th and the $j$th position[6]:

$$P_{ij} |\ldots, i : \alpha_i, \ldots, j : \alpha_j, \ldots\rangle = |\ldots, i : \alpha_j, \ldots, j : \alpha_i, \ldots\rangle. \qquad (23.10)$$

In words: Particle $i$ now has the quantum numbers $\alpha_j$ and particle $j$ the quantum numbers $\alpha_i$. The transpositions are unitary in $\mathcal{H}_{(N)}$ and Hermitian, because of $P_{ij}^2 = 1$:

$$P_{ij}^{-1} = P_{ij}^\dagger = P_{ij}. \qquad (23.11)$$

---

[5]Since the spin of both quantum objects is $\frac{1}{2}$, we omit it below and write for short $|i : \pm\frac{1}{2}\rangle$ instead of $|i : \frac{1}{2}, \pm\frac{1}{2}\rangle$, and similarly for the total state, $|1 : \frac{1}{2}; 2 : -\frac{1}{2}\rangle$ instead of $|1 : \frac{1}{2}, \frac{1}{2}; 2 : \frac{1}{2}, -\frac{1}{2}\rangle$.
[6]We note that $P_{ij}$ is not a projection operator. The letter $P$ is derived from 'permutation'.

We stated above that experimentally measurable quantities must not depend on the method of numbering. For matrix elements of an observable $A_N$ with an allowed state $|\varphi\rangle$, this requirement can be formulated as follows:

$$\langle\varphi|\, A_N\,|\varphi\rangle \overset{!}{=} \langle P_{ij}\varphi|\, A_N\,|P_{ij}\varphi\rangle = \langle\varphi|\, P_{ij}^\dagger A_N P_{ij}\,|\varphi\rangle \quad \text{for all pairs } i,j, \quad (23.12)$$

which implies[7] that:

$$A_N = P_{ij}^\dagger A_N P_{ij} \quad \text{or} \quad [A_N, P_{ij}] = 0 \quad \text{for all pairs } i,j. \quad (23.13)$$

In particular, of course, this must apply to the Hamiltonian $H_N$ (otherwise we would have found a tool to distinguish the particles after all):

$$[H_N, P_{ij}] = 0 \quad \text{for all pairs } i,j. \quad (23.14)$$

We now ask for the eigenvalues $\eta_{ij}$ of $P_{ij}$:

$$P_{ij}\,|\psi\rangle = \eta_{ij}\,|\psi\rangle. \quad (23.15)$$

Because of $P_{ij}^2 = 1$, we have $\eta_{ij}^2 = 1$; thus there are two possible eigenvalues $\eta_{ij} = \pm 1$. Now, if $|\psi\rangle$ is an eigenvector of *all* transpositions,[8] then the eigenvalues $\eta_{ij}$ are the same for all pairs $i,j$. This holds because due to e.g. $P_{ni} P_{mj} P_{nm} P_{ni} P_{mj} = P_{ij}$ (see exercises), the equation $\eta_{ij} = \eta_{ni}^2 \eta_{mj}^2 \eta_{nm} = \eta_{nm} = \eta$ applies for all pairs $i,j$ and all pairs $n,m$. Generalizing (23.6), we can therefore distinguish two cases:

$$\eta = +1 : \text{totally symmetric state } \left|\psi_N^{(+)}\right\rangle \leftrightarrow P_{ij}\left|\psi_N^{(+)}\right\rangle = \left|\psi_N^{(+)}\right\rangle \;\forall\; (i,j)$$

$$\eta = -1 : \text{totally antisymmetric state } \left|\psi_N^{(-)}\right\rangle \leftrightarrow P_{ij}\left|\psi_N^{(-)}\right\rangle = -\left|\psi_N^{(-)}\right\rangle \;\forall\; (i,j).$$
$$(23.16)$$

Hence, the states of a quantum system of identical objects are either symmetric or antisymmetric with regard to the interchange of two indices. Other possibilities do not exist.

This means, *inter alia*, that identical particles do not populate the whole Hilbert space $\mathcal{H}_{(N)}$, but only the subspaces $\mathcal{H}_{(N)}^{(+)}$ (symmetric states) and $\mathcal{H}_{(N)}^{(-)}$ (antisymmetric states). These two subspaces are mutually orthogonal:

$$\left\langle\psi_N^{(+)}\,\Big|\,\psi_N^{(-)}\right\rangle = \left\langle\psi_N^{(+)}\,\Big|\,P_{ij}^\dagger P_{ij}\,\Big|\,\psi_N^{(-)}\right\rangle = -\left\langle\psi_N^{(+)}\,\Big|\,\psi_N^{(-)}\right\rangle = 0, \quad (23.17)$$

---

[7] In general, transpositions do not commute, $P_{23} P_{12} \neq P_{12} P_{23}$.

[8] The $N$-particle states (23.3) do not satisfy this condition; we still have to construct the said eigenvectors.

and, due to (23.13), there is no observable $A_N$ which mediates between the two spaces:

$$\left\langle \psi_N^{(+)} \right| A_N \left| \psi_N^{(-)} \right\rangle = \left\langle \psi_N^{(+)} \right| P_{ij}^\dagger A_N P_{ij} \left| \psi_N^{(-)} \right\rangle = - \left\langle \psi_N^{(+)} \right| A_N \left| \psi_N^{(-)} \right\rangle = 0. \quad (23.18)$$

Thus, the two subspaces $\mathcal{H}_{(N)}^{(+)}$ and $\mathcal{H}_{(N)}^{(-)}$ are strictly separated.[9]

Now what are the allowed states? We start from the $N$-particle product states (23.3) that are *not* totally symmetric or antisymmetric. We can construct symmetric or antisymmetric states from them by suitable superpositions. For this purpose, we introduce two new operators, namely the *symmetrization operator* $S_N^{(+)}$ and the *antisymmetrization* operator $S_N^{(-)}$:

$$S_N^{(\pm)} = \frac{1}{N!} \sum_P (\pm 1)^p \, P \quad (23.19)$$

where the sum runs over all the $N!$ permutations $P$ of the $N$-tuple.[10] The number $p$ denotes the number of transpositions from which $P$ is constructed.[11] The states

$$\left| \varphi_N^{(\pm)} \right\rangle = |1 : \alpha_1, 2 : \alpha_2, \ldots N : \alpha_N\rangle^{(\pm)} = S_N^{(\pm)} |1 : \alpha_1, 2 : \alpha_2, \ldots N : \alpha_N\rangle \quad (23.20)$$

are elements of $\mathcal{H}_{(N)}^{(\pm)}$, and it holds that

$$P_{ij} \left| \varphi_N^{(\pm)} \right\rangle = \pm \left| \varphi_N^{(\pm)} \right\rangle; \quad P \left| \varphi_N^{(\pm)} \right\rangle = (\pm 1)^p \left| \varphi_N^{(\pm)} \right\rangle. \quad (23.21)$$

This leads immediately to

$$S_N^{(\pm)} \left| \varphi_N^{(\pm)} \right\rangle = \frac{1}{N!} \sum_P (\pm 1)^p \, P \left| \varphi_N^{(\pm)} \right\rangle = \left| \varphi_N^{(\pm)} \right\rangle. \quad (23.22)$$

It follows that the (anti)symmetrized product states $\left| \varphi_N^{(\pm)} \right\rangle$ form an orthogonal basis of their subspaces $\mathcal{H}_{(N)}^{(+)}$ and $\mathcal{H}_{(N)}^{(-)}$. However, they are not normalized. The normalized states are

$$|1 : \alpha_1, 2 : \alpha_2, \ldots N : \alpha_N\rangle_{norm}^{(-)} = \frac{1}{\sqrt{N!}} \sum_P (-1)^p \, P |1 : \alpha_1, 2 : \alpha_2, \ldots N : \alpha_N\rangle \quad (23.23)$$

---

[9]This illustrates with an example the remark of Chap. 14, Vol. 1 that not every vector in the Hilbert space necessarily corresponds to a physically realizable state (i.e. that a superselection rule exits).

[10]Example for a 3-tuple: (1, 2, 3), (1, 3, 2), (2, 1, 3), (2, 3, 1), (3, 1, 2), (3, 2, 1).

[11]$p$ is even for a cyclic permutation and odd for a non-cyclic one.

and

$$|1 : \alpha_1, 2 : \alpha_2, \ldots N : \alpha_N\rangle_{norm}^{(+)} = \sqrt{\frac{N_1! N_2! \ldots}{N!}} \sum_{P'} P' |1 : \alpha_1, 2 : \alpha_2, \ldots N : \alpha_N\rangle,$$

(23.24)

where $N_i$ is the multiplicity of $\alpha_i$ and the sum extends only over those permutations $P'$ which lead to different states (whose number is $\frac{N!}{N_1! N_2! \ldots}$). The completeness relation then reads:

$$\sum_{\alpha_1 \ldots \alpha_N} |1 : \alpha_1, 2 : \alpha_2, \ldots N : \alpha_N\rangle_{norm}^{(\pm)} \langle 1 : \alpha_1, 2 : \alpha_2, \ldots N : \alpha_N|_{norm}^{(\pm)} = 1. \quad (23.25)$$

By the way, we see that the states $|1 : \alpha_1, 2 : \alpha_2, \ldots N : \alpha_N\rangle^{(\pm)}$ do not factorize—they are entangled. This means that the individual identical quantum objects do not have individually assignable properties, which is just the basic requirement of this section.[12]

## 23.3 The Pauli Exclusion Principle

If one looks closely at the state $|1 : \alpha_1, 2 : \alpha_2, \ldots N : \alpha_N\rangle^{(-)}$, it can be seen that it vanishes if two sets of quantum numbers are equal, i.e. $\alpha_k = \alpha_l$ with $k \neq l$. Thus we have found the *Pauli (exclusion) principle*; it states that two identical particles cannot agree on all their quantum numbers in a totally antisymmetric state. This is shown perhaps even more clearly in the equivalent notation

$$|1 : \alpha_1, 2 : \alpha_2, \ldots N : \alpha_N\rangle_{norm}^{(-)} = \frac{1}{\sqrt{N!}} \begin{vmatrix} |1 : \alpha_1\rangle & |2 : \alpha_1\rangle & \ldots & |N : \alpha_1\rangle \\ |1 : \alpha_2\rangle & |2 : \alpha_2\rangle & \ldots & |N : \alpha_2\rangle \\ \vdots & \vdots & \vdots & \vdots \\ |1 : \alpha_N\rangle & |2 : \alpha_N\rangle & \ldots & |N : \alpha_N\rangle \end{vmatrix}.$$

(23.26)

This determinant, formed of single-particle states, is called the *Slater determinant*. We see directly that for $\alpha_k = \alpha_l$ with $k \neq l$, two rows of this determinant are equal and therefore it is zero.

The particles which are described by a totally antisymmetric state are called *fermions*; they have half-integer spins and obey Fermi-Dirac statistics. In contrast, *bosons* are described by totally symmetric states; they have integer spins and obey Bose-Einstein statistics. Unlike fermions, bosons are *not* subject to the Pauli prin-

---

[12] Another way to formulate states of many quantum objects is provided by the so-called second quantization, see Appendix W, Vol. 2. This method works with creation and annihilation operators, and in this regard is somewhat similar to the algebraic treatment of the angular momentum or the harmonic oscillator.

ciple, which means that bosons can agree in all quantum numbers, i.e. more than one can occupy precisely the same state. As already mentioned in Chap. 16, all the elementary particles belong to one of these two particle classes.[13] Incidentally, this connection between spin and statistics cannot be derived from what we have said here, but it follows from the so-called *spin-statistics theorem*.

The Pauli principle is of fundamental importance for the structure of matter, since it ensures that atoms cannot collapse: Each state $(n, l, m_l)$ can be occupied by only two electrons, which differ by their spin orientations. If there are more electrons, they must populate higher levels, i.e. outer shells; as is well known, the periodic table is based on this principle. Similarly, in astronomy, the Pauli principle explains why old stars (with the exception of black holes) do not collapse under the weight of their own gravity: The fermions must occupy different states, thus creating a back pressure which prevents further collapse.

## 23.4  The Helium Atom

The helium atom is a prime example of the application of the ideas discussed in this chapter. The problem cannot be solved exactly, so that we have to introduce some approximations, but we will see that even with this approximate method, the indistinguishability of the two electrons leads to classically inexplicable effects (summarized under the term *exchange energy*).

We neglect the motion of the nucleus,[14] which is about 8000 times heavier than the electrons, and place our coordinate origin in the nucleus, which has the atomic number $Z = 2$. In addition, we neglect spin-dependent interactions, in contrast to our treatment of the hydrogen atom in Chap. 19. The spin will be considered later, but only for the classification of the electrons. Thus, the single-particle states can be represented as $|n_1 l_1 m_1\rangle$.

The total Hamiltonian reads

$$H = H_1 + H_2 + V_{1,2} \tag{23.27}$$

with

$$H_i = \frac{\mathbf{p}_i^2}{2m} - \frac{Ze^2}{4\pi\varepsilon_0} \frac{1}{r_i} \; ; \; V_{1,2} = \frac{e^2}{4\pi\varepsilon_0} \frac{1}{|\mathbf{r}_1 - \mathbf{r}_2|}. \tag{23.28}$$

$Ze$ is the nuclear charge, with $Z = 2$ for helium. The interaction $V_{1,2}$ describes the electron-electron interaction, i.e. the mutual electrostatic repulsion of the two electrons.

---

[13]In theoretical solid-state physics, one considers so-called anyons (not to be confused with anions). These are quasiparticles in two dimensions which are neither bosons nor fermions.

[14]In the equations, therefore, the mass of the electrons and not the reduced mass appears.

We treat the problem as follows: We consider $V_{1,2}$ as a 'small' perturbation potential in the sense of the perturbation theory of Chap. 19. Accordingly, we neglect $V_{1,2}$ in the first step and look for the states which the (identical) electrons can occupy generally. Then we take $V_{1,2}$ into account and calculate the first energy correction as a perturbation.

### 23.4.1 Spectrum Without $V_{1,2}$

If we 'turn off' the electron-electron interaction as a start, then each electron can occupy the hydrogen eigenstates, as discussed in Chap. 17. We call them position states in the following (which is somewhat imprecise, since we write them in the abstract representation), in order to distinguish them readily from the spin states. The single-particle product states are given by

$$|1 : \alpha_1\rangle |2 : \alpha_2\rangle ; \quad \alpha_i = n_i l_i m_i \tag{23.29}$$

and the total state reads[15]:

$$|1 : \alpha_1\rangle |2 : \alpha_2\rangle \equiv |1 : n_1 l_1 m_1, 2 : n_2 l_2 m_2\rangle . \tag{23.30}$$

For the unperturbed energies, we have

$$E_{n_1 n_2}^{(0)} = -Z^2 R_\infty \frac{1}{n_1^2} - Z^2 R_\infty \frac{1}{n_2^2}, \tag{23.31}$$

with[16]

$$R_\infty = \frac{m e^4}{2 \hbar^2 (4\pi \varepsilon_0)^2}. \tag{23.32}$$

In accordance with these considerations, the ground state of helium is given by $E_{11}^{(0)} = -8 \cdot 13.6\,\text{eV} = -108.8\,\text{eV}$. But since the experimental value is $-78.975\,\text{eV}$, we obviously have to improve our method. This is done below by means of perturbation theory.

---

[15]Since we have only two particles, we could use the shorter notation, familiar from Chap. 20, in which the state of the first or second particle is listed in first or second place, i.e. $|1 : \alpha_1\rangle |2 : \alpha_2\rangle \equiv |n_1 l_1 m_1, n_2 l_2 m_2\rangle$. Nevertheless, we choose the slightly more cumbersome version, since it is unambiguous.

[16]The mass effect could be taken into account here by

$$R_{He} = R_\infty \left(1 + \frac{m}{m_{He}}\right)^{-1}.$$

But first, we want to get an overview of the states, whereby we consider the symmetrization postulate for identical particles. The (anti)symmetrized position states are

$$|n_1l_1m_1, n_2l_2m_2\rangle^{(\pm)} = C\{|1 : n_1l_1m_1, 2 : n_2l_2m_2\rangle \pm |1 : n_2l_2m_2, 2 : n_1l_1m_1;\rangle\}$$
(23.33)

with the normalization constant

$$C = \begin{cases} \frac{1}{\sqrt{2}} \\ \frac{1}{2} \end{cases} \text{ for } \begin{cases} (n_1l_1m_1) \neq (n_2l_2m_2) \\ (n_1l_1m_1) = (n_2l_2m_2) \end{cases}$$
(23.34)

In the next step, we take into account also the spins of the two electrons. Two spins of $\frac{1}{2}$ can be added to give a total spin of $S = 1$ or $S = 0$; in (23.8) and (23.9), we have written down the corresponding spin states $|S, m_S\rangle$. For a total spin $S = 1$, they form a triplet $|1, m_S\rangle$; for $S = 0$, a singlet $|0, 0\rangle$. For historical reasons, these are called *orthohelium* (triplet) and *parahelium* (singlet). The spin part in orthohelium is symmetric, so that the position state must be antisymmetric according to the Pauli principle, and *vice versa* for parahelium. With (23.33), the possible states for orthohelium are:

$$|n_1l_1m_1, n_2l_2m_2\rangle^{(-)} |1, 1\rangle$$
$$|n_1l_1m_1, n_2l_2m_2\rangle^{(-)} |1, 0\rangle$$
$$|n_1l_1m_1, n_2l_2m_2\rangle^{(-)} |1, -1\rangle ;$$
(23.35)

and for parahelium, they are:

$$|n_1l_1m_1, n_2l_2m_2\rangle^{(+)} |0, 0\rangle .$$
(23.36)

We see that for orthohelium, it must hold that $(n_1l_1m_1) \neq (n_2l_2m_2)$, while for parahelium, the spin states must be different. This means that the undisturbed ground state $|1 : n_1l_1m_1, 2 : n_2l_2m_2\rangle = |1 : 1, 0, 0; 2 : 1, 0, 0\rangle$ can be occupied only by parahelium. Neglecting the terms caused by $V_{1,2}$, the helium spectrum has the structure shown in Fig. 23.2. In this spectrum, the boundary to the continuum is a limiting point for the bound energy levels; this is only implied in the figure.

Obviously, there are discrete levels in the continuum. This is due to the following: To raise both electrons from the ground state into the first excited state, we need $81.6\,\text{eV} = (108.8 - 68) \cdot 2\,\text{eV}$ of energy. On the other hand, the ionization energy (one of the electrons is transferred into the continuum) is $54.4\,\text{eV} = \frac{1}{2} \cdot 108.8\,\text{eV}$. The doubly-excited state therefore does not necessarily decay to the ground state or another bound state, but can also lead to a state of a singly-ionized helium plus a free electron (*autoionization*). It follows that in *all* the discrete states below the ionization limit, one electron is in the one-particle ground state.

**Fig. 23.2** Helium spectrum without electron-electron interaction (not to scale). The yellow part indicates the continuous spectrum

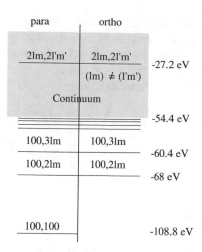

### 23.4.2 Spectrum with $V_{1,2}$ (Perturbation Theory)

Having established the 'rough' helium spectrum, we calculate by means of the tools of Chap. 19 the corrections due to the electron-electron interaction $V_{1,2}$, i.e. in first-order perturbation theory. We have for the ground state:

$$E^{(1)}_{100;100} = \langle 1, 0, 0; 1, 0, 0|^{(+)} \ V_{1,2} \ |1, 0, 0; 1, 0, 0\rangle^{(+)}$$

$$= \frac{e^2}{4\pi\varepsilon_0} \int d^3r_1 d^3r_2 \left|\psi_{1,0,0}(\mathbf{r}_1)\right|^2 \left|\psi_{1,0,0}(\mathbf{r}_2)\right|^2 \frac{1}{|\mathbf{r}_1 - \mathbf{r}_2|} \tag{23.37}$$

with the position functions[17]

$$\psi_{1,0,0}(\mathbf{r}) = \frac{1}{\sqrt{4\pi}} \left(\frac{Z}{a_0}\right)^{3/2} e^{-\frac{Zr}{a_0}}. \tag{23.38}$$

After some computations, the result reads

$$E^{(1)}_{1,0,0;1,0,0} = \frac{5}{4} Z R_\infty. \tag{23.39}$$

For $Z = 2$, it follows that $E^{(1)}_{1,0,0;1,0,0} \approx \frac{5}{2} \cdot 13.6\,\text{eV} = 34\,\text{eV}$, so that we obtain the result for the ground state energy $E_{1,0,0;1,0,0}$:

$$E_{1,0,0;1,0,0} \approx E^{(0)}_{11} + E^{(1)}_{1,0,0;1,0,0} = -108.8\,\text{eV} + 34\,\text{eV} = -74.8\,\text{eV}, \tag{23.40}$$

---

[17]cf. Chap. 17 and Appendix B, Vol. 2.

and thus we have finally obtained a value which is after all in the neighborhood of the experimentally-measured energy of $-78.975\,\text{eV}$.

Next, let us look at the perturbative correction for states with $n_1 = 1$ and $n_2 = n \geq 2$. Since one electron must be in the state $|1, 0, 0\rangle$ (one-particle ground state), we have (see exercises)

$$
\begin{aligned}
E^{(1)}_{1,0,0;nlm} &= \langle 1, 0, 0; nlm|^{(\pm)}\; V_{1,2}\,|1, 0, 0; nlm\rangle^{(\pm)} \\
&= \frac{e^2}{4\pi\varepsilon_0} \int d^3r_1 d^3r_2 \frac{\left|\psi_{1,0,0}\,(\mathbf{r}_1)\,\psi_{nlm}\,(\mathbf{r}_2) \pm \psi_{nlm}\,(\mathbf{r}_1)\,\psi_{1,0,0}\,(\mathbf{r}_2)\right|^2}{2\,|\mathbf{r}_1 - \mathbf{r}_2|} \\
&= C_{nl} \pm A_{nl},
\end{aligned}
\tag{23.41}
$$

with the *Coulomb energy*

$$
C_{nl} = \frac{e^2}{4\pi\varepsilon_0} \int d^3r_1 d^3r_2 \frac{\left|\psi_{1,0,0}\,(\mathbf{r}_1)\,\psi_{nlm}\,(\mathbf{r}_2)\right|^2}{|\mathbf{r}_1 - \mathbf{r}_2|}
\tag{23.42}
$$

and the *exchange energy*

$$
A_{nl} = \frac{e^2}{4\pi\varepsilon_0} \int d^3r_1 d^3r_2 \frac{\psi_{1,0,0}\,(\mathbf{r}_1)\,\psi_{nlm}\,(\mathbf{r}_2)\,\psi^*_{nlm}\,(\mathbf{r}_1)\,\psi^*_{1,0,0}\,(\mathbf{r}_2)}{|\mathbf{r}_1 - \mathbf{r}_2|}.
\tag{23.43}
$$

The exchange energy is due to the Pauli principle and is a purely quantum-mechanical effect, which is not explainable classically.[18] The corrections to the energy are given by

$$
E_{12} = E^{(0)}_{12} + C_{2l} \pm A_{2l}; \; + \text{parahelium}, \; - \text{orthohelium}.
\tag{23.44}
$$

$C_{nl}$ is positive, which can be seen directly; $A_{nl}$ is also positive, as the calculation shows. The detailed calculations can be found in Appendix N, Vol. 2. The result now reads

$$
\begin{aligned}
C_{20} &= \frac{e^2}{4\pi\varepsilon_0}\frac{17}{81}\frac{Z}{a_0}; \quad C_{21} = \frac{e^2}{4\pi\varepsilon_0}\frac{59}{243}\frac{Z}{a_0} \\
A_{20} &= \frac{e^2}{4\pi\varepsilon_0}\frac{16}{729}\frac{Z}{a_0}; \quad A_{21} = \frac{e^2}{4\pi\varepsilon_0}\frac{112}{6561}\frac{Z}{a_0}.
\end{aligned}
\tag{23.45}
$$

Numerically,[19] we have for $Z = 2$

---

[18]The Coulomb energy would also have the same form for non-identical particles.

[19]In some textbooks, a few incorrect values for $C_{2l}$ and $A_{2l}$ are quoted. We mention this here not in order to find fault with other textbooks; in fact, there are always some mistakes in any longer text (in this one as well), in spite of the most careful editing. The remark is aimed rather at making it clear that learning is a process for which each individual is responsible for his or her own progress. A certain critical distance should be maintained towards every textbook; no single one is absolutely correct. Thus, look not only into *one* book as an aid to learning, but always use several!

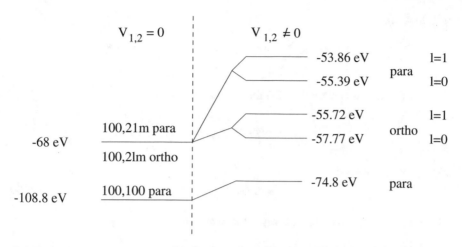

**Fig. 23.3** The lowest discrete levels of the helium spectrum with and without electron-electron interactions (not to scale)

$$E_{12}^{(0)} \approx -68.0 \text{ eV} \; ; \; C_{20} \approx 11.42 \text{ eV}; \; C_{21} \approx 13.21 \text{ eV}$$
$$A_{20} \approx 1.19 \text{ eV}; \; A_{21} \approx 0.93 \text{ eV}. \tag{23.46}$$

With these values, the energy levels of parahelium are found to be $-55.39$ eV and $-53.86$ eV; and those of orthohelium are $-57.77$ eV and $-55.72$ eV. Figure 23.3 illustrates the situation. We see that the degeneracy is removed (strictly speaking, only partially, because the $m$-degeneracy remains).

## 23.5 The Ritz Method

The (Rayleigh-) Ritz method is a general,[20] very simple, elegant and—properly applied—very effective method of approximation for the energy of the ground state.

We have a Hamiltonian with

$$H \, |\varphi_n\rangle = E_n \, |\varphi_n\rangle . \tag{23.47}$$

An arbitrary state $|\psi\rangle$ (not necessarily normalized) can be expanded in terms of a CONS $\{|\varphi_n\rangle\}$

$$|\psi\rangle = \sum_n c_n \, |\varphi_n\rangle . \tag{23.48}$$

---

[20]i.e. not limited to identical quantum objects.

It follows that

$$\langle \psi | H | \psi \rangle = \sum_{n,m} c_n^* c_m \langle \varphi_n | H | \varphi_m \rangle = \sum_n |c_n|^2 E_n. \tag{23.49}$$

We now perform an estimation. On the one hand, we have

$$\langle \psi | \psi \rangle = \sum_n |c_n|^2, \tag{23.50}$$

and on the other hand

$$E_n \geq E_0 \text{ for } n > 0 \tag{23.51}$$

where $E_0$ is the ground-state energy. This leads to

$$\langle \psi | H | \psi \rangle = \sum_n |c_n|^2 E_n \geq \sum_n |c_n|^2 E_0 = E_0 \langle \psi | \psi \rangle, \tag{23.52}$$

or[21]

$$E_0 = \inf_{\psi} \frac{\langle \psi | H | \psi \rangle}{\langle \psi | \psi \rangle}. \tag{23.53}$$

This means that we have to choose reasonable states with one or several parameters and vary them so that the right side is a minimum. If we find values that are lower than the experimental value, this is not a failure of the variational principle, but rather evidence that the Hamiltonian chosen does not correctly describe the problem and needs to be improved. The method can thus provide important information even if it does not 'work'.

As an example, we consider the helium atom. We assume the following trial function:

$$\psi(\mathbf{r}_1, \mathbf{r}_2) = e^{-\frac{\zeta}{a_0} r_1} e^{-\frac{\zeta}{a_0} r_2}. \tag{23.54}$$

It is the product of single-particle functions as found in Chap. 17. A relatively unimportant difference is the lack of normalization,[22] but an essential difference is the use of $\zeta$ instead of $Z$; this parameter is chosen in such a way that our value in (23.53) becomes minimal. With $H = H_1 + H_2 + V_{1,2}$, we find after some calculations

$$\frac{\langle \psi | H_1 + H_2 | \psi \rangle}{\langle \psi | \psi \rangle} = R_\infty \left( 2\zeta^2 - 8\zeta \right), \tag{23.55}$$

and (see also (23.39))

$$\frac{\langle \psi | V_{1,2} | \psi \rangle}{\langle \psi | \psi \rangle} = R_\infty \frac{5}{4} \zeta. \tag{23.56}$$

---

[21]inf means *infimum*, the greatest lower bound.

[22]This shortcoming is cured by the process itself, as stated above.

This leads to

$$\frac{\langle \psi | H | \psi \rangle}{\langle \psi | \psi \rangle} = R_\infty \left( 2\zeta^2 - \frac{27}{4}\zeta \right). \tag{23.57}$$

The right-hand side shows a minimum with respect to $\zeta$ at $\zeta = \frac{27}{16} = 2 - \frac{5}{16}$. One interprets this equation to mean that the two electrons shield each other to some extent from the nuclear charge, and accordingly experience a smaller charge. The quantity $\zeta$ is called the *effective charge number*. The numerical value for the ground-state energy is

$$E_0 \leq -R_\infty \frac{(27)^2}{128} \approx -77.5 \text{ eV}. \tag{23.58}$$

This value is already relatively close to the experimental value of $E_0 = -78.975 \text{ eV}$.

## 23.6   How Far does the Pauli Principle Reach?

In this section, we consider the question of whether we always have to apply the symmetrization postulate. Does the Pauli principle not include *all* identical particles in the universe? Why can we consider the properties of a system that consists of only one or a few quantum objects? Obviously, we need not take into account all identical quantum properties of our universe in most of our considerations.

We consider two electrons in two spatial regions, region $A$ with $|\Phi\rangle$, and region $B$ with $|\Psi\rangle$, where the regions are so far apart that there is virtually no overlap of the two wavefunctions; see Fig. 23.4. This condition, which will apply in the following, is the salient point of our discussion. We can write it as $|\langle \Phi | \Psi \rangle| = \varepsilon \ll 1$. In addition, we want to measure a two-particle observable, which has eigenfunctions in $A$ with $|\varphi_n\rangle$ and in $B$ with $|\psi_m\rangle$, where it holds that $|\langle \varphi_n | \psi_m \rangle| = \eta \ll 1$. These eigenfunctions each form a CONS.

**Fig. 23.4** Probability density for two electrons in two separate spatial regions, $A$ and $B$

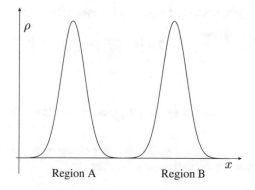

Region A          Region B

To keep the following considerations clear, we consider only the limiting case of $\varepsilon, \eta \to 0$. Thus we can assume for simplicity that all the scalar products of states in $A$ with states in $B$ vanish, e.g.

$$\langle \Phi \,|\Psi \rangle \approx 0; \quad \langle \varphi_n \,|\psi_m \rangle \approx 0. \tag{23.59}$$

### 23.6.1   Distinguishable Quantum Objects

Let us first assume that the electrons are *distinguishable*. The product state reads

$$|\Phi\Psi\rangle := |1 : \Phi\rangle \,|2 : \Psi\rangle. \tag{23.60}$$

We want to measure

$$|\varphi_n\psi_m\rangle := |1 : \varphi_n\rangle \,|2 : \psi_m\rangle. \tag{23.61}$$

We have

$$\langle \varphi_n\psi_m \,|\Phi\Psi\rangle = \langle 1 : \varphi_n \,|1 : \Phi\rangle \,\langle 2 : \psi_m \,|2 : \Psi\rangle. \tag{23.62}$$

The probability $w_{nm}$ to find this state in a measurement is given as usual by the squared value

$$w_{nm} = |\langle \varphi_n\psi_m \,|\Phi\Psi\rangle|^2. \tag{23.63}$$

If we are interested only in electron 1, we can average over the variables of electron 2. Then the probability $w_n$ of measuring the state $|\varphi_n\rangle$ is given by

$$w_n = \sum_m |\langle \varphi_n\psi_m \,|\Phi\Psi\rangle|^2 = \sum_m |\langle 1 : \varphi_n \,|1 : \Phi\rangle \,\langle 2 : \psi_m \,|2 : \Psi\rangle|^2$$
$$= |\langle 1 : \varphi_n \,|1 : \Phi\rangle|^2 \sum_m |\langle 2 : \psi_m \,|2 : \Psi\rangle|^2 = |\langle 1 : \varphi_n \,|1 : \Phi\rangle|^2, \tag{23.64}$$

where the last equation follows because of

$$\sum_m |\langle 2 : \psi_m \,|2 : \Psi\rangle|^2 = \sum_m \langle 2 : \Psi \,|2 : \psi_m\rangle \,\langle 2 : \psi_m \,|2 : \Psi\rangle = 1. \tag{23.65}$$

### 23.6.2   Identical Quantum Objects

Let us now assume that the electrons are *indistinguishable*. Then we have to start from the antisymmetric states

$$|\Phi\Psi\rangle^{(-)} := \frac{|1 : \Phi\rangle \,|2 : \Psi\rangle - |1 : \Phi\rangle \,|2 : \Psi\rangle}{\sqrt{2}} \tag{23.66}$$

and

$$|\varphi_n\psi_m\rangle^{(-)} := \frac{|1:\varphi_n\rangle\,|2:\psi_m\rangle - |1:\psi_m\rangle\,|2:\varphi_n\rangle}{\sqrt{2}}. \tag{23.67}$$

This leads to

$$^{(-)}\langle\varphi_n\psi_m\,|\Phi\Psi\rangle^{(-)} \approx \frac{\langle 1:\varphi_n|\,1:\Phi\rangle\,\langle 2:\psi_m|\,2:\Psi\rangle + \langle 2:\varphi_n|\,2:\Phi\rangle\,\langle 1:\psi_m|\,1:\Psi\rangle}{2} \tag{23.68}$$

where we have assumed, following our above assumptions about the spatial distribution of the two electrons, that 'mixed terms' according to (23.59) do not have to be taken into account, i.e.

$$\langle 1:\varphi_n|\,1:\Psi\rangle, \quad \langle 2:\psi_m|\,2:\Phi\rangle, \quad \langle 2:\varphi_n|\,2:\Psi\rangle, \quad \langle 1:\psi_m|\,1:\Phi\rangle \approx 0. \tag{23.69}$$

With this, it follows that

$$w_{nm} = \left|{}^{(-)}\langle\varphi_n\psi_m\,|\Phi\Psi\rangle^{(-)}\right|^2. \tag{23.70}$$

If we are interested only in the electron in region 1, we can average over region 2 and obtain (see the exercises):

$$w_n = \sum_m \left|{}^{(-)}\langle\varphi_n\psi_m\,|\Phi\Psi\rangle^{(-)}\right|^2$$
$$= |\langle 1:\varphi_n|\,1:\Phi\rangle|^2 = |\langle 2:\varphi_n|\,2:\Phi\rangle|^2 = |\langle\varphi_n|\,\Phi\rangle|^2. \tag{23.71}$$

In other words, under the assumptions made above, the measurement result for 1 is independent of 2—we need not worry about 2 or the Pauli principle.

## 23.7 Exercises

1. Two identical quantum objects are in the states $|\alpha_1\rangle$ and $|\alpha_2\rangle$. Show that the total state must be symmetric or antisymmetric,

$$|\psi_\pm\rangle = \frac{|1:\alpha_1, 2:\alpha_2\rangle \pm |1:\alpha_2, 2:\alpha_1\rangle}{\sqrt{2}}. \tag{23.72}$$

2. Two identical particles are in the states $|a\rangle$ and $|b\rangle$. What is the correct expression for the total state $|\psi\rangle$?
3. Let $|\varphi\rangle = |1:\alpha_1, 2:\alpha_2, 3:\alpha_3\rangle$. Determine $P_{12}P_{23}|\varphi\rangle$ and $P_{23}P_{12}|\varphi\rangle$. Under what conditions do $P_{12}$ and $P_{23}$ commute?
4. Write down explicitly the normalized states $|1:\alpha_1, 2:\alpha_2, \ldots, N:\alpha_N\rangle^{(\pm)}_{\text{norm}}$ for 2 and 3 particles.

5. Given 3 identical particles; to save paperwork, we denote the product states simply by $|1, 2, 3\rangle$ instead of by $|1 : \alpha_1, 2 : \alpha_2, 3 : \alpha_3\rangle$; $|1 : \alpha_2, 2 : \alpha_1, 3 : \alpha_3\rangle$ is then $|2, 1, 3\rangle$, etc.

   (a) Write down all 6 product states.
   (b) Show explicitly that for the total (anti)symmetrical state, $P_{12} |\psi\rangle^\pm = \eta_{12} |\psi\rangle^\pm$. Determine $\eta_{12}$.
   (c) Given the state $|\varphi\rangle = |1, 2, 3\rangle - |1, 3, 2\rangle + |2, 1, 3\rangle - |2, 3, 1\rangle + |3, 1, 2\rangle - |3, 2, 1\rangle$, show explicitly that $P_{12} |\varphi\rangle$ cannot be written as $c |\varphi\rangle$.

6. Show explicitly that $P_{ni} P_{mj} P_{nm} P_{ni} P_{mj} = P_{ij}$.
7. Show that

$$E^{(1)}_{100;nlm} = \frac{e^2}{4\pi\varepsilon_0} \int d^3 r_1 d^3 r_2 \frac{|\psi_{100}(\mathbf{r}_1) \psi_{nlm}(\mathbf{r}_2) \pm \psi_{nlm}(\mathbf{r}_1) \psi_{100}(\mathbf{r}_2)|^2}{2 |\mathbf{r}_1 - \mathbf{r}_2|} = C_{nl} \pm A_{nl}.$$
$$(23.73)$$

8. Prove (23.71), i.e.

$$w_n = \sum_m \left|^{(-)} \langle \varphi_n \psi_m | \Phi \Psi \rangle^{(-)}\right|^2 = |\langle \varphi_n | \Phi \rangle|^2. \qquad (23.74)$$

# Chapter 24
# Decoherence

The theory of decoherence solves some significant problems associated with the measurement process in quantum mechanics. The basic idea is to take into account the effect of the environment on a quantum system.

As we have seen in the preceding chapters, quantum mechanics is a theory that can predict the outcome of measurements with great accuracy—fine for all practical purposes, *fapp*. However, fundamental questions of interpretation remain open, and this is particularly the case when we are dealing with the concept of 'measurement', as we have repeatedly seen. In our formulation so far, we have on the one hand the deterministic SEq, on the other hand the measurement process which introduces probabilities into the theory—how do these two aspects fit together? Another ambiguity: Just previous to the time of a measurement, a quantum-mechanical system is in general a superposition of different states—but due to the measurement, *one* of the states is selected out of this superposition; we do not obtain any sort of 'superimposed' states as the result of a measurement. How does this 'collapse' of the wavefunction take place, and on which time scale does it occur?

This is essentially the already often-mentioned *quantum mechanical measurement problem*. At least a part of the unclear issues is answered by the theory of *decoherence*, which we now wish to discuss briefly. The basic idea is that one takes into account the interactions of a quantum system with its environment. Indeed, the term 'isolated system' is *per se* an idealization that never can be realized, strictly speaking, except possibly for the entire universe. In experimental practice, it may require a very elaborate procedure to isolate quantum systems even approximately or to separate them sufficiently from their surroundings.[1]

---

[1] Indeed, the experimental challenges are enormous if one tries to isolate certain quantum objects from their environment. Serge Haroche and David Wineland developed new ground-breaking experimental methods, making it possible to measure non-destructively those quantum objects and to control them, which was previously thought to be impossible. In 2012, they were awarded the Nobel Prize in physics. As the BBC wrote (http://www.bbc.co.uk/news/science-environment-19879890 (November 2012)), "But for physicists, the import of the pair's techniques is outlined in a layman's summary on the Nobel site: they preserve the delicate quantum-mechanical states

© Springer Nature Switzerland AG 2018
J. Pade, *Quantum Mechanics for Pedestrians 2*, Undergraduate Lecture
Notes in Physics, https://doi.org/10.1007/978-3-030-00467-5_24

**Fig. 24.1** System $\mathcal{S}$, measuring device $\mathcal{M}$ and environment $\mathcal{U}$

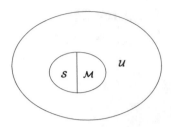

We must therefore consider not only the measured quantum object $\mathcal{S}$ and the measuring apparatus[2] $\mathcal{M}$, but in addition the two must be understood as an *open system* interacting with its environment $\mathcal{U}$; see Fig. 24.1.[3] It turns out that the influence of the environment actually destroys superpositions and makes (relative) phases unobservable. In this way, decoherence explains the non-occurrence of macroscopic superpositions, discussed e.g. in the example of Schrödinger's cat.

In the following, we want to illustrate the basic idea by using a simple example, before we present a slightly more formal approach.

## 24.1  A Simple Example

To conceive an intuitive idea of decoherence, we discuss a very simplified example. We assume a material object, whose dimensions may be microscopic (e.g. an electron) or macroscopic (e.g. a grain of sand), and which can be in two states $|z_1\rangle$ and $|z_2\rangle$, with sharp energies $E_1$ and $E_2$. The time evolution of the states is then given by $|z_n\rangle \rightarrow e^{-iE_n t/\hbar}|z_n\rangle$. The initial state $|z\rangle$ is supposed to be a superposition of the two states, i.e. $|z\rangle = c_1|z_1\rangle + c_2|z_2\rangle$. The unitary time evolution of this state is then given by[4]:

$$|z\rangle \rightarrow e^{-iE_1 t/\hbar}c_1|z_1\rangle + e^{-iE_2 t/\hbar}c_2|z_2\rangle. \tag{24.1}$$

---

of the photons and ions—states that theorists had for decades hoped to measure in the laboratory, putting the ideas of quantum mechanics on a solid experimental footing. Those include the slippery quantum-mechanical ideas of entanglement—the seemingly ethereal connection between two distant particles . . . and of decoherence, in which the quantum nature of a particle slowly slips away through its interactions with other matter."

[2] Of course, $\mathcal{M}$ is not confined to the usual equipment of the physics laboratory. In the case of e.g. Schrödinger's cat, the cat is the measuring apparatus with the pointer states 'dead' and 'alive', which measure the state of the radioactive atom.

[3] Some remarks on the concepts of 'open', 'isolated', etc. can be found in Appendix S, Vol. 1.

[4] Note: For a grain of sand this is a *cat-state*. Today, this term is understood as the superposition of two quantum states that are macroscopically distinguishable. Originally, the term referred to an entangled state between a macrostate (cat) and a microstate (radioactive nucleus). Schrödinger cat states are 'normal' quantum mechanical states which for instance can be entangled, see e.g. C. Wang et al. 'A Schrödinger cat living in two boxes, *Science* 352, 1087 (2016), https://doi.org/10.1126/Science.aaf2941.

**Table 24.1** Frequencies $\omega$ for different distances and masses

| $\omega$ | $\Delta z = 1\,\mathrm{nm}$ | $\Delta z = 1\,\mathrm{m}$ |
|---|---|---|
| $m = 10^{-30}\,\mathrm{kg}$ | $10^{-4}\,\mathrm{s}^{-1}$ | $10^{5}\,\mathrm{s}^{-1}$ |
| $m = 10^{-3}\,\mathrm{kg}$ | $10^{23}\,\mathrm{s}^{-1}$ | $10^{32}\,\mathrm{s}^{-1}$ |

We can factor out one of the two exponential terms (which results in a physically uninteresting global phase as overall factor), for instance:

$$|z\rangle \rightarrow e^{-iE_1 t/\hbar}\left[c_1\,|z_1\rangle + e^{-i\omega t}c_2\,|z_2\rangle\right], \tag{24.2}$$

with $\omega = (E_2 - E_1)/\hbar$. Of course, this also works with $E_2$ instead of $E_1$. To get an idea of the order of magnitude of $\omega$ and the period of oscillation, we assume that the object is in the earth's gravitational field. Then we have $\omega = \frac{mg\Delta z}{\hbar}$. For an electron and a separation of the wave packets of $\Delta z = 1\,\mathrm{nm}$ or $\Delta z = 1\,\mathrm{m}$, we obtain $\omega = 10^{-4}$ or $10^{5}\,\mathrm{s}^{-1}$; for a mass of $1\,\mathrm{g}$, these values are $\omega = 10^{23}$ or $10^{32}\mathrm{s}^{-1}$, see Table 24.1. To compare: $10^{-22}\,\mathrm{s}$ is the time which light requires to 'pass through' an atomic nucleus.

This means that for macroscopic masses and distances, the phase $\omega t$ in (24.2) rotates so fast that during the measurement one records only its average. In other words, when phases are changing so rapidly, only *one* of two states $|z_1\rangle$ and $|z_2\rangle$ can be detected. Which one of the two states this actually applies to remains an open question—not only with this simple heuristic reasoning, but also within the scope of the approaches to decoherence which we discuss in the following.

We can also treat the problem using the density matrix. We have from (24.1):

$$\rho = \begin{pmatrix} |c_1|^2 & c_1 c_2^* e^{i\omega t} \\ c_1^* c_2 e^{-i\omega t} & |c_2|^2 \end{pmatrix}. \tag{24.3}$$

Again, we see a rapid 'flickering' in the superpositions or coherences. If we want to carry out a measurement, we must remember that *every* measurement requires a finite amount of time $T$, although this may seem to our everyday understanding vanishingly small (e.g. $10^{-12}\,\mathrm{s}$). Thus, we have to average the density matrix over the measuring time $T$ (of course, this is again a heuristic argument) and obtain (see exercises):

$$\frac{1}{T}\int_0^T \rho\,dt = \begin{pmatrix} |c_1|^2 & c_1 c_2^* \cdot s(T) \\ c_1^* c_2 \cdot s^*(T) & |c_2|^2 \end{pmatrix} \tag{24.4}$$

with

$$s(T) = e^{i\omega T/2}\frac{\sin \omega T/2}{\omega T/2} \underset{\omega T \to \infty}{\rightarrow} 0. \tag{24.5}$$

Hence, for sufficiently long averaging times, we can write

$$\rho \approx \begin{pmatrix} |c_1|^2 & 0 \\ 0 & |c_2|^2 \end{pmatrix}. \tag{24.6}$$

For e.g. $\omega = 10^{32}\,\text{s}^{-1}$ and a measurement lasting $T = 10^{-16}\,\text{s}$, we have $|s(T)|$ $\sim 10^{-16}$; under these circumstances, one effectively cannot see the coherences or superpositions. This means that the off-diagonal elements of the density matrix, i.e. the superpositions, disappear very quickly—in this way, a statistical mixture emerges out of a pure state. Which of the two $c_i$ is finally selected by the measuring process cannot be said at this point.

As we pointed out in the introduction, we have to work with open systems, i.e. to take into account the influence of the environment. Here, we have modelled this effect by considering the finite time resolution of the measuring apparatus. We did not try to describe the effects of the environment as realistically as possible. The fact that we still obtain a result such as (24.6) indicates that decoherence in reality does not depend on specific interactions or conditions, but rather that it is robust with regard to them and constitutes a universal phenomenon.

## 24.2  Decoherence

In this section, we describe the process of measurement on a very schematic level. One reason is that we have considered only Hamiltonians that do not depend on time. But here, we consider *open* systems interacting with their environment; they of course experience an evolution in time (e.g. due to the fact that the measuring device adjusts itself to a new value), and we have no conceptual tools[5] at hand in this regard. A further justification for the following simple approach (which is representative of the standard scenario) is due to the aforementioned universality of decoherence— the details do not matter here if one wants to explain the phenomenon in a more qualitative than quantitative manner. And finally, the basic idea of decoherence may be worked out better by means of systems which are as simple as possible (i.e. toy models). The following considerations are therefore characterized in some places by plausibility arguments rather than strict mathematics, but this does not diminish their general validity.

First, we repeat an example treated in Chap. 20, as shown in Fig. 24.2.

The photon can occupy the state $|H\rangle$ or $|V\rangle$ (horizontal or vertical). We describe the detectors (i.e. the measuring apparatus) by the ket $|10\rangle$ or $|01\rangle$, if DH or DV clicks, respectively. If both detectors are inactive, we have $|00\rangle$. With this notation we can write the process in short form as:

---

[5] Such as time-ordering operators etc.

**Fig. 24.2** A photon is incident on a beam splitter BS and is detected in one of the detectors DH or DV

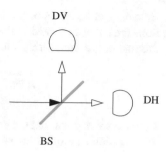

$$|H\rangle\,|00\rangle \to \frac{1+i}{2}\,[|H\rangle + i\,|V\rangle]\,|00\rangle \to \frac{1+i}{2}\,[|H\rangle\,|10\rangle + i\,|V\rangle\,|01\rangle]. \quad (24.7)$$

In words: The photon in the state $|H\rangle$ propagates towards the beam splitter, no detector is activated and the total state is factorizable. This is true even after it passes the first beam splitter. Eventually, the photon will hit and activate one of the two detectors with 50 % probability, e.g. if it is in the state $|H\rangle$, it will activate detector DH. Obviously, the final state in (24.7) is entangled. With regard to the following considerations, we denote $[|H\rangle + i\,|V\rangle]\,|00\rangle$ as the initial state and $[|H\rangle\,|10\rangle + i\,|V\rangle\,|01\rangle]$ as the final state, which means that we confine ourselves to the process $[|H\rangle + i\,|V\rangle]\,|00\rangle \to [|H\rangle\,|10\rangle + i\,|V\rangle\,|01\rangle]$.

We now generalize this example. To this end, we assume a quantum system $\mathcal{S}$ which has the states $|S_m\rangle$. $\mathcal{S}$ is to be measured by a measuring apparatus $\mathcal{M}$ with the states $|M_k\rangle$. For simplicity and without loss of generality, we assume that $\mathcal{S}$ and $\mathcal{M}$ have the same number $M$ of states; typically, $M$ has values of less than $10^4$.

At the beginning of our observations, the system is in the state $|S_m\rangle$ and the measuring apparatus is in its initial state $|M_0\rangle$. An ideal (recoilless) measurement then eventually causes the measuring apparatus to indicate the state of $\mathcal{S}$:

$$|S_m\rangle\,|M_0\rangle \to |S_m\rangle\,|M_m\rangle. \quad (24.8)$$

If the initial state of $\mathcal{S}$ is a superposition, i.e. $\sum s_m\,|S_m\rangle$, then it follows due to the linearity of the dynamics[6] that:

$$\sum_m s_m\,|S_m\rangle\,|M_0\rangle \to \sum_m s_m\,|S_m\rangle\,|M_m\rangle. \quad (24.9)$$

We see that we have on the right-hand side an entangled state whose density operator is given by

$$\rho = \sum_{mm'} s_m s_{m'}^*\,|S_m\rangle\,|M_m\rangle\,\langle M_{m'}|\,\langle S_{m'}|. \quad (24.10)$$

---

[6] Just as in the example of the beam splitter considered above.

Evidently, coherences occur, i.e. terms with $m \neq m'$. On the other hand, superpositions are never observed in the macroscopic domain, and the density operator for a mixture of non-interfering states would be

$$\rho = \sum_m |s_m|^2 \, |S_m\rangle \, |M_m\rangle \, \langle M_m| \, \langle S_m| \,. \tag{24.11}$$

How does it happen that the coherences disappear? If one considers only $S$ and $\mathcal{M}$, this question cannot be answered.

### 24.2.1  The Effect of the Environment I

The dilemma can be resolved only if—and this is the essential idea of decoherence—the environment $\mathcal{U}$ is included in the considerations. The states of $\mathcal{U}$ are $|U_n\rangle$. Their number is $N$, where $N$ is a very large number, of order $10^{20}$, $10^{30}$ or larger and in any case substantially greater than $M$.

Initially, we assume a factorized state of the form:

$$|\psi\rangle_0 = \sum_{m=1}^{M} s_m \, |S_m\rangle \, |M_0\rangle \sum_{n=1}^{N} u_n \, |U_n\rangle \,. \tag{24.12}$$

In the course of time, the system, measuring apparatus and environment interact and the state changes to:

$$|\psi\rangle = \sum_{m,n=1}^{M,N} c_{mn} \, |S_m\rangle \, |M_m\rangle \, |U_n\rangle \,. \tag{24.13}$$

The coupling is in the factors $c_{mn}$ which do not generally factorize in the form $c_{mn} = a_m b_n$. In other words, the state (24.13) is entangled, and this with respect to the $(S, \mathcal{M})$ states on the one hand and the environmental states on the other hand.

Thus, the essential mechanism here is not the direct effect of the environment on $(S, \mathcal{M})$, which would possibly change the states of $S$ or $\mathcal{M}$ (noise). Rather, we assume that these states remain unchanged. Instead, the fact that the evolution of $(S, \mathcal{M})$ leaves its mark on the environment due to the entanglement—in other words, that information about $(S, \mathcal{M})$ 'seeps out' to the environment, is decisive.

The density operator for the state (24.13) as an element of $\mathcal{H}_S \otimes \mathcal{H}_M \otimes \mathcal{H}_U$ is

$$\rho = \sum_{m,n,m',n'} c_{mn} c_{m'n'}^* \, |S_m\rangle \, |M_m\rangle \, |U_n\rangle \, \langle S_{m'}| \, \langle M_{m'}| \, \langle U_{n'}| \,. \tag{24.14}$$

We note that this density operator describes a pure state and that, at this point, we have the *total* information about the triplet $(S, \mathcal{M},$ and $\mathcal{U})$ at our disposal.

Crucial for the following argument is the *entanglement* between $(\mathcal{S}, \mathcal{M})$ and the environment $\mathcal{U}$; thus, the state (24.13) must not factorize (e.g. as denoted by $c_{mn} = a_m b_n$). Without this entanglement, there is no decoherence; the environment *must* be included in the considerations. This naturally raises quite difficult questions (how far does the environment extend?), but on a *fapp* level, we need only the argument that the environment has very many degrees of freedom; whether there are $10^{20}$ or $10^{30}$ is irrelevant.

The coherences (i.e. superpositions) occurring in the density operator (24.14) can be eliminated by averaging over the environment variables. Since there is an enormous number of environmental variables or states interacting with $(\mathcal{S}, \mathcal{M})$, we see—as in the ideal gas[7]—not every single contribution, but a mean value; this is analogous to the averaging over the phase that we performed in the above 'grain of sand' example. The averaging over environmental states (trace over $\mathcal{U}$, here also called *trace over unobservables*) leads in a first step to the reduced density operator

$$\rho_{S,red} = tr_U(\rho) = \sum_{k=1}^{N} \langle U_k | \rho | U_k \rangle$$

$$= \sum_{k=1}^{N} \sum_{m,n,m',n'} c_{mn} c_{m'n'}^* |S_m\rangle |M_m\rangle \, \delta_{nk} \langle S_{m'}| \langle M_{m'}| \, \delta_{n'k}$$

$$= \sum_{m,m'} \sum_{k=1}^{N} c_{mk} c_{m'k}^* |S_m\rangle |M_m\rangle \langle S_{m'}| \langle M_{m'}|. \tag{24.15}$$

In the representation as density matrix (of dimension $M \times M$), we can write this as in Chap. 22 as the product of an $M \times N$ matrix $C$ with its adjoint:

$$\rho_{S,red} = CC^\dagger \quad \text{with} \quad C = (c_{mn}), \tag{24.16}$$

and the matrix elements of $\rho_{S,red}$ are then the dot product of the $N$-dimensional $m$th row of $C$ with the $m'$th column of $C^\dagger$ (this is shown explicitly in the exercises).

In a second step, we want to estimate the order of magnitude of the matrix elements of $\rho_{S,red}$. We start with the diagonal elements that must be positive, being absolute squared values. Because of the normalization (or $tr\rho = 1$), they are $1/M$ on average.

On this basis, also the off-diagonal terms are on average $1/M$, so that we can write this common factor in front of the matrix. The diagonal terms within the matrix then are of order 1. Concerning the off-diagonal elements, we have to sum over generally both positive and negative contributions (from real and imaginary terms). For very large $N$ (as said, we have for macroscopic environments values of $N \approx 10^{20}$ or more), these contributions will cancel out on average; if we can assume a normal distribution, the deviation from the mean value zero is on the order of in the magnitude of the

---

[7]There, we do not know the phase-space coordinates of $10^{23}$ particles (more precisely, we neither can know them nor want to know them).

relative statistical error, i.e. $\sim \frac{1}{\sqrt{N}}$. Overall, the estimate of the order of magnitude of the matrix elements reads:

$$(\rho_{S,\text{red}})_{ij} = O\left(\frac{1}{M}\right)\left(\delta_{ij} + O\left(\frac{1}{\sqrt{N}}\right)\right). \qquad (24.17)$$

A more detailed analysis is given in the exercises.

For coupling to a macroscopic environment, we have the result that the off-diagonal elements of the reduced density matrix disappear, so to speak for statistical reasons alone[8]; to a good approximation, it holds that:

$$\rho_{S,\text{red}} = \sum_m p_m \left|S_m\right\rangle \left|M_m\right\rangle \left\langle S_m\right| \left\langle M_m\right|; \quad p_m = \sum_{k=1}^{N} |c_{mk}|^2. \qquad (24.18)$$

In other words, due to the influence of the environment, the coherences seem to have disappeared from the system; the $p_m$ are the measurement probabilities of the *individual* configurations.

### 24.2.2  Simplified Description

We can also simplify by assuming that after the measurement, we consider *one* state of the environment for each setting of $(\mathcal{S}, \mathcal{M})$, and write the total state correspondingly instead of (24.13) as:

$$|\psi\rangle = \sum_{m=1}^{M} d_m \left|S_m\right\rangle \left|M_m\right\rangle \left|U_m\right\rangle. \qquad (24.19)$$

Then the density operator reads

$$\rho = \sum_{m,m'} d_m d_{m'}^* \left|S_m\right\rangle \left|M_m\right\rangle \left|U_m\right\rangle \left\langle S_{m'}\right| \left\langle M_{m'}\right| \left\langle U_{m'}\right| \qquad (24.20)$$

and for the reduced density operator, it follows for the moment:

$$\rho_{S,\text{red}} = tr_U(\rho) = \sum_{k=1}^{M} \sum_{m,m'} d_m d_{m'}^* \left|S_m\right\rangle \left|M_m\right\rangle \left\langle U_k\right| U_m\rangle \left\langle U_{m'}\right| U_k\rangle \left\langle S_{m'}\right| \left\langle M_{m'}\right|.$$
$$(24.21)$$

Since the states $|U_m\rangle$ come from a high-dimensional space (as we said above, dimension $10^{20}$ or more), two arbitrarily chosen states are orthogonal to each other with a high probability (this is a similar argument to the one leading to the estimate (24.17));

---

[8]One can imagine that the system itself has lost the information about certain interference terms because it has migrated into the environment. Therefore, the coupling to a *great many* degrees of freedom is substantial—then the process is virtually irreversible.

actually we have $\langle U_k | U_m \rangle \approx \delta_{mk}$. This means that we arrive at the same result as above:

$$\rho_{S,\text{red}} = \sum_m p_m |S_m\rangle |M_m\rangle \langle S_m| \langle M_m|; \quad p_m = |d_m|^2 . \tag{24.22}$$

Thus, the system looks *fapp* like a mixture.

### 24.2.3  The Effect of the Environment II

The influence of the environment is not limited to receiving information about the system. It also structures the possible results and resolves ambiguities. To consider an example, we assume a spin-1/2 system $S$ in the state

$$|\alpha\rangle = \frac{|z+\rangle + |z-\rangle}{\sqrt{2}}, \tag{24.23}$$

where $|z+\rangle$ and $|z-\rangle$ are the eigenstates of $\sigma_z$ with the eigenvalues $+1$ and $-1$.[9] $S$ interacts with $\mathcal{M}$, which in our example is a two-dimensional system, also. After some time, the state evolves to:

$$|\psi\rangle = \frac{|z+\rangle |m_{z+}\rangle + |z-\rangle |m_{z-}\rangle}{\sqrt{2}}. \tag{24.24}$$

We can think of $|m_{z+}\rangle$ and $|m_{z-}\rangle$ as the two eigenstates of a pointer variable (pointer observable) $M_z$ of $\mathcal{M}$.

The state $|\psi\rangle$ has the form of a *biorthonormal decomposition*.[10] Such decompositions are not unique if the squared values of the coefficients are all equal. This is the case here, and indeed we can represent $|\psi\rangle$ e.g. by the following biorthonormal decomposition with respect to $x$ (the example is considered in the exercises):

$$|\psi\rangle = \frac{|x+\rangle |m_{x+}\rangle + |x-\rangle |m_{x-}\rangle}{\sqrt{2}} \tag{24.25}$$

where the $|m_{x\pm}\rangle$ are eigenstates of a pointer variable $M_x$ which are related to the $|m_{z\pm}\rangle$ by

---

[9]In vector notation, we have $|z+\rangle \cong \begin{pmatrix} 1 \\ 0 \end{pmatrix}$ and $|z-\rangle \cong \begin{pmatrix} 0 \\ 1 \end{pmatrix}$. Correspondingly, $|\alpha\rangle$ is the 'up' state of $\sigma_x$.

[10]Each state $|\psi\rangle$ of the total system $S + \mathcal{M}$ can be represented in the form

$$|\psi\rangle = \sum c_i |u_i\rangle |v_i\rangle$$

where $\{|u_i\rangle\}$ and $\{|v_i\rangle\}$ are CONS in $\mathcal{H}_S$ and $\mathcal{H}_{\mathcal{M}}$ (Schmidt decomposition, biorthogonal decomposition). The decomposition is unique iff the coefficients $|c_i|^2$ are all different.

$$|m_{x+}\rangle = \frac{|m_{z+}\rangle + |m_{z-}\rangle}{\sqrt{2}}; \ |m_{x-}\rangle = \frac{-|m_{z+}\rangle + |m_{z-}\rangle}{\sqrt{2}}. \qquad (24.26)$$

We note that (24.24) and (24.25) are two different biorthonormal decompositions of the same state $|\psi\rangle$. Consequently, certain entangled states of $S + M$ can represent both a correlation between $\sigma_z$ values and the values of an observable $M_z$ of $M$, and simultaneously a corresponding correlation between $\sigma_x$ values and the values of an observable $M_x$ von $M$. Of course, this is unsatisfactory, if (as in our example) $M_x$ does not commute with $M_z$. One way out of this dilemma is the inclusion of the environment.

For this, we consider in our model in addition a third, likewise two-dimensional system $U$ (our model environment), which interacts with $M$. In the course of the measurement, the state changes over time and reads:

$$|\Psi\rangle = \frac{|z+\rangle \, |m_{z+}\rangle \, |u_{z+}\rangle + |z-\rangle \, |m_{z-}\rangle \, |u_{z-}\rangle}{\sqrt{2}}, \qquad (24.27)$$

where $|u_{z+}\rangle$ and $|u_{z-}\rangle$ are two orthogonal states which span $\mathcal{H}_{U}$. We have three components, $S + M + U$, and accordingly a triorthonormal decomposition. For this we can use another decomposition theorem (theorem of triorthogonal decompositions[11]) which ensures the uniqueness of the state (24.27) when the three pairs of states $\{|z+\rangle, |z-\rangle\}$, $\{|m_{z-}\rangle, |m_{z+}\rangle\}$ and $\{|u_{z+}\rangle, |u_{z-}\rangle\}$ are orthogonal in their respective Hilbert spaces, independently of the expansion coefficients. We have therefore, in contrast to the biorthonormal decomposition, the result that $M_x$ and $M_z$ cannot be measured simultaneously, or

$$|\Psi\rangle \neq \frac{|x+\rangle \, |m_{x+}\rangle \, |u_{x+}\rangle + |x-\rangle \, |m_{x-}\rangle \, |u_{x-}\rangle}{\sqrt{2}}. \qquad (24.28)$$

The disturbing ambiguity appearing in (24.24) and (24.25) is therefore removed by the inclusion of the model environment (technically: tri- instead of biorthogonal decompositions).

We can, on the basis of our model system, summarize our considerations as follows: We include the environment by extending the system $(S, M)$ to the system $(S, M, U)$, where we assume that there is an interaction $W$ between $M$ and $U$, although at this point it is unknown.[12] $W$ commutes with a $M$ observable (the pointer variable[13]); we call it $M_z$. Under these conditions, the system $(S, M)$ will be perfectly correlated in *one* product basis (in the example $\{|z+\rangle \, |m_{z+}\rangle, |z-\rangle \, |m_{z-}\rangle\}$). Hence, $\{|m_{z+}\rangle, |m_{z-}\rangle\}$ is the pointer basis of $M$, which appears in the course of time in the diagonal of the reduced density matrix that is obtained by tracing out over the

---

[11]The theorem of triorthogonal decompositions can be generalized to vectors that are linearly independent (and not necessarily orthogonal), as well as to $n$ systems (instead of 3).

[12]Of course, one knows these interactions in model computations.

[13]This pointer variable contains the possible states of the measuring apparatus.

environmental modes. Measurements of other spin directions are impossible. The pointer observable is thus actually determined by the interaction with $\mathcal{U}$, because this 'monitoring' of $\mathcal{M}$ by $\mathcal{U}$ leads to an almost immediate decoherence between different pointer states. So we have an effective 'collapse' of the total state into a tensor product of a pointer state and a correlated eigenstate of $S$.

### 24.2.4 Interim Review

Following the model concepts which we have just outlined, the major mechanism is that information about the system $S$ (and the measuring apparatus $\mathcal{M}$) finds its way into the environment $\mathcal{U}$. On the one hand, the enormous number of degrees of freedom of the environment ensures thereby that this process is virtually irreversible. On the other hand, the environment specifies the pointer basis via its interaction with $\mathcal{M}$ and in this way prevents ambiguities and contradictions. For this reason, we speak of *measurement by the environment* (environmental monitoring)[14]—in accordance with these ideas, macroscopic properties are created by the environment.

Of course, the dividing lines between $S$, $\mathcal{M}$ and $\mathcal{U}$ are not sharply defined and are to a certain extent arbitrary. At first glance, this is similar to the problem of demarcation between quantum mechanics and classical mechanics (the Heisenberg cut), discussed e.g. in Chap. 14, Vol. 1, which arises when one postulates that the measuring apparatus must obey the rules of classical mechanics. But there is the fundamental difference that in the decoherence theory, $S$, $\mathcal{M}$ and $\mathcal{U}$, i.e. *all* systems, obey the rules of quantum mechanics—we no longer have to make use of classical mechanics, but have a coherent and consistent representation within the framework of quantum mechanics alone. In other words, the problems discussed in Chap. 14, Vol. 1, regarding the boundary between quantum mechanics and classical mechanics have been resolved by the introduction of decoherence.

Finally, a note with regard to the measuring apparatus. It plays only the role of an agent which shows us macroscopically the microscopic result. But even without measuring apparatus, entanglement of the system states with those of the environment occurs, and hence decoherence. Since measurement by the environment is always present (unless the system were indeed isolated), we can for the sake of simplicity dispense with the explicit consideration of the measuring apparatus in the frame of certain considerations (or ascribe it either to the system or to the environment), and restrict ourselves to the pair $(S, \mathcal{U})$.[15] In this way, the state $|\psi\rangle_0 = \sum_{m=1}^{M} s_m |S_m\rangle \sum_{n=1}^{N} u_n |U_n\rangle$ becomes $|\psi\rangle = \sum_{m,n=1}^{M,N} c_{mn} |S_m\rangle |U_n\rangle$, and by decoherence effects this leads to the reduced density operator $\rho_{S,red} = \sum_m p_m |S_m\rangle \langle S_m|$ with $p_m = \sum_{k=1}^{N} |c_{mk}|^2$.

---

[14]Other designations are *environmentally induced decoherence* or *einselection*, an abbreviation of *environment-induced selection*.

[15]One can guarantee the uniqueness (or tri- instead of biorthogonal decomposition) by (formally) splitting the environment into two or more systems.

## 24.2.5   *Formal Treatment*

To give a rough outline of the formal treatment, we start with the total Hamiltonian (for clarity, we dispense here with the measuring apparatus):

$$H = H_S + H_U + H_{US}, \tag{24.29}$$

where $H_{US}$ describes the interaction between $U$ and $S$. In this context, the environment is often referred to as a '(thermal) bath' or 'reservoir' (based on the concepts of thermodynamics).

The time evolution of the total density operator reads

$$\rho(t) = \hat{U}(t)\,\rho(0)\,\hat{U}^\dagger(t); \ \ \hat{U}(t) = e^{-i\frac{Ht}{\hbar}}, \tag{24.30}$$

and the reduced density operator is given by the trace over the degrees of freedom of the environment:

$$\rho_S(t) = tr_U\left[\hat{U}(t)\,\rho(0)\,\hat{U}^\dagger(t)\right]. \tag{24.31}$$

We assume that initially, system and environment are not entangled, i.e.

$$\rho(0) = \rho_S(0) \otimes \rho_U(0). \tag{24.32}$$

Furthermore, we assume that we know the orthogonal basis states $|n\rangle$ of the environment (which of course, strictly speaking, is the case only if we define a model environment). Since we do not know in which exact state of superposition $\mathcal{U}$ is initially, we assume a statistical mixture:

$$\rho_U(0) = \sum_n p_n\,|n\rangle\,\langle n|\,; \ \ \sum_n p_n = 1. \tag{24.33}$$

Then we have in $S$ the reduced density operator

$$
\begin{aligned}
\rho_S(t) &= \sum_m \langle m|\left[\hat{U}(t)\,\rho_S(0) \otimes \sum_n p_n\,|n\rangle\,\langle n|\,\hat{U}^\dagger(t)\right]|m\rangle \\
&= \sum_{m,n} \sqrt{p_n}\,\langle m|\,\hat{U}(t)\,|n\rangle\,\rho_S(0)\,\sqrt{p_n}\,\langle n|\,\hat{U}^\dagger(t)\,|m\rangle \\
&= \sum_{i=(m,n)} A_i(t)\,\rho_S(0)\,A_i^\dagger(t); \ \ A_{i=(m,n)}(t) = \sqrt{p_n}\,\langle m|\,\hat{U}(t)\,|n\rangle. 
\end{aligned} \tag{24.34}
$$

The operators $A_i(t)$ act in the Hilbert space of $S$. Due to the unitarity of the propagator, they satisfy (see the exercises):

$$\sum_{i=(m,n)} A_i^\dagger(t)\,A_i(t) = 1. \tag{24.35}$$

With the equation $\rho_S(t) = \sum_i A_i(t) \rho_S(0) A_i^\dagger(t)$, the problem is formally solved. An explicit calculation of $A_i(t)$ and therefore of $\rho_S(t)$, of course, demands specific assumptions about the system and the environment and requires considerable calculation. Therefore, we leave the discussion of the formal treatment with these remarks.

## 24.3 Time Scales, Universality

Decoherence remained unnoticed for quite a while (see also the 'Historical side note' below). Partly responsible for this is the extremely high speed with which it proceeds; cf. the above example of the 'grain of sand'. Theoretical and experimental work suggests that for macroscopic conditions, the time scales are of order $10^{-20}$, $10^{-30}$, $10^{-40}$ s or even higher orders of magnitude.

As pointed out above, these values are obtained from various model calculations. In these, micro- and macroscopic objects are placed in different model environments. Let us for example assume that the initial state of the object is a superposition of two states (given e.g. by two bell curves) at the positions $x$ and $x'$ (the locations of the peaks of the bell curves). Then, under suitable assumptions, one can describe [16] the evolution by a density matrix of the form $\rho(x, x', t) = \rho(x, x', 0) e^{-\Lambda t(x-x')^2}$. Clearly, this expression eventually becomes diagonal, $\rho(x, x', t) \to \rho(x, x, t) \delta_{xx'}$. The localization rate $\Lambda > 0$ is a measure of the speed at which this process carries through. Numerical values of the model calculations are given in Table 24.2. In this case, the dust particle has a diameter of $10^{-5}$ m, and the (large) molecule is $10^{-8}$ m in diameter.

We see that for dust particles in the air, even an overlap of $x - x' = 10^{-10}$ m has decayed in a time of about $10^{-20}$ s. [17] Therefore, if Schrödinger's cat were in a superposition state between death and life, it would last at most a period of perhaps the order of the Planck time. Even molecules are already on the border between quantum mechanics and classical mechanics, and also in microscopic bodies, coherences may fade away in $10^{-12}$ s or similarly short times. These short decoherence times give the impression of a jump or collapse. In fact, the momentum is continuous, so it is only an 'apparent' collapse.

All these considerations are quite general; at least in the macroscopic world, decoherence has a certain universality. Studies show that decoherence is the only relevant dynamics on extremely short time scales. Details of the system or environment do not play a role; the result is insensitive to them. Hence, decoherence appears as a universal phenomenon of macroscopic superpositions.

---

[16]E. Joos and H.D. Zeh, 'The Emergence of Classical Properties Through Interaction with the Environment', *Z. Phys.* B59 223-243 (1985); M. Tegmark, 'Apparent Wave Function Collapse Caused by Scattering', *Found. Phys. Lett.* 6, 571–590 (1993).

[17]The specific values—whether $10^{17}$ or $10^{19}$—do not matter; just the orders of magnitude are relevant.

**Table 24.2** Localization rate $\Lambda$ (unit $m^{-2}s^{-1}$) for different objects and environments

| Model environment\object | Electron | Molecule | Dust particle | Bowling ball |
|---|---|---|---|---|
| Cosmic background radiation | $10^{-6}$ | $10^{-8}$ | $10^{10}$ | $10^{21}$ |
| Sunlight on earth | $10^5$ | $10^{17}$ | $10^{24}$ | $10^{32}$ |
| Vacuum ($10^3$ particles/cm$^3$) | $10^{22}$ | $10^{21}$ | $10^{27}$ | $10^{35}$ |
| Air molecules (s.t.p.) | $10^{35}$ | $10^{34}$ | $10^{40}$ | $10^{49}$ |

## 24.4  Decoherence-Free Subspaces, Basis

It is clear that the very short decoherence times (together with the universal validity of decoherence) may be 'deadly' for quantum computers and other applications that operate on the basis of the superposition principle of quantum mechanics. The decoherence time itself is not crucial, but rather the ratio of 'switching time' of quantum gates (see Chap. 26) to the decoherence time, i.e. the number $N_{op}$ of possible operations during the lifetime of the system. For $N_{op}$ of e.g. the systems quantum dots/MRI/ion traps, currently (2011), numbers of the order of $10^3/10^7/10^{10}$ have been reported or considered possible.

There are several strategies as to how to escape the problem of short decoherence times; a keyword in this context is 'decoherence-free subspaces'.[18] Intuitively explained, these are subspaces of the state space to whose individual states the environment reacts identically. Accordingly, the phase relationships in superpositions will remain unchanged under the influence of the environment, and the coherences can survive.

We consider a simple model: We have two states $|0\rangle$ and $|1\rangle$; the influence of the environment lies in the fact that it adds to each state a random phase: $|0\rangle \rightarrow e^{i\varphi_0}|0\rangle$ and $|1\rangle \rightarrow e^{i\varphi_1}|1\rangle$ (for the sake of simplicity, we omit the measuring apparatus). If we first consider a simple superposition, we obtain

$$|0\rangle + |1\rangle \rightarrow e^{i\varphi_0}|0\rangle + e^{i\varphi_1}|1\rangle = e^{i\varphi_0}\left[|0\rangle + e^{i(\varphi_1-\varphi_0)}|1\rangle\right], \qquad (24.36)$$

and (averaged over the difference between the uncorrelated random phases) the coherences disappear, similar to the example of the 'grain of sand' discussed above. But things look different if we consider two quantum objects in the (entangled) states

$$|\psi^{\pm}\rangle = \frac{|01\rangle \pm |10\rangle}{\sqrt{2}}. \qquad (24.37)$$

---

[18]Other methods to protect decoherence are to rely on the quantum Zeno effect (cf. Appendix L, Vol. 1), or on a special quantum measuring technique (called *weak measurement*); see Sabrina Maniscalco et al. 'Protecting entanglement via the quantum Zeno effect', *Phys. Rev. Lett.* 100, 090503 (2008), or Yong-Su Kim et al. 'Protecting entanglement from decoherence using weak measurement and quantum measurement reversal', Nature Physics (2011), https://doi.org/10.1038/nphys2178. See also H. Le Jeannic et al. 'Slowing Quantum Decoherence by Squeezing in Phase Space', *Phys. Rev. Lett.* 120, 073603 (Feb 2018).

The coupling to our model environment then leads to

$$\left|\psi^{\pm}\right\rangle \rightarrow e^{i(\varphi_0+\varphi_1)}\frac{|01\rangle \pm |10\rangle}{\sqrt{2}}. \tag{24.38}$$

Since the global phase is unobservable, the initial state will in fact be conserved—even if we take the environment into account.

Hence, if there is no entanglement in the overall state (24.13) and, consequently, no coherences in the full density matrix (24.14), then the environment cannot distinguish between the individual $S$ states, and we have no decoherence—in other words, $S$ behaves as an isolated system. In accordance with the simple examples just treated, we can therefore conclude that a subspace is decoherence-free if the environment cannot distinguish between its components.

However, this conclusion holds in the above example only if we restrict the discussion to the basis $\{|0\rangle, |1\rangle\}$. In another basis, such as

$$|\pm\rangle = \frac{|0\rangle \pm |1\rangle}{\sqrt{2}}, \tag{24.39}$$

we have e.g.

$$\left|\psi^+\right\rangle = \frac{|++\rangle - |--\rangle}{\sqrt{2}}. \tag{24.40}$$

When now the environment adds random phases to each state, that is $|+\rangle \rightarrow e^{i\varphi_+}|+\rangle$ and $|-\rangle \rightarrow e^{i\varphi_-}|-\rangle$, it follows (see exercises) that:

$$\left|\psi^+\right\rangle \rightarrow e^{2i\varphi_+}\frac{|++\rangle - e^{2i(\varphi_--\varphi_+)}|--\rangle}{\sqrt{2}}, \tag{24.41}$$

and clearly, this state is not decoherence-free for $\varphi_- - \varphi_+ \neq 2m\pi$.

In order to obtain a unique situation, we can consider e.g. the measuring apparatus or formally split the environment into two or more parts, as we have outlined above. With three systems such as $S$, $\mathcal{M}$ and $\mathcal{U}$, we can then perform a triorthogonal decomposition, whereby a unique decomposition (or pointer basis) can be achieved due to the interaction of the environment with the measuring apparatus.

## 24.5 Historical Side Note

For decades, the attitude of the 'old' Copenhagen school was authoritative; it claimed that the physical analysis of the measurement process in quantum-mechanical terms would be a pointless undertaking (see also Chap. 28). However, many people found it quite unsatisfactory to 'split' the world into a quantum realm dominated by the SEq and a separate realm of classical instruments. Where and by which criteria should one draw the line? A quote[19]: "The principle of superposition was suspended by 'decree' in the classical domain. This point of view—known as the Copenhagen Interpretation (CI)—has kept many a physicist out of despair. On the other hand, as long as a compelling reason for the quantum-classical border could not be found, the CI universe would be governed by two sets of laws, with poorly defined domains of jurisdiction. This fact has kept many a student, not to mention their teachers, in despair."

That the 'thought control' of the Copenhagen interpretation actually dominated quantum physics in the past to a great extent was experienced by H. Dieter Zeh, one of the first protagonists of the idea of decoherence, among others. He published his ideas in 1970 in a paper 'On the interpretation of measurement in quantum theory'. Before that, he had submitted an earlier version of the paper to the renowned physics journal *Nuovo Cimento*. That version was rejected because of the devastating judgment of the referee: "The paper is completely senseless. It is clear that the author has not fully understood the problem and the previous contributions in the field".[20] We see that even in the natural sciences, the 'right thing' can prevail more readily if it is generally accepted (and understood).

In the meantime, the theory of decoherence is considered an important element that can contribute to the explanation of the measurement problem.

## 24.6   Conclusions

Decoherence is a purely quantum-mechanical phenomenon that classical mechanics cannot even begin to explain. It is caused by the interaction of an open system with the environment, which thereby absorbs information about the system—the quantum nature of the system 'leaks out', so to speak, into the environment.[21] Since the environment in general has very many degrees of freedom among which the information about the system is distributed, the process is virtually irreversible; we cannot know the state of all environmental degrees of freedom (if we could, we would see superpositions in the macroscopic domain, also).

This information transfer becomes apparent in *entangled* states, which are a purely quantum-mechanical phenomenon. As soon as enough information about the entanglement has found its way into the environment, so that it can distinguish between

---

[19]W.H. Zurek, 'Decoherence, einselection and the quantum origins of the classical', quant-ph/0105127 v2, 11.7.2002

[20]Quoted in E. Joos, 'Elements of environmental decoherence', quant-ph/9908008 v1, 2.8.1999. In retrospect, it is clear that it was the referee who did not understand the problem fully.

[21]In J. Samuel, 'Gravity and decoherence: the double slit experiment revisited', *Classical and Quantum Gravity* 35 045004, https://doi.org/10.1088/1361-6382/aaa313 (Jan 2018) the view is represented that gravity is responsible for decoherence.

system states, the states can no longer interfere. Thus, entanglement as a *nonlocal* phenomenon of quantum mechanics leads to the *local* classical properties.[22]

It is only in isolated systems that superpositions and entanglement can be maintained over extended periods. For open systems, with their inevitable and uncontrollable interactions with their environment, decoherence acts on extremely short time scales. Thus it is understandable that one does not see superpositions, although they are allowed by quantum mechanics, under macroscopic conditions (not even for very short times in everyday terms), and why classical objects always maintain their familiar properties or, strictly speaking, *seem to have* them.[23]

Following the ideas outlined above, we can assume that the pointer variables (i.e. the display states) of the measuring apparatus $\mathcal{M}$ are determined by the interaction of $\mathcal{M}$ with $\mathcal{U}$. At least this assumption applies in simple model systems. In any case, it is certain that we do not need classical mechanics to describe the measurement process; by virtue of this point alone, decoherence contributes significantly to the internal consistency of quantum mechanics.

In principle, decoherence has the same effect as the postulate of the reduction of the wavefunction (wavefunction collapse), but in contrast, it describes a physically (and mathematically) plausible process. So we can dispense with the collapse postulate (see Chap. 14, Vol. 1, postulate 2.3) and hence the ominous distinction between the two dynamics (SEq and measurement); also the discussion about the boundary between quantum mechanics and classical mechanics may be placed on a different basis. For practical purposes, so to say as a working tool, we can of course continue to use the state reduction concept—we know now that it is simply an abbreviation for the process described above.

Given the dependence on the environment, one speaks of environment-induced decoherence; more vivid terms are *measurement by the environment* or *environment as a witness*.

Although decoherence provides a catchy explanation of why the world around us appears so classical, it cannot solve the measurement problem in its entirety. We note that decoherence does not provide a mechanism for the actual collapse, but rather a mechanism for the *appearance* of the collapse. In addition, decoherence does not explain why in a particular experiment precisely one out of many possibilities for the measured result is realized (i.e. that one which is actually measured). In addition, one would also like to know e.g. why certain superpositions are not observed (superselection rules[24]); is that due to the mechanism of decoherence? A further problem is provided by quantum cosmology, which attempts to describe quantum states of the entire universe; of course, there is after all no environment in this situation which could produce decoherence. Finally, one can argue that decoherence

---

[22] See e.g. J. Richens et al. 'Entanglement is Necessary for Emergent Classicality in All Physical Theories', *Phys. Rev. Lett.* 119, 080503 (Aug 2017).

[23] "The unresolved problem today is rather the classical physics itself. How is it possible that there is after all something like the familiar ideal world of classical physics?" Peter Mittelstaedt, 'Quantum Mechanics at the End of the 20th Century', *Physikalische Blätter* 56 (2000), No. 12, p. 65.

[24] For example, there are no interferences between fermions and bosons.

only claims irreversibility, when in fact all time evolutions are reversible in theory. That would of course be different if we could show that the correlations lost between the environment and the system can, even in principle, never be recovered. In short, decoherence answers some questions regarding the measurement process, but leaves others open. In essence, we still do not know how the transition from 'possible' to 'factual' really happens.

Be that as it may be—decoherence is at least a very common approach these days (if not one of the most common approaches) in quantum mechanics, which is also due to its physically and mathematically compelling content. Obviously, the opinion has found more and more followers in recent years that with decoherence, a step has been made in the right direction concerning the treatment of the measurement problem.

Among other things, decoherence has led to the very significant finding that quantum mechanics must be considered not only in isolated, but also in open systems. This entails the understanding that superpositions of states, such as those used for example in quantum computers, are very fragile constructs under normal conditions, and can be kept 'alive' only when the system is sufficiently isolated from its environment.[25] Thus, decoherence is a great, if not *the* great obstacle to the construction of a quantum computer. Such systems have to be very carefully shielded from their environments.

## 24.7   Exercises

1. Given the density matrix

$$\rho = \begin{pmatrix} |c_1|^2 & c_1 c_2^* e^{i\omega t} \\ c_1^* c_2 e^{-i\omega t} & |c_2|^2 \end{pmatrix};$$

(24.42)

calculate $\frac{1}{T} \int_0^T \rho \, dt$.

2. Consider the reduced density matrix $\rho_{S,red} = CC^\dagger$ of (24.16), where $C$ is given as an $M \times N$-matrix:

$$C = (c_{mn}) = \begin{pmatrix} c_{11} & c_{12} & \cdots & c_{1N} \\ c_{21} & c_{22} & \cdots & c_{2N} \\ \vdots & \vdots & \ddots & \vdots \\ c_{M1} & c_{M2} & \cdots & c_{MN} \end{pmatrix}.$$

(24.43)

---

[25] "The fundamental limitation to an observer's ability is of a different nature: he must restrict his observations to a finite part of the Universe. Insurmountable difficulties do not arise from having to handle large, complicated systems; rather, they appear with limited and not perfectly isolated systems." A. Peres, *Quantum Theory*, p. 346.

Hence, the system has $M$ states, and the environment has $N$. Estimate the order of magnitude of the elements of $\rho_{S,\text{red}}$.

3. Calculate explicitly the eigenvalues of the density matrix

$$\rho = \begin{pmatrix} |c_1|^2 & c_1 c_2^* \\ c_1^* c_2 & |c_2|^2 \end{pmatrix} \tag{24.44}$$

with $|c_1| + |c_2|^2 = 1$.

4. We consider two quantum objects with $\mathcal{H} = \mathcal{H}_1 \otimes \mathcal{H}_1$. The CONS $\{|0\rangle, |1\rangle\}$ is a basis of $\mathcal{H}_1$.

   (a) Show that the states

   $$|\pm\rangle = \frac{|0\rangle \pm |1\rangle}{\sqrt{2}} \tag{24.45}$$

   are also a CONS in $\mathcal{H}_1$.

   (b) Write down the states

   $$|\psi^\pm\rangle = \frac{|01\rangle \pm |10\rangle}{\sqrt{2}} \tag{24.46}$$

   in the basis $\{|+\rangle, |-\rangle\}$.

   (c) As assumed in the text, the effect of the environment is to add to each basis state a corresponding random phase. How are the new states $|\psi^\pm\rangle$ formulated?

5. Show that

$$\sum_{i=(m,n)} A_i^\dagger (t) A_i (t) = 1; \quad A_{i=(m,n)} (t) = \sqrt{p_n} \langle m| \hat{U} (t) |n\rangle. \tag{24.47}$$

see (24.34).

6. Two quantum objects each have a two-dimensional Hilbert space with the orthonormal basis vectors $|0\rangle$ and $|1\rangle$. They are in the ground state:

$$|\psi\rangle = c_0 |0\rangle |0\rangle + c_1 |1\rangle |1\rangle. \tag{24.48}$$

We now perform a change of basis via

$$|0\rangle = a_{11} |+\rangle + a_{12} |-\rangle; \quad |1\rangle = a_{21} |+\rangle + a_{22} |-\rangle, \tag{24.49}$$

where $|+\rangle$ and $|-\rangle$ are also an orthonormal basis. Under which conditions does $|\psi\rangle = d_+ |+\rangle |+\rangle + d_- |-\rangle |-\rangle$ hold?

# Chapter 25
# Scattering

Scattering theory is an important and very well elaborated branch of quantum mechanics; we consider some of its basics here.

Scattering is of crucial importance for investigating the structure of matter. It is no coincidence that the perhaps most expensive experiment on earth is the Large Hadron Collider (CERN), where the analysis of high-energy scattering processes has given information about the Higgs particle, whose existence was until 2012 only postulated.

We find the beginning and, at the same time, the basic idea of all scattering experiments in the classical experiment of Rutherford, who in 1911 passed $\alpha$ particles through a gold foil. It was found that the majority of the $\alpha$ particles passed the gold foil unimpeded, but a few of them were very strongly deflected or scattered. Thus, the atomic model of Thompson could be disproved; it postulated that the electrons were stuck in the atoms like raisins in the dough of a positively-charged background. It turned out that in contrast, the atom must be nearly empty, the positive charge being concentrated in a tiny region called the atomic nucleus.

Up to now, we have treated scattering by means of very simple examples in Chap. 15 (piecewise-constant potentials). Of course, there are far more advanced formulations; indeed, scattering is a very comprehensive and thoroughly elaborated branch of quantum mechanics which naturally is due to its importance in physics. Here, we can only sketch some of the basics.

In the following, we will consider the simplest case, namely *elastic scattering*. These are scattering processes where the sum of the kinetic energies of the scattering partners are the same before and after the collision, and where the scattering partners themselves remain the same before and after the collision. Thus, energy transfers e.g. into rotational or vibrational energy are excluded, and we have no recombinations such as $AB + CD \rightarrow AC + BD$ or the like. In short, the scattering partners are structureless.

Throughout, we assume an infinitely heavy scattering center, i.e. the difference between mass and reduced mass is neglected. In Appendix O, Vol. 2 are some remarks on the scattering of identical particles, where we use relative and center-of-mass

© Springer Nature Switzerland AG 2018

J. Pade, *Quantum Mechanics for Pedestrians 2*, Undergraduate Lecture Notes in Physics, https://doi.org/10.1007/978-3-030-00467-5_25

coordinates. Some information about the functions occurring (e.g. Bessel functions) and the decomposition of waves in terms of angular momentum components are collected in Appendix B, Vol. 2.

## 25.1   Basic Idea; Scattering Cross Section

We first formulate some of the basic elements of scattering in classical mechanics, and then transfer these terms to quantum mechanics.

### 25.1.1   Classical Mechanics

Particles are incident along the $z$ axis and are scattered at the scattering center (scatterer), cf. Fig. 25.1. The scattered particles can be detected at certain angles; these *scattering angles* are, as usual, compactly denoted by $\Omega = (\vartheta, \varphi)$ (see Chap. 16 and Appendix D, Vol. 1). The interaction between scatterer and particles is assumed to be of sufficiently short range so that the particles are asymptotically free (i.e. for large $|z|$).[1] The basic idea is now that we let many particles impinge on the scattering center and measure the number of scattered particles at all possible angles. It is unavoidable that the incident particles in such experiments have a certain spread perpendicular to the $z$ axis, i.e. the incident particle beam cannot be perfectly collimated.

We place the detector at a distance $r$ from the scattering center and at a given solid angle $\Omega\,(\vartheta, \varphi)$. The flux density (current density) of the incident particles[2] is given by $j_{in} = \frac{\text{number incoming}}{\text{area}\cdot\text{time}}$. We can measure the number of scattered particles $i_{scatt} = \frac{\text{number scattered}}{\text{solid angle}\cdot\text{time}}$ (which is the way these quantities are actually measured in the experiment). These two quantities are proportional to each other, $i_{scatt} \sim j_{in}$. For the (generally angle-dependent) constant of proportionality, a special designation is conventional: it is called the *differential cross section* or *scattering cross section*, and is denoted by $\frac{d\sigma}{d\Omega}$ (note that $d\Omega = \sin\vartheta\,d\vartheta\,d\varphi$). Thus we have as the definition of the scattering cross section:

$$\frac{d\sigma}{d\Omega} = \frac{i_{scatt}}{j_{in}}. \tag{25.1}$$

This can be rewritten using the current density $j_{scatt}\,(r, \vartheta, \varphi)$ of the scattered particles, which for geometrical reasons[3] vanishes asymptotically proportionally to $\frac{1}{r^2}$, namely $i_{scatt} = r^2 j_{scatt}$. It follows that

---

[1] We make this assumption for simplicity; the situation for long-range potentials can also be treated.

[2] The number $dN$ of particles passing in a time $dt$ through the area $d\mathbf{A}$ is related to the current density $\mathbf{j}$ by $dN = \mathbf{j} \cdot d\mathbf{A}\,dt$.

[3] The surface area of a sphere of radius $r$ is $4\pi r^2$.

**Fig. 25.1** Scattering of a classical particle

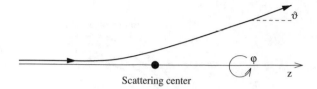

Scattering center

$$\frac{d\sigma}{d\Omega} = \frac{r^2 \; j_{\text{scatt}} \, (r, \vartheta, \varphi)}{j_{\text{in}}}. \tag{25.2}$$

Evidently, the scattering cross section has the dimension of an area—hence the term 'cross section'. This also applies to the integral of the scattering cross section over all angles, which is called the *total (scattering) cross section* or *bulk cross section $\sigma$*:

$$\sigma = \int \frac{d\sigma}{d\Omega} d\Omega. \tag{25.3}$$

An illustrative example: For a hard sphere of radius $R$, the total scattering cross section is equal to the great circle area, i.e. $\sigma = \pi R^2$.

### 25.1.2 Quantum Mechanics

We now want to transfer these concepts to quantum mechanics, confining ourselves to short-range potentials,

$$\lim_{r \to \infty} r V(\mathbf{r}) = 0. \tag{25.4}$$

First, we know that strictly speaking, we have to represent all quantum objects in the form of wave packets; for example, the incident part of the wavefunction as a suitable superposition of plane waves:

$$\psi_{\text{in}} (\mathbf{r}, t) = \int d^3 k \; \hat{\psi} (\mathbf{k}) \; e^{i(\mathbf{kr} - \omega t)}. \tag{25.5}$$

On the other hand, due to the linearity of the SEq, we can first consider separately individual wave components and then superimpose them; we went through these procedures already in Chap. 15 (potential step). In principle, we want to proceed here in the same manner.[4] Hence, the following considerations are based on the (improper or unphysical) states of sharp momentum; we keep in mind that we can

---

[4] We repeat the already familiar note: We refer to $e^{ikz}$ etc. as waves, since we take tacitly into account the factor $e^{-i\omega t}$.

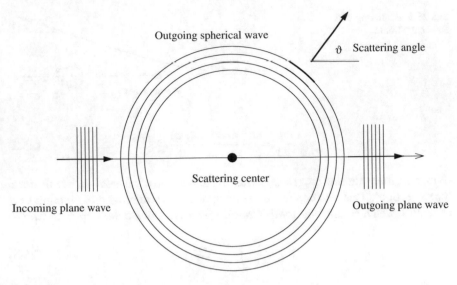

**Fig. 25.2**  Quantum-mechanical scattering

compose these individual solutions into the total (and physical) solution (however, we will not carry this out explicitly).[5]

Accordingly, for the incoming part we use the *ansatz*:

$$\varphi_{\text{in}}(\mathbf{r}) = e^{ikz}. \tag{25.6}$$

The scattered wave $\varphi_{\text{scatt}}(\mathbf{r})$ has only outgoing components (moving away from the scattering center). Now one can decompose a plane wave asymptotically into the sum of an incoming $\left(\frac{e^{-ikr}}{r}\right)$ and an outgoing $\left(\frac{e^{ikr}}{r}\right)$ spherical wave; thus, we can write asymptotically for the scattered wave:

$$\varphi_{\text{scatt}}(\mathbf{r}) \underset{r \to \infty}{\to} f(\vartheta, \varphi) \frac{e^{ikr}}{r}. \tag{25.7}$$

The function $f(\vartheta, \varphi)$ is called the *scattering amplitude*; it contains all the information about the scattering process that is available to us. The situation is shown schematically in Fig. 25.2 (where the circles should be actually modified depending on the angle proportional to $f(\vartheta, \varphi)$).

In this way, we have defined in our context the *stationary scattering problem*: It involves the solution of the stationary SEq

---

[5]One uses wave packets, among other things, for the explicit graphical representation of scattering processes, as we have done in Chap. 15. For reasons of space, we omit this here and refer to other textbooks on quantum mechanics.

$$E\varphi\left(\mathbf{r}\right) = \left[-\frac{\hbar^2}{2m}\nabla^2 + V\left(r\right)\right]\varphi\left(\mathbf{r}\right) \tag{25.8}$$

with $E > 0$, where the wavefunction contains incoming and scattered parts:

$$\varphi\left(\mathbf{r}\right) = \varphi_{\text{in}}\left(\mathbf{r}\right) + \varphi_{\text{scatt}}\left(\mathbf{r}\right) \tag{25.9}$$

to which (25.6) and (25.7) apply.

Finally, we establish the connection to the scattering cross section (25.2). The well-known definition of the current (or particle flux) density:

$$\mathbf{j} = \frac{\hbar}{2mi}\left(\psi^*\nabla\psi - \psi\nabla\psi^*\right) \tag{25.10}$$

leads to

$$\mathbf{j}_{\text{in}} = \frac{\hbar k}{m}\mathbf{e}_z; \tag{25.11}$$

and for the scattered wave, it follows asymptotically (see exercises) that:

$$\mathbf{j}_{\text{scatt}} \underset{r\to\infty}{\to} \frac{\hbar k}{m}\left|f\left(\vartheta,\varphi\right)\right|^2\frac{\mathbf{e}_r}{r^2}. \tag{25.12}$$

We insert this in (25.2) and obtain for the differential scattering cross section

$$\frac{d\sigma}{d\Omega} = \left|f\left(\vartheta,\varphi\right)\right|^2; \tag{25.13}$$

and for the total cross section

$$\sigma = \int \frac{d\sigma}{d\Omega}d\Omega = \int_0^{2\pi}d\varphi \int_0^{\pi}d\vartheta\,\sin\vartheta\,\left|f\left(\vartheta,\varphi\right)\right|^2. \tag{25.14}$$

We see that $\left|f\left(\vartheta,\varphi\right)\right|^2$ is the central quantity in our scattering experiments. On the one hand, it can be experimentally measured; on the other hand, it is determined by the potential $V\left(\mathbf{r}\right)$. For a given potential, the scattering amplitude can be always determined uniquely, either analytically (exact or in approximation) or numerically. The inverse problem, however, has no unique solution—a direct extrapolation from the (measured) scattering cross section to the potential is not possible.

## 25.2 The Partial-Wave Method

The partial-wave method is one of the standard procedures of scattering theory. We have already learned its essential ingredients in Chaps. 16 and 17 and can now apply them. In Appendix B, Vol. 2, there are some comments on Legendre polynomials, spherical harmonics, Bessel functions and the partial-wave decomposition of plane waves and spherical waves.

For simplicity, we consider a spherically symmetric potential[6]:

$$V = V(r).$$                                                                  (25.15)

The basic idea is that for spherically symmetric potentials, the operators $H$, $l^2$ and $l_z$ commute, and thus an expansion in terms of spherical harmonics makes sense (multipole expansion, cf. Chap. 16). Because of rotational invariance about the $z$ axis, the formulations are independent of the azimuthal angle $\varphi$. Accordingly, we expand the wavefunction $\varphi(\mathbf{r})$ as described in Chap. 17 in terms of $Y_l^0(\vartheta)$ (and not of $Y_l^m(\vartheta, \varphi)$) or of Legendre polynomials $P_l(\cos \vartheta) = \sqrt{\frac{4\pi}{2l+1}} Y_l^0(\vartheta)$:

$$\varphi(\mathbf{r}) = \sum_{l=0}^{\infty} \frac{u_l(r)}{r} P_l(\cos \vartheta).$$      (25.16)

After some rearranging, we obtain the radial equation

$$\frac{d^2 u_l(r)}{dr^2} + \left(k^2 - v_{\text{eff}}(r)\right) u_l(r) = 0,$$            (25.17)

with

$$k^2 = \frac{2m}{\hbar^2} E$$                                                  (25.18)

and the effective potential[7]:

$$v_{\text{eff}} = \frac{2m}{\hbar^2} V_{\text{eff}} = \frac{2m}{\hbar^2} \left[V(r) + V_{\text{centrifugal}}\right] = \frac{2m}{\hbar^2} \left[V(r) + \frac{\hbar^2 l(l+1)}{2\mu r^2}\right].$$  (25.19)

At the origin, the solutions of (25.17) must fulfill the condition $u_l(r) \underset{r \to 0}{\sim} r^{l+1}$. Due to $\lim_{r \to \infty} r V(r) = 0$ (25.4), the wavefunction must go over asymptotically to a free solution at large distances.[8]

---

[6]For general potentials $V(\mathbf{r})$, we have no single radial equation as in the case $V(r)$, but rather systems of coupled radial equations, which couple eigenfunctions of different angular momenta. The reason is that we need to expand the potential $V$ and the wavefunction $\varphi$ in terms of the angular momentum (multipole expansion). In the product $V\varphi$, total angular momenta occur according to the laws of angular momentum addition. If we sort according to these total angular momenta, we obtain coupled systems of radial equations (see exercises).

[7]We note again that we assume an infinitely heavy scattering center, so that the mass of the scattered quantum object and not the reduced mass enters.

[8]These two conditions are necessary for the square integrability of the wavefunction, see Chap. 17.

Now we have the general relation (multipole expansion of a plane wave):

$$e^{ikz} = \sum_{l=0}^{\infty} (2l + 1)\, i^l\, j_l\, (kr)\, P_l\, (\cos \vartheta)\,, \tag{25.20}$$

where $j_l\,(kr)$ are specific and important functions of mathematical physics, the *spherical Bessel functions* (see Appendix B, Vol. 2).[9] They satisfy

$$j_l\,(kr) \underset{r \to 0}{\sim} (kr)^{l+1}\,; \quad j_l\,(kr) \underset{r \to \infty}{\sim} \frac{\sin\left(kr - \frac{l\pi}{2}\right)}{kr}. \tag{25.21}$$

As can be seen in the last equation, and because of

$$\frac{\sin\left(kr - \frac{l\pi}{2}\right)}{kr} = \frac{(-i)^l\, e^{ikr} - i^l e^{-ikr}}{2ikr}\,, \tag{25.22}$$

one can regard a plane wave asymptotically as the sum of incoming and outgoing spherical waves.

Since asymptotically the radial function $u_l\,(r)$ describes free behavior, it will be given by essentially the corresponding Bessel function at large $r$, apart from a possible phase shift $\delta_l$, in which the effect of the potential manifests itself:

$$u_l\,(r) \underset{r \to \infty}{\sim} c_l \sin\left(kr - \frac{l\pi}{2} + \delta_l\right). \tag{25.23}$$

Now we assemble everything. On the one hand, we have asymptotically[10]:

$$\varphi_{asy}\,(\mathbf{r}) = e^{ikz} + f\,(\vartheta)\, \frac{e^{ikr}}{r} = \sum_{l=0}^{\infty} \left[ (2l + 1)\, i^l\, j_l\,(kr) + f_l\,(\vartheta)\, \frac{e^{ikr}}{r} \right] P_l\,(\cos \vartheta)\,;$$
$$\tag{25.24}$$

and on the other hand

$$\varphi_{asy}\,(\mathbf{r}) = \sum_{l=0}^{\infty} \frac{u_l\,(r)}{r}\, P_l\,(\cos \vartheta)\,, \tag{25.25}$$

where for $j_l\,(kr)$ and $u_l\,(r)$, we still have to insert the asymptotic behavior. The comparison of these two formulations leads to:

$$f_l\,(\vartheta) = \frac{(2l + 1)}{2ik}\left(e^{2i\delta_l} - 1\right) = \frac{(2l + 1)}{k}\, e^{i\delta_l} \sin \delta_l \tag{25.26}$$

---

[9]They can be calculated e.g. by means of the recursion relation $j_l\,(x) = (-1)^l\, x^l \left(\frac{1}{x}\frac{d}{dx}\right)^l \frac{\sin x}{x}$.
[10]Note that due to the spherical symmetry of the potential, $f\,(\vartheta)$ is independent of $\varphi$. For general potentials, we have $f = f\,(\vartheta, \varphi)$.

and

$$f(\vartheta) = \sum_{l=0}^{\infty} f_l(\vartheta)\, P_l(\cos\vartheta) = \frac{1}{k}\sum_{l=0}^{\infty}(2l+1)\,e^{i\delta_l}\sin\delta_l\, P_l(\cos\vartheta). \qquad (25.27)$$

The calculation can be found in the exercises. The total scattering cross section is calculated according to (25.14) and with $\int d\Omega\, P_l(\cos\vartheta)\, P_{l'}(\cos\vartheta) = \frac{4\pi}{2l+1}\delta_{ll'}$; the result reads

$$\sigma = \frac{4\pi}{k^2}\sum_{l=0}^{\infty}(2l+1)\sin^2\delta_l. \qquad (25.28)$$

Because of $P_l(1) = 1$, the comparison of (25.27) and (25.28) leads immediately to the *optical theorem*:

$$\sigma = \mathrm{Im}\, f(0). \qquad (25.29)$$

It can be shown that this relation follows from the conservation of the probability current density. The calculation can be found for example in Schwabl, p. 332, where also the more or less intuitive reasoning is given that the interference of the scattered wave with the incoming wave along the $z$ direction leads to a reduction of the probability current density in the forward direction. The interference term, which is proportional to $f(0)$, yields by definition precisely the total scattering cross section.

At first glance, it is perhaps not quite clear what is the advantage of the partial-wave analysis—instead of *one* wavefunction $\varphi(\mathbf{r})$, one has to calculate in principle *infinitely many* wavefunctions $u_l(r)$. First, an advantage is certainly the fact that the functions $u_l(r)$ obey *ordinary* differential equations, which are analytically and numerically much simpler than partial differential equations. Furthermore, under appropriate circumstances, one in fact needs only a few scattering phases—and then the partial-wave analysis is suitable as a practical method.

Let us assume in this context for simplicity that the scattering potential vanishes for $r > a$. Then, only the centrifugal term of the effective potential (25.19) acts in this domain, see Fig. 25.3. For a quantum object incident with the energy $E$, the classical turning point[11] is given by

$$E = \frac{\hbar^2 k^2}{2m} = V_{\text{centrifugal}} = \frac{\hbar^2 l(l+1)}{2mr^2} \rightarrow r_{\text{return}} = \frac{\sqrt{l(l+1)}}{k}. \qquad (25.30)$$

For $r < r_{\text{return}}$ (classically forbidden domain), the wavefunction decays exponentially. Hence, if the turning point is outside the range of the potential, $r_{\text{return}} > a$, the quantum object sees (almost) nothing of the potential. Consequently, scattering occurs only for $r_{\text{return}} < a$, i.e. for $\sqrt{l(l+1)} \approx l < ka$. For very short-range potentials and/or low energies, it is therefore sufficient to calculate only a few scattering phases.

---

[11]The point where $E = V$; a classical object must reverse at this point, i.e. it is reflected. See also Chap. 15.

**Fig. 25.3** Turning point for
a short-range potential.
Classically, a particle with
the energy given 'sees' no
distinction between the
different potential forms
beyond the reversal point

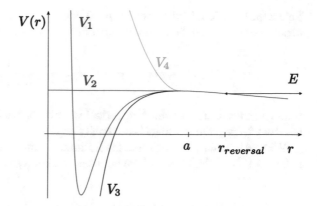

In particular, this is the case for example for the Bose-Einstein condensation, which is known to occur only at extremely low temperatures (or energies). Here, consideration of the scattering contribution of only the angular momentum $l = 0$ is often sufficient, the so-called *s-wave scattering* (for the nomenclature, see Chap. 17).

## 25.3 Integral Equations, Born Approximation

Apart from the formulation of the stationary scattering problem as in (25.8) and (25.9), there is a representation in terms of an *integral equation*. This equation is interesting in itself and offers in addition the starting point for an approximation procedure to the scattering problem, the *Born approximation*.

The method uses *Green's functions*; see also Appendix H, Vol. 1. To make clear the basic idea, we start from a differential equation of the form

$$\left(\nabla_{\mathbf{r}}^2 + k^2\right) \varphi\left(\mathbf{r}\right) = \rho\left(\mathbf{r}\right). \tag{25.31}$$

Instead of the operator $\nabla_{\mathbf{r}}^2 + k^2$, any other operator may occur; the only thing that matters is that the homogeneous equation is linear. Its solution $\varphi_0\left(\mathbf{r}\right)$ is known. Now we assume that we know the solution of this differential equation for the case that the inhomogeneity is a delta function. This solution is called a *Green's function*, and it is

$$\left(\nabla_{\mathbf{r}}^2 + k^2\right) G\left(\mathbf{r} - \mathbf{r}'; k\right) = \delta\left(\mathbf{r} - \mathbf{r}'\right). \tag{25.32}$$

Then we can describe the inhomogeneity $\rho\left(\mathbf{r}\right)$ as a superposition of corresponding delta functions, and because of the linearity of the differential equation, this transfers to the solutions of the differential equation. In other words, we have:

$$\varphi\left(\mathbf{r}\right) = \varphi_0\left(\mathbf{r}\right) + \int d^3r' G\left(\mathbf{r} - \mathbf{r}'; k\right) \rho\left(\mathbf{r}'\right). \tag{25.33}$$

This is the basic idea behind the Green's functions. The specific form of $G\left(\mathbf{r} - \mathbf{r}'; k\right)$ depends of course on the particular operator,; in our example, with $\nabla_{\mathbf{r}}^2 + k^2$, it is

$$G\left(\mathbf{r} - \mathbf{r}'; k\right) = -\frac{1}{4\pi}\frac{e^{ik|\mathbf{r}-\mathbf{r}'|}}{|\mathbf{r} - \mathbf{r}'|}. \tag{25.34}$$

Intuitively, these are outgoing[12] spherical waves propagating from the center $\mathbf{r}' = \mathbf{r}$, i.e. from every point of the inhomogeneity $\rho\left(\mathbf{r}'\right)$.[13]

We now transfer these results to the stationary scattering problem with an arbitrary potential $V\left(\mathbf{r}\right)$, which we write in the usual notation $k^2 = \frac{2m}{\hbar^2} E$ as

$$\left(\nabla^2 + k^2\right)\psi\left(\mathbf{r}\right) = v\left(\mathbf{r}\right)\psi\left(\mathbf{r}\right); \quad \psi\left(\mathbf{r}\right) \underset{r\to\infty}{\to} e^{i\mathbf{k}\cdot\mathbf{r}} + f\left(\vartheta, \varphi\right)\frac{e^{ikr}}{r}, \tag{25.35}$$

where we choose for the incident wave $\varphi_0\left(\mathbf{r}\right) = e^{i\mathbf{k}\cdot\mathbf{r}}$ and use the abbreviation $v\left(\mathbf{r}\right) = \frac{2m}{\hbar^2} V\left(\mathbf{r}\right)$. We consider the right-hand side of the SEq as a formal inhomogeneity and have as its solution (also formal):

$$\psi\left(\mathbf{r}\right) = e^{i\mathbf{k}\cdot\mathbf{r}} - \frac{1}{4\pi}\int d^3 r' \frac{e^{ik|\mathbf{r}-\mathbf{r}'|}}{|\mathbf{r} - \mathbf{r}'|} v\left(\mathbf{r}'\right)\psi\left(\mathbf{r}'\right). \tag{25.36}$$

This equation, also called the *Lippmann-Schwinger equation*, is equivalent to the SEq (25.35), including the boundary conditions (because of the form of the Green's function as outgoing spherical waves). Thus, it is a more compact representation of the stationary scattering problem, and perhaps also physically more transparent.[14]

To get more information, we consider the asymptotic behavior. With the approximation $|\mathbf{r} - \mathbf{r}'| \underset{r\to\infty}{\to} r - \hat{\mathbf{r}}\cdot\mathbf{r}' + \cdots$, it follows (see exercises) that:

$$\frac{e^{ik|\mathbf{r}-\mathbf{r}'|}}{|\mathbf{r} - \mathbf{r}'|} \underset{r\to\infty}{\to} \frac{e^{ikr-ik\hat{r}\cdot\mathbf{r}'}}{r} + \cdots = \frac{e^{ikr}}{r}e^{-ik\hat{r}\cdot\mathbf{r}'} + \cdots = \frac{e^{ikr}}{r}e^{-i\mathbf{k}'\cdot\mathbf{r}'} + \cdots \tag{25.37}$$

where we have introduced the momentum vector $\mathbf{k}' = k\hat{\mathbf{r}}$, pointing in the direction of the scattered quantum object. We insert this into (25.36) and find:

$$\psi\left(\mathbf{r}\right) \underset{r\to\infty}{\to} e^{i\mathbf{k}\cdot\mathbf{r}} - \frac{1}{4\pi}\frac{e^{ikr}}{r}\int d^3 r' e^{-i\mathbf{k}'\cdot\mathbf{r}'} v\left(\mathbf{r}'\right)\psi\left(\mathbf{r}'\right). \tag{25.38}$$

This gives the following integral representation for the scattering amplitude:

---

[12] In addition, there are solutions in the form of incoming spherical waves (mathematically on equal footing). But we can neglect them for physical reasons, because we want to describe scattering processes.

[13] This is none other than a somewhat technical formulation of Huygens' principle.

[14] Of course it is not an exact *explicit* solution—this does not exist in general.

$$f\left(\vartheta, \varphi\right) = -\frac{1}{4\pi} \int d^3r' e^{-i\mathbf{k}'\cdot\mathbf{r}'} v\left(\mathbf{r}'\right) \psi\left(\mathbf{r}'\right). \tag{25.39}$$

With the abstract representation of the plane wave,[15] $e^{i\mathbf{k}\cdot\mathbf{r}} \to |\mathbf{k}\rangle$, the abstract representation of the scattering amplitude in the bracket formalism is

$$f\left(\vartheta, \varphi\right) = -\frac{1}{4\pi} \langle\mathbf{k}'| v |\psi\rangle = -\frac{m}{2\pi\hbar^2} \langle\mathbf{k}'| V |\psi\rangle; \tag{25.40}$$

and for the total wave, with (25.36), we have

$$|\psi\rangle = |\psi_0\rangle + Gv |\psi\rangle. \tag{25.41}$$

Assuming that $Gv |\psi\rangle$ is a small term[16] compared to $|\psi_0\rangle$, we can start an iterative solution method by inserting in each case the lower approximation instead of the state $|\psi\rangle$ on the right-hand side:

$$|\psi\rangle_0 = |\mathbf{k}\rangle$$
$$|\psi\rangle_1 = |\mathbf{k}\rangle + Gv |\psi\rangle_0$$
$$\cdots$$
$$|\psi\rangle_{n+1} = |\mathbf{k}\rangle + Gv |\psi\rangle_n. \tag{25.42}$$

Inserting into the scattering amplitude, this leads to

$$f\left(\vartheta, \varphi\right) = -\frac{1}{4\pi} \langle\mathbf{k}'| v |\psi\rangle = -\frac{1}{4\pi} \langle\mathbf{k}'| v \{1 + Gv + GvGv + \cdots\} |\mathbf{k}\rangle \tag{25.43}$$

or

$$f\left(\vartheta, \varphi\right) = -\frac{1}{4\pi} \langle\mathbf{k}'| v + vGv + vGvGv + \cdots |\mathbf{k}\rangle. \tag{25.44}$$

Of course, it must be checked in each individual case whether this *Born series for the scattering amplitude* converges (or at least converges asymptotically).

We will not deal with this difficult matter any further, but terminate the series after the first term. Using such an approximation, we obtain for the scattering amplitude the (first) *Born approximation*:

$$f_{\text{Born}}\left(\vartheta, \varphi\right) = -\frac{1}{4\pi} \langle\mathbf{k}'| v |\mathbf{k}\rangle. \tag{25.45}$$

In the position representation and with $\mathbf{q} = \mathbf{k} - \mathbf{k}'$, this expression reads

---

[15]In order to simplify the notation, we omit here the normalization factor $(2\pi)^{-3/2}$, cf. Chap. 12, Vol. 1.

[16]More precisely: $\|Gv |\psi\rangle\| \ll \||\psi_0\rangle\|$.

$$f_{\text{Born}}\left(\vartheta, \varphi\right) = -\frac{m}{2\pi\hbar^2} \int d^3 r' V\left(\mathbf{r}'\right) e^{i\mathbf{q}\cdot\mathbf{r}'}. \tag{25.46}$$

The vector $\hbar\mathbf{q}$ is the *momentum transfer* and indicates the exchange of momentum from the incoming wave ($\mathbf{k}$) to the scattered wave ($\mathbf{k}'$). We see that in this approximation, the scattering amplitude is simply the Fourier transform (with respect to $\mathbf{q}$) of the potential.

We want to transform the last expression explicitly for our problem of elastic scattering on a central potential. Elastic scattering means $|\mathbf{k}| = |\mathbf{k}'|$, and for a wave incident along the $z$ axis, it follows that $\mathbf{k} \cdot \mathbf{k}' = k^2 \cos\vartheta$, and thus

$$q = \left|\mathbf{k} - \mathbf{k}'\right| = 2k \sin\frac{\vartheta}{2}. \tag{25.47}$$

For the scattering amplitude of a central potential in the Born approximation, we obtain[17]:

$$\begin{aligned}
f_{\text{Born}}\left(\vartheta\right) &= -\frac{m}{2\pi\hbar^2} \int_0^\infty dr'\, r'^2 V\left(r'\right) \int_0^\pi d\vartheta' \sin\vartheta'\, e^{iqr'\cos\vartheta'} \int_0^{2\pi} d\varphi' \\
&= -\frac{m}{\hbar^2} \int_0^\infty dr'\, r'^2 V\left(r'\right) \int_{-1}^1 d\cos\vartheta'\, e^{iqr'\cos\vartheta'} \\
&= -\frac{2m}{q\hbar^2} \int_0^\infty dr'\, r' V\left(r'\right) \sin qr'. \tag{25.48}
\end{aligned}$$

As an example, the Born approximation for the Yukawa and the Coulomb potentials is found in the exercises.

## 25.4  Exercises

1. Show that:
$$\left|\mathbf{r} - \mathbf{r}'\right| \underset{r\to\infty}{\to} r - \hat{\mathbf{r}} \cdot \mathbf{r}'. \tag{25.49}$$

2. Prove that
$$\frac{e^{ik|\mathbf{r}-\mathbf{r}'|}}{|\mathbf{r}-\mathbf{r}'|} \underset{r\to\infty}{\to} \frac{e^{ikr}}{r} e^{-i\mathbf{k}'\cdot\mathbf{r}'}. \tag{25.50}$$

3. Calculate explicitly the asymptotic form of the current density for the scattered wave.

4. Determine the general relation between scattering amplitude and scattering phases.

---

[17]For the integration, we choose the $\mathbf{q}$ axis as $z$ axis; the integration then runs over the spherical coordinates ($r', \vartheta', \varphi'$).

5. Determine the radial equations for a general potential $V(\mathbf{r})$.
6. The Yukawa potential (also called the screened Coulomb potential) has the form

$$V(r) = V_0 \frac{e^{-r/a}}{r}; \quad a > 0. \tag{25.51}$$

The range of the potential is of order $a$. Determine the scattering amplitude for the potential in the Born approximation. The Coulomb potential follows for $a \to \infty$ (infinite range of the Coulomb potential). Calculate also in this case the scattering cross section (*Rutherford scattering cross section*).
7. In this exercise, we address the transformation between the abstract representation and the position representation. We recall that this topic is discussed in more detail in Chap. 12, Vol. 1.

(a) Transform the equation

$$|\psi\rangle = |\psi_0\rangle + Gv|\psi\rangle \tag{25.52}$$

into the position representation.
(b) Write the right-hand side of the following equation:

$$f_{\text{Born}}(\vartheta, \varphi) = -\frac{1}{4\pi} \langle \mathbf{k}' | v | \mathbf{k} \rangle \tag{25.53}$$

explicitly in the position representation.

# Chapter 26
# Quantum Information

Quantum information is one of the modern applications of quantum mechanics. Apart from quantum teleportation, we consider the fundamentals of quantum computers and the algorithms of Deutsch, Grover and Shor.

Quantum information (QI) means the transfer and processing of information, as far it is specifically quantum mechanical and not classical. In other words, quantum-mechanical principles such as superposition and entanglement of states play a central role in QI.

Hence in QI, we are dealing not with the fresh discovery of new principles of quantum mechanics, but rather with the new application of known relationships—QI was always implicit in quantum mechanics.[1] It is just the way of looking at things which has changed in the last two or three decades, probably because some concepts that had long been handled rather gingerly (entanglement, nonlocality, etc.) have proven their theoretical and practical significance.

With *quantum cryptography*, we already addressed a subtopic of quantum information in Chap. 10, Vol. 1. Two further topics that we outline below are *quantum teleportation* and *quantum computation*. But first, we show that there is no general quantum copier.

## 26.1 No-Cloning Theorem (Quantum Copier)

The *no-cloning theorem* states that it is impossible to duplicate *arbitrary* quantum-mechanical states. Thus, it is *not* possible to observe a state $|z\rangle$ non-destructively by producing an arbitrary number of (identical) copies of $|z\rangle$ and measuring this ensemble in leisurely fashion, as we will now demonstrate. We assume without loss of generality that all states are normalized.

---

[1]However, this is somewhat concealed by the fact that it makes use of a peculiar notation that (naturally) is oriented more to the needs of information processing than to those of theoretical physics.

© Springer Nature Switzerland AG 2018
J. Pade, *Quantum Mechanics for Pedestrians 2*, Undergraduate Lecture
Notes in Physics, https://doi.org/10.1007/978-3-030-00467-5_26

We have an *unknown* state $|a\rangle$ which we want to duplicate. The system to which we want to transfer the copy is $|\varphi\rangle$; it plays the role of the blank sheet for copying and must be suitable, of course, to accept the copy. For instance, if we want to copy an unknown state $|a\rangle$ with spin 1, then $|\varphi\rangle$ must also allow spin 1. Thus, we want to transfer the product state[2] $|a \otimes \varphi\rangle$ by a unitary transformation $U$ into the clone $|a \otimes a\rangle$. We do not have to go into details about $U$; it is enough to know that $U$ must be independent of the state to be copied. We have:

$$U\,|a \otimes \varphi\rangle = |a \otimes a\rangle\,. \tag{26.1}$$

Now, a copier should be able to duplicate not only a single state (which here is $|a\rangle$); we need to have at least a second original state $|b\rangle$ which we can copy onto our blank sheet:

$$U\,|b \otimes \varphi\rangle = |b \otimes b\rangle\,. \tag{26.2}$$

We multiply the adjoint of the second equation into the first equation:

$$\langle b \otimes \varphi|\,U^\dagger U\,|a \otimes \varphi\rangle = \langle b \otimes b|\,a \otimes a\rangle\,. \tag{26.3}$$

We obtain

$$\langle b \otimes \varphi|\,a \otimes \varphi\rangle = \langle b \otimes b|\,a \otimes a\rangle\,, \tag{26.4}$$

or, with $\langle\varphi|\,\varphi\rangle = 1$,

$$\langle b|\,a\rangle = \langle b|\,a\rangle^2\,. \tag{26.5}$$

It follows that either $\langle b|\,a\rangle = 0$, or $\langle b|\,a\rangle = 1$. So we have either $|b\rangle \perp |a\rangle$ or $|b\rangle = |a\rangle$, which means that we cannot clone other states (recall that we assume normalized states).

Hence, strictly speaking, the name *no-cloning theorem* is not quite correct, because one can copy a state and the states orthogonal to it—but only those; all other states cannot be copied. If we know that *all* the states to be measured are parallel or orthogonal to a known state $|a\rangle$, we can make arbitrarily many copies of each state. This explains in retrospect why in quantum cryptography one uses *two different* orientations for the measurement of linear polarization: to spoil the possibility of reliably copying the states.

Another exceptional case in which copies of quantum states are possible occurs when there is a classical (i.e. non-quantum-mechanical) sub-step in the information processing. This of course can be copied perfectly.

Apart from these exceptional cases, the no-cloning theorem applies globally—there is no universal copier for pure quantum states.

---

[2]For reasons of greater clarity, we use here the detailed notation with $\otimes$, i.e. $|a \otimes b\rangle$.

## 26.2 Quantum Cryptography

We have already discussed this topic in Chap. 10, Vol. 1. Without going into detail, we want just briefly to mention here that there are protocols that work with entangled photons (e.g. the E91 protocol) and thereby provide an increase in security.

## 26.3 Quantum Teleportation

Teleportation is the (hypothetical) process whereby matter is transported from point $A$ to point $B$ without traversing the intervening space physically. This procedure, so much appreciated by sci-fi authors,[3] has in fact little in common with quantum teleportation, since in the latter, it is not the body, but rather its state—or more precisely, the quantum state—which is teleported. The tools are entangled states; in a certain part of the process, the information must be transmitted via a classical channel. We discuss the subject on the basis of quantum objects that can exist in two states, which we call $|0\rangle$ and $|1\rangle$.[4] The two states are normalized and mutually orthogonal.

Here we meet up again with Alice and Bob from Chap. 10, Vol. 1. The starting point is as follows: Alice wants to inform Bob of the state of a quantum object $Q1$, such as

$$|\varphi\rangle_1 = c\,|0\rangle_1 + d\,|1\rangle_1\,, \tag{26.6}$$

but without sending him the quantum object itself. Alice herself does not know the state and therefore cannot measure it reliably, since a single measurement gives no information about the constants $c$ and $d$ and will in general change the state (26.6). Preparing an ensemble by copying $Q1$ would indeed allow for the measurement of $c$ and $d$ with arbitrary precision, but it is prohibited according to the no-cloning theorem, as we have just seen. What to do?

The solution is achieved with a pair of entangled quantum objects $QA$ and $QB$, transported to Alice and Bob (see Fig. 26.1) and containing the information about their status. To be concrete, the state of the entangled quantum objects $QA$ and $QB$ is

---

[3]Teleportation was made popular especially by 'Star Trek'. Apparently it was introduced in the series mainly for cost reasons—it simply would have been much more expensive to film / animate landings of spacecrafts on alien planets. "Beam me up, Scotty!"

[4]Instead of $|0\rangle$ and $|1\rangle$, we could choose other designations such as $|h\rangle$ and $|v\rangle$. But since $|0\rangle$ and $|1\rangle$ are used in quantum information exclusively, we adopt this notation. It should be mentioned in any case that $|0\rangle$ is *not* the zero vector (and of course, not the ground state of the harmonic oscillator). For concrete calculations, we use the representation

$$|0\rangle \cong \begin{pmatrix} 1 \\ 0 \end{pmatrix}; \quad |1\rangle \cong \begin{pmatrix} 0 \\ 1 \end{pmatrix}.$$

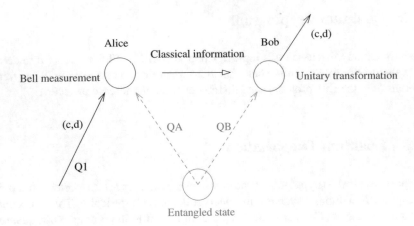

**Fig. 26.1** Schematics of quantum teleportation

$$|\Phi^-\rangle_{AB} = \frac{1}{\sqrt{2}}\left[|01\rangle_{AB} - |10\rangle_{AB}\right]. \tag{26.7}$$

The total state $|\Psi\rangle_{1AB}$ of the three quantum objects is given by:

$$
\begin{aligned}
|\Psi\rangle_{1AB} &= |\varphi\rangle_1 \otimes |\Phi^-\rangle_{AB} \\
&= \frac{1}{\sqrt{2}}\left[c\,|001\rangle_{1AB} + d\,|101\rangle_{1AB} - c\,|010\rangle_{1AB} - d\,|110\rangle_{1AB}\right]. \tag{26.8}
\end{aligned}
$$

Alice now measures the quantum objects $Q1$ and $QA$ (which are *not* entangled), using as a basis the Bell states, which we met up with already, in Chap. 20:

$$|\Phi^\pm\rangle_{1A} = \frac{1}{\sqrt{2}}\left[|01\rangle_{1A} \pm |10\rangle_{1A}\right]$$

$$|\Psi^\pm\rangle_{1A} = \frac{1}{\sqrt{2}}\left[|00\rangle_{1A} \pm |11\rangle_{1A}\right], \tag{26.9}$$

with the inversions

$$|01\rangle_{1A} = \frac{1}{\sqrt{2}}\left[|\Phi^+\rangle_{1A} + |\Phi^-\rangle_{1A}\right]; \quad |10\rangle_{1A} = \frac{1}{\sqrt{2}}\left[|\Phi^+\rangle_{1A} - |\Phi^-\rangle_{1A}\right]$$

$$|00\rangle_{1A} = \frac{1}{\sqrt{2}}\left[|\Psi^+\rangle_{1A} + |\Psi^-\rangle_{1A}\right]; \quad |11\rangle_{1A} = \frac{1}{\sqrt{2}}\left[|\Psi^+\rangle_{1A} - |\Psi^-\rangle_{1A}\right]. \tag{26.10}$$

We insert these inversions into the total state (26.8). It follows that:

$$|\Psi\rangle_{1AB} = \frac{1}{2}\left[\begin{array}{l} |\Psi^+\rangle_{1A}\,(-d\,|0\rangle_B + c\,|1\rangle_B) + |\Psi^-\rangle_{1A}\,(d\,|0\rangle_B + c\,|1\rangle_B) \\ +\,|\Phi^+\rangle_{1A}\,(-c\,|0\rangle_B + d\,|1\rangle_B) - |\Phi^-\rangle_{1A}\,(c\,|0\rangle_B + d\,|1\rangle_B) \end{array}\right]. \tag{26.11}$$

We can rewrite this with the help of the state $|\varphi\rangle_B = c\,|0\rangle_B + d\,|1\rangle_B$ and the (unitary) Pauli matrices (see the exercises):

$$|\Psi\rangle_{1AB} = \frac{1}{2}\left[\begin{array}{l} -|\Phi^-\rangle_{1A}\,|\varphi\rangle_B - |\Phi^+\rangle_{1A}\,\sigma_z\,|\varphi\rangle_B \\ +\,|\Psi^+\rangle_{1A}\,i\sigma_y\,|\varphi\rangle_B + |\Psi^-\rangle_{1A}\,\sigma_x\,|\varphi\rangle_B \end{array}\right]. \tag{26.12}$$

When Alice carries out her measurement, one of the four summands in the brackets is filtered out of the total state $|\Psi\rangle_{1AB}$, depending on which state $Q1QA$ she projects. Alice now tells Bob via a classical method (telephone, etc.) which state she has measured, so that Bob knows which unitary transformation he has to apply for preserving the original condition. In this way, the state $|\varphi\rangle_1$ of the quantum object 1 (but not $Q1$ itself!) has been teleported without ever measuring $Q1$ directly. Bob knows that $QB$ now has precisely the unknown state that $Q1$ had before (i.e. 26.6). A few remarks are in order:

1. The coefficients $c$ and $d$ are not measured. They are unknown for $|\varphi\rangle_1$, and likewise, after the teleportation, for $|\varphi\rangle_B$.
2. This is not copying, since the state $|\varphi\rangle_1$ is destroyed by Alice's measurement. So there is no contradiction with the no-cloning theorem.
3. Bob can prepare $|\varphi\rangle_B$ as soon as he receives the result of Alice's measurement. The transmission of this information is done via a classical channel, i.e. with a speed equal to or less than the speed of light. So there is no instantaneous non-local information transfer.
4. Quantum teleportation *never* involves the transport of matter.
5. Experimental realizations of quantum teleportation have been carried out since 1994. Outside the lab and over longer distances, they were performed e.g. in 2004 (in Vienna, over a distance of 600 m, using a fiber-optic cable in a sewer tunnel crossing the Danube River).[5] Methods for the teleportation of a beam of light including its temporal correlations have been proposed for instance in 2009.[6]

---

[5] Quantum teleportation is not restricted to the range of some few centimeters or meters; for 'records' see e.g. Xiao-Song Ma et al., 'Quantum teleportation over 143 kms using active feed-forward', *Nature* 489, 269–273 (2012), or Ji-Gang Ren et al., Ground-to-satellite quantum teleportation, *Nature* 549, 70–73 (2017) where quantum teleportation of independent single-photon qubits from a ground observatory to a low-Earth-orbit satellite over distances of up to 1400 km is reported.

[6] C. Noh et al., 'Quantum Teleportation of the Temporal Fluctuations of Light', *Phys. Rev. Lett.* 102, 230501 (2009).

## 26.4   The Quantum Computer

A quantum computer (QC) operates under the laws of quantum mechanics. In particular, it uses superposition and entanglement of states—principles which do not exist in classical mechanics. Due to this, a QC can (could)[7] carry out a large number of parallel operations (*quantum parallelism*), and solve suitable problems much faster than a classical computer. Although in a measurement, only *one* state is observed, it is nevertheless possible in certain cases to extract global information.

The topic experienced a big boost in 1994, when Peter W. Shor showed that a quantum computer can factorize large integers much faster[8] than a classical computer. However, it appears today (2018) that there is still a long road to a generally-applicable QC. The main problem is decoherence. The processes in a QC are essentially unitary transformations which must not be disturbed (or at least only in a manageable way) or even destroyed due to uncontrolled interactions with the environment.

All in all, this is a very active research area with a correspondingly extensive literature. We shall outline below a few basic ideas.

### 26.4.1   Qubits, Registers (Basic Concepts)

In classical information theory, the basic unit is the *bit* which can have one of *two* values. The bit is stored in a system that can assume only two states, and the system is *either* in the one *or* in the other state.[9] We call these states $|0\rangle$ and $|1\rangle$ once again.

In contrast, a quantum-mechanical system can be in a superposition $|z\rangle$ of the two states:

$$|z\rangle = a\,|0\rangle + b\,|1\rangle\,; \quad |a|^2 + |b|^2 = 1 \tag{26.13}$$

This system stores not a *bit*, but rather a *quantum bit*, a *qubit* for short.[10] A qubit can be implemented by any system with two states—such as a spin-1/2 quantum object (spin up or spin down), a polarized photon (vertical or horizontal), an atom (excited or not), and so on.

The initially indeterminate value of the system or the qubit is determined by a measurement. We obtain with a probability of $|a|^2$ the state $|0\rangle$, and with $|b|^2$ the state $|1\rangle$. This fact in itself is not remarkably useful. In other words, it is *not* the case

---

[7]The conditional 'could' is more appropriate insofar that until now (2018), a generally working, large QC exists only on paper.

[8]Under appropriate circumstances, the computation time of the classic computer grows exponentially with the number of digits of $N$, that of the quantum computer only polynomially. See the later section on the Shor algorithm.

[9]Possible realizations are familiar examples such as the ubiquitous coin with heads and tails, a switch (on/off), etc. Of course, all bi-stable systems are suitable in principle.

[10]Analogously, a linear combination of three states is called a *qutrit*, $a\,|0\rangle + b\,|1\rangle + c\,|2\rangle$.

that we can extract all the information—after the measurement, the system is either in the state $|0\rangle$ or in the state $|1\rangle$. In this sense, the information content of a qubit equals that of a classical bit. However, the superposition principle allows a certain parallelism in the calculations, as we shall see shortly.

Just as in a classical computer, one combines several qubits to give *registers*.[11] A quantum register of $n$ qubits (register of size $n$ or length $n$) is a state of the $2^n$-dimensional product space (Hilbert space); for its basis, we can choose the product states. For a register $|a\rangle \otimes |b\rangle = |ab\rangle$ of two qubits, a possible basis is e.g. $\{|00\rangle, |01\rangle, |10\rangle, |11\rangle\}$. If we understand this as a binary notation,[12] then the states are given in decimal notation by $\{|0\rangle, |1\rangle, |2\rangle, |3\rangle\}$. Unlike a classical bit, which can take on only *one* of these values, the qubit states can be superposed. We point out that a register can be entangled.

**Example**: Suppose we want to store a number between 0 and 7 in a register. For this we need 3 bits (in general, we need $n$ bits to store one of the $2^n$ numbers between 0 and $2^{n-1}$). A classical register would thus store *one* of the following configurations

$$
\begin{array}{llll}
0 = (000) & 1 = (001) & 2 = (010) & 3 = (011) \\
4 = (100) & 5 = (101) & 6 = (110) & 7 = (111) \,.
\end{array}
\tag{26.14}
$$

A system of 3 qubits can occupy the following product states[13]:

$$
\begin{array}{llll}
0 : |000\rangle & 1 : |001\rangle & 2 : |010\rangle & 3 : |011\rangle \\
4 : |100\rangle & 5 : |101\rangle & 6 : |110\rangle & 7 : |111\rangle \,.
\end{array}
\tag{26.15}
$$

Because we can generate superpositions like

$$
|q\rangle = \sum_{x,y,z \in \{0,1\}} c_{xyz} |xyz\rangle \,,
\tag{26.16}
$$

one could conclude that the state vector allows us to store the $2^3 = 8$ numbers $c_{xyz}$ at once, and generally $2^N$ numbers with $N$ qubits.[14] But a measurement gives of course only one of the basis states. Thus, with the coefficients $c_{xyz}$, we have a remarkable virtual information store at our disposal, but we cannot read it out directly from the system. A measurement yields *one* of the numbers 0 to 7, and not all 8 in one sweep.

**A remark on notation**: As already indicated, there are different notations for qubits. Let us take as an example three qubits, which are all in the same superposition state[15];

---

[11]For the sake of simplicity, we assume that the quantum objects are distinguishable.

[12]So we have e.g. $10 \,\hat{=}\, 1 \cdot 2^1 + 0 \cdot 2^0 = 2$ or $1101 \,\hat{=}\, 1 \cdot 2^3 + 1 \cdot 2^2 + 0 \cdot 2^1 + 1 \cdot 2^0 = 13$.

[13]We use the abbreviation

$$|a\rangle \otimes |b\rangle \otimes |c\rangle = |abc\rangle$$

[14]For $N = 100$, we would have $2^N \approx 1.27 \cdot 10^{30}$.

[15]For the sake of clarity, we leave off indexing: $|0\rangle \, |0\rangle \, |0\rangle \equiv |000\rangle \equiv |0_1 0_2 0_3\rangle$, and so on.

$$|z\rangle = \frac{|0\rangle + |1\rangle}{\sqrt{2}} \otimes \frac{|0\rangle + |1\rangle}{\sqrt{2}} \otimes \frac{|0\rangle + |1\rangle}{\sqrt{2}}. \tag{26.17}$$

Performing the multiplications yields:

$$|z\rangle = \frac{|000\rangle + |001\rangle + |010\rangle + |011\rangle + |100\rangle + |101\rangle + |110\rangle + |111\rangle}{2\sqrt{2}}. \tag{26.18}$$

This reads in decimal notation:

$$|z\rangle = \frac{|0\rangle + |1\rangle + |2\rangle + |3\rangle + |4\rangle + |5\rangle + |6\rangle + |7\rangle}{2\sqrt{2}} = \frac{1}{2\sqrt{2}} \sum_{k=0}^{7} |k\rangle. \tag{26.19}$$

The three notations (26.17)–(26.19) all denote exactly the same facts. In general, we have for a product of $n$ qubits

$$|z\rangle = \frac{|0\rangle + |1\rangle}{\sqrt{2}} \otimes \frac{|0\rangle + |1\rangle}{\sqrt{2}} \otimes \cdots \otimes \frac{|0\rangle + |1\rangle}{\sqrt{2}} = \frac{1}{2^{n/2}} \sum_{k=0}^{2^n-1} |k\rangle. \tag{26.20}$$

If it is not clear whether a ket is meant in decimal or binary notation, then $|0\rangle$ and $|1\rangle$ may be ambiguous, since it is not clear from the outset whether they are states of a single qubit or a register. This must be determined from context.

### 26.4.2   Quantum Gates and Quantum Computers

The manipulation of the registers is achieved by *gates*. A quantum gate is a device that acts on certain qubits of a register by means of a specific unitary operation. This means in the context of the following considerations that we can equate a gate with a unitary transformation. A *quantum network* or *quantum circuit* consists of gates which are interconnected in a certain way and which act in a specified time sequence.[16] The gates are connected by *quantum wires*, i.e. by ideal, loss-free and error-free connecting links.

A *quantum computer* is a quantum circuit which changes an input state according to a quantum algorithm[17] and yields the result as output or final state. It is important that the computation be reversible, as it involves only unitary transformations (=gates). The final state is measured as usual (by a projective measurement).

---

[16]The size of the network corresponds to the number of gates.

[17]An algorithm is a procedure for solving a problem (in finitely many steps).

**Fig. 26.2** Symbolic representation of the Hadamard gate and two different symbolic representations of the phase shift gate

The charm of the quantum gates is, *inter alia*, that only *three* of them are needed to perform all sorts of computational operations, namely two 1-qubit gates and a 2-qubit gate.[18]

### 26.4.2.1  1-Qubit Gates

The first 1-qubit gate that we will consider is the *Hadamard gate* or the Hadamard transformation (see also Appendix P, Vol. 2)[19]:

$$H \cong \frac{1}{\sqrt{2}} \begin{pmatrix} 1 & 1 \\ 1 & -1 \end{pmatrix}. \tag{26.21}$$

We obtain (see the exercises):

$$|q\rangle \to H\,|q\rangle = \frac{|1 - q\rangle + (-1)^q\,|q\rangle}{\sqrt{2}}; \quad q \in \{0, 1\}. \tag{26.22}$$

The second 1-qubit gate is the *phase shift gate* (phase shift, phase gate)[20]:

$$\Phi_\varphi \cong \begin{pmatrix} 1 & 0 \\ 0 & e^{i\varphi} \end{pmatrix}, \tag{26.23}$$

or

$$|q\rangle \to \Phi_\varphi\,|q\rangle = e^{iq\varphi}\,|q\rangle; \quad q \in \{0, 1\}. \tag{26.24}$$

These two components, schematically shown in Fig. 26.2, can be combined e.g. in such a way that they transform the state $|0\rangle$ into the general state of one bit (in other words, with these two gates we can construct any unitary 1-qubit operation). We have for example (see the exercises):

---

[18] We note that for this purpose there are other equivalent ways to choose three different 1 and 2-qubit gates. In addition, one can represent all the operations with *Toffoli gates*, which are 3-qubit gates.

[19] $H$ here always means the Hadamard transformation. Since the Hamiltonian does not occur in this chapter, there is no risk of confusion.

[20] Also called 'rotation'.

**Fig. 26.3** Symbolic
representation of the
transformation (26.25)

$$|0\rangle \rightarrow \Phi_\varphi H \Phi_\vartheta H |0\rangle = e^{i\vartheta/2}\left(\cos\frac{\vartheta}{2}|0\rangle - ie^{i\varphi}\sin\frac{\vartheta}{2}|1\rangle\right). \tag{26.25}$$

Figure 26.3 shows the symbolic representation of the transformation (26.25).

#### 26.4.2.2   2-Qubit Gate

Now one cannot perform all necessary operations with 1-bit gates. On the one hand, this is due to the fact that one wants to generate entangled states, and this requires at least two bits. The other reason is that certain classical operations are not reversible and therefore cannot be directly translated in terms of quantum information,[21] but this is achieved by using gates that process two or more bits.

We therefore introduce *two* qubits, one ($|p\rangle$, control) of length $n$ and another ($|q\rangle$, target) of length $m$. The control bit remains unchanged, but the target bit is changed by the unitary transformation, in a way which is determined by the control bit[22]:

$$\begin{bmatrix} |p\rangle \\ |q\rangle \end{bmatrix} \rightarrow \begin{bmatrix} |p\rangle \\ |q \oplus f(p)\rangle \end{bmatrix}, \tag{26.26}$$

where $\oplus$ in quantum information always denotes the addition modulo $2^n$ and not (as in vector spaces) the direct sum (see Appendix C, Vol. 2).[23]

A simple and important example is the *controlled NOT gate* (CNOT gate, controlled not, CNOT, also called measuring gate) with $n = m = 1$, $|p, q\rangle \rightarrow |p, q \oplus p\rangle$. The symbolic representation is given in Fig. 26.4. Here, the second qubit $|q\rangle$ is changed when the first qubit $|p\rangle$ is in the state $|1\rangle$, otherwise nothing happens; in detail, this reads: $|00\rangle \rightarrow |00\rangle$, $|01\rangle \rightarrow |01\rangle$, $|10\rangle \rightarrow |11\rangle$, $|11\rangle \rightarrow |10\rangle$. We see in the target bit the (not uniquely reversible) XOR structure, but together with the control bit, a uniquely reversible unitary transformation results; it reads in matrix form:

---

[21] As an example, we consider the mod2 sum (=exclusive OR = XOR) $p \oplus q$ with $p, q \in \{0, 1\}$. Obviously this is not a reversible mapping, since we have $0 \oplus 0 = 1 \oplus 1$ and $0 \oplus 1 = 1 \oplus 0$. Similarly, the traditional gates AND and OR are not unitary and therefore are not directly eligible for quantum applications.

[22] This gate is also called a controlled-U gate.

[23] When using the notation $a \oplus b$, the information about $n$ has to come from somewhere else.

**Fig. 26.4** Symbolic representation of the CNOT gate

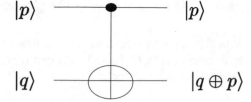

$$C \cong \begin{pmatrix} 1 & 0 & 0 & 0 \\ 0 & 1 & 0 & 0 \\ 0 & 0 & 0 & 1 \\ 0 & 0 & 1 & 0 \end{pmatrix}. \tag{26.27}$$

If the target bit is in the state $|0\rangle$, we have

$$|p\rangle |0\rangle \to |p\rangle |p\rangle ; \quad p \in \{0, 1\}. \tag{26.28}$$

This looks at first glance like a copier—but we know that copiers do not exist. In fact, the CNOT gate works as a copier only for $|0\rangle |0\rangle$ and $|1\rangle |0\rangle$.[24] Arbitrary states are not copied, but are entangled. To see this, we assume that the control bit is a superposition:

$$|p\rangle = a |0\rangle + b |1\rangle ; \quad a, b \neq 0. \tag{26.29}$$

It follows that:

$$|p\rangle |0\rangle = (a |0\rangle + b |1\rangle) |0\rangle \to a |0\rangle |0\rangle + b |1\rangle |1\rangle , \tag{26.30}$$

and that is not a copy of the state $|p\rangle$, but rather an entangled state.

Another important operation is the *kickback*, with $m = 1$, while $n$ is not specified. The otherwise arbitrary function $f$ can take on the two values 0 and 1. For $|q\rangle$, we choose the superposition $|q\rangle = \frac{|0\rangle - |1\rangle}{\sqrt{2}}$. Then we have, because of $|0 \oplus f(p)\rangle = |f(p)\rangle$:

$$|p\rangle \frac{|0\rangle - |1\rangle}{\sqrt{2}} \to |p\rangle \frac{|f(p)\rangle - |1 \oplus f(p)\rangle}{\sqrt{2}} = \begin{cases} |p\rangle \frac{|0\rangle - |1\rangle}{\sqrt{2}} \\ |p\rangle \frac{|1\rangle - |0\rangle}{\sqrt{2}} \end{cases} \quad \text{for } f(p) = \begin{cases} 0 \\ 1 \end{cases} \tag{26.31}$$

or briefly

$$|p\rangle \frac{|0\rangle - |1\rangle}{\sqrt{2}} \to |p\rangle (-1)^{f(p)} \frac{|0\rangle - |1\rangle}{\sqrt{2}}. \tag{26.32}$$

The modular addition in the second register leaves it unchanged, apart from the change of sign controlled by $p$. As the overall result, we have $|p\rangle \to (-1)^{f(p)} |p\rangle$.

---

[24]Copying two (orthogonal) states is indeed allowed; see the section 'quantum copier'.

### 26.4.3   The Basic Idea of the Quantum Computer

Now that we have discussed some fundamental functions, we briefly describe the basic idea of the QC. There is an input register of $N$ qubits, which are stored in a special state $|\psi\rangle$, namely in

$$|\psi\rangle = \frac{1}{\sqrt{2^N}} \sum_{n=1}^{N} |\varphi_n\rangle .  \qquad (26.33)$$

The states $|\varphi_n\rangle$ are the state of the product basis; for the case $N = 3$, they are given in (26.15). This input state is now changed into a final state by a controlled sequence of unitary (i.e. reversible) transformations such as $H$.

We describe this by constructing the tensor product $|\Psi\rangle$ of $|\psi\rangle$ with the state $|\chi\rangle$ of an output register of $2^M$ qubits:

$$|\Psi\rangle = |\psi \otimes \chi\rangle = \frac{1}{\sqrt{2^N}} \sum_{n} |\varphi_n \otimes \chi\rangle .  \qquad (26.34)$$

The QC is thus essentially a unitary (total) operator $U$ which transforms the system into the entangled state

$$|\Psi\rangle \rightarrow |\Psi'\rangle = U\,|\Psi\rangle = \frac{1}{\sqrt{2^N}} \sum_{n} |\varphi_n \otimes f\,(\varphi_n)\rangle .  \qquad (26.35)$$

Both registers together now simultaneously contain $2^{N+M}$ values of the pair $(\varphi, f(\varphi))$.

The result of computations is read out by an (irreversible) measurement process. We thus have available significant virtual information, but cannot obtain it directly from the system, since the measurement returns only *one* couple $|\varphi_k \otimes f\,(\varphi_k)\rangle$.

However, it is possible to get more information from the state (26.35). There are several methods of doing this; we will consider the two simplest in more detail in the following, namely the algorithms of Deutsch and of Grover. After that, we give a brief comment on the Shor algorithm; more details on this topic are found in Appendix S, Vol. 2.

### 26.4.4   The Deutsch Algorithm

This section has the purpose of illustrating the principle of a quantum computation by means of a toy example. It is a black box (also called *oracle*). We know only that it calculates a (Boolean) function $f:\{0, 1\} \rightarrow \{0, 1, \}$, but not which of the four possibilities:

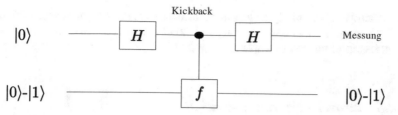

**Fig. 26.5** Symbolic representation of the Deutsch algorithm

$$\begin{array}{ll}
(1)\ f\,(0) = 1; & f(1) = 0 \\
(2)\ f\,(0) = 0; & f(1) = 1 \\
(3)\ f\,(0) = 1; & f(1) = 1 \\
(4)\ f\,(0) = 0; & f(1) = 0
\end{array} \tag{26.36}$$

is actually realized. One would like to know if it is one of the last two possibilities or not—in other words, if $f(0)$ and $f(1)$ are different or not.

Without quantum mechanics, we simply have to measure $f(0)$ and $f(1)$ to find the answer. Obviously, we need two measurement procedures. With quantum mechanics, we need only *one* measuring process.

For this purpose, we use the experimental setup shown in Fig. 26.5. The salient point is the use of the kickback transformation (26.32). We start with the state $|0\rangle\,\frac{|0\rangle-|1\rangle}{\sqrt{2}}$ and transform as follows:

$$|0\rangle\,\frac{|0\rangle - |1\rangle}{\sqrt{2}} \underset{\text{Hadamard}}{\rightarrow} \frac{|0\rangle + |1\rangle}{\sqrt{2}}\,\frac{|0\rangle - |1\rangle}{\sqrt{2}}$$

$$\underset{\text{kickback}}{\rightarrow} \frac{(-1)^{f(0)}\,|0\rangle + (-1)^{f(1)}\,|1\rangle}{\sqrt{2}}\,\frac{|0\rangle - |1\rangle}{\sqrt{2}}. \tag{26.37}$$

The second bit has done its duty in the transfer of $f$ and we omit it now. It follows for the second Hadamard gate that:

$$\frac{(-1)^{f(0)}\,|0\rangle + (-1)^{f(1)}\,|1\rangle}{\sqrt{2}}$$

$$\underset{\text{Hadamard}}{\rightarrow} \frac{\left[(-1)^{f(0)} + (-1)^{f(1)}\right]|0\rangle + \left[(-1)^{f(0)} - (-1)^{f(1)}\right]|1\rangle}{2}. \tag{26.38}$$

It is immediately obvious that for $f(0) = f(1)$, we measure $|0\rangle$; otherwise, $|1\rangle$. Thus, quantum mechanics answers the question (i.e. $f(0) = f(1)$—yes or no?) with *one* measurement, while the classical approach requires *two*.

This example is an improved version (1998) of the first quantum algorithm which was presented by David Deutsch in 1985. It shows that quantum calculations can run much faster than classical ones, provided one asks the right questions. In this toy example, we gain only a factor of 2. For other algorithms, however, the calculating time depends polynomially on a system variable (e.g. the size of the system) for the quantum computer and exponentially for the classical computer, as is the case for the extension of this toy example to $n$ inputs $|0\rangle$.

## 26.4.5   Grover's Search Algorithm

This algorithm, proposed by Lov K. Grover in 1996, performs a search in an unstructured data base. The problem can be thought of as the inverse phone book problem, i.e. looking for a name if only the phone number is known (where the phone book with $N$ entries is arranged alphabetically, of course). Classically, one has to check each entry one by one; to find the right name, one has to make $\frac{N}{2}$ attempts on the average. In contrast, the Grover algorithm needs $\sim \sqrt{N}$ steps.

We assume that each number appears only once and that there are $2^n$ entries. We can describe the phone book with respect to our search by a function $f(k)$ that vanishes for all arguments except the desired one ($\kappa$):

$$f(k) = \delta_{k\kappa}; \quad k = 0, 1, \ldots, N - 1; \quad N = 2^n; \quad 0 \le \kappa \le N - 1. \tag{26.39}$$

$f(k)$ is again a black box (oracle). The position $\kappa$ is not known and has to be identified.

We use two registers and the kickback discussed above. The first register is of length $N$, the second is in the state $\frac{|0\rangle - |1\rangle}{\sqrt{2}}$. Since this is also the output state of the second register for the kickback, we consider in the following only the first register. Here, it holds that:

$$|k\rangle \to (-1)^{f(k)} |k\rangle, \tag{26.40}$$

where $\{|k\rangle\}$ is a CONS of dimension $N$. Because of (26.39), this means that all the states remain unchanged, except for the desired state $|\kappa\rangle$, where we have $|\kappa\rangle \to -|\kappa\rangle$. Due to this, the transformation can be written as (see the exercises):

$$U_\kappa = 1 - 2 |\kappa\rangle \langle \kappa|. \tag{26.41}$$

Note that the last equation is just a different notation for the kickback. It does not mean that we know the value of $\kappa$ at this point.

The initial state for the algorithm is a normalized equally-weighted superposition of all states

$$|s\rangle = \frac{1}{\sqrt{N}} \sum_{k=0}^{N-1} |k\rangle; \quad \langle s | s \rangle = 1. \tag{26.42}$$

Using this state, we define an operator

$$U_s = 2 |s\rangle \langle s| - 1. \tag{26.43}$$

The Grover algorithm consists of a (repeated) application of $U_s U_\kappa$ to the initial state. We want to interpret this algorithm in the following in a geometrical way; the algebraic point of view can be found in Appendix R, Vol. 2.

The geometrical analogy is based on the fact that the two vectors $|\kappa\rangle$ and $|s\rangle$ define a plane. The two vectors are normalized, but due to

$$\langle \kappa \,|s\rangle = \frac{1}{\sqrt{N}} \tag{26.44}$$

they are not orthogonal (although 'almost' orthogonal for large $N$). We denote the vector lying in this plane which is orthogonal to $|\kappa\rangle$ as $|k'\rangle$. The angle between $|s\rangle$ and $|k'\rangle$ is $\varphi$. Because of $\langle \kappa \,|s\rangle = \frac{1}{\sqrt{N}} = \cos\left(\frac{\pi}{2} - \varphi\right) = \sin\varphi$, we have:

$$\varphi = \arcsin\frac{1}{\sqrt{N}}. \tag{26.45}$$

The two operators $U_\kappa$ and $U_s$ have a simple geometrical interpretation—they describe reflections (see the exercises). $U_\kappa$ leaves the components of $|s\rangle$ which are orthogonal to $|\kappa\rangle$ unchanged, and reverses the sign of the $|\kappa\rangle$-component of $|s\rangle$. In the end, it is therefore a reflection around $|k'\rangle$. Analogously, the operator $U_s = 2\,|s\rangle\,\langle s|-1$ produces a reflection in $|s\rangle$. The transformed vectors remain in the plane spanned by $|\kappa\rangle$ and $|s\rangle$.

If we apply $U_\kappa$ to $|s\rangle$, we get a vector $|s'\rangle$ reflected at $|k'\rangle$, and this vector is transformed into $|s''\rangle$ by $U_s$. In sum we have the result that $U_s U_\kappa$ turns the vector $|s\rangle$ by the angle $2\varphi$ into the direction of $|\kappa\rangle$; see Fig. 26.6. By repeated application of $U_s U_\kappa$, we can rotate the initial state $|s\rangle$ closer and closer to $|\kappa\rangle$. This means that with each step, the relative amplitude of the $|\kappa\rangle$ component of $|s\rangle$ increases, and after a suitable number of steps a measurement we will almost certainly obtain the desired state $|\kappa\rangle$.

**Fig. 26.6** Geometrical interpretation of the Grover algorithm

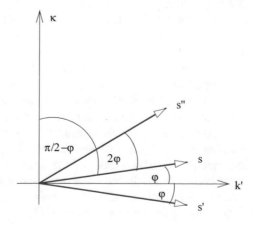

This behavior is called *amplitude amplification*. A superposition of states is thus changed by this unitary transformation in such a way that the amplitude of the state sought is particularly large. This state is then measured with a particularly high probability, especially if the amplification process is applied several times or iteratively; while wrong answers will cancel out. By the way, amplitude amplification is used also in Shor's algorithm.

The appropriate number of steps is given by the consideration that $\frac{\pi}{2} - \varphi$ must be in the vicinity of an integer multiple of $2\varphi$, i.e.

$$\frac{\pi}{2} - \varphi \approx 2\varphi \cdot m \ \text{ or } \ m \approx \frac{\pi}{4\varphi} - \frac{1}{2}. \tag{26.46}$$

With (26.45), we have for sufficiently large $N$:

$$m \approx \frac{\pi}{4}\sqrt{N}. \tag{26.47}$$

This also follows from the algebraic considerations (see Appendix R, Vol. 2), which lead to the formulation

$$\langle \kappa | \, (U_s U_\kappa)^m \, |s\rangle = \sin (2m + 1)\,\varphi. \tag{26.48}$$

Thus we need $O(\sqrt{N})$ steps to answer (almost with certainty) the initial question. With this small number of steps (classical methods require $O(N)$ steps), the Grover algorithm is optimal in the sense that there is no algorithm that requires fewer than $O(\sqrt{N})$ steps.

### 26.4.6   Shor's Algorithm

This algorithm, proposed by Peter W. Shor in 1994, can be used for the factorization of very large numbers into prime factors. The problem is that while it is trivial to multiply two numbers, no matter how large, it is very time-consuming and far from trivial to find the prime factors of a given very large number. It is quickly computed that $179424673 \cdot 373587883$ gives the number $N = 67030883744037259$; but is is very hard to find the factors of a given $N$ (try it yourself with $268898680104636581$ or $170699960169639253$, remembering that these numbers are small in this context). This fact is used for encryption (RSA algorithm). The basic idea is essentially that only Alice (sender) and Bob (receiver) know the prime factorization of a very large number $N = N_1 \cdot N_2$, whereas $N$ may be known to the public. The security of the system depends crucially on the fact that the factorization of $N$ takes so much time even with the fastest computers that it is a de facto insolvable problem (the computation times would be of the order of thousands or millions of years or more, depending on the number of digits of $N$).

The Shor algorithm is now a way to perform exactly this factorization relatively quickly. The algorithm is divided into a classical and a quantum-mechanical part. Similar to Grover's algorithm, it works on the basis of amplitude amplification. One possibility is to compute a Fourier transform of the state $|\Psi'\rangle$ as in

$$|\Psi\rangle \rightarrow |\Psi'\rangle = U|\Psi\rangle = \frac{1}{\sqrt{2^N}} \sum_n |\varphi_n \otimes f(\varphi_n)\rangle. \tag{26.49}$$

From the spectrum, one can infer the period of $f(\varphi_n)$, and from this period finally the prime factors. An algorithm running on a classical computer needs $O(N)$ steps to determine the period, while this figure is $O(\ln^3 N)$ for a quantum computer. It is this tremendous reduction in computation time which makes the Shor algorithm such a valuable tool. Since the detailed calculation is quite lengthy, we relegate it to Appendix S, Vol. 2.

### 26.4.7  On The Construction of Real Quantum Computers

As said above, quantum computing is a very active research area with a wide variety of issues, both in hardware and in software. The attempt to give an overview of the current state of affairs must be incomplete and will be quickly obsolete. Thus, a few remarks will be enough.

In the last years, there were quite often reports on quantum computers operating with different numbers of qubits. But apparently, most of them were special machines with very limited possibilities, designed for particular problems, and not every researcher was convinced that these devices kept their promises.

Note that there are marked and big differences between today's QC's and 'normal' PC's. For example, current QC's can usually not be programmed and reprogrammed, as is the case with PC's. Indeed, the first reprogrammable QC (operating with 5 qubits) was announced in 2016.[25] In addition, today's QC's are not cute devices to be comfortably placed on your lap. Most of them are big machines and have be cooled down elaborately to quite low temperatures (a few mK). In 2012 a QC was announced which was functional at room temperature, but it was operating with just 2 qubits, see P.C. Maurer et al., Room-Temperature Quantum Bit Memory Exceeding One Second, *Science* 336, 1283–1286, DOI: 10.1126/science.1220513 (08 june 2012).

The progress in the development of quantum computers is not as fast as hoped, although worldwide not only academic institutions, but also big players of the IT-military-industrial complex are engaged in research, as Google, Microsoft, IBM, NASA, to name a few. Despite all the efforts of the last years, the 'general-purpose QC for everybody' seems to be still far away, and the question remains whether it will be possible one day to produce 'real' quantum computers, perhaps even in

---

[25]S. Debnath et al., 'Demonstration of a small programmable quantum computer with atomic qubits'; *Nature* 536, 63–66 (04 August 2016); doi:10.1038/nature18648.

mass production. Opinions are divided on this issue, ranging from very pessimistic to very optimistic. The main problem is decoherence. That is, the quantum parallelism requires a unitary evolution, and this implies that uncontrollable interactions with the environment must be eliminated.

The experimental difficulties are obvious. A large number of quantum gates has to be 'wired' to each other and must be insulated from the environment as well as possible. Of course, such interactions cannot be eliminated completely. The task is therefore to minimize the perturbations induced by the environment and to offset the unavoidable errors by suitable correction algorithms. Thus, one needs good ideas on how to correct errors during a calculation, and how to restore superposition states.

It seems that quantum computers will not replace the classical computer by any means - and they will probably look very different. Which 'hardware' will ultimately prevail is not yet clear at the moment. Candidates for qubits presently include, among others, ion traps, molecular nuclear spins, entangled atoms, quantum dots in semiconductors, or spins of single atoms embedded in a semiconductor. More recent developments are photon–photon quantum gates, gates between a flying optical photon and a single trapped atom, and silicon quantum gates.[26]

Also on the theoretical side there are as yet unanswered questions. For example: Is there a general class of tasks which a quantum computer can solve better than a classical computer—or is it just a question of individual cases (such as the Shor algorithm), which have been found, at least so far, more or less by chance?

Concerning practical applications, there are a few calculations of difficult physical problems using QC's, for instance the first high-energy physics simulation on a quantum computer (2015, creation of pairs of particles and antiparticles) or real-time dynamics of lattice gauge theories.[27]

A further field of activity is the simulation of QC's (i.e., quantum simulators) with the help of supercomputers. The aim is to develop and to test algorithms suitable for 'real' QC's. In 2017, the Jülich Supercomputing Centre has announced a new world record by simulating a QC with 46 qubits.[28]

For those who want to gain experience with quantum devices, there are some interactive home pages. Among others, IBM provides a page where one can do own calculations, for instance write and run quantum algorithms on a real QC.[29]

---

[26]See e.g. D. M. Zajac et al., 'Quantum CNOT Gate for Spins in Silicon', *Science* 07 December 2017. DOI: 10.1126/science.aap5965.

[27]E.A. Martinez et al., 'Real-time dynamics of lattice gauge theories with a few-qubit quantum computer', *Nature* 534, 516–519 (23 June 2016), doi:10.1038/nature18318.

[28]http://www.fz-juelich.de/SharedDocs/Pressemitteilungen/UK/EN/2017/2017-12-15-world-record-juelich-researchers-simulate-quantum-computer.html?nn=897918.

[29]https://www.research.ibm.com/ibm-q/ as of december 2017.

## 26.5 Exercises

1. Above, it was proposed that you yourself try to find the prime factorization of 268898680104636581 and 170699960169639253. Did you find it?
2. Pauli matrices and qubits:

   (a) How do the Pauli matrices act on the qubit states $|0\rangle$ and $|1\rangle$?
   (b) How do the Pauli matrices act on the qubit state $|\varphi\rangle = c\,|0\rangle + d\,|1\rangle$?

3. Calculate the full expression containing $N$ terms:

$$|z\rangle = \frac{|0\rangle - |1\rangle}{\sqrt{2}} \otimes \frac{|0\rangle - |1\rangle}{\sqrt{2}} \otimes \cdots \otimes \frac{|0\rangle - |1\rangle}{\sqrt{2}}. \tag{26.50}$$

4. Show that:

$$|q\rangle \to H\,|q\rangle = \frac{|1 - q\rangle + (-1)^q\,|q\rangle}{\sqrt{2}}; \quad q \in \{0, 1\}, \tag{26.51}$$

   where $H$ is the Hadamard matrix.
5. Calculate explicitly

$$\Phi_\varphi H \Phi_\vartheta H \tag{26.52}$$

   where $H$ is the Hadamard transformation and $\Phi$ the phase shifter.
6. Kickback and Grover's algorithm: Given that:

$$f(k) = \delta_{k\kappa}; \quad k = 0, 1, \ldots, d - 1; \quad d = 2^n; \quad 0 \le \kappa \le d - 1. \tag{26.53}$$

   The effect of the kickback may be written as:

$$|k\rangle \to (-1)^{f(k)}\,|k\rangle \quad \text{or} \quad U_\kappa\,|k\rangle = (-1)^{f(k)}\,|k\rangle \tag{26.54}$$

   where $\{|k\rangle\}$ is a CONS. Show that

$$U_\kappa = 1 - 2\,|\kappa\rangle\,\langle\kappa|. \tag{26.55}$$

7. Given the normalized states $|x\rangle$ and $|y\rangle$, with $\langle x\,|y\rangle = 0$; show that the operator $U = 2\,|x\rangle\,\langle x| - 1$ describes a reflection at $|x\rangle$ and $-U$ a reflection at $|y\rangle$.
8. Given the normalized state

$$|\psi\rangle = \sum_{n=1}^{N} c_n\,|\varphi_n\rangle \quad \text{with} \quad \langle\varphi_n\,|\varphi_m\rangle = \delta_{nm}. \tag{26.56}$$

   The probability of measuring the state $|\varphi_k\rangle$ is thus given by $|c_k|^2$. We selectively amplify the amplitude $c_m \ne 0$ by the following unitary transformation $U$ (see Fig. 26.7):

**Fig. 26.7** Effect of
$-U = 1 - 2 |x\rangle \langle x|$ on a
general state

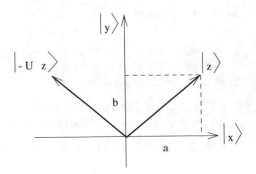

$$U : c_n \to \alpha c_n \text{ for } n \neq m; \quad c_m \to \beta c_m \text{ for } n = m \qquad (26.57)$$

with suitably chosen $\alpha, \beta$.

(a) How are $\alpha$ and $\beta$ connected?
(b) How do the measurement probabilities behave under a $k$-fold iteration of $U$?
(c) Specialize to the case of an initially uniform distribution $c_n = \frac{1}{\sqrt{N}}$ and $\alpha = \frac{1}{4}$.
    How often does one have to iterate in order to measure the state $m$ with a
    probability of $w > 1 - 10^{-6}$ (assuming $N \gg 1$)?

# Chapter 27
# Is Quantum Mechanics Complete?

We delve further into the question of whether quantum mechanics is complete. As the Kochen–Specker theorem and the GHZ states show, a realistic representation of quantum mechanics is compatible neither with non-contextuality nor with locality.

In the context of the EPR paradox and Bell's inequality (see Chap. 20), we came across the question of whether quantum mechanics as a physical theory is complete.

If we assume that it is, and trust the formalism of quantum mechanics developed thus far, then we must also accept e.g. that objective chance exists, and that properties are not necessarily fixed a priori, but are only 'created' by a measurement.[1] This contradicts classical physics, where properties are presumed to exist independently of measurements (pre-existence), and where measurement does not mean the creation of a property, but rather the reduction of our ignorance about this property.

In contrast, if we do not accept this contradiction, and therefore (or for other reasons) take the view that quantum mechanics is *in*complete, we have to introduce the additional variables that make it a complete theory. These postulated other quantities are usually called *hidden variables* (HV). They would ensure that all aspects of probability can be removed from quantum mechanics (at least in principle), and that its predictions generally are deterministic. Hence, one speaks also of *cryptodeterminism* and the sub-quantum world.

So if we want to postulate that a realistic view of the world as is commonly accepted in classical physics also applies to quantum mechanics, then this means that (1) quantum states refer to individual systems, not just to the results of repeated

---

[1]In any case, this is apparently the prevailing view, e.g.: "Values cannot be ascribed to observables prior to measurement; such values are only the outcomes of measurement." K. Gottfried and T.-M. Yan, *Quantum Mechanics: Fundamentals*, 2nd Edition, 2003, p. 42.

© Springer Nature Switzerland AG 2018
J. Pade, *Quantum Mechanics for Pedestrians 2*, Undergraduate Lecture
Notes in Physics, https://doi.org/10.1007/978-3-030-00467-5_27

measurements,[2] and in particular that (2) a measurement determines the value of a physical quantity which that quantity had immediately before and independently of the measurement.

For example, if a circularly-polarized photon impinges on an analyzer sensitive to linear polarization, we obtain with probability 1/2 either a horizontally or vertically linear-polarized photon; more cannot be said according to the rules of quantum mechanics. If we now demand that it must be certain *before* the measurement whether the photon will be polarized vertically or horizontally, we cannot avoid introducing additional variables which contain this information.

The key question is the following[3]: Let us suppose that a quantum system is in the state $|\psi\rangle$. Does then every observable $A$ have an objectively pre-existing value $A(\psi, \lambda_1, \lambda_2, \ldots)$, determined by $|\psi\rangle$ and a set of hidden variables $(\lambda_1, \lambda_2, \ldots)$? If we could answer this question with 'yes', then the values of all observables would be elements of physical reality; we would have a realistic theory.

But our considerations based on Bell's inequality have shown that naive realism collides with quantum mechanics. As we have seen, it is an experimentally testable and confirmed statement that realism and locality are not at the same time compatible with quantum mechanics.

Before we take up this issue again, we will examine another combination of conditions: Is it compatible with quantum mechanics that all properties of a quantum system (a) are defined at all times (value-definiteness) and in addition (b) do not depend on the context of the measurement (non-contextuality)? The answer is 'no', as we will see in the following on the basis of the Kochen–Specker theorem.[4]

Finally a point of clarification: The fact that we take up the question of hidden variables again and in a wider context here does not mean that we aim to introduce them (through the back door, so to speak) at the end of this book. In fact, according to present knowledge, hidden variables are fighting rather a losing battle. The question of interest is instead: Why does the introduction of hidden variables fail? In trying to answer this question, we can learn more about the way quantum mechanics 'works'.

## 27.1  The Kochen–Specker Theorem

It is a seemingly innocuous assumption that everything that exists in the physical world is 'really there' and, furthermore, exists independently of our measurements. We substantiate this idea in two terms or conditions:

---

[2]However, also in approaches based on objective chance or tending towards the many-worlds interpretation, one assumes that states refer to individual systems and not merely to ensembles. The crucial element for the following considerations is requirement (2).

[3]As a Hamlet-style question, so to speak: 'To be' or 'to be found'?

[4]Contextuality states that the measurement outcome of an observable depends on the set of compatible observables that are measured at the same time. Thus, nonlocality can be considered as a reflection of contextuality in spatially separated systems.

1. *Value-definiteness* (VD, also called pre-assigned initial values): All properties of a quantum system are defined at all times, even when the system is for example in a superposition state.
2. *Non-contextuality* (NC): The properties of a quantum system do not depend on which quantities are measured in an experiment. They are thus independent of the measurement context, i.e. they are non-contextual.

The Kochen–Specker theorem (KST) shows that in quantum mechanics, these two demands for value definiteness and non-contextuality cannot be fulfilled simultaneously. It follows that there cannot exist realistic non-contextual models with hidden variables in quantum mechanics.

Basically, this theorem from 1967 is a mathematical result about the nature of Hilbert spaces; it can be reduced to the purely geometric problem that it is not possible to color the surface of a three-dimensional sphere in a certain way.[5]

### 27.1.1 Value Function

In order to quantify the concepts, we introduce a value function $V_{|\psi\rangle}(A)$. It denotes the value of the physical quantity $A$ when the system is in state $|\psi\rangle$. If $|\psi\rangle$ is an eigenvector of $A$ with eigenvalue $a_n$, then we can assume $V_{|\psi\rangle}(A) = a_n$. But if $|\psi\rangle$ is not an eigenvector, we need to ask for additional properties of $V$ in order to arrive at a reasonable statement. It seems natural and intuitively obvious to require

$$V_{|\psi\rangle}(F(A)) = F\left(V_{|\psi\rangle}(A)\right) \tag{27.1}$$

Therefore, for each condition of a quantum system, the value function of the function of a physical quantity equals the function of the value function—or in brief, the value of a function equals the function of the value. A simple example: The value of $L_x$ is $m\hbar$; then the value of $L_x^2$ equals the square of the value of $L_x$, i.e., it equals $(m\hbar)^2$ and not $(m\hbar)^{3/2}$ or the like.

Requiring (27.1) has the consequence that if $[A, B] = 0$, it holds that:

$$V_{|\psi\rangle}(A + B) = V_{|\psi\rangle}(A) + V_{|\psi\rangle}(B) : \text{sum rule}$$
$$V_{|\psi\rangle}(A \cdot B) = V_{|\psi\rangle}(A) \cdot V_{|\psi\rangle}(B) : \text{product rule,} \tag{27.2}$$

and that in addition, $V_{|\psi\rangle}(1) = 1$. For the proofs see the exercises. Note that $[A, B] = 0$ is a precondition for (27.2). With noncommuting observables, there is generally no consistent way to assign values. As an example, we consider $A = \sigma_x$, $B = \sigma_y$; the assigned value is an eigenvalue of each operator. The eigenvalues

---

[5]In this context, Gleasons's theorem is of interest (see Appendix T, Vol. 2). It deals in fact with the question of how to introduce probabilities into quantum mechanics, but also refers to a contradiction in the assignment of properties of a quantum system. This contradiction is addressed by the Kochen–Specker theorem.

of $\sigma_x$ and $\sigma_y$ are $\pm 1$, but those of $\sigma_x + \sigma_y$ are $\pm\sqrt{2}$, so that the requirement $V_{|\psi\rangle}(A + B) = V_{|\psi\rangle}(A) + V_{|\psi\rangle}(B)$ cannot be fulfilled.

With the value function, the above assumptions can now be rewritten as follows:

1. VD: Each set of physical properties which is represented by corresponding operators $A$, $B$, $C$, $\ldots$ in $\mathcal{H}$, has well-defined values $V_{|\psi\rangle}(A)$, $V_{|\psi\rangle}(B)$, $V_{|\psi\rangle}(C)$, $\ldots$
2. NC: For commuting operators $A$, $B$, the rules $V_{|\psi\rangle}(A + B) = V_{|\psi\rangle}(A) + V_{|\psi\rangle}(B)$ and $V_{|\psi\rangle}(AB) = V_{|\psi\rangle}(A) V_{|\psi\rangle}(B)$ apply.

As we have already discussed in Chap. 13, Vol. 1, the question of whether a quantum system has a property or not may be conveniently formulated by means of projection operators. We briefly review: The basis of the $N$-dimensional Hilbert space is the CONS $\{|a_n\rangle, n = 1, 2, \ldots\}$. Because of the completeness of this basis, the projection operators $P_{|a_n\rangle} = |a_n\rangle \langle a_n|$ satisfy

$$\sum_n P_{|a_n\rangle} = 1. \tag{27.3}$$

The projection operators act on the basis states according to

$$P_{|a_n\rangle} |a_m\rangle = |a_n\rangle \langle a_n| a_m\rangle = \delta_{nm} \cdot |a_m\rangle. \tag{27.4}$$

Hence, $|a_n\rangle$ is an eigenvector of $P_{|a_n\rangle}$ with eigenvalue 1; all other vectors $|a_m\rangle$ with $n \neq m$ are eigenvectors of $P_{|a_n\rangle}$ with eigenvalues 0.

The CONS $\{|a_n\rangle\}$ can be seen as the eigenvectors of an operator $A = \sum_n a_n |a_n\rangle \langle a_n| = \sum_n a_n P_{|a_n\rangle}$ acting in $\mathcal{H}$ (spectral representation), where the eigenvalue equation is $A |a_n\rangle = a_n |a_n\rangle$. Thus, we can understand the spectral operators (projection operators) as a representation of yes-no observables, i.e. as a response to the question as to whether a quantum-mechanical system has a property $a_n$ (1, yes) or not (0, no) with respect to the physical quantity $A$.

### 27.1.2  From the Value Function to Coloring

We now take advantage of this connection between projection operators $P$ and properties. With the product rule and because of $P^2 = P$, it generally follows that $V_{|\psi\rangle}(P^2) = V_{|\psi\rangle}^2(P) = V_{|\psi\rangle}(P)$, and thus

$$V_{|\psi\rangle}(P) = 0 \text{ or } 1. \tag{27.5}$$

If we think of $P$ as a statement, then the value function gives an assignment of 'true' (equal to 1) or 'false' (equal to 0) in the state $|\psi\rangle$.

For the spectral operators $P_{|a_n\rangle}$, we have[6]:

---

[6]We note that these projectors commute.

$$P_{|a_i\rangle} P_{|a_j\rangle} = \delta_{ij} P_{|a_i\rangle}; \quad \sum_n P_{|a_n\rangle} = 1. \tag{27.6}$$

Because of the sum rule, exactly one of the values of the set $\left\{ V_{|\psi\rangle} \left( P_{|a_i\rangle} \right) \right\}$ is 1 (this statement is true), let us say $V_{|\psi\rangle} \left( P_{|a_m\rangle} \right)$, and all others are 0 (these statements are false). We therefore find for the CONS $\{|a_n\rangle\}$ the results:

$$\sum_n V_{|\psi\rangle} \left( P_{|a_n\rangle} \right) = 1 \quad \text{with} \quad V_{|\psi\rangle} \left( P_{|a_n\rangle} \right) = 1 \quad \text{or} \quad 0. \tag{27.7}$$

This means in other words that the operator $A$ has a well-defined value, namely the eigenvalue $a_m$, for which $V_{|\psi\rangle} \left( P_{|a_m\rangle} \right) = 1$. At this point, we do not know $m$ or which eigenvalue this is; it is sufficient that there is exactly one.

One can render these formulations a bit more intuitive. Namely, we color in such a way that the basis vector $|a_m\rangle$ with $V_{|\psi\rangle} \left( P_{|a_m\rangle} \right) = 1$ becomes black and all others $|a_n\rangle$ with $V_{|\psi\rangle} \left( P_{|a_n\rangle} \right) = 0, n \neq m$ become white. Then the two assumptions can be written as

1. NC: Given a CONS, the vector with $V_{|\psi\rangle} = 1$ is colored black and the other vectors white ($V_{|\psi\rangle} = 0$).[7]
2. VD: This coloring process must be performed for *all* basis systems (CONS) of the Hilbert space.

If condition 2 were not fulfilled, then not all the properties of the quantum system would be well defined.

We note that the physical quantity $A$ has a value independent of the measurement. Thus, if a vector is colored black in a certain basis and also appears in a different basis, it is black there, too (non-contextuality).

### 27.1.3 Coloring

The KST now states that the last two assumptions cannot be met in Hilbert spaces (i.e. in quantum mechanics) of dimension $\geq 3$. One can divide the proof into several steps. First, one proves that the existence of a value function for a Hilbert space of dimension $N$ implies that there is a value function for all spaces with dimension less than $N$. Next, one proves that the existence of a value function for a complex Hilbert space implies that there also a value function for the real Hilbert space of the same dimension. For simplicity, we accept these results. Finally, one must still show that there are no such value functions in the three-dimensional real Euclidean space.

Thus one has boiled down the initial question to a three-dimensional geometrical problem. The question is now whether one can mark *all* basis systems (more precisely, all CONS) in a 3-dimensional real space so that one vector is always black ($V_{|\psi\rangle} = 1$)

---

[7]Strictly speaking, the statement does not apply to the states $|a_n\rangle$, but to the corresponding rays; the phase factors cancel each other in the expression $|a_n\rangle \langle a_n|$. For reasons of clarity, we accept this imprecision.

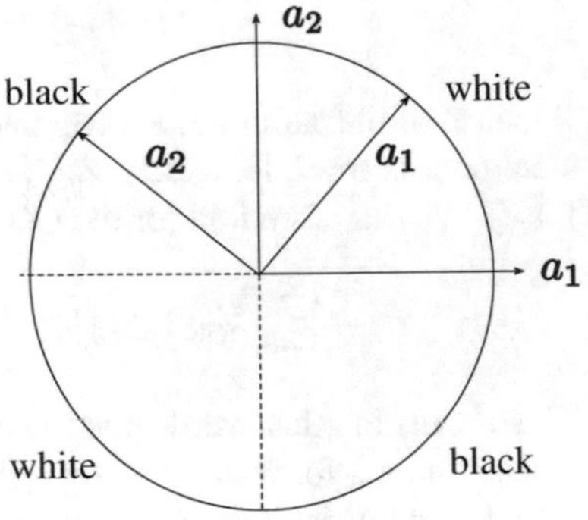

**Fig. 27.1** Coloring for
dim = 2

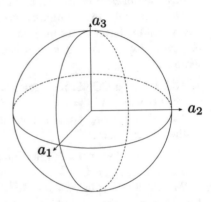

**Fig. 27.2** Coloring for
dim = 3

and the other two are white ($V_{|\psi\rangle} = 0$). Or perhaps more intuitively: Can a spherical surface be colored by means of orthogonal point triples, where one point is black and the two others white?

In fact, for dim = 2 (a circular surface), the assumptions of NC and VD are easily met, for example by coloring the four quadrants (or the corresponding circular segments) alternately black and white; see Fig. 27.1.

But for dim = 3, things look different Fig. 27.2. While we can color the equatorial plane as in two dimensions, the additional third dimension makes a consistent coloring of the entire surface impossible. To prove this, we need to find only *one* suitable CONS which cannot be colored accordingly. In other words, we need only to falsify the statements (VD + NC) once, and the simplest example will suffice. But this is surprisingly complicated, given the clarity of the question. In 1967, Kochen and Specker needed 117 vectors to demonstrate the theorem named after them.

We demonstrate their approach explicitly using the example of a set of vectors which is at the moment probably the smallest known set. It was published 1997 by A. Cabello and comprises 18 vectors in an albeit *four-dimensional* Hilbert space.[8]

---

[8]It is the set of these 18 basic vectors (or yes-no-tests) in a four-dimensional space with which recently the first experimental implementation of a Kochen–Specker set was performed; see

**Table 27.1** Four orthogonal vectors (i.e., one column) form nine different bases

| $|0,0,0,1\rangle$ | $|0,0,0,1\rangle$ | $|1,-1,1,-1\rangle$ | $|1,-1,1,-1\rangle$ | $|0,0,1,0\rangle$ | $|1,-1,-1,1\rangle$ | $|1,1,-1,1\rangle$ | $|1,1,-1,1\rangle$ | $|1,1,1,-1\rangle$ |
|---|---|---|---|---|---|---|---|---|
| $|0,0,1,0\rangle$ | $|0,1,0,0\rangle$ | $|1,-1,-1,1\rangle$ | $|1,1,1,1\rangle$ | $|0,1,0,0\rangle$ | $|1,1,1,1\rangle$ | $|1,1,1,-1\rangle$ | $|-1,1,1,1\rangle$ | $|-1,1,1,1\rangle$ |
| $|1,1,0,0\rangle$ | $|1,0,1,0\rangle$ | $|1,1,0,0\rangle$ | $|1,0,-1,0\rangle$ | $|1,0,0,1\rangle$ | $|1,0,0,-1\rangle$ | $|1,-1,0,0\rangle$ | $|1,0,1,0\rangle$ | $|1,0,0,1\rangle$ |
| $|1,-1,0,0\rangle$ | $|1,0,-1,0\rangle$ | $|0,0,1,1\rangle$ | $|0,1,0,-1\rangle$ | $|1,0,0,-1\rangle$ | $|0,1,-1,0\rangle$ | $|0,0,1,1\rangle$ | $|0,1,0,-1\rangle$ | $|0,1,-1,0\rangle$ |

They are listed in Table 27.1, where each vector occurs twice. Each of the nine columns of the table represents a CONS.

At this point, the reasoning is very simple: For each CONS of 4 vectors (i.e. for each column in the table), one vector has to be colored black, so it has the value 1. Since in the table each vector occurs exactly twice, and the values assigned to the vectors are either 0 or 1, the sum of these values over the entire table is always an *even* number. On the other hand, the sum of these values in each column must be 1 (only one vector is black, the other three are white), so that the sum over the entire table must be 9, which is *not* even.

We have thus shown that quantum mechanics (which operates in Hilbert spaces) is not compatible with the requirements of value definiteness and non-contextuality. So there is no realistic non-contextual hidden-variables theory.

## 27.1.4 Interim Review: The Kochen–Specker Theorem

The KST shows that there is a contradiction between quantum mechanics and the pair value definiteness/non-contextuality (which is in essence due to the fact that quantum mechanics operates in a Hilbert space). Logically, we must abandon one of the two assumptions, or both. But at present it is not clear where the correct path may lead.

As in Bell's inequality, the KST is independent of the physics of the quantum systems (because it ultimately is a statement about Hilbert spaces). Its role in the discussion of hidden variables is based on the following points:

1. The KST has nothing to do with the uncertainty principle, etc.; it is based on the vector-space structure of the state space.
2. The KST requires only a *finite* set of *discrete* commuting observables, and thus avoids the problems that arise when considering a continuum of quantum-mechanical statements.
3. In contrast to Bell's inequality, the KST has nothing to do with statistical correlations of an ensemble. It compares the results of various measurements that can be performed on a *single* system. Assumptions about locality and separability are

Vincenzo D'Ambrosio et al., 'Experimental Implementation of a Kochen–Specker Set of quantum Tests', *Phys. Rev.* X 3, 011012 (2013).

not needed. In this way, it is rather similar to the GHZ considerations discussed below.

Finally, we add a few remarks:

1. The theorem shows that there is no consistent value assignment for a sufficiently large but finite set of observables. The possibility of such an assignment disappears somewhere between the consideration of *one* and of *all* the observables. (We emphasize again that one cannot attribute the values 0 or 1 to all properties). How to construct the smallest non-colorable configuration is an open problem (i.e. finding the largest set of observables for which one can still make a value assignment).
2. Numbers: While in the original work of Kochen and Specker, 117 rays were used for $\mathbb{R}^3$, later on configurations of 33 and 31 rays were found. For $\mathbb{R}^4$, there exist configurations between 33 rays and the 18 given explicitly above. More numbers: in $\mathbb{R}^5$, 29; in $\mathbb{R}^6$, 31; in $\mathbb{R}^7$, 34; and in $\mathbb{R}^8$, 36 rays are the current minimal numbers.
3. The KST does not generally exclude hidden variables, but only those that are not contextual.
4. Experimentally, the KST has been confirmed impeccably.[9] Thus it was shown that the measurement of a property of a quantum system (two laser-cooled calcium ions in an electromagnetic trap) depends on other measurements on the system. By the way, techniques were used in this experiment that were originally developed for building a quantum computer.

For a recent review of the topic see e.g. D. Rajan, M. Visser, Kochen–Specker theorem revisited, arXiv:1708.01380v1 [quant-ph] (4.8.2017).

## 27.2   GHZ States

We have seen in Chap. 20 on the basis of Bell's inequality that quantum mechanics is incompatible with a local-realistic hidden variable theory. Now, one might see a certain disadvantage of Bell's argument in the fact that it is based on a statistical treatment, i.e. that it requires validation by means of an ensemble. But there is a possibility to test the compatibility of local-realistic theories with quantum mechanics which is independent of Bell's inequality and does without this statistical argument. It was proposed by Greenberg, Horne and Zeilinger[10] (GHZ) in 1989 and it involves entangled states of *three* quantum objects (GHZ states).[11]

---

[9]G. Kirchmair et al., 'State-independent experimental test of quantum contextuality', *Nature* 460, 494–497 (2009).

[10]In fact, in this paper *four* spin-1/2 systems are used. The simplification to the three quantum objects considered here was introduced some time later by Mermin.

[11]An attempt at a treatment of this topic suitable for schools was given for example by : 'EPR Paradoxon in school—Absolute and relative, and Bertelsmann's socks', by K. Jaeckel and J. Pade, in: H. Fischler (Ed.), *Quantum physics in school*, IPN 133, (1992) (text in German).

**Fig. 27.3** GHZ display for three photons

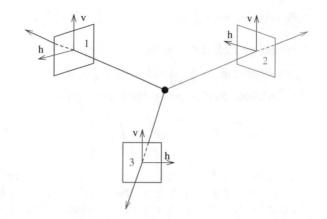

Using GHZ states, it can be shown in a way independent of Bell's inequality that quantum mechanics and local realism are not compatible. The GHZ argument has the advantage that it does not require the measurement of a whole ensemble to determine probabilities, because it does not involve statistical correlations, but rather a perfect anti-correlation. Four measurements suffice, while for the verification of Bell's inequality one needs a large number of measurements to obtain reasonable statistics.

We discuss the situation for the case that a total system decays from a certain initial state into three photons. The three photons move in a plane towards three observers, separated in each case by an angular distance of 120°; see Fig. 27.3. They can measure the following states of polarization: The linear polarizations $|h\rangle$ / $|v\rangle$, the states $|h'\rangle$ / $|v'\rangle$ rotated by 45°, and the circular polarizations $|r\rangle$ / $|l\rangle$. Thus, we have:

$$|h'\rangle = \frac{|h\rangle + |v\rangle}{\sqrt{2}} \; ; \; |v'\rangle = \frac{-|h\rangle + |v\rangle}{\sqrt{2}} \; ; \; |r\rangle = \frac{|h\rangle + i\,|v\rangle}{\sqrt{2}} \; ; \; |l\rangle = \frac{|h\rangle - i\,|v\rangle}{\sqrt{2}}.$$

$$(27.8)$$

The measurements are carried out simultaneously at the three stations, so that a 'communication' among the photons could occur only superluminally.

The system is prepared in such a way that it is in a special entangled overall state (the GHZ state) prior to measurement[12]:

$$|\psi\rangle = \frac{|h, h, h\rangle + |v, v, v\rangle}{\sqrt{2}}.$$

$$(27.9)$$

Obviously, this state is invariant under any permutation of the three observers.

We will use the *polarization operators* already introduced in Chap. 4, Vol. 1:

---

[12]Since we have only three observers, we use the shorthand notation $|\psi\rangle = \frac{|h,h,h\rangle + |v,v,v\rangle}{\sqrt{2}} \equiv \frac{|1:h,2:h,3:h\rangle + |1:v,2:v,3:v\rangle}{\sqrt{2}}$.

$$P_L = |h\rangle\langle h| - |v\rangle\langle v| \; ; \; P_{L'} = |h'\rangle\langle h'| - |v'\rangle\langle v'| \; ; \; P_C = |r\rangle\langle r| - |l\rangle\langle l|. \quad (27.10)$$

(The indices $L$ and $C$ signify, of course, longitudinal and circular). In the usual representation $|h\rangle \cong \begin{pmatrix} 1 \\ 0 \end{pmatrix}$ and $|v\rangle \cong \begin{pmatrix} 0 \\ 1 \end{pmatrix}$, one immediately sees the connection of these polarization operators to the Pauli matrices:

$$P_L = \begin{pmatrix} 1 & 0 \\ 0 & -1 \end{pmatrix} = \sigma_z ; P_{L'} = \begin{pmatrix} 0 & 1 \\ 1 & 0 \end{pmatrix} = \sigma_x ; P_C = \begin{pmatrix} 0 & -i \\ i & 0 \end{pmatrix} = \sigma_y. \quad (27.11)$$

Accordingly, the $LL'C$ measurement is also called a $zxy$ measurement.

The three observers measure the polarizations, namely two observers measure the circular ($|r\rangle$ / $|l\rangle$) and one observer the rotated linear polarization $(|h'\rangle$ / $|v'\rangle)$. We consider as an example the measurement of $(|r\rangle$ / $|l\rangle)_1$ $(|r\rangle$ / $|l\rangle)_2$ $(|h'\rangle$ / $|v'\rangle)_3$, or for short the $CCL'$ measurement. With the help of the inverse transforms of (27.8) (see exercises), we can write the state (27.9) in this case (see exercises) as:

$$|\psi\rangle_{CCL'} = \frac{|r,l,h'\rangle + |l,r,h'\rangle - |r,r,v'\rangle - |l,l,v'\rangle}{2}$$

$$|\psi\rangle_{L'CC} = \frac{|h',r,l\rangle + |h'l,r\rangle - |v'r,r\rangle - |v'l,l\rangle}{2} \quad (27.12)$$

$$|\psi\rangle_{CL'C} = \frac{|l,h',r\rangle + |r,h',l\rangle - |r,v',r\rangle - |l,v',l\rangle}{2},$$

where by the last two states follow by cyclic permutation.

In all cases, we can predict with certainty the outcome for the third photon if the results for two of the photons are known. If e.g. in the $CCL'$ measurement (27.12) photon 1 and 2 are right-handed circularly polarized (i.e. they are in the state $|r\rangle$), then photon 3 is in the state $|v'\rangle$ with certainty, without the need of a further measurement. Consequently, the local realism sees elements of reality in the individual measurement results.

We assign values to these elements which we call $L_i' = \pm 1$ for $h'/v'$ polarizations and $C_i = \pm 1$ for $r/l$ polarizations, $i = 1, 2, 3$. (These are the eigenvalues of the polarization operators in (27.11)). We choose $+1$ for $h'$ and $r$ and $-1$ for $v'$ and $l$. Hence, we assign the value $(+1)(-1)(+1) = -1$ to the state $|r,l,h'\rangle$. In this way we obtain for (27.12) and its cyclic permutations the relations

$$C_1 C_2 L_3' = -1 ; C_1 L_2' C_3 = -1 ; L_1' C_2 C_3 = -1. \quad (27.13)$$

As a fourth measurement, we consider the case that all three observers measure the linear rotated polarization (for short, an $L'L'L'$ measurement). Here, the conversion of (27.9) leads to the state (see exercises):

$$|\psi\rangle_{L'L'L'} = \frac{|h', h', h'\rangle + |h', v', v'\rangle + |v', h', v'\rangle + |v', v', h'\rangle}{2}. \quad (27.14)$$

The argument against local realism now runs as follows: Because of the locality, each measurement result $h'/v'$ of a photon is independent of the measurements for the other two photons; this applies correspondingly to the values $L'_i$ and $C_i$. Because of $C_i^2 = (\pm 1)^2 = +1$, we can with (27.13) write $L'_1 L'_2 L'_3 = (L'_1 C_2 C_3)(C_1 L'_2 C_3)(C_1 C_2 L'_3)$ and thus obtain

$$L'_1 L'_2 L'_3 = -1. \quad (27.15)$$

This being the case, for the local realism only the following $h'/v'$ measurements are possible: $v'v'v'$, $h'h'v'$, $h'v'h'$, $v'h'h'$. In other words, an *odd* number of photons is in the state $v'$.

But as we see from (27.14), according to quantum mechanics the possible results are of the form $h'h'h'$, $h'v'v'$, $v'h'v'$, $v'v'h'$; therefore, an *even* number of photons is in the state $v'$, and we have

$$L'_1 L'_2 L'_3 = 1. \quad (27.16)$$

What is the reason for this contradiction between (27.15) and (27.16)? The essential point is this: The assumption that e.g. the two terms $Z_1$ occurring in (27.13) are identical, is wrong. In fact, these values are not fixed from the outset, but are contextual, i.e. they depend on which other variables are measured simultaneously.[13] We can therefore not assume $C_i^2 = (\pm 1)^2 = +1$ and, consequently, cannot derive the (27.15).

As always, measurements have the last word. The corresponding experiment was performed for the first time in the year 2000.[14] It clearly demonstrated that the quantum-mechanical result is correct. Local-realistic hidden variables have no place in quantum mechanics.

**Interim review: GHZ**

With the help of GHZ states, it can be shown in a way independent of Bell's inequality that quantum mechanics and local realism are not compatible. The GHZ argument has the advantage that it does not require the measurement of an ensemble to determine probabilities, because it does not make use of statistical correlations, but instead of a perfect anti-correlation. Four measurements suffice, while it requires a large number of measurements for reasonable statistics in order to prove Bell's inequality.

---

[13] As in the case of Bell's inequality, the problem is that one cannot measure the six quantities $L'_1, L'_2, L'_3, C_1, C_2, C_3$ simultaneously; these are the eigenvalues of operators that do not all commute with each other. The measurement of the six variables is counterfactual: in one experiment, one cannot measure more than three of them.

[14] Jian-Wei Pan et al., 'Experimental test of quantum nonlocality in three-photon Greenberger–Horne–Zeilinger entanglement', *Nature*, Vol. 403, pp. 515–519 (2000).

## 27.3  Discussion and Outlook

As we have seen, a realistic representation of quantum mechanics is compatible neither with non-contextuality nor with locality, two properties which in the context of a
realistic approach are generally taken for granted. This leaves us with the possibilities
of a realistic contextual nonlocal description, or of simply non-realistic theories.

In recent years, the debate has concentrated on the pair realism/locality. Should
we abandon locality, or instead the notion of physical reality—or both concepts?
This question cannot be answered solely by means of logic.

In 2003, Anthony Leggett[15] explored one of the options, considering a certain
class of physically plausible theories which are nonlocal but realistic (called *crypto-
nonlocal* by Leggett). He noted that these theories are incompatible with quantum
mechanics, and expressed this in terms of new inequalities. The inequalities were
investigated experimentally in 2006 and 2007 and confirmed,[16] which would imply
that one should rather question realism instead of locality.

However, these considerations are not without controversy. One objection is, for
example, that the violation of Leggett's inequality just means that realism and a
certain type of nonlocality are incompatible, while there are other types of nonlocality that are not addressed by Leggett's inequality.[17] And Leggett himself acknowl-
edged[18] in 2008 that certain local elements have some influence in his problem:
"A critic may argue that ... we have in effect smuggled the concept of locality back
in again." Perhaps, he continues, the message is that although the concept of local
realism is clearly defined, it might not be a particularly useful exercise to analyze
this concept separately in terms of its two main components.

In any case, it is currently not clear in which direction to travel. In addition, still
other conditions, currently generally taken for granted, could be violated.[19] We have
for example always tacitly assumed causality, i.e. the fact that an event cannot be
influenced by other events that lie in the future (the arrow of time). This assumption is
made explicitly in the *objective local theories* (OLT), based on the three postulates of
locality, realism, and induction (causality). It can be shown quite generally that these

---

[15] A.J. Leggett, 'Nonlocal hidden-variable theories and quantum mechanics: an incompatibility
theorem', *Found. Phys.* 33 (2003) 1469–1493. was awarded the 2003 Nobel Prize for his work in
the field of superfluidity.

[16] S. Gröblacher et al., 'An experimental test of non-local realism', *Nature* 446 (2007), pp. 871–875.

[17] M. Socolovsky, 'Quantum mechanics and Leggett's inequalities', *Int. J. Theor. Phys.* 48 (2009),
pp. 3303–3311.

[18] A.J. Leggett, 'Realism and the physical world', *Rep. Prog. Phys.* 71 (2008), 022001.

[19] In principle, one cannot exclude for example that the rules of ordinary logic do not apply and/or
need to be expanded in the realm of quantum mechanics—rules which are tacitly applied in the
derivation of Bell's inequality and the other arguments. Here, Gödel's theorem comes into play,
according to which, roughly speaking, any theory that is proposed as the basis for mathematics,
including logic, is necessarily inadequate, incomplete or contradictory. It is not clear at this point,
however, what should be changed in conventional logic in order that Bell's inequalities not be
violated by quantum mechanics.

theories are incompatible with quantum mechanics.[20] In principle, we can also (or perhaps we must?) sacrifice the arrow of time. Thus, a measurement of an entangled quantum object at a given time would determine the properties of the other one at the moment of its emission from the same source, hence in the past.[21]

The idea of giving up the familiar notion of 'cause and effect' is more than unusual, of course.[22] It would certainly be very difficult to continue doing physics the way we are accustomed without it. However, there are very reputable and respected physicists who not only think that this sacrifice is possible, but expect that the next major revolution in physics will force us to do just that.[23]

Be that as it may—we can state that we must take leave of one or several plausible ideas in order to maintain the classical notion of realism.[24] For this reason, the expectation is often expressed that any future extension of quantum mechanics, compatible with experiments, must give up certain features of realism.[25]

Of course there are alternatives allowing one to avoid the whole discussion about Realism and Co.: For example, one can argue that quantum mechanics is just a set of calculation rules for the determination of measured values and has no intrinsic meaning beyond that. We will meet up with more viewpoints in Chap. 28, but we can already state here that in all cases, the results of the discussion about hidden variables have no direct influence on the practical usefulness of quantum mechanics. In view of this, one may ask of course why this topic should be of general interest. The answer is perhaps more a matter of personal preference; but it must be emphasized that this

---

[20]This is done by comparing experimental results with an extension of Bell's inequality, the *CSCH inequality*, proposed in 1969 by Clauser, Horne, Shimony and Holt.

[21]In connection with delayed-choice experiments (see Appendix M, Vol. 1) also, the idea of a time-reversed effect is discussed. In fact, the fundamental laws of physics are time-reversal invariant, i.e. time-symmetrically causal, and do not reflect the time-asymmetrical idea of cause and effect.

[22]Though the idea that events obey a definite causal order is deeply rooted in our understanding of the world, causal order needs not be a required property of nature. For instance, it was recently shown that in quantum mechanics, there are correlations that cannot be understood in terms of definite causal order; see Ognyan Oreshkov et al., 'Quantum correlations with no causal order', Nature Comm. 3, 1092 (2012), doi: 10.1038/ncomms2076.

[23]"I believe that in our present picture of physical reality, especially regarding the nature of time, a huge upheaval is imminent, it may be even greater than the revolution that has already been triggered by relativity theory and quantum mechanics." Roger Penrose, British mathematician and physicist, in *New Mind. The emperor's new clothes or the debate over artificial intelligence, consciousness and the laws of nature.*

[24]In Appendix U, Vol. 2, some quotes from philosophers, artists, etc. are compiled; they show illustratively that *the* classical notion of 'reality' does not exist and has never existed in the past.

[25]A certain skepticism about the concept of 'reality' is in the tradition of modern science. More than 200 years ago, Georg Christoph Lichtenberg stated in one of his physics lectures: "We care little about whether the bodies have an objective reality apart from us or not. It would always be possible that at least some would not have one. We have to imagine the things; the idea does not depend on us, but of those things that make an impression on us, the impression cannot act on us as in another way than our abilities admit. At least is that what we feel of the bodies apart from us, not always objectively real." Gottlieb Gamauf, in *Physics lectures, from the memoirs of Gottlieb Gamauf.*

debate has allowed us to look deeper into the mysteries of quantum mechanics.[26] Whether the hope will be fulfilled that this may contribute towards leading quantum information[27] from its current state of basic research to a full-blown technical revolution remains to be seen.

In any case, the debate has forced us to question our notions of 'self-evident' and to obtain in this way new insights into the world. And that is one of the main tasks of science.

## 27.4 Exercises

1. A system is in the polarization state $|r\rangle$. Using $w_P = tr\,(\rho P)$, calculate the probability of measuring the system in the state $|h\rangle$.

2. A mixture is described by $\rho = \sum p_n \,|\varphi_n\rangle \,\langle\varphi_n|$, where $\{|\varphi_n\rangle\}$ is a CONS. Using $w_P = tr\,(\rho P)$, calculate the probability of measuring the system in the state $|\varphi_N\rangle$.

3. The value function $V_{|\psi\rangle}$ is defined by $V_{|\psi\rangle}(F\,(A)) = F\left(V_{|\psi\rangle}(A)\right)$.

   (a) Prove for $[A,\,B] = 0$ the sum rule $V_{|\psi\rangle}(A + B) = V_{|\psi\rangle}(A) + V_{|\psi\rangle}(B)$.
   (b) Prove for $[A,\,B] = 0$ the product rule $V_{|\psi\rangle}(A \cdot B) = V_{|\psi\rangle}(A) \cdot V_{|\psi\rangle}(B)$.
   (c) Show that $V_{|\psi\rangle}(1) = 1$.

4. Given the polarization operators $P_L$, $P_{L'}$ and $P_C$ (or the corresponding Pauli matrices, see (27.11)):

   (a) Determine (once more) their eigenvalues and eigenvectors.
   (b) Express the eigenvectors of $P_C$ and $P_{L'}$ in terms of those of $P_L$.

5. Given the GHZ state

$$|\psi\rangle_\pm = \frac{|h, h, h\rangle \pm |v, v, v\rangle}{\sqrt{2}} \tag{27.17}$$

corresponding to an $LLL$ measurement; rewrite this for a $CCL'$ measurement (plus $CL'C$ and $L'CC$) (27.12) and for an $L'L'L'$ measurement (27.14).

---

[26]The topic is far from complete and the subject of current research. This is shown by conferences (e.g. 'The Nature of Quantum Reality', One-Day Conference, 10th June 2017, St Cross College, University of Oxford), by reviews (e.g. the very readable article by Z. Merali, 'Quantum physics: What is really real?', *Nature* 521, 278–280, (May 2015) doi:10.1038/521278a), and by a number of scientific papers (e.g. G.C. Krizek, 'The conception of reality in Quantum Mechanics', arXiv:1708.02148v1 [quant-ph] (Aug 2017)).

[27]"The development of quantum mechanics early in the twentieth century obliged physicists to change radically the concepts they used to describe the world. The main ingredient of the first quantum revolution, wave-particle duality, has led to inventions such as the transistor and the laser that are at the root of the information society. Thanks to ideas developed by Albert Einstein and John S. Bell, another essential quantum ingredient, entanglement, is now leading us through the conceptual beginnings of a second quantum revolution—this time based on quantum information." Alain Aspect, 'Quantum mechanics: To be or not to be local', *Nature* 446, pp. 866–867 (19th April, 2007).

6. The following combinations of the polarization operators (27.10) are given:

$$Q_1 = P_{1L'} P_{2C} P_{3C}; \quad Q_2 = P_{1C} P_{2L'} P_{3C}$$
$$Q_3 = P_{1C} P_{2C} P_{3L'}; \quad Q = P_{1L'} P_{2L'} P_{3L'}. \tag{27.18}$$

The numerical index denotes the space in which the particular polarization operator acts. We use in the following the fact that operators from different spaces commute, e.g. $P_{1L'} P_{2C} = P_{2C} P_{1L'}$. In addition, we have $P_{nL'} P_{nC} = -P_{nC} P_{nL'}$ as well as $P_{nC}^2 = P_{nL'}^2 = 1$.

(a) Show that the three operators $Q_i$ have the eigenvalues $\pm 1$.
(b) Show that the three operators $Q_i$ commute pairwise.
(c) Show that the states

$$|\psi\rangle_\pm = \frac{|h, h, h\rangle \pm |v, v, v\rangle}{\sqrt{2}} \tag{27.19}$$

are common eigenstates of the three operators $Q_i$ with the eigenvalues $\mp 1$, as well as eigenstates of the operator $Q$ with the eigenvalues $\pm 1$.

# Chapter 28
# Interpretations of Quantum Mechanics

The formalism of quantum mechanics is unambiguous. But the question remains open as to what it 'really' means. In this chapter, we outline some of the more popular interpretations of quantum mechanics.

The great importance of quantum mechanics for our current view of the physical world is undisputed. In the previous chapters, we have gotten an idea of how coherent it is as a theory, of how powerful it is in the treatment of various practical issues, from the hydrogen atom to the quantum computer, and what insight it provides into problems that were long considered purely philosophical.[1] Undoubtedly, quantum mechanics has profoundly changed our worldview.

At the same time, it raises unresolved central epistemological problems, such as the question of the existence of objective chance. It is typical of quantum mechanics that the formal apparatus is unique, but not its meaning. For example, let us consider the state vector, whose mathematical formulation and whose relation to experimental variables are precisely defined. But what does it mean? Does the state vector describe the physical reality of an *individual* quantum system? Or has it nothing to do with an individual system, but is applicable only to an *ensemble*? Or is it simply an indisputable *calculation recipe* which allows us to determine the probability of an experimental result?

Here, the *interpretations of quantum mechanics* appear on the scene. Interpretation essentially means explanation, clarification and giving meaning to the formalism. The aim of an interpretation of quantum mechanics is a better understanding of

---

[1] For a long time, the study of the foundations of quantum mechanics was considered a humanistic rather than a scientific activity. However, the search for better explanatory models is not necessarily a glass bead game, but, on the contrary, it can have very practical consequences, such as Bell's inequality, entanglement, decoherence, quantum computers etc.

© Springer Nature Switzerland AG 2018
J. Pade, *Quantum Mechanics for Pedestrians 2*, Undergraduate Lecture Notes in Physics, https://doi.org/10.1007/978-3-030-00467-5_28

the terminology, namely in the sense of a physical picture of the world which is as consistent as possible.[2]

It is characteristic of different interpretations that they make the *same* predictions for measurements, as they refer to the *same* formal apparatus. Therefore, one cannot differentiate one of them from another on the basis of experimental results, and thus could argue all day about them without reaching a convincing conclusion. In contrast, different theories lead in general to different predictions or experimental results that allow a decision as to which theory is more appropriate. However, the situation is currently such that the proposed changes to quantum mechanics lead to such subtle effects that they cannot (yet) be detected. Therefore, one also usually subsumes under 'interpretations' in addition those approaches that are, strictly speaking, new theories.

As we shall see, the interpretations of quantum mechanics provide some very different answers to the open questions. These different points of view have, as we want to emphasize again, no effect on the *practical* application of quantum mechanics; it is excellently *fapp* (one of the best validated physical theories known) and in this respect there is no fundamental disagreement. The dispute concerns only the ontological significance of quantum mechanics, i.e. what it 'really' means. As the various interpretations cannot be distinguished experimentally, it is largely a matter of faith, conviction or taste which interpretation one prefers (and this is sometimes expressed with missionary zeal).[3]

Especially when teaching quantum mechanics, one is often asked questions of meaning by lay people; it is not for nothing that 'bizarre' is a regularly recurring adjective when popular scientific descriptions of quantum effects which are so strange to our everyday experience are under discussion.[4] Hence it is of great interest not just for professionals, but also (in an appropriate form) for physically less-trained people, to know the conceptions of the world in which quantum mechanics is embedded, and to appreciate the differences between these explanations; in short, to learn the state of our physical worldview. That the general public is very open to these issues is shown for example by the widespread interest in the many-worlds theory, ideas about entanglement, considerations of the relationship between consciousness and measurement and the like. Of course, one must clearly distance oneself from

---

[2]Since time immemorial, philosophy has concerned itself, in one form or another, with interpretations of our world. Here are two opposing voices from the nineteenth century:

"No, just the facts do not exist, only interpretations. We can not state a fact »itself«: perhaps it is nonsense to want such a thing. »It's all subjective« you say: but even this is interpretation, the »subject« is not given, but something added and invented." Friedrich Nietzsche, in *Legacy*, KSA 12.

"The philosophers have only interpreted the world differently; it is important to change it." Karl Marx, in *Theses on Feuerbach*.

[3]The situation is confusing even for physicists: "Quantum theory was split up into dialects. Different people describe the same experiences in remarkably different languages. This is confusing even to physicists." David Finkelstein (born 1929), American theoretical physicist.

[4]The special theory of relativity is also suspect to our common sense. But there are no controversies within physics concerning the explanation and visualization of the Twin paradox and Co. This is different in quantum mechanics; here, there exist a number of distinct explanations.

any esotericism; but with appropriately popularized descriptions of the problems of quantum mechanics and the way in which different interpretations address them, interest and motivation can be established to a high degree.

## 28.1 Preliminary Remarks

### 28.1.1 Problematic Issues

In Chap. 14 (Vol. 1), we have identified some 'difficult' concepts of quantum mechanics—for example, the special role of measurement in quantum mechanics, the occurrence of probabilities, the collapse of the wavefunction, the relationship between classical mechanics and quantum mechanics (Heisenberg cut). They are all related to the measurement process and are therefore not entirely independent. In Vol. 2, we have learned more of the conceptual peculiarities of quantum mechanics. Of central importance is *entanglement*. On the one hand, it led us to the phenomenon of decoherence (via the reduced density operator), on the other hand to the question of the validity of local realism (via Bell's inequality). We recapitulate briefly the state of the most important problem areas in the following:

1. The Kochen–Specker theorem shows that in quantum mechanics, not *all* properties can be fixed prior to measurement. The classical idea—that measuring means finding out fixed properties as part of reality—cannot be maintained in quantum mechanics;there are situations in which quantum systems do not possess objectively-determined physical characteristics.[5]

   Bell's inequality and the GHZ states show that quantum mechanics cannot be simultaneously local and realistic. Whether just one of these properties is not satisfied (and if so which one), or both, is still unclear. In any case, it is obvious that the worldview of quantum mechanics differs in very essential respects from that of classical mechanics, and thus also of our everyday understanding. Indeed, according to current knowledge there is no getting around this fact.[6]

2. The concept of the collapse of the wavefunction denotes the fact that there are seemingly two different time evolutions: On the one hand the process defined by the SEq, which is deterministic, unitary and reversible, and on the other hand the rapidly changing, non-deterministic, non-unitary and irreversible course of a measurement, where the only distinction is that between 'before' and 'after'.

---

[5]We repeat the remark that these conclusions are based on the fact that quantum mechanics takes place in a Hilbert space.

[6]The belief that quantum mechanics implies a drastic break with classical physics was also expressed earlier (albeit partly for reasons other than those listed here): "Against all reactionary efforts . . . I am certain that the statistical nature of the Psi-function and thus the laws of nature . . . will determine the style of the laws, at least for several centuries . . . Dreaming of a way back, back to the classic style of Newton-Maxwell . . . seems to me hopeless, absurd." Wolfgang Pauli, Nobel Prize 1945, writing in 1952.

As we have seen, the collapse can be explained by the mechanism of decoherence—but only insofar as, due to the flow of information into the environment, superpositions (apparently) collapse very rapidly (which is why we see no macroscopic superpositions). However, decoherence does not describe how the selection of the actual measured value comes about. For this reason, decoherence provides a very plausible explanation of the state reduction, but, by strict logic, it is not the only possible description.

3. Assuming that the measurement apparatus obeys the rules of classical mechanics, one is confronted with the problem of drawing the line between classical mechanics and quantum mechanics. If one describes quantum systems as *open* systems, however, one stays within the validity of quantum mechanics, i.e. one does not require the 'services' of classical mechanics. Thus, decoherence provides some clarification also in this respect.

But e.g. the assumption that quantum mechanics is valid for our entire universe shows that this does not mean that all questions of demarcation have been resolved. Thus, there must be a SEq for the whole universe (even if we do not know the Hamiltonian). Then the temporal evolution of the universe would be deterministic, which raises many questions; among others, the question of free will. One might, of course, assume that our universe is an open system—but what would its environment then be?

These considerations are based on the belief that the present form of quantum mechanics is essentially valid. But of course it is not impossible that in fact it still has certain shortcomings, in one way or another.[7] The question arises as to whether a single consistent theory that can explain the world in fact exists, including all observable phenomena (Theory of Everything, TOE); or if certain aspects may be described only by certain theories which are mutually exclusive.[8]

4. The fact that quantum mechanics, even with complete knowledge of the state, generally provides only probability statements, can be interpreted from a number of very different positions. We consider three of them. One view assumes that quantum mechanics is complete and reflects a fundamental limitation of our knowledge of nature, so that we must content ourselves with probability statements—its probabilistic aspect is natural (objective chance). Another view also assumes the completeness of quantum mechanics, but rejects the existence of objective chance; according to this view, probabilities arise due to the limited perspective of the observer. Finally, there is the view that the present theory is not complete and should be complemented by introducing hidden variables, the consideration of which would allow us to predict results exactly, as we are accustomed from classical mechanics.

The formalism of quantum mechanics (i.e. the postulates introduced in Chap. 14,

---

[7]In this connection, we recall that there is still no satisfactory unification of general relativity theory and quantum mechanics.

[8]A very critical attitude towards the reductionism of modern science is shown by e.g. Robert B. Laughlin (Nobel prize in physics 1998) in his book *A different universe—Reinventing physics from the bottom down* (2006).

Vol. 1) does not permit us, as noted above, to decide directly which of these (and other) points of view is the correct one. The view adopted by a majority is that there are no hidden variables; only one of about a dozen now current interpretations (namely the Bohm interpretation) makes use of them. The reason is, of course, that the considerations discussed so far have substantially sharpened the constraints on hidden variables. They may be neither non-contextual-realistic nor local-realistic—and thus are in this sense as far from classical mechanics as is quantum mechanics itself. The situation is different regarding the question of whether the probabilities are based on objective chance in quantum mechanics. Here, various models of explanation are currently under discussion.

With these questions, issues are touched that go far beyond the physical framework of quantum-mechanical physics in the strict sense—determinism, causality, verifiability, reality, locality and separability. An interpretation of quantum mechanics is expected to respond to these problem areas and to provide coherent answers to the open (physical) questions. In spite of that, one can of course take the view that quantum mechanics is only a collection of rules and calculational prescriptions (albeit very well-functioning) for solving specific problems, and that the questions raised do not affect our obtaining practical results, and are therefore uninteresting.[9] This *pragmatic* or *instrumental* view works perfectly in practice, but it is for many people a very unsatisfactory and unacceptable idea that such a fundamental theory as quantum mechanics should be only some sort of physical cookbook.[10] In this sense, *realistic* views[11] of quantum mechanics do not regard it exclusively as a calculation scheme, but assume that it provides, at least partially, a faithful representation

---

[9]The old dispute between the 'pragmatist' Bohr and the 'realist' Einstein is still alive, just with different actors, for instance Stephen Hawking and Roger Penrose. Their positions are found in S. Hawking and R. Penrose, *The Nature of Space and Time*, Princeton University Press Princeton, (1996), namely:

S. Hawking: "Although I'm regarded as a dangerous radical by particle physicists for proposing that there may be loss of quantum coherence, I'm definitely a conservative compared to Roger. I take the positivist viewpoint that a physical theory is just a mathematical model and that it is meaningless to ask whether it corresponds to reality. All that one can ask is that its predictions should be in agreement with observation. I think Roger is a Platonist at heart but he must answer for himself." (pp. 3–4)

R. Penrose: "At the beginning of this debate, Stephen said that he thinks that he is a positivist, whereas I am a Platonist. I am happy with him being a positivist, but I think that the crucial point here is, rather, that I am a realist. Also, if one compares this debate with the famous debate of Bohr and Einstein, some seventy years ago, I should think that Stephen plays the role of Bohr, whereas I play Einstein's role! For Einstein argued that there should exist something like a real world, not necessarily represented by a wave function, whereas Bohr stressed that the wave function doesn't describe a 'real' microworld but only 'knowledge' that is useful for making predictions." (pp. 134–135)

[10]With the predominant recipe "Shut up and calculate!".

[11]In the debate about quantum mechanics, the adjective *realistic* has two meanings:

(a) If we require, as discussed in Chap. 27 that properties are pre-existent in quantum mechanics also, this means that (i) quantum states refer to individual systems, not just to the results of repeated measurements, and that (ii) a measurement determines the value of a physical quantity which it had immediately before and independently of the measurement.

of reality.[12] There are different views, which correspondingly show up as different interpretations.

## 28.1.2  Difficulties in the Representation of Interpretations

There are some complicating factors in the presentation of the interpretations of quantum mechanics:

1. There is quite simply a goodly number of different interpretations, so that an overview must be limited to a relatively short description of some few selected interpretations, in order not to get out of hand.
2. The formal requirements of the various interpretations are very different and, for some interpretations, go beyond the scope of our considerations. We therefore confine ourselves in all cases to working out the basic idea in a language-oriented representation. More detailed comments on certain interpretations are given in Appendix V, Vol. 2.
3. Another difficulty lies in a certain terminological fuzziness. A comparison of relevant sources in the literature quickly shows that many terms are not very precisely defined, and accordingly the descriptions of different interpretations do not always coincide. This terminological confusion goes so far that even the exact meanings of some of the concepts involved are unclear, or the same concepts are used with different meanings. We cite in this connection Peres (Quantum Theory. Concepts and Methods, pp. 23):

> The experts disagree on what is meant by "Copenhagen interpretation". Ballentine gives this name to the claim that "a pure state provides a complete and exhaustive description of a single system" The latter approach is called by Stapp the "absolute-$\psi$ interpretation". Stapp insists that"critics often confuse the Copenhagen interpretation, which is basically pragmatic, with the diametrally-opposed absolute-$\psi$ interpretation … In the Copenhagen interpretation, the notion of absolute wave function representing the world itself is unequivocally rejected". There is therefore no real conflict between Ballentine and Stapp, except that one of them calls Copenhagen interpretation what the other considers as the exact opposite of the Copenhagen interpretation.

---

(b) A realistic interpretation, on the other hand, may be based only on requirement (i), so that quantum states refer (in a not specifically detailed manner) to individual systems, not just to the result of repeated measurements.

[12]It would not be the first time in the history of science that the mathematical formalism turns out to be more than just a clever computational recipe. Those who wanted it that way could for example initially regard Kepler's laws as pure arithmetic, simply allowing the determination of the orbits of the planets more precisely than other rules. The realization that Kepler's laws indeed give a picture of reality superior to the notions existing until then had first to overcome much opposition. (We recall that Galileo Galilei was rehabilitated only in 1992 by the Catholic Church.) Another example of this phenomenon is Planck's constant, which was at first, even for Planck himself, only a pragmatic mathematical trick, introduced ad hoc in order to master convergence problems.

## 28.2 Some Interpretations in Short Form

As we said above, there are around a dozen current interpretations (some of which are also split up into different subversions), of which we present here some of the most notable in short form, by and large in the order of their dates of appearance.[13]

### 28.2.1 Copenhagen Interpretation(s)

This is a collective term for several interpretations which are not just slightly different, but sometimes even contradict each other. The first forms date from the 1920s and are due to Bohr in Copenhagen and to Heisenberg (who originated the baptismal name 'Copenhagen'); even they differed e.g. in terms of a realistic explanation. Meanwhile, the differentiation has gone further, so that the term 'Copenhagen interpretation' is very blurry—in fact it should appear in the plural in the heading of this paragraph. A contributing factor to the confusion is that different authors have different opinions as to what actually constitutes the Copenhagen interpretation, as briefly described above.

The Copenhagen interpretations essentially agree that the state vector gives the best knowledge of the system. This means that measurement results are objectively random and that behind the state vector there is no deeper reality, e.g. in the form of hidden variables; there *is no reason* why in a measurement, a specific result emerges and not another one of several possible outcomes. Similarly, the Copenhagen variants agree that quantum-mechanical statements refer to *individual* systems, and that it does not make sense to assign properties to an *unobserved* system.

Concerning other issues, the individual Copenhagen variants differ considerably in some aspects; for example, the wavefunction is in one variant only a tool for calculating probabilities, in another one an 'element of reality'.[14] We want to look at two versions in detail, which are usually called the minimal and the standard interpretations (one could also speak of the 'older' and 'newer' Copenhagen

---

[13]"Whoever merely accumulates observations and experiments, seems to me like someone who keeps a register of the pieces which two chess players lift and put down or take away; someone who notices the moves they make has taken a large step forward; it will cost him much time to determine the laws of motion precisely, and much time will pass until he guesses the intention behind why all these movements are made, and that everything is done to make the king a prisoner. Without this kind of hypotheses, nothing can be accomplished. The question of whether they are useful has something absurd in itself: Because we want in the end to explain the phenomena in nature, and such a hypothesis is indeed nothing more than such a bold statement; it immediately is thrown in disarray when the phenomena contradict it. Also, the question of whether the false hypotheses can have their uses is answered at once by itself. It is not for everyone to strike immediately the best." Georg Christoph Lichtenberg, *The Waste Books*, Vol. J (1521).

[14]To illustrate the bandwidth of what lies behind the term 'Copenhagen interpretation': The interpretation called the 'Participatory Anthropic Principle' also sees itself in the tradition of the Copenhagen interpretation; here it is assumed that observation by a conscious observer is responsible for the collapse of the wavefunction.

interpretations, or in the sense of the above quotation, of the 'Copenhagen interpretation' and the'absolute-$\psi$ interpretation').

## Minimal Interpretation

The oldest interpretation of quantum mechanics is, strictly speaking, none at all, but rather an anti-interpretation, because it is purely instrumental and dispenses almost completely with the attempt to find an inner meaning. It was largely shaped by the pragmatic Bohr, who saw in the wavefunction essentially a mathematical tool for calculating probabilities and values of measurement results. Distinctive to this interpretation is the separation of each measurement into a quantum-mechanical part (the observed, measured system) and another part obeying the laws of classical physics (the measurement apparatus, the observing system). Only the classical part is real; the objects described by quantum mechanics (electrons, atoms, etc.) do not really exist. More specifically: statements about such objects which go beyond the predictions of experimental results etc. may not and should not be made. We recall Bohr's remark, already quoted in Chap. 14, Vol. 1, in which he pointedly summarizes his position thus: 'There is no quantum world. There is only an abstract quantum-physical description. It is wrong to think that the task of physics is to find out how nature is. Physics concerns itself with what we can say about nature'.

Of course there must be a boundary between the domain described by quantum mechanics as a pure construct of thought, and the real world of the measuring apparatus (Heisenberg cut). This division has proved itself extraordinarily in practice, as we have emphasized several times, but is conceptually unsatisfying and leaves some questions unanswered—such as where exactly this boundary is located between quantum mechanics and classical mechanics. With today's experimental techniques (manipulation of individual atoms, etc.), one can no longer simply hide behind the statement that measurement devices are just to a tremendous extent larger than the quantum systems studied.[15]

The minimal interpretation may also be referred to as the 'older' Copenhagen interpretation; at least it displays essential features of that interpretation. As a further feature of the older Copenhagen interpretation, one often mentions the aspects of complementarity, uncertainty and the correspondence principle.

In regard to the history of science, it is quite interesting that the 'older' Copenhagen interpretation enjoyed a kind of monopoly position for a long time; this certainly had to do with the undisputed authority of Bohr. In any case, new ideas, not fitting the minimal interpretation, experienced difficulties in getting attention for many years; an example, at least at its beginnings, is the concept of decoherence (see Chap. 24).

---

[15]Indeed, *quantumness* is a current topic, i.e. the study of the question of to what extent a given system is quantum in nature. See e.g. E. Kot et al., 'Breakdown of the Classical Description of a Local System', *Phys. Rev. Lett.* 108, 233601 (2012); P. Kurzynski et al., 'Experimental undecidability of macroscopic quantumness', arXiv: 1111.2696v2 (2012); C. Marletto & V. Vedral, 'Witnessing the quantumness of a system by observing only its classical features', *npj Quantum Information* 3, Article number 41, doi:10.1038/s41534-017-0040-4 (Oct 2017).

**Standard Interpretation**

The very pragmatic attitude of Bohr was not shared to a similar extent by all others. Especially Dirac and von Neumann originated that version of the Copenhagen interpretation which today is called the orthodox[16] or canonical interpretation, or *standard interpretation*.[17] In contrast to Bohr's dictum, this interpretation assumes that there *is* a quantum world, that individual quantum objects such as atoms and electrons exist, i.e. they are real. The mathematical foundation of this interpretation is found in Chap. 14, Vol. 1 in the form of postulates that describe the transformation of the properties of quantum objects into measurable quantities. A state vector of the Hilbert space (or the wavefunction) provides a complete description of a real, existing individual system. Therefore we also speak of a realistic (as opposed to an instrumentalist or pragmatic) interpretation. This does not mean that we assume that *each* element of the theory has a complete correspondence in the real world; in general one assumes e.g. that the wavefunction has no real counterpart, but rather it simply provides the maximum information about a single system.[18]

In contrast to the minimal interpretation, one can also attempt to include the measurement process into quantum mechanics. To that end, the concept of 'collapse of the wave function' is introduced, without its however being clear how this collapse takes place in detail. For this dilemma, the concept of decoherence seems to show at least a partial way out, as we have seen in Chap. 24. Whether one would denote the standard interpretation with or without decoherence as a Copenhagen interpretation is again mostly a matter of taste. After all, some protagonists of decoherence see themselves in their own words 'in the tradition of the Copenhagen interpretation'. In any case, the standard interpretation, complemented by decoherence, is a currently widely-accepted interpretation of quantum mechanics.

## 28.2.2 Ensemble Interpretation

This interpretation came about in the early days of quantum mechanics along with the Copenhagen interpretation, and as a reply and alternative to it; among its supporters were Einstein and Langevin. According to this interpretation, also called the statistical interpretation, the wavefunction is an abstract mathematical entity and does not apply to a single system, but rather refers to an ensemble of identically prepared systems. Where this limitation originates is not stated. In principle, this interpretation does not immediately rule out the possibility that the measured quantity has a well-defined

---

[16]Some authors denote as *orthodox* the view that an observable has no definite value if the state (or ray) is not in an eigenspace of the observable.

[17]We note that some authors denote every formulation which involves no hidden variables or extensions of the SEq as a 'standard interpretation'.

[18]This is similar to the vector potential in electrodynamics. The attitude that the vector potential is a purely mathematical entity however is in conflict with the (quantum-mechanical) Aharonov–Bohm effect, in which an electron is influenced by the vector potential instead of by the magnetic field itself.

value for each member of the ensemble (in the sense of hidden variables or pre-existing values). However, this notion received a major blow from Bell's theorem and the experiments of Aspect. Therefore, a newer version excludes the question of the determinism of physical quantities, which of course limits its value for making meaningful statements.

### 28.2.3   Bohm's Interpretation

This interpretation, developed in 1952 by David Bohm, is based on the 'pilot' or 'guiding' wave theory, which Louis de Broglie originated in 1927 (which is why it is also called de Broglie–Bohm theory). The wavefunction here plays two roles: On the one hand, its squared value provides information on the most probable position of the particle; on the other hand, it affects the coordinates of the particle in the form of a 'quantum potential'. The physical state of a particle is completely determined by the combination of the wavefunction *and* the particle position. Both the wavefunction and the particle coordinates are regarded as real; however, the latter are unobservable and are thus the hidden variables in this interpretation.

The probabilistic character of quantum results is attributed to our ignorance of the hidden variables, i.e. to the factual impossibility of determining the initial values of all particle coordinates. In this interpretation, there is no collapse, and the particles move along well-defined trajectories. Measurement means, therefore, only the reduction of our ignorance about the system, not the generation of the measured values themselves.

A major difference from classical physics is that this interpretation is nonlocal.[19] and that there are accordingly instantaneous interactions. If we make changes, for example, to a particle in a many-particle system, then the total wavefunction changes instantaneously, and with it also the quantum potential and thus the trajectories of all the other particles.

In Appendix V, Vol. 2, a brief description of the mathematical approach of the Bohmian interpretation is given. We point out that the Bohmian and the Copenhagen interpretations make the same predictions, so that experiments cannot decide between the two approaches. On the other hand, this also means that the Bohmian interpretation also contains the same problematic superpositions.

### 28.2.4   Many-Worlds Interpretation

The many-worlds interpretation dates back to Hugh Everett (1957). It is an example of how to rigorously keep the mathematical part of quantum mechanics and yet make other statements about reality.

---

[19]This is insofar striking, as Bohm's ideas really were aimed at the elimination of the non-locality e.g. in the EPR experiment through hidden variables. But, as he says himself: "If the price of avoiding non-locality is to make an intuitive explanation impossible, one has to ask whether the cost is too great." David Bohm et al., *Phys. Rep.* 144, 321 (1987).

The basic idea is to assume that it makes sense to speak of the state vector of the *entire* universe; it is real and evolves deterministically (and reversibly) according to the SEq. In contrast to the (older) Copenhagen interpretation, there is neither a collapse of the wavefunction nor is there, in addition to quantum mechanics, a classical domain describing the measurement apparatus. Instead, it is assumed that in a measurement, or in each physical interaction, *all* physically possible events are realized. This is accomplished by the splitting of the entire universe into many parallel worlds, whereby in each parallel world exactly one of the possible outcomes is realized. Accordingly, we must therefore imagine an ongoing birth of many parallel worlds.

As a result of decoherence effects,[20] *macroscopically* different parallel universes develop independently. This is also true for the observers—in each parallel universe, there is a separate observer, not perceiving the other observers (his 'parallel clones' or 'parallel egos'). *Microscopically* different parallel universes, however, can interact, and an observer can interpret this as interference effects. Thus, the collapse of the wavefunction is a process noticed only by the respective observer, and it bears the character of objective chance, since it is not objectively predictable by that observer. However, the entire universe is, as we said above, strictly deterministic, and on this scale there is no objective chance.[21]

It goes without saying that the many-worlds interpretation was enthusiastically taken up by science fiction and fantasy fans.[22] In physics it is very controversial. Some see in it the solution to (almost) all problems, while for others it has the status of silly esoteric paraphernalia—both of these attitudes can be associated with the names of renowned physicists. After all, this interpretation solves the measurement problem without any modification of the formalism of quantum mechanics. As such, it enjoys great popularity with quantum cosmologists. For example, C.J. Isham writes (Lectures on Quantum Theory, Mathematical and Structural Foundations, pp. 183): 'Indeed, it is rather difficult to think of *any* interpretation of quantum cosmology that does not invoke this view[23] in one way or another. Thus 'post-Everett' schemes[24] have become almost obligatory for those working in the physics of the very early universe.'

Some further comments on the many-worlds interpretation can be found in Appendix V, Vol. 2.

---

[20]Today, one can argue in this way; in 1957, the term 'decoherence' was not yet known.

[21]The idea of parallel or multiple universes is not a peculiarity of quantum mechanics; in general relativity and string theory, there are several models of multi- and pluriverses (keywords e.g. 'infinite space', 'bubbles' or 'eternal inflation', 'nested multiverses in black holes').

[22]A short version, with a wink of an eye: "There are indeed such things as parallel universes, although parallel is hardly the right word—universes swoop and spiral around one another like some mad weaving machine or a squadron of Yossarians with middle-ear trouble. And they branch. But, and this is important, not all the time. The universe doesn't much care if you tread on a butterfly. There are plenty more butterflies" Terry Pratchett, in *Lords and Ladies*.

[23]I.e. the many-worlds interpretation.

[24]'Post-Everett' essentially refers to the assumption that there are no external state reductions and thus the time-dependent SEq always applies.

## 28.2.5   Consistent-Histories Interpretation

This interpretation, which summarizes and generalizes in a sense the Copenhagen and Many-worlds interpretations, was introduced in 1984 by R. Griffiths. The term 'history' here means simply an ordered sequence of physical events—it is about the construction of inherently consistent processes. A history $H_i$ is a set (a sequence) of statements $A_{i,j}$, each at a time $t_{i,j}$, in the form $H_i = (A_{i,1}, A_{i,2}, \ldots, A_{i,n})$. A physical process can be generally described by a number of different histories, which are combined into a history family. The core of the interpretation is a consistency criterion by which it can be checked whether the probability of the history family equals the sum of the probabilities of the individual histories, and thus the additive law of classical probabilities is satisfied.[25] Such history families are called consistent. An observer is not needed in this interpretation, but, on the other hand, here also the measurement problem is not solved in the end. Some further comments can be found in Appendix V, Vol. 2.

## 28.2.6   Collapse Theories

The first collapse theory (also called dynamical reduction theory) was developed 1984/5 by Ghirardi, Rimini and Weber and is named GRW theory after the authors. In the meantime, several variants have appeared. All have in common that the SEq is extended by non-linear and/or stochastic additional terms[26]; in this sense, they are therefore actually more likely to be seen as new theories than as interpretations. The additional terms are constructed in such a way that they 'disentangle' entangled states of large systems, and that a system collapses spontaneously into a spatially-localized state (spontaneous localization, dynamic collapse). Of course, the additional terms are adjusted so that these effects are very small in isolated microscopic systems, but very large and pronounced in macroscopic systems. Hence, these are realistic theories without hidden variables; a special observer is not needed. Because of the additional terms, one can in principle experimentally detect deviations from the usual quantum mechanics, even though this is not yet possible with current technology. Some further remarks on the GRW formalism can be found in Appendix V, Vol. 2.

A kind of mass-bounded mechanism was suggested by Roger Penrose. Accordingly, there is a gravitational effect on the mass of the quantum system that causes the collapse of the wavefunction. The heavier the system, the stronger and faster is the effect of gravity, so that for macroscopic systems, the collapse takes place almost immediately.

---

[25]The focus is thus on the history of a system and not on the value of an observable at a particular time.

[26]GRW proposed a stochastic additional term. Nonlinear terms can, for example, be based on gravity in the context of general relativity.

### 28.2.7 Other Interpretations

There are a number of other interpretations of quantum mechanics[27] which arose mainly in the last 30 years. A compact and brief overview, partly in tabular format, is provided by the Wikipedia article 'Interpretations of quantum mechanics'.[28] It contains also an extensive index, encompassing original articles, secondary literature, textbooks and web addresses. There are corresponding articles in some other languages, most of them also very instructive.

Finally, we want to refer briefly to the question of whether measurement has something to do with *consciousness*. This is a very controversial subject that is often classified as untrustworthy and highly speculative. Nevertheless, we want to summarize briefly some of the positions held.

In principle, these approaches differ in terms of the assumption of whether consciousness can be described in physical terms or not. The second group includes, for example, an early approach of von Neumann, according to which the human consciousness cannot be described by physics; but even so, as the ultimate measurement apparatus, consciousness transforms the possible into the factual on perceiving the result of a measurement. Of course, the introduction of consciousness as a nonphysical category, which nevertheless significantly affects physical results, is very unsatisfactory from a physical standpoint. We have only to think of quantum cosmology, which starts from quantum states of the entire universe; to maintain the range of validity of the deterministic SEq, we should therefore demand a 'universal consciousness'.

The many-minds interpretation, a variant of the many-worlds interpretation, also introduces consciousness as a new category. There is however no real, observer-independent splitting into parallel universes, but rather it is the brain or the consciousness of the individual observer which causes this splitting due to its self-awareness. There are different ideas about how this mechanism works; for example, J.B. Hartle explains (*Am. J. Phys.* 36 (1968) 704): "This 'reduction of the wave packet' does take place in the consciousness of the observer, not because of any unique physical process which takes place there, but only because the state is a construct of the observer and not an objective property of the physical system."

---

[27] Here one may also refer to *quantum logic*. This approach was suggested for the first time in 1936 by Birkhoff and von Neumann. By and large, it is about a modification of classical logic and its adaptation to the structure of Hilbert space. Classically we have, for example, yes-no statements with which, however, we cannot describe adequately the behavior of non-commuting variables such as position and momentum (is 'wrong' the same as 'not true'?). The basic idea may perhaps be described as follows: The classical view of an event is a subset of a total set (Abelian or commutative, distributive); in the view of quantum mechanics, an event is instead regarded as a subspace of a Hilbert space (non-Abelian, non-distributive).

[28] http://en.wikipedia.org/wiki/Interpretations_of_quantum_mechanics (accessed Dec 2017). Of course, there are always new papers on existing interpretations (e.g. D.H. Mahler et al., 'Experimental nonlocal and surreal Bohmian trajectories', *Science Advances* 2, doi10.1126/sciadv.150146619 (Feb 2016)) as well as new approaches and considerations (e.g. R. E. Kastner et al., Taking Heisenberg's Potentia Seriously , arXiv:1709.03595v4 (Oct 2017)).

However, if one takes the position that consciousness can be described in physical terms, one has to somehow make sure that the brain is 'disentangled', i.e. that it is not found in superposition states. This can be achieved either by means of a kind of many-minds interpretation, or by a corresponding collapse theory (e.g. with suitable non-linear terms), as it was postulated in an early form by Wigner (1961).[29] A different approach is offered by decoherence considerations, with the argument that decoherence effects (e.g. in nerve signal propagation) proceed so rapidly that the brain (considered as a quantum-mechanical system) is never perceptibly in a superposition state.

## 28.3   Conclusion

Central concepts of quantum mechanics such as probability, entanglement, state reduction, and measurement link it to fundamental epistemological categories such as deterministic versus random chance, or realism, non-locality, positivism versus subjectivism. In these aspects, quantum mechanics, although it functions excellently fapp, leaves open many questions.[30] To answer these as consistently as possible on a level beyond the mere application of the formalism is the aim of the various interpretations of quantum mechanics.

Obviously, however, none of the currently-discussed approaches can be considered as *the* solution. For each interpretation, there are pros and cons; none of them can prevail against its rivals on the basis of generally convincing objective reasons. In other words, which interpretation one prefers is more a question of taste or gut instinct than of logic.[31] Experimentally, one cannot necessarily expect a clarification. In fact, in recent years, certain issues have been decided in the laboratory (keyword 'Bell's inequality'), so that in the context of modern quantum mechanics the slogan 'practical metaphysics' or 'experimental philosophy' was coined. But since most current interpretations are not experimentally falsifiable, partly in principle, partly just not yet,[32] it will be difficult to bring about a decision in this way.

A fascinating aspect of this situation is, among others, how well quantum mechanics 'works' in spite of this interpretative fog and mist; in terms of its practical

---

[29]This approach has been quite popular at times and was then sometimes referred to as the 'standard interpretation'. We see again how diffuse the terminology can be in this area.

[30]Quite apart from further implications such as that 'entanglement' is in direct contradiction to the analytical, reductionist approach of occidental science.

[31]A non-representative survey among participants of the conference 'Quantum Physics and the Nature of Reality' (July 2011) revealed, among other things, that 42 % favor the 'Copenhagen interpretation', 24 % an 'information-based or information-theoretical interpretation' and, after all, 18 % the many-worlds theory of Everett. Unfortunately, the initiators of the survey left out the position of 'shut up and calculate' which ignores the issues of interpretation and aims at the utility of quantum mechanics in concrete applications M. Schlosshauer et al., 'A Snapshot of Foundational Attitudes Toward Quantum Mechanics', arXiv:1301.1069 (2013).

[32]We remember that it is not possible to justify the validity of a theory from a finite set of experiments, i.e. to verify it. Only its falsification is unique.

applications, there is virtually no dissent. Although, from this perspective, the discussion of interpretations may seem unnecessary, it belongs for many physicists to the self-image of physics to provide more than a collection of formal rules, but instead to aim at the most faithful possible representation of reality in the sense of a closed physical worldview. Insofar, such discussions as those about the interpretation of quantum mechanics are not simply meaningless for science, but rather they help to identify more sharply the problems and open issues. In addition, for example, the discussion about the term 'entanglement' shows that such 'soft' and non-formal debates can lead to very concrete results, for example the quantum computer.

To be fair, we must state that today we cannot foresee with certainty which (if any) of the current interpretations will prevail. Of course, quantum mechanics will (and must) continue to develop; just think of the missing link between quantum mechanics and gravity (or general relativity). But to what extent this development will help to clarify the open epistemological questions remains to be seen.

In any case, the situation is currently such that, as we pointed out in Chap. 27, even the foundations of physics are no longer taboo, and it is being discussed, among other things, whether we need a major revision including tacitly made assumptions such as logic or causality. It would not be a particularly great surprise if necessary extensions of quantum mechanics should prove to be extremely counter-intuitive.

Be that as it may—important issues are still unclear and open at present; no one knows on what journey quantum mechanics will yet take us. Only one thing is certain: it remains exciting and fascinating—the suspense continues.

# Appendix A
# Abbreviations and Notations

For a better overview, we collect here some abbreviations and specific notation.

**Abbreviations**

| | |
|---|---|
| ala | Algebraic approach |
| ana | Analytical approach |
| ClM | Classical mechanics |
| CONS | Complete orthonormal system |
| CSCO | Complete system of commuting observables |
| DEq | Differential equation |
| EPR | Einstein-Podolsky-Rosen paradox |
| $fapp$ | 'Fine for all practical purposes' |
| MZI | Mach–Zehnder interferometer |
| ONS | Orthonormal system |
| PBS | Polarizing beam splitter |
| QC | Quantum computer |
| QM | Quantum mechanics |
| QZE | Quantum Zeno effect |
| SEq | Schrödinger equation |

**Operators**

There are several different notations for an operator which is associated with a physical quantity $A$; among others: (1) $A$, that is the symbol itself, (2) $\hat{A}$, notation with hat (3) $\mathcal{A}$, calligraphic typeface, (4) $A_{op}$, notation with index. It must be clear from the context what is meant in each case.

For special quantities such as the position $x$, one also finds the uppercase notation $X$ for the corresponding operator.

**Many-Particle States**

For two quantum objects, the position gives the object number, if nothing is otherwise specified:

$$|nm\rangle = |n_1 m_2\rangle; \tag{A.1}$$

© Springer Nature Switzerland AG 2018
J. Pade, *Quantum Mechanics for Pedestrians 2*, Undergraduate Lecture
Notes in Physics, https://doi.org/10.1007/978-3-030-00467-5

$n$ and $m$ each stand for a single or for several quantum numbers.

With more than two quantum objects (object 1 with quantum numbers $\alpha_1$, object 2 with quantum numbers $\alpha_2$)..., we generally use the following notation:

$$|1 : \alpha_1, 2 : \alpha_2, \ldots, n : \alpha_n\rangle. \tag{A.2}$$

It is more transparent than the equivalent notation

$$\left|\varphi_{\alpha_1}^{(1)} \varphi_{\alpha_2}^{(2)} \ldots \varphi_{\alpha_n}^{(n)}\right\rangle. \tag{A.3}$$

Interchanging the quantum numbers (e.g. those of object 1 and 2) looks like this:

$$|1 : \alpha_2, 2 : \alpha_1, \ldots, n : \alpha_n\rangle \tag{A.4}$$

instead of

$$\left|\varphi_{\alpha_2}^{(1)} \varphi_{\alpha_1}^{(2)} \ldots \varphi_{\alpha_n}^{(n)}\right\rangle. \tag{A.5}$$

**The Hamiltonian and the Hadamard transformation**

We denote the Hamiltonian by $H$. With reference to questions of quantum information, especially in Chap. 27, Vol. 2, $H$ stands for the Hadamard transformation.

**Perturbation Calculations**

To denote Hamiltonians and states in perturbation theory, we use a superscript index in parentheses, indicating the perturbation order:

$$H^{(0)}; \ \left|\varphi^{(1)}\right\rangle \ \text{etc.} \tag{A.6}$$

**Tracing Out**

The reduced density operator, arising through tracing out all degrees of freedom $\neq k$, is denoted by a superscript index in parentheses:

$$\rho^{(k)} \tag{A.7}$$

**Vector Spaces**

We denote a vector space by $\mathcal{V}$, a Hilbert space by $\mathcal{H}$.

Based on the notation $\mathbb{R}^3$ or $\mathbb{C}^3$ for the three-dimensional real or complex space, we select the following notation, if necessary, for a more precise specification of Hilbert spaces:

$$\begin{aligned} &\qquad\qquad d = \text{dimension} \\ \mathcal{H}_{n(m)}^d \ \text{with} \ &n = \text{number of the corresponding quantum object} \\ &m = \text{total number of quantum objects.} \end{aligned} \tag{A.8}$$

# Appendix B
# Special Functions

We compile here some material for important special functions of quantum mechanics.

## B.1 Spherical Harmonics

The general form of the *spherical harmonics* $Y_l^m(\vartheta, \varphi)$[1] is

$$Y_l^m(\vartheta, \varphi) = f_l^m(\vartheta) \, e^{im\varphi} = (-1)^{\frac{m+|m|}{2}} \left[ \frac{2l+1}{4\pi} \frac{(l-|m|)!}{(l+|m|)!} \right]^{1/2} P_l^{|m|} (\cos \vartheta) \, e^{im\varphi},$$

(B.1)

where $P_l^m$ are the *associated Legendre functions*:

$$P_l^m(x) = \frac{(-1)^m}{2^l l!} \left(1 - x^2\right)^{\frac{m}{2}} \frac{d^{l+m}}{dx^{l+m}} \left(x^2 - 1\right)^l.$$

(B.2)

These are solutions of the differential equation

$$\left(1 - x^2\right) \frac{d^2 g(x)}{dx^2} - 2x \frac{dg(x)}{dx} + \left[ l(l+1) - \frac{m^2}{1 - x^2} \right] g(x) = 0.$$

(B.3)

In particular, we have for $m = 0$ the *Legendre polynomials* $P_l (\cos \vartheta)$:

$$Y_l^0(\vartheta, \varphi) = \sqrt{\frac{2l+1}{4\pi}} P_l (\cos \vartheta).$$

(B.4)

The spherical harmonics form a CONS, they are complete:

---

[1] Also called spherical functions, surface spherical harmonics, Laplace spherical harmonics, Laplace spherical functions or the like.

© Springer Nature Switzerland AG 2018
J. Pade, *Quantum Mechanics for Pedestrians 2*, Undergraduate Lecture Notes in Physics, https://doi.org/10.1007/978-3-030-00467-5

$$\sum_{l=0}^{\infty} \sum_{m=-l}^{l} Y_l^m (\vartheta, \varphi) \, Y_l^{m*} (\vartheta', \varphi') = \frac{\delta (\vartheta - \vartheta') \, \delta (\varphi - \varphi')}{\sin \vartheta} \tag{B.5}$$

and orthonormal

$$\int_0^{\pi} d\vartheta \sin \vartheta \int_0^{2\pi} d\varphi Y_l^{m*} (\vartheta, \varphi) \, Y_{l'}^{m'} (\vartheta, \varphi) = \delta_{ll'} \delta_{mm'}. \tag{B.6}$$

With the notation $\Omega = (\vartheta, \varphi)$ for the solid angle and $d\Omega = \sin \vartheta \, d\vartheta \, d\varphi$ (also written $d^2 \hat{r}$ or $d\hat{r}$), the orthogonality relation is written as:

$$\int Y_l^{m*} (\vartheta, \varphi) \, Y_{l'}^{m'} (\vartheta, \varphi) \, d\Omega = \delta_{ll'} \delta_{mm'}. \tag{B.7}$$

Thus, for the Legendre polynomials:

$$\int d\Omega \, P_l (\cos \vartheta) \, P_{l'} (\cos \vartheta) = \frac{4\pi}{2l + 1} \delta_{ll'}. \tag{B.8}$$

The addition theorem of the spherical harmonics reads

$$\frac{2l + 1}{4\pi} P_l (\cos \alpha) = \sum_{m=-l}^{l} Y_l^{m*} (\vartheta_1, \varphi_1) \, Y_l^m (\vartheta_2, \varphi_2), \tag{B.9}$$

where $\alpha$ is the angle between the directions $(\vartheta_1, \varphi_1)$ and $(\vartheta_2, \varphi_2)$.

The product of two spherical harmonics is given by[2]:

$$Y_{l_1}^{m_1} (\vartheta, \varphi) \, Y_{l_2}^{m_2} (\vartheta, \varphi) = \sqrt{\frac{(2l_1 + 1) (2l_2 + 1)}{4\pi}}$$

$$\times \sum_{L=|l_1-l_2|}^{l_1+l_2} \sqrt{\frac{1}{2L + 1}} \, \langle l_1 l_2 0 0 \, | L 0 \rangle$$

$$\times \langle l_1 l_2 m_1 m_2 \, | L M \rangle \, Y_L^M (\hat{\mathbf{r}}); \quad M = m_1 + m_2. \tag{B.10}$$

Finally, we give explicitly the first few spherical harmonics:

---

[2]For the proof, one uses properties of the rotation matrices which are related to the spherical harmonics by $D_{m0}^{(l)} (\vartheta, \varphi) = \sqrt{\frac{4\pi}{2l+1}} Y_l^{m*} (\vartheta, \varphi)$. For the Clebsch–Gordan coefficients see Chap. 16.

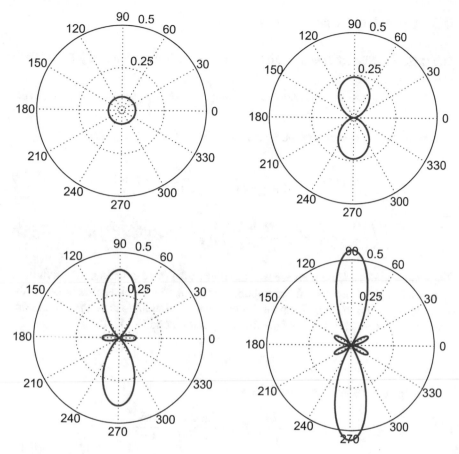

**Fig. B.1** Polar diagram of $\left|Y_l^0\right|^2$ for $l = 0, 1, 2, 3$

$$Y_0^0 = \frac{1}{\sqrt{4\pi}}; \quad Y_1^0 = \sqrt{\frac{3}{4\pi}} \cos \vartheta; \quad Y_1^1 = -\sqrt{\frac{3}{8\pi}} \sin \vartheta e^{i\varphi}$$

$$Y_2^0 = \sqrt{\frac{5}{16\pi}} \left(3 \cos^2 \vartheta - 1\right); \quad Y_2^1 = -\sqrt{\frac{15}{8\pi}} \sin \vartheta \cos \vartheta e^{i\varphi}; \quad Y_2^2 = \sqrt{\frac{15}{32\pi}} \sin^2 \vartheta e^{2i\varphi}$$

$$Y_3^0 = \sqrt{\frac{7}{16\pi}} \left(5 \cos^3 \vartheta - 3 \cos \vartheta\right); \quad Y_3^1 = -\sqrt{\frac{21}{64\pi}} \sin \vartheta \left(5 \cos^2 \vartheta - 1\right) e^{i\varphi}$$

$$Y_3^2 = \sqrt{\frac{105}{32\pi}} \sin^2 \vartheta \cos \vartheta e^{2i\varphi}; \quad Y_3^3 = -\sqrt{\frac{35}{64\pi}} \sin^3 \vartheta e^{3i\varphi}. \tag{B.11}$$

The graphical representation of some spherical harmonics is shown in Fig. B.1.

## B.2   Spherical Bessel Functions

The stationary SEq for free quantum objects in spherical coordinates (cf. Chap. 17) reads:

$$E\psi = -\frac{\hbar^2}{2m}\nabla^2\psi = -\frac{\hbar^2}{2m}\left(\frac{1}{r}\frac{\partial^2}{\partial r^2}r - \frac{\mathbf{l}^2}{\hbar^2 r^2}\right)\psi. \qquad (B.12)$$

Correspondingly, we can use the *ansatz* $\psi(\mathbf{r}) = y_l(r)\,Y_l^m(\vartheta, \varphi)$ and obtain

$$Ey_l(r) = -\frac{\hbar^2}{2m}\left(\frac{1}{r}\frac{\partial^2}{\partial r^2}r - \frac{l(l+1)}{r^2}\right)y_l(r), \qquad (B.13)$$

or

$$\left(\frac{1}{r}\frac{\partial^2}{\partial r^2}r + k^2 - \frac{l(l+1)}{r^2}\right)y_l(r) = 0; \quad k^2 = \frac{2m}{\hbar^2}E. \qquad (B.14)$$

The *spherical Bessel functions* are special solutions of these equations. The solutions which are regular at $r = 0$ are called proper spherical Bessel functions $j_l$; the irregular solutions are the Neumann functions $n_l$. Combinations of these functions are the Hankel functions $h_l^{(\pm)}$ of the first $(+)$ and second $(-)$ kind:

$$h_l^{(\pm)} = n_l \pm i j_l. \qquad (B.15)$$

The functions with $l = 0$ and $l = 1$ are:

$$j_0 = \frac{\sin kr}{kr}; \quad j_1 = \frac{\sin kr}{(kr)^2} - \frac{\cos kr}{kr}$$

$$(B.16)$$

$$n_0 = \frac{\cos kr}{kr}; \quad n_1 = \frac{\cos kr}{(kr)^2} + \frac{\sin kr}{kr}.$$

Functions with higher indices can be computed recursively; with $x = kr$, it follows for instance:

$$(2l+1)f_l = x(f_{l+1} + f_{l-1}); \quad l \neq 0 \qquad (B.17)$$

or

$$f_l = -x^{l-1}\frac{d}{dx}\left(\frac{f_{l-1}}{x^{l-1}}\right) = \left[x^l\left(-\frac{1}{x}\frac{d}{dx}\right)^l\right]f_0. \qquad (B.18)$$

Here, $f_l = c_1 j_l + c_2 n_l$ is an arbitrary linear combination of $j_l$ and $n_l$.

Their behavior at the origin is given by

$$j_l(x) \sim \frac{x^l}{(2l+1)!!} \left[ 1 + O\left(x^2\right) \right]$$

$$n_l(x) \sim \frac{(2l+1)!!}{(2l+1)} \left(\frac{1}{x}\right)^{l+1} \left[ 1 + O\left(x^2\right) \right] \quad ; \quad x \to 0, \qquad (B.19)$$

and the asymptotic forms are

$$j_l(x) \sim \frac{1}{x} \sin\left(x - \frac{l\pi}{2}\right) \left[ 1 + O\left(\frac{1}{x}\right) \right]$$

$$n_l(x) \sim \frac{1}{x} \cos\left(x - \frac{l\pi}{2}\right) \left[ 1 + O\left(\frac{1}{x}\right) \right] \quad ; \quad x \to \infty. \qquad (B.20)$$

The spherical Bessel functions play an important role in (among others) scattering theory, since under certain conditions they constitute the asymptotic solutions. We have e.g. for the outgoing scattered wave:

$$\psi_{\text{out}} \to h_l^{(+)}(x) = n_l(x) + i j_l(x) \to \frac{e^{ix}}{x}. \qquad (B.21)$$

Due to the relation

$$j_l(x) = \sqrt{\frac{\pi}{2x}} J_{l+\frac{1}{2}}(x)$$

$$n_l(x) = (-1)^l \sqrt{\frac{\pi}{2x}} J_{-l-\frac{1}{2}}(x), \qquad (B.22)$$

where $J_\nu(x)$ are the ordinary Bessel functions of order $\nu$, the spherical Bessel functions are also called 'half-integer Bessel functions' or 'small Bessel functions'.

## B.3 Eigenfunctions of the Hydrogen Atom

The potential is given by

$$V(r) = -\frac{1}{4\pi\varepsilon_0} \frac{Ze^2}{r} \qquad (B.23)$$

where $Ze$ is the nuclear charge and $e$ the electronic charge. The eigenfunctions are

$$\psi_{nlm}(r) = R_{nl}(r) Y_l^m(\vartheta, \varphi). \qquad (B.24)$$

Here, $n = 1, 2, \ldots$ is the principal quantum number, $l$ and $m$ determine the angular momentum.

The *radial functions* are given by:

$$R_{nl} = \sqrt{\frac{(n-l-1)!\,(2\kappa)^3}{2n\,((n+l)!)^3}}\,(2\kappa r)^l\,e^{-\kappa r}L_{n+l}^{2l+1}\,(2\kappa r) \qquad \text{(B.25)}$$

with

$$\kappa = \frac{Z}{na_0}. \qquad \text{(B.26)}$$

The radius $a_0$ is given by ($\mu$ = reduced mass):

$$a = \frac{a_0}{Z} = \frac{\hbar^2}{Z\mu e^2};\; a_0 \approx \text{Bohr radius}\, a_B = \frac{\hbar^2}{me^2}. \qquad \text{(B.27)}$$

The functions $L_{n+l}^{2l+1}\,(y)$ are the *associated Laguerre polynomials*; they can be calculated from

$$L_r^s\,(y) = \left(-\frac{\mathrm{d}}{\mathrm{d}y}\right)^s e^y \left(\frac{\mathrm{d}}{\mathrm{d}y}\right)^r e^{-y} y^r. \qquad \text{(B.28)}$$

The first radial functions are:

K-shell, s-orbital: $R_{10}\,(r) = 2\left(\frac{Z}{a_0}\right)^{\frac{3}{2}} e^{-\frac{Zr}{a_0}}$

L-shell, s-orbital: $R_{20}\,(r) = 2\cdot\left(\frac{Z}{2a_0}\right)^{\frac{3}{2}}\left(1 - \frac{Zr}{2a_0}\right)e^{-\frac{Zr}{2a_0}}$

L-shell, p-orbital: $R_{21}\,(r) = \frac{1}{\sqrt{3}}\cdot\left(\frac{Z}{2a_0}\right)^{\frac{3}{2}}\frac{Zr}{a_0}e^{-\frac{Zr}{2a_0}}$

M-shell, s-orbital: $R_{30}\,(r) = 2\cdot\left(\frac{Z}{2a_0}\right)^{\frac{3}{2}}\left(1 - \frac{2Zr}{3a_0} + \frac{2Z^2r^2}{27a_0^2}\right)e^{-\frac{Zr}{3a_0}}$ $\qquad$ (B.29)

M-shell, p-orbital: $R_{31}\,(r) = \frac{4\sqrt{2}}{3}\cdot\left(\frac{Z}{3a_0}\right)^{\frac{3}{2}}\frac{Zr}{a_0}\left(1 - \frac{Zr}{6a_0}\right)e^{-\frac{Zr}{3a_0}}$

M-shell, d-orbital: $R_{32}\,(r) = \frac{2\sqrt{2}}{27\sqrt{5}}\cdot\left(\frac{Z}{3a_0}\right)^{\frac{3}{2}}\left(\frac{Zr}{a_0}\right)^2 e^{-\frac{Zr}{3a_0}}.$

A graphical representation of some radial functions is shown in Figs. B.2 and B.3. For the mean values in the state $\psi_{nlm}$, we have

$$\left\langle\frac{1}{r}\right\rangle = \frac{Z}{a_0 n^2};\; \left\langle\frac{1}{r^2}\right\rangle = \frac{Z^2}{a_0^2 n^3}\frac{1}{l+\frac{1}{2}};\; \left\langle\frac{1}{r^3}\right\rangle = \frac{Z^3}{a_0^3 n^3}\frac{1}{l\left(l+\frac{1}{2}\right)(l+1)}$$

$$\langle r\rangle = \frac{1}{2}\left[3n^2 - l\,(l+1)\right]\frac{a_0}{Z};\; \langle r^2\rangle = \frac{1}{2}\left[5n^2 + 1 - 3l\,(l+1)\right]n^2\frac{a_0^2}{Z^2},$$
$$\text{(B.30)}$$

and for $s > -2l - 1$, we find the recursion relation

$$\frac{s+1}{n^2}\langle r^s\rangle - (2s+1)\frac{a_0}{Z}\langle r^{s-1}\rangle + \frac{s}{4}\left[(2l+1)^2 - s^2\right]\frac{a_0^2}{Z^2}\langle r^{s-2}\rangle = 0. \qquad \text{(B.31)}$$

**Fig. B.2** The radial functions $R_{10}$ (*red*), $R_{20}$ (*green*) and $R_{21}$ (*blue*). Not normalized; $x = \frac{Zr}{a_0}$

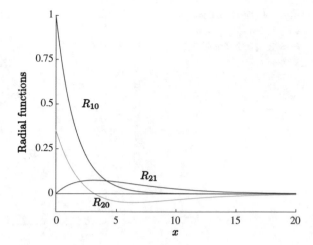

**Fig. B.3** The radial functions $R_{30}$ (*red*), $R_{31}$ (*green*) and $R_{32}$ (*blue*). Not normalized; $x = \frac{Zr}{a_0}$

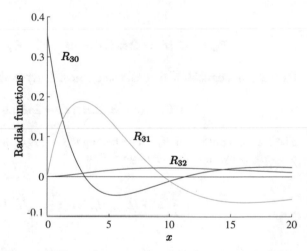

## B.4 Hermite Polynomials

The eigenfunctions of the harmonic oscillator can be written as

$$\psi_n(x) = \left(\frac{m\omega}{\hbar\pi}\right)^{1/4} \frac{1}{\sqrt{n! \cdot 2^n}} \cdot e^{-\frac{x^2}{2}} \cdot H_n(x), \tag{B.32}$$

where the *Hermite polynomials* are defined by

$$H_n(x) = e^{\frac{x^2}{2}} \left(x - \frac{d}{dx}\right)^n e^{-\frac{x^2}{2}}. \tag{B.33}$$

The first few Hermite polynomials are

**Fig. B.4** The functions $e^{-\frac{x^2}{2}} H_n(x)/\sqrt{n!2^n}$ for $n = 0, 1, 2, 3$ (*black, red, green, blue*)

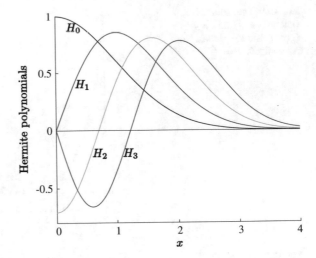

$$H_0 = 1; \; H_1 = 2x; \; H_2 = 4x^2 - 2; \; H_3 = 8x^3 - 12x. \tag{B.34}$$

Further polynomials are best calculated recursively from

$$H_{n+1}(x) = 2xH_n(x) - 2nH_{n-1}(x). \tag{B.35}$$

The Hermite polynomials, which belong to the important class of *orthogonal polynomials*, obey the orthogonality relation:

$$\int_{-\infty}^{\infty} e^{-\frac{x^2}{2}} H_n(x)H_m(x)dx = \sqrt{\pi}n!2^n \delta_{nm}. \tag{B.36}$$

A graphical representation of some Hermite polynomials is found in Fig. B.4.

## B.5 Waves

A *plane wave*[3] travelling in the $z$ direction (i.e. $\mathbf{k} = (0, 0, k)$) can be decomposed into partial waves:

$$e^{ikz} = \sum_{l=0}^{\infty} (2l + 1) i^l j_l(kr) P_l(\cos \vartheta). \tag{B.37}$$

---

[3]Wave, because one considers explicitly only the spatial part of $e^{i(kz-\omega t)}$, adding tacitly the term $e^{-i\omega t}$.

Here, spherical coordinates are assumed.

Generalizing, we have for $\mathbf{r} \to (r, \vartheta_r, \varphi_r)$ and $\mathbf{k} \to (k, \vartheta_k, \varphi_k)$ the representation

$$e^{i\mathbf{k}\cdot\mathbf{r}} = 4\pi \sum_{l=0}^{\infty} \sum_{m=-l}^{l} i^l j_l(kr) Y_l^{m*}(\vartheta_k, \varphi_k) Y_l^m(\vartheta_r, \varphi_r). \qquad (B.38)$$

With the help of the addition theorem (B.9), we can also write:

$$e^{i\mathbf{k}\cdot\mathbf{r}} = \sum_{l=0}^{\infty} (2l+1) i^l j_l(kr) P_l(\cos\alpha), \qquad (B.39)$$

where $\alpha$ is the angle between the directions $(\vartheta_k, \varphi_k)$ and $(\vartheta_r, \varphi_r)$.

For *outgoing spherical waves*, we have (for incoming waves correspondingly $e^{-i\cdots}$):

$$\frac{e^{ik|\mathbf{r}_1-\mathbf{r}_2|}}{|\mathbf{r}_1 - \mathbf{r}_2|} = k \sum_{l=0}^{\infty} (2l+1) i^l j_l(kr_<) h_l^{(+)}(kr_>) P_l(\cos\alpha), \qquad (B.40)$$

where $\alpha$ is the angle between the directions of $\mathbf{r}_1$ and $\mathbf{r}_2$, and the abbreviations are defined as $r_< = \min(r_1, r_2)$ and $r_> = \max(r_1, r_2)$. In particular, for $k = 0$ we have

$$\frac{1}{|\mathbf{r}_1 - \mathbf{r}_2|} = \sum_{l=0}^{\infty} \frac{r_<^l}{r_>^{l+1}} P_l(\cos\alpha) = \sum_{l,m} \frac{r_<^l}{r_>^{l+1}} \frac{4\pi}{2l+1} Y_l^m(\vartheta_1, \varphi_1) Y_l^{m*}(\vartheta_2, \varphi_2). \qquad (B.41)$$

# Appendix C
# Tensor Product

We discuss here some properties of the *tensor product* of vector spaces. In order to make it more familiar, we will write down some results in explicit form (and in doing so, we will see that this notation is quite clumsy).

## C.1 Direct Product

The tensor product connects two or more vector spaces to form a common vector space, also called the *product space*. Assuming two vector spaces $V_1$ (dimension $N$) and $V_2$ (dimension $M$), we write for the vector product $V$

$$V = V_1 \otimes V_2 \tag{C.1}$$

If the spaces $V_1$ and $V_2$ have the bases $\{|n\rangle_1\}$ and $\{|m\rangle_2\}$, the basis of the product space is the set of all pairs $\{|n\rangle_1 \otimes |m\rangle_2\}$. Thus, the dimension of the product space is $N \cdot M$, and a general state vector has the form

$$|\psi\rangle = \sum_{n,m} c_{nm} |n\rangle_1 \otimes |m\rangle_2, \tag{C.2}$$

where we write also $|n_1 \otimes m_2\rangle$ or simply $|n\rangle_1 |m\rangle_2$, $|n_1\rangle |m_2\rangle$ or $|n_1 m_2\rangle$ instead of $|n\rangle_1 \otimes |m\rangle_2$. If the meaning is clear from context, we will omit the indices.

Here, we give an example in component representation: Given that

$$|u\rangle \cong \begin{pmatrix} u_1 \\ u_2 \\ u_3 \end{pmatrix}; \quad |v\rangle \cong \begin{pmatrix} v_1 \\ v_2 \end{pmatrix}; \tag{C.3}$$

© Springer Nature Switzerland AG 2018
J. Pade, *Quantum Mechanics for Pedestrians 2*, Undergraduate Lecture
Notes in Physics, https://doi.org/10.1007/978-3-030-00467-5

then we have[4]:

$$|w\rangle = |u\rangle \otimes |v\rangle \cong \begin{pmatrix} u_1 \\ u_2 \\ u_3 \end{pmatrix} \otimes \begin{pmatrix} v_1 \\ v_2 \end{pmatrix} = \begin{pmatrix} u_1 v_1 \\ u_1 v_2 \\ u_2 v_1 \\ u_2 v_2 \\ u_3 v_1 \\ u_3 v_2 \end{pmatrix}. \tag{C.4}$$

## C.2 Direct Sum of Vector Spaces

We mention this term briefly since it is sometimes confused with the direct product. For the vector spaces $\mathcal{V}_1$ (dimension $N$) and $\mathcal{V}_2$ (dimension $M$), the direct sum is written as:

$$\mathcal{V} = \mathcal{V}_1 \oplus \mathcal{V}_2, \tag{C.5}$$

where the space has dimension $N + M$. If $\mathcal{V}_1$ and $\mathcal{V}_2$ have the bases $\{|n\rangle_1\}$ and $\{|m\rangle_2\}$, then the basis of the sum space is the set $\{|1\rangle_1, |2\rangle_2, \ldots |1\rangle_2, |2\rangle_2, \ldots\}$.

Example in component representation: we have

$$|w\rangle = |u\rangle \oplus |v\rangle \cong \begin{pmatrix} u_1 \\ u_2 \\ u_3 \end{pmatrix} \oplus \begin{pmatrix} v_1 \\ v_2 \end{pmatrix} = \begin{pmatrix} u_1 \\ u_2 \\ u_3 \\ v_1 \\ v_2 \end{pmatrix}. \tag{C.6}$$

## C.3 Properties of the Tensor Product

Tensor products can be carried out multiply; an example is:

$$\mathcal{V} = \mathcal{V}_1 \otimes \mathcal{V}_2 \otimes \cdots \otimes \mathcal{V}_n = \bigotimes_{l=1}^{n} \mathcal{V}_l. \tag{C.7}$$

For identical spaces ($\mathcal{V}_1 = \mathcal{V}_2 = \cdots$), one can write this as *tensor power* $\mathcal{V} = \mathcal{V}_1^{\otimes N}$ or shortly $\mathcal{V}_1^N$.

We assume for the following that the operator $U$ acts only in space 1 and $V$ only in space 2.

A tensor product of operators acts on a tensor product of vectors in each space separately:

---

[4]Rule: The right index changes the fastest.

$$(U \otimes V)(|u\rangle \oplus |v\rangle) = U|u\rangle \oplus V|v\rangle. \tag{C.8}$$

In contrast to proper operator products, the order is not changed in the adjoint

$$(U \otimes V)^\dagger = U^\dagger \otimes V^\dagger; \quad (|u\rangle \oplus |v\rangle)^\dagger = \langle u| \otimes \langle v|. \tag{C.9}$$

We have for example

$$\{(U_1 U_2 U_3) \otimes (V_1 V_2)\}^\dagger = (U_1 U_2 U_3)^\dagger \otimes (V_1 V_2)^\dagger = \left(U_3^\dagger U_2^\dagger U_1^\dagger\right) \otimes \left(V_2^\dagger V_1^\dagger\right). \tag{C.10}$$

## C.4 Examples

### C.4.1 General Examples

Given two matrices $A$ and $B$ with

$$A = \begin{pmatrix} 1 & 2 & 3 \\ 4 & 5 & 6 \\ 7 & 8 & 9 \end{pmatrix}; \quad B = \begin{pmatrix} a & b \\ c & d \end{pmatrix} \tag{C.11}$$

Then it follows for $A \otimes B$:

$$A \otimes B = \begin{pmatrix} 1B & 2B & 3B \\ 4B & 5B & 6B \\ 7B & 8B & 9B \end{pmatrix} = \begin{pmatrix} a & b & 2a & 2b & 3a & 3b \\ c & d & 2c & 2d & 3c & 3d \\ 4a & 4b & 5a & 5b & 6a & 6h \\ 4c & 4d & 5c & 5d & 6c & 6d \\ 7a & 7b & 8a & 8b & 9a & 9b \\ 7c & 7d & 8c & 8d & 9c & 9d \end{pmatrix} \tag{C.12}$$

and for $A \oplus B$

$$A \oplus B = \begin{pmatrix} A & 0 \\ 0 & B \end{pmatrix} = \begin{pmatrix} 1 & 2 & 3 & 0 & 0 \\ 4 & 5 & 6 & 0 & 0 \\ 7 & 8 & 9 & 0 & 0 \\ 0 & 0 & 0 & a & b \\ 0 & 0 & 0 & c & d \end{pmatrix}. \tag{C.13}$$

### C.4.2 Example with Reference to Chap. 20

We start with

$$|h\rangle \cong \begin{pmatrix} 1 \\ 0 \end{pmatrix}; \quad |v\rangle \cong \begin{pmatrix} 0 \\ 1 \end{pmatrix} \tag{C.14}$$

Then it follows with $|hh\rangle \equiv |h\rangle \otimes |h\rangle$

$$|hh\rangle \cong \begin{pmatrix} 1 \\ 0 \\ 0 \\ 0 \end{pmatrix}, \quad |hv\rangle \cong \begin{pmatrix} 0 \\ 1 \\ 0 \\ 0 \end{pmatrix}, \quad |vh\rangle \cong \begin{pmatrix} 0 \\ 0 \\ 1 \\ 0 \end{pmatrix}, \quad |vv\rangle \cong \begin{pmatrix} 0 \\ 0 \\ 0 \\ 1 \end{pmatrix}$$

and

$$|\Phi\rangle = \frac{|hv\rangle - |vh\rangle}{\sqrt{2}} \cong \frac{1}{\sqrt{2}} \begin{pmatrix} 0 \\ 1 \\ -1 \\ 0 \end{pmatrix}. \tag{C.15}$$

The measurement of the first component of this state with respect to horizontal polarization is described by

$$(|h_1\rangle \otimes I_2)\,(|h_1\rangle \otimes I_2)^\dagger = (|h_1\rangle \otimes I_2)\,(\langle h_1| \otimes I_2), \tag{C.16}$$

where $I_2$ is the one-operator in space 2 (for the sake of clarity, we use indices). We have

$$|h_1\rangle \otimes I_2 \cong \begin{pmatrix} 1 \\ 0 \end{pmatrix} \otimes \begin{pmatrix} 1 & 0 \\ 0 & 1 \end{pmatrix} = \begin{pmatrix} 1 & 0 \\ 0 & 1 \\ 0 & 0 \\ 0 & 0 \end{pmatrix}, \tag{C.17}$$

and it follows that

$$(|h_1\rangle \otimes I_2)\,(\langle h_1| \otimes I_2) \cong \begin{pmatrix} 1 & 0 \\ 0 & 1 \\ 0 & 0 \\ 0 & 0 \end{pmatrix} \begin{pmatrix} 1 & 0 & 0 & 0 \\ 0 & 1 & 0 & 0 \end{pmatrix} = \begin{pmatrix} 1 & 0 & 0 & 0 \\ 0 & 1 & 0 & 0 \\ 0 & 0 & 0 & 0 \\ 0 & 0 & 0 & 0 \end{pmatrix} \tag{C.18}$$

This leads to

$$(|h_1\rangle \otimes I_2)\,(\langle h_1| \otimes I_2)\,|\Phi\rangle \cong \frac{1}{\sqrt{2}} \begin{pmatrix} 1 & 0 & 0 & 0 \\ 0 & 1 & 0 & 0 \\ 0 & 0 & 0 & 0 \\ 0 & 0 & 0 & 0 \end{pmatrix} \begin{pmatrix} 0 \\ 1 \\ -1 \\ 0 \end{pmatrix} = \frac{1}{\sqrt{2}} \begin{pmatrix} 0 \\ 1 \\ 0 \\ 0 \end{pmatrix} \tag{C.19}$$

as is required.

The same procedure in a more compact notation reads:

$$
\begin{aligned}
(|h_1\rangle \otimes I_2)\,(\langle h_1| \otimes I_2)\,|\Phi\rangle &= (|h_1\rangle \otimes I_2)\,(\langle h_1| \otimes I_2)\,\frac{|h_1 v_2\rangle - |v_1 h_2\rangle}{\sqrt{2}} \\
&= (|h_1\rangle \otimes I_2)\,\frac{\langle h_1| \otimes I_2\,|h_1 v_2\rangle - \langle h_1| \otimes I_2\,|v_1 h_2\rangle}{\sqrt{2}} \\
&= (|h_1\rangle \otimes I_2)\,\frac{|v_2\rangle}{\sqrt{2}} = \frac{|h_1\rangle \otimes |v_2\rangle}{\sqrt{2}} = \frac{|h_1\rangle\,|v_2\rangle}{\sqrt{2}}.
\end{aligned}
\tag{C.20}
$$

# Appendix D
# Wave Packets

## D.1 General Remarks

A plane wave is not a physically realizable state: it is infinitely extended and has the same magnitude in all places and at all times (squared amplitude value). Mathematically, this is expressed by the fact that it is not square integrable. But because of the linearity of quantum mechanics, we can superpose individual waves in such a way that physically 'reasonable' expressions arise (keyword: Fourier transformation). In addition, we can construct these superpositions in such a way that they have (at least approximately) a well-defined momentum, like classical particles. We want to discuss in the following some of the characteristics of these *wave packets*.[5]

### D.1.1 One-Dimensional Wave Packet

A one-dimensional wave packet generally has the form[6]:

$$\psi(x, t) = \frac{1}{\sqrt{2\pi}} \int_{-\infty}^{\infty} A(k) e^{i(kx - \omega t)} dk, \qquad (D.1)$$

where the amplitude function $A(k)$ is usually centered around a value $K$ and has a pronounced maximum at $K$; see Fig. D.1. As an example, we can imagine a bell curve (Gaussian curve), centered at $K$, in which case $\psi(x, t)$ may be explicitly represented (see below). But what information can be obtained in the general case?

---

[5]We have already discussed some of the properties of wave packets in Chap. 15.
[6]We extract the factor $\frac{1}{\sqrt{2\pi}}$ from the integral in order to write the Fourier transform as usual.

© Springer Nature Switzerland AG 2018
J. Pade, *Quantum Mechanics for Pedestrians 2*, Undergraduate Lecture Notes in Physics, https://doi.org/10.1007/978-3-030-00467-5

**Fig. D.1** Schematic
representation of the
amplitude function $|A(k)|$

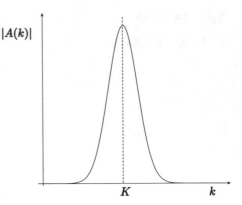

To answer this question, we first write $A(k) = |A(k)| e^{i\varphi(k)}$ and obtain

$$\psi(x, t) = \frac{1}{\sqrt{2\pi}} \int_{-\infty}^{\infty} |A(k)| e^{i(kx - \omega t + \varphi(k))} dk = \frac{1}{\sqrt{2\pi}} \int_{-\infty}^{\infty} |A(k)| e^{i\Phi(k)} dk. \quad (D.2)$$

The magnitude of the integral now depends on how fast $e^{i\Phi(k)}$ oscillates in the neighborhood of $K$ (only there is $|A(k)|$ significantly different from zero)—the faster the oscillations, the smaller becomes the integral (or its value). In general, we obtain the largest contribution if $\Phi(k)$ does not vary in the neighborhood of $K$, i.e. if

$$\frac{d\Phi(k)}{dk} = \frac{d(kx - \omega t + \varphi(k))}{dk} = x - \frac{d\omega}{dk}t + \frac{d\varphi(k)}{dk} = 0 \text{ for } k = K. \quad (D.3)$$

With $\omega = \frac{\hbar k^2}{2m}$, it follows that:

$$x - \frac{\hbar k}{m}t + \frac{d\varphi(k)}{dk} = 0; \quad (D.4)$$

or with $k = K$,

$$x = \frac{\hbar K}{m}t - \frac{d\varphi(k)}{dk}\Big/_{k=K} = v_g t - \frac{d\varphi(k)}{dk}\Big/_{k=K}. \quad (D.5)$$

The *group velocity* $v_g = \frac{d\omega}{dk}$ $(k = K)$ denotes the propagation velocity of the wave packet, while the *phase velocity* $v_{ph} = \frac{\omega}{k}$ denotes the propagation velocity of the individual partial waves (with fixed $k$). Generally, it holds that $v_g \neq v_{ph}$ (the wave packet deforms in the course of time, e.g. it diverges); for $v_g = v_{ph}$, one speaks of a dispersion-free wave (e.g. electromagnetic waves in a vacuum). The concept of group velocity, moreover, makes sense only if the superposition still has a recognizable coherence, i.e. it is not fragmented.

To obtain more information about the behavior of the wave packet, we expand $\omega$ in the neighborhood of $k = K$:

$$\omega(k) = \frac{\hbar k^2}{2m} = \frac{\hbar(k - K + K)^2}{2m} = \frac{\hbar K^2}{2m} + \frac{\hbar(k - K)K}{m} + \frac{\hbar(k - K)^2}{2m} \quad \text{(D.6)}$$

and obtain with $\Omega = \frac{\hbar K^2}{2m}$ and the group velocity $v_g = \frac{d\omega}{dk} = \frac{\hbar K}{m}$ for the phase $\Phi(k)$:

$$\begin{aligned}
\Phi(k) &= kx - \omega(k)t + \varphi(k) \\
&= kx - \left[\frac{\hbar K^2}{2m} + \frac{\hbar k K}{m} - \frac{\hbar K^2}{m} + \frac{\hbar(k - K)^2}{2m}\right]t + \varphi(k) \\
&= kx + \Omega t - v_g k t - \frac{\hbar(k - K)^2}{2m}t + \varphi(k) \\
&= k\left(x - v_g t\right) + \Omega t - \frac{\hbar(k - K)^2}{2m}t + \varphi(k). \quad \text{(D.7)}
\end{aligned}$$

If we can neglect the quadratic term $(k - K)^2$ (for instance, if $c(k)$ is very narrowly concentrated about $K$), we obtain with (D.2) (see the exercises):

$$\psi(x, t) = \frac{1}{\sqrt{2\pi}} e^{i\Omega t} \int_{-\infty}^{\infty} A(k) e^{ik(x - v_g t)} dk = e^{i\Omega t} \psi\left(x - v_g t, 0\right). \quad \text{(D.8)}$$

This means that, under these conditions, the wavefunction moves unchanged to the right (for $v_g > 0$) with the velocity $v_g$. The approximation is valid for

$$\frac{\hbar(k - K)^2}{2m}t \ll 1 \text{ or } t \ll \frac{2m}{\hbar(k - K)^2}. \quad \text{(D.9)}$$

If this inequality is satisfied, the wave packet (almost) does not disperse.

## D.1.2  Example: Bell Curve

A normalized $k$ distribution in the form of a bell curve is given by:

$$A(k) = \left(\frac{b_0^2}{\pi}\right)^{1/4} \exp\left(-\frac{b_0^2}{2}(k - K)^2\right). \quad \text{(D.10)}$$

Its maximum is at $K$; its width in momentum space is $\Delta k = \frac{2}{b_0}$ (the two turning points of $A(k)$ are at $k = K \pm \frac{1}{b_0}$). Thus, the corresponding wave packet is

$$\psi(x, t) = \frac{1}{\sqrt{2\pi}} \left(\frac{b_0^2}{\pi}\right)^{1/4} \int_{-\infty}^{\infty} e^{-\frac{b_0^2}{2}(k-K)^2} e^{i(kx-\omega t)} dk \tag{D.11}$$

with $\omega = \frac{\hbar k^2}{2m}$. This expression allows for a closed solution (calculation by variable substitution and integration in the complex plane; for details, see relevant textbooks). The result reads:

$$\psi(x, t) = \frac{1}{\sqrt{2}N(t)} \left(\frac{b_0^2}{\pi}\right)^{1/4} \exp\left(E^2(x, t) - \frac{1}{2}b_0^2 K^2\right) \tag{D.12}$$

with

$$N(t) = \sqrt{\frac{b_0^2}{2} + \frac{i\hbar t}{2m}}; \quad E(x, t) = \frac{Kb_0^2 + ix}{2N(t)}. \tag{D.13}$$

We are interested especially in the 'size' of $\psi(x, t)$, i.e. the squared amplitude value. From (D.12), it follows that:

$$|\psi(x, t)|^2 = \frac{1}{\sqrt{\pi}b(t)} \exp\left(-\frac{\left(x - \frac{\hbar K}{m}t\right)^2}{b^2(t)}\right) \tag{D.14}$$

with

$$b(t) = \sqrt{b_0^2 + \left(\frac{\hbar t}{mb_0}\right)^2}. \tag{D.15}$$

As we see from (D.14), $|\psi(x, t)|^2$ is very small for values that are far from $x - \frac{\hbar K}{m}t = 0$. Also from (D.14), we see directly that the group velocity is

$$v_g = \frac{d\omega}{dk} = \frac{\hbar K}{m}. \tag{D.16}$$

This illustrates the above considerations.

By the way, the results of this section are found also in the discussion of free motion (see Chap. 5, Vol. 1).

### D.1.3 Many-Dimensional Wave Packet

The generalization from one- to $n$-dimensional wave packets offers no fundamental surprises. We have

$$\psi(\mathbf{x}, t) = \frac{1}{(2\pi)^{n/2}} \int\limits_{-\infty}^{\infty} A(\mathbf{k})\, e^{i(\mathbf{kx}-\omega t)} d^n k, \qquad (D.17)$$

where the amplitude function $c(\mathbf{k})$ is centered about a maximum at $\mathbf{k} = \mathbf{K}$. The group velocity is given by

$$\mathbf{v}_g(k) = \nabla_k \omega(k)\, /_{\mathbf{k=K}}. \qquad (D.18)$$

## D.2 Potential Step and Wave Packet

As an example of an application, we consider scattering by a one-dimensional potential step. The potential is

$$V = \begin{cases} V_0 \\ 0 \end{cases} \quad \text{for} \quad \begin{matrix} x < 0, & \text{region 2} \\ x > 0, & \text{region 1.} \end{matrix} \qquad (D.19)$$

A quantum object is incident from the right onto the potential step; see Fig. D.2.
    With the usual abbreviations

$$k^2 = \frac{2mE}{\hbar^2}; \; \kappa^2 = \frac{2m}{\hbar^2}(V_0 - E) \text{ for } V_0 > E; \; k'^2 = \frac{2m}{\hbar^2}(E - V_0) \text{ for } V_0 < E$$
$$(D.20)$$

and

**Fig. D.2** Situation at the potential step for $E > V_0$ (*above*) and $E < V_0$ (*below*). The *horizontal lines* indicate a wave, the *curved line* an exponential decay

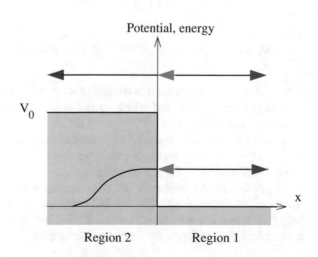

$$
\gamma(k) = \begin{cases} \kappa = \sqrt{k_0^2 - k^2} \\ -ik' = -i\sqrt{k^2 - k_0^2} \end{cases} \quad \text{for} \quad \begin{matrix} E < V_0 \\ E > V_0 \end{matrix} \text{ or } \begin{matrix} k < k_0 \\ k > k_0 \end{matrix}; \; k_0 = \sqrt{\frac{2m}{\hbar^2} V_0},
$$

(D.21)

a partial solution for fixed $k$, as shown in Chap. 15, is given by:

$$
\varphi_1 = e^{-ikx} + \frac{ik + \gamma}{ik - \gamma} e^{ikx}; \quad \varphi_2 = \frac{2ik}{ik - \gamma} e^{\gamma x}
$$

(D.22)

with $k > 0$, $\kappa > 0$.

From this, we obtain the total solution by integration over the continuous index $k > 0$:

$$
\Psi_1(x,t) = \int_0^\infty c(k) \left( e^{-ikx} + \frac{ik+\gamma}{ik-\gamma} e^{ikx} \right) e^{-i\omega t} dk; \; x > 0
$$
$$
\Psi_2(x,t) = \int_0^\infty c(k) \frac{2ik}{ik-\gamma} e^{\gamma x} e^{-i\omega t} dk; \; x < 0 \qquad \text{with } \omega = \frac{\hbar k^2}{2m}. \text{ (D.23)}
$$

$c(k)$ is a function of $k$ which, as discussed above, is nonzero only in a neighborhood of a certain momentum $K$ (as e.g. in the bell curve).

With the definition of $\gamma$, it follows that

$$
\Psi_1(x,t) = \int_0^\infty c(k) e^{-i(kx+\omega t)} dk - \int_0^{k_0} c(k) e^{i(kx-\omega t) + 2i \arctan k/\kappa} dk
$$
$$
+ \int_{k_0}^\infty c(k) \frac{k-k'}{k+k'} e^{i(kx-\omega t)} dk
$$
$$
\Psi_2(x,t) = \int_0^{k_0} c(k) \frac{2ik}{ik-\kappa} e^{\kappa x} e^{-i\omega t} dk + \int_{k_0}^\infty c(k) \frac{2k}{k+k'} e^{-i(k'x+\omega t)} dk. \tag{D.24}
$$

We see that $\Psi_1(x,t)$ contains two types of waves: On the one hand, there are waves travelling from right to left which in our model concept represent the incoming quantum object ($\varphi_{\text{ein}} \sim e^{-i(kx+\omega t)}$); on the other hand there are waves travelling from left to right that represent the reflected quantum object ($\varphi_{\text{refl}} \sim e^{i(kx-\omega t)}$). $\Psi_2(x,t)$ for $k < k_0$ is the exponentially-damped term ($\sim e^{\kappa x}$), and for $k > k_0$, it is the transmitted part of the wavefunction ($\varphi_{\text{trans}} \sim e^{-i(k'x+\omega t)}$).[7]

Depending on the choice of $c(k)$, one can create very complicated wavefunctions. To allow the comparison with classical behavior, we again choose $c(k)$ in such a way that it is centered around a value $K$ (i.e. we have $\Delta k \ll K$) and has a pronounced maximum there. Furthermore, we restrict the discussion to two cases: (1)

---

[7]We recall that the wavefunction does not describe the object itself, but allows for the calculation of its position probability.

The maximum of $c(k)$ is at $K < k_0$; outside of the interval $(0, k_0)$, $c(k)$ vanishes. (2) The maximum of $c(k)$ is at $K > k_0$; outside of the interval $(k_0, \infty)$, $c(k)$ vanishes.

**Case 1**: For the energy of all the partial waves (and therefore for the total wave), we take $E < V_0$. From the point of view of classical physics, this case corresponds to an object which is incident first from right to left with velocity $v$ at the potential step, then is reflected and travels back at the same velocity from left to right. The quantum-mechanical behavior differs from this in two (causally related) points: First, the wavefunction penetrates into the classically forbidden region 2 (i.e. $x < 0$); on the other hand (and as a result of this intrusion), the reflected wave experiences a phase delay.

To see this more closely, we consider the wavefunction. Because of $c(k) = 0$ for $k > k_0$, we have:

$$\Psi_1(x, t) = \int_0^\infty c(k)\, e^{-i(kx + \omega t)}\, dk - \int_0^{k_0} c(k)\, e^{i(kx - \omega t) + 2i\,\arctan k/\kappa}\, dk$$

$$\Psi_2(x, t) = \int_0^{k_0} c(k)\, \frac{2ik}{ik - \kappa}\, e^{\kappa x}\, e^{-i\omega t}\, dk. \tag{D.25}$$

Incoming $(\sim e^{-i(kx + \omega t)})$ and reflected $(\sim e^{i(kx - \omega t)})$ wave components have the same absolute value of their amplitudes; the group velocity for both components of the wave packet is $\left|v_g\right| = \frac{d\omega}{dk}\,/_{k=K} = \frac{\hbar K}{m}$. Thus far, the quantum-mechanical behavior corresponds to the classical solution. The main difference lies in the fact that the wavefunction does not vanish identically for $x < 0$. Intuitively, this means that the quantum object penetrates into the classically forbidden region (essentially the beginning of the tunnel effect). This leads to a ($k$-dependent) phase shift $\tau$ of the reflected partial wave:

$$\tau = \frac{2}{\omega} \arctan \frac{k}{\kappa} = \frac{2m}{\hbar k} \arctan \frac{k}{\kappa}. \tag{D.26}$$

**Case 2**: For the energy of all the partial waves (and therefore the total wave), we take $E > V_0$. From the point of view of classical physics, this case corresponds to an object which is first incident from right to left with velocity $v$ onto the potential step, and from there continues to travel in the same direction, but at a lower speed, $v'$. The quantum-mechanical behavior differs in one respect: it includes a reflection at the step.

We can see this directly from the wavefunction:

$$\Psi_1(x, t) = \int_0^\infty c(k)\, e^{-i(kx + \omega t)}\, dk + \int_{k_0}^\infty c(k)\, \frac{k - k'}{k + k'}\, e^{i(kx - \omega t)}\, dk$$

$$\Psi_2(x, t) = \int_{k_0}^{\infty} c(k) \frac{2k}{k + k'} e^{-i(k'x + \omega t)} dk. \tag{D.27}$$

The amplitudes of the reflected wave components vanish with increasing energy $E$, proportionally to $\frac{V_0}{4E}$. The reflection takes place (in contrast to case 1) with no phase delay, i.e. instantaneously. The group velocity of the transmitted component is obtained from the stationarity of the total phase:

$$0 = \frac{d(k'x + \omega t)}{dk} = \frac{k}{k'} x + \frac{\hbar k}{m} t = \frac{k}{k'} \left( x + \frac{\hbar k'}{m} t \right), \tag{D.28}$$

giving $v'_g = \frac{\hbar}{m}\sqrt{K^2 - k_0^2}$, compared to $v_g = \frac{\hbar K}{m}$ for the incoming component. Hence, as in the classical case, the group velocity of the transmitted component is lower as that of the incoming component.

We note that the wavefunction *always* has a reflected part under these circumstances (albeit possibly a very small one). Intuitively speaking, quantum mechanics thus allows for the scattering of a cannon ball by a snowflake (as the historical example goes). This contradicts our everyday experience, but it seems less strange when one thinks not of matter, but of light. We consider the propagation of light in a non-absorbing medium with a variable refractive index. In case 1, we have a change from a real refractive index (region 1) to an imaginary one (region 2) and correspondingly total reflection. In case 2, we have a sudden change of the value of the real refractive index, which always causes a partial reflection of the light.

In determining the group velocity, we have tacitly assumed that the maxima of the functions $F(k) = \frac{k-k'}{k+k'} c(k)$ and $G(k) = \frac{2k}{k+k'} c(k)$ are located approximately at $k = K$. We want to check this briefly. With

$$\frac{2k}{k + k'} = \frac{k - k'}{k + k'} + 1; \quad k' = \sqrt{k^2 - k_0^2}; \quad \frac{dk'}{dk} = \frac{k}{k'}; \quad \frac{d}{dk} \frac{2k}{k + k'} = \frac{2}{k'} \frac{k' - k}{k + k'}, \tag{D.29}$$

we obtain for the position of the maxima the conditional equations

$$\frac{d}{dk} c(k) \frac{k - k'}{k + k'} = c^{(1)} \frac{k - k'}{k + k'} + c \frac{2}{k'} \frac{k' - k}{k + k'} = 0$$
$$\frac{d}{dk} c(k) \frac{2k}{k + k'} = c^{(1)} \frac{2k}{k + k'} + c \frac{2}{k'} \frac{k' - k}{k + k'} = 0 \tag{D.30}$$

or

$$k'c^{(1)} - 2c = 0; \quad k'kc^{(1)} + c(k' - k) = 0. \tag{D.31}$$

As a typical distribution, we insert a bell curve (D.10) for $c(k)$; with $c^{(1)} = -b_0^2(k - K)c$, it follows that

$$k'b_0^2 (k - K) + 2 = 0; \quad k'kb_0^2 (k - K) + k - k' = 0. \tag{D.32}$$

Instead of looking for exact (as possible) solutions of these equations, we use approximations according to our more qualitative approach. We use the fact that the width of the distribution in momentum space is given by $\Delta k = \frac{2}{b_0}$. Thus, the last two equations can be written as

$$k - K = -\frac{2}{k'b_0^2} = -\frac{1}{2k'} (\Delta k)^2; \quad k - K = \frac{k' - k}{k'kb_0^2} = -\frac{k - k'}{4k'k} (\Delta k)^2 \tag{D.33}$$

or

$$k = K \left[ 1 - \frac{(\Delta k)^2}{2k'K} \right]; \quad k = K \left[ 1 - \frac{(\Delta k)^2}{2k'K} \frac{k - k'}{2k} \right]. \tag{D.34}$$

Due to $\Delta k \ll K$ and $k_0 < k$, we can replace $k$ approximately by $K$ and $k'$ by $K' = \sqrt{K^2 - k_0^2}$, obtaining for the position of the maxima:

$$k = K \left[ 1 - \frac{(\Delta k)^2}{2K'K} \right]; \quad k = K \left[ 1 - \frac{(\Delta k)^2}{2K'K} \frac{K - K'}{2K} \right]. \tag{D.35}$$

We see directly that for sufficiently narrow distributions (essentially $(\Delta k)^2 \ll K'K$), the maxima of the two distributions are at $k \approx K$, and thus our discussion concerning the group velocity was consistent.

## D.3   Exercises

1. The function $\psi (x, t)$ is given as

$$\psi (x, t) = \frac{1}{\sqrt{2\pi}} \int_{-\infty}^{\infty} A (k) e^{ik(x - v_g t)} dk. \tag{D.36}$$

Show that:

$$\psi (x, t) = \psi (x - v_g t, 0) . \tag{D.37}$$

Solution: We have

$$\psi (x, 0) = \frac{1}{\sqrt{2\pi}} \int_{-\infty}^{\infty} A (k) e^{ikx} dk. \tag{D.38}$$

It follows that

$$\psi\left(x, t\right) = \frac{1}{\sqrt{2\pi}} \int_{-\infty}^{\infty} A\left(k\right) e^{ik\left(x - v_g t\right)} dk \underset{y = x - v_g t}{=} \frac{1}{\sqrt{2\pi}} \int_{-\infty}^{\infty} A\left(k\right) e^{iky} dk$$

$$= \psi\left(y, 0\right) = \psi\left(x - v_g t, 0\right). \tag{D.39}$$

2. For which times can one neglect the broadening of the Gaussian wave packet (D.12)?

   Solution: The width of the distribution in position space ($\Delta x = \sqrt{2} b(t)$) is given by (D.15), i.e. by

$$b(t) = \sqrt{b_0^2 + \left(\frac{\hbar t}{m b_0}\right)^2} \approx b_0 \left[1 + \frac{1}{2} \left(\frac{\hbar t}{m b_0^2}\right)^2\right]. \tag{D.40}$$

The packet has practically not broadened for times

$$t \ll \sqrt{2} \frac{m b_0^2}{\hbar}. \tag{D.41}$$

Because of $\Delta k = \frac{2}{b_0}$, it follows that

$$t \ll \frac{4\sqrt{2} m}{\hbar \left(\Delta k\right)^2} \tag{D.42}$$

in accordance with (D.9).

3. The relativistic energy-momentum relation is given by

$$E^2 = m_0^2 c^4 + c^2 p^2. \tag{D.43}$$

Determine the group velocity and the phase velocity $v_g$ and $v_{ph}$. Show that $v_g v_{ph} = c^2$. Which velocity is greater than $c$?

Solution: We have:

$$v_g = \frac{dE}{dp} = \frac{c^2 p}{E}; \quad v_{ph} = \frac{E}{p}. \tag{D.44}$$

For the product, it follows immediately that:

$$v_g v_{ph} = \frac{c^2 p}{E} \frac{E}{p} = c^2. \tag{D.45}$$

For the phase velocity, we find

$$v_{ph}^2 = \frac{m_0^2 c^4 + c^2 p^2}{p^2} = c^2 + \frac{m_0^2 c^4}{p^2} \geq c^2. \tag{D.46}$$

# Appendix E
# Laboratory System, Center-of-Mass System

The hydrogen atom is a concrete example of the general case of a two-body problem, where two interacting masses or quantum objects are considered without external forces. The total energy of this system is composed of the kinetic energies of the two bodies and the potential energy, i.e. the interaction energy $V$ between them:

$$E = E_{\text{kin}} + E_{\text{pot}} = \frac{\mathbf{p}_1^2}{2m_1} + \frac{\mathbf{p}_2^2}{2m_2} + V. \tag{E.1}$$

We assume in the following that the potential depends only on the relative distance $\mathbf{r}_1 - \mathbf{r}_2$, i.e. on $V = V(\mathbf{r}_1 - \mathbf{r}_2)$. Under this assumption, one can reduce the problem to an equivalent one-body problem; specialization to the Coulomb interaction of point charges leads in the quantum-mechanical treatment to the well-known form of the hydrogen spectrum.

## E.1 The Equivalent One-Body Problem

For a simpler description of the problem, we introduce new coordinates, namely *center-of-mass coordinates* and *relative coordinates*:

$$\mathbf{R} = \frac{m_1 \mathbf{r}_1 + m_2 \mathbf{r}_2}{m_1 + m_2} \text{ and } \mathbf{r} = \mathbf{r}_1 - \mathbf{r}_2$$
$$\mathbf{R} = (X, Y, Z) \text{ and } \mathbf{r} = (x, y, z) \tag{E.2}$$

as well as the *total mass* and the *reduced mass*:

$$M = m_1 + m_2 \text{ and } \mu = \frac{m_1 m_2}{m_1 + m_2}. \tag{E.3}$$

One can perform this transformation classically and then change to quantum mechanics, or proceed *vice versa*. In the following, both approaches are discussed.

© Springer Nature Switzerland AG 2018
J. Pade, *Quantum Mechanics for Pedestrians 2*, Undergraduate Lecture
Notes in Physics, https://doi.org/10.1007/978-3-030-00467-5

## E.2    Transformation Laboratory System → Center-of-Mass System

### E.2.1    First Transformation, Then Transition to Quantum Mechanics

The inverse transformations to (E.2) read:

$$\mathbf{r}_1 = \mathbf{R} + \frac{m_2}{M}\mathbf{r} \text{ and } \mathbf{r}_2 = \mathbf{R} - \frac{m_1}{M}\mathbf{r}. \tag{E.4}$$

With $\mathbf{p}_1 = m_1\dot{\mathbf{r}}_1$ etc., taking derivatives with respect to time leads to:

$$\mathbf{p}_1 = m_1\dot{\mathbf{R}} + \mu\dot{\mathbf{r}} \text{ and } \mathbf{p}_2 = m_2\dot{\mathbf{R}} - \mu\dot{\mathbf{r}}. \tag{E.5}$$

For the kinetic energy, we find:

$$E_{\text{kin}} = \frac{\mathbf{p}_1^2}{2m_1} + \frac{\mathbf{p}_2^2}{2m_2} = \frac{1}{2}M\dot{\mathbf{R}}^2 + \frac{1}{2}\mu\dot{\mathbf{r}}^2 = \frac{\mathbf{P}^2}{2M} + \frac{\mathbf{p}^2}{2\mu} \tag{E.6}$$

with the *center-of-mass momentum* and the *relative momentum*

$$\mathbf{P} = M\dot{\mathbf{R}} \text{ and } \mathbf{p} = \mu\dot{\mathbf{r}} \tag{E.7}$$

The total energy is thus

$$E = E_{\text{kin}} + E_{\text{pot}} = \frac{\mathbf{P}^2}{2M} + \frac{\mathbf{p}^2}{2\mu} + V(\mathbf{r}). \tag{E.8}$$

We now go into the *center-of-mass system*, where $\mathbf{P} = 0$; it follows that:

$$E = \frac{\mathbf{p}^2}{2\mu} + V(\mathbf{r}) \text{ in the center-of-mass system.} \tag{E.9}$$

This problem depends only on the relative coordinate; it is called the (classical) one-body problem. If we now proceed to quantum mechanics, we obtain from the last equation in the usual way, i.e. setting $\mathbf{p} = \frac{\hbar}{i}\nabla$, the Hamiltonian of the relative motion

$$H = -\frac{\hbar^2}{2\mu}\nabla^2 + V(\mathbf{r}). \tag{E.10}$$

## E.2.2 First Transition to Quantum Mechanics, Then Transformation

We start with (E.1) and obtain

$$H = -\frac{\hbar^2}{2m_1} \nabla_1^2 - \frac{\hbar^2}{2m_2} \nabla_2^2 + V(\mathbf{r}_1 - \mathbf{r}_2). \tag{E.11}$$

Now we have to rearrange the nabla operators $\nabla_1$ and $\nabla_2$ using

$$\nabla_n = \left( \frac{\partial}{\partial x_n}, \frac{\partial}{\partial y_n}, \frac{\partial}{\partial z_n} \right) \tag{E.12}$$

by means of the variable transformation (E.2) to give the nabla operators $\nabla_{\mathbf{R}}$ and $\nabla_{\mathbf{r}}$. We have (chain rule)

$$\frac{\partial}{\partial x_1} = \frac{\partial X}{\partial x_1}\frac{\partial}{\partial X} + \frac{\partial Y}{\partial x_1}\frac{\partial}{\partial Y} + \frac{\partial Z}{\partial x_1}\frac{\partial}{\partial Z} + \frac{\partial x}{\partial x_1}\frac{\partial}{\partial x} + \frac{\partial y}{\partial x_1}\frac{\partial}{\partial y} + \frac{\partial z}{\partial x_1}\frac{\partial}{\partial z}. \tag{E.13}$$

With

$$\frac{\partial X}{\partial x_1} = \frac{m_1}{M} \quad \text{and} \quad \frac{\partial x}{\partial x_1} = 1, \tag{E.14}$$

it follows that

$$\frac{\partial}{\partial x_1} = \frac{m_1}{M}\frac{\partial}{\partial X} + \frac{\partial}{\partial x}, \tag{E.15}$$

and similarly for the other expressions.[8] Overall, we find:

$$\nabla_1 = \nabla_{\mathbf{r}} + \frac{m_1}{M}\nabla_{\mathbf{R}} \quad \text{and} \quad \nabla_2 = -\nabla_{\mathbf{r}} + \frac{m_2}{M}\nabla_{\mathbf{R}}.$$

Inserting into the Hamiltonian (E.11) yields:

$$H = -\frac{\hbar^2}{2M}\nabla_{\mathbf{R}}^2 - \frac{\hbar^2}{2\mu}\nabla_{\mathbf{r}}^2 + V(\mathbf{r}). \tag{E.16}$$

The time-dependent Schrödinger equation

$$i\hbar\dot{\Psi} = H\Psi \quad \text{with} \quad \Psi = \Psi(\mathbf{R}, \mathbf{r}, t) \tag{E.17}$$

---

[8]One can write this and similar conversions compactly using the (mathematically very sloppy) notation $\nabla_1 = \frac{\partial}{\partial \mathbf{r}_1}$ (this does not mean that one 'divides' by the vector $\mathbf{r}_1$; it is only a different notation for the nabla operator). Then it follows e.g. that $\frac{\partial}{\partial \mathbf{r}_1} = \frac{\partial \mathbf{r}}{\partial \mathbf{r}_1}\frac{\partial}{\partial \mathbf{r}} + \frac{\partial \mathbf{R}}{\partial \mathbf{r}_1}\frac{\partial}{\partial \mathbf{R}}$, etc.; transformations of this kind can thus be performed quite cleverly and with little paperwork. But one has to know what this notation means and how to deal with it.

yields with the usual separation *ansatz*

$$\Psi(\mathbf{R}, \mathbf{r}, t) = e^{-i\omega t} \Phi(\mathbf{R}, \mathbf{r}) \text{ with } E_{\text{total}} = \hbar\omega \tag{E.18}$$

the stationary total Schrödinger equation

$$E_{\text{total}}\Phi(\mathbf{R}, \mathbf{r}) = H\Phi(\mathbf{R}, \mathbf{r}) = \left(-\frac{\hbar^2}{2M}\nabla_{\mathbf{R}}^2 - \frac{\hbar^2}{2\mu}\nabla_{\mathbf{r}}^2 + V(\mathbf{r})\right)\Phi(\mathbf{R}, \mathbf{r}). \tag{E.19}$$

The coordinate $\mathbf{R}$ appears only in the first term on the right-hand side; thus, a separation *ansatz* makes sense:

$$\Phi(\mathbf{R}, \mathbf{r}) = g(\mathbf{R})\psi(\mathbf{r}). \tag{E.20}$$

Inserting and applying the known argumentation gives with

$$E_{\text{total}} = E_{\mathbf{R}} + E_{\mathbf{r}} \tag{E.21}$$

finally a split of the equation into two equations, namely:

1. an equation for the center-of-mass, in fact a free motion:

$$-\frac{\hbar^2}{2M}\nabla_{\mathbf{R}}^2 g(\mathbf{R}) = E_{\mathbf{R}} g(\mathbf{R}) ; \tag{E.22}$$

2. an equation for the relative motion (which contains the interaction between the two quantum objects):

$$-\frac{\hbar^2}{2\mu}\nabla_{\mathbf{r}}^2 \psi(\mathbf{r}) + V(\mathbf{r}) \psi(\mathbf{r}) = E_{\mathbf{r}} \psi(\mathbf{r}). \tag{E.23}$$

Thus we have again (E.10). In other words, no matter whether we proceed first classically or first quantum mechanically, we end up (as we must) with the same result, i.e. (E.23). This is the *equivalent one-body problem*. It differs basically only in one point from the problem of a quantum object of mass $m$ in a potential $V$, namely in the occurrence of the reduced mass $\mu$ instead of the mass $m$.

# Appendix F
# Analytic Treatment of the Hydrogen Atom

In this section, we consider the detailed derivation of the solution of the radial equation of the hydrogen atom by a power series approach.

## F.1   Nonrelativistic Case: Schrödinger equation

We start from the radial equation in the form (17.25), namely:

$$\frac{d^2 u_{nl}(r)}{dr^2} - \left( \frac{l(l+1)}{r^2} - \frac{2\mu\gamma}{\hbar^2 r} + \frac{2\mu}{\hbar^2} |E_n| \right) u_{nl}(r) = 0. \tag{F.1}$$

For the meaning of the individual quantities, see Chap. 17

### Simplification of the Constants

In (F.1), the appearance of five constants ($l$, $\mu$, $\gamma$, $\hbar$, and $E_n$) is annoying. For simplicity, we transform to a new variable $\rho$:

$$u_{nl}(r) = S_{nl}(\xi r) = S_{nl}(\rho) \tag{F.2}$$

where we choose $\rho$ or $\xi$ suitably. Inserting in the radial equation yields:

$$\frac{d^2 S_{nl}(\rho)}{d\rho^2} - \left( \frac{l(l+1)}{\rho^2} - \frac{2\mu\gamma}{\hbar^2 \xi \rho} + \frac{2\mu}{\hbar^2 \xi^2} |E_n| \right) S_{nl}(\rho) = 0. \tag{F.3}$$

We choose

$$\xi = \sqrt{\frac{8\mu |E_n|}{\hbar^2}} \tag{F.4}$$

and obtain with the abbreviation

© Springer Nature Switzerland AG 2018
J. Pade, *Quantum Mechanics for Pedestrians 2*, Undergraduate Lecture
Notes in Physics, https://doi.org/10.1007/978-3-030-00467-5

$$c = \gamma \sqrt{\frac{\mu}{2\hbar^2 |E_n|}} \tag{F.5}$$

the following equation for $S_{nl}(\rho)$, with only two constants:

$$\frac{d^2 S_{nl}(\rho)}{d\rho^2} - \left( \frac{l(l+1)}{\rho^2} - \frac{c}{\rho} + \frac{1}{4} \right) S_{nl}(\rho) = 0. \tag{F.6}$$

**Separating the Behavior for $r \to 0, r \to \infty$**

For $\rho \to \infty$ (i.e. $r \to \infty$), this equation becomes

$$\frac{d^2 S_{nl}(\rho)}{d\rho^2} - \frac{1}{4} S_{nl}(\rho) = 0 \tag{F.7}$$

with the solution[9]:

$$S_{nl}(\rho) = e^{-\rho/2}, \tag{F.8}$$

and for $\rho \to 0$, we obtain approximately

$$\frac{d^2 S_{nl}(\rho)}{d\rho^2} - \frac{l(l+1)}{\rho^2} S_{nl}(\rho) = 0 \tag{F.9}$$

with the solutions

$$S_{nl}(\rho) = \begin{cases} \rho^{l+1}: \text{regular solution} \\ \rho^{-l}: \text{irregular solution} \end{cases} \tag{F.10}$$

The irregular solution has a pole at zero; we exclude it as unphysical. We consider the behavior for $r \to 0$ and for $r \to \infty$ and obtain as our *ansatz*:

$$S_{nl}(\rho) = \rho^{l+1} e^{-\rho/2} f_{nl}(\rho). \tag{F.11}$$

Inserting this into (F.6) yields

$$\rho \frac{d^2 f_{nl}(\rho)}{d\rho^2} + [2(l+1) - \rho] \frac{d f_{nl}(\rho)}{d\rho} + [c - l - 1] f_{nl}(\rho) = 0. \tag{F.12}$$

**Solution by Power Series**

To solve this differential equation, we use a power series *ansatz* of the form

---

[9]We note that the exact solution of an asymptotically-approximated differential equation need not be identical with the asymptotic solution of the exact differential equation. However, that does not matter here, as we are seeking only a clever *ansatz* to simplify the problem, but not (at this point) the exact solution.

$$f_{nl}(\rho) = \sum_{k=0}^{\infty} a_k \rho^k. \tag{F.13}$$

Inserting gives[10]:

$$0 = \sum_{k=0}^{\infty} k(k+1)a_{k+1}\rho^k + 2(l+1)\sum_{k=0}^{\infty}(k+1)a_{k+1}\rho^k$$

$$- \sum_{k=0}^{\infty} ka_k\rho^k + [c-l-1]\sum_{k=0}^{\infty} a_k\rho^k. \tag{F.14}$$

Comparing the coefficients of like powers leads to

$$k(k+1)a_{k+1} + 2(l+1)(k+1)a_{k+1} - ka_k + [c-l-1]a_k = 0 \tag{F.15}$$

or

$$a_{k+1} = \frac{k+l+1-c}{(k+1)(k+2l+2)}a_k. \tag{F.16}$$

Using this relation, we can thus recursively compute all the coefficients of the power series, if we specify $a_0$ (the zero-th term is determined only by the normalization of the total wavefunction). However, it is still unclear what the radius of convergence of the power series (F.13) is and whether the solution is physically acceptable. To answer these questions, we first note that

$$\frac{a_{k+1}}{a_k} \xrightarrow[k\to\infty]{} \frac{1}{k+1}. \tag{F.17}$$

By the ratio test, the series converges (because of $\lim_{k\to\infty} \frac{a_{k+1}}{a_k} = \lim_{k\to\infty} \frac{1}{k+1} = 0 < 1$). That is, the power series has the same convergence behavior as $e^\rho$ (because of $e^\rho = \sum c_k \rho^k = \sum \rho^k/k!$ and $c_{k+1}/c_k = 1/(k+1)$). However, although the power series converges (the radius of convergence is in fact $\infty$), it is not acceptable for physical reasons. This is because, as we can see from (F.11), the asymptotic behavior of the radial function $R_{nl} = u_{nl(r)}/r$ would be given in this case by $\rho^l e^{\rho/2}$ and, accordingly, $R_{nl}$ would not be square integrable.

We examine this in more detail. In addition to the function

$$f_{nl}(\rho) = \sum_{k=0}^{\infty} a_k \rho^k, \tag{F.18}$$

---

[10]Here, we use rearrangements such as

$$\sum_{k=0}^{\infty} ka_k\rho^{k-1} = \sum_{k=1}^{\infty} ka_k\rho^{k-1} = \sum_{k=0}^{\infty}(k+1)a_{k+1}\rho^k.$$

we consider the 'comparison function'

$$e^{\lambda\rho} = \sum_{k=0}^{\infty} b_k \rho^k = \sum_{k=0}^{\infty} \frac{\lambda^k}{k!} \rho^k \quad \text{with } 0 < \lambda < 1. \tag{F.19}$$

Now there is a $K$ such that for $k \geq K$, we have:

$$\frac{a_{k+1}}{a_k} > \frac{b_{k+1}}{b_k} > 0. \tag{F.20}$$

One can easily see why, since due to (F.16),

$$\frac{a_{k+1}}{a_k} = \frac{k + l + 1 - c}{(k+1)(k+2l+2)} > \frac{b_{k+1}}{b_k} = \frac{\lambda}{k+1} > 0. \tag{F.21}$$

Solving this equation for $k$ yields

$$K > \frac{(l+1)(2\lambda - 1) + c}{1 - \lambda} > c - l - 1. \tag{F.22}$$

Clearly, this condition is the most stringent ($K$ is largest) if we restrict ourselves to the range $1/2 < \lambda < 1$; this we take to be the case from now on. We split:

$$f_{nl}(\rho) = \sum_{k=0}^{K-1} a_k \rho^k f(\rho) + \sum_{k=K}^{\infty} a_k \rho^k = P(\rho) + \sum_{k=K}^{\infty} a_k \rho^k$$

$$= P(\rho) + \sum_{m=0}^{\infty} a_{K+m} \rho^{K+m} \tag{F.23}$$

and

$$e^{\lambda\rho} = \sum_{k=0}^{K-1} b_k \rho^k + \sum_{k=K}^{\infty} b_k \rho^k = Q(\rho) + \sum_{k=K}^{\infty} b_k \rho^k = Q(\rho) + \sum_{m=0}^{\infty} b_{K+m} \rho^{K+m}, \tag{F.24}$$

and note that:

$$a_{K+m} = a_K \prod_{l=0}^{m-1} \frac{a_{K+l+1}}{a_{K+l}} > a_K \prod_{l=0}^{m-1} \frac{b_{K+l+1}}{b_{K+l}} = a_K \frac{b_{K+m}}{b_K}. \tag{F.25}$$

Now we can estimate as follows:

$$f_{nl}(\rho) - P(\rho) = \sum_{m=0}^{\infty} a_{K+m} \rho^{K+m} > \frac{a_K}{b_K} \sum_{m=0}^{\infty} b_{K+m} \rho^{K+m} = \frac{a_K}{b_K} [e^{\lambda\rho} - Q(\rho)]. \tag{F.26}$$

This inequality holds for all $\rho$, i.e. also for $\rho \to \infty$. But in that case, we can neglect the polynomials $P$ and $Q$ in comparison with the corresponding function, so that we obtain:

$$f_{nl}(\rho) \geq \frac{a_K}{b_K} e^{\lambda \rho}, \tag{F.27}$$

and we have finally

$$S_{nl}(\rho) = \rho^{l+1} e^{-\rho/2} f_{nl}(\rho) \geq \frac{a_K}{b_K} \rho^{l+1} e^{(\lambda - 1/2)\rho}. \tag{F.28}$$

Because of $1/2 < \lambda < 1$, we have $R_{nl} = u_{nl(r)}/r \underset{r \to \infty}{\to} \infty$, and therefore the radial function would not be square integrable and would not be physically meaningful. In other words, the power series (F.18) always gives a physically meaningless result.

There is only one way out of this situation, namely if $f_{nl}(\rho)$ in (F.13) is *not* an infinite power series, but a rather polynomial.[11] For some natural number $m$, it must therefore hold that $a_m = 0$, because then the radial function behaves for large $r$ essentially as (a polynomial in $r$ times $e^{-\rho/2}$), and thus is square integrable. So we have to require that the numerator in (F.16) vanishes—in other words, it must hold that $c \in \mathbb{N}$. This is exactly why we rename $c$ to $n$ and call this number the *principal quantum number*. In addition, from (F.16), it follows for $l = n - 1 - k$ that for a given $n$, the quantum number $l$ can have only the values

$$l = 0, 1, \ldots, n - 2, n - 1. \tag{F.29}$$

With (F.5), we thus have the identity:

$$n = \gamma \sqrt{\frac{\mu}{2\hbar^2 |E_n|}}; n \in \mathbb{N}, \tag{F.30}$$

and solving for $|E_n|$ yields

$$|E_n| = \frac{\mu \gamma^2}{2\hbar^2} \frac{1}{n^2}. \tag{F.31}$$

Hence, the energy spectrum is discrete for negative energies $E$, i.e. for bound states.

## F.2  Relativistic Case: Dirac equation

The Dirac equation describes the Hydrogen spectrum with considerably more precision than the Schrödinger equation. Historically, the good agreement (fine structure etc.) with the experimental results was an important contribution to the triumph of the Dirac equation and the underlying ideas.

---

[11] As it turns out, these polynomials are the associated Laguerre polynomials; see Chap. 17.

In the following, we will sketch the proceeding in an abbreviated manner, leaving some steps to the reader. We refer to Sect. 16.5 (addition of angular momenta), Chap. 17 (Hydrogen atom) and Sect. 19.3 (Hydrogen: fine structure). Note that in this section we write $m_0$ for the mass of the electron in order to avoid confusion with the $z$-component $m$ of the angular momentum.

## F.2.1   From 4-Spinor to 2-Spinors

We start with the Dirac equation in the form

$$i\hbar \frac{\partial}{\partial t} \psi = c\boldsymbol{\alpha} \left(\mathbf{p} - q\mathbf{A}\right) \psi + q\Phi\psi + m_0 c^2 \beta\psi \tag{F.32}$$

where $m_0$ is the rest mass of the electron. Considering an electron in a Coulomb potential, we have

$$\mathbf{A} = 0 \; ; q\Phi = -\frac{1}{4\pi\varepsilon_0} \frac{Ze^2}{r} = V\left(r\right) \tag{F.33}$$

where $Z$ is the proton number and $e$ the charge of the electron. The Hamilton operator reads

$$H = c\boldsymbol{\alpha}\mathbf{p} + V\left(r\right) + m_0 c^2 \beta. \tag{F.34}$$

The method to calculate the eigenvalues is in parts similar to the non-relativistic case though there are fundamental differences. For instance, the nonrelativistic state is 1-dimensional, whereas in the Dirac case, the state is a 4-spinor. In addition, in the Schrödinger case as considered in Chap. 17, we have the orbital angular momentum $\mathbf{l}$ only, whereas here we have to take into account the spin $\mathbf{s}$ in addition. The total angular momentum is given by $\mathbf{j} = \mathbf{l} + \mathbf{s}$.

Since we consider a central field, the total angular momentum $\mathbf{j}$ commutes with the Hamiltonian (F.34), so we can construct eigenfunctions of simultaneously $H$, $j^2$ and $j_z$. In addition, the Hamiltonian (F.34) is invariant against space reflection given by (see Appendix U, Vol. 1).

$$P = \beta P^{(\mathbf{x})} \; ; \; P^{(\mathbf{x})} : \mathbf{x} \to -\mathbf{x}. \tag{F.35}$$

We know that $P$ commutes with $H$, that $P^2 = 1$ and that, correspondingly, $P$ has the eigenvalues $\pm 1$. In other words, $H$ allows for eigenfunctions with defined parity, even or odd.

Let us go now into details. We are searching solutions for the stationary equation with $E > 0$:

$$E\Psi = c\,\boldsymbol{\alpha} \cdot \mathbf{p}\Psi + V\left(r\right)\Psi + m_0 c^2 \beta\Psi. \tag{F.36}$$

Since the problem has spherical symmetry, a description in terms of spherical coordinates will be favourable. In view of the block structure of the matrices $\alpha$ and $\beta$, we write the 4-spinor $\Psi$ as

$$\Psi = \begin{pmatrix} \Phi \\ X \end{pmatrix} \tag{F.37}$$

where $\Phi$ and $X$ are 2-spinors. Thus, in the standard representation of $\alpha$ and $\beta$ follows

$$E \begin{pmatrix} \Phi \\ X \end{pmatrix} = c \begin{pmatrix} 0 & \sigma \\ \sigma & 0 \end{pmatrix} \mathbf{p} \begin{pmatrix} \Phi \\ X \end{pmatrix} + V(r) \begin{pmatrix} \Phi \\ X \end{pmatrix} + m_0 c^2 \begin{pmatrix} 1 & 0 \\ 0 & -1 \end{pmatrix} \begin{pmatrix} \Phi \\ X \end{pmatrix} \tag{F.38}$$

or

$$\begin{aligned} E\Phi &= c\,\boldsymbol{\sigma}\cdot \mathbf{p} X + V(r)\,\Phi + m_0 c^2 \Phi \\ EX &= c\,\boldsymbol{\sigma}\cdot \mathbf{p}\Phi + V(r) X - m_0 c^2 X \end{aligned} \tag{F.39}$$

In order that the 4-spinor $\Psi$ has a defined parity, the 2-spinors have to fulfill

$$\beta P_0 \begin{pmatrix} \Phi(\mathbf{r}) \\ X(\mathbf{r}) \end{pmatrix} = \beta \begin{pmatrix} \Phi(-\mathbf{r}) \\ X(-\mathbf{r}) \end{pmatrix} = \begin{pmatrix} \Phi(-\mathbf{r}) \\ -X(-\mathbf{r}) \end{pmatrix} \stackrel{!}{=} \pm \begin{pmatrix} \Phi(\mathbf{r}) \\ X(\mathbf{r}) \end{pmatrix} \tag{F.40}$$

or explicitly

$$\begin{pmatrix} \Phi(-\mathbf{r}) \\ X(-\mathbf{r}) \end{pmatrix} \stackrel{!}{=} \begin{pmatrix} \Phi(\mathbf{r}) \\ -X(\mathbf{r}) \end{pmatrix} \quad \text{or} \quad \begin{pmatrix} \Phi(-\mathbf{r}) \\ X(-\mathbf{r}) \end{pmatrix} \stackrel{!}{=} \begin{pmatrix} -\Phi(\mathbf{r}) \\ X(\mathbf{r}) \end{pmatrix}. \tag{F.41}$$

## F.2.2  Angular Part of the 2-Spinors

The two 2-spinors $\Phi$ and $X$ can be expressed as linear combinations of eigenfunctions of simultaneously $\mathbf{j}^2$, $j_z$, $\mathbf{l}^2$ and $\mathbf{s}^2$ which is more easily seen by converting the term $\boldsymbol{\sigma}\mathbf{p}$ in a suitable form. We have[12]

$$\boldsymbol{\sigma}\cdot\mathbf{p} = \left( \frac{1}{r^2} \boldsymbol{\sigma}\cdot\mathbf{r}\ \boldsymbol{\sigma}\cdot\mathbf{r} \right) \boldsymbol{\sigma}\cdot\mathbf{p} = \frac{1}{r^2}\boldsymbol{\sigma}\cdot\mathbf{r}\ (\boldsymbol{\sigma}\cdot\mathbf{r}\ \boldsymbol{\sigma}\cdot\mathbf{p}) = \frac{1}{r^2}\boldsymbol{\sigma}\cdot\mathbf{r}\ (\mathbf{r}\cdot\mathbf{p} + i\,\boldsymbol{\sigma}\cdot\mathbf{l}) \tag{F.42}$$

or with $\hat{\mathbf{r}} = \frac{\mathbf{r}}{r}$

$$\boldsymbol{\sigma}\cdot\mathbf{p} = \boldsymbol{\sigma}\cdot\hat{\mathbf{r}} \left( \hat{\mathbf{r}}\cdot\mathbf{p} + \frac{i}{r}\boldsymbol{\sigma}\cdot\mathbf{l} \right). \tag{F.43}$$

Due to $\mathbf{j}^2 = (\mathbf{l}+\mathbf{s})^2 = \mathbf{l}^2 + 2\mathbf{s}\cdot\mathbf{l} + \mathbf{s}^2$, we have[13]

---

[12]We make use of $\boldsymbol{\sigma}\cdot\mathbf{a}\ \boldsymbol{\sigma}\cdot\mathbf{b} = \mathbf{a}\cdot\mathbf{b} + i\,\boldsymbol{\sigma}\cdot[\mathbf{a}\times\mathbf{b}]$ which means, inter alia, $\boldsymbol{\sigma}\cdot\hat{\mathbf{r}}\ \boldsymbol{\sigma}\cdot\hat{\mathbf{r}} = 1$. As usually, $\mathbf{r}\times\mathbf{p} = \mathbf{l}$.

[13]Remember $\mathbf{s} = \frac{\hbar}{2}\boldsymbol{\sigma}$.

$$2\mathbf{s} \cdot \mathbf{l} = \mathbf{j}^2 - \mathbf{l}^2 - \mathbf{s}^2. \tag{F.44}$$

The corresponding two-dimensional eigenspinors $\varphi_{jm_j}^{(+)}$ and $\varphi_{jm_j}^{(-)}$ (i.e., eigenfunctions of $\mathbf{j}^2$, $j_z$, $\mathbf{l}^2$ and $\mathbf{s}^2$) were derived in Sect. 16.5; they read in position representation (here, we note the angular part only)

$$\varphi_{jm_j}^{(+)} = \begin{pmatrix} \sqrt{\frac{l+m_j+1/2}{2l+1}}\, Y_l^{m_j-1/2} \\ \sqrt{\frac{l-m_j+1/2}{2l+1}}\, Y_l^{m_j+1/2} \end{pmatrix} \text{ for } j = l+\frac{1}{2} \; ; \; \varphi_{jm_j}^{(-)} = \begin{pmatrix} \sqrt{\frac{l-m_j+1/2}{2l+1}}\, Y_l^{m_j-1/2} \\ -\sqrt{\frac{l+m_j+1/2}{2l+1}}\, Y_l^{m_j+1/2} \end{pmatrix} \text{ for } j = l-\frac{1}{2}. \tag{F.45}$$

In comparison with Sect. 16.5, we have changed the sign for $\varphi_{jm_j}^{(-)}$ in order to arrive at the simple formulation (F.48).

The result may be written compactly as

$$\varphi_{jm_j}^{(\pm)} = \begin{pmatrix} \sqrt{\frac{l\pm m_j+1/2}{2l+1}}\, Y_l^{m_j-1/2} \\ \pm\sqrt{\frac{l\mp m_j+1/2}{2l+1}}\, Y_l^{m_j+1/2} \end{pmatrix} \text{ for } j = l \pm \frac{1}{2} \tag{F.46}$$

where the $Y_m^l$ are spherical functions. We have $l = 0, 1, 2, \ldots$, whereby $\varphi_{jm_j}^{(-)}$ exists only for $l > 0$. Note that for a given $j$, the two spinors have opposite parity,[14] since their $l$-values differ by one.

The eigenequations are

$$\begin{aligned}
\mathbf{j}^2 \varphi_{jm_j}^{(\pm)} &= \hbar^2 j\, (j+1)\, \varphi_{jm_j}^{(\pm)} \\
\mathbf{l}^2 \varphi_{jm_j}^{(\pm)} &= \hbar^2 l\, (l+1)\, \varphi_{jm_j}^{(\pm)} \\
j_z \varphi_{jm_j}^{(\pm)} &= \hbar m_j \varphi_{jm_j}^{(\pm)} \\
\mathbf{s}^2 \varphi_{jm_j}^{(\pm)} &= \tfrac{3}{4}\hbar^2 \varphi_{jm_j}^{(\pm)}.
\end{aligned} \tag{F.47}$$

We note in passing that

$$\varphi_{jm_j}^{(+)} = \boldsymbol{\sigma} \cdot \hat{\mathbf{r}}\; \varphi_{jm_j}^{(-)} \tag{F.48}$$

which again shows that $\varphi_{jm_j}^{(+)}$ and $\varphi_{jm_j}^{(-)}$ have opposite parity.[15]

### F.2.3   From 2-Spinors to 4-Spinor

Thus, the general expression for the four-spinor $\Psi$ for given values of $j$ and $m_j$ reads[16]

---

[14] Remind that the parity of $Y_l^m$ is given by $(-1)^l$.

[15] Note that in spherical coordinates, $\boldsymbol{\sigma} \cdot \hat{\mathbf{r}}$ only contains the angle variables $\vartheta$ and $\varphi$.

[16] The factors $i$ and $\frac{1}{r}$ are chosen in order to simplify the following calculations.

$$\Psi_{jm_j} = \frac{1}{r} \begin{pmatrix} iG_j^{(+)}(r)\,\varphi_{jm_j}^{(+)} + iG_j^{(-)}(r)\,\varphi_{jm_j}^{(-)} \\ F_j^{(+)}(r)\,\varphi_{jm_j}^{(-)} + F_j^{(-)}(r)\,\varphi_{jm_j}^{(+)} \end{pmatrix}. \tag{F.49}$$

We can split this expression into two solutions with defined parity. Due to the (F.40) and (F.48), they read

$$\Psi_{jm_j}^{(+)} = \frac{1}{r} \begin{pmatrix} iG_j^{(+)}(r)\,\varphi_{jm_j}^{(+)} \\ F_j^{(+)}(r)\,\boldsymbol{\sigma}\!\cdot\!\hat{\mathbf{r}}\,\varphi_{jm_j}^{(+)} \end{pmatrix} \;\; ; \;\; \Psi_{jm_j}^{(-)} = \frac{1}{r} \begin{pmatrix} iG_j^{(-)}(r)\,\varphi_{jm_j}^{(-)} \\ F_j^{(-)}(r)\,\boldsymbol{\sigma}\!\cdot\!\hat{\mathbf{r}}\,\varphi_{jm_j}^{(-)} \end{pmatrix} \tag{F.50}$$

which usually is written compactly as

$$\Psi_{jm_j}^l = \frac{1}{r} \begin{pmatrix} iG_{jl}(r)\,\varphi_{jm_j}^l \\ F_{jl}(r)\,\boldsymbol{\sigma}\!\cdot\!\hat{\mathbf{r}}\,\varphi_{jm_j}^l \end{pmatrix} \;\; ; \;\; j = l \pm \frac{1}{2}. \tag{F.51}$$

Note that $\Psi_{jm_j}^l$ has parity $(-1)^l$.
Inserting this expression into (F.38) results in

$$E\frac{G_{jl}}{r}\varphi_{jm_j}^l = -ic\,\boldsymbol{\sigma}\!\cdot\!\mathbf{p}\,\frac{F_{jl}}{r}\,\boldsymbol{\sigma}\!\cdot\!\hat{\mathbf{r}}\,\varphi_{jm_j}^l + V(r)\frac{G_{jl}}{r}\varphi_{jm_j}^l + m_0c^2\frac{G_{jl}}{r}\varphi_{jm_j}^l$$
$$E\frac{F_{jl}}{r}\,\boldsymbol{\sigma}\!\cdot\!\hat{\mathbf{r}}\,\varphi_{jm_j}^l = ic\,\boldsymbol{\sigma}\!\cdot\!\mathbf{p}\,\frac{G_{jl}}{r}\varphi_{jm_j}^l + V(r)\frac{F_{jl}}{r}\,\boldsymbol{\sigma}\!\cdot\!\hat{\mathbf{r}}\,\varphi_{jm_j}^l - m_0c^2\frac{F_{jl}}{r}\,\boldsymbol{\sigma}\!\cdot\!\hat{\mathbf{r}}\,\varphi_{jm_j}^l. \tag{F.52}$$

We now consider the action of the operators $\boldsymbol{\sigma}\!\cdot\!\mathbf{p}$ and $\boldsymbol{\sigma}\!\cdot\!\mathbf{p}\,\boldsymbol{\sigma}\!\cdot\!\hat{\mathbf{r}}$ on the eigenfunctions. With (F.43), we have for a general function $H(r)$

$$\boldsymbol{\sigma}\!\cdot\!\mathbf{p}\,H(r) = \boldsymbol{\sigma}\!\cdot\!\hat{\mathbf{r}}\left(\hat{\mathbf{r}}\!\cdot\!\mathbf{p} + \frac{i}{r}\boldsymbol{\sigma}\!\cdot\!\mathbf{l}\right)H(r) = \boldsymbol{\sigma}\!\cdot\!\hat{\mathbf{r}}\left(\hat{\mathbf{r}}\!\cdot\!\mathbf{p} + \frac{2i}{\hbar r}\mathbf{s}\!\cdot\!\mathbf{l}\right)H(r). \tag{F.53}$$

Using $\hat{\mathbf{r}}\cdot\mathbf{p} = \frac{\hbar}{i}\frac{\partial}{\partial r}$ and $2\mathbf{s}\cdot\mathbf{l} = \mathbf{j}^2 - \mathbf{l}^2 - \mathbf{s}^2$ yields

$$\boldsymbol{\sigma}\!\cdot\!\mathbf{p}\,H(r)\,\varphi_{jm_j}^l = \boldsymbol{\sigma}\!\cdot\!\hat{\mathbf{r}}\left(\frac{\hbar}{i}\frac{\partial}{\partial r} + \frac{i}{\hbar r}\left[\mathbf{j}^2 - \mathbf{l}^2 - \mathbf{s}^2\right]\right)H(r)\,\varphi_{jm_j}^l =$$
$$= \boldsymbol{\sigma}\!\cdot\!\hat{\mathbf{r}}\left(\frac{\hbar}{i}\frac{\partial}{\partial r} + \frac{i\hbar}{r}\left[j(j+1) - l(l+1) - \tfrac{3}{4}\right]\right)H(r)\,\varphi_{jm_j}^l. \tag{F.54}$$

With

$$j(j+1) - l(l+1) - \frac{3}{4} = \begin{cases} j - \frac{1}{2} = -1 + (j + \frac{1}{2}) \\ -j - \frac{3}{2} = -1 - (j + \frac{1}{2}) \end{cases} \text{ for } j = l \pm \frac{1}{2} \tag{F.55}$$

we arrive at

$$\boldsymbol{\sigma}\!\cdot\!\mathbf{p}\,H(r)\,\varphi_{jm_j}^l = \boldsymbol{\sigma}\!\cdot\!\hat{\mathbf{r}}\left(\frac{\hbar}{i}\frac{\partial}{\partial r} + \frac{i\hbar}{r}\left[j(j+1) - l(l+1) - \tfrac{3}{4}\right]\right)H(r)\,\varphi_{jm_j}^l =$$
$$= \boldsymbol{\sigma}\!\cdot\!\hat{\mathbf{r}}\,\frac{\hbar}{ir}\left[r\frac{\partial H(r)}{\partial r} + \left(1 \mp (j + \tfrac{1}{2})\right)H(r)\right]\varphi_{jm_j}^l \text{ for } j = l \pm \tfrac{1}{2}. \tag{F.56}$$

In addition, we need $\boldsymbol{\sigma}\!\cdot\!\mathbf{p}\,H(r)\,\boldsymbol{\sigma}\!\cdot\!\hat{\mathbf{r}}\,\varphi_{jm_j}^l$:

$$\boldsymbol{\sigma}{\cdot}\mathbf{p}\, H(r)\, \boldsymbol{\sigma}{\cdot}\hat{\mathbf{r}}\, \varphi^l_{jm_j} = \boldsymbol{\sigma}{\cdot}\mathbf{p}\, \boldsymbol{\sigma}{\cdot}\hat{\mathbf{r}}\, H(r)\varphi^l_{jm_j} =$$
$$= \left[\mathbf{p}{\cdot}\hat{\mathbf{r}} + i\boldsymbol{\sigma}\left(\mathbf{p}\times\hat{\mathbf{r}}\right)\right] H(r)\varphi^l_{jm_j} = \left(-\tfrac{2i\hbar}{r} + \hat{\mathbf{r}}\cdot\mathbf{p} - \tfrac{i}{r}\,\boldsymbol{\sigma}{\cdot}\mathbf{L}\right) H(r)\varphi^l_{jm_j}. \tag{F.57}$$

Evaluating $\hat{\mathbf{r}}\cdot\mathbf{p}$ and $\tfrac{i}{r}\,\boldsymbol{\sigma}{\cdot}\mathbf{L}$ yields

$$\boldsymbol{\sigma}{\cdot}\mathbf{p}\, H(r)\, \boldsymbol{\sigma}{\cdot}\hat{\mathbf{r}}\, \varphi^l_{jm_j} = \left(-\tfrac{2i\hbar}{r} + \tfrac{\hbar}{i}\tfrac{\partial}{\partial r} - \tfrac{i\hbar}{r}\left[j\,(j+1) - l\,(l+1) - \tfrac{3}{4}\right]\right) H(r)\varphi^l_{jm_j} =$$
$$= \left(-\tfrac{2i\hbar}{r} + \tfrac{\hbar}{i}\tfrac{\partial}{\partial r} + \tfrac{i\hbar}{r}\left[1 \mp (j+\tfrac{1}{2})\right]\right) H(r)\varphi^l_{jm_j} \tag{F.58}$$

or

$$\boldsymbol{\sigma}{\cdot}\mathbf{p}\, H(r)\boldsymbol{\sigma}{\cdot}\hat{\mathbf{r}}\, \varphi^l_{jm_j} = \frac{\hbar}{ir}\left[r\frac{\partial}{\partial r} + 1 \pm \left(j+\frac{1}{2}\right)\right] H(r)\varphi^l_{jm_j} \quad \text{for } j = l\pm\frac{1}{2}. \tag{F.59}$$

Inserting the results in (F.52) yields finally

$$-\tfrac{1}{c\hbar}\left[E + \tfrac{1}{4\pi\varepsilon_0}\tfrac{Ze^2}{r} - m_0c^2\right] G_{jl} = \tfrac{\partial}{\partial r} F_{jl} \pm \tfrac{j+\frac{1}{2}}{r} F_{jl}$$
$$\tfrac{1}{c\hbar}\left[E + \tfrac{1}{4\pi\varepsilon_0}\tfrac{Ze^2}{r} + m_0c^2\right] F_{jl} = \tfrac{\partial}{\partial r} G_{jl} \mp \tfrac{j+\frac{1}{2}}{r} G_{jl}. \tag{F.60}$$

## F.2.4  Coupled Radial Equations, Solution

To get rid of the many constants, we insert the fine-structure constant[17] $\alpha$

$$\alpha = \frac{1}{4\pi\varepsilon_0}\frac{e^2}{c\hbar} \tag{F.61}$$

and introduce the abbreviations

$$k = \pm\left(j+\frac{1}{2}\right) \;;\; \alpha_1 = \frac{m_0c^2 + E}{c\hbar} \;;\; \alpha_2 = \frac{m_0c^2 - E}{c\hbar} \;;\; \tau = \sqrt{\alpha_1\alpha_2} \;;\; \rho = \tau r \;;\; \gamma = Z\alpha. \tag{F.62}$$

Thus, (F.60) are written as

$$\left(\tfrac{d}{d\rho} + \tfrac{k}{\rho}\right) F_{jl} - \left(\tfrac{\alpha_2}{\tau} - \tfrac{\gamma}{\rho}\right) G_{jl} = 0$$
$$\left(\tfrac{d}{d\rho} - \tfrac{k}{\rho}\right) G_{jl} - \left(\tfrac{\alpha_1}{\tau} + \tfrac{\gamma}{\rho}\right) F_{jl} = 0. \tag{F.63}$$

From now on, the method is essentially the same as in the case of the Hydrogen atom of the Schrödinger equation. We just sketch the essential steps of the approach,

---

[17]Do not confuse the fine structure constant $\alpha$ and the Dirac matrices $\boldsymbol{\alpha}$.

leaving the detailed calculation to the reader (it is just too comprehensive to be performed in an appendix).

First, one can show that the physically meaningful solutions of (F.63) have an asymptotical behavior proportional to $e^{-\rho}$. In view of this, one separates this behavior by choosing

$$F_{jl}(\rho) = e^{-\rho} f(\rho) \; ; \; G_{jl}(\rho) = e^{-\rho} g(\rho). \tag{F.64}$$

The solutions of the resulting differential equations for $f$ and $g$ are searched in form of $\rho^s$ times power series. Demanding regularity of the functions at $\rho = 0$ leads to $s = \sqrt{k^2 - \gamma^2}$. Furthermore, comparison of the coefficients of equal powers in the power series yields recursion relations between these coefficients. The determination of the coefficients leads to power series with an asymptotic behavior $\sim e^{2\rho}$ which would according to (F.64) yield non-normalizable functions. This unphysical behavior can only be avoided, if the power series stop at a certain power $N$, i.e., if they are polynomials.[18] The termination condition reads

$$E = m_0 c^2 \left[ 1 + \frac{\gamma^2}{(s+N)^2} \right]^{-\frac{1}{2}}. \tag{F.65}$$

With the *main quantum number n*

$$n = N + j + \frac{1}{2} \tag{F.66}$$

the energy levels for Coulomb interaction are given by:

$$E_{nj} = m_0 c^2 \left[ 1 + \left( \frac{Z\alpha}{n - (j + \frac{1}{2}) + \sqrt{(j + \frac{1}{2})^2 - (Z\alpha)^2}} \right)^2 \right]^{-\frac{1}{2}} \tag{F.67}$$

with

$$n = 1, 2, \ldots, \infty \; ; \; 0 < j + \frac{1}{2} \le n \; ; \; 0 \le l \le n - 1 \, , \, j = l \pm \frac{1}{2}. \tag{F.68}$$

For a discussion of this spectrum see Sect. 19.3, Vol. 2.

## F.3 Exercises and Solutions

1. Calculate $\mathbf{r} \cdot \mathbf{p}$, $\mathbf{p} \cdot \mathbf{r}$ and $\boldsymbol{\sigma} \cdot \mathbf{p}$
2. Prove (F.48), i.e.

---

[18]$N$ is called radial quantum number.

$$\varphi_{jm_j}^{(+)} = \boldsymbol{\sigma} \cdot \hat{\mathbf{r}} \, \varphi_{jm_j}^{(-)}. \tag{F.69}$$

Solution: We have

$$\hat{\mathbf{r}} = \frac{1}{r}(x, y, z) = \left( -\frac{1}{2}\sqrt{\frac{8\pi}{3}}\left(Y_1^1 - Y_1^{-1}\right), \, -\frac{1}{2i}\sqrt{\frac{8\pi}{3}}\left(Y_1^1 + Y_1^{-1}\right), \, \sqrt{\frac{4\pi}{3}}Y_1^0 \right). \tag{F.70}$$

It follows

$$\boldsymbol{\sigma} \cdot \hat{\mathbf{r}} = \frac{1}{r}\left(\sigma_x x + \sigma_y y + \sigma_z z\right) = \frac{1}{r}\left[ \begin{pmatrix} 0 & x \\ x & 0 \end{pmatrix} + \begin{pmatrix} 0 & -iy \\ iy & 0 \end{pmatrix} + \begin{pmatrix} z & 0 \\ 0 & -z \end{pmatrix} \right]$$

$$= \frac{1}{r}\begin{pmatrix} z & x - iy \\ x + iy & -z \end{pmatrix} \tag{F.71}$$

or

$$\boldsymbol{\sigma} \cdot \hat{\mathbf{r}} = \begin{pmatrix} \sqrt{\frac{4\pi}{3}}Y_1^0 & \sqrt{\frac{8\pi}{3}}Y_1^{-1} \\ -\sqrt{\frac{8\pi}{3}}Y_1^1 & -\sqrt{\frac{4\pi}{3}}Y_1^0 \end{pmatrix} = \sqrt{\frac{4\pi}{3}}\begin{pmatrix} Y_1^0 & \sqrt{2}Y_1^{-1} \\ -\sqrt{2}Y_1^1 & -Y_1^0 \end{pmatrix}. \tag{F.72}$$

# Appendix G
# The Lenz Vector

In this section, we want to derive the spectrum of the hydrogen atom in an algebraic manner; the analytic derivation can be found in Chap. 17 and Appendix F, Vol. 2. Here, we use the fact that the *Lenz vector*[19] is a constant of the motion (and thus a conserved quantity); this additional constant is, moreover, responsible for the high degree of degeneracy in the energy spectrum of the hydrogen atom.

The eigenfunctions of the hydrogen atom can be represented algebraically as well; this is given in some textbooks on quantum mechanics (see e.g. Schwabl, Annex C., p. 400).

## G.1   In Classical Mechanics

In classical mechanics, the Lenz vector is defined as:

$$\mathbf{\Lambda}_{CIM} = \frac{1}{m\gamma} \left( \mathbf{L} \times \mathbf{p} \right) + \frac{\mathbf{r}}{r}. \tag{G.1}$$

For the motion of a particle in a Coulomb field (or a Kepler field) $V(r) = -\gamma/r$, it is, in addition to the energy and the angular momentum, a further conserved quantity. Its magnitude is equal to the eccentricity of the elliptical orbit. Its conservation means that this orbital ellipse is not rotating, so there is no perihelion motion. For other potentials, this is generally[20] not the case; there, one finds *rosette orbits*.

---

[19] Also called the Laplace–Runge–Lenz vector, Runge–Lenz vector, etc. and, especially in quantum mechanics, the Runge–Lenz–Pauli operator.

[20] Aside from the Coulomb potential, only the potential of the harmonic oscillator, $V \sim r^2$, leads to closed elliptical orbits.

© Springer Nature Switzerland AG 2018
J. Pade, *Quantum Mechanics for Pedestrians 2*, Undergraduate Lecture Notes in Physics, https://doi.org/10.1007/978-3-030-00467-5

## G.2   In Quantum Mechanics

For the translation into quantum mechanics, we have to symmetrize:

$$\mathbf{\Lambda} = \frac{1}{2m\gamma} \left( \mathbf{L} \times \mathbf{p} - \mathbf{p} \times \mathbf{L} \right) + \frac{\mathbf{r}}{r}. \tag{G.2}$$

The Hamiltonian reads

$$H = \frac{p^2}{2m} - \frac{\gamma}{r}. \tag{G.3}$$

$\mathbf{\Lambda}$ is a Hermitian vector operator that commutes with $H$ (i.e. it represents a vectorial conserved quantity) and is orthogonal[21] to $\mathbf{L}$:

$$\mathbf{\Lambda} = \mathbf{\Lambda}^\dagger; \quad [\mathbf{\Lambda}, H] = 0; \quad \mathbf{\Lambda} \cdot \mathbf{L} = \mathbf{L} \cdot \mathbf{\Lambda} = 0. \tag{G.4}$$

In principle, it is technically rather easy to prove these and the other statements that follow, but often it is also a lengthy procedure. Therefore, the proofs are left to the exercises.

For $\mathbf{\Lambda}^2$, we obtain[22]:

$$\mathbf{\Lambda}^2 = \frac{2H}{m\gamma^2} \left( \mathbf{L}^2 + \hbar^2 \right) + 1. \tag{G.5}$$

We restrict ourselves to negative energies $-|E|$ (and thus to bound states; in principle, the reasoning could be extended to scattering states). With the rescaling[23]

$$\mathbf{R} = \sqrt{\frac{m\gamma^2}{2\,|E|}} \, \mathbf{\Lambda}, \tag{G.6}$$

it follows that

$$\mathbf{R}^2 + \mathbf{L}^2 + \hbar^2 = -\frac{m\gamma^2}{2E}. \tag{G.7}$$

Finally, we introduce two generalized angular-momentum operators:

$$\mathbf{J}_1 = \frac{1}{2} \left( \mathbf{L} + \mathbf{R} \right); \quad \mathbf{J}_2 = \frac{1}{2} \left( \mathbf{L} - \mathbf{R} \right). \tag{G.8}$$

---

[21] $\mathbf{\Lambda}$ is a polar vector, $\mathbf{L}$ an axial vector.

[22] We recall that for a central potential, $[H, \mathbf{L}^2] = 0$.

[23] The vector operator $\mathbf{R}$ introduced here is not to be confused with the center-of-mass vector.

They satisfy the equations

$$[\mathbf{J}_1, \mathbf{J}_2] = 0; \quad \mathbf{J}_1^2 = \mathbf{J}_2^2. \tag{G.9}$$

With (G.8) , we can write (G.7) in the form:

$$2\left(\mathbf{J}_1^2 + \mathbf{J}_2^2\right) + \hbar^2 = -\frac{m\gamma^2}{2E}, \tag{G.10}$$

and because of (G.9), it follows that

$$4\mathbf{J}_1^2 + \hbar^2 = -\frac{m\gamma^2}{2E}. \tag{G.11}$$

Because $\mathbf{J}_1$ is a generalized angular-momentum operator, its eigenvalues have the form $\hbar^2 j \, (j + 1)$, where $j$ is a positive integer or half-integer. (This operator has a different symmetry from the angular-momentum operators considered previously; therefore its eigenvalues can take on both integer and half-integer values.)

Hence, it follows that

$$4\hbar^2 j \, (j + 1) + \hbar^2 = \frac{m\gamma^2}{2\,|E|} \tag{G.12}$$

or

$$E = -\frac{m\gamma^2}{2\hbar^2\,(2j + 1)^2}; \quad j = 0, \frac{1}{2}, 1, \frac{3}{2}, \ldots \tag{G.13}$$

If we identify the numbers $2j + 1$ with the principal quantum number $n$, we obtain the the familiar form of the energy levels of the hydrogen atom.

## G.3   General Theorems on Vector Operators

For the manipulations in the exercises, we compile here a few facts about commutators and vector operators:

### G.3.1   General Commutator Relations

We need, among others, the general commutator relations

$$[A, [B, C]] + [B, [C, A]] + [C, [A, B]] = 0 \tag{G.14}$$

(Jacobi identity) and

$$[A, BC] = [A, B] C + B [A, C]$$
$$[AB, C] = A [B, C] + [A, C] B. \tag{G.15}$$

## G.3.2  Vector Operators

An operator is a vector operator iff

$$\left[L_j, A_k\right] = i\hbar \sum_m \varepsilon_{jkm} A_m; \quad \left[\mathbf{L}^2, \mathbf{A}^2\right] = 0. \tag{G.16}$$

Here, $\mathbf{L}$ is the angular-momentum operator and $\varepsilon_{jkm}$ is the Levi-Civita tensor (Levi-Civita symbol, permutation tensor, antisymmetric symbol, alternating symbol; see Appendix F, Vol. 1).[24]

For two vector operators $\mathbf{B}$ and $\mathbf{C}$, it holds that:

$$[\mathbf{L}, \mathbf{B} \cdot \mathbf{C}] = 0. \tag{G.17}$$

This is due to

$$[L_i, \mathbf{B} \cdot \mathbf{C}] = \sum_j \left[L_i, B_j C_j\right] = \sum_j \left[L_i, B_j\right] C_j + B_j \left[L_i, C_j\right]$$
$$= i\hbar \sum_j \sum_m \varepsilon_{ijm} \left(B_m C_j + B_j C_m\right) = 0. \tag{G.18}$$

The last step follows because of $\varepsilon_{1jm} = -\varepsilon_{1mj}$:

$$\sum_{jm} \varepsilon_{ijm} \left(B_m C_j + B_j C_m\right) = \sum_{jm} B_m C_j \left(\varepsilon_{ijm} + \varepsilon_{imj}\right) = 0. \tag{G.19}$$

In addition, for a vector operator, we have:

$$\mathbf{A} \times \mathbf{L} = -\mathbf{L} \times \mathbf{A} + 2i\hbar\mathbf{A}; \quad \mathbf{L} \times \mathbf{A} = -\mathbf{A} \times \mathbf{L} + 2i\hbar\mathbf{A}. \tag{G.20}$$

---

[24]Typically, in this context, the Einstein summation convention is used, according to which one sums over repeated indices without noting this explicitly. Since we rarely use the Leci-Civita tensor, we write out the summation sign (cf. Appendix F, Vol. 1). Instead of using the usual notation, for example

$$\varepsilon_{ijk}\varepsilon_{mnk} = \delta_{im}\delta_{jn} - \delta_{in}\delta_{jm},$$

we write

$$\sum_k \varepsilon_{ijk}\varepsilon_{mnk} = \delta_{im}\delta_{jn} - \delta_{in}\delta_{jm}.$$

Proof:

$$(\mathbf{A} \times \mathbf{L})_i = \sum_{jk} \varepsilon_{ijk} A_j L_k = \sum_{jk} \varepsilon_{ijk} \left( L_k A_j - \hbar \sum_m \varepsilon_{jkm} A_m \right)$$
$$= \sum_{jk} \varepsilon_{ijk} L_k A_j - \sum_{jk} \varepsilon_{ijk} \sum_m \varepsilon_{jkm} A_m = -(\mathbf{L} \times \mathbf{A}) + 2i\hbar\mathbf{A},$$

$$(G.21)$$

due to

$$\sum_{jk} \varepsilon_{ijk} \varepsilon_{jkm} = -2\delta_{mi}. \qquad (G.22)$$

Furthermore, vector operators satisfy generally the equation

$$\mathbf{A} \cdot (\mathbf{B} \times \mathbf{C}) = (\mathbf{A} \times \mathbf{B}) \cdot \mathbf{C}. \qquad (G.23)$$

In particular, for the momentum and the position, we have

$$\mathbf{r} \cdot \mathbf{L} = \mathbf{L} \cdot \mathbf{r} = 0; \quad \mathbf{p} \cdot \mathbf{L} = \mathbf{L} \cdot \mathbf{p} = 0. \qquad (G.24)$$

## G.4   Exercises

1. Show that $\mathbf{\Lambda}$ is a Hermitian vector operator which commutes with $H = \frac{\mathbf{p}^2}{2m} - \frac{\gamma}{r}$ and satisfies the equation $\mathbf{\Lambda} \cdot \mathbf{L} = \mathbf{L} \cdot \mathbf{\Lambda} = 0$.
   Solution:

   (a) For the Hermiticity, we need only consider the term $(\mathbf{L} \times \mathbf{p} - \mathbf{p} \times \mathbf{L})$, which, due to the symmetrization (and because $\mathbf{L}$ and $\mathbf{p}$ are Hermitian), is automatically Hermitian. We see this explicitly for e.g. the $x$ or 1 component:

   $$\begin{aligned}(\mathbf{L} \times \mathbf{p} - \mathbf{p} \times \mathbf{L})_1 &= L_2 p_3 - L_3 p_2 - p_2 L_3 + p_3 L_2 \\ (\mathbf{L} \times \mathbf{p} - \mathbf{p} \times \mathbf{L})_1^\dagger &= p_3 L_2 - p_2 L_3 - L_3 p_2 + L_2 p_3 \\ &= (\mathbf{L} \times \mathbf{p} - \mathbf{p} \times \mathbf{L})_1. \end{aligned} \qquad (G.25)$$

   (b) Concerning the question of the vector operator, we realize first that

   $$\mathbf{L} \times \mathbf{p} - \mathbf{p} \times \mathbf{L} = \frac{i}{\hbar} [\mathbf{L}^2, \mathbf{p}]. \qquad (G.26)$$

   We calculate this following the example of the $x$ or 1 component:

   $$[\mathbf{L}^2, p_1] = \sum_i [L_i^2, p_1] = \sum_i L_i [L_i, p_1] + [L_i, p_1] L_i \qquad (G.27)$$

where we have used (G.15). Since the momentum is a vector operator, we can use (G.16) and obtain explicitly

$$
\begin{aligned}
\left[\mathbf{L}^2, p_1\right] &= \sum_i \sum_m L_i i\hbar\varepsilon_{i1m}\, p_m \; + \; i\hbar\varepsilon_{i1m}\, p_m L_i \\
&= L_2 i\hbar\varepsilon_{213}\, p_3 + i\hbar\varepsilon_{213}\, p_3 L_2 + L_3 i\hbar\varepsilon_{312}\, p_2 + i\hbar\varepsilon_{312}\, p_2 L_3.
\end{aligned}
\tag{G.28}
$$

With the corresponding values for the Levi-Civita tensor, it follows that

$$
\begin{aligned}
\left[\mathbf{L}^2, p_1\right] &= i\hbar\left(-L_2 p_3 - p_3 L_2 + L_3 p_2 + p_2 L_3\right) \\
&= -i\hbar\left(\mathbf{L} \times \mathbf{p} - \mathbf{p} \times \mathbf{L}\right)_1
\end{aligned}
\tag{G.29}
$$

and thus (G.26). Hence, we can write the Lenz vector $\mathbf{\Lambda} = \frac{1}{2m\gamma}\left(\mathbf{L} \times \mathbf{p} - \mathbf{p} \times \mathbf{L}\right) + \hat{\mathbf{r}}$ also as

$$
\mathbf{\Lambda} = \frac{i}{2m\gamma\hbar}\left[\mathbf{L}^2, \mathbf{p}\right] + \hat{\mathbf{r}}.
\tag{G.30}
$$

Now we can treat the question of the vector operator. We have to show that:

$$
\left[L_j, \Lambda_k\right] = i\hbar \sum_m \varepsilon_{jkm} \Lambda_m.
\tag{G.31}
$$

Since $\hat{\mathbf{r}}$ is a vector operator, we can confine ourselves to $\left[\mathbf{L}^2, \mathbf{p}\right]$, i.e. we have to show that

$$
\left[L_j, \left[\mathbf{L}^2, p_k\right]\right] = i\hbar \sum_m \varepsilon_{jkm} \left[\mathbf{L}^2, p_m\right].
\tag{G.32}
$$

With the Jacobi identity (G.14), we obtain

$$
\left[\mathbf{L}^2, \left[L_j, p_k\right]\right] = i\hbar \sum_m \varepsilon_{jkm} \left[\mathbf{L}^2, p_m\right].
\tag{G.33}
$$

Since the momentum is a vector operator, we find with $\left[L_j, p_k\right] = i\hbar \sum_m \varepsilon_{jkm} p_m$ from the last equation that:

$$
\left[\mathbf{L}^2, i\hbar \sum_m \varepsilon_{jkm} p_m\right] = i\hbar \sum_m \varepsilon_{jkm} \left[\mathbf{L}^2, p_m\right],
\tag{G.34}
$$

with which the vector operator character of $\mathbf{\Lambda}$ is proven.

(c) The question of the commutator is next: We have to show that

$$[H, \Lambda] = \left[H, \frac{i}{2m\gamma\hbar}[L^2, \mathbf{p}] + \hat{\mathbf{r}}\right] = 0. \qquad (G.35)$$

We will do this again step by step and in detail. First, we have

$$[H, \Lambda] = \frac{i}{2m\gamma\hbar}[H, [L^2, \mathbf{p}]] + [H, \hat{\mathbf{r}}]$$

$$= -\frac{i}{2m\gamma\hbar}[L^2, [\mathbf{p}, H]] + [H, \hat{\mathbf{r}}] \qquad (G.36)$$

where we have used the fact that for a central potential, $[H, L^2] = 0$. We insert $H = \frac{p^2}{2m} - \frac{\gamma}{r}$ and obtain

$$[H, \Lambda] = \frac{i}{2m\hbar}\left[L^2, \left[\mathbf{p}, \frac{1}{r}\right]\right] + \frac{1}{2m}[\mathbf{p}^2, \hat{\mathbf{r}}]. \qquad (G.37)$$

Calculating the commutator $\left[\mathbf{p}, \frac{1}{r}\right]$ yields

$$\left[\mathbf{p}, \frac{1}{r}\right] = -\frac{\hbar}{i}\frac{\hat{\mathbf{r}}}{r^2}. \qquad (G.38)$$

It follows that

$$[H, \Lambda] = -\frac{1}{2m}\left[L^2, \frac{\hat{\mathbf{r}}}{r^2}\right] + \frac{1}{2m}[\mathbf{p}^2, \hat{\mathbf{r}}]. \qquad (G.39)$$

We know that $L^2$ contains derivatives only with respect to the angles; therefore we can write

$$[H, \Lambda] = \frac{1}{2m}[\mathbf{p}^2, \hat{\mathbf{r}}] - \frac{1}{2m}\left[\frac{L^2}{r^2}, \hat{\mathbf{r}}\right]. \qquad (G.40)$$

From the representation of the Laplacian in spherical coordinates, we obtain:

$$\mathbf{p}^2 = p_r^2 + \frac{L^2}{r^2}. \qquad (G.41)$$

It follows that

$$[H, \Lambda] = \frac{1}{2m}\left[p_r^2 + \frac{L^2}{r^2}, \hat{\mathbf{r}}\right] - \frac{1}{2m}\left[\frac{L^2}{r^2}, \hat{\mathbf{r}}\right] = \frac{1}{2m}[p_r^2, \hat{\mathbf{r}}] = 0. \quad (G.42)$$

The last equals sign applies since $p_r$ contains only derivatives with respect to $r$, while in $\hat{\mathbf{r}}$, only angles occur.

(d) Finally, we show that $\mathbf{\Lambda} \cdot \mathbf{L} = \mathbf{L} \cdot \mathbf{\Lambda} = 0$. We start with $\mathbf{\Lambda} = \frac{1}{2m\gamma}$ $(\mathbf{L} \times \mathbf{p} - \mathbf{p} \times \mathbf{L}) + \frac{\mathbf{r}}{r}$ and find:

$$\mathbf{\Lambda} \cdot L = \frac{1}{2m\gamma} (\mathbf{L} \times \mathbf{p} - \mathbf{p} \times \mathbf{L}) \cdot \mathbf{L} + \hat{\mathbf{r}} \cdot \mathbf{L}. \tag{G.43}$$

From Chap. 16, we know that $\hat{\mathbf{r}} \cdot \mathbf{L} = \mathbf{L} \cdot \hat{\mathbf{r}} = 0$. (Hint: Write the operator equation out in coordinates and examine it in detail). So we have to show that

$$(\mathbf{L} \times \mathbf{p} - \mathbf{p} \times \mathbf{L}) \cdot \mathbf{L} = \mathbf{L} \cdot (\mathbf{L} \times \mathbf{p} - \mathbf{p} \times \mathbf{L}) = 0. \tag{G.44}$$

We consider here only $\mathbf{L} \cdot (\mathbf{L} \times \mathbf{p} - \mathbf{p} \times \mathbf{L})$; the treatment of the other term proceeds analogously. First, we rewrite using (G.20) and obtain

$$\mathbf{L} \cdot (\mathbf{L} \times \mathbf{p} - \mathbf{p} \times \mathbf{L}) = \mathbf{L} \cdot (2\mathbf{L} \times \mathbf{p} - 2i\hbar\mathbf{p}.) \tag{G.45}$$

We can rearrange this using (G.23):

$$\mathbf{L} \cdot (2\mathbf{L} \times \mathbf{p} - 2i\hbar\mathbf{p}) = 2(\mathbf{L} \times \mathbf{L}) \cdot \mathbf{p} - 2i\hbar\mathbf{L} \cdot \mathbf{p} = 0 \tag{G.46}$$

because of $\mathbf{L} \cdot \mathbf{p} = 0$.

For practice, we perform the calculation once more using the Levi-Civita tensor:

$$(\mathbf{L} \times \mathbf{p} - \mathbf{p} \times \mathbf{L}) \cdot \mathbf{L} = \sum_{ijk} \varepsilon_{ijk} \left( L_j p_k - p_j L_k \right) L_i. \tag{G.47}$$

Here, we insert

$$\left[ L_j, p_k \right] = i\hbar \sum_m \varepsilon_{jkm} p_m \tag{G.48}$$

and obtain

$$(\mathbf{L} \times \mathbf{p} - \mathbf{p} \times \mathbf{L}) \cdot \mathbf{L} = \sum_{ijk} \varepsilon_{ijk} \left[ i\hbar \sum_m \varepsilon_{jkm} p_m - p_k L_j - p_j L_k \right] L_i$$

$$= i\hbar \sum_{ijkm} \varepsilon_{ijk}\varepsilon_{jkm} p_m L_i - \sum_{ijk} \varepsilon_{ijk} \left( p_k L_j + p_j L_k \right) L_i. \tag{G.49}$$

We rewrite the last summand:

$$\sum_{ijk} \varepsilon_{ijk} \left( p_k L_j + p_j L_k \right) L_i = \sum_{ijk} \left( \varepsilon_{ijk} + \varepsilon_{ikj} \right) p_k L_j L_k L_i = 0$$

$$\text{due to } \varepsilon_{ijk} = -\varepsilon_{ikj}. \tag{G.50}$$

For the first summand, we have using (G.22):

$$\sum_{ijkm} \varepsilon_{ijk} \varepsilon_{jkm} p_m L_i = -2 \sum_{im} \delta_{mi} p_m L_i$$

$$= -2 \sum_{i} p_i L_i = -2 \mathbf{p} \cdot \mathbf{L} = 0. \tag{G.51}$$

2. Prove the equation

$$\mathbf{\Lambda}^2 = \frac{2H}{m\gamma^2} \left( \mathbf{L}^2 + \hbar^2 \right) + 1. \tag{G.52}$$

Solution: With the abbreviation

$$g = \frac{1}{2m\gamma}, \tag{G.53}$$

we can write the square of the Lenz vector

$$\mathbf{\Lambda} = \frac{1}{2m\gamma} \left( \mathbf{L} \times \mathbf{p} - \mathbf{p} \times \mathbf{L} \right) + \frac{\mathbf{r}}{r} \tag{G.54}$$

as

$$\mathbf{\Lambda}^2 = g^2 \left( \mathbf{L} \times \mathbf{p} - \mathbf{p} \times \mathbf{L} \right) \left( \mathbf{L} \times \mathbf{p} - \mathbf{p} \times \mathbf{L} \right)$$

$$+ g \left[ \left( \mathbf{L} \times \mathbf{p} - \mathbf{p} \times \mathbf{L} \right) \frac{\mathbf{r}}{r} + \frac{\mathbf{r}}{r} \left( \mathbf{L} \times \mathbf{p} - \mathbf{p} \times \mathbf{L} \right) \right] + 1. \tag{G.55}$$

We treat the terms $\sim g^2$ and $\sim g$ separately.

(a) The terms $\sim g^2$ are

$$\left( \mathbf{L} \times \mathbf{p} \right) \left( \mathbf{L} \times \mathbf{p} \right) - \left( \mathbf{p} \times \mathbf{L} \right) \left( \mathbf{L} \times \mathbf{p} \right) - \left( \mathbf{L} \times \mathbf{p} \right) \left( \mathbf{p} \times \mathbf{L} \right) + \left( \mathbf{p} \times \mathbf{L} \right) \left( \mathbf{p} \times \mathbf{L} \right)$$

$$= p^2 L^2 - \left( -p^2 L^2 - 4\hbar^2 p^2 \right) - \left( -p^2 L^2 \right) + p^2 L^2 = 4p^2 L^2 + 4\hbar^2 p^2. \tag{G.56}$$

To realize this, we first consider the last term:

$$\left( \mathbf{p} \times \mathbf{L} \right) \left( \mathbf{p} \times \mathbf{L} \right) = \sum_{i} \sum_{jk,mn} \varepsilon_{ijk} p_j L_k \varepsilon_{imn} p_m L_n$$

$$= \sum_{jk} p_j L_k p_j L_k - p_j L_k p_k L_j$$

$$= \sum_{jk} p_j \left( p_j L_k - i\hbar \sum_m \varepsilon_{jkm} p_m \right) L_k - p_j \left( \mathbf{L} \cdot \mathbf{p} \right) L_j$$

$$= \sum_{jk} \left( p_j^2 L_k^2 - i\hbar p_j \sum_m \varepsilon_{jkm} p_m L_k \right)$$

$$= \mathbf{p}^2 \mathbf{L}^2 - i\hbar \sum_k \left( \sum_{jm} \varepsilon_{jkm} p_j p_m \right) L_k$$

$$= \mathbf{p}^2 \mathbf{L}^2 + i\hbar \sum_k (\mathbf{p} \times \mathbf{p})_k L_k = \mathbf{p}^2 \mathbf{L}^2 = p^2 L^2, \quad \text{(G.57)}$$

where we have used $\mathbf{L} \cdot \mathbf{p} = 0$.
With this equation and with (G.20) and (G.23), the other terms can be calculated. We have

$$(\mathbf{L} \times \mathbf{p})(\mathbf{p} \times \mathbf{L}) = (-\mathbf{p} \times \mathbf{L} + 2i\hbar\mathbf{p})(\mathbf{p} \times \mathbf{L})$$
$$= -(\mathbf{p} \times \mathbf{L})(\mathbf{p} \times \mathbf{L}) + 2i\hbar\mathbf{p}(\mathbf{p} \times \mathbf{L}) = -p^2 L^2 + 2i\hbar (\mathbf{p} \times \mathbf{p})\mathbf{L} = -p^2 L^2$$
$$\text{(G.58)}$$

and

$$(\mathbf{L} \times \mathbf{p})(\mathbf{L} \times \mathbf{p}) = (\mathbf{L} \times \mathbf{p})(-\mathbf{p} \times \mathbf{L} + 2i\hbar\mathbf{p})$$
$$= -(\mathbf{L} \times \mathbf{p})(\mathbf{p} \times \mathbf{L}) + 2i\hbar(\mathbf{L} \times \mathbf{p})\mathbf{p} = -(\mathbf{L} \times \mathbf{p})(\mathbf{p} \times \mathbf{L}) = p^2 L^2$$
$$\text{(G.59)}$$

and

$$(\mathbf{p} \times \mathbf{L})(\mathbf{L} \times \mathbf{p}) = (-\mathbf{L} \times \mathbf{p} + 2i\hbar\mathbf{p})(\mathbf{L} \times \mathbf{p})$$
$$= -(\mathbf{L} \times \mathbf{p})(\mathbf{L} \times \mathbf{p}) + 2i\hbar\mathbf{p}(\mathbf{L} \times \mathbf{p}) = -p^2 L^2 + 2i\hbar\mathbf{p}(-\mathbf{p} \times \mathbf{L} + 2i\hbar\mathbf{p})$$
$$= -p^2 L^2 - 2i\hbar\mathbf{p}(\mathbf{p} \times \mathbf{L}) - 4\hbar^2 \mathbf{p}^2 = -p^2 L^2 - 4\hbar^2 p^2. \quad \text{(G.60)}$$

(b) The terms $\sim g$ are

$$(\mathbf{L} \times \mathbf{p} - \mathbf{p} \times \mathbf{L})\frac{\mathbf{r}}{r} + \frac{\mathbf{r}}{r}(\mathbf{L} \times \mathbf{p} - \mathbf{p} \times \mathbf{L})$$
$$= (\mathbf{L} \times \mathbf{p})\frac{\mathbf{r}}{r} - (\mathbf{p} \times \mathbf{L})\frac{\mathbf{r}}{r} + \frac{\mathbf{r}}{r}(\mathbf{L} \times \mathbf{p}) - \frac{\mathbf{r}}{r}(\mathbf{p} \times \mathbf{L}). \quad \text{(G.61)}$$

Because of (G.23), it follows that

$$(\mathbf{L} \times \mathbf{p})\mathbf{r} = \mathbf{L}(\mathbf{p} \times \mathbf{r}) = -L^2. \quad \text{(G.62)}$$

Since $\mathbf{L}$ contains only derivatives with respect to angles, we have

$$(\mathbf{L} \times \mathbf{p})\frac{\mathbf{r}}{r} = \mathbf{L}\left(\mathbf{p} \times \frac{\mathbf{r}}{r}\right) = -\frac{L^2}{r}. \quad \text{(G.63)}$$

Furthermore, we have:

$$(\mathbf{p} \times \mathbf{L}) \frac{\mathbf{r}}{r} = (-\mathbf{L} \times \mathbf{p} + 2i\hbar\mathbf{p}) \frac{\mathbf{r}}{r} = \frac{L^2}{r} + 2i\hbar\mathbf{p}\frac{\mathbf{r}}{r}. \tag{G.64}$$

For the last term here, it holds that:

$$2i\hbar\mathbf{p}\frac{\mathbf{r}}{r} = 2i\hbar\frac{\hbar}{i}\nabla\frac{\mathbf{r}}{r} = 2i\hbar\frac{\hbar}{i}\left(\frac{3}{r} - \frac{\mathbf{r}\cdot\mathbf{r}}{r^3} + \frac{\mathbf{r}}{r}\nabla\right) = 2\hbar^2\frac{2}{r} + 2i\hbar\frac{\mathbf{r}}{r}\mathbf{p}. \tag{G.65}$$

Moreover, we have

$$\frac{\mathbf{r}}{r}(\mathbf{p} \times \mathbf{L}) = \frac{1}{r}(\mathbf{r} \times \mathbf{p})\mathbf{L} = \frac{L^2}{r} \tag{G.66}$$

and

$$\frac{\mathbf{r}}{r}(\mathbf{L} \times \mathbf{p}) = \frac{\mathbf{r}}{r}(-\mathbf{p} \times \mathbf{L} + 2i\hbar\mathbf{p}) = -\frac{L^2}{r} + 2i\hbar\frac{\mathbf{r}}{r}\mathbf{p}. \tag{G.67}$$

Hence, it follows for the terms $\sim g$:

$$(\mathbf{L} \times \mathbf{p} - \mathbf{p} \times \mathbf{L})\frac{\mathbf{r}}{r} + \frac{\mathbf{r}}{r}(\mathbf{L} \times \mathbf{p} - \mathbf{p} \times \mathbf{L})$$

$$= -\frac{L^2}{r} - \left[\frac{L^2}{r} + 2\hbar^2\frac{2}{r} + 2i\hbar\frac{\mathbf{r}}{r}\mathbf{p}\right] + \left[-\frac{L^2}{r} + 2i\hbar\frac{\mathbf{r}}{r}\mathbf{p}\right] - \left[\frac{L^2}{r}\right]$$

$$= -\frac{4L^2}{r} - \frac{4\hbar^2}{r}. \tag{G.68}$$

(c) In sum, we have:

$$\Lambda^2 = g^2(\mathbf{L} \times \mathbf{p} - \mathbf{p} \times \mathbf{L})(\mathbf{L} \times \mathbf{p} - \mathbf{p} \times \mathbf{L})$$

$$+ g(\mathbf{L} \times \mathbf{p} - \mathbf{p} \times \mathbf{L})\frac{\mathbf{r}}{r} + g\frac{\mathbf{r}}{r}(\mathbf{L} \times \mathbf{p} - \mathbf{p} \times \mathbf{L}) + 1$$

$$= g^2\{4p^2L^2 + 4\hbar^2p^2\} + g\left\{-\frac{4L^2}{r} - \frac{4\hbar^2}{r}\right\} + 1$$

$$= 4g^2p^2L^2 + 4\hbar^2g^2p^2 - g\frac{4L^2}{r} - g\frac{4\hbar^2}{r} + 1$$

$$= 4\left(g^2p^2 - g\frac{1}{r}\right)L^2 + 4\hbar^2\left(g^2p^2 - g\frac{1}{r}\right) + 1. \tag{G.69}$$

With

$$\frac{p^2}{2m} - \frac{\gamma}{r} = H; \quad p^2 = 2mH + 2m\frac{\gamma}{r} \tag{G.70}$$

it follows that

$$\Lambda^2 = 4\left(g^2 p^2 - g\frac{1}{r}\right)\left(L^2 + \hbar^2\right) + 1$$

$$= 4\left(g^2 2mH + g^2 2m\frac{\gamma}{r} - g\frac{1}{r}\right)\left(L^2 + \hbar^2\right) + 1. \tag{G.71}$$

In order that the terms $\sim 1/r$ vanish, we must require:

$$g^2 2m\gamma - g = 0; \quad g = \frac{1}{2m\gamma}, \tag{G.72}$$

which is in fact the case; we thus finally obtain

$$\Lambda^2 = \frac{2}{m\gamma^2} H\left(L^2 + \hbar^2\right) + 1. \tag{G.73}$$

(d) Prove the two equations

$$[\mathbf{J}_1, \mathbf{J}_2] = 0; \quad \mathbf{J}_1^2 = \mathbf{J}_2^2 \tag{G.74}$$

with $\mathbf{J}_1 = \frac{1}{2}(\mathbf{L} + \mathbf{R})$ and $\mathbf{J}_2 = \frac{1}{2}(\mathbf{L} - \mathbf{R})$ and $R = \sqrt{\frac{m\gamma^2}{2|E|}}\mathbf{\Lambda}$.
Solution: We have initially

$$[\mathbf{J}_1, \mathbf{J}_2] = \mathbf{J}_1\mathbf{J}_2 - \mathbf{J}_2\mathbf{J}_1 = \frac{1}{2}(\mathbf{L} + \mathbf{R})\frac{1}{2}(\mathbf{L} - \mathbf{R}) - \frac{1}{2}(\mathbf{L} - \mathbf{R})\frac{1}{2}(\mathbf{L} + \mathbf{R})$$

$$= \frac{1}{4}\left\{\mathbf{L}^2 - \mathbf{L}\mathbf{R} + \mathbf{R}\mathbf{L} + \mathbf{R}^2\right\} - \frac{1}{4}\left\{\mathbf{L}^2 + \mathbf{L}\mathbf{R} - \mathbf{R}\mathbf{L} + \mathbf{R}^2\right\} = 0. \tag{G.75}$$

For the last step, we have used $\mathbf{L}\mathbf{R} = \mathbf{R}\mathbf{L} = 0$.
Concerning the proof of the second equation, we note that:

$$\mathbf{J}_1^2 = \frac{1}{4}\left(\mathbf{L}^2 + \mathbf{L}\mathbf{R} + \mathbf{R}\mathbf{L} + \mathbf{R}^2\right) = \frac{1}{4}\left(\mathbf{L}^2 + \mathbf{R}^2\right)$$

$$\mathbf{J}_2^2 = \frac{1}{4}\left(\mathbf{L}^2 - \mathbf{L}\mathbf{R} - \mathbf{R}\mathbf{L} + \mathbf{R}^2\right) = \frac{1}{4}\left(\mathbf{L}^2 + \mathbf{R}^2\right). \tag{G.76}$$

We have again used $\mathbf{L}\mathbf{R} = \mathbf{R}\mathbf{L} = 0$.

3. Show that $\mathbf{J}_1$ (and therefore also $\mathbf{J}_2$) is a generalized angular-momentum operator. Solution: We consider $[J_{1x}, J_{1y}]$ in more detail and infer the other relations from cyclic permutations. We have

$$
\begin{aligned}
[J_{1x}, J_{1y}] &= \frac{1}{2}(L_x + R_x)\frac{1}{2}(L_y + R_y) - \frac{1}{2}(L_y + R_y)\frac{1}{2}(L_x + R_x) \\
&= \frac{1}{4}\{L_xL_y + L_xR_y + R_xL_y + R_xR_y\} \\
&\quad - \frac{1}{4}\{L_yL_x + L_yR_x + R_yL_x + R_yR_x\} \\
&= \frac{1}{4}\{[L_x, L_y] + [L_x, R_y] + [R_x, L_y] + [R_x, R_y]\}. \quad \text{(G.77)}
\end{aligned}
$$

For the individual commutators, it holds that:

$$
\begin{aligned}
[L_x, L_y] &= i\hbar L_z; \quad [R_x, R_y] = i\hbar L_z \\
[L_x, R_y] &= i\hbar R_z; \quad [R_x, L_y] = i\hbar R_z,
\end{aligned} \quad \text{(G.78)}
$$

and therefore it follows that

$$
[J_{1x}, J_{1y}] = \frac{i\hbar}{2}(L_z + R_z) = i\hbar J_{1z} \quad \text{(G.79)}
$$

and the corresponding other relations follow by cyclic permutation.

# Appendix H
# Perturbative Calculation of the Hydrogen Atom

In this section, we want to outline the perturbation calculation for

$$H = H^{(0)} + W_{mp} + W_{ls} + W_D; \quad H^{(0)} = \frac{\mathbf{p}^2}{2m} - \frac{\gamma}{r}; \quad \gamma = \frac{Ze^2}{4\pi\varepsilon_0} \tag{H.1}$$

in a little more detail than was given in Chap. 19. We have:

$$W_{mp} = -\frac{\mathbf{p}^4}{8m^3 c^2}$$

$$W_{ls} = \frac{1}{2m^2 c^2} \frac{1}{r} \frac{dV(r)}{dr} \mathbf{l} \cdot \mathbf{s} = \frac{1}{2m^2 c^2} \frac{\gamma}{r^3} \mathbf{l} \cdot \mathbf{s} \tag{H.2}$$

$$W_D = \frac{\hbar^2}{8m^2 c^2} \nabla^2 V(r) = \frac{\pi \hbar^2 \gamma}{2m^2 c^2} \delta(\mathbf{r}).$$

With the orthonormal states $\left|n; j, m_j; l\right\rangle$, we obtain the energy corrections as

$$\left\langle n; j', m_j'; l' \right| W \left| n; j, m_j; l \right\rangle. \tag{H.3}$$

For brevity we use in the following the notation

$$\langle A \rangle := \left\langle n; j', m_j'; l \right| A \left| n; j, m_j; l \right\rangle. \tag{H.4}$$

## H.1 Calculation of the Matrix Elements

### H.1.1 Matrix Elements of $W_{mp}$

Because of $\frac{\mathbf{p}^2}{2m} = H^{(0)} + \frac{\gamma}{r}$, we have

© Springer Nature Switzerland AG 2018
J. Pade, *Quantum Mechanics for Pedestrians 2*, Undergraduate Lecture
Notes in Physics, https://doi.org/10.1007/978-3-030-00467-5

$$\langle W_{mp} \rangle = \left\langle -\frac{\mathbf{p}^4}{8m^3c^2} \right\rangle = \left\langle -\frac{1}{2mc^2} \left( H^{(0)} + \frac{\gamma}{r} \right)^2 \right\rangle$$

$$= -\frac{1}{2mc^2} \left\langle \left( H^{(0)} \right)^2 + H^{(0)}\frac{\gamma}{r} + \frac{\gamma}{r}H^{(0)} + \frac{\gamma^2}{r^2} \right\rangle. \tag{H.5}$$

Since $H^0$ is Hermitian, it follows that

$$\langle W_{mp} \rangle = -\frac{1}{2mc^2} \left\langle \left( E_n^{(0)} \right)^2 + 2E_n^{(0)}\frac{\gamma}{r} + \frac{\gamma^2}{r^2} \right\rangle. \tag{H.6}$$

This leads to

$$\langle W_{mp} \rangle = -\frac{1}{2mc^2} \left\{ \left( E_n^{(0)} \right)^2 + 2E_n^{(0)}\gamma\left\langle \frac{1}{r} \right\rangle + \gamma^2\left\langle \frac{1}{r^2} \right\rangle \right\} \delta_{j'j}\delta_{m'_j m_j}. \tag{H.7}$$

### H.1.2   Matrix Elements of $W_{ls}$

We have:

$$\langle W_{ls} \rangle = \left\langle \frac{1}{2m^2c^2}\frac{\gamma}{r^3}\mathbf{l} \cdot \mathbf{s} \right\rangle = \frac{\gamma}{2m^2c^2}\left\langle \frac{1}{r^3}\mathbf{l} \cdot \mathbf{s} \right\rangle. \tag{H.8}$$

Making use of

$$\mathbf{j}^2 = \mathbf{l}^2 + 2\mathbf{l} \cdot \mathbf{s} + \mathbf{s}^2 \text{ or } \mathbf{l} \cdot \mathbf{s} = \frac{1}{2}\left( \mathbf{j}^2 - \mathbf{l}^2 - \mathbf{s}^2 \right), \tag{H.9}$$

we obtain:

$$\langle W_{ls} \rangle = \frac{\gamma\hbar^2}{2m^2c^2}\frac{1}{2}\left[ j(j+1) - l(l+1) - \frac{3}{4} \right]\left\langle \frac{1}{r^3} \right\rangle \delta_{j'j}\delta_{m'_j m_j}. \tag{H.10}$$

For the case $l = 0$ (where the term $\left\langle \frac{1}{r^3} \right\rangle$ does not exist), we find no contribution, since then $j(j+1) - l(l+1) - \frac{3}{4} = 0$.

### H.1.3   Matrix Elements of $W_D$

We have:

$$\langle W_D \rangle = \left\langle \frac{\pi\hbar^2\gamma}{2m^2c^2}\delta(\mathbf{r}) \right\rangle = \frac{\pi\hbar^2\gamma}{2m^2c^2}\langle \delta(\mathbf{r}) \rangle. \tag{H.11}$$

Due to the delta function, this term involves only $s$ orbitals, because only for these is $\psi(0) \neq 0$ (because of $R_{nl} \sim r^l$ for small $r$). Thus, we obtain the contribution:

$$\langle W_D \rangle = \frac{\pi \hbar^2 \gamma}{2m^2 c^2} |R_{n0}(0)|^2 \, \delta_{j'j} \delta_{m'_j m_j} \tag{H.12}$$

for $l = 0$ only.

## H.2 Fine Structure Corrections

Summed up: All the corrections are diagonal in $j$ and $m_j$. We add them and obtain as the total correction to the energy:

$$E_n^{(1)} = \left[ \begin{array}{c} -\frac{1}{2mc^2} \left\{ \left( E_n^{(0)} \right)^2 + 2E_n^{(0)} \gamma \langle \frac{1}{r} \rangle + \gamma^2 \langle \frac{1}{r^2} \rangle \right\} \\ + \frac{\gamma \hbar^2}{2m^2 c^2} \frac{1}{2} \left[ j(j+1) - l(l+1) - \frac{3}{4} \right] \langle \frac{1}{r^3} \rangle + \frac{\pi \hbar^2 \gamma}{2m^2 c^2} |R_{n0}(0)|^2 \delta_{l,0} \end{array} \right] \delta_{j'j} \delta_{m'_j m_j}. \tag{H.13}$$

Finally, we have to calculate the three matrix elements

$$\left\langle \frac{1}{r^a} \right\rangle = \langle n; j, m_j; l | \frac{1}{r^a} | n; j, m_j; l \rangle \sim \int R_{nl}^2 \frac{1}{r^a} r^2 dr; \quad a = 1, 2, 3. \tag{H.14}$$

We take the result from Appendix B, Vol. 2, 'Special functions':

$$\left\langle \frac{1}{r} \right\rangle = \frac{Z}{a_0 n^2}; \quad \left\langle \frac{1}{r^2} \right\rangle = \frac{Z^2}{a_0^2 n^3} \frac{1}{l + \frac{1}{2}}; \quad \left\langle \frac{1}{r^3} \right\rangle = \frac{Z^3}{a_0^3 n^3} \frac{1}{l \left( l + \frac{1}{2} \right) (l + 1)}. \tag{H.15}$$

# Appendix I
# The Production of Entangled Photons

Entangled photons are of great importance for experiments in basic research as well as for practical applications, from quantum cryptography to quantum teleportation to quantum computing. In the following, we outline three experimental production methods. In each case, photon pairs are generated which are entangled with respect to their polarization, i.e. they are in a Bell state as $|\Psi\rangle = \frac{|hv\rangle - |vh\rangle}{\sqrt{2}}$.

## I.1 Atomic Sources

In this method, appropriate atoms such as calcium or mercury are excited so that they emit two polarization-entangled photons on returning to their ground states. Calcium, for example, has in its ground state (1) two electrons in the 4s shell. By irradiation with UV light, an excited state (3, both electrons in the 4p state) is populated; it decays through a cascade via an intermediate state (2),[25] see Fig. I.1. In the first step $3 \rightarrow 2$, a photon with $\lambda = 551.3$ nm is emitted; in the second step, $2 \rightarrow 1$, a photon is emitted with $\lambda = 422.7$ nm.

It is essential that the intermediate level be degenerate with respect to the magnetic quantum number $m$ (that is, with respect to the polarization). Assuming that the two photons are emitted along the quantization axis, the $m = 0$ state does not contribute to the optical transition. Thus, if we consider photons which are emitted in opposite directions, their state of polarization is given by

$$|\psi\rangle = \frac{|r\rangle\,|r\rangle + |l\rangle\,|l\rangle}{\sqrt{2}} = \frac{|rr\rangle + |ll\rangle}{\sqrt{2}} = \frac{|hh\rangle - |vv\rangle}{\sqrt{2}}. \tag{I.1}$$

We see that the linear polarization direction is not specified, but that the two photons are polarized parallel to each other.

---

[25] An electron that is in the highest level (3) cannot go directly into the ground state (1), because then angular momentum would not be conserved.

© Springer Nature Switzerland AG 2018

J. Pade, *Quantum Mechanics for Pedestrians 2*, Undergraduate Lecture Notes in Physics, https://doi.org/10.1007/978-3-030-00467-5

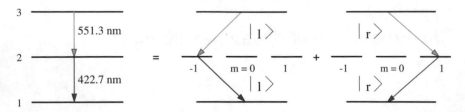

**Fig. I.1**  Principle of the cascade transition in calcium

This was the first method that was reliable enough to be used in experiments. However, it has serious drawbacks—among other things, the fact that the photons are not emitted in fixed directions, but rather are randomly distributed over the whole solid angle, so that the method is not very efficient.

## I.2  Parametric Fluorescence

In this process, one slices a photon so to speak into two daughter photons with longer wavelengths. The 'scissors' is an optically active (non-linear) crystal such as barium borate. It converts a UV photon into two red photons, commonly referred to as *signal* and *idler*, that are emitted in certain directions due to the conservation of momentum and energy; cf. Fig. I.2. Because of this conversion to lower photon energies, the process is also called 'parametric down conversion' (PDC).

Due to the polarization of the daughter photons, one can generally distinguish two types: Type I fluorescence (orthogonal polarizations), and type II fluorescence (parallel polarizations). We will hereafter consider only type II fluorescence, as only in this case can entanglement be obtained directly. In addition, we confine ourselves to the case that both daughter photons have the same energy (degenerate fluorescence).

In contrast to the two-photon cascade emission, here the two daughter photons are generated simultaneously. Above all, conservation of momentum ensures a restriction of the possible directions of emission of the signal and idler photons; namely, one can

**Fig. I.2**  Scheme of
parametric fluorescence

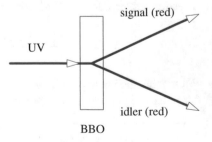

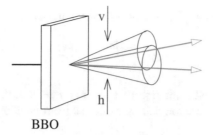

BBO

**Fig. I.3** Parametric fluorescence: photons with horizontal (*bottom*) and vertical (*top*) polarization are emitted along the surfaces of two cones. A pair of photons which is emitted along the section lines of the cones is polarization-entangled

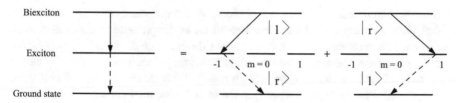

**Fig. I.4** Generation of an entangled photon pair via an exciton cascade process

obtain emission of the photons along two conical surfaces. In the case of degenerate fluorescence, these emission cones have the same opening angles; see Fig. I.3.

The opening angle of the cones depends on the angle between the optical axis of the crystal and the direction of the incident light beam. In particular, one can arrange this so that the two cones intersect. If a photon is emitted in one of these intersecting directions, then the other photon is emitted in the other direction. The polarization state of this photon pair is entangled; it is given by:

$$|\psi\rangle = \frac{|hv\rangle + |vh\rangle}{\sqrt{2}}. \tag{I.2}$$

Parametric fluorescence is currently the most common method for generating entangled photons. However, the yield is limited to a few percent.

## I.3 Semiconductor Sources

Here, there are several methods. One is to produce a biexciton, i.e. a state of two bound excitons.[26] Under appropriate conditions, the biexciton decays in a cascade process that is quite similar to the atomic decay described above; see Fig. I.4.[27]

---

[26] An exciton is a bound state between an electron and a hole in a semiconductor or insulator.

[27] In other words, the photon pair is produced by the radiative decay of two electron-hole pairs.

This results in an entangled pair of photons in the state

$$|\psi\rangle = \frac{|hh\rangle + |vv\rangle}{\sqrt{2}}. \qquad (I.3)$$

There is no fundamental limitation of the yield here, and the mechanism is quite effective. However, the photon pairs are emitted isotropically, in all directions.

## I.4  Concluding Remarks

Given the importance of entangled photons, it is understandable that researchers are actively working on reducing the limitations of the procedures described. With parametric fluorescence, one can e.g. enhance the entrance channel by 'amplifying' the UV pulses in a resonator. Thereby, ultrashort light pulses in very rapid succession are produced, with which one can obtain much higher emission rates of entangled photons, and even more pairs which are entangled with one another.[28]

With semiconductor sources, for example, a method was developed that allows for effective collection of the photon pairs generated; keyword 'photonic molecule'. Using this method, an efficiency of 80–90 % should be achievable, i.e. 8–9 photon pairs per 10 excitation pulses.[29] By means of quantum points it is possible to produce entangled photons so to say off the production line.[30]

Finally, we note that apart from photons/polarization, one can of course entangle other quantum objects/properties. An example is the spin entanglement of atoms. Here, a suitable diatomic molecule with zero spin is excited so highly that it dissociates, i.e. it decays into two atoms, each with spin 1/2. This atomic pair is spin entangled, where the entanglement of course refers to the $z$ components of the spins.

---

[28] R. Krischek et al., 'Ultraviolet enhancement cavity for ultrafast nonlinear optics and high-rate multiphoton entanglement experiments.' *Nature Photonics* 4, 170–173 (2010).

[29] A. Dousse et al., 'Ultrabright source of entangled photon pairs.' *Nature* 466, 217 (2010).

[30] See I. Schwartz et al., Deterministic generation of a cluster state of entangled photons, *Science*, online: 8. September 2016; http://dx.doi.org/10.1126/science.aah4758.

# Appendix J
# The Hardy Experiment

## J.1 The Experiment

Interaction-free measurement and quantum entanglement—these two elements of quantum mechanics are combined in Hardy's experiment,[31] shown schematically in Fig. J.1. The point is, *inter alia*, that we can entangle two quantum objects in an 'interaction-free' manner[32] here.

An electron-positron pair (or another particle-antiparticle pair) enters two Mach–Zehnder interferometers[33] (MZI) at the same time, the positron into the upper and the electron into the lower MZI. We denote the properties of the positron and the electron in the following as 1 and 2. If positron and electron meet at the point WW, they annihilate with certainty, $e^+e^- \rightarrow 2\gamma$.

In both MZI's, the first beam splitter BS and the mirror M are fixed. In each case, the second beam splitter, i.e. BS1 or BS2, can be moved out of/into the beam path. At WW, the arms of the two MZI's cross. Here, by an appropriate arrangement of the arms, we can choose either that the electron and the positron (a) do not meet and thus are not annihilated; or (b) do meet and thus are annihilated.

We consider first the situation in which the destructive interaction is switched on and both beam splitters BSi are in place. As described in Chap. 6, Vol. 1, each MZI is adjusted in such a way that a quantum object entering horizontally or vertically is always detected at the horizontal or vertical detector (the 'bright' detectors). If the other ('dark') detectors register a quantum, we know that some disturbing object is

---

[31]Lucien Hardy, 'Quantum mechanics, local realistic theories, and Lorentz-invariant realistic theories.' *Phys. Rev. Lett.* 68, 2981 (1992).

[32]For the limitations on the term 'interaction-free', see Chap. 6, Vol. 1.

[33]Of course, the components are made in such a way that they in fact perform as beam splitters and mirrors for the electrons and positrons. By the way, recent experiments were carried out with two photons, where the pair annihilation is replaced by a process of destructive interference. Here, a special type of quantum measurement was used (called *weak measurement*); see J.S. Lundeen & A.M. Steinberg, 'Experimental joint weak measurement on a photon pair as a probe of Hardy's paradox', arXiv:0810.4229 (2008).

© Springer Nature Switzerland AG 2018
J. Pade, *Quantum Mechanics for Pedestrians 2*, Undergraduate Lecture
Notes in Physics, https://doi.org/10.1007/978-3-030-00467-5

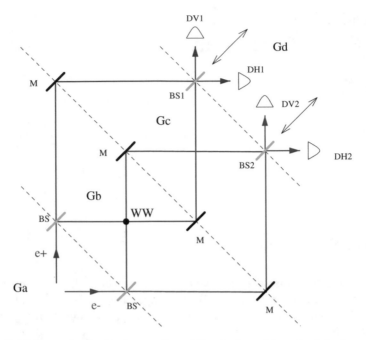

**Fig. J.1** Hardy's experiment. The beam splitters BS and mirrors M are fixed, the beam splitters BSi can be moved in/out of the ray path. Electrons (*blue*) and positrons (*red*) are detected in the detectors DVi and DHi

in the beam path. That means in Hardy's experiment that a click from a dark detector indicates that the quantum object must have 'probed' the arm with WW. When both dark detectors register a count, i.e. DH1 and DV2, the electron and the positron must have simultaneously 'probed' the WW point and thus should have annihilated with certainty. But that would contradict the result that they were both detected by the two dark detectors, which in fact is observed in a certain percentage of cases. From a classical point of view, this is a contradiction, known as Hardy's paradox.

We now describe the reactions of the detectors as a function of whether we turn the annihilation interaction at WW on or off, and whether the beam splitters BS1, 2 are in the path or not. We give the calculation further below, and just anticipate the result here.

For simplicity, we consider for the moment only the case that the two movable detectors are in the path of the setup (J.15). The probabilities for the activation of the two detectors are given in Table J.1:

For inactivated annihilation ($\alpha = 1$), we find again the result of Chap. 6, Vol. 1, indicating that the setup reproduces the initial state. Thus, detectors (DV1, DH2) are 'bright' and always respond at the same time.

For activated annihilation ($\alpha = 0$), a quarter of the positron-electron pairs annihilates, and three quarters can be detected in the detectors. The pairs landing in (DH1, DH2), (DH1, DV2) and (DV1, DV2) cannot have interacted directly by means of

**Table J.1** Probabilities for the simultaneous activation of two detectors

| Detectors | $\alpha = 1$ | $\alpha = 0$ |
|---|---|---|
| DH1,DH2 | 0 | 1/16 |
| DH1,DV2 | 0 | 1/16 |
| DV1,DH2 | 1 | 9/16 |
| DV1,DV2 | 0 | 1/16 |

For $\alpha = 0$ or 1, the annihilation process is activated or deactivated

the pair annihilation, as they would then have been annihilated; but they must have 'realized' the pair annihilation, since otherwise neither one nor two 'dark' detectors would have responded. In particular, in one sixteenth of the cases, both dark detectors respond. This means that both the electron and the positron have 'realized' that a disturbing object is in their beam paths—and this, of course, could be only the other partner. Consequently, $e^+$ and $e^-$ must have been at the same point (WW) in some way, without having destroyed each other.

## J.2 Calculation of the Probabilities

For the calculation, we adopt the notation of Chap. 6, Vol. 1, namely that $|H\rangle$ and $|V\rangle$ denote horizontal and vertical propagation channels. Moreover, we derived there for the operators $T$ (beam splitter) and $S$ (mirror) the expressions:

$$T = \frac{1+i}{2} [1 + i |H\rangle \langle V| + i |V\rangle \langle H|]$$
$$S = - |H\rangle \langle V| - |V\rangle \langle H| . \tag{J.1}$$

The annihilation interaction at point WW can be effective only if in the first MZI, a horizontal state is present, and in the second MZI, a vertical state. We can, therefore, represent this by an operator $W_\alpha$ with the following properties:

$$W_\alpha |H_1 H_2\rangle = |H_1 H_2\rangle ; \quad W_\alpha |H_1 V_2\rangle = \alpha |H_1 V_2\rangle$$
$$W_\alpha |V_1 H_2\rangle = |V_1 H_2\rangle ; \quad W_\alpha |V_1 V_2\rangle = |V_1 V_2\rangle$$
$$\alpha = \begin{cases} 1 \\ 0 \end{cases} \text{ for } \begin{matrix} \text{inactivated} \\ \text{activated} \end{matrix} \text{ annihilation interaction.} \tag{J.2}$$

Since $\{|H\rangle, |V\rangle\}$ is a CONS, we can write

$$W_\alpha = |H_1 H_2\rangle \langle H_1 H_2| + \alpha |H_1 V_2\rangle \langle H_1 V_2| + |V_1 H_2\rangle \langle V_1 H_2| + |V_1 V_2\rangle \langle V_1 V_2|$$
$$= 1 + (\alpha - 1) |H_1 V_2\rangle \langle H_1 V_2| . \tag{J.3}$$

This enables us to formulate the effect of the setup shown in Fig. J.1. We employ here the method used in Chap. 6, Vol. 1, namely dividing the setup into four main regions: In the first region Ga, we have the incoming state $|z_a\rangle$; in Gb the state $|z_b\rangle = T_1 T_2 |z_a\rangle$ transformed by the beam splitter, which possibly is modified by the annihilation interaction $W_\alpha$.[34] This state $|z_b\rangle = W_\alpha T_1 T_2 |z_a\rangle$ is reflected by the mirrors, so that in Gc, we have $|z_c\rangle = S_1 S_2 W_\alpha T_1 T_2 |z_1\rangle$. Finally, in Gd, following the beam splitters BS1, 2, we have the final state $|z_d\rangle$. We denote the operator representing the double MZI by $M$ and distinguish three cases:

1. no beam splitters BS1,2 inserted          $M(0) = S_1 S_2 W_\alpha T_1 T_2$
2. one beam splitter BS1,2 inserted    $M(1, i) = T_i S_1 S_2 W_\alpha T_1 T_2;\ i = 1, 2$   (J.4)
3. both beam splitters BS1,2 inserted       $M(2) = T_1 T_2 S_1 S_2 W_\alpha T_1 T_2$.

In the following, we calculate $M$ in the matrix representation. We start from

$$|H\rangle \cong \begin{pmatrix} 1 \\ 0 \end{pmatrix};\ \ |V\rangle \cong \begin{pmatrix} 0 \\ 1 \end{pmatrix} \tag{J.5}$$

and obtain (cf. Chap. 6, Vol. 1):

$$T \cong \frac{1+i}{2} \begin{pmatrix} 1 & i \\ i & 1 \end{pmatrix};\ \ S \cong -\begin{pmatrix} 0 & 1 \\ 1 & 0 \end{pmatrix}. \tag{J.6}$$

For the direct product of two beam splitters or mirrors, it follows that:

$$T_1 T_2 \cong \frac{i}{2} \begin{pmatrix} 1 & i & i & -1 \\ i & 1 & -1 & i \\ i & -1 & 1 & i \\ -1 & i & i & 1 \end{pmatrix};\ \ S_1 S_2 \cong \begin{pmatrix} 0 & 0 & 0 & 1 \\ 0 & 0 & 1 & 0 \\ 0 & 1 & 0 & 0 \\ 1 & 0 & 0 & 0 \end{pmatrix}. \tag{J.7}$$

If we use only one beam splitter, the matrices ($E$ is the $2 \times 2$ unit matrix) are given by:

$$T_1 E_2 = \frac{(1+i)}{2} \begin{pmatrix} 1 & 0 & i & 0 \\ 0 & 1 & 0 & i \\ i & 0 & 1 & 0 \\ 0 & i & 0 & 1 \end{pmatrix};\ \ E_1 T_2 = \frac{(1+i)}{2} \begin{pmatrix} 1 & i & 0 & 0 \\ i & 1 & 0 & 0 \\ 0 & 0 & 1 & i \\ 0 & 0 & i & 1 \end{pmatrix}. \tag{J.8}$$

The annihilation interaction is then represented by

---

[34]We leave off the explicit specification of the direct product, and write simply $T_1 T_2$ instead of $T_1 \otimes T_2$.

$$W_\alpha \cong \begin{pmatrix} 1 & 0 & 0 & 0 \\ 0 & \alpha & 0 & 0 \\ 0 & 0 & 1 & 0 \\ 0 & 0 & 0 & 1 \end{pmatrix}. \tag{J.9}$$

Now we can calculate the operators $M$. We have for $M\,(0)$:

$$M\,(0) \cong \frac{i}{2} \begin{pmatrix} -1 & i & i & 1 \\ i & -1 & 1 & i \\ \alpha i & \alpha & -\alpha & \alpha i \\ 1 & i & i & -1 \end{pmatrix}, \tag{J.10}$$

and for $M\,(1, i) = T_i M\,(0)$:

$$M\,(1, 1) \cong \frac{i}{2} \frac{(1 + i)}{2} \begin{pmatrix} -1 - a & i\,(1 + a) & i\,(1 - a) & 1 - a \\ 2i & -2 & 0 & 0 \\ -i\,(1 - a) & -1 + a & -1 - a & i\,(1 + a) \\ 0 & 0 & 2i & -2 \end{pmatrix}, \tag{J.11}$$

as well as

$$M\,(1, 2) \cong \frac{i}{2} \frac{(1 + i)}{2} \begin{pmatrix} -2 & 0 & 2i & 0 \\ 0 & -2 & 0 & 2i \\ i\,(\alpha + 1) & \alpha - 1 & -\alpha - 1 & i\,(\alpha - 1) \\ -\alpha + 1 & i\,(\alpha + 1) & i\,(-\alpha + 1) & -\alpha - 1 \end{pmatrix}. \tag{J.12}$$

For $M\,(2)$, it follows that:

$$M\,(2) \cong -\frac{1}{4} \begin{pmatrix} -3 - \alpha & i\,(\alpha - 1) & i\,(1 - \alpha) & 1 - \alpha \\ i\,(\alpha - 1) & -3 - \alpha & \alpha - 1 & i\,(1 - \alpha) \\ i\,(1 - \alpha) & \alpha - 1 & -3 - \alpha & i\,(\alpha - 1) \\ 1 - \alpha & i\,(\alpha - 1) & i\,(1 - \alpha) & -3 - \alpha \end{pmatrix}. \tag{J.13}$$

We now specialize, as shown in Fig. J.1, to the non-entangled initial state:

$$|v_1 h_2\rangle \cong \begin{pmatrix} 0 \\ 0 \\ 1 \\ 0 \end{pmatrix} \tag{J.14}$$

and obtain as final states $|z_d\rangle$:

$$|z_d(0)\rangle = \tfrac{i}{2} \begin{pmatrix} i \\ 1 \\ -\alpha \\ i \end{pmatrix}; \quad |z_d(1,1)\rangle = \tfrac{i}{2}\tfrac{(1+i)}{2} \begin{pmatrix} i(1-\alpha) \\ 0 \\ -1-\alpha \\ 2i \end{pmatrix}$$

$$|z_d(1,2)\rangle = \tfrac{i}{2}\tfrac{(1+i)}{2} \begin{pmatrix} 2i \\ 0 \\ -1-\alpha \\ i(1-\alpha) \end{pmatrix}; \quad |z_d(2)\rangle = \tfrac{1}{4} \begin{pmatrix} i(\alpha-1) \\ 1-\alpha \\ 3+\alpha \\ i(\alpha-1) \end{pmatrix}. \tag{J.15}$$

We wish now to discuss these expressions briefly. We consider first the detection probabilities and then the entanglement status of the $e^+e^-$ pairs detected.

### Detection Probabilities

If the final state is given e.g. as $|h_1 h_2\rangle$, then both detectors DH1 and DH2 are activated. As usual, the detection probabilities are given by the sum of the squared values of the corresponding coefficients in (J.15). The results are collected in Table J.2.

We first leave the pair annihilation switched off ($\alpha = 1$). We see a uniform distribution of the results if both beam splitters are withdrawn; if both are in place, however, the setup reproduces the initial state $|v_1 h_2\rangle$.

Now we let the pair annihilation act ($\alpha = 0$). In this case, one quarter of the positron-electron pairs will always annihilate, regardless of whether there are beam splitters in the beam path or not. The remaining 75 % of the pairs, which are detected by the detectors, cannot have interacted *directly* by pair annihilation, as they would have been annihilated in that case, and not detected.

We repeat our discussion on the configuration with two beam splitters BSi inserted (i.e. the final state is $|z_d(2)\rangle$). Without pair annihilation (i.e. for $\alpha = 1$), only the detector pair DV1 and DH2 responds. Thus, if a 'dark' detector responds we know that some 'obstacle' must have been in the path; but this can only be the pair annihilation at WW, i.e. $\alpha = 0$. Hence, if *both* dark detectors DH1 and DV2 respond, both the electron and the positron must have 'remarked' that a disturbing object is in the beam path, i.e. the other partner. Consequently, $e^+$ and $e^-$ must have met somehow at WW without having annihilated each other—Hardy's paradox.

**Table J.2** Probabilities for the simultaneous response of two detectors

| Detectors | $|z_d(0)\rangle$ | $|z_d(1,1)\rangle$ | $|z_d(1,2)\rangle$ | $|z_d(2)\rangle$ |
|---|---|---|---|---|
| $|h_1 h_2\rangle \hateq (\text{DH1, DH2})$ | $\frac{1}{4}$ | $\frac{(1-\alpha)^2}{8}$ | $\frac{1}{2}$ | $\frac{(1-\alpha)^2}{16}$ |
| $|h_1 v_2\rangle \hateq (\text{DH1, DV2})$ | $\frac{1}{4}$ | $0$ | $0$ | $\frac{(1-\alpha)^2}{16}$ |
| $|v_1 h_2\rangle \hateq (\text{DV1, DH2})$ | $\frac{\alpha^2}{4}$ | $\frac{(1+\alpha)^2}{8}$ | $\frac{(1+\alpha)^2}{8}$ | $\frac{(3+\alpha)^2}{16}$ |
| $|v_1 v_2\rangle \hateq (\text{DV1, DV2})$ | $\frac{1}{4}$ | $\frac{1}{2}$ | $\frac{(1-\alpha)^2}{8}$ | $\frac{(1-\alpha)^2}{16}$ |

For $\alpha = 0$ or 1, the annihilation process is switched on or off

## Entanglement

We first recall the simple criterion, formulated in Chap. 20, as to whether a four-dimensional vector (as the direct product of two 2-dimensional vectors) represents an entangled state: It must fulfill $c_1 c_4 \neq c_2 c_3$, where $c_i$ are the four components of the corresponding vector.

Applying this criterion to the states (J.15), we obtain Table J.3. As we can read off immediately, the final states are always factorizable for $\alpha = 1$ (no annihilation), i.e. they are not entangled. However, they are *always* entangled for activated pair annihilation ($\alpha = 0$), independently of whether BS1 and/or BS2 are in the beam path or not.

This is interesting because the initial state (J.14) is not entangled. Hence, we have the result for activated pair annihilation that a non-entangled state becomes entangled, and this in an interaction-free manner, in the sense that the electron and the positron cannot have interacted directly via pair annihilation.

By the way, the converse is not true: If we send an entangled initial state through the setup with two beam splitters BSi, it will not be dis-entangled, but rather it remains entangled. An initial state of the form $\sim |h_1 h_2\rangle + |v_1 v_2\rangle$ is for example converted into a final state of the same form.

**Table J.3** Entanglement criterion for the final states (J.15)

| | Without BSi's | With BS1 | With BS2 | With both BSi's |
|---|---|---|---|---|
| $c_1 c_4 - c_2 c_3$ | $i^2 + \alpha$ | $2i^2 (1 - \alpha)$ | $2i^2 (1 - \alpha)$ | $i^2 (\alpha - 1) - (1 - \alpha)(3 - \alpha)$ |

# Appendix K
# Set-Theoretical Derivation of the Bell Inequality

Bell's inequality, considered in Chap. 20, may also be derived using the calculus of set theory.

Given a total set $U$, from which we single out three subsets $a$, $b$ and $c$; we label the respective (absolute) complements by $\neg a$, $\neg b$ and $\neg c$. We can combine and intersect these sets; we can form e.g. $a \cap b$.

We start with the equation

$$a \cap b = (a \cap b) \cap (c \cup \neg c). \tag{K.1}$$

With the distributive property of sets:

$$A \cap (B \cap C) = (A \cap B) \cup (A \cap C), \tag{K.2}$$

it follows for $A = a \cap b$, $B = c$ and $C = \neg c$ that:

$$a \cap b = (a \cap b) \cap (c \cap \neg c) = ((a \cap b) \cap c) \cup ((a \cap b) \cap \neg c). \tag{K.3}$$

The relations

$$(a \cap b) \cap c \subseteq a \cap c; \quad (a \cap b) \cap \neg c \subseteq b \cap \neg c \tag{K.4}$$

lead to the inequality

$$a \cap b \subseteq (a \cap c) \cup (b \cap \neg c). \tag{K.5}$$

For the numbers (occurrence frequencies) of the elements, it holds correspondingly that

$$n(a, b) \leq n(a, c) + n(b, \neg c). \tag{K.6}$$

© Springer Nature Switzerland AG 2018
J. Pade, *Quantum Mechanics for Pedestrians 2*, Undergraduate Lecture Notes in Physics, https://doi.org/10.1007/978-3-030-00467-5

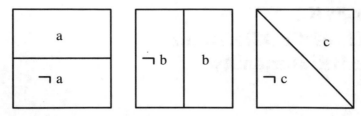

**Fig. K.1**  Example of the partition of a set according to the properties $a$, $b$ and $c$

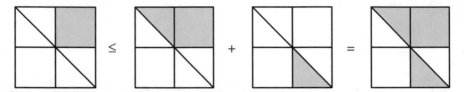

**Fig. K.2**  Graphical representation of inequality (K.6) $n(a, b) \leq n(a, c) + n(b, \neg c)$ for the partition as shown in Fig. K.1

We illustrate this inequality by means of the example in Fig. K.1. Then we can represent the inequality (K.6) graphically as shown in Fig. K.2.

# Appendix L
# The Special Galilei Transformation

Restricting ourselves to non-relativistic conditions, we consider here inertial frames related by a Galilei transformation[35] $\mathbf{r}' = \mathbf{r} + \mathbf{v}t$ with constant velocity $\mathbf{v}$, see Fig. L.1. Unlike rotations and translations, the kinetic energy is not invariant in this transformation. Nevertheless, because we are dealing with inertial frames, the form of the laws of physics, specifically the SEq, must stay the same under the transformation (shape invariance).[36] This is the case, as we show explicitly in the following. To avoid ambiguities, we denote the position and momentum operators by $\mathbf{X}$ and $\mathbf{P}$, and their eigenvalues by $\mathbf{r}$ or $\mathbf{x}$ and $\mathbf{p}$.

## L.1 Special Galilei Transformation

### L.1.1 Abstract Notation

We have two inertial frames $S$ and $S'$ which are related by $\mathbf{r}' = \mathbf{r} + \mathbf{v}t$. Since the transformation is continuous, it is represented by a unitary operator

$$U(\mathbf{v}) = e^{-i\mathbf{v}\mathbf{G}/\hbar} \tag{L.1}$$

with a Hermitian[37] operator $\mathbf{G}$. To determine $\mathbf{G}$, we translate the classical relations

$$\mathbf{r}' = \mathbf{r} + t\mathbf{v}; \ \mathbf{p}' = \mathbf{p} + m\mathbf{v} \tag{L.2}$$

---

[35] We perform here an active coordinate transformation (the system is shifted, the coordinate system remains); it is also called a *boost*. In a passive transformation, the system remains, while the coordinate system is shifted.

[36] Otherwise, the inertial systems would not be equivalent.

[37] The minus sign in the exponent is purely conventional; $\mathbf{G}$ as in 'Galileo'.

© Springer Nature Switzerland AG 2018
J. Pade, *Quantum Mechanics for Pedestrians 2*, Undergraduate Lecture Notes in Physics, https://doi.org/10.1007/978-3-030-00467-5

**Fig. L.1** Galilei
transformation

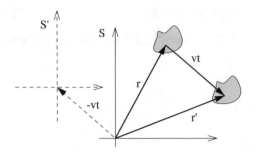

by means of

$$\mathbf{X}' = U^{-1}(\mathbf{v})\mathbf{X}U(\mathbf{v}) \overset{!}{=} \mathbf{X} + t\mathbf{v}; \quad \mathbf{P}' = U^{-1}(\mathbf{v})\mathbf{P}U(\mathbf{v}) \overset{!}{=} \mathbf{P} + m\mathbf{v} \qquad (\text{L.3})$$

into quantum mechanics. For sufficiently small, but otherwise arbitrary $\mathbf{v}$, we obtain
for $\mathbf{X}$ (with the infinitesimal approximation $e^{-i\mathbf{v}\mathbf{G}/\hbar} \approx 1 - i\mathbf{v}\mathbf{G}/\hbar$) the following
expression:

$$(1 + i\,(\mathbf{v}\mathbf{G})\,/\hbar)\,\mathbf{X}(1 - i\,(\mathbf{v}\mathbf{G})\,/\hbar) = \mathbf{X} + t\mathbf{v},$$
$$\text{or } \mathbf{X} + i\,(\mathbf{v}\mathbf{G})\,/\hbar\mathbf{X} - \mathbf{X}i\,(\mathbf{v}\mathbf{G})\,/\hbar = \mathbf{X} + t\mathbf{v}, \qquad (\text{L.4})$$
$$\text{or } \frac{i}{\hbar}\,((\mathbf{v}\mathbf{G})\,\mathbf{X} - \mathbf{X}\,(\mathbf{v}\mathbf{G})) = t\mathbf{v},$$

and analogously for $\mathbf{P}$. Since the last equation must hold for any velocities, it follows
that

$$\frac{i}{\hbar}\left(G_iX_j - X_jG_i\right) = t\delta_{ij}; \text{ analogously } \frac{i}{\hbar}\left(G_iP_j - P_jG_i\right) = m\delta_{ij}. \qquad (\text{L.5})$$

To specify the exact form of $\mathbf{G} = \mathbf{G}\,(\mathbf{X}, \mathbf{P})$, we use a result from Chap. 21, namely

$$\left[X_i, g(P_j)\right] = i\hbar\frac{\partial g(P_j)}{\partial P_j}\delta_{ij}; \quad \left[P_i, f(X_j)\right] = -i\hbar\frac{\partial f(X_j)}{\partial X_j}\delta_{ij}. \qquad (\text{L.6})$$

It follows that

$$\frac{\partial G_i}{\partial P_j} = t\delta_{ij}; \quad \frac{\partial G_i}{\partial X_j} = -m\delta_{ij} \qquad (\text{L.7})$$

and we can take as solution

$$G_i = tP_i - mX_i \text{ or } \mathbf{G} = t\mathbf{P} - m\mathbf{X}. \qquad (\text{L.8})$$

Hence, the unitary transformation reads

$$U(\mathbf{v}) = e^{-i\mathbf{v}\mathbf{G}/\hbar} = e^{-i\mathbf{v}(t\mathbf{P}-m\mathbf{X})/\hbar}. \qquad (\text{L.9})$$

One has to be a little careful with this operator, as the two operators $X$ and $P$ appearing in the exponent do not commute, $[X_i, P_j] = i\hbar\delta_{ij}$. Here, we will exploit a transformation[38] which we proved in Chap. 13, Vol. 1: Two operators $A$ and $B$, both of which commute with their commutator $[A, B]$, satisfy

$$e^A e^B = e^{A+B} e^{\frac{1}{2}[A,B]}.$$ 
(L.10)

It follows (see exercises) that:

$$U(\mathbf{v}) = e^{-i\mathbf{v}(t\mathbf{P}-m\mathbf{X})/\hbar} = \begin{cases} e^{im\mathbf{v}\mathbf{X}/\hbar} e^{-i\mathbf{v}\mathbf{P}t/\hbar} e^{-i\frac{mv^2}{2\hbar}t} \\ e^{-i\mathbf{v}\mathbf{P}t/\hbar} e^{im\mathbf{v}\mathbf{X}/\hbar} e^{i\frac{mv^2}{2\hbar}t} \end{cases}$$
(L.11)

How are the states in $S$ and $S'$ related? We have

$$\left|\psi'(t)\right\rangle = U(\mathbf{v})\left|\psi(t)\right\rangle.$$
(L.12)

In the position representation, it follows that

$$\langle\mathbf{q}\left|\psi'(t)\right\rangle = \langle\mathbf{q}|\,U(\mathbf{v})\left|\psi(t)\right\rangle.$$
(L.13)

Because of $\langle\mathbf{q}|\,U(\mathbf{v}) = \left(U^\dagger(\mathbf{v})\,|\mathbf{q}\rangle\right)^\dagger$, we consider first $U^\dagger(\mathbf{v})\,|\mathbf{q}\rangle$[39]:

$$U^\dagger(\mathbf{v})\,|\mathbf{q}\rangle = e^{-i\frac{mv^2}{2\hbar}t} e^{-im\mathbf{v}\mathbf{X}/\hbar} e^{i\mathbf{v}\mathbf{P}t/\hbar}\,|\mathbf{q}\rangle$$

$$= e^{-i\frac{mv^2}{2\hbar}t} e^{-im\mathbf{v}\mathbf{X}/\hbar}\,|\mathbf{q}-\mathbf{v}t\rangle = e^{-i\frac{mv^2}{2\hbar}t} e^{-im\mathbf{v}(\mathbf{q}-\mathbf{v}t)/\hbar}\,|\mathbf{q}-\mathbf{v}t\rangle.$$
(L.14)

It follows that:

$$\langle\mathbf{q}\left|\psi'(t)\right\rangle = \langle\mathbf{q}|\,U(\mathbf{v})\left|\psi(t)\right\rangle = e^{i\frac{mv^2}{2\hbar}t} e^{im\mathbf{v}(\mathbf{q}-\mathbf{v}t)/\hbar} \langle\mathbf{q}-\mathbf{v}t\left|\psi(t)\right\rangle$$
(L.15)

or

$$\psi'(\mathbf{q}, t) = e^{-i\frac{mv^2}{2\hbar}t} e^{im\mathbf{v}\mathbf{q}/\hbar}\psi(\mathbf{q}-\mathbf{v}t, t).$$
(L.16)

Finally, we set $\mathbf{q} = \mathbf{r}' = \mathbf{r} + \mathbf{v}t$ and obtain

$$\psi'(\mathbf{r}', t) = e^{-i\frac{mv^2}{2\hbar}t} e^{im\mathbf{v}\mathbf{r}'/\hbar}\psi(\mathbf{r}, t) = e^{i\frac{mv^2}{2\hbar}t} e^{im\mathbf{v}\mathbf{r}/\hbar}\psi(\mathbf{r}, t).$$
(L.17)

---

[38]The Baker–Campbell–Hausdorff formula, see the exercises for Chap. 13, Vol. 1.
[39]Note the sign: $e^{\pm i\mathbf{a}\mathbf{p}/\hbar}\,|\mathbf{r}\rangle = |\mathbf{r}\mp\mathbf{a}\rangle$.

### L.1.2  Position Representation

For comparison with the abstract approach of the last paragraph, we consider the effect of the transformation $\mathbf{r}' = \mathbf{r} + \mathbf{v}t$ by starting directly from the position representation. The SEq reads

$$i\hbar \frac{\partial}{\partial t} \psi(\mathbf{r}, t) = \left[ -\frac{\hbar^2}{2m} \nabla_{\mathbf{r}}^2 + V(\mathbf{r}, t) \right] \psi(\mathbf{r}, t). \qquad (L.18)$$

The potential may be time dependent. To distinguish between the derivatives in the two inertial systems, we use subscript indices in the derivative operators, e.g. $\nabla_{\mathbf{r}'} = \left( \frac{\partial}{\partial x'}, \frac{\partial}{\partial y'}, \frac{\partial}{\partial z'} \right)$. For the wavefunction and spatial derivative with $\mathbf{r}' = \mathbf{r} + t\mathbf{v}$, it clearly holds that

$$\psi(\mathbf{r}, t) = \psi(\mathbf{r}' - \mathbf{v}t, t) \, ; \; \nabla_{\mathbf{r}} \psi(\mathbf{r}, t) = \nabla_{\mathbf{r}'} \psi(\mathbf{r}' - \mathbf{v}t, t). \qquad (L.19)$$

The time derivatives in the two reference systems are related by[40]:

$$\frac{\partial}{\partial t} \psi(\mathbf{r}, t) = \frac{\partial}{\partial t} \psi(\mathbf{r}' - \mathbf{v}t, t) + \mathbf{v} \cdot \nabla_{\mathbf{r}'} \psi(\mathbf{r}' - \mathbf{v}t, t). \qquad (L.20)$$

We insert all the intermediate results into the SEq (L.18) and obtain

$$i\hbar \frac{\partial}{\partial t} \psi(\mathbf{r}' - \mathbf{v}t, t) + i\hbar \mathbf{v} \cdot \nabla_{\mathbf{r}'} \psi(\mathbf{r}' - \mathbf{v}t, t)$$
$$= \left[ -\frac{\hbar^2}{2m} \nabla_{\mathbf{r}'}^2 + V(\mathbf{r}' - \mathbf{v}t, t) \right] \psi(\mathbf{r}' - \mathbf{v}t, t). \qquad (L.21)$$

Although this representation is correct, the occurrence of the argument $\mathbf{r}' - \mathbf{v}t$ and of the term $\nabla_{\mathbf{r}'} \psi$ on the left-hand side is annoying. A remedy is provided by the unitary transformation (see exercises):

$$\psi(\mathbf{r}, t) = \psi(\mathbf{r}' - \mathbf{v}t, t) = e^{i\left(-m\mathbf{v}\mathbf{r}' + mv^2 t/2\right)/\hbar} \psi'(\mathbf{r}', t). \qquad (L.22)$$

It leads to

$$i\hbar \frac{\partial}{\partial t} \psi'(\mathbf{r}', t) = \left[ -\frac{\hbar^2}{2m} \nabla_{\mathbf{r}'}^2 + V'(\mathbf{r}', t) \right] \psi'(\mathbf{r}', t) \, ; \; V'(\mathbf{r}', t) := V(\mathbf{r}' - \mathbf{v}t, t),$$
$$(L.23)$$

i.e. the familiar form of the SEq, however with a modified potential. Only the free SEq is completely invariant under Galilei transformations, and not just form invariant.

Inverting the unitary transformation (L.22) yields

---

[40]On the left side, the variable $\mathbf{r}$ is fixed in the derivative with respect to $t$. Therefore, also $\mathbf{r}' - \mathbf{v}t$ must be constant, and thus the term $-\mathbf{v}\nabla_{\mathbf{r}'} \psi(\mathbf{r}' - \mathbf{v}t, t)$ must be subtracted on the right side.

$$\psi'\left(\mathbf{r}', t\right) = e^{i\left(m\mathbf{v}\mathbf{r}' - mv^2 t/2\right)/\hbar}\psi\left(\mathbf{r}, t\right),\tag{L.24}$$

i.e. the same result as (L.17). As an example, the transformation of a plane wave is considered in the exercises.

### L.1.3 Several Quantum Objects

We start with a closed system of two quantum objects. The interaction depends only on the relative distance $\mathbf{r}$; external interactions do not exist. We can thus formulate the SEq using the relative coordinate $\mathbf{r}$ and the center-of-mass coordinate $\mathbf{R}$ (see Appendix E, Vol. 2):

$$i\hbar\partial_t \Psi\left(\mathbf{R}, \mathbf{r}, t\right) = \left[-\frac{\hbar^2}{2M}\nabla_{\mathbf{R}}^2 - \frac{\hbar^2}{2\mu}\nabla_{\mathbf{r}}^2 + V\left(\mathbf{r}\right)\right]\Psi\left(\mathbf{R}, \mathbf{r}, t\right)\tag{L.25}$$

where $M$ is the total mass and $\mu$ the reduced mass.

A Galilei boost leaves the relative coordinate unchanged, while for the center-of-mass coordinate, we see that

$$\mathbf{R}' = \mathbf{R} + \mathbf{v}t.\tag{L.26}$$

In the system $S'$, $\Psi\left(\mathbf{R}, \mathbf{r}, t\right)$ becomes $\Psi\left(\mathbf{R}' - \mathbf{v}t, \mathbf{r}, t\right)$, and we obtain with the transformation

$$\Psi\left(\mathbf{R}' - \mathbf{v}t, \mathbf{r}, t\right) = e^{i\left(-M\mathbf{v}\mathbf{R}' + Mv^2 t/2\right)/\hbar}\Psi'\left(\mathbf{R}', \mathbf{r}, t\right)\tag{L.27}$$

the new SEq

$$i\hbar\partial_t \Psi'\left(\mathbf{R}' - \mathbf{v}t, \mathbf{r}, t\right) = \left[-\frac{\hbar^2}{2M}\nabla_{\mathbf{R}'}^2 - \frac{\hbar^2}{2\mu}\nabla_{\mathbf{r}}^2 + V\left(\mathbf{r}\right)\right]\Psi'\left(\mathbf{R}' - \mathbf{v}t, \mathbf{r}, t\right).\tag{L.28}$$

For $N$ quantum objects, the considerations proceed analogously. The transformation for the position and momentum operators of the quantum object with index $n$ is

$$U^{-1}(\mathbf{v})\mathbf{X}_n U(\mathbf{v}) = \mathbf{X}_n + t\mathbf{v}; \quad U^{-1}(\mathbf{v})\mathbf{P}_n U(\mathbf{v}) = \mathbf{P}_n + m\mathbf{v}.\tag{L.29}$$

Since the operators for different quantum objects commute, we have

$$U(\mathbf{v}) = \prod_n e^{-i\mathbf{v}(t\mathbf{P}_n - m\mathbf{X}_n)/\hbar}\tag{L.30}$$

and it follows that

$$\Psi'\left(\mathbf{R}', \mathbf{r}_1, \ldots, \mathbf{r}_N, t\right) = e^{iM\mathbf{v}\mathbf{R}'/\hbar} e^{-i\frac{Mv^2}{2\hbar}t} \Psi\left(\mathbf{R}', \mathbf{r}_1 - \mathbf{v}t, \ldots, \mathbf{r}_N - \mathbf{v}t, t\right). \quad \text{(L.31)}$$

## L.2 The Special Galilei Transformation and Kinetic Energy

In this section, we consider the most general form of a Hamiltonian which is compatible with the special Galilei transformation. For this purpose, we start from the transformation (L.9), where we confine ourselves to $t = 0$ for the sake of simplicity.

We recapitulate briefly: The requirement

$$U^{-1}(\mathbf{v})PU(\mathbf{v}) \stackrel{!}{=} \mathbf{P} + m\mathbf{v} \quad \text{(L.32)}$$

leads with $U(\mathbf{v}) = e^{-i\mathbf{v}\mathbf{G}/\hbar}$ and $\mathbf{G} = t\mathbf{P} - m\mathbf{X}$, and at $t = 0$, to

$$U(\mathbf{v}) = e^{i\mathbf{v}m\mathbf{X}/\hbar}. \quad \text{(L.33)}$$

For our further considerations, we define a velocity operator $\dot{\mathbf{X}}$ by the equation $\dot{\mathbf{X}} = \frac{i}{\hbar}[H, \mathbf{X}]$. The idea is that the kinetic energy $E_{\text{kin}}$ is determined by $E_{\text{kin}} = \frac{m\dot{\mathbf{r}}^2}{2}$. We have up to now assumed that this is the same as $\frac{\mathbf{P}^2}{2m}$, but we will soon see that this is true only under certain conditions.

Concerning the transformation behavior of the velocity operator, we assume because of (L.3) that

$$U^{-1}(\mathbf{v})\dot{\mathbf{X}}U(\mathbf{v}) \stackrel{!}{=} \dot{\mathbf{X}} + \mathbf{v}. \quad \text{(L.34)}$$

From (L.32) and (L.34), it follows that:

$$U^{-1}(\mathbf{v})\left(m\dot{\mathbf{X}} - \mathbf{P}\right)U(\mathbf{v}) = m\dot{\mathbf{X}} - \mathbf{P}. \quad \text{(L.35)}$$

This means that the operator $m\dot{\mathbf{X}} - P$ commutes with $U$ and therefore also with $\mathbf{X}$ (because of $U(\mathbf{v}) = e^{-i\mathbf{v}m\mathbf{X}/\hbar}$). Therefore, we can write:

$$m\dot{\mathbf{X}} - \mathbf{P} = \mathbf{f}(\mathbf{X}). \quad \text{(L.36)}$$

The relevant question is whether we can always eliminate the function $\mathbf{f}(\mathbf{X})$ by a unitary transformation. This is in fact the case in one dimension, but not in three dimensions, as we now show.

## L.2.1   One-Dimensional Case

Generally, one can eliminate the function $f(X)$ by a unitary transformation only in the case of *one* dimension. Here is how: Let $F(x)$ be the anti-derivative of $f(x)$, i.e. $F'(X) = f(X)$. Then we consider the unitary transformation[41]:

$$S = e^{i\frac{F(X)}{\hbar}}. \tag{L.37}$$

Apparently, $X$ remains unchanged under the transformation $X' = S^{-1}XS$, i.e. $X' = X$. To calculate $P'$, we use $[P, f(X)] = -i\hbar\frac{df(X)}{dX}$ and obtain initially

$$S^{-1}PS - P = S^{-1}(PS - SP) = -i\hbar S^{-1}\frac{dS(X)}{dX} = -i\hbar S^{-1}\frac{i}{\hbar}f(X)S = f(X) \tag{L.38}$$

and from this, with $P' = S^{-1}PS = P + f(X)$ and (L.36), finally

$$m\dot{X}' = m\dot{X} = P + f(X) = P'. \tag{L.39}$$

In short: One can always assume that $P = m\dot{X}$; this choice is unitarily equivalent to all other choices.

With this result, we now want to determine the most general form that a Hamiltonian can take which is compatible with Galilei transformations. For this, we define the operator of the kinetic energy as

$$K = \frac{m\dot{X}^2}{2} = \frac{P^2}{2m}. \tag{L.40}$$

It follows that[42]:

$$[K, X] = \frac{1}{2m}\left[P^2, X\right] = -\frac{i\hbar}{m}P. \tag{L.41}$$

With

$$P = m\dot{X} = m\frac{i}{\hbar}[H, X], \tag{L.42}$$

we obtain

$$[H - K, X] = 0. \tag{L.43}$$

This means that $H - K$ is a function only of $X$, which is usually referred to as $V(X)$. Hence, the general form of a Hamiltonian which is compatible with the Galilean transformation reads in the one-dimensional case:

---

[41] Such a transformation is also called a *local gauge transformation*.
[42] Here, we again use $[x, f(p)] = i\hbar\frac{\partial f(p)}{\partial p}$.

$$H = K + V(X) = \frac{P^2}{2m} + V(X). \tag{L.44}$$

Galilei invariance manifests itself by the fact that the *form* of $H$ is unchanged: If $H$ is a function of $X$ and $P$, then the transformed Hamiltonian $H_v$ is the *same* function of $X' = X$ and $P' = P - mv$:

$$H = \frac{P^2}{2m} + V(X) \rightarrow$$
$$H' = \frac{P'^2}{2m} + V(X') = \frac{P^2}{2m} - vP + \tfrac{1}{2}mv^2 + V(X'). \tag{L.45}$$

## L.2.2   Three-Dimensional Case

We start again with (L.36):

$$m\dot{\mathbf{X}} - \mathbf{P} = \mathbf{f}(\mathbf{X}). \tag{L.46}$$

In general, there is no transformation which causes the function $\mathbf{f}(\mathbf{x})$ to vanish. In fact, one would have to find a unitary transformation

$$S = e^{i\frac{F(\mathbf{X})}{\hbar}} \tag{L.47}$$

which satisfies

$$\mathbf{f}(\mathbf{X}) = \nabla_{\mathbf{X}} F(\mathbf{X}), \tag{L.48}$$

which is possible only for $\nabla_{\mathbf{X}} \times \mathbf{f}(\mathbf{X}) = 0$.
From (L.36), the commutation relation

$$\left[\dot{X}_i, X_j\right] = -\frac{i\hbar}{m}\delta_{ij} \tag{L.49}$$

follows. The kinetic energy is defined by

$$K = \frac{m\dot{\mathbf{X}}^2}{2} = \frac{(\mathbf{P} - \mathbf{f}(\mathbf{X}))^2}{2m}. \tag{L.50}$$

As mentioned above, we cannot assume $\mathbf{f}(\mathbf{X}) = 0$ or $m\dot{\mathbf{X}} = P$.
For the commutator $[K, X_i]$, we have:

$$[K, X_i] = \frac{m}{2}\sum_j \left[\dot{X}_j^2, X_i\right] = \frac{m}{2}\sum_j \left(\dot{X}_j \left[\dot{X}_j, X_i\right] + \left[\dot{X}_j, X_i\right]\dot{X}_j\right) = -i\hbar\dot{X}_i. \tag{L.51}$$

Comparing this with $[H, X_i] = -i\hbar\dot{X}_i$, we can conclude that

$$[H - K, X_i] = 0. \tag{L.52}$$

This means that $H - K$ can only be a function of $\mathbf{X}$, so that we have $H = K + V(\mathbf{X})$. Thus, the most general form of a Hamiltonian which is compatible with Galilei invariance in three dimensions is:

$$H = \frac{1}{2m}(\mathbf{P} - \mathbf{f}(\mathbf{X}))^2 + V(\mathbf{X}). \tag{L.53}$$

Note the difference between $\frac{\mathbf{P}}{m}$ and $\dot{\mathbf{X}}$; the kinetic energy reads

$$K = \frac{m\dot{\mathbf{X}}^2}{2} \neq \frac{\mathbf{P}^2}{2m}. \tag{L.54}$$

In electrodynamics, the Hamiltonian is

$$H_{cl} = \frac{1}{2m}(\mathbf{P} - q\mathbf{A})^2 + q\varphi, \tag{L.55}$$

where $\mathbf{A}$ is the vector potential and $\varphi$ the scalar potential. Thus, the elimination of $\mathbf{f}(\mathbf{X})$ works only if $\mathbf{A}$ can be represented as a gradient, or for

$$\mathbf{B} = \nabla \times \mathbf{A} = 0, \tag{L.56}$$

i.e. for vanishing magnetic field.

## L.3 Exercises

1. Given the transformation
$$U(\mathbf{v}) = e^{-i\mathbf{v}(t\mathbf{P} - m\mathbf{X})/\hbar}; \tag{L.57}$$

show that:

$$U(\mathbf{v}) = e^{-i\mathbf{v}(t\mathbf{P} - m\mathbf{X})/\hbar} = \begin{cases} e^{im\mathbf{v}\mathbf{X}/\hbar}e^{-i\mathbf{v}\mathbf{P}t/\hbar}e^{-i\frac{mv^2}{2\hbar}t} \\ e^{-i\mathbf{v}\mathbf{P}t/\hbar}e^{im\mathbf{v}\mathbf{X}/\hbar}e^{i\frac{mv^2}{2\hbar}t} \end{cases}. \tag{L.58}$$

Solution: For two operators $A$ and $B$ with $[A, B] = c$ ($c$ is a complex number), it holds that $e^A e^B = e^{A+B}e^{\frac{1}{2}[A,B]}$. We consider first

$$A = \frac{im\mathbf{v}\mathbf{X}}{\hbar}; \quad B = -\frac{it\mathbf{v}\mathbf{P}}{\hbar}. \tag{L.59}$$

We calculate the commutator $[A, B]$ using $\left[X_i, P_j\right] = i\hbar\delta_{ij}$:

$$[A, B] = \frac{mt}{\hbar^2} \left( (\mathbf{vX}) (\mathbf{vP}) - (\mathbf{vP}) (\mathbf{vX}) \right) = \frac{mt}{\hbar^2} \sum_{i,j} v_i v_j \left( X_i P_j - P_j X_i \right)$$

$$= \frac{mt}{\hbar^2} \sum_{i,j} v_i v_j i \hbar \delta_{ij} = i \frac{mt}{\hbar} \mathbf{v}^2. \tag{L.60}$$

With $e^{A+B} = e^{-\frac{1}{2}[A,B]} e^A e^B$, it follows that

$$U(\mathbf{v}) = e^{-i\mathbf{v}(t\mathbf{P}-m\mathbf{X})/\hbar} = e^{-i\frac{mt}{2\hbar}\mathbf{v}^2} e^{\frac{im\mathbf{vX}}{\hbar}} e^{-\frac{it\mathbf{vP}}{\hbar}}. \tag{L.61}$$

We write

$$B = \frac{im\mathbf{vX}}{\hbar}; \quad A = -\frac{it\mathbf{vP}}{\hbar}. \tag{L.62}$$

This leads to

$$[A, B] = -i\frac{mt}{\hbar}\mathbf{v}^2 \tag{L.63}$$

and therefore

$$U(\mathbf{v}) = e^{-i\mathbf{v}(m\mathbf{X}-t\mathbf{P})/\hbar} = e^{i\frac{mt}{2\hbar}\mathbf{v}^2} e^{-\frac{it\mathbf{vP}}{\hbar}} e^{\frac{im\mathbf{vX}}{\hbar}}. \tag{L.64}$$

2. Show the form invariance of the SEq under Galilei transformations.
   Solution: If we insert $\psi(\mathbf{r}, t) = \psi(\mathbf{r}' - \mathbf{v}t, t)$ into the SEq, we obtain (L.21):

$$i\hbar\frac{\partial}{\partial t}\psi\left(\mathbf{r}' - \mathbf{v}t, t\right) + i\hbar\mathbf{v} \cdot \nabla_{\mathbf{r}'}\psi\left(\mathbf{r}' - \mathbf{v}t, t\right)$$

$$= \left[-\frac{\hbar^2}{2m}\nabla_{\mathbf{r}'}^2 + V\left(\mathbf{r}' - \mathbf{v}t, t\right)\right]\psi\left(\mathbf{r}' - \mathbf{v}t, t\right). \tag{L.65}$$

We want to have $\mathbf{r}'$ instead of the argument $\mathbf{r}' - \mathbf{v}t$; in addition, we want to eliminate the term $\nabla_{\mathbf{r}'}\psi$ by a transformation. To this end, we apply the *ansatz*

$$\psi\left(\mathbf{r}' - \mathbf{v}t, t\right) = e^{iT}\varphi\left(\mathbf{r}', t\right), \tag{L.66}$$

with $T = T\left(\mathbf{r}', t\right)$ (we note that we have $\varphi$ and not $\psi$ on the right side). For the left-hand side, it follows that:

$$i\hbar\frac{\partial}{\partial t}\psi + i\hbar\mathbf{v} \cdot \nabla_{\mathbf{r}'}\psi$$
$$= i\hbar\left[e^{iT}\dot{\varphi} + i\dot{T}e^{iT}\varphi\right] + i\hbar\mathbf{v}\cdot\left[e^{iT}\nabla_{\mathbf{r}'}\varphi + i\left(\nabla_{\mathbf{r}'}T\right)e^{iT}\varphi\right]. \tag{L.67}$$

and for the Laplacian on the right-hand side,

$$\nabla_{\mathbf{r}'}^2 \psi = \nabla_{\mathbf{r}'} \left( e^{iT} \nabla_{\mathbf{r}'} \varphi \right) + i \nabla_{\mathbf{r}'} \left( \nabla_{\mathbf{r}'} T \right) e^{iT} \varphi$$

$$= e^{iT} \nabla_{\mathbf{r}'}^2 \varphi + 2i \left( \nabla_{\mathbf{r}'} T \right) e^{iT} \left( \nabla_{\mathbf{r}'} \varphi \right) + i \left( \nabla_{\mathbf{r}'}^2 T \right) e^{iT} \varphi - \left( \nabla_{\mathbf{r}'} T \right)^2 e^{iT} \varphi. \tag{L.68}$$

Inserting into (L.65) yields

$$i\hbar \left[ \dot{\varphi} + i \dot{T} \varphi \right] + i\hbar \mathbf{v} \cdot \left[ \nabla_{\mathbf{r}'} \varphi + i \left( \nabla_{\mathbf{r}'} T \right) \varphi \right]$$

$$= -\frac{\hbar^2}{2m} \left[ \nabla_{\mathbf{r}'}^2 \varphi + 2i \left( \nabla_{\mathbf{r}'} T \right) \left( \nabla_{\mathbf{r}'} \varphi \right) + i \left( \nabla_{\mathbf{r}'}^2 T \right) \varphi - \left( \nabla_{\mathbf{r}'} T \right)^2 \varphi \right]$$
$$+ V \left( \mathbf{r}' - \mathbf{v}t, t \right) \varphi. \tag{L.69}$$

If we want to obtain as our result the SEq for $\varphi$:

$$i\hbar \dot{\varphi} = -\frac{\hbar^2}{2m} \nabla_{\mathbf{r}'}^2 \varphi + V \left( \mathbf{r}' - \mathbf{v}t, t \right) \varphi, \tag{L.70}$$

then the other terms must cancel. Sorting by coefficients of $\varphi$ and $\nabla_{\mathbf{r}'} \varphi$, we initially have

$$\varphi \Rightarrow -\hbar \dot{T} - \hbar \mathbf{v} \cdot \nabla_{\mathbf{r}'} T = -\frac{\hbar^2}{2m} i \left( \nabla_{\mathbf{r}'}^2 T \right) + \frac{\hbar^2}{2m} \left( \nabla_{\mathbf{r}'} T \right)^2$$
$$\nabla_{\mathbf{r}'} \varphi \Rightarrow i\hbar \mathbf{v} = -\frac{\hbar^2}{2m} 2i \nabla_{\mathbf{r}'} T. \tag{L.71}$$

From the second equation, it follows that

$$\frac{\hbar}{m} \nabla_{\mathbf{r}'} T = -\mathbf{v} \text{ or } T = \frac{m}{\hbar} \left( -\mathbf{v} \cdot \mathbf{r} + F(t) \right) \tag{L.72}$$

with a still-to-be-determined function $F(t)$. Inserting $T$ into the first equation gives

$$\dot{F} = \frac{1}{2} v^2. \tag{L.73}$$

Hence it follows that

$$T = \frac{m}{\hbar} \left( -\mathbf{v}\mathbf{r} + \frac{v^2}{2} \right) \tag{L.74}$$

and we obtain finally the relation

$$\psi \left( \mathbf{r}' - \mathbf{v}t, t \right) = e^{i \left( -m\mathbf{v}\mathbf{r}' + mv^2 t/2 \right)/\hbar} \varphi \left( \mathbf{r}', t \right). \tag{L.75}$$

The wavefunction $\varphi \left( \mathbf{r}', t \right)$, defined in such a manner in the system $S'$, satisfies the SEq:

$$i\hbar\dot{\varphi}\left(\mathbf{r}',t\right) = -\frac{\hbar^2}{2m}\nabla_{\mathbf{r}'}^2\varphi\left(\mathbf{r}',t\right) + V\left(\mathbf{r}' - \mathbf{v}t,t\right)\varphi\left(\mathbf{r}',t\right). \tag{L.76}$$

If we finally define

$$V\left(\mathbf{r}' - \mathbf{v}t,t\right) := V'\left(\mathbf{r}',t\right), \tag{L.77}$$

then the SEq is written in the system $S'$ as:

$$i\hbar\dot{\varphi}\left(\mathbf{r}',t\right) = -\frac{\hbar^2}{2m}\nabla_{\mathbf{r}'}^2\varphi\left(\mathbf{r}',t\right) + V'\left(\mathbf{r}',t\right)\varphi\left(\mathbf{r}',t\right). \tag{L.78}$$

3. What happens to a plane wave under a Galilei boost?
   Solution: We have in $S$

$$\varphi(\mathbf{r},t) = \varphi_0 e^{i(\mathbf{kr}-\omega t)}. \tag{L.79}$$

With the transformation

$$\psi'\left(\mathbf{r}',t\right) = e^{-i\frac{mv^2}{2\hbar}t}e^{imvr'/\hbar}\psi\left(\mathbf{r},t\right) = e^{i\frac{mv^2}{2\hbar}t}e^{imvr/\hbar}\psi\left(\mathbf{r},t\right) \tag{L.80}$$

we obtain for the transformed wave function:

$$\varphi'(\mathbf{r}',t) = \varphi_0 e^{i(\mathbf{kr}-\omega t)+i\left(mvr'-mv^2t/2\right)/\hbar}. \tag{L.81}$$

We consider the exponent (without the factor $i$). We have:

$$\begin{aligned}
\mathbf{kr} - \omega t + \frac{m}{\hbar}\mathbf{vr}' - \frac{m}{2\hbar}v^2t &= \mathbf{kr}' - \mathbf{kv}t - \frac{\hbar k^2}{2m}t + \frac{m}{\hbar}\mathbf{vr}' - \frac{m}{2\hbar}v^2t \\
&= \mathbf{kr}' + \frac{m}{\hbar}\mathbf{vr}' - \frac{\hbar t}{2m}\left(k^2 + \frac{2m}{\hbar}\mathbf{kv} + \frac{m^2}{\hbar^2}v^2\right) \\
&= \left(\mathbf{k} + \frac{m\mathbf{v}}{\hbar}\right)\mathbf{r}' - \frac{\hbar t}{2m}\left(\mathbf{k} + \frac{m\mathbf{v}}{\hbar}\right)^2, \tag{L.82}
\end{aligned}$$

and it follows that

$$\varphi'(\mathbf{r}',t) = \varphi_0 e^{i(\mathbf{k}'\mathbf{r}'-\omega't)}; \quad \mathbf{k}' = \mathbf{k} + \frac{m\mathbf{v}}{\hbar}, \omega' = \frac{\hbar k'^2}{2m}. \tag{L.83}$$

# Appendix M
# Kramers' Theorem

The theorem of Kramers deals with the consequences which time-reversal invariance has for the energy levels of a system of $N$ quantum objects with spin $\frac{1}{2}$.

We know that for the time-reversal operator $\mathcal{T}$, it holds that $\mathcal{T}^2 = c$, with $|c| = 1$. To determine the constant $c$, we consider the commutator $[\mathcal{T}^2, \mathcal{T}]$. On the one hand, we have:

$$[\mathcal{T}^2, \mathcal{T}] = \mathcal{T}[\mathcal{T}, \mathcal{T}] - [\mathcal{T}, \mathcal{T}]\mathcal{T} = 0, \tag{M.1}$$

and on the other hand,

$$[\mathcal{T}^2, \mathcal{T}] = [c, \mathcal{T}] = c\mathcal{T} - \mathcal{T}c = c\mathcal{T} - c^*\mathcal{T}. \tag{M.2}$$

It follows directly that $c - c^* = 0$, and because of $|c| = 1$, we find $c = \pm 1$. One can show that for integer spin, $\mathcal{T}^2 = 1$, and for half-integer spin, $\mathcal{T}^2 = -1$.

We apply these facts to a system of $N$ quantum objects with spin $\frac{1}{2}$. On time reversal, each spin must be considered; consequently we obtain in this case $\mathcal{T}^2 = (-1)^N$. Now let $\{|E_n; N\rangle\}$ be a CONS of this system. If we assume that $H$ is invariant under time reversal, then also $\mathcal{T}|E_n; N\rangle$ must be an eigenstate of $H$ for the energy $E_n$. If, on the other hand, $\mathcal{T}|E_n; N\rangle$ is linearly dependent on $|E_n; N\rangle$ (and this *must* be the case if there is no degeneracy), then it must hold that $\mathcal{T}|E_n; N\rangle = \lambda|E_n; N\rangle$, and it follows that:

$$\mathcal{T}^2|E_n; N\rangle = \mathcal{T}\lambda|E_n; N\rangle = \lambda^*\mathcal{T}|E_n; N\rangle = \lambda^*\lambda|E_n; N\rangle = (-1)^N|E_n; N\rangle. \tag{M.3}$$

However, this is not satisfied for odd $N$ because of $\lambda^*\lambda > 0$. Hence, it follows that the states $|E_n; N\rangle$ are degenerate for odd $N$. This is the statement of the *theorem of Kramers*: If the Hamiltonian of a system of $N$ quantum objects with spin $\frac{1}{2}$ is invariant under time reversal, then for odd $N$, all stationary states are degenerate (Kramers' degeneracy). It can be shown that the degree of degeneracy is even.

© Springer Nature Switzerland AG 2018
J. Pade, *Quantum Mechanics for Pedestrians 2*, Undergraduate Lecture Notes in Physics, https://doi.org/10.1007/978-3-030-00467-5

# Appendix N
# Coulomb Energy and Exchange Energy in the Helium Atom

In this appendix, we consider in more detail the calculation of the Coulomb energy and exchange energy as addressed in Chap. 23.

The Coulomb energy is given by

$$C_{nl} = \frac{e^2}{4\pi\varepsilon_0} \int d^3r_1 d^3r_2 \frac{|\psi_{100}(\mathbf{r}_1)\,\psi_{nlm}(\mathbf{r}_2)|^2}{|\mathbf{r}_1 - \mathbf{r}_2|}, \tag{N.1}$$

and the exchange energy by

$$A_{nl} = \frac{e^2}{4\pi\varepsilon_0} \int d^3r_1 d^3r_2 \frac{\psi_{100}(\mathbf{r}_1)\,\psi_{nlm}(\mathbf{r}_2)\,\psi^*_{nlm}(\mathbf{r}_1)\,\psi^*_{100}(\mathbf{r}_2)}{|\mathbf{r}_1 - \mathbf{r}_2|}. \tag{N.2}$$

The wavefunction reads

$$\psi_{nlm}(r) = R_{nl}(r)\,Y_l^m(\vartheta, \varphi). \tag{N.3}$$

It follows that

$$C_{nl} = \frac{e^2}{4\pi\varepsilon_0}\frac{1}{4\pi} \int d^3r_1 d^3r_2 \frac{R_{10}^2(r_1)\,R_{nl}^2(r_2)\,|Y_l^m(\vartheta_2,\varphi_2)|^2}{|\mathbf{r}_1 - \mathbf{r}_2|} \tag{N.4}$$

and

$$A_{nl} = \frac{e^2}{4\pi\varepsilon_0}\frac{1}{4\pi} \int d^3r_1 d^3r_2$$
$$\frac{R_{10}(r_1)\,R_{nl}(r_1)\,Y_l^{m*}(\vartheta_1,\varphi_1)\,Y_l^m(\vartheta_2,\varphi_2)\,R_{10}(r_2)\,R_{nl}(r_2)}{|\mathbf{r}_1 - \mathbf{r}_2|}. \tag{N.5}$$

We have (see Appendix B, Vol. 2):

J. Pade, *Quantum Mechanics for Pedestrians 2*, Undergraduate Lecture Notes in Physics, https://doi.org/10.1007/978-3-030-00467-5

$$\frac{1}{|\mathbf{r}_1 - \mathbf{r}_2|} = \sum_{L,M} \frac{r_<^L}{r_>^{L+1}} \frac{4\pi}{2L+1} Y_L^M(\vartheta_1, \varphi_1) Y_L^{M*}(\vartheta_2, \varphi_2). \tag{N.6}$$

Here, $r_<$ and $r_>$ denote the smaller and larger values of $r_1$ and $r_2$.

It follows that:

$$C_{nl} = \frac{e^2}{4\pi\varepsilon_0} \frac{1}{4\pi} \sum_{L,M} \frac{4\pi}{2L+1} \int d^3r_1 d^3r_2 R_{10}^2(r_1) R_{nl}^2(r_2) Y_l^m(\vartheta_2, \varphi_2) Y_l^{m*}(\vartheta_2, \varphi_2)$$

$$\times \frac{r_<^L}{r_>^{L+1}} Y_L^M(\vartheta_1, \varphi_1) Y_L^{M*}(\vartheta_2, \varphi_2) \tag{N.7}$$

and

$$A_{nl} = \frac{e^2}{4\pi\varepsilon_0} \frac{1}{4\pi} \sum_{L,M} \frac{4\pi}{2L+1} \int d^3r_1 d^3r_2 R_{10}(r_1) R_{nl}(r_1) Y_l^{m*}(\vartheta_1, \varphi_1) Y_l^m(\vartheta_2, \varphi_2)$$

$$\times \frac{r_<^L}{r_>^{L+1}} R_{10}(r_2) R_{nl}(r_2) Y_L^M(\vartheta_1, \varphi_1) Y_L^{M*}(\vartheta_2, \varphi_2). \tag{N.8}$$

The angular part of the term $C_{nl}$ is given by

$$C_{nl} \rightarrow \sum_{L,M} \frac{4\pi}{2L+1} \int d\Omega_1 d\Omega_2 Y_l^m(\vartheta_2, \varphi_2) Y_l^{m*}(\vartheta_2, \varphi_2) Y_L^M(\vartheta_1, \varphi_1) Y_L^{M*}(\vartheta_2, \varphi_2)$$

$$= \sum_{L,M} \frac{4\pi}{2L+1} \int Y_L^M(\vartheta_1, \varphi_1) d\Omega_1 Y_l^m(\vartheta_2, \varphi_2) Y_l^{m*}(\vartheta_2, \varphi_2) Y_L^{M*}(\vartheta_2, \varphi_2) d\Omega_2$$

$$= \sqrt{4\pi} \sum_{L,M} \frac{4\pi}{2L+1} \int Y_0^{0*}(\vartheta_1, \varphi_1) Y_L^M(\vartheta_1, \varphi_1) d\Omega_1 Y_l^m(\vartheta_2, \varphi_2) Y_l^{m*}(\vartheta_2, \varphi_2)$$

$$\times Y_L^{M*}(\vartheta_2, \varphi_2) d\Omega_2$$

$$= \sum_{L,M} \frac{4\pi}{2L+1} \delta_{L0}\delta_{M0} \int Y_l^m(\vartheta_2, \varphi_2) Y_l^{m*}(\vartheta_2, \varphi_2) d\Omega_2 = 4\pi\delta_{L0}\delta_{M0}, \tag{N.9}$$

and for $A_{nl}$ by

$$A_{nl} \rightarrow \sum_{L,M} \frac{4\pi}{2L+1} \int d\Omega_1 d\Omega_2 Y_l^{m*}(\vartheta_1, \varphi_1) Y_l^m(\vartheta_2, \varphi_2) Y_L^M(\vartheta_1, \varphi_1) Y_L^{M*}(\vartheta_2, \varphi_2)$$

$$= \sum_{L,M} \frac{4\pi}{2L+1} \int Y_l^{m*}(\vartheta_1, \varphi_1) Y_L^M(\vartheta_1, \varphi_1) d\Omega_1 Y_l^m(\vartheta_2, \varphi_2) Y_L^{M*}(\vartheta_2, \varphi_2) d\Omega_2$$

$$= \sum_{L,M} \frac{4\pi}{2L+1} \delta_{lL}\delta_{mM} = \frac{4\pi}{2l+1}. \tag{N.10}$$

This gives

$$
\begin{aligned}
C_{nl} &= \frac{e^2}{4\pi\varepsilon_0} \int r_1^2 dr_1 r_2^2 dr_2 R_{10}^2 (r_1) R_{nl}^2 (r_2) \frac{1}{r_>} \\
&= \frac{e^2}{4\pi\varepsilon_0} \int r_1^2 dr_1 R_{10}^2 (r_1) \left[ \frac{1}{r_1} \int_0^{r_1} r_2^2 R_{nl}^2 (r_2) \, dr_2 + \int_{r_1}^{\infty} r_2 R_{nl}^2 (r_2) \, dr_2 \right]
\end{aligned}
$$

$$(N.11)$$

and

$$
\begin{aligned}
A_{nl} &= \frac{e^2}{4\pi\varepsilon_0} \frac{1}{2l+1} \int r_1^2 dr_1 r_2^2 dr_2 R_{10} (r_1) R_{nl} (r_1) R_{10} (r_2) R_{nl} (r_2) \frac{r_<^l}{r_>^{l+1}} \\
&= \frac{e^2}{4\pi\varepsilon_0} \frac{1}{2l+1} \int r_1^2 dr_1 R_{10} (r_1) R_{nl} (r_1) \\
&\quad \times \left[ \frac{1}{r_1^{l+1}} \int_0^{r_1} r_2^{2+l} R_{10} (r_2) R_{nl} (r_2) \, dr_2 + r_1^l \int_{r_1}^{\infty} r_2^{1-l} R_{10} (r_2) R_{nl} (r_2) \, dr_2 \right].
\end{aligned}
$$

$$(N.12)$$

We need the expressions for $n = 2$ and $l = 0, 1$; the radial functions (for helium, we have $Z = 2$) are given by:

$$
R_{10} (r) = 2 \left( \frac{Z}{a_0} \right)^{\frac{3}{2}} e^{-\frac{Zr}{a_0}}
$$

$$
R_{20} (r) = 2 \left( \frac{Z}{2a_0} \right)^{\frac{3}{2}} \left( 1 - \frac{Zr}{2a_0} \right) e^{-\frac{Zr}{2a_0}}
$$

$$(N.13)$$

$$
R_{21} (r) = \frac{1}{\sqrt{3}} \left( \frac{Z}{2a_0} \right)^{\frac{3}{2}} \frac{Zr}{a_0} e^{-\frac{Zr}{2a_0}}.
$$

After inserting the radial functions, we have to calculate the integrals in (N.11) and (N.12). One can do this by hand (essentially by partial integrations of the form $\int x^n e^{-x} = -x^n e^{-x} + n \int x^{n-1} e^{-x}$), or with the help of a computational program such as *Maple* or *Mathematica*. The result is in either case:

$$
C_{20} = \frac{e^2}{4\pi\varepsilon_0} \frac{17}{81} \frac{Z}{a_0}; \quad C_{21} = \frac{e^2}{4\pi\varepsilon_0} \frac{59}{243} \frac{Z}{a_0}
$$

$$(N.14)$$

and

$$
A_{20} = \frac{e^2}{4\pi\varepsilon_0} \frac{16}{729} \frac{Z}{a_0}; \quad A_{21} = \frac{e^2}{4\pi\varepsilon_0} \frac{112}{6561} \frac{Z}{a_0}.
$$

$$(N.15)$$

# Appendix O
# The Scattering of Identical Particles

As might be expected, quantum scattering of identical objects[43] has certain peculiarities, which we briefly describe here.

We consider the scattering of two (identical) quantum objects in their center-of-mass system, i.e. we use center-of-mass coordinates $\mathbf{R}$ and relative coordinates $\mathbf{r}$:

$$\mathbf{R} = \frac{\mathbf{r}_1 + \mathbf{r}_2}{2}; \quad \mathbf{r} = \mathbf{r}_1 - \mathbf{r}_2. \tag{O.1}$$

Interchanging the quantum objects results in $\mathbf{R} \to \mathbf{R}$ and $\mathbf{r} \to -\mathbf{r}$.

Now we let the two quantum objects scatter. There are two possibilities, as outlined in Fig. O.1, which we can formally describe by the transition $(r, \vartheta, \varphi) \to (r, \pi - \vartheta, \varphi + \pi)$.

We begin with the case that the two quantum objects are in principle distinguishable, although the two detectors are insensitive to this difference.[44] Then the differential scattering cross section for one quantum object to be scattered at the angle $(\vartheta, \varphi)$ is equal to the scattering cross section of the relative-coordinate particle into the same direction, i.e.

$$\frac{d\sigma^{(1)}}{d\Omega} = |f(\vartheta, \varphi)|^2. \tag{O.2}$$

The scattering cross section for quantum object 2 to be scattered at the same angle is obtained by the transformation $\mathbf{r} \to -\mathbf{r}$ or $(\vartheta, \varphi) \to (\pi - \vartheta, \varphi + \pi)$:

$$\frac{d\sigma^{(2)}}{d\Omega} = |f(\pi - \vartheta, \varphi + \pi)|^2. \tag{O.3}$$

The total differential cross section ($=$counting rate of the detector at an angle $(\vartheta, \varphi)$) is the sum:

---

[43] Although the term 'identical particles' is familiar, we instead prefer to use the term 'quantum object' in this chapter.

[44] For example, the scattering of an electron and a muon, or of two isotopes such as $^{12}C$ and $^{13}C$.

© Springer Nature Switzerland AG 2018
J. Pade, *Quantum Mechanics for Pedestrians 2*, Undergraduate Lecture
Notes in Physics, https://doi.org/10.1007/978-3-030-00467-5

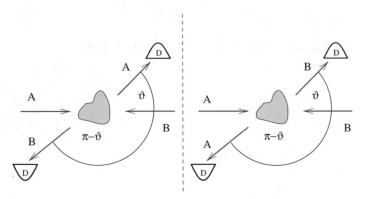

**Fig. O.1** Scattering of two objects under the angles $\vartheta$ and $\pi - \vartheta$ (D = detector)

$$\frac{d\sigma}{d\Omega} = \frac{d\sigma^{(1)}}{d\Omega} + \frac{d\sigma^{(2)}}{d\Omega} = |f(\vartheta, \varphi)|^2 + |f(\pi - \vartheta, \varphi + \pi)|^2. \tag{O.4}$$

It is crucial that we *in principle* can distinguish between the two quantum objects, regardless of whether or not we do so.

Now we suppose that the two quantum objects are identical. Then the two scattering processes in Fig. O.1 are indistinguishable even in principle. Consequently, we must perform a symmetrization of the total wavefunction (integer spin) or an antisymmetrization (half-integer spin), as described in Chap. 23.

Since the spin part is symmetric for bosons, the position part must also be symmetrized. We consider in the following two bosons with spin 0. In the case of fermions, the two quantum objects are supposed to have spin $\frac{1}{2}$. If the two fermions are in the singlet state, the spin part is antisymmetric (see Chap. 23); therefore, the spatial part must be symmetrical. In contrast, when the two fermions are in the triplet state, the spin part is symmetric and the spatial part must be antisymmetric. The bottom line is that the differential scattering cross section for a symmetrized (upper sign) or an antisymmetrized (lower sign) spatial wavefunction reads

$$\frac{d\sigma}{d\Omega} = |f(\vartheta, \varphi) \pm f(\pi - \vartheta, \varphi + \pi)|^2. \tag{O.5}$$

In other words, we have to add the (properly symmetrized) amplitudes and not their absolute squares, as we have always done for those systems where we could not decide which path the quantum objects have taken (double slit, interaction-free quantum measurement, etc.).

For the bosons, it follows for the differential cross section (we restrict ourselves to central forces, which eliminates the dependence on $\varphi$):

$$\frac{d\sigma}{d\Omega} = |f(\vartheta) + f(\pi - \vartheta)|^2$$
$$= |f(\vartheta)|^2 + |f(\pi - \vartheta)|^2 + 2\operatorname{Re}\left[f(\vartheta) f^*(\pi - \vartheta)\right]. \qquad (O.6)$$

For the two fermions with spin $\frac{1}{2}$, we distinguish between the singlet case:

$$\frac{d\sigma}{d\Omega} = |f_s(\vartheta, \varphi) + f_s(\pi - \vartheta, \varphi + \pi)|^2 ; \qquad (O.7)$$

and the triplet case:

$$\frac{d\sigma}{d\Omega} = |f_t(\vartheta, \varphi) - f_t(\pi - \vartheta, \varphi + \pi)|^2 . \qquad (O.8)$$

These two cases are distinguishable in principle; for an equal distribution of the states (triplet to singlet ratio $= 3/1$), the spin-insensitive detector thus measures the differential scattering cross-section

$$\frac{d\sigma}{d\Omega} = \frac{1}{4} |f_s(\vartheta, \varphi) + f_s(\pi - \vartheta, \varphi + \pi)|^2$$
$$+ \frac{3}{4} |f_t(\vartheta, \varphi) - f_t(\pi - \vartheta, \varphi + \pi)|^2 . \qquad (O.9)$$

For spin-independent central forces, we find:

$$f_s(\vartheta, \varphi) = f_t(\vartheta, \varphi) = f(\vartheta), \qquad (O.10)$$

and it follows that

$$\frac{d\sigma}{d\Omega} = |f(\vartheta)|^2 + |f(\pi - \vartheta)|^2 - \operatorname{Re}\left[f(\vartheta) f^*(\pi - \vartheta)\right]. \qquad (O.11)$$

We see the familiar distinction between distinguishable (O.4) and indistinguishable terms, namely the occurrence of interference.

Generalization of (O.6) and (O.11) for arbitrary integer and half-integer angular momenta $j$ leads to:

$$\frac{d\sigma}{d\Omega} = |f(\vartheta)|^2 + |f(\pi - \vartheta)|^2 + 2\frac{(-1)^{2j}}{2j+1} \operatorname{Re}\left[f(\vartheta) f^*(\pi - \vartheta)\right]. \qquad (O.12)$$

One sees nicely how the interferences diminish with increasing angular momentum.

# Appendix P
# The Hadamard Transformation

We will discuss here briefly the relation between the Hadamard transformation and the Mach–Zehnder interferometer (MZI, Fig. P.1) or beam splitter (cf. Chap. 6, Vol. 1).

## P.1    The MZI and the Hadamard Transformation

Mach–Zehnder interferometers are used in many different contexts, among others also in introductions to quantum information. If the details of the setup are not of interest, we can combine the effects of the two mirrors ($S$) and of the beam splitters ($T$). In this way, we can represent a MZI (i.e. $TST$) as a combination of two 'effective' beam splitters, i.e. as $H^2$ (here, $H$ means Hadamard and *not* Hamiltonian), with

$$H^2 = TST = 1, \tag{P.1}$$

where, in the matrix representation, we have[45]:

$$T = \frac{(1+i)}{2} \begin{pmatrix} 1 & i \\ i & 1 \end{pmatrix}; \quad S = \begin{pmatrix} 0 & -1 \\ -1 & 0 \end{pmatrix}. \tag{P.2}$$

To determine $H$, we represent the effect of the mirrors as the product of two matrices $P$ and $Q$, i.e. $S = QP$, where we have to choose the matrices in such a way that $TQ = PT$ holds. In this case, we have:

$$H^2 = TST = TQPT = PTPT \text{ or } H = PT \tag{P.3}$$

---

[45]We omit here the distinction between an operator and its representation, i.e. between $=$ and $\cong$.

© Springer Nature Switzerland AG 2018
J. Pade, *Quantum Mechanics for Pedestrians 2*, Undergraduate Lecture
Notes in Physics, https://doi.org/10.1007/978-3-030-00467-5

**Fig. P.1** The Mach–Zehnder interferometer. M = mirror, BS = beam splitter, D = detector

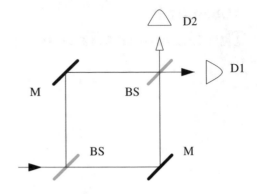

With the matrices

$$P = -\frac{i}{\sqrt{2}} \begin{pmatrix} 1 & 1 \\ i & -i \end{pmatrix}; \; Q = P^T = -\frac{i}{\sqrt{2}} \begin{pmatrix} 1 & i \\ 1 & -i \end{pmatrix} \tag{P.4}$$

it follows that

$$H = PT = \frac{1}{\sqrt{2}} \begin{pmatrix} 1 & 1 \\ 1 & -1 \end{pmatrix}. \tag{P.5}$$

The unitary matrix $H$ is called the *Hadamard matrix*[46]; with its help we can write the effect of the MZI (or the corresponding operator) as

$$H^2 = \frac{1}{2} \begin{pmatrix} 1 & 1 \\ 1 & -1 \end{pmatrix} \begin{pmatrix} 1 & 1 \\ 1 & -1 \end{pmatrix} = \begin{pmatrix} 1 & 0 \\ 0 & 1 \end{pmatrix}, \tag{P.6}$$

as it indeed must be.

The Hadamard matrix is related to the Pauli matrices by

$$H = \frac{\sigma_x + \sigma_z}{\sqrt{2}}. \tag{P.7}$$

## P.2   The Beam Splitter and the Hadamard Transformation

To realize the Hadamard transformation *directly* by a beam splitter (BS) (i.e. without mirrors), we have to perform phase shifts. A possible setup is shown in Fig. P.2:

---

[46]This is a two-dimensional special case. A Hadamard matrix $H_n$ of order $n$ is an $n \times n$-matrix with elements $+1$ and $-1$, which satisfies $H_n H_n^T = n E_n$, where $E_n$ is the $n$-dimensional unit matrix. A generalized Hadamard matrix can contain arbitrary elements and satisfies the equation $H_n H_n^\dagger = n E_n$.

**Fig. P.2** Beam splitter with
phase shifters

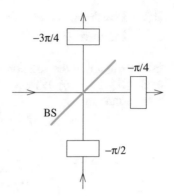

Phase shifts of $-\pi/2$ in each case at the two vertical beams and in addition $-\pi/4$ in
each case at the two outgoing beams.

The effect of the beam splitter can be written as (cf. Chap. 6, Vol. 1):

$$|H\rangle \rightarrow \frac{(1+i)}{2} [|H\rangle + i\,|V\rangle]$$
$$|V\rangle \rightarrow \frac{(1+i)}{2} [|V\rangle + i\,|H\rangle]. \tag{P.8}$$

If an incoming vertical photon impinges upon a beam splitter plus phase shifter, we
have (with $\frac{(1+i)}{2} = \frac{e^{i\pi/4}}{\sqrt{2}}$ and $i = e^{i\pi/2}$):

$$|V\rangle \underset{\text{phase}}{\rightarrow} e^{-i\pi/2} |V\rangle \underset{\text{BS}}{\rightarrow} e^{-i\pi/2} \frac{e^{i\pi/4}}{\sqrt{2}} \left[|V\rangle + e^{i\pi/2} |H\rangle\right]$$

$$\underset{\text{phase}}{\rightarrow} e^{-i\pi/2} \frac{e^{i\pi/4}}{\sqrt{2}} \left[e^{-3i\pi/4} |V\rangle + e^{i\pi/2} e^{-i\pi/4} |H\rangle\right], \tag{P.9}$$

and for a horizontal photon:

$$|H\rangle \underset{\text{BS}}{\rightarrow} \frac{e^{i\pi/4}}{\sqrt{2}} \left[|H\rangle + e^{i\pi/2} |V\rangle\right] \underset{\text{phase}}{\rightarrow} \frac{e^{i\pi/4}}{\sqrt{2}} \left[e^{-i\pi/4} |H\rangle + e^{i\pi/2} e^{-3i\pi/4} |V\rangle\right]; \tag{P.10}$$

or, summarizing,

$$|H\rangle \rightarrow \frac{1}{\sqrt{2}} [|H\rangle + |V\rangle]; \quad |V\rangle \rightarrow \frac{1}{\sqrt{2}} [|H\rangle - |V\rangle] \tag{P.11}$$

as required.

## P.3 The Hadamard Transformation and Quantum Information

In order to apply the notation usual in quantum information, we make the replacement $|H\rangle \rightarrow |0\rangle$, $|V\rangle \rightarrow |1\rangle$. Hence, we find with (P.5) for the Hadamard transformation:

$$|0\rangle \rightarrow \frac{|0\rangle + |1\rangle}{\sqrt{2}}; \quad |1\rangle \rightarrow \frac{|0\rangle - |1\rangle}{\sqrt{2}}, \tag{P.12}$$

or, in short form:

$$|x\rangle \underset{H}{\rightarrow} \frac{|1 - x\rangle + (-1)^x |x\rangle}{\sqrt{2}}; \quad x = 0, 1. \tag{P.13}$$

Especially in quantum information, the Hadamard transformation is very frequently used (see Chap. 26); its specific symbol or abbreviation is $-\boxed{H}-$.

As an example of an application, we consider the preparation of special states. We assume three quantum objects, each with states $|0\rangle$ and $|1\rangle$. Initially, all three objects are in the state $|0\rangle$; thus, we have for the ground state $|0\rangle\,|0\rangle\,|0\rangle = |000\rangle$. We apply a Hadamard transformation to each individual state and obtain

$$|0\rangle\,|0\rangle\,|0\rangle \rightarrow \frac{|0\rangle + |1\rangle}{\sqrt{2}} \frac{|0\rangle + |1\rangle}{\sqrt{2}} \frac{|0\rangle + |1\rangle}{\sqrt{2}}$$

$$= \frac{|0\rangle|0\rangle|0\rangle + |0\rangle|0\rangle|1\rangle + |0\rangle|1\rangle|0\rangle + |1\rangle|0\rangle|0\rangle + |0\rangle|1\rangle|1\rangle + |1\rangle|0\rangle|1\rangle + |1\rangle|1\rangle|0\rangle + |1\rangle|1\rangle|1\rangle}{2\sqrt{2}}. \tag{P.14}$$

We see that we obtain a linear combination of *all possible* states through the application of the Hadamard transformation. In short, we can write:

$$|000\rangle \rightarrow \frac{|000\rangle + |001\rangle + |010\rangle + |100\rangle + |011\rangle + |101\rangle + |110\rangle + |111\rangle}{2\sqrt{2}}. \tag{P.15}$$

If we assume this to be the binary representation of numbers, we obtain in the decimal representation

$$|000\rangle \underset{\text{binary}}{\rightarrow} \frac{|000\rangle + |001\rangle + |010\rangle + |011\rangle + |100\rangle + |101\rangle + |110\rangle + |111\rangle}{2\sqrt{2}}$$

$$|0\rangle \underset{\text{decimal}}{\rightarrow} \frac{|0\rangle + |1\rangle + |2\rangle + |3\rangle + |4\rangle + |5\rangle + |6\rangle + |7\rangle}{2\sqrt{2}}. \tag{P.16}$$

In general, we start from $n$ states $|0\rangle$. If we apply a Hadamard transformation to each individual state, we can generate all the possible states of the total system. This means in decimal notation:

$$|0\rangle \underset{\text{decimal}}{\rightarrow} \sum_{m=0}^{2^n-1} |m\rangle. \tag{P.17}$$

If we start from an initial configuration in which the quantum objects are in the state $|0\rangle$ or $|1\rangle$, we obtain again a linear combination of all possible total states; however, the signs are different.

# Appendix Q
# From the Interferometer to the Computer

We want to show here that the Mach–Zehnder interferometer, introduced in Chap. 6, Vol. 1, is basically one of the essential building blocks for a quantum computer as discussed in Chap. 26. We follow in part the presentation given in arXiv:0011013.

**The Mach–Zehnder Interferometer**

We assume a setup which is slightly modified compared to Chap. 6, Vol. 1. First, we add a phase shifter in the upper and the lower beam paths, and secondly, we use Hadamard beam splitters (H-BS, i.e. conventional beam splitters with phase shifters added[47]); see Fig. Q.1. This setup corresponds to the overall transformation $M_C$:

$$M_C = \frac{1}{\sqrt{2}} \begin{pmatrix} 1 & 1 \\ 1 & -1 \end{pmatrix} \begin{pmatrix} e^{i\alpha_1} & 0 \\ 0 & 1 \end{pmatrix} \begin{pmatrix} 0 & -1 \\ -1 & 0 \end{pmatrix} \begin{pmatrix} e^{i\alpha_0} & 0 \\ 0 & 1 \end{pmatrix} \frac{1}{\sqrt{2}} \begin{pmatrix} 1 & 1 \\ 1 & -1 \end{pmatrix}, \quad (Q.1)$$

or, more compactly (see the exercises):

$$M_C = -e^{i\frac{(\alpha_0 + \alpha_1)}{2}} \begin{pmatrix} \cos \frac{(\alpha_0 - \alpha_1)}{2} & i \sin \frac{(\alpha_0 - \alpha_1)}{2} \\ -i \sin \frac{(\alpha_0 - \alpha_1)}{2} & -\cos \frac{(\alpha_0 - \alpha_1)}{2} \end{pmatrix}. \quad (Q.2)$$

In order to facilitate comparison with the usual notation for quantum computers, we use $|0\rangle$ and $|1\rangle$ in this appendix instead of $|H\rangle$ and $|V\rangle$. For e.g. the initial state $|0\rangle$, we can write explicitly:

$$|0\rangle \rightarrow -e^{i\frac{(\alpha_0 + \alpha_1)}{2}} \left[ \cos \frac{(\alpha_0 - \alpha_1)}{2} |0\rangle + i \sin \frac{(\alpha_0 - \alpha_1)}{2} |1\rangle \right]$$
$$= -e^{i\frac{((\alpha_0 + \alpha_1)}{2})} \left[ \cos \frac{\alpha}{2} |0\rangle - i \sin \frac{\alpha}{2} |1\rangle \right] ; \ \alpha = \alpha_1 - \alpha_0. \quad (Q.3)$$

---

[47]The standard beam splitter (see Chap. 6, Vol. 1) corresponds to the transformation $\frac{(1+i)}{2} \begin{pmatrix} 1 & i \\ i & 1 \end{pmatrix}$, while the Hadamard beam splitter is described by $\frac{1}{\sqrt{2}} \begin{pmatrix} 1 & 1 \\ 1 & -1 \end{pmatrix}$; see Appendix P, Vol. 2.

© Springer Nature Switzerland AG 2018
J. Pade, *Quantum Mechanics for Pedestrians 2*, Undergraduate Lecture Notes in Physics, https://doi.org/10.1007/978-3-030-00467-5

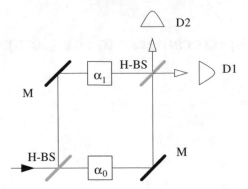

**Fig. Q.1** A Mach–Zehnder interferometer, with phase shifters and Hadamard beam splitters

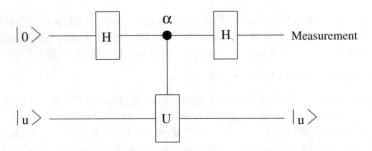

**Fig. Q.2** Graphical representation of the transformation (Q.5)

### Network

We restrict ourselves in the following discussion to the consideration of the input state $|0\rangle$. Then we can write the setup of Fig. Q.1 in diagram form as a network with three quantum logic gates, namely as (see the exercises):

$$-\boxed{H}-\overset{\varphi}{\bullet}-\boxed{H}- \tag{Q.4}$$

We generalize this special process of phase shifting somewhat by describing it as the application of a unitary operator $U$ with $U|u\rangle = e^{i\alpha}|u\rangle$ in a CNOT gate as shown in Fig. Q.2. This structure is one of the basic building blocks of quantum algorithms. An input state $|0\rangle$ is transformed as follows (see the exercises):

$$|u\rangle\,|0\rangle \overset{H}{\to} |u\rangle \frac{|0\rangle + |1\rangle}{\sqrt{2}} \overset{c\text{-}U}{\to} |u\rangle \frac{|0\rangle + e^{i\alpha}|1\rangle}{\sqrt{2}} \overset{H}{\to} |u\rangle\, e^{i\frac{\alpha}{2}}\left[\cos\frac{\alpha}{2}\,|0\rangle - i\sin\frac{\alpha}{2}\,|1\rangle\right].$$
$$\tag{Q.5}$$

If we choose for example

$$\alpha_0 = \pm\pi; \ \alpha_1 = \alpha + \alpha_0, \tag{Q.6}$$

then this is (referred to the input state $|0\rangle$) the *exact* simulation of the Mach–Zehnder setup (Q.1), as the comparison of (Q.3) and (Q.5) shows. Note that the qubit $|u\rangle$ is not changed.

### Extensions

Instead of the phase shift, we can insert any other unitary transformation for $U$. For example, we can choose a transformation of the controlled-$U$ type for $f : \{0, 1\}^n \rightarrow \{0, 1\}^m$:

$$|x\rangle \, |y\rangle \rightarrow |x\rangle \, |[y + f(x)] \bmod 2^m\rangle. \tag{Q.7}$$

If we let the initial state be a superposition of all states $|y\rangle$,

$$|u\rangle = \frac{1}{2^{m/2}} \sum_{y=0}^{2^m - 1} e^{-\frac{2\pi i y}{2^m}} |y\rangle, \tag{Q.8}$$

then we obtain initially

$$|x\rangle \, |u\rangle = \frac{1}{2^{m/2}} \sum_{y=0}^{2^m - 1} e^{-\frac{2\pi i y}{2^m}} |x\rangle \, |y\rangle \rightarrow \frac{1}{2^{m/2}} \sum_{y=0}^{2^m - 1} e^{-\frac{2\pi i y}{2^m}} |x\rangle \, |[y + f(x)] \bmod 2^m\rangle.$$
$$\tag{Q.9}$$

We rearrange this expression by expanding the exponent by $f(x)$ and subsequently renaming the summation index:

$$\frac{1}{2^{m/2}} \sum_{y=0}^{2^m - 1} e^{-\frac{2\pi i y}{2^m}} |x\rangle \, |[y + f(x)] \bmod 2^m\rangle$$

$$= \frac{1}{2^{m/2}} e^{2\pi i \frac{f(x)}{2^m}} \sum_{y=0}^{2^m - 1} e^{-2\pi i \frac{y + f(x)}{2^m}} |x\rangle \, |[y + f(x)] \bmod 2^m\rangle$$

$$= \frac{1}{2^{m/2}} e^{2\pi i \frac{f(x)}{2^m}} \sum_{y=0}^{2^m - 1} e^{-\frac{2\pi i y}{2^m}} |x\rangle \, |y\rangle, \tag{Q.10}$$

where we have used (periodicity of the complex $e$-function, $e^{ix} = e^{ix + 2\pi i m}$):

$$\sum_{y=0}^{2^m - 1} e^{-2\pi i \frac{y + f(x)}{2^m}} |[y + f(x)] \bmod 2^m\rangle$$

$$\underset{z = y + f(x)}{=} \sum_{z=0}^{2^m - 1} e^{-\frac{2\pi i z}{2^m}} |z\rangle = \sum_{y=0}^{2^m - 1} e^{-\frac{2\pi i y}{2^m}} |y\rangle. \tag{Q.11}$$

Hence, (Q.10) becomes

$$|x\rangle\,|u\rangle \to e^{2\pi i \frac{f(x)}{2^m}}\,|x\rangle\,|u\rangle\,. \tag{Q.12}$$

In particular, for $m = 1$ we obtain

$$|x\rangle\,|u\rangle \to e^{\pi i f(x)}\,|x\rangle\,|u\rangle = (-1)^{f(x)}\,|x\rangle\,|u\rangle\,. \tag{Q.13}$$

**Exercises**

1. Determine explicitly

$$M_C = \frac{1}{\sqrt{2}}\begin{pmatrix}1 & 1\\ 1 & -1\end{pmatrix}\begin{pmatrix}e^{i\alpha_1} & 0\\ 0 & 1\end{pmatrix}\begin{pmatrix}0 & -1\\ -1 & 0\end{pmatrix}\begin{pmatrix}e^{i\alpha_0} & 0\\ 0 & 1\end{pmatrix}\frac{1}{\sqrt{2}}\begin{pmatrix}1 & 1\\ 1 & -1\end{pmatrix}. \tag{Q.14}$$

Solution: We start with

$$M_C = \frac{1}{2}\begin{pmatrix}1 & 1\\ 1 & -1\end{pmatrix}\begin{pmatrix}0 & -e^{i\alpha_1}\\ -e^{i\alpha_0} & 0\end{pmatrix}\begin{pmatrix}1 & 1\\ 1 & -1\end{pmatrix}, \tag{Q.15}$$

from which it follows that

$$M_C = -\frac{1}{2}\begin{pmatrix}e^{i\alpha_0} + e^{i\alpha_1} & e^{i\alpha_0} - e^{i\alpha_1}\\ -e^{i\alpha_0} + e^{i\alpha_1} & -e^{i\alpha_0} - e^{i\alpha_1}\end{pmatrix}. \tag{Q.16}$$

We transform with

$$M_C = -\frac{e^{i(\alpha_0+\alpha_1)/2}}{2}\begin{pmatrix}e^{i\frac{\alpha_0-\alpha_1}{2}} + e^{-i\frac{\alpha_0-\alpha_1}{2}} & e^{i\frac{\alpha_0-\alpha_1}{2}} - e^{-i\frac{\alpha_0-\alpha_1}{2}}\\ -e^{i\frac{\alpha_0-\alpha_1}{2}} + e^{-i\frac{\alpha_0-\alpha_1}{2}} & -e^{i\frac{\alpha_0-\alpha_1}{2}} - e^{-i\frac{\alpha_0-\alpha_1}{2}}\end{pmatrix} \tag{Q.17}$$

and obtain finally

$$M_C = -e^{i\frac{(\alpha_0+\alpha_1)}{2}}\begin{pmatrix}\cos\frac{(\alpha_0-\alpha_1)}{2} & i\sin\frac{(\alpha_0-\alpha_1)}{2}\\ -i\sin\frac{(\alpha_0-\alpha_1)}{2} & -\cos\frac{(\alpha_0-\alpha_1)}{2}\end{pmatrix}. \tag{Q.18}$$

2. Determine the transformation corresponding to Fig. Q.1, if standard beam splitters are used instead of the Hadamard beam splitters.
   Solution: The transformation reads

$$M = \frac{(1+i)}{2}\begin{pmatrix}1 & i\\ i & 1\end{pmatrix}\begin{pmatrix}e^{i\alpha_1} & 0\\ 0 & 1\end{pmatrix}\begin{pmatrix}0 & -1\\ -1 & 0\end{pmatrix}\begin{pmatrix}e^{i\alpha_0} & 0\\ 0 & 1\end{pmatrix}\frac{(1+i)}{2}\begin{pmatrix}1 & i\\ i & 1\end{pmatrix}. \tag{Q.19}$$

It follows that:

$$M = -\frac{i}{2}\begin{pmatrix}1 & i\\ i & 1\end{pmatrix}\begin{pmatrix}0 & e^{i\alpha_1}\\ e^{i\alpha_0} & 0\end{pmatrix}\begin{pmatrix}1 & i\\ i & 1\end{pmatrix}, \tag{Q.20}$$

or

$$M = -\frac{i}{2} \begin{pmatrix} ie^{i\alpha_1} + ie^{i\alpha_0} & e^{i\alpha_1} - e^{i\alpha_0} \\ -e^{i\alpha_1} + e^{i\alpha_0} & ie^{i\alpha_1} + ie^{i\alpha_0} \end{pmatrix}. \tag{Q.21}$$

3. Determine the transformation corresponding to the network (Q.4) and show that it agrees for the input state $|0\rangle$ with the transformation (Q.2) up to a global phase.
   Solution: The network (Q.4) corresponds to the transformation

$$\frac{1}{\sqrt{2}} \begin{pmatrix} 1 & 1 \\ 1 & -1 \end{pmatrix} \begin{pmatrix} 1 & 0 \\ 0 & e^{i\alpha} \end{pmatrix} \frac{1}{\sqrt{2}} \begin{pmatrix} 1 & 1 \\ 1 & -1 \end{pmatrix} = \frac{1}{2} \begin{pmatrix} 1 + e^{i\alpha} & 1 - e^{i\alpha} \\ 1 - e^{i\alpha} & 1 + e^{i\alpha} \end{pmatrix}$$

$$= \frac{1}{2} e^{i\alpha/2} \begin{pmatrix} e^{-i\alpha/2} + e^{i\alpha/2} & e^{-i\alpha/2} - e^{i\alpha/2} \\ e^{-i\alpha/2} - e^{i\alpha/2} & e^{-i\alpha/2} + e^{i\alpha/2} \end{pmatrix} = e^{i\alpha/2} \begin{pmatrix} \cos\frac{\alpha}{2} & -i\sin\frac{\alpha}{2} \\ -i\sin\frac{\alpha}{2} & \cos\frac{\alpha}{2} \end{pmatrix}. \tag{Q.22}$$

Hence, the initial state $|0\rangle$ is transformed to $e^{i\alpha/2} \left[ \cos\frac{\alpha}{2} |0\rangle - i\sin\frac{\alpha}{2} |1\rangle \right]$, which is identical to (Q.3) up to a global phase.

4. Derive (Q.5). How do the phases $\alpha$, $\alpha_0$ and $\alpha_1$ have to be chosen so that the MZI of Fig. Q.1 and the network of Fig. Q.2 work identically?
   Solution: The action of the Hadamard transformation is

$$|0\rangle \overset{H}{\to} \frac{|0\rangle + |1\rangle}{\sqrt{2}}; \quad |1\rangle \overset{H}{\to} \frac{|0\rangle - |1\rangle}{\sqrt{2}}, \tag{Q.23}$$

and the c-U transformation is

$$|0\rangle \overset{\text{c-U}}{\to} |0\rangle ; \quad |1\rangle \overset{\text{c-U}}{\to} e^{i\alpha} |1\rangle . \tag{Q.24}$$

Then it follows first of all that

$$|u\rangle |0\rangle \overset{H}{\to} |u\rangle \frac{|0\rangle + |1\rangle}{\sqrt{2}} \overset{\text{c-U}}{\to} |u\rangle \frac{|0\rangle + e^{i\alpha} |1\rangle}{\sqrt{2}}. \tag{Q.25}$$

The last Hadamard transformation acts as follows:

$$\frac{|0\rangle + e^{i\alpha} |1\rangle}{\sqrt{2}} \overset{H}{\to} \frac{|0\rangle + |1\rangle + e^{i\alpha} [|0\rangle - |1\rangle]}{2} = \frac{[1 + e^{i\alpha}] |0\rangle + [1 - e^{i\alpha}] |1\rangle}{2}$$

$$= \frac{e^{i\alpha/2} [e^{-i\alpha/2} + e^{i\alpha/2}] |0\rangle + e^{i\alpha/2} [e^{-i\alpha/2} - e^{i\alpha/2}] |1\rangle}{2} = e^{i\alpha/2} \left[ \cos\frac{\alpha}{2} |0\rangle - i\sin\frac{\alpha}{2} |1\rangle \right]. \tag{Q.26}$$

Summarized, this means that for the network,

$$|u\rangle |0\rangle \overset{H,\text{c-U},H}{\to} |u\rangle e^{i\alpha/2} \left[ \cos\frac{\alpha}{2} |0\rangle - i\sin\frac{\alpha}{2} |1\rangle \right]; \tag{Q.27}$$

and for the MZI,

$$|0\rangle \rightarrow -e^{i\frac{(\alpha_0 + \alpha_1)}{2}}\left[\cos \frac{(\alpha_1 - \alpha_0)}{2}|0\rangle - i\sin \frac{(\alpha_1 - \alpha_0)}{2}|1\rangle\right]. \quad (Q.28)$$

The square brackets are identical for $\alpha = \alpha_1 - \alpha_0$ and the phase factor for $\alpha = \alpha_0 + \alpha_1 \pm 2\pi$. Hence the two setups operate identically for

$$\alpha_0 = \pm\pi; \quad \alpha_1 = \alpha + \alpha_0. \quad (Q.29)$$

# Appendix R
# The Grover Algorithm, Algebraically

The geometric treatment of the Grover algorithm is found in Chap. 26. To avoid back-and-forth browsing, we list here once more the essential elements from that chapter.

We assume a function $f(k)$ which vanishes for all arguments with the exception of one (the one being sought):

$$f(k) = \delta_{k\kappa}; \ k = 0, 1, \ldots, N - 1; \ N = 2^n; \ 0 \leq \kappa \leq N - 1. \tag{R.1}$$

We use a kickback and obtain the mapping

$$|k\rangle \to (-1)^{f(k)} |k\rangle, \tag{R.2}$$

where $\{|k\rangle\}$ is a CONS of dimension $N$. Because of (R.1), this means that all states remain unchanged, except for the state being sought, $|\kappa\rangle$, where $|\kappa\rangle \to -|\kappa\rangle$ holds. Therefore, the mapping (R.2) can also be written as

$$U_\kappa = 1 - 2 |\kappa\rangle \langle\kappa|. \tag{R.3}$$

The initial state for the algorithm is a normalized and equally-weighted superposition of all the states:

$$|s\rangle = \frac{1}{\sqrt{N}} \sum_{k=0}^{N-1} |k\rangle; \ \langle s | s \rangle = 1. \tag{R.4}$$

With this state, we define the operator

$$U_s = 2 |s\rangle \langle s| - 1. \tag{R.5}$$

Grover's algorithm is now a (repeated) application of $U_s U_\kappa$ to the initial state.

We calculate explicitly the first iteration. We apply $U_\kappa$ to $|s\rangle$:

© Springer Nature Switzerland AG 2018
J. Pade, *Quantum Mechanics for Pedestrians 2*, Undergraduate Lecture Notes in Physics, https://doi.org/10.1007/978-3-030-00467-5

$$U_\kappa \left| s \right\rangle = (1 - 2 \left| \kappa \right\rangle \left\langle \kappa \right|) \left| s \right\rangle = \left| s \right\rangle - 2 \left| \kappa \right\rangle \left\langle \kappa \left| s \right\rangle \right. = \left| s \right\rangle - \frac{2}{\sqrt{N}} \left| \kappa \right\rangle . \quad \text{(R.6)}$$

Here, we have used

$$\left\langle \kappa \left| s \right\rangle \right. = \frac{1}{\sqrt{N}}. \quad \text{(R.7)}$$

In the next step, we consider $U_s$ in addition:

$$U_s U_\kappa \left| s \right\rangle = U_s \left( \left| s \right\rangle - \frac{2}{\sqrt{N}} \left| \kappa \right\rangle \right) = (2 \left| s \right\rangle \left\langle s \right| - 1) \left( \left| s \right\rangle - \frac{2}{\sqrt{N}} \left| \kappa \right\rangle \right)$$

$$= (2 \left| s \right\rangle \left\langle s \right| - 1) \left| s \right\rangle - \frac{2}{\sqrt{N}} (2 \left| s \right\rangle \left\langle s \right| - 1) \left| \kappa \right\rangle = \frac{N - 4}{N} \left| s \right\rangle + \frac{2}{\sqrt{N}} \left| \kappa \right\rangle . \quad \text{(R.8)}$$

At the outset, we had $\left\langle \kappa \left| s \right\rangle \right. = 1 / \sqrt{N}$; applying $U_s U_\kappa$ yields

$$\left\langle \kappa \right| U_s U_\kappa \left| s \right\rangle = \left\langle \kappa \right| \left( \frac{N - 4}{N} \left| s \right\rangle + \frac{2}{\sqrt{N}} \left| \kappa \right\rangle \right)$$

$$= \frac{N - 4}{N \sqrt{N}} + \frac{2}{\sqrt{N}} = \frac{3N - 4}{N \sqrt{N}} = \frac{3}{\sqrt{N}} \left( 1 - \frac{4}{3N} \right) . \quad \text{(R.9)}$$

The absolute square of the amplitude value has increased in the second step by a factor of about 9:

$$\left| \left\langle \kappa \left| s \right\rangle \right. \right|^2 = \frac{1}{N}; \quad \left| \left\langle \kappa \right| U_s U_\kappa \left| s \right\rangle \right|^2 = \frac{9}{N} \left( 1 - \frac{4}{3N} \right)^2 \approx \frac{9}{N}. \quad \text{(R.10)}$$

For the calculation of $\left\langle \kappa \right| (U_s U_\kappa)^m \left| s \right\rangle$ for arbitrary $m$, we first examine the influence of the operators on linear combinations of $\left| \kappa \right\rangle$ and $\left| s \right\rangle$. We have:

$$U_\kappa (a \left| \kappa \right\rangle + b \left| s \right\rangle) = (1 - 2 \left| \kappa \right\rangle \left\langle \kappa \right|) (a \left| \kappa \right\rangle + b \left| s \right\rangle)$$
$$U_s (a \left| \kappa \right\rangle + b \left| s \right\rangle) = (2 \left| s \right\rangle \left\langle s \right| - 1) (a \left| \kappa \right\rangle + b \left| s \right\rangle) . \quad \text{(R.11)}$$

Expanding gives:

$$U_\kappa (a \left| \kappa \right\rangle + b \left| s \right\rangle) = a \left| \kappa \right\rangle + b \left| s \right\rangle - 2a \left| \kappa \right\rangle \left\langle \kappa \left| \kappa \right\rangle \right. - 2b \left| \kappa \right\rangle \left\langle \kappa \left| s \right\rangle \right.$$
$$U_s (a \left| \kappa \right\rangle + b \left| s \right\rangle) = 2a \left| s \right\rangle \left\langle s \left| \kappa \right\rangle \right. + 2b \left| s \right\rangle \left\langle s \left| s \right\rangle \right. - a \left| \kappa \right\rangle - b \left| s \right\rangle . \quad \text{(R.12)}$$

With $\left\langle \kappa \left| \kappa \right\rangle \right. = 1$ and $\left\langle \kappa \left| s \right\rangle \right. = 1 / \sqrt{N}$, it follows that

$$U_\kappa (a \left| \kappa \right\rangle + b \left| s \right\rangle) = a \left| \kappa \right\rangle + b \left| s \right\rangle - 2a \left| \kappa \right\rangle - 2b \left| \kappa \right\rangle \frac{1}{\sqrt{N}}$$
$$U_s (a \left| \kappa \right\rangle + b \left| s \right\rangle) = 2a \left| s \right\rangle \frac{1}{\sqrt{N}} + 2b \left| s \right\rangle - a \left| \kappa \right\rangle - b \left| s \right\rangle , \quad \text{(R.13)}$$

and this leads to

$$U_\kappa \left( a \, |\kappa\rangle + b \, |s\rangle \right) = -a \, |\kappa\rangle + b \, |s\rangle - \frac{2b}{\sqrt{N}} \, |\kappa\rangle$$
$$U_s \left( a \, |\kappa\rangle + b \, |s\rangle \right) = \frac{2a}{\sqrt{N}} \, |s\rangle + b \, |s\rangle - a \, |\kappa\rangle . \tag{R.14}$$

Writing this in matrix form, we obtain

$$U_\kappa : \begin{pmatrix} a \\ b \end{pmatrix} \to \begin{pmatrix} -a - \frac{2b}{\sqrt{N}} \\ b \end{pmatrix} = \begin{pmatrix} -1 & -\frac{2}{\sqrt{N}} \\ 0 & 1 \end{pmatrix} \begin{pmatrix} a \\ b \end{pmatrix}$$

$$U_s : \begin{pmatrix} a \\ b \end{pmatrix} \to \begin{pmatrix} -a \\ \frac{2a}{\sqrt{N}} + b \end{pmatrix} = \begin{pmatrix} -1 & 0 \\ \frac{2}{\sqrt{N}} & 1 \end{pmatrix} \begin{pmatrix} a \\ b \end{pmatrix} . \tag{R.15}$$

It follows[48] that

$$U_s U_\kappa = \begin{pmatrix} -1 & 0 \\ \frac{2}{\sqrt{N}} & 1 \end{pmatrix} \begin{pmatrix} -1 & -\frac{2}{\sqrt{N}} \\ 0 & 1 \end{pmatrix} = \begin{pmatrix} 1 & \frac{2}{\sqrt{N}} \\ -\frac{2}{\sqrt{N}} & 1 - \frac{4}{N} \end{pmatrix} , \tag{R.16}$$

or

$$U_s U_\kappa : \begin{pmatrix} a \\ b \end{pmatrix} \to \begin{pmatrix} a + \frac{2}{\sqrt{N}} b \\ -\frac{2}{\sqrt{N}} a + \left( 1 - \frac{4}{N} \right) b \end{pmatrix} . \tag{R.17}$$

In order to calculate $(U_s U_\kappa)^m$, we first diagonalize $U_s U_\kappa$. We have (see the exercises):

$$U_s U_\kappa = \begin{pmatrix} 1 & \frac{2}{\sqrt{N}} \\ -\frac{2}{\sqrt{N}} & 1 - \frac{4}{N} \end{pmatrix} = M \begin{pmatrix} e^{2i\varphi} & 0 \\ 0 & e^{-2i\varphi} \end{pmatrix} M^{-1} \tag{R.18}$$

with

$$\varphi = \arcsin \frac{1}{\sqrt{N}} \tag{R.19}$$

and

$$M = \begin{pmatrix} -i & i \\ e^{i\varphi} & e^{-i\varphi} \end{pmatrix} ; \quad M^{-1} = \frac{1}{2i \cos \varphi} \begin{pmatrix} -e^{-i\varphi} & i \\ e^{i\varphi} & i \end{pmatrix} . \tag{R.20}$$

It follows that:

$$(U_s U_\kappa)^m = M \begin{pmatrix} e^{2im\varphi} & 0 \\ 0 & e^{-2im\varphi} \end{pmatrix} M^{-1} \tag{R.21}$$

A small calculation (see the exercises) shows

$$(U_s U_\kappa)^m = \frac{1}{\cos \varphi} \begin{pmatrix} \cos (2m - 1) \varphi & \sin 2m\varphi \\ -\sin 2m\varphi & \cos (2m + 1) \varphi \end{pmatrix} , \tag{R.22}$$

---

[48] For simplicity, we do not distinguish in this section between an operator and its matrix representation, i.e. we always write $=$ instead of $\cong$.

or

$$(U_s U_\kappa)^m : \begin{pmatrix} a \\ b \end{pmatrix} \to \frac{1}{\cos\varphi} \begin{pmatrix} a\cos(2m-1)\,\varphi + b\sin 2m\varphi \\ -a\sin 2m\varphi + b\cos(2m+1)\,\varphi \end{pmatrix}, \qquad (R.23)$$

or

$$(U_s U_\kappa)^m \,(a\,|\kappa\rangle + b\,|s\rangle) = \frac{1}{\cos\varphi}\big([a\cos(2m-1)\,\varphi + b\sin 2m\varphi]\,|\kappa\rangle$$
$$+ [-a\sin 2m\varphi + b\cos(2m+1)\,\varphi]\,|s\rangle\,). \quad (R.24)$$

We then find from this:

$$(U_s U_\kappa)^m \,|s\rangle = \frac{1}{\cos\varphi}\,(\sin 2m\varphi \cdot |\kappa\rangle + \cos(2m+1)\,\varphi \cdot |s\rangle), \qquad (R.25)$$

and therefore (see the exercises),

$$\langle\kappa|\,(U_s U_\kappa)^m \,|s\rangle = \frac{1}{\cos\varphi}\,\langle\kappa|\,(\sin 2m\varphi \cdot |\kappa\rangle + \cos(2m+1)\,\varphi \cdot |s\rangle)$$
$$= \frac{1}{\cos\varphi}\left(\sin 2m\varphi + \frac{1}{\sqrt{N}}\cos(2m+1)\,\varphi\right) = \sin(2m+1)\,\varphi.$$
$$(R.26)$$

As a test, we check this result for $m = 1$. We find:

$$\langle\kappa|\,(U_s U_\kappa)\,|s\rangle = \sin 3\varphi = 3\sin\varphi - 4\sin^3\varphi$$
$$= 3\frac{1}{\sqrt{N}} - 4\frac{1}{N\sqrt{N}} = \frac{3N-4}{N\sqrt{N}} \qquad (R.27)$$

as expected; see (R.9).

**Exercises**

1. Show that:

$$\begin{pmatrix} 1 & \frac{2}{\sqrt{N}} \\ -\frac{2}{\sqrt{N}} & 1-\frac{4}{N} \end{pmatrix} = M \begin{pmatrix} e^{2i\varphi} & 0 \\ 0 & e^{-2i\varphi} \end{pmatrix} M^{-1} \qquad (R.28)$$

with $\varphi = \arcsin\frac{1}{\sqrt{N}}$ and

$$M = \begin{pmatrix} -i & i \\ e^{i\varphi} & e^{-i\varphi} \end{pmatrix}; \quad M^{-1} = \frac{1}{ie^{-i\varphi} + ie^{i\varphi}} \begin{pmatrix} -e^{-i\varphi} & i \\ e^{i\varphi} & i \end{pmatrix}. \qquad (R.29)$$

Solution: We have

$$M \begin{pmatrix} e^{2i\varphi} & 0 \\ 0 & e^{-2i\varphi} \end{pmatrix} M^{-1} = \frac{1}{ie^{-i\varphi} + ie^{i\varphi}} \begin{pmatrix} -i & i \\ e^{i\varphi} & e^{-i\varphi} \end{pmatrix} \begin{pmatrix} e^{2i\varphi} & 0 \\ 0 & e^{-2i\varphi} \end{pmatrix} \begin{pmatrix} -e^{-i\varphi} & i \\ e^{i\varphi} & i \end{pmatrix}$$

$$= \frac{1}{ie^{-i\varphi} + ie^{i\varphi}} \begin{pmatrix} ie^{i\varphi} + ie^{-i\varphi} & e^{2i\varphi} - e^{-2i\varphi} \\ -e^{2i\varphi} + e^{-2i\varphi} & ie^{3i\varphi} + ie^{-3i\varphi} \end{pmatrix}$$

$$= \begin{pmatrix} 1 & \frac{1}{i}\left( e^{i\varphi} - e^{-i\varphi} \right) \\ -\frac{1}{i}\left( e^{i\varphi} - e^{-i\varphi} \right) & e^{2i\varphi} + e^{-2i\varphi} - 1 \end{pmatrix}. \tag{R.30}$$

With

$$e^{i\varphi} = \cos \arcsin \frac{1}{\sqrt{N}} + i \sin \arcsin \frac{1}{\sqrt{N}} = \sqrt{1 - \frac{1}{N}} + i\frac{1}{\sqrt{N}};$$

$$e^{2i\varphi} = 1 - \frac{2}{N} + 2i\frac{1}{\sqrt{N}}\sqrt{1 - \frac{1}{N}}, \tag{R.31}$$

it follows finally that

$$M \begin{pmatrix} e^{2i\varphi} & 0 \\ 0 & e^{-2i\varphi} \end{pmatrix} M^{-1} = \begin{pmatrix} 1 & \frac{1}{i}2i\frac{1}{\sqrt{N}} \\ -\frac{1}{i}2i\frac{1}{\sqrt{N}} & 2 - \frac{4}{N} - 1 \end{pmatrix} = \begin{pmatrix} 1 & \frac{2}{\sqrt{N}} \\ -\frac{2}{\sqrt{N}} & 1 - \frac{4}{N} \end{pmatrix}. \tag{R.32}$$

2. Show that

$$(U_s U_\kappa)^m = \frac{1}{\cos \varphi} \begin{pmatrix} \cos (2m - 1) \varphi & \sin 2m\varphi \\ -\sin 2m\varphi & \cos (2m + 1) \varphi \end{pmatrix}. \tag{R.33}$$

Solution: We start from

$$(U_s U_\kappa)^m = M \begin{pmatrix} e^{2im\varphi} & 0 \\ 0 & e^{-2im\varphi} \end{pmatrix} M^{-1} \tag{R.34}$$

with

$$M = \begin{pmatrix} -i & i \\ e^{i\varphi} & e^{-i\varphi} \end{pmatrix}; \quad M^{-1} = \frac{1}{2i \cos \varphi} \begin{pmatrix} -e^{-i\varphi} & i \\ e^{i\varphi} & i \end{pmatrix}. \tag{R.35}$$

It follows that:

$$(U_s U_\kappa)^m = \frac{1}{2i \cos \varphi} \begin{pmatrix} -i & i \\ e^{i\varphi} & e^{-i\varphi} \end{pmatrix} \begin{pmatrix} e^{2im\varphi} & 0 \\ 0 & e^{-2im\varphi} \end{pmatrix} \begin{pmatrix} -e^{-i\varphi} & i \\ e^{i\varphi} & i \end{pmatrix}$$

$$= \frac{1}{2i \cos \varphi} \begin{pmatrix} ie^{i(2m-1)\varphi} + ie^{-i(2m-1)\varphi} & e^{2im\varphi} - e^{-2im\varphi} \\ -e^{2im\varphi} + e^{-2im\varphi} & ie^{i(2m+1)\varphi} + ie^{-i(2m+1)\varphi} \end{pmatrix}$$

$$= \frac{1}{\cos \varphi} \begin{pmatrix} \cos (2m - 1) \varphi & \sin 2m\varphi \\ -\sin 2m\varphi & \cos (2m + 1) \varphi \end{pmatrix}. \tag{R.36}$$

3. Given that

$$\langle \kappa | (U_s U_\kappa)^m | s \rangle = \frac{1}{\cos \varphi} \left( \sin 2m\varphi + \frac{1}{\sqrt{N}} \cos (2m + 1) \varphi \right) ; \qquad (R.37)$$

show that

$$\langle \kappa | (U_s U_\kappa)^m | s \rangle = \sin (2m + 1) \varphi \qquad (R.38)$$

holds.

Solution: With $\sin \varphi = \frac{1}{\sqrt{N}}$ and the relevant theorems for the trigonometric functions, we find:

$$\frac{1}{\cos \varphi} \left( \sin 2m\varphi + \frac{1}{\sqrt{N}} \cos (2m + 1) \varphi \right)$$

$$= \frac{1}{\cos \varphi} (\sin 2m\varphi + \sin \varphi \cos (2m + 1) \varphi)$$

$$= \frac{1}{\cos \varphi} \frac{\sin (2m\varphi) + \sin ((2m + 2) \varphi)}{2}$$

$$= \frac{1}{\cos \varphi} \sin \frac{2m\varphi + (2m + 2) \varphi}{2} \cdot \cos \frac{2m\varphi - (2m + 2) \varphi}{2}$$

$$= \frac{1}{\cos \varphi} \sin (2m + 1) \varphi \cdot \cos \varphi = \sin (2m + 1) \varphi. \qquad (R.39)$$

# Appendix S
# Shor Algorithm

Shor's algorithm serves to decompose very large numbers into their prime factors. We first discuss the classical and then the quantum-mechanical part of the algorithm.

## S.1 Classical Part

Given a number $N$ (odd, not prime) whose prime factorization is to be determined, $N = p_1^{\alpha_1} p_2^{\alpha_2} \ldots p_m^{\alpha_m}$. To accomplish this, we choose at random a number $a$ with $2 \leq a \leq N - 1$, which is relatively prime to $N$ (otherwise we would have found a divisor of $N$), i.e. $gcd\,(a, N) = 1$.[49] We consider

$$a^j \bmod N; \quad j = 0, 1, 2, \ldots \tag{S.1}$$

Beginning with a value $j_p$ of $j$ which depends on $a$ and $N$, $a^j \bmod N$ is periodic with the *order* $r$ (also called the *period*)[50]:

$$a^{j+r} \bmod N = a^j \bmod N \tag{S.2}$$

where $r$ is the smallest number for which this equation is satisfied. Table S.1 shows some examples.

As can be seen, the period can be even or odd; $j_p$ can be 0, but also assumes other values.

With the help of the period $r$ we can determine factors of $N$. This works as follows: We assume that we have determined the period $r$ (actually, this is the business of quantum mechanics) and that it meets the following conditions:

---

[49] $gcd$ is an acronym for 'greatest common divisor'. There are very effective methods for determining the $gcd$.

[50] Some remarks on modular arithmetic are to be found below.

© Springer Nature Switzerland AG 2018
J. Pade, *Quantum Mechanics for Pedestrians 2*, Undergraduate Lecture Notes in Physics, https://doi.org/10.1007/978-3-030-00467-5

**Table S.1** Some examples for $a^j \bmod N$

| $j =$ | 0 | 1 | 2 | 3 | 4 | 5 | 6 | ... | $r$ | $j_p$ |
|---|---|---|---|---|---|---|---|---|---|---|
| $a = 8, N = 21$ | 1 | 8 | 1 | 8 | 1 | 8 | 1 | ... | 2 | 0 |
| $a = 13, N = 35$ | 1 | 13 | 29 | 27 | 1 | 13 | 29 | ... | 4 | 0 |
| $a = 19, N = 35$ | 1 | 19 | 11 | 34 | 16 | 24 | 1 | ... | 6 | 0 |
| $a = 4, N = 63$ | 1 | 4 | 16 | 1 | 4 | 16 | 1 | ... | 3 | 0 |
| $a = 6, N = 63$ | 1 | 6 | 36 | 27 | 36 | 27 | 36 | ... | 2 | 2 |
| $a = 7, N = 63$ | 1 | 7 | 49 | 28 | 7 | 49 | 28 | ... | 3 | 1 |

$$(1) \; r \text{ is even}; \; (2) \; a^{\frac{r}{2}} \bmod N \neq 1. \tag{S.3}$$

As one can show, the inequality $w \geq 1 - \frac{1}{2^{(m-1)}}$ holds for the probability $w$ that these two conditions are satisfied for a number $N = p_1^{\alpha_1} p_2^{\alpha_2} \ldots p_m^{\alpha_m}$. If the conditions (S.3) are not met, we select a different $a$ and start the process again. Numerical examples are given in the exercises.

Under the assumptions of (S.3), one can show that the two expressions

$$d_\pm = gcd\left(a^{\frac{r}{2}} \pm 1, N\right) \tag{S.4}$$

yield factors of $N$. One divides $N$ by these factors and applies, if necessary, the same procedures at $\frac{N}{d_- d_+}$ to identify any other factors.

An example: Let $N = 21$; we choose $a = 8$. Then we have $r = 2$ (see the above table). With the two expressions

$$d_- = gcd\,(7, 21) = 7; \; d_+ = gcd\,(9, 21) = 3 \tag{S.5}$$

we have found the two factors of $21 = 3 \cdot 7$. Other examples are given below and in the exercises.

## S.2 Quantum-Mechanical Part

The quantum-mechanical part is confined to the determination of the period for given $N$ and $a$. We set up two registers, the first register (argument register) of length $m$, and the second (function register) of length $L$. Usually one chooses $N^2 \leq M = 2^m < 2N^2$ and $L \gtrsim \log_2 N$.

In the following, we illustrate each step with the specific example $N = 35, a = 13$ and[51] $r = 4$. For this example, $j_p = 0$, and we restrict the general formalism to this case (which implies $a^r \bmod N = 1$). The extension to $j_p > 0$ is simple, but it leads to

---

[51] In a 'real' problem, of course, one does not know the period; it is a number being sought.

more paperwork. Similarly, in the examples explicitly worked out, we choose for $M$ not the just as the 'usual' value denoted by $M = 2048 = 2^{11}$, but confine ourselves for clarity[52] to $M = 128 = 2^7$.

**First step**: We prepare the state $|0\rangle \otimes |0\rangle \equiv |0\rangle |0\rangle$ (i.e. argument register $\otimes$ function register) and then apply the Hadamard transformation $H_2^{\otimes m}$ to the argument register; this yields the superposition (see Chap. 26):

$$|\varphi_1\rangle = \frac{1}{\sqrt{M}} \sum_{j=0}^{M-1} |j\rangle |0\rangle . \tag{S.6}$$

This state (a product state) contains all $M = 2^m$ numbers from 0 to $M - 1$ at the same time (it must of course be guaranteed that $M > r$ holds).

In our example with $N = 35$, $a = 13$ and $M = 128$, this means that:

$$|\varphi_1\rangle = \frac{1}{\sqrt{128}} (|0\rangle + |1\rangle + \cdots |127\rangle) |0\rangle . \tag{S.7}$$

**Second step**: We modify the function register by the unitary transformation[53] $|j\rangle |0\rangle \rightarrow |j\rangle |a^j \bmod N\rangle$. Then we obtain the total state (it is obviously an entangled state):

$$|\varphi_2\rangle = \frac{1}{\sqrt{M}} \sum_{j=0}^{M-1} |j\rangle |a^j \bmod N\rangle . \tag{S.8}$$

In this way, the quantum computer has calculated *in one passage* the possible values of $a^j \bmod N$ (quantum parallelism). This is the step that gives significant time savings compared to traditional methods.

We still have to read out the relevant information, i.e. the period $r$, but the essential part of the computational work is done. In order to see this more clearly, we re-sort. We know that $a^j \bmod N$ is periodic with period $r$; we can therefore write

$$j = J + kr; \quad J = 0, 1, 2, \ldots, k = 0, 1, 2, \ldots \tag{S.9}$$

The numbers $J$ are called the offset. In our example ($N = 35$, $a = 13$, $r = 4$)), the offset can take on the four values $J = 0, 1, 2, 3$.

Thus we can write:

$$a^j \bmod N = a^{J+kr} \bmod N = a^J \bmod N \text{ independently of } k, \tag{S.10}$$

---

[52]If one applies the Shor algorithm to numbers that are suitable for encryption purposes, one encounters of course some more pitfalls than are seen here. But since we just want to outline the principle of the method here, toy examples will do nicely.

[53]The exact form of the transformation is not important. It is enough to know that we can construct it by suitable combinations of $H$, $\Phi$ and $C$ (Hadamard, phase, CNOT), and that it is reversible.

and it follows that

$$|\varphi_2\rangle = \frac{1}{\sqrt{M}} \sum_J \sum_{k=0}^{s-1} |J + kr\rangle \, |a^J \bmod N\rangle \text{ with } s = 1 + \left[\frac{M-1-J}{r}\right], \quad (S.11)$$

where $\left[\frac{p}{q}\right]$ denotes the integer portion of $\frac{p}{q}$ (also called floor or integer part). The outer sum runs over all possible offsets[54] $J$. Thus, we have a *superposition* of states of the argument register as 'prefactor' of a possible value of the function register.

In our example, $r = 4$, and we obtain for (S.8) the expression:

$$|\varphi_2\rangle = \frac{1}{\sqrt{128}} \sum_{j=0}^{127} |j\rangle \, |13^j \bmod 35\rangle$$

$$= \frac{|0\rangle\,|1\rangle + |1\rangle\,|13\rangle + |2\rangle\,|29\rangle + |3\rangle\,|27\rangle + |4\rangle\,|1\rangle + \cdots}{\sqrt{128}}. \quad (S.12)$$

Re-sorting yields the formulations corresponding to (S.11):

$$|\varphi_2\rangle = \frac{|0\rangle + |4\rangle + |8\rangle + \cdots + |124\rangle}{\sqrt{128}} \, |1\rangle + \frac{|1\rangle + |5\rangle + |9\rangle + \cdots + |125\rangle}{\sqrt{128}} \, |13\rangle$$

$$+ \frac{|2\rangle + |6\rangle + |10\rangle + \cdots + |126\rangle}{\sqrt{128}} \, |29\rangle + \frac{|3\rangle + |7\rangle + |11\rangle + \cdots + |127\rangle}{\sqrt{128}} \, |27\rangle. \quad (S.13)$$

With $M = 128$ and $r = 4$, it follows that $s = \left[\frac{128}{r}\right] = 32$. The crucial point here is that *all* superpositions are 4-periodic—only their offsets are different.

At this point, one could determine the period by measuring often enough the argument register and the function register, and sorting the results according to the different measured values. However, our toy example with $r = 4$ is insofar somewhat misleading in that it is very transparent due to the small numbers used. In 'real' problems, one is dealing with very large numbers, and accordingly the periods can be very large.[55] It would be an important aid if each function register had some superposition of the *same* states (apart from phases); in (S.11), the function registers differ due to the different offsets. This crucial simplification is achieved with the quantum Fourier transform (QFT, see Appendix H, Vol. 1, 'Fourier transforms and the delta function').

---

[54]The length $L$ of the function register must be greater than or equal to the number of different offsets; therefore, we have the condition $L \gtrsim \log_2 N$.

[55]Even toy examples with 'small' numbers can lead to remarkable period lengths. As an example, we consider the prime factorization of $N = 2149841$. For $a = 3$, we obtain the period $r = 213330$. With $gcd\left(3^{106665} - 1, 2149841\right) = 131$ and $gcd\left(3^{106665} + 1, 2149841\right) = 16411$, the factorization $N = 2149841 = 131 \cdot 16411$ follows.

**Third step**: We apply the QFT (acting on the argument register):

$$U_{QFT} = \frac{1}{\sqrt{M}} \sum_{n,l=0}^{M-1} e^{2\pi i \frac{nl}{M}} |l\rangle \langle n| \tag{S.14}$$

to $|\varphi_2\rangle$ in (S.11) and obtain (due to $\langle n| j \rangle = \delta_{jn}$) the relationship:

$$\begin{aligned}
|\varphi_3\rangle &= U_{QFT} |\varphi_2\rangle \\
&= \frac{1}{\sqrt{M}} \sum_{n,l=0}^{M-1} e^{2\pi i \frac{nl}{M}} |l\rangle \langle n| \frac{1}{\sqrt{M}} \sum_{J} \sum_{k=0}^{s-1} |J + kr\rangle |a^J \bmod N\rangle \\
&= \frac{1}{M} \sum_{J} \sum_{k=0}^{s-1} \sum_{n,l=0}^{M-1} e^{2\pi i \frac{nl}{M}} |l\rangle \, \delta_{n,J+kr} \, |a^J \bmod N\rangle \\
&= \frac{1}{M} \sum_{J} \sum_{l=0}^{M-1} \sum_{k=0}^{s-1} e^{2\pi i \frac{(J+kr)l}{M}} |l\rangle \, |a^J \bmod N\rangle .
\end{aligned} \tag{S.15}$$

We have thus achieved our goal that the same states always occur in the argument registers, independently of the offset; in the expression (S.11) for $|\varphi_2\rangle$, this was still not the case. In order to formulate this more clearly, we use the fact that $|a^J \bmod N\rangle$ in (S.15) does not depend on $k$, and we thus can carry out the sum over $k$. With (see the exercises for the calculation):

$$c(J, r, s; l) := \sum_{k=0}^{s-1} e^{2\pi i \frac{(J+kr)l}{M}} = \begin{cases} e^{2\pi i \frac{Jl}{M}} e^{\pi i \frac{rl(s-1)}{M}} \frac{\sin \pi \frac{rl}{M} s}{\sin \pi \frac{rl}{M}} & ; \quad \frac{rl}{M} \notin \mathbb{N}_0 \\ e^{2\pi i \frac{Jl}{M}} \cdot s & \quad \frac{rl}{M} \in \mathbb{N}_0 \end{cases} \tag{S.16}$$

we can write (S.15) as

$$|\varphi_3\rangle = \frac{1}{M} \sum_{J} \sum_{l=0}^{M-1} c(J, r, s; l) |l\rangle \, |a^J \bmod N\rangle ; \quad s = 1 + \left[\frac{M-1-J}{r}\right] . \tag{S.17}$$

The crucial point is that $|c(J, r, s; l)|$, according to (S.16), is not only *independent* of $J$ (i.e. of the function register), but also has *distinct maxima* for particular values of $l$, namely for $l$ values which are multiples of $\frac{M}{r}$ (or at least approximate multiples). In other words, the QFT (S.14) causes an *amplitude amplification*, which leads to the result that measurements yield mainly $l$ values with $l \approx n \cdot \frac{M}{r}$ with a high probability.

We illustrate these findings by means of the example $M = 128$, $r = 4$. It follows first of all that

$$s = 1 + \left[\frac{128-1-J}{4}\right] = 1 + 31 + \left[\frac{3-J}{4}\right] = 32 \text{ due to } J = 0, 1, 2, 3,$$

$$\tag{S.18}$$

and from this (note: $0 \leq l \leq M - 1$) with (S.16):

$$|c(J, r, s; l)| = \left| \frac{\sin \pi l}{\sin \frac{\pi l}{32}} \right| = \begin{cases} 32 \\ 0 \end{cases} \text{ for } \begin{matrix} l = 0, 32, 64, 96 \\ \text{otherwise.} \end{matrix} \qquad (S.19)$$

Accordingly, we see in the argument register only states $|n \cdot 32\rangle$; all others vanish exactly. In this transparent example, we see immediately that the period is $r = 4$; we can formally determine it by $\frac{rl}{M} \in \mathbb{N}_0$ or $r = \frac{M}{l} n_l$. Since 32 is the greatest common divisor of $(32, 64, 96)$, the period is given by $\frac{128}{32} = 4$.

A similar behavior always results if the period $r$ is a power of 2 (because of $M = 2^m$, $M$ is also a power of two). Only the states $|n \cdot \frac{M}{r}\rangle$ with $n \in \mathbb{N}_0$ survive, all others vanish exactly. If $r$ is not a power of two, the 'unwanted' states do not disappear exactly, but their amplitudes are much smaller than those of the desired states. As an example, we choose $M = 128$ and $r = 6$. In this case, we find:

$$s = 1 + \left[ \frac{128 - 1 - J}{6} \right] = 22 + \left[ \frac{1 - J}{6} \right] = \begin{cases} 22 \text{ for } J = 0, 1 \\ 21 \text{ for } J = 2, 3, 4, 5 \end{cases} \qquad (S.20)$$

and thus

$$|c(J, r, s; l)| = \begin{cases} \left| \frac{\sin \pi \frac{3l}{64} s}{\sin \pi \frac{3l}{64}} \right| ; & \frac{3l}{64} \notin \mathbb{N}_0 \\ s & \frac{3l}{64} \in \mathbb{N}_0 \end{cases} ; \; l = 0, 1, 2, \ldots, 127. \qquad (S.21)$$

Obviously, $\frac{3l}{64}$ is an integer only for $l = 0$ and $l = 64$, but this holds true approximately also for the values $l = 21$ and $22$, $42$ and $43$, $85$ and $86$, $106$ and $107$. This manifests itself in the fact that the corresponding coefficients $|c(J, r, s; l)|$ are relatively large; see Fig. S.1. Accordingly, when measuring one obtains one of these states with a high probability. Thus, we have here another non-trivial example of amplitude amplification, as discussed in Chap. 26. From Fig. S.1, the period 6 can be read out directly; formally, we can calculate it as $r \approx n \cdot \frac{M}{l}$. A more detailed analysis of this example is found in the exercises.

We summarize: *Before* the QFT, the states of the argument register depend on the function register, as is seen in the example of (S.13). *After* the QFT, the argument registers contain superpositions of the *same* states, independently of the function register; cf. (S.15). In addition, the QFT causes the measurement probabilities for states in the argument registers to become unequally distributed, as in (S.13), but with pronounced maxima at $l \approx n\frac{M}{r}$ with $n = 0, 1, 2, \ldots$. In other words, with a high probability (with certainty for $r = 2^N$), *each* measurement yields one of those states in the argument register from which the period can be determined directly. In this manner, the quantum computer determines the periodicity of a function in a few steps.

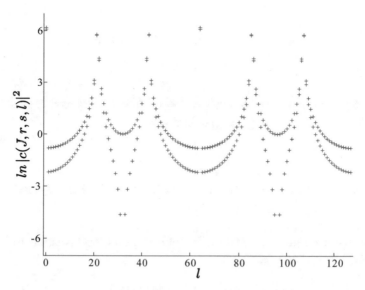

**Fig. S.1** The coefficient (S.21) for $M = 128$ and $r = 6$. *Red crosses* for $s = 22$, *blue crosses* for $s = 21$

As we said above, this works with very high probability. Should it not work in a particular case (or if the conditions (S.3) are not satisfied), one chooses another number $a$ and lets the algorithm run through once more.

## S.3   Supplement on Modular Arithmetic

We present here some information on modular arithmetic. We have

$$a = p \cdot n + r \rightarrow a \bmod n = r \text{ or } a \bmod n = a - \left[\frac{a}{n}\right] n, \qquad (S.22)$$

where all numbers are $\in \mathbb{N}_0$; $\left[\frac{a}{n}\right]$ is the integer part. Example: $a = 51, n = 7$.

$$51 \bmod 7 = (7 \cdot 7 + 2) \bmod 7 = 2$$
$$51 \bmod 7 = 51 - \left[\frac{51}{7}\right] \cdot 7 = 51 - 7 \cdot 7 = 2. \qquad (S.23)$$

Modular addition and multiplication are defined by

$$(a + b) \bmod n = (a \bmod n + b \bmod n) \bmod n \qquad (S.24)$$

and
$$ab \bmod n = [(a \bmod n) \cdot (b \bmod n)] \bmod n. \tag{S.25}$$

Example:

$$(52 + 34) \bmod 7 = 86 \bmod 7 = 2 = (52 \bmod 7 + 34 \bmod 7) \bmod 7$$
$$= (3 + 6) \bmod 7 = 2 \tag{S.26}$$

and

$$(52 \cdot 34) \bmod 7 = 1768 \bmod 7 = 4 = (52 \bmod 7 \cdot 34 \bmod 7) \bmod 7$$
$$= (3 \cdot 6) \bmod 7 = 4. \tag{S.27}$$

If the calculations refer to only *one* $n$, the following shorthand notation is often used in quantum information applications:

$$a \oplus b := (a + b) \bmod n \tag{S.28}$$

where here the symbol $\oplus$ of course does *not* denote the direct sum of vectors.
    Because of (S.25), powers can be calculated recursively:

$$a^{r+1} \bmod N = [(a^r \bmod N) \cdot (a \bmod N)] \bmod N. \tag{S.29}$$

## S.4  Exercises

1. Given two integers $n$ and $a$, calculate the period $r = a \bmod n$, and with (S.4) the factors of $n$.

    (a) $n = 35$ and $a = 13$
        Solution: We have

| | | | |
|---|---|---|---|
| $13^0$ | $= 1$ | $= 0 \cdot 35 + 1$ | $= 1 \ (\bmod 35)$ |
| $13^1$ | $= 13$ | $= 0 \cdot 35 + 13$ | $= 13 \ (\bmod 35)$ |
| $13^2$ | $= 169$ | $= 4 \cdot 35 + 29$ | $= 29 \ (\bmod 35)$ |
| $13^3$ | $= 2197$ | $= 62 \cdot 35 + 27$ | $= 27 \ (\bmod 35)$ |
| $13^4$ | $= 28561$ | $= 816 \cdot 35 + 1$ | $= 1 \ (\bmod 35)$ |
| $13^5$ | $= 371293$ | $= 10608 \cdot 35 + 13$ | $= 13 \ (\bmod 35)$ |

Hence, the period is $r = 4$. It follows that $a^{r/2} = 13^2 = 169$ and thus

$$\gcd(168, 35) = 7; \ \gcd(170, 35) = 5; \ 7 \cdot 5 = 35 \tag{S.30}$$

**Table S.2** Solution of exercise 2 (a), $N = 21$

| $a$ | $r$ | $a^{r/2}$ | $gcd(a^{r/2} - 1, 15)$ | $gcd(a^{r/2} + 1, 15)$ |
|---|---|---|---|---|
| 2 | 6 | $2^3 = 8$ | $(7, 21) = 7$ | $(9, 21) = 3$ |
| 4 | 3 | | | |
| 5 | 6 | $5^3 = 125$ | $(124, 21) = 1$ | $(126, 21) = 21$ |
| 8 | 2 | $8^1 = 8$ | $(7, 21) = 7$ | $(9, 21) = 3$ |
| 10 | 6 | $10^3 = 1000$ | $(999, 21) = 3$ | $(1001, 21) = 7$ |
| 11 | 6 | $11^3 = 1331$ | $(1330, 21) = 7$ | $(1332, 21) = 3$ |
| 13 | 2 | $13^1 = 13$ | $(12, 21) = 3$ | $(14, 21) = 7$ |
| 16 | 3 | | | |
| 17 | 6 | $17^3 = 4913$ | $(4912, 21) = 1$ | $(4914, 21) = 21$ |
| 19 | 6 | $19^3 = 6859$ | $(6858, 21) = 3$ | $(6860, 21) = 7$ |
| 20 | 2 | $20^1 = 20$ | $(19, 21) = 1$ | $(21, 21) = 21$ |

5 blanks, 6 hits, $w = \frac{6}{11}$

(b) $n = 437$ and $a = 94$

Solution: The period is $r = 22$ (calculation best by computer). It follows that

$$gcd\left(94^{11} - 1, 437\right) = 23; \quad gcd\left(94^{11} + 1, 437\right) = 19; \quad 23 \cdot 19 = 437. \tag{S.31}$$

2. We denote by $w$ the probability that the conditions (S.3) are satisfied for the number $N = p_1^{\alpha_1} p_2^{\alpha_2} \dots p_m^{\alpha_m}$ (i.e. $r$ is even and $a^{\frac{r}{2}} \bmod N \neq 1$). Then the inequality $w \geq 1 - \frac{1}{2^{m-1}}$ applies. Particularly for $m = 2$ (the 'hardest' case), it follows that $w \geq 1 - \frac{1}{2^{m-1}} = \frac{1}{2}$. Check the findings for $N = 21$ and $N = 33$.

   The solution is found in Table S.2. As is seen, we have 5 blanks, 6 hits, $w = \frac{6}{11}$.
   The solution is found in Table S.3. As is seen, we have 9 blanks, 10 hits, $w = \frac{10}{19}$.

3. For which values of $l$ is the sum

$$A = \left| \sum_{k=0}^{s-1} e^{2\pi i \frac{krl}{M}} \right| ; \quad l \in \mathbb{N}_0; \; r, M \in \mathbb{N} \tag{S.32}$$

maximal?

Solution: For $\frac{rl}{M} \in \mathbb{N}_0$, we have for each summand $e^{2\pi i \frac{krl}{M}} = 1$; it follows $A = s$. For $\frac{rl}{M} \notin \mathbb{N}_0$, each summand has magnitude 1, but there are different phases in dependence on $k$, such that the absolute value of the sum *must* be smaller than $s$.

4. Calculate

$$c(J, r, s; l) = \sum_{k=0}^{s-1} e^{2\pi i \frac{(J+kr)l}{M}}. \tag{S.33}$$

Solution: We have

**Table S.3** Solution of exercise 2 (b), $N = 33$

| $a$ | $r$ | $a^{r/2}$ | $gcd(a^{r/2} - 1, 15)$ | $gcd(a^{r/2} + 1, 15)$ |
|---|---|---|---|---|
| 2 | 10 | $2^5 = 32$ | $(31, 33) = 1$ | $(33, 33) = 33$ |
| 4 | 5 | | | |
| 5 | 10 | $5^5 = 3125$ | $(3124, 33) = 11$ | $(3126, 33) = 3$ |
| 7 | 10 | $7^5 = 16807$ | $(16806, 33) = 3$ | $(16808, 33) = 11$ |
| 8 | 10 | $8^5 = 32768$ | $(32767, 33) = 1$ | $(32769, 33) = 33$ |
| 10 | 2 | $10^1 = 10$ | $(9, 33) = 3$ | $(11, 33) = 11$ |
| 13 | 10 | $13^5 = 371293$ | $(371292, 33) = 3$ | $(371294, 33) = 11$ |
| 14 | 10 | $14^5 = 537824$ | $(537823, 33) = 11$ | $(537825, 33) = 3$ |
| 16 | 5 | | | |
| 17 | 10 | $17^5 = 1419857$ | $(1419856, 33) = 1$ | $(1419858, 33) = 33$ |
| 19 | 10 | $19^5 = 2476099$ | $(2476098, 33) = 3$ | $(2476100, 33) = 11$ |
| 20 | 10 | $20^5 = 3200000$ | $(3199999, 33) = 11$ | $(3200001, 33) = 3$ |
| 23 | 2 | $23^1 = 23$ | $(22, 33) = 11$ | $(24, 33) = 3$ |
| 25 | 5 | | | |
| 26 | 10 | $26^5 = 11881376$ | $(11881375, 33) = 11$ | $(11881377, 33) = 3$ |
| 28 | 10 | $28^5 = 17210368$ | $(17210367, 33) = 3$ | $(17210369, 33) = 11$ |
| 29 | 10 | $29^5 = 20511149$ | $(20511148, 33) = 1$ | $(20511150, 33) = 33$ |
| 31 | 5 | | | |
| 32 | 2 | $32^1 = 32$ | $(31, 33) = 1$ | $(33, 33) = 33$ |

9 blanks, 10 hits, $w = \frac{10}{19}$

$$c(J, r, s; l) = \sum_{k=0}^{s-1} e^{2\pi i \frac{(J+kr)l}{M}} = e^{2\pi i \frac{Jl}{M}} \sum_{k=0}^{s-1} \left( e^{2\pi i \frac{rl}{M}} \right)^k = e^{2\pi i \frac{Jl}{M}} \frac{1 - e^{2\pi i \frac{rl}{M} s}}{1 - e^{2\pi i \frac{rl}{M}}}.$$
$$(S.34)$$

With

$$e^{2\pi i \frac{Jl}{M}} \frac{1 - e^{2\pi i \frac{rl}{M} s}}{1 - e^{2\pi i \frac{rl}{M}}} = e^{2\pi i \frac{Jl}{M}} \frac{e^{\pi i \frac{rl}{M} s}}{e^{\pi i \frac{rl}{M}}} \frac{e^{-\pi i \frac{rl}{M} s} - e^{\pi i \frac{rl}{M} s}}{e^{-\pi i \frac{rl}{M}} - e^{\pi i \frac{rl}{M}}}$$

$$= e^{2\pi i \frac{Jl}{M}} e^{\pi i \frac{rl}{M}(s-1)} \frac{\sin \pi \frac{rl}{M} s}{\sin \pi \frac{rl}{M}} = e^{\pi i l \frac{2J + r(s-1)}{M}} \frac{\sin \pi \frac{rl}{M} s}{\sin \pi \frac{rl}{M}}, \tag{S.35}$$

it follows that

$$c(J, r, s; l) = \begin{cases} e^{\pi i l \frac{2J + r(s-1)}{M}} \dfrac{\sin \pi \frac{rl}{M} s}{\sin \pi \frac{rl}{M}} & ; \quad \dfrac{rl}{M} \notin \mathbb{N}_0 \\ e^{\pi i l \frac{2J}{M}} \cdot s & \quad \dfrac{rl}{M} \in \mathbb{N}_0. \end{cases} \tag{S.36}$$

The lower expression follows from the upper one by L'Hôpital's rule.
For the absolute value, we have the simple relationship

$$|c(J, r, s; l)| = \begin{cases} \left| \dfrac{\sin \pi \frac{rl}{M} s}{\sin \pi \frac{rl}{M}} \right| & ; \quad \dfrac{rl}{M} \notin \mathbb{N}_0 \\ s & \quad \dfrac{rl}{M} \in \mathbb{N}_0. \end{cases} \tag{S.37}$$

Note: From the previous problem, we know that $\left| \dfrac{\sin \pi \frac{rl}{M} s}{\sin \pi \frac{rl}{M}} \right| \le s$. This will be shown in the next exercise by a different route.

5. Prove the inequality

$$|\sin(ny)| \le n |\sin y| ; \quad n = 1, 2, 3... \tag{S.38}$$

Solution: We use induction. For $n = 1$, the inequality is evidently satisfied. We assume that it is also satisfied for $n$. Then we have for $n + 1$ according to the addition theorems for trigonometric functions:

$$|\sin((n + 1)y)| = |\sin(ny) \cos y + \cos(ny) \sin y| . \tag{S.39}$$

We estimate the right side:

$$|\sin(ny) \cos y + \cos(ny) \sin y| \le |\sin(ny) \cos y| + |\cos(ny) \sin y| . \tag{S.40}$$

For the first term on the right, we insert $|\sin(ny)| \le n |\sin y|$; it follows that

$$|\sin((n + 1)y)| \le n |\sin y| |\cos y| + |\cos(ny)| |\sin y| . \tag{S.41}$$

On the right side, we factor out $|\sin y|$ and use $|\cos y| \le 1$, $|\cos(ny)| \le 1$. We then have

$$|\sin((n + 1)y)| \le (n + 1) |\sin y| , \tag{S.42}$$

and thus the statement is proven.

6. Formulate the single steps of the Shor algorithm explicitly for $N = 35$ and $a = 13$. Choose as in the text $M = 2^m = 128$.

Solution: From the first step, we obtain

$$|\varphi_1\rangle = \frac{1}{\sqrt{128}} \left( |0\rangle + |1\rangle + \cdots |127\rangle \right) |0\rangle . \tag{S.43}$$

Then we transform unitarily and find

$$|\varphi_2\rangle = \frac{1}{\sqrt{128}} \sum_{j=0}^{127} |j\rangle \left| 13^j \bmod 35 \right\rangle . \tag{S.44}$$

Because of

$$13^0 \bmod 35 = 1; \quad 13^1 \bmod 35 = 13; \\ 13^2 \bmod 35 = 29; \quad 13^3 \bmod 35 = 27; \quad 13^4 \bmod 35 = 1, \tag{S.45}$$

it follows that $r = 4$. Thus we obtain in detail

$$|\varphi_2\rangle = \frac{|0\rangle |1\rangle + |1\rangle |13\rangle + |2\rangle |29\rangle + |3\rangle |27\rangle + |4\rangle |1\rangle + \cdots}{\sqrt{128}} . \tag{S.46}$$

Re-sorting gives

$$|\varphi_2\rangle = \frac{|0\rangle + |4\rangle + |8\rangle + \cdots + |124\rangle}{\sqrt{128}} |1\rangle + \frac{|1\rangle + |5\rangle + |9\rangle + \cdots + |125\rangle}{\sqrt{128}} |13\rangle \\ + \frac{|2\rangle + |6\rangle + |10\rangle + \cdots + |126\rangle}{\sqrt{128}} |29\rangle + \frac{|3\rangle + |7\rangle + |11\rangle + \cdots + |127\rangle}{\sqrt{128}} |27\rangle . \tag{S.47}$$

The QFT acting on the argument register

$$U_{QFT} = \frac{1}{\sqrt{128}} \sum_{l,k=0}^{127} e^{2\pi i \frac{lk}{128}} |k\rangle \langle l| \tag{S.48}$$

leads to the state $|\varphi_3\rangle$ (written in detail, i.e. without using $c (J, r, s; m)$):

$$|\varphi_3\rangle = U_{QFT} |\varphi_2\rangle$$

$$= \frac{1}{128} \left( \sum_{k=0}^{127} e^{2\pi i \frac{0k}{128}} |k\rangle + \sum_{k=0}^{127} e^{2\pi i \frac{4k}{128}} |k\rangle + \sum_{k=0}^{127} e^{2\pi i \frac{8k}{128}} |k\rangle + \cdots \right.$$

$$\left. + \sum_{k=0}^{127} e^{2\pi i \frac{124k}{128}} |k\rangle \right) |1\rangle$$

$$+ \frac{1}{128} \left( \sum_{k=0}^{127} e^{2\pi i \frac{k}{128}} |k\rangle + \sum_{k=0}^{127} e^{2\pi i \frac{5k}{128}} |k\rangle + \sum_{k=0}^{127} e^{2\pi i \frac{9k}{128}} |k\rangle + \cdots$$

$$+ \sum_{k=0}^{127} e^{2\pi i \frac{125k}{128}} |k\rangle \Bigg) |13\rangle$$

$$+ \frac{1}{128} \Bigg( \sum_{k=0}^{127} e^{2\pi i \frac{2k}{128}} |k\rangle + \sum_{k=0}^{127} e^{2\pi i \frac{6k}{128}} |k\rangle + \sum_{k=0}^{127} e^{2\pi i \frac{10k}{128}} |k\rangle + \cdots$$

$$+ \sum_{k=0}^{127} e^{2\pi i \frac{126k}{128}} |k\rangle \Bigg) |29\rangle$$

$$+ \frac{1}{128} \Bigg( \sum_{k=0}^{127} e^{2\pi i \frac{3k}{128}} |k\rangle + \sum_{k=0}^{127} e^{2\pi i \frac{7k}{128}} |k\rangle + \sum_{k=0}^{127} e^{2\pi i \frac{11k}{128}} |k\rangle + \cdots$$

$$+ \sum_{k=0}^{127} e^{2\pi i \frac{127k}{128}} |k\rangle \Bigg) |27\rangle ; \tag{S.49}$$

and re-sorting gives

$$|\varphi_3\rangle = \frac{1}{128} \sum_{k=0}^{127} \left[ e^{2\pi i \frac{0k}{128}} + e^{2\pi i \frac{4k}{128}} + e^{2\pi i \frac{8k}{128}} + \ldots + e^{2\pi i \frac{124k}{128}} \right] |k\rangle |1\rangle$$

$$+ \frac{1}{128} \sum_{k=0}^{127} e^{2\pi i \frac{k}{128}} \left[ e^{2\pi i \frac{0k}{128}} + e^{2\pi i \frac{4k}{128}} + e^{2\pi i \frac{8k}{128}} + \ldots + e^{2\pi i \frac{124k}{128}} \right] |k\rangle |13\rangle$$

$$+ \frac{1}{128} \sum_{k=0}^{127} e^{2\pi i \frac{2k}{128}} \left[ e^{2\pi i \frac{0k}{128}} + e^{2\pi i \frac{4k}{128}} + e^{2\pi i \frac{8k}{128}} + \ldots + e^{2\pi i \frac{124k}{128}} \right] |k\rangle |29\rangle$$

$$+ \frac{1}{128} \sum_{k=0}^{127} e^{2\pi i \frac{3k}{128}} \left[ e^{2\pi i \frac{0k}{128}} + e^{2\pi i \frac{4k}{128}} + e^{2\pi i \frac{8k}{128}} + \ldots + e^{2\pi i \frac{124k}{128}} \right] |k\rangle |27\rangle . \tag{S.50}$$

The square brackets are nonzero only for $k = 0, 32, 64, 96$ (note that $k \leq 127$), since (see also (S.35)) we have:

$$\sum_{l=0}^{31} e^{2\pi i \frac{4lk}{128}} = e^{\pi i k \frac{31}{32}} \cdot \frac{\sin \pi k}{\sin \frac{\pi k}{32}}$$

$$= 32 \cdot \left( \delta_{k,0} + \delta_{k,32} + \delta_{k,64} + \delta_{k,96} \right) . \tag{S.51}$$

Thus it follows that

$$|\varphi_3\rangle = \frac{1}{4} \sum_{k=0}^{127} e^{\pi i k \frac{31}{32}} \cdot \left( \delta_{k,0} + \delta_{k,32} + \delta_{k,64} + \delta_{k,96} \right) |k\rangle |1\rangle$$

$$+ \frac{1}{4} \sum_{k=0}^{127} e^{2\pi i \frac{k}{128}} \cdot \left( \delta_{k,0} + \delta_{k,32} + \delta_{k,64} + \delta_{k,96} \right) |k\rangle \, |13\rangle$$

$$+ \frac{1}{4} \sum_{k=0}^{127} e^{2\pi i \frac{2k}{128}} \cdot \left( \delta_{k,0} + \delta_{k,32} + \delta_{k,64} + \delta_{k,96} \right) |k\rangle \, |29\rangle$$

$$+ \frac{1}{4} \sum_{k=0}^{127} e^{2\pi i \frac{3k}{128}} \cdot \left( \delta_{k,0} + \delta_{k,32} + \delta_{k,64} + \delta_{k,96} \right) |k\rangle \, |27\rangle , \qquad \text{(S.52)}$$

or, explicitly,

$$|\varphi_3\rangle = \frac{|0\rangle + |32\rangle + |64\rangle + |96\rangle}{4} |1\rangle$$

$$+ \frac{|0\rangle + i \, |32\rangle - |64\rangle - i \, |96\rangle}{4} |13\rangle$$

$$+ \frac{|0\rangle - |32\rangle + |64\rangle - |96\rangle}{4} |29\rangle$$

$$+ \frac{|0\rangle - i \, |32\rangle - |64\rangle + i \, |96\rangle}{4} |27\rangle . \qquad \text{(S.53)}$$

Each measurement of the argument register yields one of the values 0, 32, 64, 96, and this independently of the value of the function register. The greatest common divisor of 32, 64, 96 is 32; hence the period is $r = \frac{128}{32} = 4$.

7. Work through the Shor algorithm for $N = 35$ and $a = 19$. Choose again $M = 128$. Solution: In the first step, we have

$$|\varphi_1\rangle = \frac{1}{\sqrt{128}} \sum_{j=0}^{127} |j\rangle \, |0\rangle , \qquad \text{(S.54)}$$

and in the second:

$$|\varphi_2\rangle = \frac{1}{\sqrt{128}} \sum_{j=0}^{127} |j\rangle \, \left| 19^j \bmod 35 \right\rangle . \qquad \text{(S.55)}$$

The period is $r = 6$ due to

| | | | |
|---|---|---|---|
| $19^0 \bmod 35 = 1$ | $19^1 \bmod 35 = 19$ | $19^2 \bmod 35 = 11$ | $19^3 \bmod 35 = 34$ |
| $19^4 \bmod 35 = 16$ | $19^5 \bmod 35 = 24$ | $19^6 \bmod 35 = 1$ | $19^7 \bmod 35 = 19$ |

Re-sorting yields

$$|\varphi_2\rangle = \frac{1}{\sqrt{128}} \sum_J \sum_{k=0}^{s-1} |J + 6k\rangle \left|19^J \bmod 35\right\rangle \quad \text{with } s = 1 + \left[\frac{127 - J}{6}\right].$$
(S.56)

The offset can take on the values $J = 0, 1, 2, 3, 4, 5$; correspondingly, we have $s = 22$ for $J = 0, 1$ and $s = 21$ for $J = 2, 3, 4, 5$.

In the third step, we transform the state $|\varphi_2\rangle$ via QFT to the state $|\varphi_3\rangle$:

$$|\varphi_3\rangle = \frac{1}{128} \sum_J \sum_{l=0}^{127} c\,(J, 6, s; l)\, |l\rangle \left|19^J \bmod 35\right\rangle$$
(S.57)

with

$$|c\,(J, 6, s; l)| = \begin{cases} \frac{\sin \pi \frac{3l}{64} s}{\sin \pi \frac{3l}{64}} & \frac{3l}{64} \notin \mathbb{N}_0 \\ s & \frac{3l}{64} \in \mathbb{N}_0 \end{cases} \ ; \ s = \begin{cases} 22 \\ 21 \end{cases} \text{for } J = \begin{matrix} 0, 1 \\ 2, 3, 4, 5 \end{matrix}.$$
(S.58)

We obtain particularly large values if $\frac{3l}{64}$ is (at least approximately) an integer. We obtain an integer for $l = 0$ and $l = 64$, but $\frac{3l}{64}$ is also approximately an integer for $l = 21$ and $22$, $42$ and $43$, $85$ and $86$, $106$ and $107$, and thus $|c\,(J, 6, s; l)|$ is also comparatively large. A numerical example for $s = 22$ illustrates this:

| $l =$ | 40 | 41 | 42 | 43 | 44 | 45 | $\cdots$ | 64 |
|---|---|---|---|---|---|---|---|---|
| $|c\,(J, 6, 22; l)|^2 =$ | 3, 4 | 10, 1 | 72, 0 | 323, 0 | 22, 4 | 8, 1 | $\cdots$ | 1024 |

If we calculate the period by $r \approx n \cdot \frac{M}{l}$, we find for the special $l$ values;

| $l =$ | 21 | 22 | 42 | 43 | 64 | 85 | 86 | 106 | 107 |
|---|---|---|---|---|---|---|---|---|---|
| $\frac{r}{n} =$ | 6, 09 | 5, 82 | 3, 05 | 2, 98 | 2 | 1, 51 | 1, 49 | 1, 21 | 1, 20 |

From the last row, $r = 6$ results, so that we have determined the period by using the measurement data.

# Appendix T
# The Gleason Theorem

Gleason's theorem addresses the question of how we can define probabilities in quantum mechanics.

We assume that a system is in state $|\psi\rangle$; the density operator is given by $\rho = |\psi\rangle\langle\psi|$. Furthermore, we want to measure a property which we represent by the projection operator $P$. We denote by $w_P$ the probability that the system in the state $|\psi\rangle$ yields the property associated with $P$ in a measurement, or 'has' this property. Then, as we have derived in Chap. 22, $w_P$ is given by

$$w_P = \langle P \rangle = tr\,(\rho P)\,. \tag{T.1}$$

The question is whether one can define the probabilities in a quite different way from (T.1). To state the question more precisely, we require as usual of the probabilities $w_P$ the following properties:

$$0 \leq w_P \leq 1 \text{ for all } P \text{ in the Hilbert space}$$
$$w_P\,(0) = 0; \;\; w_P\,(1) = 1; \;\; w_P\left(\sum_{i=1}^{\infty} P_i\right) = \sum_{i=1}^{\infty} w_P\,(P_i)\,. \tag{T.2}$$

Under these conditions, the theorem of Gleason (1957) states that on a Hilbert space of dimension $\geq 3$, the only possible probability measures are described by (T.1).[56] Since we want to use a formalism that applies for all dimensions, the restriction of the theorem to dim $\geq 3$ is irrelevant, and we can say that all possible probability measures which can be defined in $\mathcal{H}$ are generated by the density operators of pure and mixed states.

It was soon recognized (by John Bell, among others) that Gleason's theorem is in conflict with our notions of realism. This is mainly because, according to Gleason's

---

[56]Perhaps this theorem thus provides a deeper reason for the special significance of density operators in quantum mechanics.

© Springer Nature Switzerland AG 2018
J. Pade, *Quantum Mechanics for Pedestrians 2*, Undergraduate Lecture
Notes in Physics, https://doi.org/10.1007/978-3-030-00467-5

theorem, the assignment of probabilities to all possible properties in a Hilbert space must be continuous, i.e. all vectors in the space must be mapped continuously into the interval [0, 1]. On the other hand, we understand the projectors $P$ as a representation of yes-no observables, i.e. we can say of any property whether the system in fact possesses it or not. This results in a probability function that maps all $P$ to 0 or 1, i.e. a discontinuous mapping. The Kochen–Specker theorem deals with this contradiction; see Chap. 27.

# Appendix U
# What is Real? Some Quotations

The questions 'What is real? What is reality? What is the nature of our knowledge of things?'[57] have preoccupied mankind from time immemorial; there are whole libraries on the subject. In the following, we have collected some quotes which are not meant as an attempt at a systematic exposition, but are simply incidental findings. The intention is rather to illustrate different positions—sometimes with wink of the eye.

The quotes are arranged by year of birth of the author (except the last one). Since the small collection has a rather casual character, and moreover a few quotes are probably incorrectly attributed to their 'authors', we dispense with detailed references.

1. 570 BC, Xenophanes, Greek philosopher,
   "And of course the clear and certain truth no man has seen nor will there be anyone who knows about the gods and what I say about all things. For even if, in the best case, one happened to speak just of what has been brought to pass, still he himself would not know. But opinion is allotted to all."

2. 460 BC, Democritus, Greek philosopher,
   "By convention sweet and by convention bitter, by convention hot, by convention cold, by convention color; but in reality atoms and void."

3. 428 BC, Plato, Greek philosopher,
   A precise quote is missing, but since Plato's allegory of the cave is so well known and fundamental, it should be briefly summarized. Some people are living in a cave. They can look only at a wall, and behind them a fire is burning. Items that

---

[57]Occasionally, the terms 'actuality' and 'reality' are used with different meanings. A distinction may boil down to the usage that *actuality* encompasses all objectively true statements, regardless of whether they are known or apparent to us at all, while *reality* includes any statements that we believe to be true (reality is what we perceive, actuality is what truly is). Another distinction is based on etymology: reality (Latin res = thing) refers to the materiality, actuality (to act) to the aspect of interaction or cause-effect. Since we want simply to illustrate here several different points of view and have no high philosophical aspirations, we accept the usage of common language which treats reality and actuality as synonyms.

© Springer Nature Switzerland AG 2018
J. Pade, *Quantum Mechanics for Pedestrians 2*, Undergraduate Lecture
Notes in Physics, https://doi.org/10.1007/978-3-030-00467-5

are carried past between the fire and the backs of the people cast their shadows on the wall. The people now know nothing about the items themselves, but only their shadows, so they take them for the 'real' world. Just like these people in the cave, we see only a semblance of true existence (i.e. the Platonic ideals); only philosophy can lead us to a perception of the 'real'. (Perhaps Plato today would refer to the virtual world of television or computer screens instead of the shadows on a cave's wall...). The influence of Plato on philosophy was immense for many centuries. In the words of Stephen Jay Gould: "The spirit of Plato dies hard. We have been unable to escape the philosophical tradition that what we can see and measure in the world is merely the superficial and imperfect representation of an underlying reality."

4. 1303, Bridget of Sweden, Swedish mystic,
"Although a blind man does not see it, the sun still shines clearly in splendor and beauty even while he is falling down the precipice."

5. 1469, Niccolò Machiavelli, Italian politician and philosopher,
"Men in general judge more from appearances than from reality. All men have eyes, but few have the gift of penetration."
"For the great majority of mankind are satisfied with appearances, as though they were realities, and are often even more influenced by the things that seem than by those that are."

6. 1623, Blaise Pascal, French philosopher, physicist and mathematician,
"Something incomprehensible is not for that reason less real."

7. 1646, Gottfried Wilhelm Leibniz, German philosopher, scientist, diplomat, politician ('the last polymath'),
"Quite often a consideration of the nature of things is nothing but the knowledge of the nature of our minds and of these innate ideas, and there is no need to look for them outside oneself."

8. 1685, George Berkeley, Irish theologian and philosopher,
"*Esse est percipi.*" ("To be means to be perceived.") Also quoted as "*Esse est percipi vel percipere.*" ("To be means to be perceived or to perceive."). Only perceptions and perceivers really exist. The moon is not there when no one perceives it.

9. 1724, Immanuel Kant, German philosopher,
"Hitherto it has been assumed that all our knowledge must conform to objects. But all attempts to extend our knowledge of objects by establishing something in regard to them a priori, by means of concepts, have, on this assumption, ended in failure. We must therefore test whether we might not have more success in the tasks of metaphysics, if we suppose that objects must conform to our knowledge. This would agree better with what is desired, namely, that it should be possible to have knowledge of objects a priori, determining something in regard to them prior to their being given. We should then proceed precisely on the lines of Copernicus' primary hypothesis. Lacking satisfactory progress in explaining the movements of the heavenly bodies on the supposition that they all revolved around the spectator, he tested whether he might not have more success if he allowed the spectator to revolve and the stars to remain at rest."

10. 1742, Georg Christoph Lichtenberg, German writer, aphorist, mathematician and the first German professor of experimental physics,

"Euler says in his letters upon various subjects in connection with natural science (Vol. II, p. 228), that there would be thunder and lightning just as well if there were no man present whom the lightning might strike. It is a very common expression, but I must confess that it has never been easy for me to comprehend it completely. It always seems to me as if the conception *being* were something derived from our thought, and thus, if there were no longer any sentient and thinking creatures, then there would be nothing more whatever. Although this sounds simpleminded, and although I would be laughed at if I said something like that publicly, I think yet that it is one of the greatest benefits, actually one of the strangest qualities of the human spirit, to be able to make such a conjecture."

11. 1770, Georg Wilhelm Friedrich Hegel, German philosopher,

"The spiritual alone is the real."

"Reason is the conscious certainty of being all reality."

"What is rational is real; and what is real is rational."

12. 1772, Novalis (Georg Philipp Friedrich Freiherr von Hardenberg), German poet and philosopher,

"Love is the highest reality—the deepest basis of everything."

13. 1821, Fyodor Mikhailovich Dostoyevsky, Russian writer,

"But does it matter whether it was a dream or reality, if the dream made known to me the truth?"

14. 1844, Friedrich Nietzsche, German philosopher,

"No, facts are precisely what there is not, only interpretations. We cannot establish any fact 'in itself'; perhaps it is folly to want to do such a thing. 'Everything is subjective', you say; but even this is interpretation."

15. 1871, Christian Morgenstern, German writer.

"For, he reasons pointedly, that which must not, can not be."

16. 1871, Marcel Proust, French writer,

"Reality is always the bait that lures us towards something unknown along a path that we can follow only a little way."

17. 1879, Albert Einstein, German-Swiss-American physicist,

"Reality is merely an illusion, albeit a very persistent one."

"As far as the laws of mathematics refer to reality, they are not certain; and as far as they are certain, they do not refer to reality."

18. 1881, Pablo Picasso, Spanish painter,

"Everything you can imagine is real."

19. 1887, Marc Chagall, Russian-French painter,

"All our interior world is reality, and that, perhaps, more so than our apparent world."

20. 1887, Erwin Schrödinger, Austrian physicist,

"Reality is nothing more than a convenient fiction."

21. 1899, Alfred Hitchcock, British film director and producer,

"A glimpse into the world proves that horror is nothing other than reality."

22. 1889, Martin Heidegger, German philosopher,
    "Higher than reality is potentiality."
    "Being essences (or happens, occurs) as presencing."
23. 1901, Jacques Lacan, French psychoanalyst,
    "If there is a notion of the real, it is extremely complex and, because of this,
    incomprehensible; it cannot be comprehended in a way that would make an All
    out of it."
    "The obviousness of reality, the obviousness of our sight which comprehends
    and takes up the world in terms of a logic of knowledge and of an understanding
    in pictures, is becoming questionable."
24. 1903, Walker Evans, American photographer,
    "Reality is not totally real."
25. 1904, Salvador Dali, Spanish painter,
    "One day it will have to be officially admitted that what we have christened
    reality is an even greater illusion than the world of dreams."
26. 1906, Nelson Goodman, American philosopher,
    "Truth, far from being a solemn and severe master, is a docile and obedient
    servant. The scientist who supposes that he is single-mindedly dedicated to the
    search for truth deceives himself.... He seeks system, simplicity, scope; and when
    satisfied on these scores he tailors truth to fit. He as much decrees as discovers
    the laws he sets forth, as much designs as discerns the patterns he delineates."
27. 1914, Arno Schmidt, German writer,
    "The > Real World <? is, in truth, only the caricature of our great novels!"
    (Schmidt used a very personal orthography; in German, the quote reads: "Die >
    Wirkliche Welt <? : ist, in Wahrheit, nur die Karikatur unsrer Großn Romane!").
28. 1921, Paul Watzlawick, Austrian-American psychiatrist,
    "The belief that one's own view of reality is the only reality is the most dangerous
    of all delusions."
29. 1928, Philip K. Dick, American SF-writer,
    "Reality is that which, when you stop believing in it, doesn't go away."
30. 1928, Robert M. Pirsig, American writer and philosopher,
    "Laws of nature are human inventions…the world has no existence whatsoever
    outside the human imagination."
31. 1929, Audrey Hepburn, British actress and humanitarian,
    "Anyone who does not believe in miracles is not a realist."
32. 1929, Jean Baudrillard, French philosopher and sociologist,
    "For the world was not created in order to understand it. It does not care about
    knowledge. Maybe it was even created to be not understood. Knowledge is
    indeed part of the world, but only as a total illusion. That's what I find inter-
    esting, because it means that the mind is only part of a whole, and that there is
    no interpretation for this whole thing. ... Inside this world there is definitely a
    knowledge- and thought system that produces something like truth- and reality
    effects. But I think it's important that philosophy has always in mind this radical
    uncertainty and illusion. One must beware of the truth."

33. 1931, Roger Penrose, British mathematician and physicist,
    "It is my opinion that our present picture of physical reality, particularly in relation to the nature of time, is due for a grand shake up—even greater, perhaps, than that which has already been provided by present-day relativity and quantum mechanics."
34. 1935, David Mermin, American physicist,
    "We know that the moon is demonstrably not there when nobody looks."
35. 1935, Woody Allen, American screenwriter, director, actor, comedian, author,
    "I hate reality, but it's still the best place to get a good steak."
36. 1940, John Lennon, British musician,
    "Nothing is real and nothing to get hung about. Strawberry Fields forever."
37. 1944, Yves Michaud, French philosopher,
    "What we call reality is an unsatisfactory system of a small number of sensory experiences, of ill-founded beliefs and superficially perceived images."
38. 1945, Richard Tarnas, American philosopher,
    "The world is in some essential sense a construct. Human knowledge is radically interpretive. There are no perspective-independent facts. Every act of perception and cognition is contingent, mediated, situated, contextual, theory-soaked. Human language cannot establish its ground in an independent reality."
39. 2010, Christian Lange, Nils Ohlsen (eds.): *Realism—The Adventure of Reality*, exhibition catalog for the exhibition at the Kunsthalle in Emden, Germany; January through June, 2010,
    "Based on the paradigm of cultural studies of a 'radical constructivism' and on the theory of the collective reality production holds today, that reality is not objectively evident. ... One knows that the perception of people can hardly distinguish between the subjectivity in the phenomenal and objective reality. ... Obviously, the knowledge that reality can and must be continuously challenged contains just the prerequisite for the efforts to get, again and again, a new and most accurate idea of it."

# Appendix V
# Remarks on Some Interpretations of Quantum Mechanics

We consider in this appendix four interpretations in more detail than in Chap. 28, namely the Bohmian interpretation, the many-worlds interpretation, consistent histories and, finally, the Ghirardi-Rimini-Weber theory.

## V.1 Bohmian Interpretation

This is not only the currently effectively unique interpretation of quantum mechanics based on hidden variables, but surely also the most thoroughly elaborated one. The basic idea is to consider the wavefunction and the associated particle as two different, separately existing and real objects. It is also given other names such as de Broglie-Bohm theory, pilot-wave theory, Bohmian mechanics, causal interpretation, etc.

### V.1.1  Sketch of the Formalism

We begin with the SEq:

$$i\hbar\dot{\Psi} = -\frac{\hbar^2}{2m}\nabla^2\Psi + V\Psi. \tag{V.1}$$

Since the wavefunction and probability density are related by $\rho = |\Psi|^2$, we apply the *ansatz*

$$\Psi = \rho^{\frac{1}{2}}e^{i\frac{S}{\hbar}}; \ \rho, S \in \mathbb{R}, \tag{V.2}$$

so that $S$ has the dimensions of an action. We insert (V.2) into the SEq and separate with respect to real part and imaginary part. This yields:

© Springer Nature Switzerland AG 2018
J. Pade, *Quantum Mechanics for Pedestrians 2*, Undergraduate Lecture Notes in Physics, https://doi.org/10.1007/978-3-030-00467-5

$$i\hbar \frac{1}{2}\rho^{-\frac{1}{2}}\dot{\rho} = -\frac{\hbar}{2m}\left[\rho^{-\frac{1}{2}}\boldsymbol{\nabla}\rho \cdot i\boldsymbol{\nabla}S + \rho^{\frac{1}{2}}i\boldsymbol{\nabla}^2 S\right]$$

$$-\rho^{\frac{1}{2}}\dot{S} = -\frac{\hbar^2}{2m}\left[-\frac{1}{4}\rho^{-\frac{3}{2}}\left(\boldsymbol{\nabla}\rho\right)^2 + \frac{1}{2}\rho^{-\frac{1}{2}}\boldsymbol{\nabla}^2\rho - \rho^{\frac{1}{2}}\frac{1}{\hbar^2}\left(\boldsymbol{\nabla}S\right)^2\right] + V\rho^{\frac{1}{2}} \quad \text{(V.3)}$$

or

$$\dot{\rho} + \frac{1}{m}\left[\boldsymbol{\nabla}\rho \cdot \boldsymbol{\nabla}S + \rho\boldsymbol{\nabla}^2 S\right] = \dot{\rho} + \boldsymbol{\nabla}\left(\rho \cdot \frac{\boldsymbol{\nabla}S}{m}\right) = 0$$

$$\dot{S} + \frac{1}{2m}\left(\boldsymbol{\nabla}S\right)^2 + V - \frac{\hbar^2}{4m}\left[\rho^{-1}\boldsymbol{\nabla}^2\rho - \frac{1}{2}\rho^{-2}\left(\boldsymbol{\nabla}\rho\right)^2\right] = 0. \quad \text{(V.4)}$$

For $\hbar \to 0$ (i.e. in the classical limit), $S$ is a solution of the Hamilton-Jacobi equation[58]; hence, the term $\boldsymbol{\nabla}S$ can be interpreted as the momentum $\mathbf{p}$, or $\frac{\boldsymbol{\nabla}S}{m}$ as $\mathbf{v}$, i.e. as the velocity of a point particle.[59]

This interpretation can be maintained for $\hbar \neq 0$. The first equation in (V.4) then reads $\dot{\rho} + \boldsymbol{\nabla}\left(\rho \cdot \mathbf{v}\right) = 0$, which is none other than the conservation of probability. The second equation can be further understood as the Hamilton-Jacobi equation, but, in addition to the 'standard' potential $V$, it contains a *quantum potential* $W$, with

$$W = -\frac{\hbar^2}{4m}\left[\rho^{-1}\boldsymbol{\nabla}^2\rho - \frac{1}{2}\rho^{-2}\left(\boldsymbol{\nabla}\rho\right)^2\right] = -\frac{\hbar^2}{2m}\frac{\boldsymbol{\nabla}^2\rho^{\frac{1}{2}}}{\rho^{\frac{1}{2}}} = -\frac{\hbar^2}{2m}\frac{\boldsymbol{\nabla}^2|\psi|}{|\psi|}. \quad \text{(V.5)}$$

With this, (V.4) can be rewritten as

$$\dot{\rho} + \boldsymbol{\nabla}\left(\rho \cdot \frac{\boldsymbol{\nabla}S}{m}\right) = \dot{\rho} + \boldsymbol{\nabla}\left(\rho \cdot \frac{\mathbf{p}}{m}\right) = 0$$

$$\dot{S} + \frac{(\boldsymbol{\nabla}S)^2}{2m} + V + W = \dot{S} + \frac{\mathbf{p}^2}{2m} + V + W = 0. \quad \text{(V.6)}$$

Here, it is important to note that the point particle experiences not only the classical interaction $V$, but also the quantum potential $W$.

In principle, one can now obtain the Hamilton-Jacobi function $S$ by integrating equation (V.6). The speed of the point particle follows from this in terms of $\mathbf{v} = \frac{\boldsymbol{\nabla}S}{m}$, and its trajectory from $\mathbf{x} = \int \mathbf{v}\, dt$. However, we do not know the initial value; the integral over the velocity gives a family (ensemble) of possible trajectories. Below, we will go through the procedure for the motion of a free particle as an example.

But first, some general remarks about the Bohmian interpretation: We see that the wavefunction $\psi$ plays a double role. It supplies the information about the most probable position of the particle in terms of $\rho = |\Psi|^2$, and on the other hand, it affects the particle through the quantum potential $W$. In this interpretation, the physical state of a particle is completely defined not by the wavefunction alone, but by the combination of the wavefunction and the particle's position.

---

[58]This of course holds true only if the terms $\rho^{-1}\boldsymbol{\nabla}^2\rho - \frac{1}{2}\rho^{-2}\left(\boldsymbol{\nabla}\rho\right)^2$ do not vanish as $1/\hbar^2$, or vanish more slowly.

[59]Here, one in fact considers a point particle with well-defined position and velocity, and not a quantum object.

Since the wavefunction (or at least its absolute square) exerts a force on the particles by means of $W$, we have to consider $\psi$ (or at least $|\psi|$) as the mathematical representation of a real field.[60] Of course, the coordinates of the particles are considered a priori as real; but they are not observable and represent the hidden variables in this interpretation.

Thus, the particles move along well-defined trajectories, which we do not know, but about which we can make probability statements with the help of $\rho$. Only the particles are relevant to the measurement; the wavefunction acts as a (somewhat nebulous) 'guiding field'.

In this interpretation, there is no collapse due to the measurement; instead, measurement simply means reducing our ignorance—after the measurement, we know on which trajectory the particle is moving (on which it was moving even before the measurement, but then we did not know this).

We note that the interpretation is nonlocal due to the occurrence of the quantum potential. We can see this easily if we imagine making changes on just one particle in a multi-particle system. Then its wavefunction $\psi$ and quantum potential $W$ change instantaneously, and hence the trajectories of all the other particles also must change.[61]

## V.1.2   Example: Free Motion

For the sake of illustration, we consider the free motion of a particle. With the wavefunction

$$\psi = N e^{i(kx - \omega t)} \tag{V.7}$$

($N$ is a suitable normalization factor), we obtain

$$S = \hbar (kx - \omega t). \tag{V.8}$$

Evidently, the quantum potential $W$ vanishes, and the Bohmian momentum $p$ is given by

$$p = \frac{\mathrm{d}}{\mathrm{d}x} S = \hbar k. \tag{V.9}$$

With this and $p = m\dot{x}$, the family of trajectories follows:

$$x = \frac{\hbar k}{m} t + x_0, \tag{V.10}$$

---

[60] Where this field comes from and what its physical cause is are left unspecified.

[61] According to Bell, the term 'hidden variables', which is commonly used for variables that are intended to supplement the quantum-mechanical description (using wavefunctions), is misleading here, since on the contrary, the wavefunction is 'hidden'.

whereby we do not know the specific initial value $x_0$ which applies to the system under consideration. Since the probability density (in the distribution sense) is given by $|N|^2$, all initial positions are equally probable.

## V.1.3 Conclusions

The Bohmian interpretation yields the same results as the usual quantum mechanics. Moreover, it offers no advantages, either formally or computationally: although it has recourse to classical ideas, it cannot avoid phenomena such as non-locality, in contradiction to common sense. Hence a striking advantage of this approach is not apparent. In addition, there are other difficulties not mentioned here, including among others the extension to multi-particle systems or to relativistic velocities. A further main point of criticism is the asymmetry between position and momentum. Quantum mechanics can be formulated, for example, either in position or in momentum space. This is not true for the Bohmian interpretation; it depends largely on the position representation.

The attitude of the scientific community towards the Bohmian interpretation differs greatly (as with all interpretations). While many do not see it any longer as a serious explanation of quantum mechanics, others do not share this view; for example, the group 'Bohmian Mechanics' at the University of Munich.[62]

## V.2 The Many-Worlds Interpretation

The many-worlds interpretation (MWI) is based on the unmodified SEq. In contrast to the standard interpretation, it assumes that all of the changes that the SEq describes over time will be realized; it is therefore a strictly deterministic theory.

We illustrate this fact with the help of a photon of unknown polarization which is measured with respect to its horizontal/vertical polarization. The basis states of the photon are $|h\rangle$ and $|v\rangle$; the measuring apparatus has the states $|M_h\rangle$ and $|M_v\rangle$. Then the total state is:

$$|\psi\rangle = c_h |h\rangle |M_h\rangle + c_v |v\rangle |M_v\rangle ; \quad |c_h|^2 + |c_v|^2 = 1. \tag{V.11}$$

In the standard interpretation, this means that we measure the state $|i\rangle$ with a probability of $|c_i|^2$, where the measuring apparatus is in the state $|M_i\rangle$ $(i = h, v)$. The MWI, however, assumes that both terms on the right side of (V.11) describe something that really exists. So there is no state reduction and no reference to probabilities occurring in measurements. A common view is that the universe branches out into a number of different parallel worlds; in our example into two universes, one with

---

[62]http://www.mathematik.uni-muenchen.de/~bohmmech/.

$|h\rangle \, |M_h\rangle$ and the second with $|v\rangle \, |M_v\rangle$. Thus, the concept of 'measurement' plays no fundamental role, which is the reason why the MWI is currently enjoying some popularity, especially with quantum cosmologists.

However, there are some problems with the MWI connected with the concept of probability. Let us consider again the state (V.11). Both 'worlds', i.e. $|h\rangle \, |M_h\rangle$ and $|v\rangle \, |M_v\rangle$, exist after the split. What then is the significance of the coefficients $c_h$ and $c_v$? It is clear is that $|c_h|^2$ and $|c_v|^2$ cannot be interpreted as probabilities for the occurrence of one or another world, since *both* are realized. There are different explanations,[63] for example the many-minds interpretation, according to which each conscious being has available a continuum of states of consciousness, representing the branched worlds.

In science fiction, parallel universes allow for all sorts of spectacular activities. But in fact, and here the various offshoots of the MWI agree, we experience nothing of these branches (and thus cannot visit or communicate with parallel universes), because in each branch of the state vector there is a perfect correlation between our memory states and the other events which have occurred.

Another problem of the MWI is that the decomposition or branching is not always unique. We consider a system of two electrons with total spin zero. The antisymmetric (and entangled) total state is

$$|\psi\rangle = \frac{|\uparrow\rangle \, |\downarrow\rangle - |\downarrow\rangle \, |\uparrow\rangle}{\sqrt{2}}, \tag{V.12}$$

where $|\uparrow\rangle$ and $|\downarrow\rangle$ are the eigenstates (referred to the $z$ axis) for the eigenvalues $+\frac{\hbar}{2}$ and $-\frac{\hbar}{2}$. In the MWI, this state describes a split into two branches, one with $|\uparrow\rangle \, |\downarrow\rangle$ and one with $|\downarrow\rangle \, |\uparrow\rangle$.

Instead of the $z$ axis, we could just as well refer to the $x$ axis. The spin matrices are

$$S_x = \begin{pmatrix} 0 & 1 \\ 1 & 0 \end{pmatrix}; \; S_z = \begin{pmatrix} 1 & 0 \\ 0 & -1 \end{pmatrix}. \tag{V.13}$$

The eigenvectors for the eigenvalues $+\frac{\hbar}{2}$ and $-\frac{\hbar}{2}$ are given for $S_x$ by

$$|\rightarrow\rangle = \frac{1}{\sqrt{2}} \begin{pmatrix} 1 \\ 1 \end{pmatrix}; \; |\leftarrow\rangle = \frac{1}{\sqrt{2}} \begin{pmatrix} 1 \\ -1 \end{pmatrix}, \tag{V.14}$$

and for $S_z$ by

$$|\uparrow\rangle = \begin{pmatrix} 1 \\ 0 \end{pmatrix}; \; |\downarrow\rangle = \begin{pmatrix} 0 \\ 1 \end{pmatrix}. \tag{V.15}$$

---

[63] In any case, the question cannot be decided by an outside observer who can determine, so to speak, the 'weight' of the individual worlds— simply because there is not such an 'outsider's perspective' in the MWI (in other words, because for our universe, according to the common belief, there is neither an 'outside' nor a 'before', i.e. neither anteriority nor exteriority).

This us allows to convert the state (V.12), formulated in the $z$ basis, into the $x$ basis. It follows that

$$|\psi\rangle = \frac{|\leftarrow\rangle\,|\rightarrow\rangle - |\rightarrow\rangle\,|\leftarrow\rangle}{\sqrt{2}}. \tag{V.16}$$

We see that the total states (V.12) and (V.16) are the same, but not the individual states on the right-hand sides. This means that also the possible ramifications are different. The question arises: Which is the correct splitting, or what is the special basis in the Hilbert space which is related to the branching? We discussed this issue in Chap. 24 in the context of decoherence (non-uniqueness of a biorthonormal decomposition for the same coefficients). Similarly, we would have to assume, if we demand uniqueness, that by the interaction with the environment, a pointer variable is selected that determines which one of the two options (V.12) and (V.16) is realized. Apparently, however, this question can not be answered satisfactorily at present.

All in all, there are several points in the MWI which seem still not completely resolved on a quite fundamental level; perhaps the naive notion of a constantly-branching universe can be improved upon.

## V.3   Consistent Histories

### V.3.1   Definitions

The term *quantum history* essentially means 'time-ordered sequence of quantum events'. The term 'event' is quite flexible; events can be e.g. wavefunctions or properties such as position or momentum or others. We will represent each event (seen as a property of the system) by a projection operator $F$. In this section, we use subscript indices exclusively to indicate different times; for other distinctions, we choose accordingly superscript indices. For a given time sequence $t_1 < t_2 < t_f$, a history is characterized by a sequence of projectors $(F_1, F_2, \ldots, F_f)$, one projector for each time.

One can imagine a history as a stroboscopic recording of a (possibly continuous) process, but the intervals need not be constant and the events considered must not be the same at each time step. As an example, we consider a harmonic oscillator with eigenstates $\{|\varphi^i\rangle, i = 1, 2, \ldots\}$. At the first time step, we project onto the subspaces 1 and 2, at the second onto the subspace 3 and at the third, we measure whether the system is in the region $a \leq x \leq b$:

$$F_1 = |\varphi^1\rangle\langle\varphi^1| + |\varphi^2\rangle\langle\varphi^2|\,;\ F_2 = |\varphi^3\rangle\langle\varphi^3|\,;\ F_3 = X^{a,b}. \tag{V.17}$$

The crucial point is that the successive events need not be linked via the SEq, i.e. one can also incorporate e.g. stochastic evolutions. However, the SEq (or equivalently the time-evolution operator $T(t', t)$) may be used for the calculation of probabilities.

**Fig. V.1** Schematic
representation of different
histories

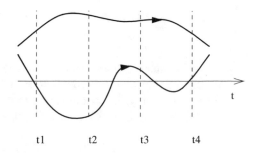

$$t1 \qquad t2 \qquad t3 \qquad t4$$

We can now define a *history Hilbert space* $\check{\mathcal{H}}$ by

$$\check{\mathcal{H}} = \mathcal{H}_1 \odot \mathcal{H}_2 \odot \cdots \mathcal{H}_f, \tag{V.18}$$

where $\mathcal{H}_j$ is a copy of the Hilbert space $\mathcal{H}$ for each time point $t_j$, which describes the system at a fixed time. Like $\otimes$, the symbol $\odot$ denotes a tensor product; the difference is that $\otimes$ couples Hilbert spaces at the same times, and $\odot$ at different times. In the space $\check{\mathcal{H}}$, the history $(F_1, F_2, \ldots, F_f)$ can be represented by the tensor product

$$Y = F_1 \odot F_2 \odot \cdots F_f. \tag{V.19}$$

This formulation means '$F_1$ at $t_1$, $F_2$ at $t_2$, ..., $F_f$ at $t_f$' ; at other times, it provides no information. Since each $F_i$ is a projector, this also applies to $Y$.

Between two points in time, there may be different histories, as indicated in Fig. V.1. One can now assign a probability to each history, so that one can distinguish between more or less probable histories. To this end, one defines an operator $K(Y)$ (chain operator):

$$K(Y) = F_f T(t_f, t_{f-1}) F_{f-1} T(t_{f-1}, t_{f-2}) \cdots T(t_1, t_0) F_0 \tag{V.20}$$

where the operators $T(t_j, t_{j'})$ are the corresponding time evolution operators which convey the system between the points $t_{j'}$ and $t_j$. The probability for the history (V.19) is determined by

$$W(Y) = tr\left(K^\dagger(Y) K(Y)\right). \tag{V.21}$$

Some properties of a history can be read directly off these formulations. For instance, $W$ vanishes for $F_f T(t_f, t_{f-1}) F_{f-1} = 0$, i.e. for a contradictory history.

As an example, we consider projections onto pure states, e.g. $|a\rangle$, $|b\rangle$ and $|c\rangle$. It follows that:

$$Y = |a_0\rangle \langle a_0| \odot |b_1\rangle \langle b_1| \odot |c_2\rangle \langle c_2|, \tag{V.22}$$

and the chain operator

$$K(Y) = |c_2\rangle \langle c_2| T (t_2, t_1) |b_1\rangle \langle b_1| T (t_1, t_0) |a_0\rangle \langle a_0|$$
$$= \langle c_2| T (t_2, t_1) |b_1\rangle \langle b_1| T (t_1, t_0) |a_0\rangle \cdot |c_2\rangle \langle a_0| \qquad (V.23)$$

is a product of complex numbers (transition amplitudes) with the operator $|c_2\rangle \langle a_0|$.

One can join different histories into a family, if they meet certain *consistency conditions*. Intuitively, the conditions mean that the probabilities of various histories are additive. For two histories $Y^a$ and $Y^b$, this is the case for

$$tr \left( K^\dagger (Y^a) K(Y^b) \right) = 0 \text{ for } a \neq b. \qquad (V.24)$$

Of course, the probabilities of all observed histories have to add up to give 1. Such a set of histories is called a *consistent family of histories* or framework. In essence, this means that a consistent family of histories consists of different, mutually exclusive histories.

The consistent histories approach now assumes that for the description of a measurement with certain results, a framework must be used in which these results are included, and that the framework must also include the measurement properties at a time before the measurement takes place. Within this consistent family, there are no contradictions, no paradoxes. These occur only when histories from different (i.e. incompatible) families are compared.

A final note: The consistency conditions (V.24) are also called *decoherence conditions*, and accordingly, the expressions 'decoherent family' or 'decoherent set' are occasionally used to denote a consistent family. The use of 'decoherence' in this context is perhaps a somewhat confusing example of language, since it is *not* directly related to the phenomenon of decoherence as a flow of information into the environment (which we treated in Chap. 24; although they may have something to do with each other). Thus, a consistent family of histories can become inconsistent for an isolated system if the environment is taken into account, and *vice versa*, an inconsistent family can become a consistent one.

## V.3.2   A Simple Example

For a very simple illustration of the basic ideas, we consider a Mach–Zehnder interferometer, where the beam splitter BS2 is mobile and can be moved out of/into the beam path; see Fig. V.2.

We denote the initial state of the photon by $|a\rangle$ (horizontal incidence), the upper and lower beam paths by $|o\rangle$ and $|u\rangle$, and the superposition state by $|s\rangle$; apart from that, we use the results of Chap. 6, Vol. 1. We write the total state as $|\varphi\rangle |AB\rangle$, i.e. as the tensor product of the Hilbert spaces of the photon $|\varphi\rangle$ and the two detectors $|AB\rangle$. If the photon triggers a detector, for example $A$, it is absorbed and vanishes; we write this as $|\varphi\rangle |AB\rangle \rightarrow |A^*B\rangle$.

**Fig. V.2** Mach–Zehnder interferometer used for the simple example illustrating consistent histories

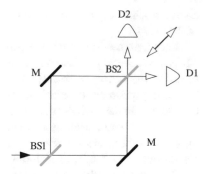

(1) We first consider the situation where the beam splitter BS2 is not in the path. The incoming photon is incident on the first beam splitter, after which it triggers either detector D1 or D2, and thus has to go through the lower or the upper arm. This we can write as

$$|a\rangle |D1D2\rangle \rightarrow \begin{pmatrix} |o\rangle |D1D2\rangle \rightarrow |D1^*D2\rangle \,;\; W = \tfrac{1}{2} \\ |u\rangle |D_1D_2\rangle \rightarrow |D1D2^*\rangle \,;\; W = \tfrac{1}{2} \end{pmatrix}. \qquad (V.25)$$

Thus, we have formulated a consistent family of histories; the probabilities of the two family members add up to 1. The two paths (and thus the displays of the detectors) are mutually exclusive.

(2) We now bring the beam splitter BS2 into the beam path. Then we have the following family of consistent histories.

$$|a\rangle |D1D2\rangle \rightarrow \begin{pmatrix} |s\rangle |D1D2\rangle \rightarrow |D1^*D2\rangle \,;\; W = 1 \\ |s\rangle |D1D2\rangle \rightarrow |D1D2^*\rangle \,;\; W = 0 \end{pmatrix}. \qquad (V.26)$$

The photon initially strikes the first beam splitter, then passes into the superposition state, meets the second beam splitter and triggers (at least in principle) D1/D2. As we have shown in Chap. 6, Vol. 1, for a horizontally incident photon, the upper history has the probability 1, the lower one 0; only D1 responds.

Thus we have two different and mutually-exclusive families of histories—no member of one family can belong to the other family. Accordingly, questions that are useful for one family can be meaningless for the other, for example the question as to which path the photon has followed.

## V.3.3  Conclusions

Consistent histories is an approach that characterizes the physics of a closed system by a discrete time series of properties that are represented by projectors in Hilbert space. The successive events need not be connected by the SEq; the evolution can also be stochastic. Consistency conditions ensure a consistent physical interpretation of the events.

Since here the SEq is not *the* general governing principle of the time evolution, but one of several possible ones, this interpretation avoids many problems that are linked to the concept of measurement—questions about the classical nature of the measuring apparatus, the mechanism of the collapse of the wavefunction, etc., do not even arise. On the other hand, this approach also cannot do without the SEq if probabilities for different histories are to be calculated. And fundamental questions remain unanswered here, also—for example whether objective chance exists or how nature selects out of the family of histories that one history which we actually measure.

There is a comprehensive primary and secondary literature on consistent histories.[64] An introduction is given by e.g. Robert B. Griffiths, *Consistent Quantum Theory* (2003). The online version of the book can be found at http://quantum.phys. cmu.edu/CQT/.

## V.4    Ghirardi-Rimini-Weber

This approach goes beyond a simple interpretation, since the dynamics is modified. The aim is to suppress superpositions of states which are located in widely-separated regions (i.e. macroscopically separated) through an additional term in the Hamiltonian. In other words, the modified dynamics conditions are constructed in such a way that they cause the transition from a pure state into a statistical mixture under macroscopic conditions. Thus it is clear that the starting point is not the SEq for pure states, but the von Neumann equation for the density operator:

$$\frac{d}{dt}\rho(t) = -\frac{i}{\hbar}[H, \rho(t)].\qquad(V.27)$$

In this equation, a term is inserted which corresponds to a localization process:

$$\frac{d}{dt}\rho = -\frac{i}{\hbar}[H, \rho] - \lambda(\rho - T[\rho]),\qquad(V.28)$$

with

$$T[\rho] = \sqrt{\frac{\alpha}{\pi}} \int_{-\infty}^{\infty} \left(e^{-\frac{\alpha}{2}(q-x)^2}\rho e^{-\frac{\alpha}{2}(q-x)^2}\right) dx,\qquad(V.29)$$

where $q$ is the position operator. The quantity $\lambda$ describes the frequency of the localization process and $\frac{1}{\sqrt{\alpha}}$ corresponds to the distance after which the linear superposition changes into a statistical mixture (localization distance).

---

[64]The approach is applied e.g. also in quantum cosmology.

Intuitively, this term means that the quantum object is subject to random localization processes around its approximate position, at random times (Poisson distribution with main frequency $\lambda$). Formally, this localization can be summarized as[65]:

$$\psi(q) \rightarrow \psi_x(q) = \frac{\varphi_x(q)}{\|\varphi_x(q)\|}; \quad \varphi_x(q) = \left(\frac{\alpha}{\pi}\right)^{\frac{3}{4}} e^{-\frac{\alpha}{2}(q-x)^2} \psi(q). \tag{V.30}$$

If $\psi(q)$ is a wave packet centered around $q_0$, then $\varphi_x(q)$ in general does not yield noteworthy contributions for $x \neq q_0$.

We mention here only that the localization process is proportional to the number of quantum objects involved—the larger the number, the faster the total system becomes localized.

The quantities $\lambda$ and $\alpha$ are not physically defined, but are freely adjustable parameters; they are chosen in such a way that they meet the predetermined requirements as well as possible. Typical values are $\lambda \approx 10^{-16}$ s$^{-1}$ and $\frac{1}{\sqrt{\alpha}} \approx 10^{-7}$ m. With these values, one obtains localization times of e.g. $10^{16}$ s ($\approx 3 \cdot 10^7$ a) for microscopic and $10^{-7}$ s for macroscopic systems.

The GRW approach thus alters standard quantum mechanics so that it yields almost unchanged results for microscopic systems, but ensures the decay of superpositions for macroscopic systems.

Among other things, a weak point of this approach can be seen in the fact that the localization process is postulated ad hoc; its physical origin (if it even exists) is not considered. So it is rather an empirically- and phenomenologically-oriented model with free model parameters which are adapted to fit measurements.

---

[65]The generalization to $N$ quantum objects in three-dimensional space, if the $i$th quantum object is located at $\mathbf{q}_i$:

$$\psi(\mathbf{q}_1, \ldots, \mathbf{q}_1) \rightarrow \psi_x(\mathbf{q}_1, \ldots, \mathbf{q}_N) = \frac{\varphi_x(\mathbf{q}_1, \ldots, \mathbf{q}_N)}{\|\varphi_x(\mathbf{q}_1, \ldots, \mathbf{q}_N)\|}$$
$$\varphi_x(\mathbf{q}_1, \ldots, \mathbf{q}_N) = \left(\frac{\alpha}{\pi}\right)^{\frac{3}{4}} e^{-\frac{\alpha}{2}(\mathbf{q}_i - \mathbf{x})^2} \psi(\mathbf{q}_1, \ldots, \mathbf{q}_N).$$

# Appendix W
# Elements of Quantum Field Theory

## W.1 Foreword

In Volume 1 we discussed some topics of relativistic quantum mechanics (RQM). In this volume we want to provide a glimpse into relativistic *quantum field theory* (QFT).

Why QFT? The subjects of quantum mechanics (QM) are characterized by three properties: (1) They are *single* particles, e.g. electrons. (2) They retain their *identity* (an electron remains an electron remains an electron) (3) They behave *non-relativistically*. RQM, which we have dealt with in Volume 1, extends the theory into the relativistic domain, thereby attempting to retain properties (1) and (2). But as we have seen with respect to property (1), there are no relativistic single-particle theories. In addition, properties (1) and (2) mean that there is no way to treat processes like the decay of elementary particles or reactions like $A + B \rightarrow C + D + E$. And finally, we have found quite problematic negative energies which can not be interpreted meaningfully in the framework of these equations (Klein–Gordon and Dirac equation). To attack all these questions, an advanced theory is required, and as we will see, QFT will resolve the issue.

Also in another regard, property (1) is a limitation, since it refers to individual particles, but not to extended objects such as, e.g. a string. How could one quantize such objects? A mass point has the parameters $x(t)$ and $\dot{x}(t)$ or $p(t)$ (here $x$ is a Lagrange coordinate, 'moving in the river with the mass point'). If we combine many mass points $x_i(t)$ to form e.g. a string, we speak no longer of individual coordinates. Instead, we have a field variable $\varphi(x, t)$, e.g. the amplitude of the string (here $x$ is an Euler coordinate, 'sitting on the bank of the river'). Thus, instead of the label $i$ we have now the label $x$, and instead of quantizing $x$ of the mass point (called 1. quantization), we now have to quantize the field $\varphi$ (called 2. quantization).[66] As the name quantum *field* theory implies, QFT is the right candidate for this task.

---

[66]The naming 1. and 2. quantization is a little bit unfortunate but established.

© Springer Nature Switzerland AG 2018
J. Pade, *Quantum Mechanics for Pedestrians 2*, Undergraduate Lecture
Notes in Physics, https://doi.org/10.1007/978-3-030-00467-5

The formalism relies very much on classical field theory (see outline in Appendix T, Vol. 1). A central term is the Lagrange density which allows for the determination of the conjugated momentum density and the Hamiltonian density. QFT starts from the equations of RQM, e.g. the Klein–Gordon and the Dirac equation. In principle one can also formulate a nonrelativistic quantum field theory, starting from the Schrödinger equation. Indeed, there are applications for this nonrelativistic field theory. Nevertheless, the name QFT means almost exclusively *relativistic* quantum field theory.

Another essential component of QFT is the formalism of the harmonic oscillator as developed in Vol. 2 Chap. 18. Actually, the quantization procedure for fields is formulated by means of creation and annihilation operators and uses terms like number operator and so on.

Among the advantages of QFT are the following: (1) Electrons and positrons have equal status; we do not have an electron sitting on an infinite sea of positrons anymore. (2) We get rid off negative energies; QFT knows positive energies only. (3) We can consider processes with say $n$ particles in the incoming channel and $m$ particles in the outgoing channel. (4) With quantum electrodynamics, we have the most stringently proven theory in physics. However, we should also mention that there exist structural problems with infinities in QFT. And on the technical level, QFT is sometimes more demanding than QM or RQM.

The content plan is as follows: First we show by means of a toy example the main ideas which are underlying the quantization of fields. This is followed by three sections in which we quantize the Klein–Gordon, Dirac and radiation field. As we will see, there occur infinities; the way to handle this problem is discussed in the section 'Operator ordering'. In the last sections, we formulate the theory for interacting fields using the example of quantum electrodynamics (QED). We define the $S$-matrix and its approximation by first order and second order terms. On the basis of first order terms, we address among others Feynman diagrams. In order to calculate the second order terms, we define Feynman propagators and present the Wick theorem. Finally, we consider the second order term of the $S$-matrix, exemplarily treating in more detail Bhabha and Møller scattering.

QFT is a very extensive topic. There are certainly quite different opinions about how to present it on a few pages as we try to do here. Be that as it may, in view of the space limitation we can not build here a royal road into the realm of quantum field theory, but can only pave the way with a few stepping stones. A lot of questions and issues cannot even be mentioned, and in the discussed topics there will be inevitably some gaps. The reader is invited to actively work on the material.

## W.2    Quantizing a Field - A Toy Example

We present a simple toy example in order to clarify the basic ideas of field quantization. The system consists of two conducting plates at $z = 0$ and $z = L$ which are infinitely extended in $x - y$-direction. Between the plates, we assume an electrical field in the form of standing waves in $z$-direction which are linearly polarized in $x$-direction. Since the plates are conducting, the electrical field vanishes at $z = 0$ and $z = L$.

We start from the source-free Maxwell equations

$$\nabla \cdot \mathbf{E}\,(\mathbf{r}, t) = 0 \; ; \; \nabla \cdot \mathbf{B}\,(\mathbf{r}, t) = 0$$
$$\nabla \times \mathbf{E}\,(\mathbf{r}, t) = -\frac{\partial}{\partial t}\mathbf{B}\,(\mathbf{r}, t) \; ; \; \nabla \times \mathbf{B}\,(\mathbf{r}, t) = \frac{1}{c^2}\frac{\partial}{\partial t}\mathbf{E}\,(\mathbf{r}, t). \tag{W.1}$$

## W.2.1 The Classical Case

According to our assumptions, the electrical field is given by

$$\mathbf{E}\,(\mathbf{r}, t) = (E_x\,(z, t)\,, 0, 0) \; \text{ with } E_x\,(0, t) = E_x\,(L, t) = 0 \tag{W.2}$$

where $E_x\,(z, t)$ may be written as

$$E_{xl}\,(z, t) = C_l \cdot q_l\,(t) \sin k_l z \; ; \; k_l = \frac{\pi}{L}l \text{ with } l = 1, 2, \ldots. \tag{W.3}$$

$C_l$ is a normalization factor, $q_l\,(t)$ the amplitude of the field (yet to be determined). Due to the last two Maxwell equations, for the magnetic field holds $\mathbf{B}\,(\mathbf{r}, t) = (0, B_y\,(z, t)\,, 0)$ with

$$B_{yl}\,(z, t) = C_l \cdot \frac{1}{c^2 k_l}\dot{q}_l\,(t) \cos k_l z. \tag{W.4}$$

Of course, the general solution for e.g. the electrical field is the linear superposition of all partial solutions (W.3), i.e., the sum over all $l$. However, in order to hold the discussion transparent, we consider for the moment only *one* partial solution for a definite $l$.

The general expression for the electromagnetic energy $H$ is given by $H = \frac{\varepsilon_0}{2}\int d^3x\,(\mathbf{E}^2 + c^2\mathbf{B})$. In our simple case, this reads

$$H_l = \frac{\varepsilon_0}{2}\int_0^L dz\,(E_{xl}^2 + c^2 B_{yl}^2). \tag{W.5}$$

Evaluation of the integral leads with $\omega_l^2 = k_l^2 c^2$ to

$$H_l = \frac{1}{2}\left[\dot{q}_l^2\,(t) + \omega_l^2 q_l^2\,(t)\right] \tag{W.6}$$

where we have fixed the normalization constant by $C_l = \sqrt{\frac{2\omega_l^2}{\varepsilon_0 L}}$. This expression is formally identically equal to the energy of a harmonic oscillator (of a fictive mass $m = 1$), if we identify $\dot{q}_l$ with $p_l$.

## W.2.2  Quantization

Quantization means, that we interpret the classical quantities $p_l$ and $q_l$ as operators[67] $\hat{p}_l$ and $\hat{q}_l$ which obey the commutation rule $[\hat{q}_l, \hat{p}_l] = i\hbar$. In this way, we arrive at the quantized version of our toy system

$$\hat{H}_l = \frac{1}{2}\left[\hat{p}_l^2(t) + \omega_l^2 \hat{q}_l^2(t)\right] \; ; \; [\hat{q}_l, \hat{p}_l] = i\hbar. \tag{W.7}$$

Of course, the fields become operators, too (hence called field operators):

$$\hat{E}_{xl}(z,t) = C_l \cdot \hat{q}_l(t)\sin k_l z \; ; \; \hat{B}_{yl}(z,t) = \frac{1}{c^2 k_l}C_l \cdot \hat{p}_l(t)\cos k_l z. \tag{W.8}$$

Note that with $\hat{p}_l$ and $\hat{q}_l$ also the fields are hermitian operators and as such observables. Let us point out, that now the field operators do not commute:

$$\left[\hat{E}_{xl}(z,t), \hat{B}_{yl}(z,t)\right] = \hat{E}_{xl}(z,t)\hat{B}_{yl}(z,t) - \hat{B}_{yl}(z,t)\hat{E}_{xl}(z,t) =$$
$$= \frac{C_l^2}{c^2 k_l}\sin k_l z \cos k_l z \left[\hat{q}_l(t) \cdot \hat{p}_l(t) - \hat{p}_l(t)\cdot\hat{q}_l(t)\right] = i\hbar\frac{C_l^2}{c^2 k_l}\sin k_l z \cos k_l z. \tag{W.9}$$

## W.2.3  Creation and Annihilation Operators, Hamiltonian

We now can apply the whole machinery of the harmonic oscillator as developed in Chap. 18, Vol. 2. We define a creation (raising) operator $a^\dagger(k_l)$ and an annihilation (lowering) operator $a(k_l)$ by (as usual, these ladder operators are written without hat)

$$a^\dagger(k_l) = \frac{1}{\sqrt{2\hbar\omega_l}}\left(\omega_l\hat{q}_l - i\,\hat{p}_l\right) \; ; \; a(k_l) = \frac{1}{\sqrt{2\hbar\omega_l}}\left(\omega_l\hat{q}_l + i\,\hat{p}_l\right)$$
$$\hat{q}_l = \sqrt{\frac{\hbar}{2\omega_l}}\left(a^\dagger(k_l) + a(k_l)\right) \; ; \; \hat{p}_l = i\sqrt{\frac{\hbar\omega_l}{2}}\left(a^\dagger(k_l) - a(k_l)\right) \tag{W.10}$$

and the commutation relations read

$$\left[a(k_l), a^\dagger(k_l)\right] = 1 \; ; \; [\hat{q}_l, \hat{p}_l] = i\hbar. \tag{W.11}$$

The field operators are given by

$$\hat{E}_{xl}(z,t) = C_l\sqrt{\frac{\hbar}{2\omega_l}}\left[a^\dagger(k_l) + a(k_l)\right]\sin k_l z \; ; \; \hat{B}_{yl}(z,t) = C_l\frac{i}{c^2 k_l}\sqrt{\frac{\hbar\omega_l}{2}}\left[a^\dagger(k_l) - a(k_l)\right]\cos k_l z. \tag{W.12}$$

With (W.7) and (W.10) we obtain the Hamiltonian $\hat{H}_l$ in the form

---

[67] In this section, we denote operators by a hat.

The quantization procedure is canonical quantization; cf. Vol. 1, App. T.3.

$$\hat{H}_l = \hbar \omega_l \frac{a^\dagger(k_l)a(k_l) + a(k_l)a^\dagger(k_l)}{2}. \tag{W.13}$$

One can show that the product $aa^\dagger$ is time-independent and therefore also $\hat{H}_l$ (see exercises). Hence, we can suppress the (uninteresting) time dependence and keep the notation $a(k_l)$.

With the help of the commutation relation (W.11), i.e., $a(k_l)a^\dagger(k_l) = 1 - a^\dagger(k_l)a(k_l)$, we rewrite $\hat{H}_l$ in (W.13) and obtain

$$\hat{H}_l = \hbar \omega_l \left[ a^\dagger(k_l)a(k_l) + \frac{1}{2} \right] \tag{W.14}$$

which expression formally equals the energy of the harmonic oscillator. In addition, we can also define a number operator $\hat{N}_l = a^\dagger(k_l)a(k_l)$ with eigenstates $|n_l\rangle$; the eigenwert equation reads $N_l |n_l\rangle = n_l |n_l\rangle$. The states $|n_l\rangle$ are orthogonal, $\langle n_l| n_j \rangle = \delta_{n_l n_j}$.

Due to

$$\hat{H}_l |n_l\rangle = \hbar \omega_l \left( \hat{N}_l + 1/2 \right) |n_l\rangle = \hbar \omega_l (n_l + 1/2) |n_l\rangle = E_{l,n_l} |n_l\rangle \tag{W.15}$$

we can write for the Hamilton function and for the energy[68]

$$\hat{H}_l = \hbar \omega_l \left( \hat{N}_l + \frac{1}{2} \right) \; ; \; E_{l,n_l} = \hbar \omega_l \left( n_l + \frac{1}{2} \right). \tag{W.16}$$

The ground state is $|0\rangle$ (the vacuum), and the ladder operators rise and lower the states, $a^\dagger(k_l) |n_l\rangle = \sqrt{n_l + 1} |n_l + 1\rangle$ and $a(k_l) |n_l\rangle = \sqrt{n_l} |n_l - 1\rangle$. A state $|n_l\rangle$ can be produced out of the vacuum by repeated application of $a^\dagger(k_l)$:

$$a^\dagger(k_l) |0\rangle = \sqrt{1} |1\rangle \; ; \; a^\dagger(k_l) |1\rangle = \left[ a^\dagger(k_l) \right]^2 |0\rangle = \sqrt{2} |1\rangle \; ; \ldots \left[ a^\dagger(k_l) \right]^m |0\rangle = \sqrt{m!} |m\rangle \tag{W.17}$$

or

$$|n_l\rangle = \frac{1}{\sqrt{n_l!}} \left[ a^\dagger(k_l) \right]^{n_l} |0\rangle. \tag{W.18}$$

Remind that the application of $a(k_l)$ to the vacuum vanishes, $a(k_l) |0\rangle = 0$.

Though the formalism is the same for the harmonic oscillator[69] and our toy example, the interpretation differs. In case of the harmonic oscillator, $E_m = \hbar \omega \left( m + \frac{1}{2} \right)$ means that the oscillator is in the $m$th level, but we now interpret $E_{l,n_l} = \hbar \omega_l \left( n_l + \frac{1}{2} \right)$ as an indication that the mode[70] is occupied by $n_l$ 'particles' of energy $\hbar \omega_l$. These

---

[68] Do not confuse energy $E_n$ and electric field $E_x$.

[69] We remark that it is not at all self-evident that we can use here the formalismus of the harmonic oscillator. It is rather one of those serendipities in physics.

[70] Here, a mode of the field is determined by its energy $\hbar \omega_l$.

'particles' are named *photons*. Note that 'photon' means just the basic unit of the energy of the electromagnetic field and not some sort of point particle running through the space. Indeed, the energy of a mode is delocalized and property of the mode.

Thus, e.g. the state $|n_l\rangle$ contains $n_l$ photons of energy $\hbar\omega_l$ - in other words, $|n_l\rangle$ is the state in which the mode is occupied by $n_l$ photons.

## W.2.4  Generalization

In our toy example, we have assumed two plates between which there are standing waves in $z$-direction which are linearly polarized in $x$-direction. We now generalize the example. We assume that we have not only two plates but a cube of length $L$. Thus, the allowed wave vectors are given by

$$\mathbf{k} = \frac{\pi}{L}\mathbf{n} \ ; \ n_x, n_y, n_z \in \mathbb{Z}. \tag{W.19}$$

In addition, we allow for arbitrary plane waves (i.e., fields with three non-vanishing components) and for arbitrary polarizations (not only in $x$-direction). Thus, a mode is determined by the numbers $(\mathbf{k}, r)$ where $r = 1, 2, 3$ indicates the polarization direction (e.g., in $x$-, $y$- and $z$-direction). Thus, we write the creation operator (formerly $a(k_l)$) as $a_r(\mathbf{k})$. In addition, instead of the partial Hamilton function $H_l$ (W.14) for one mode, we consider all modes $(\mathbf{k}, r)$. Thus, with a similar calculation as above we arrive at the Hamilton operator $\hat{H}$:

$$\hat{H} = \sum_{\mathbf{k},r} \hbar\omega_\mathbf{k} \left[ a_r^\dagger(\mathbf{k})\, a_r(\mathbf{k}) + \frac{1}{2} \right]. \tag{W.20}$$

Here, the summation runs over all allowed $\mathbf{k}$-values and all polarization directions $r$, and $\omega_\mathbf{k} = c\,|\mathbf{k}|$. Each mode behaves like a independent harmonic oscillator and can accept energy in an integer number of portions (quanta) of size $\hbar\omega_\mathbf{k}$. The commutation relation (W.11) takes on the form

$$\left[ a_r^\dagger(\mathbf{k}), a_r(\mathbf{k}) \right] = \delta_{\mathbf{k}\mathbf{k}'}\delta_{rr'}. \tag{W.21}$$

The number operator is defined by

$$N_{\mathbf{k}r} = a_r^\dagger(\mathbf{k})\, a_r(\mathbf{k}) \tag{W.22}$$

measuring the number of modes with quantum numbers $\mathbf{k}r$. We denote by $|n_{\mathbf{k}r}\rangle$ the eigenvectors of $N_{\mathbf{k}r}$, and the eigenvalue equation reads[71]

---

[71] A more detailed notation would be $|\mathbf{k}r, n_{\mathbf{k}r}\rangle$, i.e., a state with quantum numbers $\mathbf{k}r$ and the occupation number $n_{\mathbf{k}r}$. Instead, we prefer the shorter and equivalent notation $|n_{\mathbf{k}r}\rangle$.

$$N_{\mathbf{k}r} |n_{\mathbf{k}r}\rangle = n_{\mathbf{k}r} |n_{\mathbf{k}r}\rangle \tag{W.23}$$

where $n_{\mathbf{k}r}$ is the occupation number of the mode indicated by $(\mathbf{k}r)$. As in the simpler version of our example, we can adopt the interpretation that we have $n_{\mathbf{k}r}$ photons with wave vector $\mathbf{k}$ and polarization $r$ - or: the mode $(\mathbf{k}, r)$ is occupied by $n_{\mathbf{k}r}$ photons. Application of the operators $a_r^\dagger(\mathbf{k})$ and $a_r(\mathbf{k})$ increases and lowers this number by one, and we can represent the states by repeated application of the creation operator $a_r^\dagger(\mathbf{k})$:

$$|n_{\mathbf{k}r}\rangle = \frac{1}{\sqrt{n_{\mathbf{k}r}!}} \left[ a_r^\dagger(\mathbf{k}) \right]^{n_{\mathbf{k}r}} |0\rangle \tag{W.24}$$

and

$$a_r(\mathbf{k}) |0\rangle = 0. \tag{W.25}$$

A general state which comprises all modes may be written $|n_1, n_2, n_3, \ldots\rangle$, i.e., the first mode is occupied by $n_1$ photons, the second by $n_2$ and so on. In compact form, this may be written as

$$|n_1, n_2, n_3, \ldots\rangle = \prod_{\mathbf{k},r} \frac{1}{\sqrt{n_{\mathbf{k}r}!}} \left[ a_r^\dagger(\mathbf{k}) \right]^{n_{\mathbf{k}r}} |0\rangle. \tag{W.26}$$

In other words, we do not label the (identical) photons by assigning each of them an individual quantum state.[72] Instead, we count how many photons are occupying a mode.

This notation and its interpretation is known as *occupation number representation*. As seen from (W.26), it enables us to go almost without state vectors except the vacuum state $|0\rangle$. One says that the operators $a_r^\dagger(\mathbf{k})$ and $a_r(\mathbf{k})$ create and annihilate a particle (photon) with quantum numbers $(\mathbf{k}, r)$. The states $|n\rangle$ are called *number states* or *Fock states* since they live in Fock space.

### W.2.4.1  Summary of the Quantization Approach

Let us resume the approach. We started with the classical field equations and represented the solution in terms of plane waves whose amplitudes were essentially $q$ and $p = \dot{q}$. Thus, we could identify by mere inspection $q$ and $p$ as canonical variables and transform them into non-commuting operators. This means that also the fields are transformed into field operators. The energy of the system is a function of $q$ and $p$ and formally identical to that of the harmonic oscillator. Exploiting the formal analogy, we can use the formalism of the harmonic operator to construct ladder, number and energy operators though the two systems (toy example and harmonic oscillator) are physically completely different. In contrast to the harmonic oscillator, we adopt

---

[72]By the way, this would not make sense since photons are indistinguishable; see Vol. 2 Chap. 23 'Identical Particles'.

for our toy system the occupation number representation which counts how many photons are occupying a single mode.

In principle, this is the method to quantize all fields. However, the way chosen here is tailored to our toy system and we have to find a general approach by which we can identify the canonical variables. In possession of these terms, we can formulate creation and annihilation operators and all other quantities of interest. As we will see below, this tool is the Lagrange–Hamilton formalism.

### W.2.4.2   A Big Problem

So far, everything is coherent and fine - except for a 'small' problem, which in fact is an infinite one. The energy associated with the Hamiltonian (W.20) is given by

$$E = \sum_{\mathbf{k},r} \hbar\omega_{\mathbf{k}} \left( n_{\mathbf{k}r} + \frac{1}{2} \right). \tag{W.27}$$

We can split and write this as

$$E = \sum_{\mathbf{k},r} \hbar\omega_{\mathbf{k}}\, n_{\mathbf{k}r} + \sum_{\mathbf{k},r} \frac{\hbar\omega_{\mathbf{k}}}{2}. \tag{W.28}$$

Due to the second summand, i.e., the sum over the zero point energies, the energy $E$ is *always infinite*, even if all occupation numbers $n_{\mathbf{k}r}$ vanish. Let us point out in advance that this is not a speciality of our toy system. In fact, all fields to be quantized in the following (Klein–Gordon, Dirac, photons) display the same problem of infinite zero point or vacuum energy.

Of course, this is a serious problem. Can a formalism be credible in which one always has to take account of an infinitely large number - or formulated differently: in which one has to subtract an infinite number to get finite results? In general, physicists are rather nonchalant with their mathematical methods, and sometimes allow for some sloppiness for the sake of argument. But this divergence problem literally has a different order of magnitude.

We postpone the discussion and address the subject again in section 'Operator ordering'.

## W.2.5   Exercises and Solutions

1. Show that (W.3) and (W.4) satisfy the Maxwell equations (W.1).
   Solution: We have for the divergence terms

$$\nabla\cdot\mathbf{E}\,(\mathbf{r}, t) = \partial_x E_x\,(z, t) = 0 \;;\; \nabla\cdot\mathbf{B}\,(\mathbf{r}, t) = \partial_y B_y\,(z, t) = 0. \tag{W.29}$$

The rotation terms are given by $\nabla \times \mathbf{E}(\mathbf{r}, t) = (0, \partial_z E_x, 0)$ and $\nabla \times \mathbf{B}(\mathbf{r}, t) = (-\partial_z B_y, 0, 0)$. It follows

$$\partial_z E_x(z, t) = -\frac{\partial}{\partial t} B_y(z, t) \rightarrow k_m C \cdot q(t) \cos k_m z = -C \cdot \frac{1}{c^2 k_m} \ddot{q}(t) \cos k_m z \tag{W.30}$$

and

$$-\partial_z B_y(z, t) = \frac{1}{c^2} \frac{\partial}{\partial t} E_x(z, t) \rightarrow C \cdot \frac{1}{c^2} \dot{q}(t) \sin k_m z = \frac{1}{c^2} C \cdot \dot{q}(t) \sin k_m z. \tag{W.31}$$

The last equation is always satisfied, and the first for $\ddot{q}(t) = -c^2 k_m^2 q(t)$.

2. Prove (W.6)., i.e.,

$$H = \frac{1}{2}\left[\dot{q}^2(t) + \omega_m^2 q^2(t)\right]. \tag{W.32}$$

Solution: We have

$$H = \frac{\varepsilon_0}{2} \int_0^L dz \left[E_x^2 + c^2 B_y^2\right] = \frac{\varepsilon_0}{2} \int_0^L dz \left[C^2 q^2(t) \sin^2 k_m z + c^2 C^2 \frac{1}{c^4 k_m^2} \dot{q}^2(t) \cos^2 k_m z\right]. \tag{W.33}$$

This yields

$$H = \frac{\varepsilon_0}{2} C^2 \left[q^2(t) \int_0^L dz \sin^2 k_m z + \frac{1}{c^2 k_m^2} \dot{q}^2(t) \int_0^L dz \cos^2 k_m z\right]. \tag{W.34}$$

With

$$\int_0^L dz \sin^2 k_m z = \left\{\frac{z}{2} - \frac{\sin 2 k_m z}{4 k_m}\right\}_0^L = \frac{L}{2} - \frac{\sin 2\frac{\pi}{L} m L}{4 k_m} = \frac{L}{2} \tag{W.35}$$

and

$$\int_0^L dz \cos^2 k_m z = \int_0^L dz \left(1 - \sin^2 k_m z\right) = \frac{L}{2} \tag{W.36}$$

follows with $\omega_m^2 = c^2 k_m^2$

$$H = \frac{\varepsilon_0}{2} C^2 \frac{L}{2} \left[q^2(t) + \frac{1}{c^2 k_m^2} \dot{q}^2(t)\right] = \frac{\varepsilon_0 L}{2 \omega_m^2} C^2 \frac{1}{2}\left[\dot{q}^2(t) + \omega_m^2 q^2(t)\right]. \tag{W.37}$$

The choice $C^2 = \frac{2\omega_m^2}{\varepsilon_0 L}$ brings the desired result.

3. Show (W.13).

Solution: We have with (W.7) and (W.10)

$$\hat{H}_l = \frac{1}{2}\left[-\frac{\hbar\omega_l}{2}\left(a^\dagger(k_l) - a(k_l)\right)\left(a^\dagger(k_l) - a(k_l)\right) + \omega_l^2 \frac{\hbar}{2\omega_l}\left(a^\dagger(k_l) + a(k_l)\right)\left(a^\dagger(k_l) + a(k_l)\right)\right] =$$

$$= \frac{\hbar\omega_l}{4}\left[\begin{matrix}-a^\dagger(k_l)a^\dagger(k_l) + a^\dagger(k_l)a(k_l) + a(k_l)a^\dagger(k_l) - a(k_l)a(k_l) + \\ + \left(a^\dagger(k_l)a^\dagger(k_l) + a(k_l)a^\dagger(k_l) + a(k_l)a^\dagger(k_l) + a(k_l)a(k_l)\right)\end{matrix}\right] =$$

$$= \frac{\hbar\omega_l}{2}\left[a^\dagger(k_l)a(k_l) + a(k_l)a^\dagger(k_l)\right].$$

$$\text{(W.38)}$$

4. Show that $a(k_l)a^\dagger(k_l)$ is time-independent.

   Solution: Let us write for the moment $a(k_l, t)$. We invoke the definition of the time derivative in the Heisenberg picture as given by $dA/dt = i/\hbar\,[H, Aa]$. In our case this leads to

$$\frac{da(k_l, t)}{dt} = \frac{i}{\hbar}\left[\hat{H}_l, a(k_l, t)\right] = i\omega_l\left[a^\dagger(k_l, t)a(k_l, t), a(k_l, t)\right]$$

$$= -i\omega_l a(k_l, t) \; ; \quad \frac{da^\dagger(k_l, t)}{dt} = i\omega_l a^\dagger(k_l, t)$$

$$\text{(W.39)}$$

with the solutions

$$a(k_l, t) = a(k_l, 0) \cdot e^{-i\omega_l t} \; ; \; a^\dagger(k_l, t) = a^\dagger(k_l, 0) \cdot e^{i\omega_l t} \qquad \text{(W.40)}$$

and it follows

$$a(k_l, t)a^\dagger(k_l, t) = a(k_l, 0)a^\dagger(k_l, 0). \qquad \text{(W.41)}$$

Thus, we see clearly that the product $aa^\dagger$ is time-independent and therefore also $\hat{H}_l$. Hence, we switch back to the notation $a(k_l)$, suppressing the (now uninteresting) time dependence.

## W.3   Quantization of Free Fields, Introduction

In the last section, we used an approach for the quantization which was tailored to the system under consideration. We discuss now the general method (in fact, *the* general method) which is based on the Lagrange–Hamilton formalism.[73] It answers the relevant questions as, for instant, how to find those variables which are transformed into non-commuting operators, how to find the energy density (Hamiltonian density) and so on.

The Lagrangian contains the complete information about the physical system. It enables us to derive (1) the equations of motions, (2) the conjugated momentum, (3) the Hamiltonian. Moreover, it forms the basis of the canonical quantization by which the field operator and its conjugated momentum are subject to certain commutation relations.

In addition, following our approach for the toy system, we have a further point in the quantizing procedure, namely to express the field operators in terms of anni-

---

[73] In Vol. 1, App. T.3, there is a short outline of the basics of this formalism.

hilation and creation operators. To this end, we construct the free solutions and replace the expansion coefficients by operators. The final step is the formulation of the commutation relation for these annihilation and creation operators.

We summarize the *canonical quantization* procedure which we will apply in case of the Klein–Gordon field. Assume that we have a physical (classical, non-quantum) system with an appropriate Lagrangian $\mathcal{L}$ for several fields $\varphi_r, r = 1, 2, \ldots$. Then we have to carry out the following steps:

1. We calculate the conjugated momentum fields $\pi_r$ and formulate the corresponding Hamiltonian $\mathcal{H}$ density as a function of the fields $\varphi_r$ and the momentum densities $\pi_r$.

2. We consider $\varphi_r$ and $\pi_r$ as operators obeying certain commutation relations.

3. On the basis of free solutions of the system, we formulate $\varphi_r$ and $\pi_r$ in terms of annihilation and creation operators and deduce the commutation relations for these two types of operators.

However, this approach is, in a certain sense, idealized and/or of limited value. The reason is that it is based on the knowledge of $\mathcal{L}$ (or $\mathcal{H}$) of the classical system to be quantized. But there are systems with no classical $\mathcal{L}$ as for instance the spinor field of the Dirac equation. Thus, in this case, the approach can not be applied directly, and we have to look for the right Lagrangian, not to mention the appropriate commutation relations.

In the following, we will first consider the Klein–Gordon field as example for the canonical quantization. Then we will show for the Dirac system how to proceed if there is no classical Lagrangian. Finally, we compile some results for the free photon field.

Starting with this section, we will use those physical units in which $c = 1$ and $\hbar = 1$ (see Appendix B, Vol. 1, 'Natural units'). This is common practice in quantum field theory and very functional. As a consequence, several quantities are now directly equivalent. We have for instance $\mathbf{p} = \mathbf{k}$ or $E_\mathbf{k} = \omega_\mathbf{k}$, the dispersion relation may be written $E_\mathbf{k} = \sqrt{\mathbf{k}^2 + m^2}$, and for the 4-vector $p$ holds $p^0 = k^0 = E_\mathbf{k} = \omega_\mathbf{k} = E_\mathbf{p} = \omega_\mathbf{p}$.

# W.4   Quantization of Free Fields, Klein–Gordon

We now consider the real Klein–Gordon field and its quantization. This field describes electrical neutral mesons with spin zero which are equal to their antiparticles.[74] Apart from its physical importance, the Klein–Gordon field is a nice example for canonical quantization.

---

[74]Mesons which differ from their antiparticles (e.g. the $K_0$ meson with its antiparticle $\bar{K}_0$) are described by a complex Klein–Gordon field.

### W.4.1 Lagrangian, Conjugated Momentum, Poisson Brackets, Hamiltonian

The Lagrangian density $\mathcal{L}$ of the real Klein–Gordon field is given by[75]

$$\mathcal{L} = \frac{1}{2} \left(\partial_\mu \phi\right)\left(\partial^\mu \phi\right) - \frac{1}{2}m^2\phi^2. \tag{W.42}$$

The Euler–Lagrange equation reads

$$\frac{\partial \mathcal{L}}{\partial \phi} - \partial_\mu \left(\frac{\partial \mathcal{L}}{\partial \left(\partial_\mu \phi\right)}\right) = 0. \tag{W.43}$$

With

$$\frac{\partial \mathcal{L}}{\partial \phi} = -m^2 \phi \; ; \; \frac{\partial \mathcal{L}}{\partial \left(\partial_\mu \phi\right)} = \partial^\mu \phi \tag{W.44}$$

follows the Klein–Gordon equation

$$\partial_\mu \partial^\mu \phi + m^2 \phi = 0. \tag{W.45}$$

The conjugated momentum density $\pi$ is given by

$$\pi = \frac{\delta \mathcal{L}}{\delta \left(\partial_0 \phi\right)} = \partial_0 \phi \tag{W.46}$$

and the Hamiltonian density follows as

$$\mathcal{H} = \frac{1}{2}\left[\dot{\phi}^2 + (\nabla\phi)^2 + m^2\phi^2\right]. \tag{W.47}$$

### W.4.2 Canonical Quantization

The Poisson brackets for the Klein–Gordon field are given by[76] (see Appendix T, Vol. 1)

---

[75] Remarks concerning the notation: Instead of $\left(\partial_\mu \phi\right)\left(\partial^\mu \phi\right)$, some textbooks write $\left(\partial_\mu \phi\right)^2$. One finds also $\partial_\mu \phi \partial^\mu \phi$, whereby this is not meant in the sense of the product rule, i.e., $\left(\partial_\mu \phi\right)\left(\partial^\mu \phi\right) + \phi \partial_\mu \partial^\mu \phi$. In any case, the following expression is meant: $(\partial_0 \phi)\left(\partial^0 \phi\right) - (\partial_k \phi)(\partial_k \phi) = (\partial_0 \phi)^2 - (\partial_k \phi)^2 = \dot{\phi}^2 - (\nabla\phi)^2$. In these terms, (W.42) is written $\mathcal{L} = \frac{1}{2}\dot{\phi}^2 - \frac{1}{2}(\nabla\phi)^2 - \frac{1}{2}m^2\phi^2$.

[76] In order to emphasize the dimension of the argument, we write occasionally $\delta^{(3)}(\mathbf{x})$ instead of $\delta(\mathbf{x})$; analogously for other dimensions.

$$\begin{aligned} \{\phi\,(t,\mathbf{x})\,,\pi\,(t,\mathbf{x}')\}_{PB} &= \delta^{(3)}\,(\mathbf{x}-\mathbf{x}') \\ \{\phi\,(t,\mathbf{x})\,,\phi\,(t,\mathbf{x}')\}_{PB} &= 0 \\ \{\pi\,(t,\mathbf{x})\,,\pi\,(t,\mathbf{x}')\}_{PB} &= 0. \end{aligned} \qquad (\text{W.48})$$

The canonical quantization of Klein–Gordon field means that we consider the field variables $\phi\,(x)$ and $\pi\,(x)$ as operators obeying the commutation relations[77]

$$\begin{aligned} \left[\phi\,(t,\mathbf{x})\,,\dot{\phi}\,(t,\mathbf{x}')\right] &= i\delta^{(3)}\,(\mathbf{x}-\mathbf{x}') \\ \left[\phi\,(t,\mathbf{x})\,,\phi\,(t,\mathbf{x}')\right] &= \left[\dot{\phi}\,(t,\mathbf{x})\,,\dot{\phi}\,(t,\mathbf{x}')\right] = 0. \end{aligned} \qquad (\text{W.49})$$

Note the additional factor $i$ in the transition from the Poisson brackets to the commutation rules. We now expand $\phi\,(x)$ in terms of plane wave solutions of the Klein–Gordon equation (see Appendix U, Vol. 1). In the case of a finite volume $V$, they read

$$\phi\,(x) = \sum_{\mathbf{p}} \frac{1}{\sqrt{2VE_{\mathbf{p}}}} \left(a\,(\mathbf{p})\,e^{-ipx}+a^{\dagger}\,(\mathbf{p})\,e^{ipx}\right) \qquad (\text{W.50})$$

where the sum runs over all allowed discrete values of $\mathbf{p}$. The continuous solution reads[78]

$$\phi\,(x) = \frac{1}{(2\pi)^{3/2}} \int \frac{d^3p}{\sqrt{2E_{\mathbf{p}}}} \left(a\,(\mathbf{p})\,e^{-ipx}+a^{\dagger}\,(\mathbf{p})\,e^{ipx}\right). \qquad (\text{W.51})$$

Energy and momentum are related by the relativistic dispersion relation

$$E_{\mathbf{p}} = p^0 = \sqrt{\mathbf{p}^2+m^2}. \qquad (\text{W.52})$$

Inverting the continuous solution (W.51) gives for the operators $a\,(\mathbf{p})$ and $a^{\dagger}\,(\mathbf{p})$

$$\begin{aligned} a\,(\mathbf{p}) &= \frac{1}{\sqrt{(2\pi)^3 2E_{\mathbf{p}}}} \int d^3x\, e^{-i\mathbf{px}}\left(E_{\mathbf{p}}\phi\,(0,\mathbf{x})+i\dot{\phi}\,(0,\mathbf{x})\right) \\ a^{\dagger}\,(\mathbf{p}) &= \frac{1}{\sqrt{(2\pi)^3 2E_{\mathbf{p}}}} \int d^3x\, e^{i\mathbf{px}}\left(E_{\mathbf{p}}\phi\,(0,\mathbf{x})-i\dot{\phi}\,(0,\mathbf{x})\right). \end{aligned} \qquad (\text{W.53})$$

From (W.49) and (W.53) follow the commutation relations for $a\,(\mathbf{p})$ and $a^{\dagger}\,(\mathbf{p})$:

$$\left[a\,(\mathbf{p})\,,a^{\dagger}\,(\mathbf{p}')\right]=\delta\,(\mathbf{p},\mathbf{p}') \;\;;\;\; \left[a\,(\mathbf{p})\,,a\,(\mathbf{p}')\right]=\left[a^{\dagger}\,(\mathbf{p})\,,a^{\dagger}\,(\mathbf{p}')\right]=0 \qquad (\text{W.54})$$

which are again (formally) identical with the commutation relations of a harmonic oscillator[79] In other words: we can now follow the considerations from the toy

---

[77] Here the fields are considered at the same time. For different times, the fields commute.

[78] As in Appendix U, Vol. 1, we use the same symbol $a\,(\mathbf{p})$ in the discrete and the continuous case. Strictly speaking one would have to make a distinction e.g. by different names. But our approach is quite functional and the likelihood of confusion is very low.

[79] $\delta(a,b)$ is the generalized Kronecker symbol introduced in Vol. 1, Chap. 12:

**Table W.1** Table of simplest states of the Klein–Gordon field

| State | | Energy |
|---|---|---|
| Vacuum | $\lvert 0 \rangle$ | 0 |
| One particle | $a^\dagger(\mathbf{p})\lvert 0 \rangle$ | $E_\mathbf{p}$ |
| Two different particles | $a^\dagger(\mathbf{p}_1)\,a^\dagger(\mathbf{p}_2)\lvert 0 \rangle$ | $E_{\mathbf{p}_1} + E_{\mathbf{p}_2}$ |
| Two identical particles | $a^\dagger(\mathbf{p})\,a^\dagger(\mathbf{p})\lvert 0 \rangle$ | $2E_\mathbf{p}$ |

example, i.e., we can adopt the formalism of the harmonic oscillator and the occupation number representation.

One can regard the operators

$$N_\mathbf{p} = a^\dagger(\mathbf{p})\, a(\mathbf{p}) \tag{W.56}$$

as number operators with eigenvalues $0, 1, 2, \ldots$. The operators $a(\mathbf{p})$ and $a^\dagger(\mathbf{p})$ delete and create particles with momentum $p$.[80] For instance, $a^\dagger(\mathbf{p})\lvert 0 \rangle$ creates a particle with momentum $p$, and $a^\dagger(\mathbf{p}_1)\,a^\dagger(\mathbf{p}_2)\lvert 0 \rangle$ two particles with momenta $p_1$ and $p_2$. Some normalized particle states with their energy eigenvalues are given in Table W.1.

Finally, we look for the Hamiltonian. We have with (W.47)

$$H = \int \mathrm{d}^3x\, \mathcal{H}(x) = \int \mathrm{d}^3x\, \frac{1}{2}\left[\dot{\phi}^2(x) + (\nabla\phi(x))^2 + m^2\phi^2(x)\right]. \tag{W.57}$$

With (W.50) or with (W.51), this leads to

$$H = \frac{1}{2}\sum_\mathbf{p} E_\mathbf{p}\left(a^\dagger(\mathbf{p})\,a(\mathbf{p}) + a(\mathbf{p})\,a^\dagger(\mathbf{p})\right) \tag{W.58}$$

or

$$H = \frac{1}{2}\int \mathrm{d}^3p\, E_\mathbf{p}\left[a^\dagger(\mathbf{p})\,a(\mathbf{p}) + a(\mathbf{p})\,a^\dagger(\mathbf{p})\right]. \tag{W.59}$$

Two remarks:

(1) Note that there are only *positive* energies because the number operator $a^\dagger(\mathbf{p})\,a(\mathbf{p})$ has only positive eigenvalues (plus zero). In other words, we got rid of the problem with negative energies.

(2) Instead, we face another problem, known from the toy example: the energy is infinite always. By means of (W.54) we can write for example

$$\delta(a, b) = \begin{cases} \delta_{ab} \\ \delta(a - b) \end{cases} \text{ for } \begin{array}{l} a, b \text{ discrete} \\ a, b \text{ continuous} \end{array} \tag{W.55}$$

---

[80]Note that due to $p_0^2 = \mathbf{p}^2 + m^2$, it does not matter if we give the 4-momentum $p$ or the 3-momentum $\mathbf{p}$.

$$H = \sum_{\mathbf{p}} \frac{1}{2} E_{\mathbf{p}} \left( a^{\dagger} (\mathbf{p}) \, a (\mathbf{p}) + a (\mathbf{p}) \, a^{\dagger} (\mathbf{p}) \right) = \sum_{\mathbf{p}} E_{\mathbf{p}} \left( a^{\dagger} (\mathbf{p}) \, a (\mathbf{p}) + \frac{1}{2} \right) \quad \text{(W.60)}$$

This means that even if there is no particle at all (i.e., vacuum), there is a energy of $\frac{1}{2} \sum_{\mathbf{p}} E_{\mathbf{p}}$ (which we have omitted in the last table). We will find this issue also in the case of the Dirac field and the photon field.

Thus, there is a welcome property (energies are only positive) and an unwelcome property (infinite energies). We will discuss this problem below in the section 'Operator ordering'.

## W.4.3   Exercises and Solutions

1. Show

$$\frac{\partial}{\partial \left( \partial_{\mu} \phi \right)} \left( \partial_{\alpha} \phi \right) \left( \partial^{\alpha} \phi \right) = 2 \left( \partial^{\mu} \phi \right) . \quad \text{(W.61)}$$

Solution: We have

$$\frac{\partial}{\partial (\partial_{\mu} \phi)} \left( \partial_{\alpha} \phi \right) \left( \partial^{\alpha} \phi \right) = \delta_{\mu \alpha} \left( \partial^{\alpha} \phi \right) + \left( \partial_{\alpha} \phi \right) \frac{\partial}{\partial (\partial_{\mu} \phi)} \left( g^{\alpha \nu} \partial_{\nu} \phi \right) =$$
$$= \left( \partial^{\mu} \phi \right) + \left( \partial_{\alpha} \phi \right) g^{\alpha \nu} \delta_{\mu \nu} \phi = \left( \partial^{\mu} \phi \right) + \left( \partial^{\nu} \phi \right) \delta_{\mu \nu} \phi = \left( \partial^{\mu} \phi \right) + \left( \partial^{\mu} \phi \right) = 2 \left( \partial^{\mu} \phi \right) . \quad \text{(W.62)}$$

2. Prove (W.47).
   Solution:

$$\mathcal{H} = \pi \partial_{0} \phi - \mathcal{L} = \left( \partial_{0} \phi \right)^{2} - \frac{1}{2} \left( \partial_{\mu} \phi \right) \left( \partial^{\mu} \phi \right) + \frac{1}{2} m^{2} \phi^{2} =$$
$$= \left( \partial_{0} \phi \right)^{2} - \frac{1}{2} \left( \partial_{0} \phi \right) \left( \partial^{0} \phi \right) - \frac{1}{2} \left( \partial_{k} \phi \right) \left( \partial^{k} \phi \right) + \frac{1}{2} m^{2} \phi^{2} =$$
$$= \frac{1}{2} \left( \partial_{0} \phi \right) \left( \partial^{0} \phi \right) - \frac{1}{2} \left( \partial_{k} \phi \right) \left( \partial^{k} \phi \right) + \frac{1}{2} m^{2} \phi^{2} = \frac{1}{2} \left[ \left( \partial_{0} \phi \right)^{2} + \left( \partial_{k} \phi \right) \left( \partial^{k} \phi \right) + m^{2} \phi^{2} \right] =$$
$$= \frac{1}{2} \left[ \dot{\phi}^{2} + \left( \nabla \phi \right)^{2} + m^{2} \phi^{2} \right] . \quad \text{(W.63)}$$

3. Starting from $\mathcal{H}$, deduce the Klein–Gordon equation.
   Solution: We have

$$\mathcal{H} = \frac{1}{2} \pi^{2} - \frac{1}{2} \left( \partial_{k} \phi \right) \left( \partial^{k} \phi \right) + \frac{1}{2} m^{2} \phi^{2} = \frac{1}{2} \pi^{2} + \frac{1}{2} \left( \partial_{k} \phi \right) \left( \partial_{k} \phi \right) + \frac{1}{2} m^{2} \phi^{2} . \quad \text{(W.64)}$$

Note, that $\mathcal{H}$ does not depend on $\partial_{0} \phi$. With (cf. the proceeding section)[81]

---

[81] Remember

$$\frac{\delta}{\delta \varphi} = \frac{\partial}{\partial \varphi} - \partial_{\mu} \frac{\partial}{\partial \left( \partial_{\mu} \varphi \right)} \quad ; \quad \frac{\delta}{\delta \pi} = \frac{\partial}{\partial \pi} - \partial_{\mu} \frac{\partial}{\partial \left( \partial_{\mu} \pi \right)} \quad \text{(W.65)}$$

$$\dot{\phi} = \frac{\partial \mathcal{H}}{\partial \pi} - \partial_\mu \frac{\partial \mathcal{H}}{\partial \left( \partial_\mu \pi \right)} \ ; \ \dot{\pi} = -\frac{\partial \mathcal{H}}{\partial \phi} + \partial_\mu \frac{\partial \mathcal{H}}{\partial \left( \partial_\mu \phi \right)} \qquad (W.66)$$

follows

$$\dot{\phi} = \pi \ ; \ \dot{\pi} = -m^2 \phi + \partial_k \frac{\partial \mathcal{H}}{\partial (\partial_k \phi)} = -m^2 \phi + \frac{1}{2} \partial_k \frac{\partial}{\partial (\partial_k \phi)} \left( \partial_l \phi \right) \left( \partial_l \phi \right) = $$
$$= -m^2 \phi + \partial_k \left( \partial_l \phi \right) \delta_{kl} = -m^2 \phi + \partial_k \left( \partial_k \phi \right) . \qquad (W.67)$$

This leads to

$$\ddot{\phi} = \dot{\pi} = -m^2 \phi + \partial_k \left( \partial_k \phi \right) = -m^2 \phi + \nabla^2 \phi \qquad (W.68)$$

or

$$\partial_\mu \partial^\mu \phi + m^2 \phi = 0. \qquad (W.69)$$

4. Prove (W.53).
Solution: We consider the continuous solution (W.51). With $px = E_\mathbf{p} t - \mathbf{px}$ follows

$$\phi (0, \mathbf{x}) = \int \frac{d^3 p}{\sqrt{(2\pi)^3 2 E_\mathbf{p}}} \left[ a (\mathbf{p}) e^{i\mathbf{px}} + a^\dagger (\mathbf{p}) e^{-i\mathbf{px}} \right]$$
$$\dot{\phi} (0, \mathbf{x}) = \int \frac{d^3 p}{\sqrt{(2\pi)^3 2 E_\mathbf{p}}} \left[ -i E_\mathbf{p} a (\mathbf{p}) e^{i\mathbf{px}} + i E_\mathbf{p} a^\dagger (\mathbf{p}) e^{-i\mathbf{px}} \right] . \qquad (W.70)$$

This gives

$$E_\mathbf{P} \phi (0, \mathbf{x}) \pm i \dot{\phi} (0, \mathbf{x}) = \int \frac{d^3 p}{\sqrt{(2\pi)^3 2 E_\mathbf{p}}} \left\{ \begin{array}{l} E_\mathbf{P} \left[ a (\mathbf{p}) e^{i\mathbf{px}} + a^\dagger (\mathbf{p}) e^{-i\mathbf{px}} \right] \pm \\ \pm E_\mathbf{p} \left[ a (\mathbf{p}) e^{i\mathbf{px}} - a^\dagger (\mathbf{p}) e^{-i\mathbf{px}} \right] \end{array} \right\} .$$
$$(W.71)$$

The Fourier transformation of this expression gives with $\int d^3 x \ e^{i\mathbf{Px}} e^{-i\mathbf{px}} = (2\pi)^3 \delta (\mathbf{p} - \mathbf{P})$ and $E_{-\mathbf{P}} = \sqrt{(-\mathbf{P})^2 + m^2} = \sqrt{\mathbf{P}^2 + m^2} = E_\mathbf{P}$

$$\int d^3 x \ e^{i\mathbf{Px}} \left[ E_\mathbf{P} \phi (0, \mathbf{x}) \pm i \dot{\phi} (0, \mathbf{x}) \right] = \sqrt{\frac{(2\pi)^3}{2 E_\mathbf{P}}} \left\{ (E_\mathbf{P} \pm E_\mathbf{P}) a (-\mathbf{P}) + (E_\mathbf{P} \mp E_\mathbf{P}) a^\dagger (\mathbf{P}) \right\} .$$
$$(W.72)$$

Solving for $a_{-\mathbf{P}}$ and $a_\mathbf{P}^\dagger$ yields

$$\sqrt{(2\pi)^3 2 E_\mathbf{P}} a (-\mathbf{P}) = \int d^3 x \ e^{i\mathbf{Px}} \left[ E_\mathbf{P} \phi (0, \mathbf{x}) + i \dot{\phi} (0, \mathbf{x}) \right]$$
$$\sqrt{(2\pi)^3 2 E_\mathbf{P}} a^\dagger (\mathbf{P}) = \int d^3 x \ e^{i\mathbf{Px}} \left[ E_\mathbf{P} \phi (0, \mathbf{x}) - i \dot{\phi} (0, \mathbf{x}) \right] \qquad (W.73)$$

and finally

$$a\left(\mathbf{P}\right) = \frac{1}{\sqrt{(2\pi)^3 2E_{\mathbf{P}}}} \int d^3x \, e^{-i\mathbf{Px}} \left[E_{\mathbf{P}}\phi\left(0, \mathbf{x}\right) + i\dot{\phi}\left(0, \mathbf{x}\right)\right]$$

$$a^\dagger\left(\mathbf{P}\right) = \frac{1}{\sqrt{(2\pi)^3 2E_{\mathbf{P}}}} \int d^3x \, e^{i\mathbf{Px}} \left[E_{\mathbf{P}}\phi\left(0, \mathbf{x}\right) - i\dot{\phi}\left(0, \mathbf{x}\right)\right].$$

(W.74)

5. Prove (W.54).

Solution: We consider the continuous case and start from (W.53)

$$a\left(\mathbf{p}\right) = \frac{1}{\sqrt{(2\pi)^3 2E_{\mathbf{p}}}} \int d^3x \, e^{-i\mathbf{px}} \left(E_{\mathbf{p}}\phi\left(0, \mathbf{x}\right) + i\dot{\phi}\left(0, \mathbf{x}\right)\right)$$

$$a^\dagger\left(\mathbf{p}\right) = \frac{1}{\sqrt{(2\pi)^3 2E_{\mathbf{p}}}} \int d^3x \, e^{i\mathbf{px}} \left(E_{\mathbf{p}}\phi\left(0, \mathbf{x}\right) - i\dot{\phi}\left(0, \mathbf{x}\right)\right).$$

(W.75)

It follows

$$a\left(\mathbf{p}\right) a^\dagger\left(\mathbf{p}'\right) - a^\dagger\left(\mathbf{p}'\right) a\left(\mathbf{p}\right) = \frac{1}{\sqrt{(2\pi)^3 2E_{\mathbf{p}}}} \frac{1}{\sqrt{(2\pi)^3 2E_{\mathbf{p}'}}} \int d^3x \int d^3x' \, e^{-i\mathbf{px}} e^{i\mathbf{p'x'}} \, \Xi$$

(W.76)

with

$$\Xi = \left[ \begin{array}{l} \left(E_{\mathbf{p}}\phi\left(0, \mathbf{x}\right) + i\dot{\phi}\left(0, \mathbf{x}\right)\right) \left(E_{\mathbf{p}'}\phi\left(0, \mathbf{x}'\right) - i\dot{\phi}\left(0, \mathbf{x}'\right)\right) - \\ - \left(E_{\mathbf{p}'}\phi\left(0, \mathbf{x}'\right) - i\dot{\phi}\left(0, \mathbf{x}'\right)\right) \left(E_{\mathbf{p}}\phi\left(0, \mathbf{x}\right) + i\dot{\phi}\left(0, \mathbf{x}\right)\right) \end{array} \right] =$$

$$= \begin{array}{l} i\dot{\phi}\left(0, \mathbf{x}\right) E_{\mathbf{p}'}\phi\left(0, \mathbf{x}'\right) - E_{\mathbf{p}}\phi\left(0, \mathbf{x}\right) i\dot{\phi}\left(0, \mathbf{x}'\right) - \\ -E_{\mathbf{p}'}\phi\left(0, \mathbf{x}'\right) i\dot{\phi}\left(0, \mathbf{x}\right) + i\dot{\phi}\left(0, \mathbf{x}'\right) E_{\mathbf{p}}\phi\left(0, \mathbf{x}\right) \end{array} =$$

$$= iE_{\mathbf{p}'} \left[\dot{\phi}\left(0, \mathbf{x}\right), \phi\left(0, \mathbf{x}'\right)\right] + iE_{\mathbf{p}} \left[\phi\left(0, \mathbf{x}'\right), \dot{\phi}\left(0, \mathbf{x}\right)\right].$$

(W.77)

With the commutation relations for the field operators (W.49) follows

$$\Xi = iE_{\mathbf{p}'} \left[-i\delta\left(\mathbf{x} - \mathbf{x}'\right)\right] + iE_{\mathbf{p}} \left[-i\delta\left(\mathbf{x} - \mathbf{x}'\right)\right] = \left(E_{\mathbf{p}'} + E_{\mathbf{p}}\right) \delta\left(\mathbf{x} - \mathbf{x}'\right).$$

(W.78)

With this result, (W.76) reads

$$a\left(\mathbf{p}\right) a^\dagger\left(\mathbf{p}'\right) - a^\dagger\left(\mathbf{p}'\right) a\left(\mathbf{p}\right) = \frac{1}{\sqrt{(2\pi)^3 2E_{\mathbf{p}}}} \frac{1}{\sqrt{(2\pi)^3 2E_{\mathbf{p}'}}} \int d^3x \int d^3x' \, e^{-i\mathbf{px}} e^{i\mathbf{p'x'}} \left(E_{\mathbf{p}'} + E_{\mathbf{p}}\right) \delta\left(\mathbf{x} - \mathbf{x}'\right) =$$

$$= \frac{1}{\sqrt{(2\pi)^3 2E_{\mathbf{p}}}} \frac{1}{\sqrt{(2\pi)^3 2E_{\mathbf{p}'}}} \int d^3x \, e^{-i\mathbf{px}} e^{i\mathbf{p'x}} \left(E_{\mathbf{p}'} + E_{\mathbf{p}}\right) = \frac{(2\pi)^3}{\sqrt{(2\pi)^3 2E_{\mathbf{p}}}} \frac{\delta^{(3)}\left(\mathbf{p}' - \mathbf{p}\right)}{\sqrt{(2\pi)^3 2E_{\mathbf{p}'}}} \left(E_{\mathbf{p}'} + E_{\mathbf{p}}\right)$$

(W.79)

or

$$\left[a\left(\mathbf{p}\right), a^\dagger\left(\mathbf{p}'\right)\right] = \delta^{(3)}\left(\mathbf{p}' - \mathbf{p}\right).$$

(W.80)

6. Prove (W.58) and/or (W.59).

Solution: We have

$$H = \int d^3x \, \frac{1}{2} \left[\dot{\phi}^2\left(x\right) + \left(\nabla\phi\left(x\right)\right)^2 + m^2\phi^2\left(x\right)\right]$$

(W.81)

and the continuous form (W.51) of the field operator

$$\phi(x) = \frac{1}{(2\pi)^{3/2}} \int \frac{d^3 p}{\sqrt{2E_\mathbf{p}}} \left( a(\mathbf{p}) e^{-ipx} + a^\dagger(\mathbf{p}) e^{ipx} \right). \tag{W.82}$$

Consider the first summand. With $\partial_t e^{-ipx} = -ip^0 e^{-ipx} = -iE_\mathbf{p} e^{-ipx}$ follows

$$
\begin{aligned}
\int d^3x \, \dot\phi^2(x) &= \frac{1}{(2\pi)^3} \int d^3x \int \frac{d^3 p}{\sqrt{2E_\mathbf{p}}} \int \frac{d^3 p'}{\sqrt{2E_{\mathbf{p}'}}} \left[ \begin{array}{l} (-ip_0)\left(a(\mathbf{p}) e^{-ipx} - a^\dagger(\mathbf{p}) e^{ipx}\right) \cdot \\ \cdot (-ip_0')\left(a(\mathbf{p}') e^{-ip'x} - a^\dagger(\mathbf{p}') e^{ip'x}\right) \end{array} \right] = \\
&= \frac{1}{(2\pi)^3} \int \frac{d^3 p}{\sqrt{2E_\mathbf{p}}} \int \frac{d^3 p'}{\sqrt{2E_{\mathbf{p}'}}} (-p_0 p_0') \int d^3x \left[ \begin{array}{l} a(\mathbf{p}) a(\mathbf{p}') e^{-ipx} e^{-ip'x} - a^\dagger(\mathbf{p}) a(\mathbf{p}') e^{ipx} e^{-ip'x} - \\ -a(\mathbf{p}) a^\dagger(\mathbf{p}') e^{-ipx} e^{ip'x} + a^\dagger(\mathbf{p}) a^\dagger(\mathbf{p}') e^{ipx} e^{ip'x} \end{array} \right] = \\
&= \int \frac{d^3 p}{\sqrt{2E_\mathbf{p}}} \int \frac{d^3 p'}{\sqrt{2E_{\mathbf{p}'}}} \left[ \begin{array}{l} (-p_0 p_0') \left\{ a(\mathbf{p}) a(\mathbf{p}') e^{-ip_0 x} e^{-ip_0' x} + a^\dagger(\mathbf{p}) a^\dagger(\mathbf{p}') e^{ip_0 x} e^{ip_0' x} \right\} \delta(\mathbf{p}+\mathbf{p}') - \\ -(-p_0 p_0')\left\{ a^\dagger(\mathbf{p}) a(\mathbf{p}') e^{ip_0 x} e^{-ip_0' x} + a(\mathbf{p}) a^\dagger(\mathbf{p}') e^{-ip_0 x} e^{ip_0' x} \right\} \delta(\mathbf{p}-\mathbf{p}') \end{array} \right] = \\
&= \int \frac{d^3 p}{2E_\mathbf{p}} E_\mathbf{p}^2 \left[ -a(\mathbf{p}) a(-\mathbf{p}) e^{-2ip_0 x} - a^\dagger(\mathbf{p}) a^\dagger(-\mathbf{p}) e^{2ip_0 x} + a^\dagger(\mathbf{p}) a(\mathbf{p}) + a(\mathbf{p}) a^\dagger(\mathbf{p}) \right].
\end{aligned}
\tag{W.83}
$$

Note that the energy does not depend on the sign of $\mathbf{p}$ since $E_\mathbf{p}^2 = \mathbf{p}^2 + m^2$ and $E_{-\mathbf{p}}^2 = (-\mathbf{p})^2 + m^2 = E_\mathbf{p}^2$. Analogously, we arrive at

$$
\int d^3x \, (\nabla\phi(x))^2 = \int \frac{d^3 p}{2E_\mathbf{p}} \mathbf{p}^2 \left[ a(\mathbf{p}) a(-\mathbf{p}) e^{-2ip_0 x} a^\dagger(\mathbf{p}) a^\dagger(-\mathbf{p}) e^{2ip_0 x} \right.
$$
$$
\left. + a^\dagger(\mathbf{p}) a(\mathbf{p}) + a(\mathbf{p}) a^\dagger(\mathbf{p}) \right] \tag{W.84}
$$

and

$$
m^2 \int d^3x \, \phi^2(x) = \int \frac{d^3 p}{2E_\mathbf{p}} m^2 \left[ a(\mathbf{p}) a(-\mathbf{p}) e^{-2ip_0 x} + a^\dagger(\mathbf{p}) a^\dagger(-\mathbf{p}) e^{2ip_0 x} \right.
$$
$$
\left. + a^\dagger(\mathbf{p}) a(\mathbf{p}) + a(\mathbf{p}) a^\dagger(\mathbf{p}) \right]. \tag{W.85}
$$

Adding these three terms brings

$$
\begin{aligned}
H &= \int d^3x \, \frac{1}{2} \left[ \dot\phi^2(x) + (\nabla\phi(x))^2 + m^2 \phi^2(x) \right] = \\
&= \frac{1}{2} \int \frac{d^3 p}{2E_\mathbf{p}} \left[ \begin{array}{l} \left\{ -E_\mathbf{p}^2 + \mathbf{p}^2 + m^2 \right\} \left\{ a(\mathbf{p}) a(-\mathbf{p}) e^{-2ip_0 x} a^\dagger(\mathbf{p}) a^\dagger(-\mathbf{p}) e^{2ip_0 x} \right\} + \\ + \left\{ E_\mathbf{p}^2 + \mathbf{p}^2 + m^2 \right\} \left\{ a^\dagger(\mathbf{p}) a(\mathbf{p}) + a(\mathbf{p}) a^\dagger(\mathbf{p}) \right\} \end{array} \right] = \\
&= \frac{1}{2} \int \frac{d^3 p}{2E_\mathbf{p}} \left[ \left\{ E_\mathbf{p}^2 + \mathbf{p}^2 + m^2 \right\} \left\{ a^\dagger(\mathbf{p}) a(\mathbf{p}) + a(\mathbf{p}) a^\dagger(\mathbf{p}) \right\} \right] = \\
&= \frac{1}{2} \int d^3 p \, E_\mathbf{p} \left[ a^\dagger(\mathbf{p}) a(\mathbf{p}) + a(\mathbf{p}) a^\dagger(\mathbf{p}) \right].
\end{aligned}
\tag{W.86}
$$

# W.5  Quantization of Free Fields, Dirac

## W.5.1  No Classical Spinor Field

In the last section, we have applied the method of canonical quantization to the Klein–Gordon field. However, for the Dirac field, this approach does not work. This is due to the fact that the Dirac equation describes electrons and positrons[82] which have spin $\frac{1}{2}$. In other words, these particles are fermions. As such, they obey the Pauli exclusion principle stating that two fermions cannot occupy the same state.

In contrast, bosons as for instance photons can populate the same state without restriction. Thus, they reinforce one another and can produce a macroscopic field; in case of photons the electromagnetic field. This mechanism is not accessible for fermions. This means there is no macroscopic spinor field and, therefore, no corresponding $\mathcal{L}$ or $\mathcal{H}$, let alone Poisson brackets which we could second-quantize via the canonical quantization.

So we have to choose another approach. First, we assume that the Dirac equation is also the underlying equation for the quantized field, as we did in the case of the Klein–Gordon field. Next, we search for an Lagrangian $\mathcal{L}$ which reproduces the Dirac equation. Equipped with this information, we can calculate the conjugated momentum, the Hamiltonian density $\mathcal{H}$ and the Hamilton function $H$. Then, parallelizing the Klein–Gordon case, we insert into $H$ the free solutions of the Dirac equation which are composed of free waves $e^{-ikx}$ and $e^{ikx}$ with amplitudes $b$ and $d$. Again, we change these quantities into operators. However, we have no Poisson brackets and have to look for suitable commutation relations for $b$ and $d$ which are compatible with theoretical considerations and, first of all, with experimental results.

## W.5.2  Lagrangian, Conjugated Momentum, Hamiltonian

As it turns out, an appropriate Lagrangian is[83]

$$\mathcal{L} = \bar{\psi}\left(i\gamma^{\mu}\partial_{\mu} - m\right)\psi \; ; \; \psi = \psi(x). \tag{W.87}$$

The 4-spinor $\psi$ is complex and has real and imaginary parts. This means that we can regard the two fields $\psi$ and $\bar{\psi}$ as independent.

The conjugated fields for $\psi$ and $\bar{\psi}$ are given by[84]

---

[82]Of course, the DE is valid for all particles with spin $1/2$, e.g. also for muons or tauons and their antiparticles.

[83]Remember $\bar{\psi} = \psi^{\dagger}\gamma^{0}$ where $\psi^{\dagger}$ is the hermitian adjoint and $\bar{\psi}$ the (Dirac) adjoint.

[84]For the variational derivative $\frac{\delta}{\delta f}$ see Vol. 1, App. T.3.

$$\text{for } \psi\text{: } \pi_\psi = \frac{\delta \mathcal{L}}{\delta\left(\partial_0\psi\right)} = i\bar{\psi}\gamma^0 \text{ ; for } \bar{\psi}\text{: } \pi_{\bar{\psi}} = \frac{\delta \mathcal{L}}{\delta\left(\partial_0\bar{\psi}\right)} = 0. \tag{W.88}$$

The Hamiltonian density follows from $\mathcal{H} = \pi_\psi\partial_0\psi - \mathcal{L}$ as

$$\mathcal{H} = \pi_\psi\partial_0\psi - \mathcal{L} = i\bar{\psi}\gamma^0\partial_0\psi - \bar{\psi}\left(i\gamma^\mu\partial_\mu - m\right)\psi = -\bar{\psi}\left(i\gamma^k\partial_k - m\right)\psi. \tag{W.89}$$

This expression may be simplified taking into account the Dirac equation in the form $i\gamma^0\partial_0\psi = -\left(i\gamma^k\partial_k - m\right)\psi$, yielding

$$\mathcal{H} = \bar{\psi}i\gamma^0\partial_0\psi = i\psi^\dagger\partial_0\psi. \tag{W.90}$$

Finally, the Hamilton function $H$ reads

$$H = i\int \mathrm{d}^3x\ \psi^\dagger\partial_0\psi. \tag{W.91}$$

## W.5.3   The Free Solutions

We now invoke the free solutions of the Dirac equation (cf. Appendix U, Vol. 1). The continuous version reads

$$\psi(x) = \frac{1}{(2\pi)^{3/2}}\int \mathrm{d}^3p\ \sqrt{\frac{m}{E_\mathbf{p}}}\sum_r \left(b_r(\mathbf{p})\,u_r(\mathbf{p})\,e^{-ipx} + d_r^\dagger(\mathbf{p})\,w_r(\mathbf{p})\,e^{ipx}\right)$$

$$\tag{W.92}$$

and the discrete version is obtained by the change $\frac{1}{(2\pi)^{3/2}}\int \mathrm{d}^3p \rightarrow \frac{1}{\sqrt{V}}\sum_\mathbf{p}$ (cf. Appendix T, Vol. 1).[85] $r = 1, 2$ denotes the spin directions.

### W.5.3.1   Properties of the Spinors $u_r$ and $w_r$

In further considerations, we make use of the properties of the spinors $u_r$ and $w_r$ which we will discuss now.

Note that there is a (minor) difference between (W.92) and the free solutions as formulated in Appendix U, Vol. 1. There we have written $v_r$ instead of $w_r$, namely

$$\text{in Vol. 1: } \psi(x) = \frac{1}{(2\pi)^{3/2}}\int \mathrm{d}^3p\ \sqrt{\frac{m}{E_\mathbf{p}}}\sum_r \left(b_r(\mathbf{p})\,u_r(\mathbf{p})\,e^{-ipx} + d_r^\dagger(\mathbf{p})\,v_r(\mathbf{p})\,e^{ipx}\right) \tag{W.93}$$

with

---

[85] As in the Klein–Gordon case, we use for convenience the same symbols $b(\mathbf{p})$ and $d(\mathbf{p})$ in the discrete and the continuous case.

$$u_r\left(\mathbf{p}\right) = \begin{pmatrix} \left(\frac{E_{\mathbf{p}}+m}{2m}\right)^{1/2}\chi_r \\ \frac{\sigma\cdot\mathbf{p}}{(2m(m+E_{\mathbf{p}}))^{1/2}}\chi_r \end{pmatrix} \; ; \; v_r\left(\mathbf{p}\right) = \begin{pmatrix} \frac{\sigma\cdot\mathbf{p}}{(2m(m+E_{\mathbf{p}}))^{1/2}}\chi_r \\ \left(\frac{E_{\mathbf{p}}+m}{2m}\right)^{1/2}\chi_r \end{pmatrix} \tag{W.94}$$

where the $\chi_r$ are the 2-spinors

$$\chi_1 = \begin{pmatrix} 1 \\ 0 \end{pmatrix} \; ; \; \chi_2 = \begin{pmatrix} 0 \\ 1 \end{pmatrix}. \tag{W.95}$$

However, with regard to further considerations, it is advantageous (and therefore quite common) to replace $v$ by $w$ defined by

$$w_r\left(\mathbf{p}\right) = \begin{pmatrix} \frac{\sigma\cdot\mathbf{p}}{(2m(E_{\mathbf{p}}+m))^{1/2}}i\sigma_2\chi_r \\ \left(\frac{E_{\mathbf{p}}+m}{2m}\right)^{1/2}i\sigma_2\chi_r \end{pmatrix} \quad \text{or } w_1\left(\mathbf{p}\right) = v_2\left(\mathbf{p}\right) \; ; \; w_2\left(\mathbf{p}\right) = -v_1\left(\mathbf{p}\right). \tag{W.96}$$

In doing so, we will be able below to cast certain relations in a simpler manner.

For the pair $u$, $v$ we have derived in Appendix U, Vol. 1 the following orthogonality relations:

$$\bar{u}_r\left(\mathbf{p}\right)u_{r'}\left(\mathbf{p}\right) = \delta_{rr'} \; ; \; \bar{u}_r\left(\mathbf{p}\right)v_{r'}\left(\mathbf{p}\right) = 0$$
$$\bar{v}_r\left(\mathbf{p}\right)v_{r'}\left(\mathbf{p}\right) = -\delta_{rr'} \; ; \; \bar{v}_r\left(\mathbf{p}\right)u_{r'}\left(\mathbf{p}\right) = 0. \tag{W.97}$$

Using $w$ instead of $v$, these relations translate into

$$\bar{u}_r\left(\mathbf{p}\right)u_{r'}\left(\mathbf{p}\right) = \delta_{rr'} \; ; \; \bar{u}_r\left(\mathbf{p}\right)w_{r'}\left(\mathbf{p}\right) = 0$$
$$\bar{w}_r\left(\mathbf{p}\right)w_{r'}\left(\mathbf{p}\right) = -\delta_{rr'} \; ; \; \bar{w}_r\left(\mathbf{p}\right)u_{r'}\left(\mathbf{p}\right) = 0. \tag{W.98}$$

In addition we have

$$\bar{u}_r\left(\mathbf{p}\right)\gamma^0 u_{r'}\left(\mathbf{p}\right) = \frac{E_{\mathbf{p}}}{m}\delta_{rr'} \; ; \; \bar{u}_r\left(-\mathbf{p}\right)\gamma^0 w_{r'}\left(\mathbf{p}\right) = 0$$
$$\bar{w}_r\left(\mathbf{p}\right)\gamma^0 w_{r'}\left(\mathbf{p}\right) = \frac{E_{\mathbf{p}}}{m}\delta_{rr'} \; ; \; \bar{w}_r\left(-\mathbf{p}\right)\gamma^0 u_{r'}\left(\mathbf{p}\right) = 0. \tag{W.99}$$

As in the Klein–Gordon case, we adopt the free solutions (W.92) as field operators, i.e., we interpret now the 'amplitudes' $b_r\left(\mathbf{p}\right)$ and $d_r\left(\mathbf{p}\right)$ as operators. Apparently, there is a new feature in comparison with the Klein–Gordon case: we have now *two* types of operators.[86] In accordance with the considerations in Appendix U, Vol. 1 concerning the electron and its antiparticle, we regard them as acting on *electrons* and *positrons*: $b_r\left(\mathbf{p}\right)$ and $b_r^\dagger\left(\mathbf{p}\right)$ destroy and create an electron, $d_r\left(\mathbf{p}\right)$ and $d_r^\dagger\left(\mathbf{p}\right)$ destroy and create a positron. Note that at the current state of our considerations, this is a guess or assumption which has to prove to be true.

---

[86]This indicates that the particles described differ from their antiparticles.

## W.5.4 Energy

Since in the Dirac case we have no Poisson brackets from a macroscopic system, we have to proceed in another way to get information about the commutation rules for the operators $b_r$ (**p**) and $d_r$ (**p**). To this end, we first determine the Hamilton function (or energy) (W.91) in terms of these operators. On this basis, we can discuss appropriate commutation rules, i.e., an appropriate way to quantize the Dirac field.

Inserting the free solutions (W.92) into (W.91), i.e., $H = i \int d^3x \, \psi^\dagger \partial_0 \psi$, leads to the following expressions for the continuous version (see exercises):

$$H = i \int d^3x \, \psi^\dagger \partial_0 \psi = \int d^3 p \, E_\mathbf{p} \sum_r \left[ b_r^\dagger (\mathbf{p}) \, b_r (\mathbf{p}) - d_r (\mathbf{p}) \, d_r^\dagger (\mathbf{p}) \right] \quad \text{(W.100)}$$

whereas the discrete version is obtained by the replacement $\int d^3 p \rightarrow \sum_\mathbf{p}$.

## W.5.5 Interpretation of $b_r$ (p) and $d_r$ (p), Commutation Relations, Pauli Principle, Number Operator

On the basis of the expressions (W.100), we will now try to find out which commutation rules are appropriate for the Dirac system. Remember our assumption as presented above that $b_r$ (**p**) and $b_r^\dagger$ (**p**) destroy and create an electron, $d_r$ (**p**) and $d_r^\dagger$ (**p**) destroy and create a positron. This means, for example, that applying the operator $b_r^\dagger$ (**p**) to the vacuum state $|0\rangle$ creates an electron with quantum numbers $r$ and **p**; analogously with $d_r^\dagger$ (**p**) for positrons.

### W.5.5.1    Anticommutation Relations

One could postulate that the operators $b_r$ (**p**) and $d_r$ (**p**) obey similar commutation relations as in the Klein–Gordon case, i.e., $\left[ b_r (\mathbf{p}) , b_{r'}^\dagger (\mathbf{p}') \right] \sim \delta_{rr'} \delta_{\mathbf{pp}'}$. But doing so leads to an unviable theory with a lot of inconsistencies[87] which, above all, would not reproduce the fact that electrons and positrons are fermions.

As a matter of fact, one has to introduce *anticommutation relations*.[88] They read

$$\left\{ b_r (\mathbf{p}) , b_{r'}^\dagger (\mathbf{p}') \right\} = \delta_{rr'} \delta (\mathbf{p}, \mathbf{p}') \; ; \; \left\{ d_r (\mathbf{p}) , d_{r'}^\dagger (\mathbf{p}') \right\} = \delta_{rr'} \delta (\mathbf{p}, \mathbf{p}') . \quad \text{(W.101)}$$

All other anticommutators between two of these operators vanish,

---

[87]For instance, with commutation rules as with bosons, the energy would not be bounded from below.

[88]Note that this step is not mandatory or logically without alternative at this state of affairs and has to prove itself. Just a reminder: $\{a, b\} = ab + ba$.

$$\left\{ b_r\left(\mathbf{p}\right), b_{r'}\left(\mathbf{p}'\right)\right\} = \left\{ d_r\left(\mathbf{p}\right), b_{r'}\left(\mathbf{p}'\right)\right\} = \left\{ b_r\left(\mathbf{p}\right), d_{r'}^\dagger\left(\mathbf{p}'\right)\right\} = \cdots = 0. \quad \text{(W.102)}$$

### W.5.5.2 Pauli Principle

Note that the anticommutation relations (W.101) and (W.102) guarantee that we describe fermions. For instance, from (W.102) follows $\{b_r\left(\mathbf{p}\right), b_r\left(\mathbf{p}\right)\} = 2b_r\left(\mathbf{p}\right)$ $b_r\left(\mathbf{p}\right) = 0$ and analogously for $b_r^\dagger\left(\mathbf{p}\right)$, $d_r\left(\mathbf{p}\right)$ and $d_r^\dagger\left(\mathbf{p}\right)$, i.e.,

$$b_r\left(\mathbf{p}\right) b_r\left(\mathbf{p}\right) = 0 \; ; \; b_r^\dagger\left(\mathbf{p}\right) b_r^\dagger\left(\mathbf{p}\right) = 0 \; ; \; d_r\left(\mathbf{p}\right) d_r\left(\mathbf{p}\right) = 0 \; ; \; d_r^\dagger\left(\mathbf{p}\right) d_r^\dagger\left(\mathbf{p}\right) = 0.$$
$$\text{(W.103)}$$

Assume that we create e.g. an electron with quantum numbers $r\mathbf{p}$, i.e., we have $b_r^\dagger\left(\mathbf{p}\right)|0\rangle$. Applying once more $b_r^\dagger\left(\mathbf{p}\right)$ would, in the Klein–Gordon case, produce a second particle with the same quantum numbers. But in the Dirac case, this has to be forbidden due to the Pauli exclusion principle. Indeed, we have from (W.103)

$$b_r^\dagger\left(\mathbf{p}\right) b_r^\dagger\left(\mathbf{p}\right)|0\rangle = 0. \quad \text{(W.104)}$$

Thus, each state is either empty or simply occupied. Obviously, a commutation relation like $\left[ b_r^\dagger\left(\mathbf{p}\right), b_{r'}^\dagger\left(\mathbf{p}'\right)\right] \sim \delta_{rr'}\delta_{\mathbf{p}\mathbf{p}'}$ would not produce this behavior.

### W.5.5.3 Number Operator

Parallelizing the Klein–Gordon case, we define a number operator for electrons and positrons by

$$N_{e,r\mathbf{p}} = b_r^\dagger\left(\mathbf{p}\right) b_r\left(\mathbf{p}\right) \; ; \; N_{p,r\mathbf{p}} = d_r^\dagger\left(\mathbf{p}\right) d_r\left(\mathbf{p}\right) \quad \text{(W.105)}$$

which give us the number of electrons and positrons with quantum numbers $r$ and $\mathbf{p}$.

Note that electrons and positrons are now on the same level, and we do not have any more the situation of one single electron against an infinite sea of holes, i.e., positrons.

We want to reproduce the Pauli principle once again, this times by using the number operator. We apply $N_{e,r\mathbf{p}}$ onto a state $|\psi\rangle$ which consists of $n$ electrons with quantum numbers $r$ and $\mathbf{p}$, i.e., $N_{e,r\mathbf{p}}|\psi\rangle = n|\psi\rangle$. Due to the anticommutation relations, we have

$$N_{e,r\mathbf{p}} N_{e,r\mathbf{p}} = b_r^\dagger\left(\mathbf{p}\right) b_r\left(\mathbf{p}\right) b_r^\dagger\left(\mathbf{p}\right) b_r\left(\mathbf{p}\right) = b_r^\dagger\left(\mathbf{p}\right)\left[1 - b_r^\dagger\left(\mathbf{p}\right) b_r\left(\mathbf{p}\right)\right] b_r\left(\mathbf{p}\right) =$$
$$= b_r^\dagger\left(\mathbf{p}\right) b_r\left(\mathbf{p}\right) - b_r^\dagger\left(\mathbf{p}\right) b_r^\dagger\left(\mathbf{p}\right) b_r\left(\mathbf{p}\right) b_r\left(\mathbf{p}\right) = b_r^\dagger\left(\mathbf{p}\right) b_r\left(\mathbf{p}\right) = N_{e,r\mathbf{p}}$$
$$\text{(W.106)}$$

and it follows

$$N_{e,r\mathbf{p}} N_{e,r\mathbf{p}}|\psi\rangle = N_{e,r\mathbf{p}}|\psi\rangle \rightarrow n^2|\psi\rangle = n|\psi\rangle \rightarrow n = 0 \text{ or } n = 1. \quad \text{(W.107)}$$

Thus, $N_{e,r\mathbf{p}}$ has the eigenvalues 0 and 1. In other words, we have again found the Pauli principle that for fermions each state is either empty or simply occupied.

## W.5.6   Again Infinities

Since we want to write $H$ as a combination of number operators, we invoke the anticommutation rules and cast

$$H = \sum_{\mathbf{p},r} E_{\mathbf{p}} \left[ b_r^\dagger (\mathbf{p}) \, b_r (\mathbf{p}) - \mathrm{d}_r (\mathbf{p}) \, \mathrm{d}_r^\dagger (\mathbf{p}) \right] \tag{W.108}$$

into the form

$$H = \sum_{\mathbf{p},r} E_{\mathbf{p}} \left[ b_r^\dagger (\mathbf{p}) \, b_r (\mathbf{p}) + \mathrm{d}_r^\dagger (\mathbf{p}) \, \mathrm{d}_r (\mathbf{p}) - 1 \right]. \tag{W.109}$$

As in the Klein–Gordon case, we here have the problem of an infinite contribution $-\sum_{\mathbf{p},r} E_{\mathbf{p}}$ even if there are no particles.

It would be nice if we could simply neglect the infinite zero point energy to get around this conceptual difficulty and could write

$$H = \sum_{\mathbf{p},r} E_{\mathbf{p}} \left[ b_r^\dagger (\mathbf{p}) \, b_r (\mathbf{p}) + \mathrm{d}_r^\dagger (\mathbf{p}) \, \mathrm{d}_r (\mathbf{p}) \right] = \sum_{\mathbf{p},r} E_{\mathbf{p}} \left[ N_{e,r\mathbf{p}} + N_{p,r\mathbf{p}} \right]. \tag{W.110}$$

Indeed, this is the final form of the Hamilton operator. A thorough discussion of this delicate point is found below in the section 'Operator ordering'.

## W.5.7   Anticommutators for Field Operators

As said above, we cannot use the scheme of canonical quantization for the Dirac system, since there is no macroscopic spinor theory and, correspondingly, there are no Poisson brackets which we could quantize. But now, in possession of the appropriate commutators (W.101) for fermions, we can ask which relation may replace the (quantized) Poisson brackets. It is clear that there have to be differences with regard to e.g. the Klein–Gordon field, if only because we have anticommutators instead of commutators and 4-spinors instead of a scalar field. We will discuss this question only very briefly and without detailed calculations.

We sketch the approach in all brevity. One starts with the free solutions as given in (W.92), i.e.,

$$\psi (x) = \sum_{\mathbf{p},r} \sqrt{\frac{m}{V E_{\mathbf{p}}}} \left( b_r (\mathbf{p}) \, u_r (\mathbf{p}) \, e^{-ipx} + \mathrm{d}_r^\dagger (\mathbf{p}) \, w_r (\mathbf{p}) \, e^{ipx} \right) \tag{W.111}$$

for the discrete case (the continuous case runs analogously). By means of Fourier transformation, one can solve this equation for $b_r(\mathbf{p}) u_r(p)$ and $d_r^\dagger(\mathbf{p}) w_r(p)$. Invoking the anticommutation relations for $b_r(\mathbf{p})$ and $d_r(\mathbf{p})$ and the orthogonality relations for $u_r(p)$ and $w_r(p)$, one obtains after some calculations the anticommutation relations for the field operators

$$\left\{\psi_\alpha(t, \mathbf{x}), \psi_\beta^\dagger(t, \mathbf{x}')\right\} = \delta_{\alpha\beta}\delta^{(3)}(\mathbf{x} - \mathbf{x}')$$
$$\left\{\psi_\alpha(t, \mathbf{x}), \psi_\beta(t, \mathbf{x}')\right\} = \left\{\psi_\alpha^\dagger(t, \mathbf{x}), \psi_\beta^\dagger(t, \mathbf{x}')\right\} = 0$$

(W.112)

where $\alpha = 1, \ldots, 4$ and $\beta = 1, \ldots, 4$ indicate the components of the spinors. Note that these are equal-time relations as in the Klein–Gordon case. The relations are formulated in terms of the hermitian adjoint $\psi^\dagger$ and not the Dirac adjoint $\bar{\psi}$. With $\bar{\psi}$, we have for instance

$$\left\{\psi_\alpha(t, \mathbf{x}), \bar{\psi}_\beta(t, \mathbf{x}')\right\} = \gamma_{\alpha\beta}^0 \delta^{(3)}(\mathbf{x} - \mathbf{x}').$$

(W.113)

Of course, one can postulate (W.112) out of the blue and then derive the anticommutation relations (W.101) for the creation/annihilation operators. But it seems hard to see, especially for beginners, why relations (W.112) should have precisely the form they have.

## W.5.8 Conclusion

With the relations (W.101), we have quantized the Dirac field; the Hamiltonian is given in (W.110). The operators $b_r^\dagger(\mathbf{p})$ and $b_r(\mathbf{p})$ create and annihilate an electron and the operators $d_r^\dagger(\mathbf{p})$ and $d_r(\mathbf{p})$ create and annihilate a positron, both with quantum numbers $r$ and $\mathbf{p}$. The states are living in a Fock space, for instance $|0\rangle$ or $d_r^\dagger(\mathbf{p})|0\rangle$ or $b_{r_1}^\dagger(\mathbf{p}_1) d_{r_2}^\dagger(\mathbf{p}_2) d_{r_3}^\dagger(\mathbf{p}_3)|0\rangle$ (an electron with $r_1\mathbf{p}_1$, a positron with $r_2\mathbf{p}_2$, a positron with $r_3\mathbf{p}_3$). Concerning the spin, we have the following facts: $b_r^\dagger(\mathbf{p} = \mathbf{0}) / b_r(\mathbf{p} = \mathbf{0})$ with $r = 1$ ($r = 2$) creates/annihilates a stationary electron with spin $s_z = \frac{1}{2}$ ($s_z = -\frac{1}{2}$); analogously with $d_r^\dagger(\mathbf{p} = \mathbf{0}) / d_r(\mathbf{p} = \mathbf{0})$ for positrons. In general, $b_r^\dagger(\mathbf{p})$ or $d_r^\dagger(\mathbf{p})$ creates an electron or positron with momentum $\mathbf{p}$ which in its rest system has the spin $\frac{1}{2}$ and $-\frac{1}{2}$ for $r = 1$ and $r = 2$.

Some normalized particle states with their energy eigenvalues are given in Table W.2.

In sum, we have now an physical meaningful picture; $\psi(x)$ is not a state, but a field operator, creating and annihilating particles, i.e., electrons and positrons. These two types of particles are now on equal footing, and we do not need anymore an infinite sea of positrons in order to describe one electron as in the Dirac theory of Appendix U, Vol. 1. In other words, we have left the one-particle theories, and can describe arbitrary numbers of particles.

**Table W.2** Table of simplest states of the Dirac field

| State | | Energy |
|---|---|---|
| Vacuum | $\vert 0 \rangle$ | 0 |
| One electron | $b_r^\dagger (\mathbf{p}) \vert 0 \rangle$ | $E_{\mathbf{p}}$ |
| One positron | $d_r^\dagger (\mathbf{p}) \vert 0 \rangle$ | $E_{\mathbf{p}}$ |
| One electron, one positron | $b_r^\dagger (\mathbf{p}) d_r^\dagger (\mathbf{p}) \vert 0 \rangle$ | $2E_{\mathbf{p}}$ |
| Two different electrons | $b_{r'}^\dagger (\mathbf{p'}) b_r^\dagger (\mathbf{p}) \vert 0 \rangle$ | $E_{\mathbf{p'}} + E_{\mathbf{p}}; \mathbf{p'} \neq \mathbf{p}$ and /or $r' \neq r$ |
| Two identical electrons | $b_r^\dagger (\mathbf{p}) b_r^\dagger (\mathbf{p}) \vert 0 \rangle$ | 0 |

In addition, there are no problems with negative energies. The Hamiltonian (W.110) simply does not allow for them.

All fits nicely, and the only weak spot, if one may say, is the infinite contribution to the Hamiltonian in (W.109) and its negligence in (W.110). For the time being, we must accept this as the way nature works.

## W.5.9  Exercises and Solutions

1. Show that the Euler–Lagrange equations for $\psi$ and $\bar{\psi}$ reproduce the Dirac equation.

    Solution: We have

    $$\frac{\partial \mathcal{L}}{\partial \bar{\psi}} - \partial_\mu \left( \frac{\partial \mathcal{L}}{\partial \left( \partial_\mu \bar{\psi} \right)} \right) = 0 \text{ and } \frac{\partial \mathcal{L}}{\partial \psi} - \partial_\mu \left( \frac{\partial \mathcal{L}}{\partial \left( \partial_\mu \psi \right)} \right). \qquad \text{(W.114)}$$

    The Lagrangian being $\mathcal{L} = \bar{\psi} \left( i\gamma^\mu \partial_\mu - m \right) \psi$, we have

    $$\frac{\partial \mathcal{L}}{\partial \bar{\psi}} = \left( i\gamma^\mu \partial_\mu - m \right) \psi \; ; \; \frac{\partial \mathcal{L}}{\partial \left( \partial_\mu \bar{\psi} \right)} = 0 \Longrightarrow \left( i\gamma^\mu \partial_\mu - m \right) \psi = 0 \qquad \text{(W.115)}$$

    and

    $$\frac{\partial \mathcal{L}}{\partial \psi} = -m\bar{\psi} \; ; \; \frac{\partial \mathcal{L}}{\partial \left( \partial_\mu \psi \right)} = \frac{\partial \bar{\psi} i \gamma^\nu \partial_\nu \psi}{\partial \left( \partial_\mu \psi \right)} = \bar{\psi} i \gamma^\nu \delta_{\nu\mu} \Longrightarrow -m\bar{\psi} - \partial_\mu \bar{\psi} i \gamma^\mu = 0. \qquad \text{(W.116)}$$

    We see that (W.115) yields the Dirac equation and (W.116) its adjoint.

2. Write down $\psi^\dagger (x)$ and $\bar{\psi} (x)$.

    Solution: It is

$$\psi^\dagger(x) = \sum_{\mathbf{p},r=1,2} \left(\frac{m}{VE_\mathbf{p}}\right)^{1/2} \left(b_r^\dagger(\mathbf{p})\, u_r^\dagger(\mathbf{p})\, e^{ipx} + d_r(\mathbf{p})\, w_r^\dagger(\mathbf{p})\, e^{-ipx}\right).$$

(W.117)

With $\bar\psi = \psi^\dagger \gamma^0$ follows:

$$\bar\psi(x) = \psi^\dagger(x)\,\gamma^0 = \sum_{\mathbf{p},r=1,2}\left(\frac{m}{VE_\mathbf{p}}\right)^{1/2}\left(b_r^\dagger(\mathbf{p})\,u_r^\dagger(\mathbf{p})\,\gamma^0 e^{ipx} + d_r(\mathbf{p})\,w_r^\dagger(\mathbf{p})\,\gamma^0 e^{-ipx}\right) =$$

$$= \sum_{\mathbf{p},r=1,2}\left(\frac{m}{VE_\mathbf{p}}\right)^{1/2}\left(b_r^\dagger(\mathbf{p})\,\bar u_r(\mathbf{p})\,e^{ipx} + d_r(\mathbf{p})\,\bar w_r(\mathbf{p})\,e^{-ipx}\right).$$

(W.118)

3. Write down explicitly $u_r$ in (W.94) and $w_r$ in (W.96) for $r = 1, 2$.
Solution:

$$u_1(\mathbf{p}) = \left(\frac{E_\mathbf{p}+m}{2m}\right)^{1/2}\left(\begin{array}{c} \chi_1 \\ \frac{\boldsymbol\sigma\cdot\mathbf{p}}{E_\mathbf{p}+m}\chi_1 \end{array}\right) = \left(\frac{E_\mathbf{p}+m}{2m}\right)^{1/2}\left(\begin{array}{c} 1 \\ 0 \\ \frac{p_z}{E_\mathbf{p}+m} \\ \frac{p_x+ip_y}{E_\mathbf{p}+m} \end{array}\right)$$

$$u_2(\mathbf{p}) = \left(\frac{E_\mathbf{p}+m}{2m}\right)^{1/2}\left(\begin{array}{c} \chi_2 \\ \frac{\boldsymbol\sigma\cdot\mathbf{p}}{E_\mathbf{p}+m}\chi_2 \end{array}\right) = \left(\frac{E_\mathbf{p}+m}{2m}\right)^{1/2}\left(\begin{array}{c} 0 \\ 1 \\ \frac{p_x-ip_y}{E_\mathbf{p}+m} \\ \frac{-p_z}{E_\mathbf{p}+m} \end{array}\right)$$

(W.119)

$$w_1(\mathbf{p}) = -\left(\frac{E_\mathbf{p}+m}{2m}\right)^{1/2}\left(\begin{array}{c} \frac{\boldsymbol\sigma\cdot\mathbf{p}}{E_\mathbf{p}+m}i\sigma_2\chi_1 \\ i\sigma_2\chi_1 \end{array}\right) = -\left(\frac{E_\mathbf{p}+m}{2m}\right)^{1/2}\left(\begin{array}{c} \frac{-p_x+ip_y}{E_\mathbf{p}+m} \\ \frac{p_z}{E_\mathbf{p}+m} \\ 0 \\ -1 \end{array}\right)$$

$$w_2(\mathbf{p}) = -\left(\frac{E_\mathbf{p}+m}{2m}\right)^{1/2}\left(\begin{array}{c} \frac{\boldsymbol\sigma\cdot\mathbf{p}}{E_\mathbf{p}+m}i\sigma_2\chi_2 \\ i\sigma_2\chi_2 \end{array}\right) = -\left(\frac{E_\mathbf{p}+m}{2m}\right)^{1/2}\left(\begin{array}{c} \frac{p_z}{E_\mathbf{p}+m} \\ \frac{p_x+ip_y}{E_\mathbf{p}+m} \\ 1 \\ 0 \end{array}\right).$$

4. Write down explicitly $\bar w_r$ for $r = 1, 2$.
Solution: It is $\bar w_r = w_r^\dagger \gamma_0$ and

$$w_r(\mathbf{p}) = \left(\begin{array}{c} \frac{\boldsymbol\sigma\cdot\mathbf{p}}{(2m(m+E_\mathbf{p}))^{1/2}}i\sigma_2\chi_r \\ \left(\frac{E_\mathbf{p}+m}{2m}\right)^{1/2}i\sigma_2\chi_r \end{array}\right).$$

(W.120)

With $(i\sigma_2)^\dagger = -i\sigma_2$ follows

$$\bar w_r(\mathbf{p}) = \left(-\frac{\boldsymbol\sigma\cdot\mathbf{p}}{(2m(m+E_\mathbf{p}))^{1/2}}i\sigma_2\chi_r \quad \left(\frac{E_\mathbf{p}+m}{2m}\right)^{1/2}i\sigma_2\chi_r\right).$$

(W.121)

5. By means of the explicit expressions for $u_r$ and $w_r$, check equations (W.98).
   Solution: We calculate $\bar{w}_r\,(p)\,w_{r'}\,(p)$ and $\bar{w}_r\,(p)\,u_{r'}\,(p)$. We have[89]

$$
\bar{w}_r(\mathbf{p})w_{r'}(\mathbf{p}) = \left(-\frac{\sigma\cdot\mathbf{p}}{(2m(E_\mathbf{p}+m))^{1/2}}\chi_r^\dagger i\sigma_2\left(\tfrac{E_\mathbf{p}+m}{2m}\right)^{1/2}\chi_r^\dagger i\sigma_2\right)\left(\begin{array}{c}\frac{\sigma\cdot\mathbf{p}}{(2m(E_\mathbf{p}+m))^{1/2}}i\sigma_2\chi_{r'}\\[4pt]\left(\tfrac{E_\mathbf{p}+m}{2m}\right)^{1/2}i\sigma_2\chi_{r'}\end{array}\right) =
$$

$$
= -\frac{\sigma\cdot\mathbf{p}}{(2m(E_\mathbf{p}+m))^{1/2}}\frac{\sigma\cdot\mathbf{p}}{(2m(E_\mathbf{p}+m))^{1/2}}\left(\chi_r^\dagger i\sigma_2\right)(i\sigma_2\chi_{r'}) + \left(\tfrac{E_\mathbf{p}+m}{2m}\right)^{1/2}\left(\tfrac{E_\mathbf{p}+m}{2m}\right)^{1/2}\left(\chi_r^\dagger i\sigma_2\right)(i\sigma_2\chi_{r'}) =
$$

$$
= \left[-\frac{\mathbf{p}^2}{2m(E_\mathbf{p}+m)} + \frac{E_\mathbf{p}+m}{2m}\right]\chi_r^\dagger i\sigma_2 i\sigma_2\chi_{r'} = \frac{(E_\mathbf{p}+m)^2-\mathbf{p}^2}{2m(E_\mathbf{p}+m)}\chi_r^\dagger\,(-1)\,\chi_{r'} = \chi_r^\dagger\,(-1)\,\chi_{r'} = --\,\delta_{rr'}
$$
$$\tag{W.122}$$

and

$$
\bar{w}_r(\mathbf{p})u_{r'}(\mathbf{p}) = \left(-\frac{\sigma\cdot\mathbf{p}}{(2m(E_\mathbf{p}+m))^{1/2}}\chi_r^\dagger i\sigma_2\left(\tfrac{E_\mathbf{p}+m}{2m}\right)^{1/2}\chi_r^\dagger i\sigma_2\right)\left(\begin{array}{c}\left(\tfrac{E_\mathbf{p}+m}{2m}\right)^{1/2}\chi_r\\[4pt]\frac{\sigma\cdot\mathbf{p}}{(2m(m+E_\mathbf{p}))^{1/2}}\chi_r\end{array}\right) =
$$

$$
= -\frac{\sigma\cdot\mathbf{p}}{(2m(E_\mathbf{p}+m))^{1/2}}\chi_r^\dagger i\sigma_2\left(\tfrac{E_\mathbf{p}+m}{2m}\right)^{1/2}\chi_r + \left(\tfrac{E_\mathbf{p}+m}{2m}\right)^{1/2}\chi_r^\dagger i\sigma_2\frac{\sigma\cdot\mathbf{p}}{(2m(m+E_\mathbf{p}))^{1/2}}\chi_r =
$$

$$
= \frac{\sigma\cdot\mathbf{p}}{(2m(E_\mathbf{p}+m))^{1/2}}\left(\tfrac{E_\mathbf{p}+m}{2m}\right)^{1/2}\left[-\chi_r^\dagger i\sigma_2\chi_r + \chi_r^\dagger i\sigma_2\chi_r\right] = 0.
$$
$$\tag{W.123}$$

6. By means of the explicit expressions $u_r$ and $w_r$, check equations (W.99).
   Solution: By proxy, we calculate $\bar{u}_1(\mathbf{p})\gamma^0 u_{2'}(\mathbf{p})$, $\bar{u}_1(\mathbf{p})\gamma^0 u_{1'}(\mathbf{p})$ and $\bar{u}_r(-\mathbf{p})\gamma^0$ $w_{r'}(\mathbf{p})$. For $u_1^\dagger(\mathbf{p})u_2(\mathbf{p})$, we have

$$
\bar{u}_1(\mathbf{p})\gamma^0 u_{2'}(\mathbf{p}) = u_1^\dagger(\mathbf{p})u_2(\mathbf{p}) = \frac{E_\mathbf{p}+m}{2m}\left(\begin{array}{c}1\\0\\\frac{p_z}{E_\mathbf{p}+m}\\\frac{p_x-ip_y}{E_\mathbf{p}+m}\end{array}\right)^T\left(\begin{array}{c}0\\1\\\frac{p_x-ip_y}{E_\mathbf{p}+m}\\\frac{-p_z}{E_\mathbf{p}+m}\end{array}\right) =
$$

$$
= \frac{E_\mathbf{p}+m}{2m}\left(0+0+\frac{p_z}{E_\mathbf{p}+m}\frac{p_x-ip_y}{E_\mathbf{p}+m}+\frac{p_x-ip_y}{E_\mathbf{p}+m}\frac{-p_z}{E_\mathbf{p}+m}\right) =
$$

$$
= \frac{E_\mathbf{p}+m}{2m}\frac{p_z}{E_\mathbf{p}+m}\left(\frac{p_x-ip_y}{E_\mathbf{p}+m}-\frac{p_x-ip_y}{E_\mathbf{p}+m}\right) = 0.
$$
$$\tag{W.124}$$

$\bar{u}_1(\mathbf{p})\gamma^0 u_{1'}(\mathbf{p}) = u_1^\dagger(\mathbf{p})u_1(\mathbf{p})$ is given by

$$
u_1^\dagger(\mathbf{p})u_1(\mathbf{p}) = \frac{E_\mathbf{p}+m}{2m}\left(\begin{array}{c}1\\0\\\frac{p_z}{E_\mathbf{p}+m}\\\frac{p_x+ip_y}{E_\mathbf{p}+m}\end{array}\right)^T\left(\begin{array}{c}1\\0\\\frac{p_z}{E_\mathbf{p}+m}\\\frac{p_x-ip_y}{E_\mathbf{p}+m}\end{array}\right) = \frac{E_\mathbf{p}+m}{2m}\left(1+0+\frac{p_z^2}{(E_\mathbf{p}+m)^2}+\frac{p_x^2+p_y^2}{(E_\mathbf{p}+m)^2}\right) =
$$

$$
= \frac{E_\mathbf{p}+m}{2m}\left(1+\frac{E_\mathbf{p}^2-m^2}{(E_\mathbf{p}+m)^2}\right) = \frac{E_\mathbf{p}+m}{2m}\left(1+\frac{E_\mathbf{p}-m}{E_\mathbf{p}+m}\right) = \frac{E_\mathbf{p}+m}{2m}\frac{2E_\mathbf{p}}{E_\mathbf{p}+m} = \frac{E_\mathbf{p}}{m}.
$$
$$\tag{W.125}$$

---

[89]Remember $(\sigma\mathbf{a})\,(\sigma\mathbf{b}) = \mathbf{ab} + i\sigma\,(\mathbf{a}\times\mathbf{b})$.

Finally we calculate $\bar{u}_r(-\mathbf{p})\gamma^0 w_{r'}(\mathbf{p}) = u_r^\dagger(-\mathbf{p})w_{r'}(\mathbf{p})$. We have $E_{-\mathbf{p}} = \sqrt{(-\mathbf{p})^2 + m^2} = E_\mathbf{p}$. It follows

$$u_r^\dagger(-\mathbf{p})w_{r'}(\mathbf{p}) = -\left(\frac{E_\mathbf{p}+m}{2m}\right)^{1/2}\left(\frac{E_\mathbf{p}+m}{2m}\right)^{1/2}\left(\chi_r^\dagger\left(-\frac{\sigma\cdot\mathbf{p}}{E_\mathbf{p}+m}\chi_r\right)^\dagger\right)\left(\begin{array}{c}\frac{\sigma\cdot\mathbf{p}}{E_\mathbf{p}+m}i\sigma^2\chi_{r'}\\ i\sigma^2\chi_{r'}\end{array}\right) =$$

$$= -\frac{E_\mathbf{p}+m}{2m}\left[\chi_r^\dagger\frac{\sigma\cdot\mathbf{p}}{E_\mathbf{p}+m}i\sigma^2\chi_{r'} - \left(\frac{\sigma\cdot\mathbf{p}}{E_\mathbf{p}+m}\chi_r\right)^\dagger i\sigma^2\chi_{r'}\right] =$$

$$= -\frac{E_\mathbf{p}+m}{2m}\left[\chi_r^\dagger\frac{\sigma\cdot\mathbf{p}}{E_\mathbf{p}+m}i\sigma^2\chi_{r'} - \chi_r^\dagger\left(\frac{\sigma\cdot\mathbf{p}}{E_\mathbf{p}+m}\right)i\sigma^2\chi_{r'}\right] = 0.$$

$$(\text{W.126})$$

7. Prove (W.100) for the discrete case.
   Solution: With

$$H = i\int d^3x\,\psi^\dagger(x)\,\partial_0\psi(x) \qquad (\text{W.127})$$

and

$$\psi(x) = \sum_{\mathbf{p},r}\sqrt{\frac{m}{VE_\mathbf{p}}}\left(b_r(\mathbf{p})u_r(\mathbf{p})e^{-ipx} + d_r^\dagger(\mathbf{p})w_r(\mathbf{p})e^{ipx}\right)$$
$$\bar{\psi}(x) = \sum_{\mathbf{p},r}\sqrt{\frac{m}{VE_\mathbf{p}}}\left(d_r(\mathbf{p})\bar{w}_r(\mathbf{p})e^{-ipx} + b_r^\dagger(\mathbf{p})\bar{u}_r(\mathbf{p})e^{ipx}\right)$$

$$(\text{W.128})$$

we arrive at

$$H = i\int d^3x\,\bar{\psi}(x)\,\gamma^0\partial_0\psi(x) =$$
$$= i\sum_{\mathbf{p},r,\mathbf{p}',r'}\sqrt{\frac{m}{VE_\mathbf{p}}}\sqrt{\frac{m}{VE_{\mathbf{p}'}}}\int d^3x\left[d_r(\mathbf{p})\bar{w}_r(\mathbf{p})e^{-ipx} + b_r^\dagger(\mathbf{p})\bar{u}_r(\mathbf{p})e^{ipx}\right]\cdot$$
$$\cdot\gamma^0\partial_0\left[b_{r'}(\mathbf{p}')u_{r'}(\mathbf{p}')e^{-ip'x} + d_{r'}^\dagger(\mathbf{p}')w_{r'}(\mathbf{p}')e^{ip'x}\right].$$

$$(\text{W.129})$$

With

$$\partial_0 e^{ipx} = \partial_0 e^{i(E_\mathbf{p}t - \mathbf{p}\mathbf{x})} = iE_\mathbf{p}e^{ipx} \qquad (\text{W.130})$$

follows

$$H = -\sum_{\mathbf{p},r,\mathbf{p}',r'}\sqrt{\frac{m}{VE_\mathbf{p}}}\sqrt{\frac{m}{VE_{\mathbf{p}'}}}E_{\mathbf{p}'}\cdot$$
$$\cdot\int d^3x\left[\begin{pmatrix}-d_r(\mathbf{p})\bar{w}_r(\mathbf{p})\gamma^0 b_{r'}'(\mathbf{p}')u_{r'}(\mathbf{p}')e^{-ipx}e^{-ip'x} - b_r^\dagger(\mathbf{p})\bar{u}_r(\mathbf{p})\gamma^0 b_{r'}'(\mathbf{p}')u_{r'}(\mathbf{p}')e^{ipx}e^{-ip'x} + \\ +d_r(\mathbf{p})\bar{w}_r(\mathbf{p})\gamma^0 d_{r'p'}^\dagger w_{r'}(\mathbf{p}')e^{-ipx}e^{ip'x} + b_r^\dagger(\mathbf{p})\bar{u}_r(\mathbf{p})\gamma^0 d_{r'}^\dagger(\mathbf{p}')w_{r'}(\mathbf{p}')e^{ipx}e^{ip'x}\end{pmatrix}\right].$$

$$(\text{W.131})$$

Performing the $x$-integration gives (see Appendix T, Vol. 1)

$$
H = -V \sum_{\mathbf{p},r,\mathbf{p}',r'} \sqrt{\tfrac{m}{VE_{\mathbf{p}}}} \sqrt{\tfrac{m}{VE_{\mathbf{p}'}}} E_{\mathbf{p}} \left[ \begin{pmatrix} -d_r(\mathbf{p})\,\bar{w}_r(\mathbf{p})\,\gamma^0 b'_r(\mathbf{p}')\,u_{r'}(\mathbf{p})\,e^{-ip_0 x^0}\,e^{-ip'_0 x^0}\,\delta_{\mathbf{p}+\mathbf{p}',0} - \\ -b_r^\dagger(\mathbf{p})\,\bar{u}_r(\mathbf{p})\,\gamma^0 b'_r(\mathbf{p}')\,u_{r'}(\mathbf{p}')\,e^{ip_0 x^0}\,e^{-ip'_0 x^0}\,\delta_{\mathbf{p}-\mathbf{p}',0} + \\ +d_r(\mathbf{p})\,\bar{w}_r(\mathbf{p})\,\gamma^0 d_{r'}^\dagger(\mathbf{p}')\,w_{r'}(\mathbf{p}')\,e^{-ip_0 x^0}\,e^{ip'_0 x^0}\,\delta_{\mathbf{p}-\mathbf{p}',0} + \\ +b_r^\dagger(\mathbf{p})\,\bar{u}_r(\mathbf{p})\,\gamma^0 d_{r'}^\dagger(\mathbf{p}')\,w_{r'}(\mathbf{p}')\,e^{ip_0 x^0}\,e^{ip'_0 x^0}\,\delta_{\mathbf{p}+\mathbf{p}',0} \end{pmatrix} \right] =
$$

$$
= -\sum_{\mathbf{p},r,\mathbf{p}',r'} \sqrt{\tfrac{m}{E_{\mathbf{p}}}} \sqrt{\tfrac{m}{E_{\mathbf{p}'}}} E_{\mathbf{p}} \left[ \begin{pmatrix} \delta_{\mathbf{p}+\mathbf{p}',0} \begin{bmatrix} -d_r(\mathbf{p})\,\bar{w}_r(\mathbf{p})\,\gamma^0 b'_r(\mathbf{p}')\,u_{r'}(\mathbf{p}')\,e^{-ip_0 x^0}\,e^{-ip'_0 x^0} + \\ +b_r^\dagger(\mathbf{p})\,\bar{u}_r(\mathbf{p})\,\gamma^0 d_{r'}^\dagger(\mathbf{p}')\,w_{r'}(\mathbf{p}')\,e^{ip_0 x^0}\,e^{ip'_0 x^0} \end{bmatrix} + \\ +\delta_{\mathbf{p}-\mathbf{p}',0} \begin{bmatrix} -b_r^\dagger(\mathbf{p})\,\bar{u}_r(\mathbf{p})\,\gamma^0 b'_r(\mathbf{p}')\,u_{r'}(\mathbf{p}')\,e^{ip_0 x^0}\,e^{-ip'_0 x^0} + \\ +d_r(\mathbf{p})\,\bar{w}_r(\mathbf{p})\,\gamma^0 d_{r'}^\dagger(\mathbf{p}')\,w_{r'}(\mathbf{p}')\,e^{-ip_0 x^0}\,e^{ip'_0 x^0} \end{bmatrix} \end{pmatrix} \right].
$$

$$\tag{W.132}$$

Due to the Kronecker deltas we have $E_{\mathbf{p}'} = \sqrt{\mathbf{p}'^2 + m^2} = \sqrt{(\pm\mathbf{p})^2 + m^2} = E_{\mathbf{p}}$, i.e., $p_0 = E_{\mathbf{p}} = p'_0 = E_{\mathbf{p}'}$ It follows

$$
H = \begin{pmatrix} -m \sum_{\mathbf{p},r,r'} \begin{bmatrix} -d_r(\mathbf{p})\,\bar{w}_r(\mathbf{p})\,\gamma^0 b'_r(-\mathbf{p})\,u_{r'}(-\mathbf{p})\,e^{-2ip_0 x^0} + \\ +b_r^\dagger(\mathbf{p})\,\bar{u}_r(\mathbf{p})\,\gamma^0 d_{r'}^\dagger(-\mathbf{p})\,w_{r'}(-\mathbf{p})\,e^{2ip_0 x^0} \end{bmatrix} - \\ -m \sum_{\mathbf{p},r,r'} \begin{bmatrix} -b_r^\dagger(\mathbf{p})\,\bar{u}_r(\mathbf{p})\,\gamma^0 b'_r(\mathbf{p})\,u_{r'}(\mathbf{p}) + \\ +d_r(\mathbf{p})\,\bar{w}_r(\mathbf{p})\,\gamma^0 d_{r'}^\dagger(\mathbf{p})\,w_{r'}(\mathbf{p}) \end{bmatrix} \end{pmatrix}.
$$

$$\tag{W.133}$$

Due to (W.99), i.e.,

$$
\bar{u}_r(\mathbf{p})\,\gamma^0 u_{r'}(\mathbf{p}) = \tfrac{E_{\mathbf{p}}}{m}\delta_{rr'} \;\; ; \;\; \bar{u}_r(-\mathbf{p})\,\gamma^0 w_{r'}(\mathbf{p}) = 0
$$
$$
\bar{w}_r(\mathbf{p})\,\gamma^0 w_{r'}(\mathbf{p}) = \tfrac{E_{\mathbf{p}}}{m}\delta_{rr'} \;\; ; \;\; \bar{w}_r(-\mathbf{p})\,\gamma^0 u_{r'}(\mathbf{p}) = 0
$$

$$\tag{W.134}$$

follows finally

$$
H = -m \sum_{\mathbf{p},r,r'} \left[ -b_r^\dagger(\mathbf{p})\,\bar{u}_r(\mathbf{p})\,\gamma^0 b_{r'}(\mathbf{p}')\,u_{r'}(\mathbf{p}) + d_r(\mathbf{p})\,\bar{w}_r(\mathbf{p})\,\gamma^0 d_{r'}^\dagger(\mathbf{p})\,w_{r'}(\mathbf{p}) \right] =
$$
$$
= -m \sum_{\mathbf{p},r} \tfrac{E_{\mathbf{p}}}{m} \left[ -b_r^\dagger(\mathbf{p})\,b_r(\mathbf{p}) + d_r(\mathbf{p})\,d_r^\dagger(\mathbf{p}) \right] = \sum_{\mathbf{p},r} E_{\mathbf{p}} \left[ b_r^\dagger(\mathbf{p})\,b_r(\mathbf{p}) - d_r(\mathbf{p})\,d_r^\dagger(\mathbf{p}) \right].
$$

$$\tag{W.135}$$

8. Prove (W.100) for the continuous case.
   Solution: We have

$$
H = i \int d^3x \, \psi^\dagger \partial_0 \psi \tag{W.136}
$$

and

$$
\psi = \sum_r \int d^3p \sqrt{\tfrac{m}{(2\pi)^3 E_p}} \left( b_r(\mathbf{p})\,u_r(\mathbf{p})\,e^{-ipx} + d_r^\dagger(\mathbf{p})\,w_r(\mathbf{p})\,e^{ipx} \right)
$$
$$
\psi^\dagger = \sum_r \int d^3p \sqrt{\tfrac{m}{(2\pi)^3 E_p}} \left( d_r(\mathbf{p})\,w_r^\dagger(\mathbf{p})\,e^{-ipx} + b_r^\dagger(\mathbf{p})\,u_r^\dagger(\mathbf{p})\,e^{ipx} \right).
$$

$$\tag{W.137}$$

With $\partial_0 e^{ipx} = ip_0 e^{ipx}$, $\partial_0 \psi$ is given by

$$\partial_0 \psi = i \sum_r \sqrt{\frac{m}{(2\pi)^3}} \int \frac{d^3 p}{\sqrt{E_\mathbf{p}}} p_0 \left[ -b_r(\mathbf{p}) u_r(\mathbf{p}) e^{-ipx} + d_r^\dagger(\mathbf{p}) w_r(\mathbf{p}) e^{ipx} \right].$$

(W.138)

It follows

$$H = -\int \frac{d^3 p}{\sqrt{E_\mathbf{p}}} \int \frac{d^3 p'}{\sqrt{E_{\mathbf{p}'}}} p_0' \sum_{r,r'} \sqrt{\frac{m}{(2\pi)^3}} \sqrt{\frac{m}{(2\pi)^3}} \int d^3 x$$
$$\left[ d_r(\mathbf{p}) w_r^\dagger(\mathbf{p}) e^{-ipx} + b_r^\dagger(\mathbf{p}) u_r^\dagger(\mathbf{p}) e^{ipx} \right] \left[ -b_{r'}(\mathbf{p}') u_{r'}(\mathbf{p}') e^{-ip'x} + d_{r'}^\dagger(\mathbf{p}') w_{r'}(\mathbf{p}') e^{ip'x} \right].$$

(W.139)

We consider the $x$-integration:

$$I = \int d^3 x \left[ d_r(\mathbf{p}) w_r^\dagger(\mathbf{p}) e^{-ipx} + b_r^\dagger(\mathbf{p}) u_r^\dagger(\mathbf{p}) e^{ipx} \right] \left[ -b_{r'}(\mathbf{p}') u_{r'}(\mathbf{p}') e^{-ip'x} + d_{r'}^\dagger(\mathbf{p}') w_{r'}(\mathbf{p}') e^{ip'x} \right] =$$
$$= \int d^3 x \left[ \begin{array}{l} \left[ -d_r(\mathbf{p}) w_r^\dagger(\mathbf{p}) b_{r'}(\mathbf{p}') u_{r'}(\mathbf{p}') e^{-i(p'+p)x} + d_r(\mathbf{p}) w_r^\dagger(\mathbf{p}) d_{r'}^\dagger(\mathbf{p}') w_{r'}(\mathbf{p}') e^{i(p'-p)x} \right] + \\ + \left[ -b_r^\dagger(\mathbf{p}) u_r^\dagger(\mathbf{p}) b_{r'}(\mathbf{p}') u_{r'}(\mathbf{p}') e^{-i(p'-p)x} + b_r^\dagger(\mathbf{p}) u_r^\dagger(\mathbf{p}) d_{r'}^\dagger(\mathbf{p}') w_{r'}(\mathbf{p}') e^{i(p'+p)x} \right] \end{array} \right].$$

(W.140)

The $x$-integration yields

$$\int d^3 x \, e^{ikx} = e^{ik_0 x^0} \int d^3 x \, e^{-i\mathbf{k}\mathbf{x}} = e^{ik_0 x^0} (2\pi)^3 \delta^3(\mathbf{k}).$$

(W.141)

It follows

$$I = (2\pi)^3 \left[ \begin{array}{l} \left[ \begin{array}{l} -d_r(\mathbf{p}) w_r^\dagger(\mathbf{p}) b_{r'}(\mathbf{p}') u_{r'}(\mathbf{p}') e^{-i(p_0'+p_0)x^0} \delta^3(\mathbf{p}'+\mathbf{p}) + \\ +d_r(\mathbf{p}) w_r^\dagger(\mathbf{p}) d_{r'}^\dagger(\mathbf{p}') w_{r'}(\mathbf{p}') e^{i(p_0'-p_0)x^0} \delta^3(\mathbf{p}'-\mathbf{p}) \end{array} \right] + \\ + \left[ \begin{array}{l} -b_r^\dagger(\mathbf{p}) u_r^\dagger(\mathbf{p}) b_{r'}(\mathbf{p}') u_{r'}(\mathbf{p}') e^{-i(p_0'-p_0)x^0} \delta^3(\mathbf{p}'-\mathbf{p}) + \\ +b_r^\dagger(\mathbf{p}) u_r^\dagger(\mathbf{p}) d_{r'}^\dagger(\mathbf{p}') w_{r'}(\mathbf{p}') e^{i(p_0'+p_0)x^0} \delta^3(\mathbf{p}'+\mathbf{p}) \end{array} \right] \end{array} \right].$$

(W.142)

The Kronecker functions give $E_{\mathbf{p}'} = \sqrt{\mathbf{p}'^2 + m^2} = \sqrt{(\pm\mathbf{p})^2 + m^2} = E_\mathbf{p}$, i.e., $p_0 = E_\mathbf{p} = p_0' = E_{\mathbf{p}'}$. Thus, we arrive at

$$I = (2\pi)^3 \left[ \begin{array}{l} \delta^3(\mathbf{p}'-\mathbf{p}) \left[ \begin{array}{l} d_r(\mathbf{p}) d_{r'}^\dagger(\mathbf{p}) w_r^\dagger(\mathbf{p}) w_{r'}(\mathbf{p}) - \\ -b_r^\dagger(\mathbf{p}) b_{r'}(\mathbf{p}) u_r^\dagger(\mathbf{p}) u_{r'}(\mathbf{p}) \end{array} \right] + \\ +\delta^3(\mathbf{p}'+\mathbf{p}) \left[ \begin{array}{l} b_r^\dagger(\mathbf{p}) d_{r'}^\dagger(-\mathbf{p}) u_r^\dagger(\mathbf{p}) w_{r'}(-\mathbf{p}) e^{2ip_0 x^0} - \\ -d_r(\mathbf{p}) b_{r'}(-\mathbf{p}) w_r^\dagger(\mathbf{p}) u_{r'}(-\mathbf{p}) e^{-2ip_0 x^0} \end{array} \right] \end{array} \right].$$

(W.143)

Inserting (W.99), i.e.,

$$u_r^\dagger(\mathbf{p}) w_{r'}(-\mathbf{p}) = w_r^\dagger(\mathbf{p}) u_{r'}(-\mathbf{p}) = 0$$

(W.144)

yields

$$I = (2\pi)^3 \delta^3(\mathbf{p}'-\mathbf{p}) \left[ \begin{array}{l} d_r(\mathbf{p}) d_{r'}^\dagger(\mathbf{p}) w_r^\dagger(\mathbf{p}) w_{r'}(\mathbf{p}) - \\ -b_r^\dagger(\mathbf{p}) b_{r'}(\mathbf{p}) u_r^\dagger(\mathbf{p}) u_{r'}(\mathbf{p}) \end{array} \right].$$

(W.145)

In addition, we use (W.98), i.e.,

$$u_r^\dagger(\mathbf{p})\, u_{r'}(\mathbf{p}) = w_r^\dagger(\mathbf{p})\, w_{r'}(\mathbf{p}) = \frac{E_\mathbf{p}}{m}\delta_{rr'} \tag{W.146}$$

and obtain

$$I = (2\pi)^3\, \delta^3\left(\mathbf{p}' - \mathbf{p}\right) \frac{E_\mathbf{p}}{m}\delta_{rr'} \left[ d_r(\mathbf{p})\, d_{r'}^\dagger(\mathbf{p}) - b_r^\dagger(\mathbf{p})\, b_{r'}(\mathbf{p}) \right]. \tag{W.147}$$

We insert this result for $I$ and obtain

$$H - \int \frac{d^3 p}{\sqrt{E_\mathbf{p}}} \int \frac{d^3 p'}{\sqrt{E_{\mathbf{p}'}}} p_0' \sum_{r,r'} \sqrt{\frac{m}{(2\pi)^3}} \sqrt{\frac{m}{(2\pi)^3}} \cdot I =$$

$$= \int \frac{d^3 p}{\sqrt{E_\mathbf{p}}} \int \frac{d^3 p'}{\sqrt{E_{\mathbf{p}'}}} p_0' \sum_r \delta^3\left(\mathbf{p}' - \mathbf{p}\right) E_\mathbf{p} \left[ b_r^\dagger(\mathbf{p})\, b_r(\mathbf{p}) - d_r(\mathbf{p})\, d_r^\dagger(\mathbf{p}) \right] =$$

$$= \int d^3 p\, E_\mathbf{p} \sum_r \left[ b_r^\dagger(\mathbf{p})\, b_r(\mathbf{p}) - d_r(\mathbf{p})\, d_r^\dagger(\mathbf{p}) \right]. \tag{W.148}$$

## W.6    Quantization of Free Fields, Photons

Essential issues are already discussed in the section 'Toy example' and in Appendix T, Vol. 1. We want to add some more results here in order to establish an uniform formalism.

A suitable Lagrangian density $\mathcal{L}$ is given by

$$\mathcal{L} = -\frac{1}{2}\left(\partial_\nu A_\mu\right)\left(\partial^\nu A^\mu\right). \tag{W.149}$$

Note that we have four fields, namely $A^\mu$, $\mu = 0, \ldots, 3$. The corresponding conjugated fields are given by

$$\pi_\mu = \frac{\delta \mathcal{L}}{\delta\left(\partial_0 A^\mu\right)} = -\partial_0 A_\mu. \tag{W.150}$$

### W.6.1    Determination of $\mathcal{H}$

In the next step, we calculate the Hamiltonian $\mathcal{H}$. We have

$$\mathcal{H} = \pi_\mu \partial^0 A^\mu - \mathcal{L} = -\left(\partial_0 A_\mu\right)\left(\partial_0 A^\mu\right) - \mathcal{L} =$$
$$= -\left(\partial_0 A_\mu\right)\left(\partial^0 A^\mu\right) + \tfrac{1}{2}\left(\partial_\nu A_\mu\right)\left(\partial^\nu A^\mu\right) =$$
$$= -\left(\partial_0 A_\mu\right)\left(\partial^0 A^\mu\right) + \tfrac{1}{2}\left[\left(\partial_0 A_\mu\right)\left(\partial^0 A^\mu\right) + \left(\partial_k A_\mu\right)\left(\partial^k A^\mu\right)\right] = \qquad \text{(W.151)}$$
$$= -\tfrac{1}{2}\left(\partial_0 A_\mu\right)\left(\partial^0 A^\mu\right) + \tfrac{1}{2}\left(\partial_k A_\mu\right)\left(\partial^k A^\mu\right).$$

With $\partial_k = -\partial^k$, this may be written as $\mathcal{H} = -\tfrac{1}{2}\left(\partial_0 A_\mu\right)\left(\partial^0 A^\mu\right) - \tfrac{1}{2}\left(\partial^k A_\mu\right)\left(\partial^k A^\mu\right)$ or

$$\mathcal{H} = -\frac{1}{2}\left(\partial^\nu A_\mu\right)\left(\partial^\nu A^\mu\right). \qquad \text{(W.152)}$$

The discrete and continuous solutions in the source-free case read (see Appendix T, Vol. 1)[90]

$$A^\mu(x) = \sum_{\mathbf{k},r} \sqrt{\frac{1}{2V\omega_\mathbf{k}}} \, \varepsilon_r^\mu(\mathbf{k}) \left[\alpha_r(\mathbf{k}) e^{-ikx} + \alpha_r^\dagger(\mathbf{k}) e^{ikx}\right]$$
$$A^\mu(x) = \sum_r \int \frac{d^3k}{\sqrt{2(2\pi)^3 \omega_\mathbf{k}}} \, \varepsilon_r^\mu(\mathbf{k}) \left[\alpha_r(\mathbf{k}) e^{-ikx} + \alpha_r^\dagger(\mathbf{k}) e^{ikx}\right]. \qquad \text{(W.153)}$$

If we choose the Coulomb gauge, then the polarization vectors have the properties

$$\mathbf{k} \cdot \boldsymbol{\varepsilon}_r(\mathbf{k}) = 0 \; ; \; \varepsilon_r^0(\mathbf{k}) = 0 \; ; \; \boldsymbol{\varepsilon}_r(\mathbf{k}) \cdot \boldsymbol{\varepsilon}_s(\mathbf{k}) = \delta_{rs}. \qquad \text{(W.154)}$$

We now interpret the amplitudes $\alpha_r(\mathbf{k})$ as operators which makes the field $A^\mu(x)$ a field operator. Instead of performing the canonical quantization procedure, we calculate the energy $H$ in terms of $\alpha_r(\mathbf{k})$ and $\alpha_r^\dagger(\mathbf{k})$ and discuss then which commutation relations are suitable. For the sake of variety, we start from the continuous solution.

We need for the (lengthy) calculation the expressions $\partial^\nu A_\mu$ and $\partial^\nu A^\mu$. With

$$\partial^\nu e^{ikx} = ik_\nu e^{ikx} \; ; \; k_\nu = \left(k^0, -\mathbf{k}\right) \; ; \; \varepsilon_{\nu r}(\mathbf{k}) = g_{\nu\mu} \varepsilon_r^\mu(\mathbf{k}) \qquad \text{(W.155)}$$

follows

$$\partial^\nu A_\mu = \sum_r \int \frac{d^3k}{\sqrt{2(2\pi)^3 \omega_\mathbf{k}}} \, \varepsilon_{\mu r}(\mathbf{k}) \, ik_\nu \left[-\alpha_r(\mathbf{k}) e^{-ikx} + \alpha_r^\dagger(\mathbf{k}) e^{ikx}\right]$$
$$\partial^\nu A^\mu = \sum_r \int \frac{d^3k}{\sqrt{2(2\pi)^3 \omega_\mathbf{k}}} \, \varepsilon_r^\mu(\mathbf{k}) \, ik_\nu \left[-\alpha_r(\mathbf{k}) e^{-ikx} + \alpha_r^\dagger(\mathbf{k}) e^{ikx}\right]. \qquad \text{(W.156)}$$

Thus, the Hamiltonian is given by

$$-2\mathcal{H} = \left(\partial^\nu A_\mu\right)\left(\partial^\nu A^\mu\right) =$$
$$= \sum_r \int \frac{d^3k}{\sqrt{2(2\pi)^3 \omega_\mathbf{k}}} \, \varepsilon_{\mu r}(\mathbf{k}) \, ik_\nu \left[-\alpha_r(\mathbf{k}) e^{-ikx} + \alpha_r^\dagger(\mathbf{k}) e^{ikx}\right] \cdot$$
$$\cdot \sum_{r'} \int \frac{d^3k'}{\sqrt{2(2\pi)^3 \omega_{\mathbf{k}'}}} \, \varepsilon_{r'}^\mu(\mathbf{k}') \, ik_\nu' \left[-\alpha_{r'}(\mathbf{k}') e^{-ik'x} + \alpha_{r'}^\dagger(\mathbf{k}') e^{ik'x}\right]. \qquad \text{(W.157)}$$

---

[90]Note $\omega_\mathbf{k} = k^0$ and $\omega_\mathbf{k}^2 = (k^0)^2 = \mathbf{k}^2$.

## W.6.2 Determination of H

For the energy follows

$$-2H = -2 \int d^3x \, \mathcal{H}(x) = \int d^3x \, \left(\partial^\nu A_\mu\right)\left(\partial^\nu A^\mu\right) =$$

$$= \int d^3x \, \sum_{rr'} \int \frac{d^3k}{\sqrt{2(2\pi)^3\omega_{\mathbf{k}}}} \int \frac{d^3k'}{\sqrt{2(2\pi)^3\omega_{\mathbf{k}'}}} \, \varepsilon_{\mu r}(\mathbf{k}) \, \varepsilon_{r'}^\mu(\mathbf{k}') \, (ik_\nu)\left(ik_\nu'\right) \cdot$$

$$\cdot \left[-\alpha_r(\mathbf{k}) \, e^{-ikx} + \alpha_r^\dagger(\mathbf{k}) \, e^{ikx}\right] \cdot \left[-\alpha_{r'}(\mathbf{k}') \, e^{-ik'x} + \alpha_{r'}^\dagger(\mathbf{k}') \, e^{ikx'}\right] =$$

$$= \sum_{rr'} \int \frac{d^3k}{\sqrt{2(2\pi)^3\omega_{\mathbf{k}}}} \int \frac{d^3k'}{\sqrt{2(2\pi)^3\omega_{\mathbf{k}'}}} \, \varepsilon_{\mu r}(\mathbf{k}) \, \varepsilon_{r'}^\mu(\mathbf{k}') \, (ik_\nu)\left(ik_\nu'\right) \cdot I\left(r, r', \mathbf{k}, \mathbf{k}'\right)$$

$$\text{(W.158)}$$

with

$$I\left(r, r', \mathbf{k}, \mathbf{k}'\right) = \int d^3x \, \left[-\alpha_r(\mathbf{k}) \, e^{-ikx} + \alpha_r^\dagger(\mathbf{k}) \, e^{ikx}\right] \cdot \left[-\alpha_{r'}(\mathbf{k}') \, e^{-ik'x} + \alpha_{r'}^\dagger(\mathbf{k}') \, e^{ikx'}\right].$$

$$\text{(W.159)}$$

With

$$\int d^3x \, e^{ipx} = e^{ip_0x_0} \int d^3x \, e^{-i\mathbf{px}} = e^{ip_0x_0} \, (2\pi)^3 \, \delta(\mathbf{p}) \qquad \text{(W.160)}$$

follows

$$I\left(r, r', \mathbf{k}, \mathbf{k}'\right) = \int d^3x \, \left[-\alpha_r(\mathbf{k}) \, e^{-ikx} + \alpha_r^\dagger(\mathbf{k}) \, e^{ikx}\right] \cdot \left[-\alpha_{r'}(\mathbf{k}') \, e^{-ik'x} + \alpha_{r'}^\dagger(\mathbf{k}') \, e^{ikx'}\right] =$$

$$= \int d^3x \cdot \left[\begin{array}{l} \alpha_r(\mathbf{k}) \, \alpha_{r'}(\mathbf{k}') \, e^{-i(k+k')x} - \alpha_r^\dagger(\mathbf{k}) \, \alpha_{r'}(\mathbf{k}') \, e^{i(k-k')x} - \\ -\alpha_r(\mathbf{k}) \, \alpha_{r'}^\dagger(\mathbf{k}') \, e^{-i(k-k')x} + \alpha_r^\dagger(\mathbf{k}) \, \alpha_{r'}^\dagger(\mathbf{k}') \, e^{i(k+k')x} \end{array}\right] =$$

$$= (2\pi)^3 \cdot \left[\begin{array}{l} \left\{\alpha_r(\mathbf{k}) \, \alpha_{r'}(\mathbf{k}') \, e^{-i(k_0+k_0')x_0} + \alpha_r^\dagger(\mathbf{k}) \, \alpha_{r'}^\dagger(\mathbf{k}') \, e^{i(k_0+k_0')x_0}\right\} \delta(\mathbf{k}+\mathbf{k}') - \\ -\left\{\alpha_r^\dagger(\mathbf{k}) \, \alpha_{r'}(\mathbf{k}') \, e^{i(k_0-k_0')x_0} + \alpha_r(\mathbf{k}) \, \alpha_{r'}^\dagger(\mathbf{k}') \, e^{-i(k_0-k_0')x_0}\right\} \delta(\mathbf{k}-\mathbf{k}') \end{array}\right].$$

$$\text{(W.161)}$$

Due to the delta functions, only the following terms survive:

$$\delta\left(\mathbf{k}+\mathbf{k}'\right) \rightarrow \mathbf{k}' = -\mathbf{k} \; ; \; k_0' = k_0 \; ; \; \delta\left(\mathbf{k}-\mathbf{k}'\right) \rightarrow \mathbf{k}' = \mathbf{k} \; ; \; k_0' = k_0. \quad \text{(W.162)}$$

This yields

$$I\left(r, r', \mathbf{k}, \mathbf{k}'\right) = (2\pi)^3 \cdot \left[\begin{array}{l} \left\{\alpha_r(\mathbf{k}) \, \alpha_{r'}(-\mathbf{k}) \, e^{-2ik_0x} + \alpha_r^\dagger(\mathbf{k}) \, \alpha_{r'}^\dagger(-\mathbf{k}) \, e^{2ik_0}\right\} \delta(\mathbf{k}+\mathbf{k}') - \\ -\left\{\alpha_r^\dagger(\mathbf{k}) \, \alpha_{r'}(\mathbf{k}) + \alpha_r(\mathbf{k}) \, \alpha_{r'}^\dagger(\mathbf{k})\right\} \delta(\mathbf{k}-\mathbf{k}') \end{array}\right].$$

$$\text{(W.163)}$$

Thus, we arrive at

$$-2H = -2 \int d^3x \, \mathcal{H}(x) =$$

$$= \sum_{rr'} \int \frac{d^3k}{\sqrt{2(2\pi)^3 \omega_k}} \int \frac{d^3k'}{\sqrt{2(2\pi)^3 \omega_{k'}}} \, \varepsilon_{\mu r}(\mathbf{k}) \, \varepsilon_{r'}^{\mu}(\mathbf{k}') \, (ik_\nu) \, (ik'_\nu) \, (2\pi)^3 \cdot$$

$$\cdot \left[ \begin{array}{l} \left\{ \alpha_r(\mathbf{k}) \, \alpha_{r'}(-\mathbf{k}) \, e^{-2ik_0x} + \alpha_r^\dagger(\mathbf{k}) \, \alpha_{r'}^\dagger(-\mathbf{k}) \, e^{2ik_0} \right\} \delta(\mathbf{k}+\mathbf{k}') - \\ - \left\{ \alpha_r^\dagger(\mathbf{k}) \, \alpha_{r'}(\mathbf{k}) + \alpha_r(\mathbf{k}) \, \alpha_{r'}^\dagger(\mathbf{k}) \right\} \delta(\mathbf{k}-\mathbf{k}') \end{array} \right] =$$

$$= \sum_{rr'} \int \frac{d^3k}{2\omega_k} \int \frac{d^3k'}{\sqrt{2(2\pi)^3 \omega_{k'}}} \left[ \begin{array}{l} \varepsilon_{\mu r}(\mathbf{k}) \, \varepsilon_{r'}^{\mu}(-\mathbf{k}) \, (ik_\nu) \, (ik'_\nu) \left\{ \begin{array}{l} \alpha_r(\mathbf{k}) \, \alpha_{r'}(-\mathbf{k}) \, e^{-2ik_0x} + \\ + \alpha_r^\dagger(\mathbf{k}) \, \alpha_{r'}^\dagger(-\mathbf{k}) \, e^{2ik_0} \end{array} \right\} \delta(\mathbf{k}+\mathbf{k}') - \\ - \varepsilon_{\mu r}(\mathbf{k}) \, \varepsilon_{r'}^{\mu}(\mathbf{k}) \, (ik_\nu) \, (ik'_\nu) \left\{ \alpha_r^\dagger(\mathbf{k}) \, \alpha_{r'}(\mathbf{k}) + \alpha_r(\mathbf{k}) \, \alpha_{r'}^\dagger(\mathbf{k}) \right\} \delta(\mathbf{k}-\mathbf{k}') \end{array} \right].$$

$$\text{(W.164)}$$

Now we consider the terms $(ik_\nu) \, (ik'_\nu) \, \delta(\mathbf{k}+\mathbf{k}')$ and $(ik_\nu) \, (ik'_\nu) \, \delta(\mathbf{k}-\mathbf{k}')$. We have (remember $k_0^2 = \mathbf{k}^2$)

$$(ik_\nu) \, (ik'_\nu) \, \delta(\mathbf{k}+\mathbf{k}') = -(k_0^2 + \mathbf{k}\mathbf{k}') \, \delta(\mathbf{k}+\mathbf{k}') = -(k_0^2 - \mathbf{k}\mathbf{k}) \, \delta(\mathbf{k}+\mathbf{k}') = 0$$

$$(ik_\nu) \, (ik'_\nu) \, \delta(\mathbf{k}-\mathbf{k}') = -(k_0^2 + \mathbf{k}\mathbf{k}) \, \delta(\mathbf{k}-\mathbf{k}') = -2\omega_k^2 \delta(\mathbf{k}-\mathbf{k}').$$

$$\text{(W.165)}$$

This yields

$$-2H = \sum_{rr'} \int \frac{d^3k}{2\omega_k} \left[ -\varepsilon_{\mu r}(\mathbf{k}) \, \varepsilon_{r'}^{\mu}(\mathbf{k}) \, (-2\omega_k^2) \left\{ \alpha_r^\dagger(\mathbf{k}) \, \alpha_{r'}(\mathbf{k}) + \alpha_r(\mathbf{k}) \, \alpha_{r'}^\dagger(\mathbf{k}) \right\} \right] =$$

$$= -\sum_{rr'} \int d^3k \, \omega_k \left[ \left( -\varepsilon_{\mu r}(\mathbf{k}) \, \varepsilon_{r'}^{\mu}(\mathbf{k}) \right) \left\{ \alpha_r^\dagger(\mathbf{k}) \, \alpha_{r'}(\mathbf{k}) + \alpha_r(\mathbf{k}) \, \alpha_{r'}^\dagger(\mathbf{k}) \right\} \right].$$

$$\text{(W.166)}$$

The product of the polarization vectors gives

$$-\varepsilon_{\mu r}(\mathbf{k}) \, \varepsilon_{r'}^{\mu}(\mathbf{k}) = -\varepsilon_{0r}(\mathbf{k}) \, \varepsilon_{r'}^{0}(\mathbf{k}) - (-\varepsilon_r(\mathbf{k}) \, \varepsilon_{r'}(\mathbf{k})) = -\varepsilon_{0r}(\mathbf{k}) \, \varepsilon_{r'}^{0}(\mathbf{k}) + \varepsilon_r(\mathbf{k}) \, \varepsilon_{r'}(\mathbf{k}).$$

$$\text{(W.167)}$$

In the source-free case, the Coulomb gauge is convenient. With (W.154) follows

$$-\varepsilon_{\mu r}(\mathbf{k}) \, \varepsilon_{r'}^{\mu}(\mathbf{k}) = \varepsilon_r(\mathbf{k}) \, \varepsilon_{r'}(\mathbf{k}) = \delta_{rr'} \qquad \text{(W.168)}$$

and we obtain

$$-2H = -\sum_{rr'} \int d^3k \, \omega_k \delta_{rr'} \left\{ \alpha_r^\dagger(\mathbf{k}) \, \alpha_{r'}(\mathbf{k}) + \alpha_r(\mathbf{k}) \, \alpha_{r'}^\dagger(\mathbf{k}) \right\}$$

$$= -\sum_r \int d^3k \, \omega_k \left\{ \alpha_r^\dagger(\mathbf{k}) \, \alpha_r(\mathbf{k}) + \alpha_r(\mathbf{k}) \, \alpha_r^\dagger(\mathbf{k}) \right\}.$$

$$\text{(W.169)}$$

Thus, the final result reads

$$H = \sum_r \int d^3k \, \omega_k \frac{\alpha_r^\dagger(\mathbf{k}) \, \alpha_r(\mathbf{k}) + \alpha_r(\mathbf{k}) \, \alpha_r^\dagger(\mathbf{k})}{2}. \qquad \text{(W.170)}$$

and in the discrete case

$$H = \sum_{\mathbf{k},r} \omega_k \left[ \frac{\alpha_r^\dagger(\mathbf{k}) \, \alpha_r(\mathbf{k}) + \alpha_r(\mathbf{k}) \, \alpha_r^\dagger(\mathbf{k})}{2} \right]. \qquad \text{(W.171)}$$

We see that we have reproduced the result found in the section 'Toy example'.

The commutation rules for the photon field are

$$\left[ \alpha_r \left( \mathbf{k} \right), \alpha_{r'}^\dagger \left( \mathbf{k'} \right) \right] = \delta_{rr'} \delta \left( \mathbf{k}, \mathbf{k'} \right). \tag{W.172}$$

This may explicitly shown by the (quite lengthy) procedure of canonical quantization. But the result seems plausible anyway, apart from the fact, that we have deduced it already in a previous section. First, a photon is a boson; so we expect a commutator, not an anticommutator. Second, the state of a photon is completely determined by its momentum and its polarization. The further results can now be formulated parallel to those of the other fields. For instance, the number operator us given by $N_{\mathbf{k}r} = a_r^\dagger \left( \mathbf{k} \right) a_r \left( \mathbf{k} \right)$ and so on.

By means of the commutation rules, we can write

$$H = \sum_{\mathbf{k},r} \omega_{\mathbf{k}} \left[ \alpha_r^\dagger \left( \mathbf{k} \right) \alpha_r \left( \mathbf{k} \right) + \frac{1}{2} \right] \tag{W.173}$$

We see that also in this case we have an infinite vacuum energy.

## W.6.3  Exercises and Solutions

1. Prove (W.150).
   Solution: We have

$$\pi_\mu = \frac{\delta \mathcal{L}}{\delta(\partial_0 A^\mu)} = -\frac{1}{2} \frac{\partial}{\partial(\partial_0 A^\mu)} \left( \partial^\kappa A^\nu \right) \left( \partial_\kappa A_\nu \right) = \\ = -\frac{1}{2} \frac{\partial}{\partial(\partial_0 A^\mu)} \left[ \left( \partial^0 A^\nu \right) \left( \partial_0 A_\nu \right) + \left( \partial^k A^\mu \right) \left( \partial_k A_\mu \right) \right]. \tag{W.174}$$

The second summand contains no derivatives with respect to $\partial_0$ or $\partial^0$ and does not contribute to the result. It follows

$$\pi_\mu = -\frac{1}{2} \frac{\partial}{\partial(\partial_0 A_\mu)} \left( \partial_0 A_\nu \right) \left( \partial^0 A^\nu \right) = \\ = -\frac{1}{2} \left[ \left( \frac{\partial}{\partial(\partial_0 A^\mu)} \left( \partial^0 A^\nu \right) \right) \left( \partial_0 A_\nu \right) + \left( \partial^0 A^\nu \right) \left( \frac{\partial}{\partial(\partial_0 A^\mu)} \left( \partial_0 A_\nu \right) \right) \right] = \tag{W.175} \\ = -\frac{1}{2} \left[ \delta_{\nu\mu} \left( \partial_0 A_\nu \right) + \left( \partial^0 A^\nu \right) \frac{\partial}{\partial(\partial_0 A^\mu)} \left( \partial_0 A_\nu \right) \right].$$

For the second summand, we use $A_\nu = g_{\nu\kappa} A^\kappa$ and obtain (remember $\partial_0 = \partial^0$)

$$\frac{\partial}{\partial(\partial_0 A^\mu)} \left( \partial_0 A_\nu \right) = \frac{\partial}{\partial(\partial_0 A^\mu)} \left( \partial_0 g_{\nu\kappa} A^\kappa \right) = \\ = \frac{\partial}{\partial(\partial_0 A^\mu)} \left[ \left( \partial_0 g_{\nu\kappa} \right) A^\kappa + g_{\nu\kappa} \partial_0 A^\kappa \right] = g_{\nu\kappa} \frac{\partial}{\partial(\partial_0 A^\mu)} \partial_0 A^\kappa = g_{\nu\kappa} \delta_{\mu\kappa} \tag{W.176}$$

(due to $\partial_0 g_{\nu\kappa} = 0$). Inserting the result gives

$$
\pi_\mu = -\tfrac{1}{2} \left[ \delta_{\nu\mu} \left( \partial_0 A_\nu \right) + \left( \partial^0 A^\nu \right) g_{\nu\kappa} \delta_{\mu\kappa} \right] = 
$$
$$
= -\tfrac{1}{2} \left[ \left( \partial_0 A_\mu \right) + \left( \partial^0 A^\nu \right) g_{\nu\mu} \right] = -\tfrac{1}{2} \left[ \left( \partial_0 A_\mu \right) + \left( \partial^0 A_\mu \right) \right] = - \left( \partial_0 A_\mu \right) .
\tag{W.177}
$$

## W.7 Operator Ordering

### W.7.1 Normal Order

The different Hamilton functions $H$ which we found for the three considered fields (Klein–Gordon, Dirac, radiation) all share the feature of an infinite vacuum energy. We recap the problem on the basis of the Klein–Gordon field. In this case, the Hamilton functions $H$ reads (discrete version)

$$
H = \frac{1}{2} \sum_\mathbf{p} E_\mathbf{p} \left( a^\dagger \left( \mathbf{p} \right) a \left( \mathbf{p} \right) + a \left( \mathbf{p} \right) a^\dagger \left( \mathbf{p} \right) \right) .
\tag{W.178}
$$

By means of the commutation relations $\left[ a \left( \mathbf{p} \right), a^\dagger \left( \mathbf{p} \right) \right] = \delta_{\mathbf{p}\mathbf{p}'}$ , this may be written as

$$
H = \sum_\mathbf{p} E_\mathbf{p} \left( a^\dagger \left( \mathbf{p} \right) a \left( \mathbf{p} \right) + \frac{1}{2} \right) .
\tag{W.179}
$$

The second summand, i.e., the sum over the zero point energies $\frac{1}{2} \sum_\mathbf{p} E_\mathbf{p}$, is *always infinite*. Of course, this is a serious problem. A way out is offered by the fact that physics may take its arguments not only from mathematics alone. One can argue as follows: In the the sum $\sum_\mathbf{p}$, there occur arbitrarily large values of $\mathbf{p}$, and thus also arbitrarily large energies. Now it is certainly not sensible (from a physical point of view, not a mathematical one) to take into account energies that may be greater than say the total energy of the universe. In other words, a cut-off at a certain ('large') $P$ is physically legitimate or even necessary, restricting the summation to $|\mathbf{p}| \leq P$. It is not required to specify the exact value of $P$; it suffices to say that it exists because then the sum over the zero point energies yields a *finite* value which as common reference point for all energies may be neglected. Thus, the world would be in order again.[91]

A formal way to handle the problem of infinities is *normal ordering* (or *Wick ordering*). The notation[92] is $\mathcal{N} [ab]$ where $a$ and $b$ are scalar field operators. Normal ordering means to rearrange a product of annihilation and creation operators in such

---

[91] There is another type of infinities in quantum electrodynamics, keyword renormalization, which also is healed by introducing a cut-off, see below.

[92] An alternative notation is enclosing the operator between double-dots, i.e., $: A :$. We prefer here $\mathcal{N} [A]$ due to its better legibility.

a way that

$$\underset{\text{all creation operators}}{\text{left hand side}} - \underset{\text{all annihilation operators}}{\text{right hand side}} \qquad \text{(W.180)}$$

thereby *neglecting* all existing commutation relations.

The rearranging depends on whether if we consider bosons (integer spin as with photons or in the Klein–Gordon case) or fermions (half-integer spin as in the Dirac case). For the purpose of a compact notation, we write in this section $\pm$ where the upper sign means bosons and the lower sign fermions. To hold things simple, we use for the commutator $[a, b]$ and the anticommutator $\{a, b\}$ the notation $[a, b]_{\mp}$, i.e., $[a, b]_- = ab - ba$ and $[a, b]_+ = ab + ba$.

Swapping a product of two bosonic operators in normal ordering does not change the sign, but it does so for fermionic operators:

$$\text{bosons: } \mathcal{N}\left[aa^\dagger\right] = a^\dagger a \text{ ; fermions: } \mathcal{N}\left[aa^\dagger\right] = -a^\dagger a$$
$$\text{in short } \mathcal{N}\left[aa^\dagger\right] = \pm a^\dagger a. \qquad \text{(W.181)}$$

For example, the operator $a^\dagger(\mathbf{p}) a(\mathbf{p})$ is already in normal order, $\mathcal{N}\left[a^\dagger(\mathbf{p}) a(\mathbf{p})\right] = a^\dagger(\mathbf{p}) a(\mathbf{p})$; the operator $a(\mathbf{p}) a^\dagger(\mathbf{p})$ reads in normal order $\mathcal{N}\left[a(\mathbf{p}) a^\dagger(\mathbf{p})\right] = \pm a^\dagger(\mathbf{p}) a(\mathbf{p})$. If there are several annihilation and creation operators, for instance $ab^\dagger c^\dagger de$, normal ordering results in $\mathcal{N}\left[ab^\dagger c^\dagger de\right] = (\pm 1)^P b^\dagger c^\dagger ade$ where $P$ gives the number of swappings. We have as an additional rule that the order within the set of destruction and the set of creation operators is unchanged from the original expression. Normal ordering is linear, i.e. ($A_i$ operators, $c_i$ complex numbers): $\mathcal{N}\left[c_1 A_1 + c_2 A_2\right] = c_1 \mathcal{N}\left[A_1\right] + c_2 \mathcal{N}\left[A_2\right]$. Note that the normal order of a product differs in general from the product of the normal orders: $\mathcal{N}\left[A_1 A_2\right] \neq \mathcal{N}\left[A_1\right] \cdot \mathcal{N}\left[A_2\right]$.

### W.7.1.1 Normal Order of Energies and Charges

**Energies** As an example, we apply the considerations[93] to the three Hamilton functions (see the corresponding sections)

$$H_{\text{Klein-Gordon}} = \tfrac{1}{2} \sum_{\mathbf{p}} E_{\mathbf{p}} \left(a^\dagger(\mathbf{p}) a(\mathbf{p}) + a(\mathbf{p}) a^\dagger(\mathbf{p})\right) = \sum_{\mathbf{p}} E_{\mathbf{p}} \left(a^\dagger(\mathbf{p}) a(\mathbf{p}) + \tfrac{1}{2}\right)$$

$$H_{\text{Dirac}} = \sum_{\mathbf{p},r} E_{\mathbf{p}} \left(b_r^\dagger(\mathbf{p}) b_r(\mathbf{p}) - d_r(\mathbf{p}) d_r^\dagger(\mathbf{p})\right) = \sum_{\mathbf{p},r} E_{\mathbf{p}} \left[b_r^\dagger(\mathbf{p}) b_r(\mathbf{p}) + d_r^\dagger(\mathbf{p}) d_r(\mathbf{p}) - 1\right]$$

$$H_{\text{Radiation}} = \tfrac{1}{2} \sum_{\mathbf{k},r} \omega_{\mathbf{k}} \left(\alpha_r^\dagger(\mathbf{k}) \alpha_r(\mathbf{k}) + \alpha_r(\mathbf{k}) \alpha_r^\dagger(\mathbf{k})\right) = \sum_{\mathbf{k},r} \omega_{\mathbf{k}} \left(\alpha_r^\dagger(\mathbf{k}) \alpha_r(\mathbf{k}) + \tfrac{1}{2}\right).$$
$$\text{(W.182)}$$

The second form of these Hamiltonians is derived by applying the appropriate (anti-)commutation relations. We see that there is in each case an infinite vacuum

---

[93] We confine ourselves to the discussion of the discrete case; the considerations for the continuous case run analogously.

energy. As stated above, the 'true' energies are given by the normal ordered expressions which implies simply neglecting the infinite vacuum energy term, retaining only the number operators. Thus we have symbolically

$$H_{\text{final}} = \mathcal{N}\left[H_{\text{before}}\right]. \tag{W.183}$$

This yields

$$H_{\text{Klein-Gordon}} = \mathcal{N}\left[\tfrac{1}{2} \sum_{\mathbf{p}} E_{\mathbf{p}} \left(a^\dagger(\mathbf{p})\, a(\mathbf{p}) + a(\mathbf{p})\, a^\dagger(\mathbf{p})\right)\right] = \sum_{\mathbf{p}} E_{\mathbf{p}} a^\dagger(\mathbf{p})\, a(\mathbf{p})$$

$$H_{\text{Dirac}} = \mathcal{N}\left[\sum_{\mathbf{p},r} E_{\mathbf{p}} \left(b_r^\dagger(\mathbf{p})\, b_r(\mathbf{p}) - d_r(\mathbf{p})\, d_r^\dagger(\mathbf{p})\right)\right] = \sum_{\mathbf{p},r} E_{\mathbf{p}} \left(b_r^\dagger(\mathbf{p})\, b_r(\mathbf{p}) + d_r^\dagger(\mathbf{p})\, d_r(\mathbf{p})\right)$$

$$H_{\text{Radiation}} = \mathcal{N}\left[\tfrac{1}{2} \sum_{\mathbf{k},r} \omega_{\mathbf{k}} \left(\alpha_r^\dagger(\mathbf{k})\, \alpha_r(\mathbf{k}) + \alpha_r(\mathbf{k})\, \alpha_r^\dagger(\mathbf{k})\right)\right] = \sum_{\mathbf{k},r} \omega_{\mathbf{k}} \alpha_r^\dagger(\mathbf{k})\, \alpha_r(\mathbf{k}). \tag{W.184}$$

By comparison with (W.182) we see that the annoying infinities have disappeared. It does not matter whether we omit the (infinite) vacuum energy or whether we normal order $H$. In other words: Normal ordering makes vacuum energy go away.

The above considerations are not restricted to $H$ but hold for all observables. Thus, in general, *any* string of operators in field theory has to be normal ordered to avoid infinities. Without this convention, results are nonsensical, in general.

Note that since *all* observables $A$ are defined a priori as normal ordered products, this fact is often not explicitly mentioned. Thus, there is often no explicit additional notation to mark the normal ordered form. For instance, we have $H = \tfrac{1}{2} \sum_{\mathbf{p}} E_{\mathbf{p}} \left(a^\dagger(\mathbf{p})\, a(\mathbf{p}) + a(\mathbf{p})\, a^\dagger(\mathbf{p})\right)$ and its normal ordered form $H = \sum_{\mathbf{p}} E_{\mathbf{p}} a^\dagger(\mathbf{p})\, a(\mathbf{p})$ In many texts, it is $H = \sum_{\mathbf{p}} E_{\mathbf{p}} a^\dagger(\mathbf{p})\, a(\mathbf{p})$ which from the start is presented as Hamilton function of the Klein–Gordon field, mostly without further comment and without indicating that it is the normal ordered form.

**Charge** In addition, we want to consider briefly the total charge $Q$ of the Dirac system. The 4-current density is given by the normal ordered expression[94]

$$j^\mu(x) = q\mathcal{N}\left[\bar{\psi}(x)\, \gamma^\mu \psi(x)\right] \tag{W.185}$$

where $q$ is the charge of the electron. $j^\mu$ fulfills the continuity equation

$$\partial_\mu j^\mu(x) = 0. \tag{W.186}$$

The operator for the total charge $\hat{Q}$ is given by

$$\hat{Q} = \int d^3x\, j^0(x) = q \int d^3x\, \mathcal{N}\left[\bar{\psi}(x)\, \gamma^\mu \psi(x)\right] =$$
$$= q \sum_{\mathbf{p},r} \left(b_r^\dagger(\mathbf{p})\, b_r(\mathbf{p}) - d_r^\dagger(\mathbf{p})\, d_r(\mathbf{p})\right) = q \sum_{\mathbf{p},r} \left(N_{e,r\mathbf{p}} - N_{p,r\mathbf{p}}\right). \tag{W.187}$$

---

[94]Remember that the fermionic 4-current (probability current) is given by $\bar{\psi}\gamma^\mu\psi$, see Appendix U, Vol. 1.

We know that the '$d$-particles', i.e., the positrons, and the '$b$-particles', i.e., the electrons, are oppositely charged. This fact is confirmed by considering the eigenvalues of $\hat{Q}$. The commutators of $\hat{Q}$ with the creation operators are given by.

$$\left[\hat{Q}, b_r^\dagger(\mathbf{p})\right] = q b_r^\dagger(\mathbf{p}) \quad ; \quad \left[\hat{Q}, d_r^\dagger(\mathbf{p})\right] = -q d_r^\dagger(\mathbf{p}). \tag{W.188}$$

Now assume that we have a state $|\Psi\rangle$ which is eigenstate of $\hat{Q}$ with the eigenvalue $Q$

$$\hat{Q}|\Psi\rangle = Q|\Psi\rangle. \tag{W.189}$$

It follows with (W.188)

$$\hat{Q}b_r^\dagger(\mathbf{p})|\Psi\rangle = \left(b_r^\dagger(\mathbf{p})\hat{Q} + q b_{r\mathbf{p}}^\dagger\right)|\Psi\rangle = \left(b_r^\dagger(\mathbf{p})Q + q b_{r\mathbf{p}}^\dagger\right)|\Psi\rangle = (Q+q)b_r^\dagger(\mathbf{p})|\Psi\rangle \tag{W.190}$$

and analogously for the other operators. All in all we have

$$\hat{Q}b_r^\dagger(\mathbf{p})|\Psi\rangle = (Q+q)b_r^\dagger(\mathbf{p})|\Psi\rangle \quad ; \quad \hat{Q}d_r^\dagger(\mathbf{p})|\Psi\rangle = (Q-q)d_r^\dagger(\mathbf{p})|\Psi\rangle$$
$$\hat{Q}b_r(\mathbf{p})|\Psi\rangle = (Q-q_0)b_r(\mathbf{p})|\Psi\rangle \quad ; \quad \hat{Q}d_r(\mathbf{p})|\Psi\rangle = (Q+q)d_r(\mathbf{p})|\Psi\rangle. \tag{W.191}$$

Thus, creating an electron or deleting a positron adds $q$ to the total charge; the vacuum has charge zero.

### W.7.1.2   Normal Order for General Field Operators

The ground state energy of the normal ordered Hamilton function $H = \sum_{\mathbf{p}} E_{\mathbf{p}} a^\dagger(\mathbf{p}) a(\mathbf{p})$ is zero due to $a(\mathbf{p})|0\rangle = 0$ (or equivalently $\langle 0|a^\dagger(\mathbf{p}) = 0$):

$$\langle 0|\mathcal{N}[H]|0\rangle = \langle 0|\sum_{\mathbf{p}} E_{\mathbf{p}} a^\dagger(\mathbf{p}) a(\mathbf{p})|0\rangle = \langle 0|\sum_{\mathbf{p}} E_{\mathbf{p}} a^\dagger(\mathbf{p})(a(\mathbf{p})|0\rangle) = 0. \tag{W.192}$$

With creation operators to the left and annihilation operators to the right, *any* normal ordered operator $A$ has a vacuum expectation value of zero. Note that this holds true even if $\langle 0|A|0\rangle \neq 0$, i.e., if the mean value does not vanish:

$$\langle 0|\mathcal{N}[A]|0\rangle = 0 \text{ always.} \tag{W.193}$$

Next we will consider the normal ordering of a product of two general field operators $A$ and $B$. To this end, we split the field operator $A$ into two parts, one containing all destruction operators $A^d$, the other containing all creation operators $A^c$, and $A = A^c + A^d$.[95] Normal ordering the product $A(x)B(y)$ yields

---

[95] In many textbooks one finds the notation $A = A^+ + A^-$, where the upper index marks the sign of energy, i.e., $A^+$ contains the destruction operators and $A^-$ the creation operators. But this notation

$$\mathcal{N}[A(x)B(y)] = \mathcal{N}\left[\left(A^d(x) + A^c(x)\right)\left(B^d(y) + B^c(y)\right)\right] =$$
$$= \mathcal{N}\left[A^d(x)B^d(y)\right] + \mathcal{N}\left[A^d(x)B^c(y)\right] + \mathcal{N}\left[A^c(x)B^d(y)\right] + \mathcal{N}\left[A^c(x)B^c(y)\right] =$$
$$= A^d(x)B^d(y) \pm B^c(y)A^d(x) + A^c(x)B^d(y) + A^c(x)B^c(y).$$

(W.194)

As is seen, the second term in the last line is the only one with changed order of the operators. So we have immediately for the difference of normal ordered product and product itself

$$\mathcal{N}[A(x)B(y)] - A(x)B(y) = \pm B^c(y)A^d(x) - A^d(x)B^c(y) = -\left[A^d(x), B^c(y)\right]_{\mp}. \quad (W.195)$$

Now the commutator for bosonic operators and the anticommutator for fermionic operators is a c-number,[96] or in other words: the difference $\mathcal{N}[A(x)B(y)] - A(x)B(y)$ contains no operators anymore. We note that this result is important in the further discussion.

### W.7.1.3   Discussion of Normal Ordering

Normal ordering is firmly established in quantum field theory. However, it appears like an arbitrary rule for quantization, a mere ad-hoc convention. The problem is clear to see: the commutation relations are suspended for this step and only for this step. As we have seen, commutation relations are in the heart of Quantum Mechanics, and normal ordering simply overrides these key elements. A theory would be highly desirable which gets along without such an artificial feature.

In classical mechanics, order does not import: $px$ equals $xp$. In general, products of classical operators can be written in many equivalent ways, and it is not automatically clear which one has to be quantized.[97] However, this is to be expected: the transition from classical mechanics to quantum mechanics *must* necessarily be ambiguous - if it were unambiguous, quantum mechanics would be superfluous.[98] There has to be something new in quantum mechanics. In first quantization, this is the change from Poisson brackets to commutator relations, and in second quantization it is (perhaps) normal ordering.

Whatsoever - the cornerstone of physics is comparison with experimental results. We can accept a physical theory if and only if it agrees with the observations. And

---

may sometimes be a little bit confusing, especially for beginners, since in $A^+$ there are the terms $\sim e^{-ikx}$, while the terms $\sim e^{ikx}$ are in $A^-$. Moreover, a few authors use the signs in the upper index in the reverse meaning. Thus, in order to avoid misunderstanding, we use $A^d$ and $A^c$ instead of $A^+$ and $A^-$.

[96]One distinguishes c-numbers (classical numbers) in contrast to q-numbers (quantum mechanical numbers, i.e., operators).

[97]In Vol. 1, we have argued that a classical operator like $xp$ should give a Hermitian quantum operator; so we introduced the Hermitian term $\frac{xp+px}{2}$. But this criterion does not work here, since the operators $aa^\dagger$ and $a^\dagger a$ both are Hermitian.

[98]The transition from classical mechanics to quantum mechanics *cannot* be unambiguous. In contrast, the transition from quantum mechanics to classical mechanics *must* be unambiguous.

the normal ordered Hamilton function results in what is actually observed: In QFT, the *normal ordered* Hamiltonian is the *observable* Hamiltonian.[99] Normal ordering gives a meaningful quantum field theory; if this were not the case, it would not persist. We must accept this as the way nature works. Perhaps a future theory will get rid of this apparent inconsistency.

## W.7.2   Time Order

*Time ordering*[100] $\mathcal{T}$ can be first understood as a practical tool to simplify the notation of complicated series as e.g. the time evolution operator $U_I(t, t_0)$ in the interaction picture. In Appendix Q, Vol. 1, we have found the representation (Dyson series)

$$U_I(t, t_0) = \sum_{n=0}^{\infty} \left(\frac{1}{i\hbar}\right)^n \int_{t_0}^{t} dt_1 \int_{t_0}^{t_1} dt_2 \dots \int_{t_0}^{t_{n-1}} dt_n \, H_I(t_1) H_I(t_2) \dots H_I(t_n).$$

(W.196)

Note that the $H_I$ at different times will not commute, $[H_I(t_1), H_I(t_2)] \neq 0$, in general. Thus, the order of time is of great importance; here we have ordered times with $t_0 \leq t_n \leq t_{n-1} \leq \dots \leq t_2 \leq t_1 \leq t$.

The calculation of this integral is cumbersome, not least because the upper limits of the integrals are all different. To circumvent this difficulty, we introduce an distribution $\Theta(t_1, t_2, \dots, t_n)$ with the properties

$$\Theta(t_1, t_2, \dots, t_n) = \begin{cases} 1 & \text{if } t_1 > t_2 > \dots > t_n \\ 0 & \text{otherwise} \end{cases}.$$

(W.197)

Then we can write

$$U_I(t, t_0) = \sum_{n=0}^{\infty} \left(\frac{1}{i\hbar}\right)^n \int_{t_0}^{t} dt_1 \dots \int_{t_0}^{t} dt_n \, \Theta(t_1, \dots, t_n) H_I(t_1) \dots H_I(t_n).$$

(W.198)

Note that in this version all integrals have as upper limit the *same* value $t$, since $\Theta(t_1, t_2, \dots, t_n)$ guarantees that the additional contributions of the integrals vanish - this is the important step. Now for $n = 2$ we have the two possibilities $t_1 > t_2$ and $t_2 > t_1$, for $n = 3$ there are six possibilities ($t_1 > t_2 > t_3, t_1 > t_3 > t_2$ and so on), and for arbitrary $n$ we have $n!$ possibilities or permutations. Thus, if we consider all permutations, we can write

---

[99]In other words: a not normal ordered Hamiltonian $\mathcal{H}_{\text{not normal}}$ does not represent an observable. In this sense, $\mathcal{H}_{\text{not normal}}$ is a Hermitian operator, but not an observable. However, in other contexts the zero point energy may be measurable, and that implies a not normal ordered Hamiltonian, as is the case e.g. in molecular vibrations.

[100]Although we need time ordering only in a later section, we consider it here in anticipation on account of its intrinsic proximity to normal ordering.

$$U_I = \sum_{n=0}^{\infty} \frac{1}{n!} \left(-\frac{i}{\hbar}\right)^n \int_{t_0}^{t} dt_1 \ldots \int_{t_0}^{t} dt_n \sum_{\pi \in S_n} \Theta\left(t_{\pi(1)}, \ldots, t_{\pi(n)}\right) H_I\left(t_{\pi(1)}\right) \ldots H_I\left(t_{\pi(n)}\right)$$

(W.199)

where $\pi \in S_n$ is one of the $n!$ permutations of the numbers $1, 2, \ldots, n$. Note that (W.199) and (W.196) are strictly identical.

We now define time ordering for a string of operators $A_i$ by

$$T\left[A_1\left(t_1\right), \ldots, A_n\left(t_n\right)\right] = \sum_{\pi \in S_n} \Theta\left(t_{\pi(1)}, \ldots, t_{\pi(n)}\right) A_1\left(t_{\pi(1)}\right), \ldots, A_n\left(t_{\pi(n)}\right).$$

(W.200)

Due to the properties of $\Theta$ as given in (W.197), time ordering picks out exactly that order of operators for which the times are ordered in the right way. Thus, $T$ guarantees that the operators act in the physical correct order and not the later one before the others.

By means of $T$, we can now write (W.199) as

$$U_I\left(t, t_0\right) = \sum_{n=0}^{\infty} \left(-\frac{i}{\hbar}\right)^n \int_{t_0}^{t} dt_1 \int_{t_0}^{t} dt_2 \ldots \int_{t_0}^{t} dt_n \, T\left[H_I\left(t_1\right) H_I\left(t_2\right) \ldots H_I\left(t_n\right)\right]$$

(W.201)

or more compactly as (*Dyson's series or expansion*)

$$U_I\left(t, t_0\right) = T \, \exp\left(-\frac{i}{\hbar} \int_{t_0}^{t} dt' \, H_I\left(t'\right)\right).$$

(W.202)

We need this equation and these considerations in the further discussion.

We will need also the time ordered product of two scalar field operators. Let $A$ and $B$ scalar fields, either bosonic (upper sign) or fermionic (lower sign). Then the time ordered product of $A$ and $B$ is defined by

$$T\left[A\left(x\right) B\left(y\right)\right] = \begin{cases} A\left(x\right) B\left(y\right) \\ \pm B\left(y\right) A\left(x\right) \end{cases} \text{for} \begin{array}{l} x^0 > y^0 \\ y^0 > x^0 \end{array}.$$

(W.203)

By means of the Heaviside function[101] $\theta$, this may be written more compactly as

$$T\left[A\left(x\right) B\left(y\right)\right] = \theta\left(x^0 - y^0\right) A\left(x\right) B\left(y\right) \pm \theta\left(y^0 - x^0\right) B\left(y\right) A\left(x\right).$$ (W.205)

Similar to the case of normal ordering, we are interested in the difference of a product of two field operators and its time ordered form. It is given by

---

[101] Remember the definition of $\theta\left(x\right)$:

$$\theta\left(x\right) = \begin{cases} 1 \\ 0 \end{cases} \text{for} \begin{array}{l} x \geq 0 \\ x < 0 \end{array}$$

(W.204)

$$T\left[A\left(x\right)B\left(y\right)\right]-A\left(x\right)B\left(y\right)=\begin{cases} A\left(x\right)B\left(y\right)-A\left(x\right)B\left(y\right)=0 & \text{for } x^0 > y^0 \\ \pm B\left(y\right)A\left(x\right)-A\left(x\right)B\left(y\right) & y^0 > x^0 \end{cases}$$

(W.206)

With $A\left(x\right)=A^d\left(x\right)+A^c\left(x\right)$, the commutator in the last line is given by

$$\pm B\left(y\right)A\left(x\right)-A\left(x\right)B\left(y\right)=-\left[A\left(x\right),B\left(y\right)\right]_{\mp}=$$
$$=-\left[A^d\left(x\right)+A^c\left(x\right),B^d\left(y\right)+B^c\left(y\right)\right]_{\mp}=-\left[A^d\left(x\right),B^c\left(y\right)\right]_{\mp}-\left[A^c\left(x\right),B^d\left(y\right)\right]_{\mp}$$

(W.207)

due to $\left[A^d\left(x\right),B^d\left(y\right)\right]_{\mp}=\left[A^c\left(x\right),B^c\left(y\right)\right]_{\mp}=0$. It follows finally

$$T\left[A\left(x\right)B\left(y\right)\right]-A\left(x\right)B\left(y\right)=\begin{cases} 0 & \text{for } x^0 > y^0 \\ -\left[A^d\left(x\right),B^c\left(y\right)\right]_{\mp}-\left[A^c\left(x\right),B^d\left(y\right)\right]_{\mp} & y^0 > x^0 \end{cases}$$

(W.208)

As in case of normal ordering, we see that the difference is made of commutators and hence contains no operators, i.e., is a $c$-number.

### W.7.3   Time Ordering and Normal Ordering

Bringing together time ordering and normal ordering[102] of two scalar fields $A$ and $B$ leads with (W.195) to

$$T\left[A\left(x\right)B\left(y\right)\right]-\mathcal{N}\left[A\left(x\right)B\left(y\right)\right]=T\left[A\left(x\right)B\left(y\right)\right]-A\left(x\right)B\left(y\right)+\left[A^d\left(x\right),B^c\left(y\right)\right]_{\mp}. \quad \text{(W.209)}$$

This yields

$$T\left[A\left(x\right)B\left(y\right)\right]=\mathcal{N}\left[A\left(x\right)B\left(y\right)\right]\begin{cases} +\left[A^d\left(x\right),B^c\left(y\right)\right]_{\mp} & \text{for } x^0 > y^0 \\ -\left[A^c\left(x\right),B^d\left(y\right)\right]_{\mp} & y^0 > x^0 \end{cases}. \quad \text{(W.210)}$$

As is seen, we can express the time ordered product of two scalar field operators by their normal ordered product plus another term which contains no operators. This result will play an important below.

---

[102] $T$ and $\mathcal{N}$ are sometimes called time ordering operator and normal ordering operator. For the sake of good order, we want to point out that this is a misnomer. An operator is an object which, when applied to a state, gives new information, as for instance the angular momentum operator. In this sense, $T$ and $\mathcal{N}$ are not operators and would be better named instructions. On the other hand, the naming is established and we will use it, too, keeping in mind the caveat.

## W.7.4   Exercises and Solutions

1. Prove (W.184).

   Solution: Exemplarily, we treat detailed the Dirac case. It is

$$
\begin{aligned}
H_{\text{Dirac}} &= \mathcal{N}\left[\sum_{\mathbf{p},r} E_{\mathbf{p}}\left(b_r^\dagger(\mathbf{p})\, b_r(\mathbf{p}) - d_r(\mathbf{p})\, d_r^\dagger(\mathbf{p})\right)\right] = \\
&= \mathcal{N}\left[\sum_{\mathbf{p},r} E_{\mathbf{p}} b_r^\dagger(\mathbf{p})\, b_r(\mathbf{p})\right] - \mathcal{N}\left[\sum_{\mathbf{p},r} E d_r(\mathbf{p})\, d_r^\dagger(\mathbf{p})\right] = \\
&= \sum_{\mathbf{p},r} E_{\mathbf{p}} \mathcal{N}\left[b_r^\dagger(\mathbf{p})\, b_r(\mathbf{p})\right] - \sum_{\mathbf{p},r} E_{\mathbf{p}} \mathcal{N}\left[d_r(\mathbf{p})\, d_r^\dagger(\mathbf{p})\right] = \quad\text{(W.211)} \\
&= \sum_{\mathbf{p},r} E_{\mathbf{p}} b_r^\dagger(\mathbf{p})\, b_r(\mathbf{p}) + \sum_{\mathbf{p},r} E_{\mathbf{p}} d_r^\dagger(\mathbf{p})\, d_r(\mathbf{p}) = \\
&= \sum_{\mathbf{p},r} E_{\mathbf{p}}\left(b_r^\dagger(\mathbf{p})\, b_r(\mathbf{p}) + d_r^\dagger(\mathbf{p})\, d_r(\mathbf{p})\right).
\end{aligned}
$$

2. Consider the Klein–Gordon field (discrete version)

$$
\phi(x) = \phi^d(x) + \phi^c(x) = \sum_{\mathbf{k}} \frac{1}{\sqrt{2VE_{\mathbf{k}}}}\left(a(\mathbf{k})\, e^{-ikx} + a^\dagger(\mathbf{k})\, e^{ikx}\right). \quad\text{(W.212)}
$$

   Calculate explicitly $\mathcal{N}[\phi(x)\,\phi(y)] - \phi(x)\,\phi(y)$.

   Solution: With (W.195) we obtain

$$
\mathcal{N}[\phi(x)\,\phi(y)] - \phi(x)\,\phi(y) = -\left[\phi^d(x),\,\phi^c(y)\right]_- . \quad\text{(W.213)}
$$

   Inserting $\phi^d$ and $\phi^c$ yields

$$
\begin{aligned}
\mathcal{N}[\phi(x)\,\phi(y)] - \phi(x)\,\phi(y) &= -\left[\sum_{\mathbf{k}} \tfrac{1}{\sqrt{2VE_{\mathbf{k}}}} a(\mathbf{k})\, e^{-ikx}, \sum_{\mathbf{k}'} \tfrac{1}{\sqrt{2VE_{\mathbf{k}'}}} a^\dagger(\mathbf{k}')\, e^{ik'y}\right]_- = \\
&- -\sum_{\mathbf{k},\mathbf{k}'} \tfrac{1}{\sqrt{2VE_{\mathbf{k}}}} \tfrac{1}{\sqrt{2VE_{\mathbf{k}'}}} e^{-ikx} e^{ik'y}\left[a(\mathbf{k}), a^\dagger(\mathbf{k})\right]_- = \\
&= -\sum_{\mathbf{k},\mathbf{k}'} \tfrac{1}{\sqrt{2VE_{\mathbf{k}}}} \tfrac{1}{\sqrt{2VE_{\mathbf{k}'}}} e^{-ikx} e^{ik'y} \delta_{\mathbf{k},\mathbf{k}'} = -\sum_{\mathbf{k}} \tfrac{1}{2VE_{\mathbf{k}}} e^{-ik(x-y)}.
\end{aligned}
$$

$$
\text{(W.214)}
$$

   As is explicitly seen, the result contains no operators.

## W.8   Interacting Fields, Quantum Electrodynamics

In order to describe interacting fields (up to now we were considering free fields only), we now combine the pieces which we have developed previously. We describe the interaction by means of the interaction picture which is based on the interaction Hamiltonian $\mathcal{H}_I$. Thereby, we confine our considerations to the interaction of fermions with spin $1/2$ and photons, i.e., to the study of *quantum electrodynamics* (QED). To formulate $\mathcal{H}_I$ we need the Lagrangians $\mathcal{L}$ of the Dirac field and of the radiation field plus a term which couples these two fields. Our focus will be on scattering. The transition amplitude from the incoming to the outgoing state is described by means of the so-called $S$-matrix which we have, finally, to approximate in a suitable manner to describe scattering processes of lowest orders.

## W.8.1 Lagrangian

In order to bring together electrons, positrons and photons, we invoke the principle of minimal coupling. Since we introduced and discussed it already in Appendix T, Vol. 1, we recap it here briefly. The approach replaces the 4-momentum $p_\mu$ by the 4-vector $p_\mu - qA_\mu$ ($q$ is the charge of the considered fermion), i.e.,[103]

$$p_\mu \to p_\mu - qA_\mu \Rightarrow i\partial_\mu \to i\partial_\mu - qA_\mu. \tag{W.215}$$

It follows for the Lagrangian

$$\mathcal{L}^{\text{Dirac}} = \bar{\psi}\left(i\gamma^\mu \partial_\mu - m\right)\psi \to \bar{\psi}\left(\gamma^\mu\left(i\partial_\mu - qA_\mu\right) - m\right)\psi = \bar{\psi}\left(\gamma^\mu i\partial_\mu - m\right)\psi - q\bar{\psi}\gamma^\mu A_\mu\psi. \tag{W.216}$$

This means that by this substitution we have the free Dirac Lagrangian plus an interaction term:

$$\mathcal{L}^{\text{Dirac}} \to \mathcal{L}^{\text{Dirac}} + \mathcal{L}^{\text{interaction}} \; ; \; \mathcal{L}^{\text{interaction}} = -q\bar{\psi}\gamma^\mu A_\mu\psi. \tag{W.217}$$

The term $q\bar{\psi}\gamma^\mu A_\mu\psi$ is the interface between fermions and photons, containing contributions of both 'worlds'.[104]

In this way, we can write the *total Lagrangian* $\mathcal{L}$ as the sum of the two free Lagrangians (Dirac and photon) and the interaction Lagrangian:

$$\mathcal{L} = \mathcal{L}^{\text{Dirac}} + \mathcal{L}^{\text{photon}} + \mathcal{L}^{\text{interaction}} =$$
$$= \bar{\psi}\left(i\gamma^\mu \partial_\mu - m\right)\psi - \tfrac{1}{2}\left(\partial^\mu A^\nu\right)\left(\partial_\mu A_\nu\right) - q\bar{\psi}\gamma^\mu A_\mu\psi. \tag{W.218}$$

Thus, $\mathcal{L}$ describes the electromagnetic interaction between electrons, positrons[105] and photons. It is the basic equation for quantum electrodynamics which is one of the best, if not the best, proven theories in physics.

One can just as well regard $q\bar{\psi}\gamma^\mu\psi$ as 4-current $j^\mu$ which enters the Lagrangian for electrodynamics in the form $\mathcal{L}^{\text{electrodynamics}} = -\tfrac{1}{2}\left(\partial^\mu A^\nu\right)\left(\partial_\mu A_\nu\right) + j^\mu A_\mu$ (see section 'Normal ordering').

---

[103] Remind $p_\mu = i\partial_\mu$, see Appendix T, Vol. 1.

[104] Note that the notation of the charge may cause some confusion since it is not standardized. Here, $q$ means the fermionic charge and the specific charge of the electron ist denoted by $q = -e$. But one finds also the notation $e$ for the general charge and $-e_0$ for the electronic charge. In other contexts, $e_0$ means a hypothetical, not observable charge of the electron, i.e., the 'bare' charge. So watch out.

[105] Or muons and tauons and their antiparticles.

## W.8.2   Conjugated Momentum, Hamiltonian

Next we search for the Hamiltonian density $\mathcal{H}$ of the Lagrangian (W.218). As is seen, in the interaction term $\bar{\psi}\gamma^\mu A_\mu \psi$ there are no time derivatives ($\partial_0 \psi$) and ($\partial_0 A^\mu$) of the fields $\psi$ and $A^\mu$. In other words, the conjugated momenta are the same as in the free case, i.e., without interaction, and the Hamiltonian for the interaction reads simply $\mathcal{H}^{\text{interaction}} = -\mathcal{L}^{\text{interaction}}$. Thus, the Hamiltonian is given by

$$\mathcal{H} = \mathcal{H}^{\text{Dirac}} + \mathcal{H}^{\text{photon}} + \mathcal{H}^{\text{interaction}} =$$
$$= \bar{\psi}\left(-i\gamma^\mu \partial_\mu + m\right)\psi - \tfrac{1}{2}\left(\partial^\mu A_\nu\right)\left(\partial^\mu A^\nu\right) + q\bar{\psi}\gamma^\mu A_\mu \psi. \tag{W.219}$$

We have to add one further step, namely normal ordering. As discussed above in Section 'Operator ordering', operators in quantum field theory have to be normal ordered to be meaningful. Thus, the Hamiltonian for the interaction (which is the term we are interested in for the following) reads in its final version

$$\mathcal{H}^{\text{interaction}} = q\mathcal{N}\left[\bar{\psi}\gamma^\mu A_\mu \psi\right]. \tag{W.220}$$

Note that $\psi$ and $A_\mu$ are *free* field operators.

## W.8.3   Interaction Picture, Time Evolution Operator

It comes as no surprise that the equation of motions for (W.218) or (W.219) cannot be solved in closed form. Instead, one invokes the interaction picture. Since this issue was introduced and discussed in Appendix Q, Vol. 1, and above in section 'Operator ordering', we recap here only the main points very briefly.

In the interaction picture, it is assumed that the Hamilton function $H$, as given for instance in the Schrödinger picture, can be written as $H = H_0 + H_1$, where $H_0$ is the free part and $H_1$ the interaction part. Usually, $H_0$ may be solved exactly. Then we define states $|\psi_I(t)\rangle$ and operators $B_I(t)$ in the interaction picture by

$$|\psi_I(t)\rangle = e^{iH_0 t}|\psi(t)\rangle \;\; ; \;\; B_I(t) = e^{iH_0 t}B e^{-iH_0 t} \tag{W.221}$$

and the time behavior of the state $|\psi_I(t)\rangle$ is given by

$$i\frac{d}{dt}|\psi_I(t)\rangle = e^{iH_0 t}H_1 e^{-iH_0 t}|\psi_I(t)\rangle = H_I(t)|\psi_I(t)\rangle. \tag{W.222}$$

The time evolution operator $U_I(t, t_0)$ makes contact between $|\psi_I(t_0)\rangle$ and $|\psi_I(t)\rangle$, i.e.,

$$|\psi_I(t)\rangle = U_I(t, t_0)|\psi_I(t_0)\rangle \tag{W.223}$$

and obeys the differential equation

$$i\frac{\mathrm{d}}{\mathrm{d}t}U_I\left(t,t_0\right) = H_I\left(t\right)U_I\left(t,t_0\right) \tag{W.224}$$

The (formal) solution may be written as

$$U_I\left(t,t_0\right) = \mathcal{T}\,\exp\left(-i\int_{t_0}^{t}\mathrm{d}t'H_I\left(t'\right)\right). \tag{W.225}$$

The knowledge of $U_I\left(t,t_0\right)$ enables us to calculate $|\psi_I\left(t\right)\rangle$ for a given $|\psi_I\left(t_0\right)\rangle$. Assume that in a certain process the system is at time $t_0$ in the initial state $|\psi_I\left(t_0\right)\rangle = |i_I\rangle$. Then we have $|\psi_I\left(t\right)\rangle = U_I\left(t,t_0\right)|\psi_I\left(t_0\right)\rangle$, see (W.223), and the probability $P_{fi}$ to find it at time $t$ in a final state $|\psi_I\left(t\right)\rangle = |f_I\rangle$ is given by

$$P_{fi} = |\langle f_I|\,U_I\left(t,t_0\right)|i_I\rangle|^2. \tag{W.226}$$

Note that transition probabilities are independent from the picture chosen. Denoting the states for $t = t_0$ and $t$ in the Schrödinger picture by $|i_S\rangle$ and $|f_S\rangle$, we have shown in Appendix Q, Vol. 1, that $\langle f_I|\,U_I\left(t,t_0\right)|i_I\rangle = \langle f_S|\,U_S\left(t,t_0\right)|i_S\rangle$ - the transition amplitudes in the Schrödinger and the interaction picture are equal.

## W.8.4   S-Operator

We now focus our interest upon scattering. In a scattering process, one can distinguish three phases and their idealization:

- Phase 1: At the initial time, the initial particles are widely separated. We idealize this by assuming $t = -\infty$ for the initial time at which we have non-interacting initial particles, i.e., *free* particles.
- Phase 2: The particles encounter each other and interact. Possibly, some of the initial particles are destroyed and new final ones are created. After that, the (new) particles run away from each other.
- Phase 3: At the final time, the final particles are again widely separated. Again, we idealize this by assuming $t = \infty$ for the final time at which we have non interacting final particles, i.e., *free* particles.

The idealization of this process is tantamount to saying that in phases 1 and 3 the interaction is switched off and, in addition, that the 'interaction time' of phase 2 is much shorter than the times needed to run from the source to the scattering center or from the scattering center to the detector. Under these assumptions, we can choose the initial and the final time as $-\infty$ and $+\infty$.

We formalize now this idealized process as seen in the interaction picture. In the beginning we have the initial state $|i\rangle = |\psi_I\left(-\infty\right)\rangle$. The time evolution operator $U_I\left(\infty,-\infty\right)$ changes this state into the final state $|\psi_I\left(\infty\right)\rangle = U_I\left(\infty,-\infty\right)|\psi_I\left(-\infty\right)\rangle$. Now let be $|f\rangle$ one of the possible final states. The transition amplitude into this certain final state[106] is given by $\langle f \,|\psi_I (\infty)\rangle =$ $\langle f|\, U_I (\infty, -\infty) \,|i\rangle$; thus, the probability to find this final state $|f\rangle$ for a given initial state $|i\rangle$ is given by $|\langle f \,|\psi_I (\infty)\rangle|^2$.

We now introduce the abbreviating notation

$$S = U_I (\infty, -\infty) \qquad\qquad (\text{W.227})$$

called *scattering operator* or scattering matrix (or simply $S$-matrix or $S$-*operator*; the letter $S$ stems from scattering, of course). Thus, we can write[107]

$$\langle f \,|\psi_I (\infty)\rangle = \langle f|\, U_I (\infty, -\infty) \,|i\rangle = \langle f|\, S \,|i\rangle = S_{fi}. \qquad (\text{W.228})$$

With the time evolution operator (W.225) we have

$$S = U_I (\infty, -\infty) = \mathcal{T} \, \exp\left(-i \int_{-\infty}^{\infty} dt H_I (t)\right) \qquad (\text{W.229})$$

and with $H_I (t) = \int d^4x \, \mathcal{H}_I(x)$

$$S = \mathcal{T}\left[e^{-i \int d^4x \, \mathcal{H}_I(x)}\right]. \qquad\qquad (\text{W.230})$$

In the case under consideration, namely quantum electrodynamics, $\mathcal{H}_I$ is given by (W.220) and we have

$$\mathcal{H}_I(x) = \mathcal{H}^{\text{interaction}} = q\mathcal{N}\left[\bar{\psi}\gamma^{\mu}A_{\mu}\psi\right]. \qquad (\text{W.231})$$

So we have solved, at least in principle, the scattering problem in quantum electrodynamics. Assume an initial state $|i\rangle$. Then the probability to find after the scattering process the final state $|f\rangle$ is given by $|\langle f|\, S \,|i\rangle|^2$ where the $S$-matrix is given by (W.230) and the interaction Hamiltonian $\mathcal{H}_I(x)$ by (W.231).

## W.8.5 Approximating S

Note that (W.230) together with (W.231) is an exact formulation without approximations. We would have finished the problem, if we could find a closed analytical evaluation of the integral $\int d^4x \, \mathcal{H}_I(x)$ for given $\mathcal{H}_I(x)$. However, such an evaluation does not exist (or pretty much never), and we have to recourse to suitable approximations.

---

[106]Remind that the states $|i\rangle$ und $|f\rangle$ are free states (eigenkets of $H_0$).

[107]Remind that transition amplitudes in the Schrödinger and in the interaction picture are equal.

The usual procedure is to expand the exponential $\mathcal{T} \exp\left[-i \int d^4x \, \mathcal{H}_I(x)\right]$ in a series of the form

$$S = \mathcal{T}\left[1 + (-i)\int d^4x \, \mathcal{H}_I(x) + \frac{(-i)^2}{2!}\int\int d^4x \, d^4y \, \mathcal{H}_I(x)\mathcal{H}_I(y) + \cdots\right].$$
(W.232)

For many applications, it is sufficient to consider only the first few terms. If $\mathcal{H}_I$ is small enough compared to the full Hamiltonian density $\mathcal{H}$ (which is indeed the case, as we will see[108]), this proceeding will give satisfying results. Thus, the remaining sections are devoted to the discussion of the two terms

$$S^{(1)} = -i\int d^4x \, [\mathcal{H}_I(x)] \;;\; S^{(2)} = -\frac{1}{2}\int\int d^4x \, d^4y \, \mathcal{T}\,[\mathcal{H}_I(x)\mathcal{H}_I(y)].$$
(W.233)

The time ordering symbol $\mathcal{T}$ for $S^{(1)}$ may be omitted, since there is only one time to consider.

As we will see, the discussion will be quite extensive, though the two terms look rather simple and innocent. We begin in the next section with $S^{(1)}$. After that, we provide some tools as contractions, propagators and the Wick theorem. They are needed for the discussion of $S^{(2)}$. This discussion is only introductory and exemplary; a thorough and detailed consideration of QED would be beyond the scope of this short introduction.

## W.9   S-Matrix, First Order

We now want to discuss the first order term $S^{(1)}$

$$S^{(1)} = -i\int d^4x \, [\mathcal{H}_I(x)] = -iq\int d^4x \, \mathcal{N}\left[\bar{\psi}A\!\!\!/\psi\right]$$
(W.234)

in some detail.

With (W.234), we have the simplest case of the scattering matrix, and it is good practice to consider the simplest case first. However, discussing $S^{(1)}$ is not only a convenient finger exercise. It also facilitates the discussion of more complex cases for several reasons. One of them is that $S^{(1)}$ encompasses eight elementary processes which are constituting the more complex cases in higher orders of $S^{(n)}$. A closer look at these elementary processes will provide simple rules which enable us to write down transition amplitudes quite easily.[109] What makes life even easier is the close

---

[108] In QED, the smallness parameter in $\mathcal{H}_I = q\mathcal{N}\left[\bar{\psi}A\!\!\!/\psi\right]$ is $|q| = |e|$ which in our natural units has the value $\sim 0.303$.

[109] The connection of transition amplitude and scattering cross section will be discussed exemplarily in the context of considering $S^{(2)}$, see below.

connection between those rules and their graphical representation in form of the so-called Feynman diagrams.

To avoid disappointment later on, we want to point out already here that none of the those eight elementary processes can occur for real particles. However, we will learn a lot from them and, as mentioned, they are building blocks of the scattering processes of higher orders.

Note that the processes can not 'really' occur in the way they are described here; but in a different way, they exist for real particles. For one thing, in the case of *external fields* apply different considerations as discussed below, for another thing the processes can exist in higher orders $S^{(n)}$. Take for instance pair annihilation. Below, we consider the 'impossible' first-order case $e^- e^+ \to \gamma$. Of course, 'real' pair annihilation exists, but as $e^- e^+ \to 2\gamma$ which is part of the second order term $S^{(2)}$.

## W.9.1 Preliminary Note: Virtual Particles

Real and virtual particles are also called on mass-shell and off mass-shell. Here some comments on their definition.

The inner product of the 4-momentum $p$ is given by $p^2 = p_\mu p^\mu = \left(p^0\right)^2 - \mathbf{p}^2 = E_\mathbf{p}^2 - \mathbf{p}^2 = m^2$. The identity $p^2 = m^2$ is also known as *mass shell condition*. It is the usual dispersion relation for relativistic particles. However, in e.g. scattering processes, there occur particles with $p^2 \neq m^2$, i.e., they exist 'off mass-shell'. These *virtual particles* can not appear in the initial and final states of real processes, but are only emitted and reabsorbed in intermediate steps. One can argue that the energy-time uncertainty relation $\Delta E \Delta t \sim \hbar$ allows for off mass-shell particles with energy $\Delta E$ provided they don't live longer than $\Delta t \sim \hbar/\Delta E$. In addition, the finite velocity $\leq c$ of those particles leads to a finite range $\Delta x \lesssim c\hbar/\Delta E$ or in natural units $\Delta x \lesssim 1/\Delta E$. In short: Virtual particles can not be measured; they exist only fleetingly.

Concerning the eight processes we will study now, this means that they cannot be realized in the described form with three on-shell particles.

## W.9.2 Field Operators

For clearer distinction, we use from now on the following notation: momentum and spin of fermions[110] are labeled by $p$ and $r$, momentum and polarization of photons are labeled by $k$ and $\lambda$.

---

[110]The considerations also apply to other fermions as muons and tauons.

The field operators are given in the sections 'Quantization of free fields' for the Dirac and the photon field. The continuous versions read

$$\psi(x) = \sum_r \int d^3 p \sqrt{\frac{m}{(2\pi)^3 E_p}} \left( b_r(\mathbf{p}) u_r(\mathbf{p}) e^{-ipx} + d_r^\dagger(\mathbf{p}) w_r(\mathbf{p}) e^{ipx} \right)$$
$$A^\mu(x) = \int \sum_\lambda d^3 k \sqrt{\frac{1}{(2\pi)^3 E_k}} \epsilon_\lambda^\mu(\mathbf{k}) \left( \alpha_\lambda(\mathbf{k}) e^{-ikx} + \alpha_\lambda^\dagger(\mathbf{k}) e^{ikx} \right). \tag{W.235}$$

The commutation relations are given by

$$\left\{ b_r(\mathbf{p}), b_{r'}^\dagger(\mathbf{p}') \right\} = \delta_{rr'} \delta(\mathbf{p} - \mathbf{p}') \quad ; \quad \left\{ d_r(\mathbf{p}), d_{r'}^\dagger(\mathbf{p}') \right\} = \delta_{rr'} \delta(\mathbf{p} - \mathbf{p}')$$
$$\left[ \alpha_\lambda(\mathbf{k}), \alpha_{\lambda'}^\dagger(\mathbf{k}') \right] = \delta_{\lambda\lambda'} \delta^3(\mathbf{k} - \mathbf{k}'). \tag{W.236}$$

All other (anti-)commutators vanish.

In view of normal ordering, it is advantageous for some considerations to aggregate the contributions of deletion and creation operators in the form

$$\psi(x) = \psi^d(x) + \psi^c(x) \quad ; \quad \bar\psi(x) = \bar\psi^d(x) + \bar\psi^c(x)$$
$$A^\mu(x) = A^{\mu d}(x) + A^{\mu c}(x). \tag{W.237}$$

Thereby, the creation and annihilation parts of the field operators are explicitly given by[111]

$$\psi^d(x) = \sum_r \int d^3 p \sqrt{\frac{m}{(2\pi)^3 E_p}} b_r(\mathbf{p}) u_r(\mathbf{p}) e^{-ipx} \; ; \; \psi^c(x) = \sum_r \int d^3 p \sqrt{\frac{m}{(2\pi)^3 E_p}} d_r^\dagger(\mathbf{p}) w_r(\mathbf{p}) e^{ipx}$$
$$\bar\psi^d(x) = \sum_r \int d^3 p \sqrt{\frac{m}{(2\pi)^3 E_p}} d_r(\mathbf{p}) \bar w_r(\mathbf{p}) e^{-ipx} \; ; \; \bar\psi^c(x) = \sum_r \int d^3 p \sqrt{\frac{m}{(2\pi)^3 E_p}} b_r^\dagger(\mathbf{p}) \bar u_r(\mathbf{p}) e^{ipx}$$
$$A^{\mu d}(x) = \sum_\lambda \int d^3 k \sqrt{\frac{1}{(2\pi)^3 E_k}} \epsilon_\lambda^\mu(\mathbf{k}) \alpha_\lambda(\mathbf{k}) e^{-ikx} \; ; \; A^{\mu c}(x) = \sum_\lambda \int d^3 k \sqrt{\frac{1}{(2\pi)^3 E_k}} \epsilon_\lambda^\mu(\mathbf{k}) \alpha_\lambda^\dagger(\mathbf{k}) e^{ikx}. \tag{W.238}$$

$\psi^d(x)$ contains all terms $\sim e^{-ipx}$ and all annihilation operators, $\psi^c(x)$ all terms $\sim e^{ipx}$ and all creation operators; analogously for $A^\mu(x)$.

The action of these operators is given by

$$\left.\begin{array}{c} \psi^d \\ \bar\psi^d \\ A^d \end{array}\right\} \text{ annihilates a(n) } \left\{\begin{array}{c} \text{electron} \\ \text{positron} \\ \text{photon} \end{array}\right. \quad ; \quad \left.\begin{array}{c} \psi^c \\ \bar\psi^c \\ A^c \end{array}\right\} \text{ creates a(n) } \left\{\begin{array}{c} \text{positron} \\ \text{electron} \\ \text{photon} \end{array}\right. \quad \text{(W.239)}$$

Thus, $\psi = \psi^d + \psi^c$ annihilates an electron and creates a positron, whereas $\bar\psi = \bar\psi^d + \bar\psi^c$ annihilates a positron and creates an electron.

---

[111] As mentioned in the section 'Operator ordering', many textbooks write $\psi^+$ for the annihilation part and $\psi^-$ for the creation part. We use $\psi^d$ and $\psi^c$ instead of $\psi^+$ and $\psi^-$, where $d$ stands for 'destroying' and $c$ für 'creating'.

**Table W.3** The eight possible processes of $\mathcal{H}_I$

| Term | Normal order | Description |
|---|---|---|
| $\bar{\psi}^d A\!\!\!/^d \psi^d$ | | $\gamma e^- e^+ \to$ vacuum |
| $\bar{\psi}^d A\!\!\!/^d \psi^c$ | $\mathcal{N}\left[\bar{\psi}^d_\alpha \gamma^\mu_{\alpha\beta} A^+_\mu \psi^c_\beta\right] = -\psi^c_\beta \bar{\psi}^d_\alpha \gamma^\mu_{\alpha\beta} A^d_\mu$ | $\gamma e^+ \to e^+$ |
| $\bar{\psi}^d A\!\!\!/^c \psi^d$ | $A^c_\mu \bar{\psi}^d \gamma^\mu \psi^d$ | $e^- e^+ \to \gamma$ |
| $\bar{\psi}^d A\!\!\!/^c \psi^c$ | $\mathcal{N}\left[\bar{\psi}^d_\alpha \gamma^\mu_{\alpha\beta} A^c_\mu \psi^c_\beta\right] = -\gamma^\mu_{\alpha\beta} A^c_\mu \psi^c_\beta \bar{\psi}^d_\alpha$ | $e^+ \to \gamma e^+$ |
| $\bar{\psi}^c A\!\!\!/^d \psi^d$ | | $\gamma e^- \to e^-$ |
| $\bar{\psi}^c A\!\!\!/^d \psi^c$ | | $\gamma \to e^- e^+$ |
| $\bar{\psi}^c A\!\!\!/^c \psi^d$ | | $e^- \to \gamma e^-$ |
| $\bar{\psi}^c A\!\!\!/^c \psi^c$ | | vacuum $\to \gamma e^- e^+$ |

## W.9.3 Eight Elementary Processes of $\mathcal{H}_I$

Let us look which processes are allowed by the interaction Hamiltonian

$$\mathcal{H}_I(x) = q\mathcal{N}\left[\bar{\psi}(x) A\!\!\!/(x) \psi(x)\right]. \qquad (W.240)$$

We insert (W.237) into this equation, expand the brackets and obtain the following $2^3 = 8$ terms

$$\mathcal{H}_I(x) = q\mathcal{N}\left[\begin{array}{l} \bar{\psi}^d A\!\!\!/^d \psi^d + \bar{\psi}^d A\!\!\!/^d \psi^c + \bar{\psi}^d A\!\!\!/^c \psi^d + \bar{\psi}^d A\!\!\!/^c \psi^c + \\ +\bar{\psi}^c A\!\!\!/^d \psi^d + \bar{\psi}^c A\!\!\!/^d \psi^c + \bar{\psi}^c A\!\!\!/^c \psi^d + \bar{\psi}^c A\!\!\!/^c \psi^c \end{array}\right]. \qquad (W.241)$$

Consider for instance the term $\bar{\psi}^c A\!\!\!/^d \psi^c$ which already is in normal order. Following the action as given in (W.239), it creates a positron ($\psi^c$), annihilates a photon ($A\!\!\!/^d$) and creates an electron ($\bar{\psi}^c$), or in short $\gamma \to e^- e^+$ (pair production). Before discussing the action of the other seven terms we note that only three terms are not in normal order (which means upper index $c$ to the left, $d$ to the right), namely $\bar{\psi}^d A\!\!\!/^d \psi^c$, $\bar{\psi}^d A\!\!\!/^c \psi^d$ and $\bar{\psi}^d A\!\!\!/^c \psi^c$:

$$\mathcal{H}_I(x) = q\left[\begin{array}{l} \bar{\psi}^d A\!\!\!/^d \psi^d + \mathcal{N}\left[\bar{\psi}^d A\!\!\!/^d \psi^c\right] + \mathcal{N}\left[\bar{\psi}^d A\!\!\!/^c \psi^d\right] + \mathcal{N}\left[\bar{\psi}^d A\!\!\!/^c \psi^c\right] + \\ +\bar{\psi}^c A\!\!\!/^d \psi^d + \bar{\psi}^c A\!\!\!/^d \psi^c + \bar{\psi}^c A\!\!\!/^c \psi^d + \bar{\psi}^c A\!\!\!/^c \psi^c \end{array}\right].$$
$$(W.242)$$

Normal ordering for instance the first of these terms, $\bar{\psi}^d A\!\!\!/^d \psi^c$, can be achieved as follows:

$$\mathcal{N}\left[\bar{\psi}^d A\!\!\!/^d \psi^c\right] = \mathcal{N}\left[\bar{\psi}^d \gamma^\mu A^d_\mu \psi^c\right] = \mathcal{N}\left[\bar{\psi}^d_\alpha \gamma^\mu_{\alpha\beta} A^d_\mu \psi^c_\beta\right] = -\psi^c_\beta \bar{\psi}^d_\alpha \gamma^\mu_{\alpha\beta} A^d_\mu \qquad (W.243)$$

where $\alpha$ and $\beta$ denote the entries of the 4-spinors $\psi^d$ and $\psi^c$; $\gamma^\mu_{\alpha\beta}$ is the element $(\alpha\beta)$ of the matrix $\gamma^\mu$. We sum up all processes in Table W.3.

Note that one of the premises of quantum field theory is fulfilled: we have different numbers and types of particles in the incoming and outgoing channel.

One can show that none of these eight processes can be realized with on-shell particles only. The discussion is much easier when using Feynman diagrams, see below and the exercises.

### W.9.4  Two Worked Out Examples

Exemplarily, we want to consider in the following two of the eight processes in more detail, namely 1) $\bar{\psi}^c A^c \psi^d$ or $e^- \to \gamma e^-$ (emission of a photon) and 2) $\bar{\psi}^c A^d \psi^c$ or $\gamma \to e^- e^+$ (pair production). Thereby, we make use of the field operators in the continuous version as given in (W.235).

#### W.9.4.1  First Example: Emission of a Photon, $e^- \to \gamma e^-$

The $S$-matrix element reads

$$S^{(1)} = -iq \int d^4x \, \bar{\psi}^c A^c \psi^d. \tag{W.244}$$

**Initial and final states** The initial and final states are given by

$$|i\rangle = b_R^\dagger (\mathbf{P}) |0\rangle \; ; \; |f\rangle = b_{R'}^\dagger (\mathbf{P}') a_\Lambda^\dagger (\mathbf{K}) |0\rangle \to \langle f| = \langle 0| a_\Lambda (\mathbf{K}) b_{R'} (\mathbf{P}') \tag{W.245}$$

i.e., an incoming electron with quantum numbers $R$ and $\mathbf{P}$ and an outgoing electron with $(R', \mathbf{P}')$ plus an outgoing photon with $(\Lambda, \mathbf{K})$.

**Matching rules** Let us first consider $\psi^d |i\rangle$. We have with continuous field operators[112]

$$\psi^d |i\rangle = \sum_r \int d^3p \, \sqrt{\tfrac{m}{(2\pi)^3 E_\mathbf{p}}} b_r (\mathbf{p}) u_r (\mathbf{p}) e^{-ipx} b_R^\dagger (\mathbf{P}) |0\rangle =$$
$$= \sum_r \int d^3p \, \sqrt{\tfrac{m}{(2\pi)^3 E_\mathbf{p}}} u_r (\mathbf{p}) e^{-ipx} \delta_{rR} \delta (\mathbf{p} - \mathbf{P}) |0\rangle = \sqrt{\tfrac{m}{(2\pi)^3 E_\mathbf{p}}} u_R (\mathbf{P}) e^{-iPx} |0\rangle \tag{W.246}$$

or in short

$$\psi^d b_R^\dagger (\mathbf{P}) |0\rangle = \sqrt{\frac{m}{(2\pi)^3 E_\mathbf{P}}} u_R (\mathbf{P}) e^{-iPx} |0\rangle. \tag{W.247}$$

The argument runs as follows: $b_R^\dagger (\mathbf{P}) |0\rangle$ creates an electron with quantum numbers $\mathbf{P}$ and $R$. The only annihilation operator $b_r (\mathbf{p})$ which can destroy this electron

---

[112]For the discrete version, replace $\sqrt{\tfrac{m}{(2\pi)^3 E_\mathbf{p}}} \to \sqrt{\tfrac{m}{V E_\mathbf{p}}}$ and $\sqrt{\tfrac{1}{2(2\pi)^3 E_\mathbf{K}}} \to \sqrt{\tfrac{1}{2V E_\mathbf{K}}}$.

**Table W.4** Incoming and outgoing contributions

| | |
|---|---|
| $\psi^d b_R^\dagger (\mathbf{P}) |0\rangle = \sqrt{\frac{m}{(2\pi)^3 E_\mathbf{P}}} u_R (\mathbf{P}) e^{-iPx} |0\rangle$ | $A^{\mu d} (x) \alpha_\Lambda^\dagger (\mathbf{K}) |0\rangle = \sqrt{\frac{1}{2(2\pi)^3 E_\mathbf{K}}} \epsilon_\Lambda^\mu (\mathbf{K}) e^{-iKx} |0\rangle$ |
| $\langle 0| \bar{\psi}^c b_R (\mathbf{P}) = \langle 0| \sqrt{\frac{m}{(2\pi)^3 E_\mathbf{P}}} \bar{u}_R (\mathbf{P}) e^{iPx}$ | $\langle 0| \alpha_\Lambda (\mathbf{K}) A^{\mu c} (x) = \langle 0| \sqrt{\frac{1}{2(2\pi)^3 E_\mathbf{K}}} \epsilon_\Lambda^\mu (\mathbf{K}) e^{iKx}$ |

has to have the same quantum numbers; if not, it acts onto the vacuum, yielding zero. If one prefers a more formal argument, one considers $b_r (\mathbf{p}) b_R^\dagger (\mathbf{P})$. Due to the anticommutation rule $\left\{ b_r (\mathbf{p}) , b_{r'}^\dagger (\mathbf{p}') \right\} = \delta_{rr'} \delta (\mathbf{p} - \mathbf{p}')$ we have

$$\left\{ b_r (\mathbf{p}) , b_R^\dagger (\mathbf{P}) \right\} = \delta_{rR} \delta (\mathbf{p} - \mathbf{P}) \rightarrow b_r (\mathbf{p}) b_R^\dagger (\mathbf{P}) = \delta_{rR} \delta (\mathbf{p} - \mathbf{P}) - b_R^\dagger (\mathbf{P}) b_r (\mathbf{p}) .$$

$$(W.248)$$

Thus, for $b_r (\mathbf{p}) b_R^\dagger (\mathbf{P}) |0\rangle$ follows

$$b_r (\mathbf{p}) b_R^\dagger (\mathbf{P}) |0\rangle = \delta_{rR} \delta (\mathbf{p} - \mathbf{P}) |0\rangle - b_R^\dagger (\mathbf{P}) b_r (\mathbf{p}) |0\rangle = \delta_{rR} \delta (\mathbf{p} - \mathbf{P}) |0\rangle \quad (W.249)$$

due to $b_r (\mathbf{p}) |0\rangle = 0$.

In other words, only those parts of the field operator $\psi^d$ survive which match the quantum numbers of the incoming particle. The same holds for outgoing particles: for $\bar{\psi}^c$ only the term $b_{R'}^\dagger (\mathbf{P}')$ contributes, and for $A^c$ only the term $\alpha_\Lambda (\mathbf{K})$. Written as a short formula or rule we have in general

**Transition amplitude** In this way we arrive at

$$\langle f| S^{(1)} |i\rangle = -iq \int d^4x \, \langle f| \bar{\psi}^c A^c \psi^d |i\rangle =$$
$$= -iq \int d^4x \, \sqrt{\frac{m}{(2\pi)^3 E_{\mathbf{P}'}}} \bar{u}_{R'} (\mathbf{P}') e^{iP'x} \sqrt{\frac{1}{2(2\pi)^3 E_\mathbf{k}}} \gamma_\mu \epsilon_\Lambda^\mu (\mathbf{K}) e^{iKx} \sqrt{\frac{m}{(2\pi)^3 E_\mathbf{P}}} u_R (\mathbf{P}) e^{-iPx} =$$
$$= -iq \sqrt{\frac{m}{(2\pi)^3 E_{\mathbf{P}'}}} \bar{u}_{R'} (\mathbf{P}') \sqrt{\frac{1}{2(2\pi)^3 E_\mathbf{k}}} \gamma_\mu \epsilon_\Lambda^\mu (\mathbf{K}) \sqrt{\frac{m}{(2\pi)^3 E_\mathbf{P}}} u_R (\mathbf{P}) \int d^4x \, e^{iP'x} e^{iKx} e^{-iPx} .$$

$$(W.250)$$

The $x$-integration[113] yields the four-dimensional delta function $(2\pi)^4 \delta^{(4)} (P' + K - P)$ and it follows

$$\langle f| S^{(1)} |i\rangle = -iq (2\pi)^4 \delta^{(4)} (P' + K - P) \cdot$$
$$\cdot \sqrt{\frac{m}{(2\pi)^3 E_{\mathbf{P}'}} \frac{1}{2(2\pi)^3 E_\mathbf{k}} \frac{m}{(2\pi)^3 E_\mathbf{P}}} \bar{u}_{R'} (\mathbf{P}') \gamma_\mu \epsilon_\Lambda^\mu (\mathbf{K}) u_R (\mathbf{P}) .$$

$$(W.251)$$

As mentioned above, all processes of first order can not occur with on-shell particles. In the case under consideration, the argument runs as follows: the four-dimensional delta function contains the conservation of the momentum and of the energy. Conservation of the momentum means $\mathbf{P} = \mathbf{P}' + \mathbf{K}$ or $\mathbf{P}' = \mathbf{P} - \mathbf{K}$. Then conservation of the energy reads

---

[113] Remember $\int d^4x \, e^{ipx} = (2\pi)^4 \delta^{(4)} (p)$.

$$E_{\mathbf{P-K}} \overset{!}{=} E_{\mathbf{P}} - E_{\mathbf{K}} \rightarrow \sqrt{(\mathbf{P}-\mathbf{K})^2 + m^2} \overset{!}{=} \sqrt{\mathbf{P}^2 + m^2} - \sqrt{\mathbf{K}^2}. \qquad \text{(W.252)}$$

Squaring both sides yields

$$\begin{aligned}
(\mathbf{P}-\mathbf{K})^2 + m^2 &\overset{!}{=} \mathbf{P}^2 + m^2 - 2\sqrt{\mathbf{K}^2}\sqrt{\mathbf{P}^2 + m^2} + \mathbf{K}^2 \\
&\rightarrow \mathbf{PK} \overset{!}{=} \sqrt{\mathbf{K}^2}\sqrt{\mathbf{P}^2 + m^2}.
\end{aligned} \qquad \text{(W.253)}$$

But this equation can not be fulfilled since $\mathbf{PK} \leq |\mathbf{P}|\,|\mathbf{K}|$ and $\sqrt{\mathbf{K}^2}\sqrt{\mathbf{P}^2 + m^2} = |\mathbf{K}|\sqrt{|\mathbf{P}|^2 + m^2} > |\mathbf{K}|\,|\mathbf{P}|$.

So we conclude that the process as described here is indeed not possible for real fermions and photons, i.e., for on-shell particles. But it can exist e.g. in the frame of a higher order $S^{(n)}$ as we will see in discussion of processes of $S^{(2)}$.

### W.9.4.2  Second Example: Pair Production, $\gamma \rightarrow e^- e^+$

The $S$-matrix element reads

$$S^{(1)} = -iq \int \mathrm{d}^4 x \; \bar{\psi}^c \slashed{A}^d \psi^c = -iq \int \mathrm{d}^4 x \; \bar{\psi}^c \gamma_\mu \psi^c A^{\mu d}. \qquad \text{(W.254)}$$

This means that the initial state is a photon; the final state consists of an electron and a positron. Thus, the initial and final states are given by

$$|i\rangle = a_\Lambda^\dagger (\mathbf{K}) |0\rangle \;\; ; \;\; |f\rangle = b_{R_1}^\dagger (\mathbf{P}_1)\, d_{R_2}^\dagger (\mathbf{P}_2) |0\rangle \rightarrow \langle f| = \langle 0|\, b_{R_1} (\mathbf{P}_1)\, d_{R_2} (\mathbf{P}_2). \qquad \text{(W.255)}$$

Following the 'matching rules', developed above, the contributions to $\langle f|\, S^{(1)} |i\rangle$ for this process are $\sqrt{\frac{1}{(2\pi)^3 E_k}}\, \epsilon_\Lambda^\mu (\mathbf{K})\, e^{-iKx}$ by the photon, whereas the outgoing electron brings $\sqrt{\frac{m}{(2\pi)^3 E_{\mathbf{P}_1}}}\, \bar{u}_{R_1} (\mathbf{P}_1)\, e^{iP_1 x}$ and the outgoing positron contributes $\sqrt{\frac{m}{(2\pi)^3 E_{\mathbf{P}_2}}}\, w_{R_2} (\mathbf{P}_2)\, e^{iP_2 x}$. It follows

$$\begin{aligned}
\langle f|\, S^{(1)} |i\rangle = &-iq\, (2\pi)^4\, \delta^{(4)} (P_1 + P_2 - K) \cdot \\
&\cdot \sqrt{\frac{m}{(2\pi)^3 E_{\mathbf{P}_1}} \frac{1}{2(2\pi)^3 E_k} \frac{m}{(2\pi)^3 E_{\mathbf{P}_2}}}\, \bar{u}_{R_1} (\mathbf{P}_1)\, \gamma_\mu \epsilon_\Lambda^\mu (\mathbf{K})\, w_{R_2} (\mathbf{P}_2).
\end{aligned} \qquad \text{(W.256)}$$

Here the delta function yields:

$$E_{\mathbf{k}} = E_{\mathbf{P}_1} + E_{\mathbf{P}_2} \;\; ; \;\; \mathbf{K} = \mathbf{P}_1 + \mathbf{P}_2. \qquad \text{(W.257)}$$

Hence

$$|\mathbf{K}| = \sqrt{\mathbf{P}_1^2 + m^2} + \sqrt{\mathbf{P}_2^2 + m^2} \;\; ; \;\; \mathbf{K} = \mathbf{P}_1 + \mathbf{P}_2. \qquad \text{(W.258)}$$

But this is a contradiction: from the first equation follows $|\mathbf{K}|^2 > (|\mathbf{P}_1| + |\mathbf{P}_2|)^2 + 2m^2$, and from the second $|\mathbf{K}|^2 \leq (|\mathbf{P}_1| + |\mathbf{P}_2|)^2$. Thus, pair production as described here cannot occur. But this is to be expected - we know, that electron-positron pair production with one photon in free space can only occur near e.g. a nucleus which receives some recoil.

## W.9.5  External Fields

As mentioned in the introduction to this section, the eight processes cannot be realized with on-shell particles. However, with external fields, the situation is different. We will very briefly sketch the reason. Assume we have a static external field $A_{\text{ext}}^\mu(\mathbf{x})$. For the sake of simplicity, we confine the considerations to a scalar potential $V(\mathbf{x})$ which can be e.g. the Coulomb potential of a nucleus.

Scattering of an electron in this external field can be described by replacing the photonic contribution $A$ by the external field, i.e., by

$$\langle f | S^{(1)} | i \rangle = -iq \int d^4x \; \langle f | \bar{\psi}^c A^c \psi^d | i \rangle \rightarrow \langle f | S^{(1)}_{\text{ext}} | i \rangle = -iq \int d^4x \; \langle f | \bar{\psi}^c A^c_{\text{ext}} \psi^d | i \rangle . \tag{W.259}$$

Assume that we have an incoming electron $(\mathbf{P}, R)$ which is scattered to $(\mathbf{P}', R')$. We compare this with the transition amplitude (W.250) for $e^- \rightarrow \gamma e^-$. We have

$$\begin{Bmatrix} \langle f | S^{(1)} | i \rangle \\ \langle f | S^{(1)}_{\text{ext}} | i \rangle \end{Bmatrix} = -iq \int d^4x \sqrt{\frac{m}{(2\pi)^3 E_{\mathbf{P}'}}} \bar{u}_{R'} (\mathbf{P}') e^{iP'x} \begin{Bmatrix} \sqrt{\frac{1}{2(2\pi)^3 E_{\mathbf{k}}}} \gamma_\mu \epsilon_\Lambda^\mu (\mathbf{K}) e^{iKx} \\ \gamma_0 V(\mathbf{x}) \end{Bmatrix} \sqrt{\frac{m}{(2\pi)^3 E_{\mathbf{P}}}} u_R (\mathbf{P}) e^{-iPx} . \tag{W.260}$$

The term $\gamma_0 V(\mathbf{x})$ has no time dependence. Hence, performing the $x$-integral gives

$$\int d^4x \; e^{iP'x} V(\mathbf{x}) e^{-iPx} = 2\pi \delta^{(1)} \left( p_0 - p_0' \right) \cdot \tilde{V} \left( \mathbf{p} - \mathbf{p}' \right) \tag{W.261}$$

where $\tilde{V}$ is the Fourier transform of $V(x)$. Now the trouble with virtual particles in $\langle f | S^{(1)} | i \rangle$ comes from the delta function $\delta^{(4)} \left( P' + K - P \right)$ in (W.251). But for the external field we have instead an one-dimensional delta function $\delta^{(1)} \left( p_0 - p_0' \right)$ which does not pose any difficulties. This is the crucial point.

Due to lack of space, we will not go into greater detail (which is one the many gaps of this text mentioned in the introduction).

**Fig. W.1** Basic lines

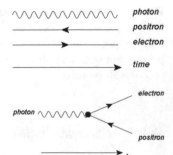

**Fig. W.2** Pair production

## W.9.6   Feynman Diagrams

By means of *Feynman diagrams*, scattering processes can be visualized in a very nice way. What is more, they also allow for formulating precisely the corresponding expressions for the transition amplitudes.

The diagrams consist of points (called vertices) and lines affixed to the vertices. Fermions are represented by a solid line with an arrow, and photons by a wavy line, see Fig. W.1. External fields are marked by a small cross as in Fig. W.4. Time is placed on the vertical or the horizontal axis; both versions are common. For particles as electrons or muons, the arrow is oriented in direction of the time; for antiparticles as positrons in opposite direction, corresponding to the conception of antiparticles in the Stückelberg-Feynman interpretation.

A further ingredient of the diagrams are vertices, i.e., points where fermions and photons interact.[114] In QED, there are always three lines attached to a vertex: one photonic line and two fermionic lines, one with the arrow toward the vertex and the other with the arrow away from the vertex. As an example, we show in Fig. W.2 the diagram for pair production. Note that in this process an electron and a positron are created (reverse direction of the positron).

The inclination of the lines has no physical meaning, as outlined in Fig. W.3. What matters is the relation to the time axis. A remark on notation: In many cases, when the situation is obvious, there is no labeling - a wavy line is a photon, and electron and positron are clearly identified by their arrows with respect to the time axis. Labeling makes sense e.g. when there are different types of particles and antiparticles, as electrons and muons.

Note that all possible processes of $S^{(1)}$ have *one* vertex. Thus, in the language of Feynman diagrams one can state 'a single vertex is not physical'.[115]

As wee will see below in more detail, Feynman diagrams are precise graphical representations of amplitudes for particle reactions. However, they do not claim to

---

[114]Note that the order of the series expansion of $S$ gives the number of vertices: diagrams for $S^{(1)}$ have one vertex, for $S^{(2)}$ two vertices and so on.

[115]With external fields as in Fig. W.4, the situation is different. However, in a certain sense, the cross is comparable to a vertex.

**Fig. W.3** Inclination of the lines does not matter. Pair production and Compton scattering (see section '$S$-matrix, 2. order')

**Fig. W.4** Electron scattering in an external field

be precise descriptions images of reality - indeed, they are rather schematic and reduced to the essentials. They show point-like objects which interact at points. But we know that there are waves which interact over a region. For example, in some Feynman diagrams, fermionic or photonic lines are drawn in such a way that they suggest instantaneous propagation (or at least faster than light). But this is an artefact, due to the simplified and schematic representation of a more complex reality. In this respect, the pictures are to be understood with caution - they represent the reality only symbolically.

But they are excellent tools concerning the mathematical formulation of the scattering processes. They greatly reduce the computations; accordingly, they are, in a certain sense, the most common method of computing amplitudes in quantum field theory.

## W.9.7  First Feynman Rules

Even for the two simple processes we have just considered we had to do some algebra, and similar calculations in higher order of approximations, i.e., $S^{(2)}$, $S^{(3)}$, ..., require an significantly higher amount of algebra. Thus, one is interested in short-cuts similar to the 'matching rules' we found in (W.4), e.g. that an incoming electron with quantum numbers[116] $(R, \mathbf{P})$ produces a factor $\sqrt{\frac{m}{(2\pi)^3 E_P}} u_R(\mathbf{P}) e^{-iPx}$.

To get more information we consider how this rules are reflected in the transition amplitudes for emission of a photon (W.251) and for pair production (W.256). We write them one above the other:

---

[116]Remember that the indication of $\mathbf{P}$ fixes the 4-vector $P$ due to $P_0 = \sqrt{\mathbf{P}^2 + m^2}$.

**Table W.5** Formal and graphical assignments to particles and vertex; Feynman rules (to be completed)

| incoming | | | outgoing | | |
|---|---|---|---|---|---|
| $e^-$ | $u_r(\mathbf{p})$ | ———→● | $e^-$ | $\bar{u}_r(\mathbf{p})$ | ●———→ |
| $e^+$ | $\bar{w}_r(\mathbf{p})$ | ←———● | $e^+$ | $w_r(\mathbf{p})$ | ●←——— |
| $\gamma$ | $\epsilon_\lambda^\mu(\mathbf{k})$ | ∿∿∿● | $\gamma$ | $\epsilon_\lambda^\mu(\mathbf{k})$ | ●∿∿∿ |
| | | vertex | ● | $-iq\gamma_\mu$ | |

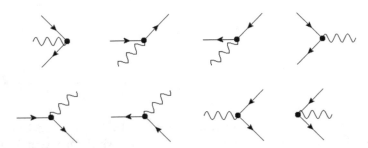

**Fig. W.5** The eight processes of $S^{(1)}$. Top: $\gamma e^- e^+ \to$ vacuum, $\gamma e^- \to e^-$, $\gamma e^+ \to e^+$, $e^- e^+ \to \gamma$. Bottom: $e^- \to \gamma e^-$, $e^+ \to \gamma e^+$, $\gamma \to e^- e^+$, vacuum $\to \gamma e^- e^+$

$$\langle f| S^{(1)} |i\rangle_{\text{emission}} = -iq\,(2\pi)^4\,\delta^{(4)}\,(P'+K-P)\cdot\sqrt{\frac{m}{(2\pi)^3 E_{P'}}\frac{1}{2(2\pi)^3 E_{\mathbf{k}}}\frac{m}{(2\pi)^3 E_P}}\,\bar{u}_{R'}(P')\,\gamma_\mu\epsilon_\Lambda^\mu(K)\,u_R(P)$$

$$\langle f| S^{(1)} |i\rangle_{\text{pair}} = -iq\,(2\pi)^4\,\delta^{(4)}\,(P_1+P_2-K)\cdot\sqrt{\frac{m}{(2\pi)^3 E_{P_1}}\frac{1}{2(2\pi)^3 E_{\mathbf{k}}}\frac{m}{(2\pi)^3 E_{P_2}}}\,\bar{u}_{R_1}(P_1)\,\gamma_\mu\epsilon_\Lambda^\mu(K)\,w_{R_2}(P_2)\,.$$

$$\text{(W.262)}$$

As is seen, the structure is very similar. From right to left we have the incoming fermion, characterized by spin vectors like $u_R(\mathbf{P})$, then the photon with polarization vector $\epsilon_\Lambda^\mu(\mathbf{K})$, and finally the outgoing fermion. It follows a square root as prefactor which contains the masses and energies of the three items. And finally we have a term which we can write as $(2\pi)^4\,\delta^{(4)}\,(P_{\text{in}} - P_{\text{out}}) \cdot (-iq\gamma_\mu)$ where $P_{\text{in}}$ and $P_{\text{out}}$ are the total momenta in the incoming and the outgoing channel. In this way, we can formulate all transition amplitudes of the eight elementary processes of Table W.3 without getting involved into long calculations doing sums and integrals.

What makes life even easier, is the connection with the Feynman diagrams. Indeed, there is a one-to-one correspondence between transition amplitude and Feynman diagram. Each term in (W.262) has a direct counterpart in the diagram. We compile the results in the following table. These assignments are part of what is called *Feynman rules* which (as enhanced version of out matching rules) are the one-to-one translation rules between the Feynman amplitudes in their mathematical form and the representation of the processes in the Feynman diagrams.

And finally, we show in Fig. W.5 the Feynman diagrams for the eight elementary processes.

## W.9.8 Exercises and Solutions

1. Show that all eight processes can not occur with real particles.
   Solution: We focus on the conservation of the 4-momentum. As is seen from Fig. W.5, the top and bottom processes are time-inverted versions of each other. If a process can not occur, the same holds for its time-reflection. So we need to consider only one row, say bottom. The last process (vacuum $\to \gamma e^- e^+$) is forbidden due to energy conservation, $E_i = 0, E_f > 0$. The first two processes in the bottom row, $e^- \to \gamma e^-$ and $e^+ \to \gamma e^+$, are equal with regard to conservation of the 4-momentum. Thus, we need to consider only one of them. This means, that there are left $e^- \to \gamma e^-$ and $\gamma \to e^- e^+$. But we have shown above that these processes cannot occur with real particles, see (W.252) and (W.257). Thus, this holds for all processes in the bottom row, and, hence, in the top row.

2. Formulate $\langle f| S^{(1)} |i\rangle$ for all eight processes of $S^{(1)}$ (cf. Fig. W.5).
   Solution: We consider $\gamma e^- \to e^-$. Incoming electron $u_R(\mathbf{P})$, photon $\epsilon_\Lambda^\mu(\mathbf{K})$, outgoing electron $u_{R'}(\mathbf{P}')$. It follows

$$\langle f| S^{(1)} |i\rangle = (2\pi)^4 \delta^{(4)} \left(P + K - P'\right) \cdot$$
$$\sqrt{\frac{m}{(2\pi)^3 E_{\mathbf{P}'}} \frac{1}{2(2\pi)^3 E_k} \frac{m}{(2\pi)^3 E_{\mathbf{P}}}} \, \bar{u}_{R'}(\mathbf{P}') \left(-iq\gamma_\mu\right) \epsilon_\Lambda^\mu(\mathbf{K}) u_R(\mathbf{P}) . \tag{W.263}$$

The other cases analogously.

# W.10 Contraction, Propagator, Wick's Theorem

The discussion of the second order term $S^{(2)} = -\frac{1}{2} \int \int d^4x \, d^4y \, T \left[\mathcal{H}_I(x)\mathcal{H}_I(y)\right]$ with $\mathcal{H}_I(x) = q\mathcal{N}\left[\bar\psi A\psi\right]$ is quite elaborate and complex compared to $S^{(1)}$. Essentially, this is due to two facts: 1) there are more terms now than in $S^{(1)}$; 2) we have to time-order a product of normal ordered strings of operators. The issue requires new tools which we will provide in this section.

We start with the contraction, i.e., the difference of a time ordered and a normal ordered product of two operators. It will turn out that this term equals essentially the propagator which in turn is closely related to the evolution of a system. Finally, Wick's theorem describes how to get rid of time ordered strings of field operators by means of contractions and (easy to handle) normal ordered products.

## W.10.1 Contraction

Given two field scalar operators $A(x)$ and $B(y)$. Then the *contraction* of $A$ and $B$, written as $\overline{A(x)B(y)}$, is defined by the difference of the time ordered and normal ordered product of $AB$:

$$\overline{A(x)B(y)} = T\left[A\left(x\right)B\left(y\right)\right] - \mathcal{N}\left[A\left(x\right)B\left(y\right)\right]. \qquad (W.264)$$

We have determined this expression in section 'Operator ordering' (see (W.210)); it is given by[117]

$$\overline{A(x)B(y)} = \begin{cases} \left[A^d\left(x\right), B^c\left(y\right)\right]_{\mp} & \text{for } \begin{matrix} x^0 > y^0 \\ y^0 > x^0 \end{matrix}. \\ -\left[A^c\left(x\right), B^d\left(y\right)\right]_{\mp} \end{cases} \qquad (W.265)$$

Note that performing the (anti)commutators 'uses up' the creation and deletion operators. Thus, the contraction $\overline{A(x)B(y)}$ contains no creation and deletion operators and is a $c$-number in this regard; we have with $\langle 0\,|0\rangle = 1$

$$\langle 0|\,\overline{A(x)B(y)}\,|0\rangle = \overline{A(x)B(y)}. \qquad (W.266)$$

Equation (W.265) may be written as[118]

$$\overline{A(x)B(y)} = \theta\left(x^0 - y^0\right)\left[A^d\left(x\right), B^c\left(y\right)\right]_{\mp} - \theta\left(y^0 - x^0\right)\left[A^c\left(x\right), B^d\left(y\right)\right]_{\mp} \qquad (W.268)$$

or in more detail:

$$\text{bosons: } \overline{A(x)B(y)} = \theta\left(x^0 - y^0\right)\left[A^d\left(x\right), B^c\left(y\right)\right] + \theta\left(y^0 - x^0\right)\left[B^d\left(y\right), A^c\left(x\right)\right]$$

$$\text{fermions: } \overline{A(x)B(y)} = \theta\left(x^0 - y^0\right)\left\{A^d\left(x\right), B^c\left(y\right)\right\} - \theta\left(y^0 - x^0\right)\left\{B^d\left(y\right), A^c\left(x\right)\right\}. \qquad (W.269)$$

We now want to form the vacuum expectation value of the contraction (W.265), i.e., $\langle 0|\,T\left[A\left(x\right)B\left(y\right)\right] - \mathcal{N}\left[A\left(x\right)B\left(y\right)\right]|0\rangle = \langle 0|\,\overline{A(x)B(y)}\,|0\rangle = \overline{A(x)B(y)}$. Since the vacuum expectation value of a normal ordered expression always vanishes, we get with (W.266)

$$\langle 0|\,T\left[A\left(x\right)B\left(y\right)\right]|0\rangle = \overline{A(x)B(y)} \qquad (W.270)$$

and with (W.265) follows

---

[117] Remember $[A, B]_- = AB - BA$ (bosons) and $[A, B]_+ = \{A, B\} = AB + BA$ (fermions).

[118] Remember the definition of the Heaviside function $\theta\left(x\right)$:

$$\theta\left(x\right) = \begin{cases} 1 & \text{for } \begin{matrix} x \geq 0 \\ x < 0 \end{matrix} \\ 0 \end{cases} \qquad (W.267)$$

$$\langle 0| \, T\left[A\left(x\right)B\left(y\right)\right]|0\rangle = \begin{cases} \left[A^d\left(x\right),B^c\left(y\right)\right]_\mp & \text{for } x^0 > y^0 \\ -\left[A^c\left(x\right),B^d\left(y\right)\right]_\mp & y^0 > x^0 \end{cases}. \qquad (W.271)$$

Remember that the commutators are $c$-numbers which do not contain any creation or annihilation operators. As we will see in the following section, they play an important role in the further discussion of QED.

Let us draw up an interim balance. We started with the definition of the contraction as $\overline{A(x)B(y)} = T\left[AB\right] - \mathcal{N}\left[AB\right]$. Conversely, we can express the time ordered product by the sum of normal ordered product plus contraction, $T\left[AB\right] = \mathcal{N}\left[AB\right] + \overline{A(x)B(y)}$. Of course, this switch only makes sense, if we can find a suitable alternative expression for the contraction. Then we would have indeed achieved our aim to replace the cumbersome time ordering by the the easy to handle normal ordering plus terms which are also rather simple to determine. We will see now that these terms are propagators.

## W.10.2 Propagators

### W.10.2.1 Green's Function, Propagator

For the sake of simplicity and brevity, we confine ourselves in the following to the Klein–Gordon field. For the Dirac field and the photon field, the results are reported at the end of this section.

Assume we have a system $S$ in some state $|\Sigma\rangle$ which we want to probe with an extra particle $p$. In an idealized process, we bring $S$ and $p$ into contact at a spacetime point $\left(y^0, \mathbf{y}\right)$ and there will be some interaction between $S$ and $p$. After that, we will remove $p$ at a spacetime point $\left(x^0, \mathbf{x}\right)$ with $x^0 > y^0$. One may ask if $S$ has remained in its previous state $|\Sigma\rangle$. The answer is given by the projection of ($p$ created at $y$) $|\Sigma\rangle$ onto ($p$ annihilated at $x$) $|\Sigma\rangle$, i.e., by the probability amplitude $\langle\Sigma|$ ($p$ annihilated at $x$) ($p$ created at $y$) $|\Sigma\rangle$. This amplitude is called Green's function or *propagator* and noted by $G^+\left(x, y\right)$. With the notation $\phi^\dagger$ and $\phi$ for the creation and annihilation operator for $p$ we can write

$$G^+\left(x, y\right) = \theta\left(x^0 - y^0\right)\,\langle\Sigma|\,\phi\left(x\right)\phi^\dagger\left(y\right)|\Sigma\rangle. \qquad (W.272)$$

The index $+$ on the left denotes that $p$ is created at $y$ *before* it is destroyed at $x$; the Heaviside function $\theta$ guarantees this behavior. $G^+$ is also called retarded propagator.

We now invoke the interpretation of Feynman-Stückelberg of antiparticles (see Appendix U, Vol. 1) which sees essentially antiparticles as particles traveling backwards in time. Thus, for the antiparticle $\bar{p}$ we have the corresponding process of creating it at $x$ and then annihilating it at $y$ with $y^0 > x^0$. Since $\phi^\dagger$ and $\phi$ are the annihilation and creation operators for $\bar{p}$, we can describe this process by

$$G^- (x, y) = \theta \left( y^0 - x^0 \right) \; \langle \Sigma | \phi^\dagger (y) \, \phi (x) | \Sigma \rangle . \tag{W.273}$$

$G^-$ is also called advanced propagator.

In QFT, we have to consider particles and antiparticles on an equal footing. It was Feynman who has pointed out that as a consequence we have to consider *both* Green's functions, i.e.,

$$\begin{aligned} G_F (x, y) &= G^+ (x, y) + G^- (x, y) = \\ = \theta \left( x^0 - y^0 \right) \; &\langle \Sigma | \phi (x) \, \phi^\dagger (y) | \Sigma \rangle + \theta \left( y^0 - x^0 \right) \; \langle \Sigma | \phi^\dagger (y) \, \phi (x) | \Sigma \rangle \quad \text{(W.274)} \\ &\rightarrow G_F (x, y) = \langle \Sigma | \mathcal{T} \left[ \phi (x) \, \phi^\dagger (y) \right] | \Sigma \rangle . \end{aligned}$$

Note the occurrence of the time ordering operator $\mathcal{T}$. As we will see, such expressions are of particular importance in QFT, and so is time ordering.

Now assume that the state $|\Sigma\rangle$ is given by the vacuum state $|0\rangle$. Then by $\phi^\dagger (y) |0\rangle$ we create a particle at $y$ and nothing else happens (there is no further interaction). By $\langle 0| \phi (x)$, this particle is destroyed at $x$. Thus, the propagator $\langle 0| \phi (x) \, \phi^\dagger (y) |0\rangle$ describes a particle which propagates from $y$ to $x$, and $\langle 0| \mathcal{T} \left[ \phi (x) \, \phi^\dagger (y) \right] |0\rangle$ describes this situation for the particle *and* the antiparticle. It is called *(free) Feynman propagator*[119] $\Delta_F (x, y)$:

$$\begin{aligned} i \Delta_F (x, y) &= \langle 0| \mathcal{T} \left[ \phi (x) \, \phi^\dagger (y) \right] |0\rangle = \\ = \theta \left( x^0 - y^0 \right) \; &\langle 0| \phi (x) \, \phi^\dagger (y) |0\rangle + \theta \left( y^0 - x^0 \right) \; \langle 0| \phi^\dagger (y) \, \phi (x) |0\rangle . \end{aligned} \tag{W.275}$$

Let us point out once more that $\theta \left( x^0 - y^0 \right) \; \langle 0| \phi (x) \, \phi^\dagger (y) |0\rangle$ describes a particle travelling from $y$ to $x$ for $x^0 > y^0$, and $\theta \left( y^0 - x^0 \right) \; \langle 0| \phi^\dagger (y) \, \phi (x) |0\rangle$ describes an antiparticle travelling from $x$ to $y$ for $y^0 > x^0$. Thus, the Feynman propagator is the transition amplitude for two processes, namely (1) a particle $p$ is created at $y$ and annihilated at $x$, (2) an antiparticle $\bar{p}$ is created at $x$ and annihilated at $y$. *Both* processes are included in the Feynman propagator; it contains both amplitudes corresponding to the possibility that we have a particle or an antiparticle.

### W.10.2.2   Connection with the Contraction

Between the Feynman propagator (W.275) and the contraction (W.265) exists a very close connection - they are identical up to a factor $i$:

$$i \Delta_F (x, y) = \langle 0| \overbracket{\phi (x) \, \phi^\dagger (y)} |0\rangle . \tag{W.276}$$

Indeed, from (W.275) we have[120]

---

[119]The factor $i$ is inserted in the definition because it makes things easier in the following. It is missing in some textbooks.

[120]Remind that the vacuum expectation value of a normal ordered product vanishes, $\langle 0| \mathcal{N} \left[ \phi (x) \, \phi^\dagger (y) \right] |0\rangle = 0$.

$$i\Delta_F(x, y) = \langle 0| \, T \left[ \phi(x) \, \phi^\dagger(y) \right] |0\rangle =$$
$$= \langle 0| \, T \left[ \phi(x) \, \phi^\dagger(y) \right] - N \left[ \phi(x) \, \phi^\dagger(y) \right] |0\rangle = \langle 0| \, \overline{\phi(x) \, \phi^\dagger(y)} \, |0\rangle. \tag{W.277}$$

In terms of bosonic deletion and creation operators, i.e., $\phi(x) = \phi^d(x) + \phi^c(x)$, $\phi^\dagger(y) = \phi^{\dagger d}(y) + \phi^{\dagger c}(y)$ we have

$$\langle 0| \, \overline{\phi(x) \, \phi^\dagger(y)} \, |0\rangle = \theta\left(x^0 - y^0\right) \langle 0| \left[ \phi^d(x), \phi^{\dagger c}(y) \right] |0\rangle + \theta\left(y^0 - x^0\right) \langle 0| \left[ \phi^{\dagger d}(y), \phi^c(x) \right] |0\rangle. \tag{W.278}$$

### W.10.2.3 Explicit Calculation of the Propagator $\Delta_F(x, y)$

How looks $\Delta_F(x, y)$ explicitly? The answer reads[121]

$$i\Delta_F(x, y) = \langle 0| \, T \left[ \phi(x) \, \phi^\dagger(y) \right] |0\rangle =$$
$$= \frac{1}{(2\pi)^3} \int \frac{d^3 p}{\sqrt{2E_p}} \left[ \theta\left(x^0 - y^0\right) e^{-ip(x-y)} + \theta\left(y^0 - x^0\right) e^{ip(x-y)} \right]. \tag{W.280}$$

Another usual formulation of the free propagator requires complex analysis which is beyond the scope of this short introduction into QFT. Thus, we report here only the result[122]; it reads

$$i\Delta_F(x, y) = \langle 0| \, T \left[ \phi(x) \, \phi^\dagger(y) \right] |0\rangle = \frac{1}{(2\pi)^4} \int d^4 p \, \frac{i}{p^2 - m^2 + i\varepsilon} e^{-ip(x-y)} \tag{W.282}$$

where $\varepsilon > 0$ is infinitesimal. With the Fourier transform

$$\Delta_F(x, y) = \frac{1}{(2\pi)^4} \int d^4 p \, \tilde{\Delta}_F(p) \, e^{-ip(x-y)} \tag{W.283}$$

we arrive at the Fourier representation $\tilde{\Delta}(p)$ of the Feynman propagator for a particle with momentum $p$ (or briefly the momentum space propagator):

$$\tilde{\Delta}_F(p) = \frac{1}{p^2 - m^2 + i\varepsilon}. \tag{W.284}$$

---

[121] Note that we have

$$\left(\partial^2 + m^2\right) \Delta(x, y) = -i\delta^4(x - y) \tag{W.279}$$

Thus, $\Delta(x, y)$ is evidently Green's function.

[122] One inserts the definition of the Heaviside function

$$\theta\left(x^0 - y^0\right) = \frac{i}{2\pi} \int_{-\infty}^{\infty} dz \, \frac{e^{-iz(x^0 - y^0)}}{z + i\varepsilon} \tag{W.281}$$

into (W.280) and obtains (W.282) after some manipulations.

One sees that there is a singularity for $p^2 = E_{\mathbf{p}}^2 - \mathbf{p}^2 = m^2$, i.e., if the usual dispersion relation for a relativistic particle is fulfilled,[123] or, in other words, if the particle is on-shell. But the Feynman propagator encompasses off-shell particles, i.e., virtual particles, which do not obey the relativistic dispersion relation. Such particles are allowed, but only for a short time, as is explained in the previous section.

### W.10.2.4 Propagators for the Klein–Gordon Field, the Dirac Field, the Radiation Field

For the Klein–Gordon field $\phi(x)$, the Dirac field $\psi(x)$ and the radiation field $A^\mu(x)$, we have the following contractions:

$$
\begin{aligned}
\overbracket{\phi(x)\,\phi^\dagger(y)} &= i\,\Delta_F(x-y) \\
\overbracket{\psi_\alpha(x)\,\bar{\psi}_\beta(y)} &= -\overbracket{\bar{\psi}_\beta(y)\,\psi_\alpha(x)} = i S_{F\alpha\beta}(x-y) \\
\overbracket{A^\mu(x)\,A^\nu(y)} &= i\mathrm{d}_F^{\mu\nu}(x-y).
\end{aligned}
\tag{W.285}
$$

All other contractions vanish, for instance $\overbracket{\psi(x)\,A^\mu(y)} = 0$ and so on. In other words: contractions vanish except when they equal a Feynman propagator.

The propagators are explicitly given by the following expressions:

$$
\begin{aligned}
\Delta_F(x-y) &= \tfrac{1}{(2\pi)^4}\int \mathrm{d}^4 p\; \tfrac{1}{p^2-m^2+i\varepsilon}e^{-ip(x-y)} \\
S_F(x-y) &= \tfrac{1}{(2\pi)^4}\int \mathrm{d}^4 p\; \tfrac{\slashed{p}+m}{p^2-m^2+i\varepsilon}e^{-ip(x-y)} \\
D_F^{\mu\nu}(x-y) &= -g^{\mu\nu}\tfrac{1}{(2\pi)^4}\int \mathrm{d}^4 k\; \tfrac{1}{k^2+i\varepsilon}e^{-ik(x-y)}.
\end{aligned}
\tag{W.286}
$$

Three remarks: (1) The Dirac case is also written as

$$
S_{F\alpha\beta}(x-y) = \frac{1}{(2\pi)^4}\int \mathrm{d}^4 p\; \frac{(\slashed{p}+m)_{\alpha\beta}}{p^2-m^2+i\varepsilon}e^{-ip(x-y)}
\tag{W.287}
$$

where $(\slashed{p}+m)_{\alpha\beta}$ is the $\alpha\beta$-entry of the matrix $\slashed{p}+m$. Note that the propagators $S_F$ and $D_F$ are matrix-valued. (2) In some textbooks, the infinitesimal character of $\varepsilon$ is stressed by writing e.g.

$$
\Delta_F(x-y) = \lim_{\varepsilon\to 0^+}\frac{1}{(2\pi)^4}\int \mathrm{d}^4 p\; \frac{1}{p^2-m^2+i\varepsilon}e^{-ip(x-y)}.
\tag{W.288}
$$

(3) In some textbooks, the fraction $\frac{\slashed{p}+m}{p^2-m^2+i\varepsilon}$ is formally reduced to $\frac{1}{\slashed{p}-m+i\varepsilon}$. Of course, this is just a symbolical notation; division by a matrix is not defined in this form.

---

[123]The infinitesimal term $i\varepsilon$ guarantees that one never hits this singularity (pole) exactly; it has to do with the integration procedures in complex analysis.

**Fig. W.6** Electron propagator $S_F(p)$ and photon propagator $D_F^{\mu\nu}(k)$ in Feynman diagrams

Finally, in momentum space the propagators read

$$
\begin{aligned}
\Delta_F(p) &= \tfrac{1}{p^2-m^2+i\varepsilon} \\
S_F(p) &= \tfrac{\not{p}+m}{p^2-m^2+i\varepsilon} \\
D_F^{\mu\nu}(k) &= -g^{\mu\nu}\tfrac{1}{k^2+i\varepsilon}.
\end{aligned}
\tag{W.289}
$$

In Feynman diagrams, propagators are 'clamped' between vertices, see Fig. W.6.

## W.10.3 Wick's Theorem

We have introduced two ordering prescriptions - normal ordering $\mathcal{N}$ and time ordering $\mathcal{T}$. Though the justification for normal ordering of operators seems perhaps opaque and not really satisfying, the mathematical procedure itself is simple and offers no difficulties. On the other hand, time ordering is not a bit opaque on the formal level, but the mathematical procedure itself is difficult and very cumbersome.

It would be a big relief and help, if we could express the demanding time ordering of a string of operators by means of the easy to handle normal ordering. In view of the fact, that these two ordering procedures are physically totally different, it comes perhaps as a surprise - but it works, as shown by *Wick's theorem*[124]: we can express time ordering of a string by a combination of normal ordered strings and contractions.

The basic idea for the theorem is time ordering of two scalar field operators $A$ and $B$ equals the sum of the normal ordered product plus the contraction, see (W.264):

$$
\mathcal{T}[AB] = \mathcal{N}[AB] + \overset{\sqcap}{AB}.
\tag{W.290}
$$

The question is what happens if we have to time-order a string of more than two field operators, i.e., $\mathcal{T}[ABCD\ldots]$. The answer is provided by Wick's theorem which we state here without proof.[125]

Wick's theorem: The time ordered product of field operators equals the sum of the normal ordered products, whereby the operators are connected by all possible contractions.[126]

---

[124]Wick, Gian Carlo, 1909–1992; Italian physicist.

[125]The theorem may be proved by induction.

[126]Remember the linearity of normal ordering, $\mathcal{N}[(A+B)C] = \mathcal{N}[AC] + \mathcal{N}[BC]$, and $\mathcal{N}[cA] = c\mathcal{N}[A]$, where $c$ is an expression which does not contain annihilation or creation operators.

$$\mathcal{T}[ABC\ldots Z] = \mathcal{N}[ABC\ldots Z + \text{all possible contractions of } ABC\ldots Z].$$
(W.291)

As an example, consider $\mathcal{T}[ABCD]$. We have

$$\mathcal{T}[ABCD] = \mathcal{N}[ABCD] +$$
$$+\mathcal{N}\left[\overset{\frown}{AB}CD\right] + \mathcal{N}\left[A\overline{BC}D\right] + \mathcal{N}\left[AB\overline{CD}\right] +$$
$$+\mathcal{N}\left[A\overset{\frown}{B}C\overset{\frown}{D}\right] + \mathcal{N}\left[\overline{AB}CD\right] + \mathcal{N}\left[AB\overset{\frown}{CD}\right] +$$
$$+\overset{\frown}{A}\overset{\frown}{B}\overset{\frown}{C}\overset{\frown}{D} + \overset{\frown}{A}\overline{BC}\overset{\frown}{D}.$$
(W.292)

So we can present the single term $\mathcal{T}[ABCD]$ by nine terms, one without, six with one and two with two contractions. This increase in terms may seem annoying, but it actually facilitates the discussion substantially, since normal ordering is a easy to perform procedure in contrast to time ordering. In addition, contractions contain no field operators and hence are $c$-numbers in this context, i.e., can be moved out of scalar products.

As an illustration, we consider the vacuum expectation value $\langle 0|\,\mathcal{T}[ABCD]\,|0\rangle$. Remember that for normal ordered strings always holds $\langle 0|\,\mathcal{N}[\ldots]\,|0\rangle = 0$. This means, that all products with a normal ordered part vanish, e.g.,

$$\langle 0|\,\mathcal{N}\left[\overset{\frown}{AB}CD\right]|0\rangle = \overset{\frown}{AB}\cdot\langle 0|\,\mathcal{N}[CD]\,|0\rangle = 0.$$
(W.293)

Thus, only the pure contraction terms survive, and due to $\langle 0|\overset{\frown}{AB}\overset{\frown}{CD}|0\rangle = \overset{\frown}{AB}\overset{\frown}{CD}\cdot$ $\langle 0|0\rangle = \overset{\frown}{AB}\overset{\frown}{CD}$ we get the final result

$$\langle 0|\,\mathcal{T}[ABCD]\,|0\rangle = \overset{\frown}{AB}\overset{\frown}{CD} + \overset{\frown}{A}\overline{BC}\overset{\frown}{D}$$
(W.294)

where on the r.h.s there are only known terms, essentially propagators.

Wick's theorem provides also an answer to the question how to treat terms like $\mathcal{T}[\mathcal{N}[AB]\,\mathcal{N}[CD]]$ (called mixed time ordered products). We state the answer without proof: A mixed time ordered product equals the sum of the normal ordered products whereby only contractions occur which connect operators from different normal ordered products.

As an example we consider $\mathcal{T}[A_1\mathcal{N}[A_2A_3A_4]]$. We have

$$\mathcal{T}[A_1\cdot\mathcal{N}[A_2A_3A_4]] =$$
$$= \mathcal{N}[A_1A_2A_3A_4] + \mathcal{N}\left[\overset{\frown}{A_1A_2}A_3A_4\right] + \mathcal{N}\left[A_1\overline{A_2A_3}A_4\right] + \mathcal{N}\left[A_1A_2\overline{A_3A_4}\right].$$
(W.295)

We see immediately, that the vacuum expectation value $\langle 0| \, T\, [A_1 \cdot \mathcal{N}\, [A_2 A_3 A_4]] \,|0\rangle$ vanishes since there are only normal ordered terms on the r.h.s.

By Wick's theorem, the difficult time ordering is replaced by the simpler normal ordering and contractions, but there may emerge a remarkable number of those terms. Consider an expression which in some sense is quite close to QED, namely $T\, [\mathcal{H}\, (x_1)\, \mathcal{H}\, (x_2)]$ with $\mathcal{H}\, (x_1) = \mathcal{N}\, [A\, (x_1)\, B\, (x_1)\, C\, (x_1)] = \mathcal{N}\, [A_1 B_1 C_1]$. We have

$$
\begin{aligned}
T\,[\mathcal{N}\,[A_1 B_1 C_1] \cdot \mathcal{N}\,[A_2 B_2 C_2]] &= \mathcal{N}\,[(A_1 B_1 C_1)\,(A_2 B_2 C_2)] + \\
+\mathcal{N}\left[\overbracket{(A_1 B_1 C_1)(A_2} B_2 C_2)\right] &+ \text{8 other terms with one contraction } + \\
+\mathcal{N}\left[\overbracket{(A_1 B_1 C_1)(A_2 B_2} C_2)\right] &+ \text{35 other terms with two contractions } + \qquad \text{(W.296)} \\
+(A_1 B_1 C_1)(A_2 B_2 C_2) &+ \text{17 other terms with three contractions.}
\end{aligned}
$$

Thus, the time ordering at the l.h.s is replaced by 64 different terms at the r.h.s, provided that all contractions exist.

Fortunately, in QED there survive only a few contractions as we know from (W.285), namely

$$
\begin{aligned}
\overbracket{\phi\,(x)\,\phi^\dagger}\,(y) &= i\,\Delta_F\,(x-y) \\
\overbracket{\psi_\alpha\,(x)\,\psi_\beta}\,(y) &= -\overbracket{\psi_\beta\,(y)\,\psi_\alpha}\,(x) = iS_{F\alpha\beta}\,(x-y) \qquad \text{(W.297)} \\
\overbracket{A^\mu\,(x)\,A^\nu}\,(y) &= iD_F^{\mu\nu}\,(x-y).
\end{aligned}
$$

All other contractions vanish as e.g. $\overbracket{\psi(x)A^\mu}(y) = 0$ or $\overbracket{\psi(x)\psi}(y) = 0$. This fact reduces the number of terms considerably, as we will see in the next section.

## W.10.4  Exercises and Solutions

1. Show (W.265).
   Solution: With

$$
T\,[A\,(x)\,B\,(y)] = \begin{cases} A\,(x)\,B\,(y) & \text{for} \quad x^0 > y^0 \\ \pm B\,(y)\,A\,(x) & \quad\quad\; y^0 > x^0 \end{cases} \qquad \text{(W.298)}
$$

we arrive at

$$
\overbracket{A(x)B}(y) = \begin{cases} A\,(x)\,B\,(y) - A\,(x)\,B\,(y) + \left[A^d\,(x)\,,\,B^c\,(y)\right]_{\mp} & \\ \pm B\,(y)\,A\,(x) - A\,(x)\,B\,(y) + \left[A^d\,(x)\,,\,B^c\,(y)\right]_{\mp} & \end{cases} \text{for} \quad \begin{matrix} x^0 > y^0 \\ y^0 > x^0 \end{matrix}.
$$

$$\text{(W.299)}$$

The case $x^0 > y^0$ is clear; for $y^0 > x^0$ we have to calculate $\pm B(y)A(x) - A(x)B(y)$. It is

$$\pm B(y)A(x) - A(x)B(y) = -[A(x)B(y) \mp B(y)A(x)] = -[A(x), B(y)]_\mp =$$
$$= -[A^d(x) + A^c(x), B^d(y) + B^c(y)]_\mp =$$
$$= -[A^d(x), B^d(y)]_\mp - [A^d(x), B^c(y)]_\mp - [A^c(x), B^d(y)]_\mp - [A^c(x), B^c(y)]_\mp =$$
$$= -[A^d(x), B^c(y)]_\mp - [A^c(x), B^d(y)]_\mp$$

(W.300)

since $[A^d(x), B^d(y)]_\mp = [A^c(x), B^c(y)]_\mp = 0$. It follows

$$\pm B(y)A(x) - A(x)B(y) + [A^d(x), B^c(y)]_\mp =$$
$$= -[A^d(x), B^c(y)]_\mp - [A^c(x), B^d(y)]_\mp + [A^d(x), B^c(y)]_\mp =. \quad (W.301)$$
$$= -[A^c(x), B^d(y)]_\mp$$

Collecting the results, we have

$$\overline{A(x)B(y)} = \begin{cases} [A^d(x), B^c(y)]_\mp & \text{for } & x^0 > y^0 \\ -[A^c(x), B^d(y)]_\mp & & y^0 > x^0 \end{cases}. \quad (W.302)$$

2. Prove (W.278).

Solution: We have to show

$$\theta(x^0 - y^0) \langle 0| \phi(x) \phi^\dagger(y) |0\rangle + \theta(y^0 - x^0) \langle 0| \phi^\dagger(y) \phi(x) |0\rangle) =$$
$$= \theta(x^0 - y^0) \langle 0| [\phi^d(x), \phi^{\dagger c}(y)] |0\rangle + \theta(y^0 - x^0) \langle 0| [\phi^{\dagger d}(y), \phi^c(x)] |0\rangle$$

(W.303)

i.e.,

$$\langle 0| \phi(x) \phi^\dagger(y) |0\rangle = \langle 0| [\phi^d(x), \phi^{\dagger c}(y)] |0\rangle \; ; \; \langle 0| \phi^\dagger(y) \phi(x) |0\rangle = \langle 0| [\phi^{\dagger d}(y), \phi^c(x)] |0\rangle. \quad (W.304)$$

We consider first $\langle 0| \phi(x) \phi^\dagger(y) |0\rangle$. With $\phi(x) = \phi^d(x) + \phi^c(x)$ and $\phi^\dagger(y) = \phi^{\dagger d}(y) + \phi^{\dagger c}(y)$ we have

$$\langle 0| \phi(x) \phi^\dagger(y) |0\rangle = \langle 0| \phi^d(x) \phi^{\dagger d}(y) |0\rangle + \langle 0| \phi^d(x) \phi^{\dagger c}(y) |0\rangle$$
$$+ \langle 0| \phi^c(x) \phi^{\dagger d}(y) |0\rangle + \langle 0| \phi^c(x) \phi^{\dagger c}(y) |0\rangle. \quad (W.305)$$

A destruction (or annihilation) operator applied to the vacuum yields zero, $\phi^{\dagger d}(y) |0\rangle$; thus, the first and the third term on the r.h.s vanish. Likewise, $\langle 0| \phi^c(x)$ vanishes. So we have in a first step

$$\langle 0| \phi(x) \phi^\dagger(y) |0\rangle = \langle 0| \phi^d(x) \phi^{\dagger c}(y) |0\rangle. \quad (W.306)$$

On the r.h.s, we subtract $0 = \langle 0| \phi^{\dagger c}(y) \phi^d(x) |0\rangle$ and obtain

$$\langle 0| \phi(x) \phi^\dagger(y) |0\rangle = \langle 0| \phi^d(x) \phi^{\dagger c}(y) |0\rangle - \langle 0| \phi^{\dagger c}(y) \phi^d(x) |0\rangle = \langle 0| \left[ \phi^d(x), \phi^{\dagger c}(y) \right] |0\rangle .$$
(W.307)

In a similar way, we arrive at

$$\langle 0| \phi^\dagger(y) \phi(x) |0\rangle = \langle 0| \left[ \phi^{\dagger d}(y), \phi^c(x) \right] |0\rangle$$
(W.308)

which proves the assertion.

2. Prove (W.280).

Solution: The Klein–Gordon free solution reads (cf. Appendix U, Vol. 1)

$$\phi^\dagger(y) = \frac{1}{(2\pi)^{3/2}} \int \frac{d^3 p}{\sqrt{2E_\mathbf{p}}} \left[ a_\mathbf{p}^\dagger e^{ipy} + a_\mathbf{p} e^{-ipy} \right].$$
(W.309)

With $a_\mathbf{p} |0\rangle = 0$ and $a_\mathbf{p}^\dagger |0\rangle = |\mathbf{p}\rangle$ follows

$$\phi^\dagger(y) |0\rangle = \frac{1}{(2\pi)^{3/2}} \int \frac{d^3 p}{\sqrt{2E_\mathbf{p}}} |\mathbf{p}\rangle e^{ipy}.$$
(W.310)

Taking the Hermitian adjoint $\left( \phi^\dagger(y) |0\rangle \right)^\dagger = \langle 0| \phi(y)$ brings (with $y \to x$ and $\mathbf{p} \to \mathbf{p}'$)

$$\langle 0| \phi(x) = \frac{1}{(2\pi)^{3/2}} \int \frac{d^3 p'}{\sqrt{2E_{\mathbf{p}'}}} \langle \mathbf{p}'| e^{-ip'x}.$$
(W.311)

Multiplying the last two expressions leads to

$$\langle 0| \phi(x) \phi^\dagger(y) |0\rangle = \frac{1}{(2\pi)^3} \int \frac{d^3 p d^3 p'}{\sqrt{2E_\mathbf{p}}\sqrt{2E_{\mathbf{p}'}}} \langle \mathbf{p}'| \mathbf{p}\rangle e^{ip(y-x)}.$$
(W.312)

As is usual, we assume the momentum states as orthonormalized, i.e., $\langle \mathbf{p}'| \mathbf{p}\rangle = \delta(\mathbf{p}' - \mathbf{p})$. Thus, we have

$$\langle 0| \phi(x) \phi^\dagger(y) |0\rangle = \frac{1}{(2\pi)^3} \int \frac{d^3 p}{\sqrt{2E_\mathbf{p}}} e^{ip(y-x)}.$$
(W.313)

In a similar way, we have

$$\langle 0| \phi^\dagger(y) \phi(x) |0\rangle = \frac{1}{(2\pi)^3} \int \frac{d^3 p}{\sqrt{2E_\mathbf{p}}} e^{ip(x-y)}.$$
(W.314)

Due to

$$i\Delta_F(x, y) = \langle 0| \, T\left[\phi(x)\,\phi^\dagger(y)\right]|0\rangle =$$
$$= \theta\left(x^0 - y^0\right)\,\langle 0|\,\phi(x)\,\phi^\dagger(y)\,|0\rangle + \theta\left(y^0 - x^0\right)\,\langle 0|\,\phi^\dagger(y)\,\phi(x)\,|0\rangle$$

$$(W.315)$$

we arrive at the result[127]

$$i\Delta_F(x, y) = \frac{1}{(2\pi)^3}\int \frac{d^3p}{\sqrt{2E_{\mathbf{p}}}}\left[\theta\left(x^0 - y^0\right)\,e^{-ip(x-y)} + \theta\left(y^0 - x^0\right)\,e^{ip(x-y)}\right].$$

$$(W.316)$$

## W.11  S-Matrix, 2. Order, General

We now focus on the discussion of the second order term of the $S$-matrix $S^{(2)} = (-i)^2\frac{1}{2!}\int\int d^4x\,d^4y\,T\left[\mathcal{H}_I(x)\,\mathcal{H}_I(y)\right]$ with $\mathcal{H}_I(x) = q\mathcal{N}\left[\bar\psi A\psi\right]$. First we transform $T\left[\mathcal{H}_I(x)\,\mathcal{H}_I(y)\right]$ by means of Wick's theorem. Then we give an overview of the possible physical processes. In the next section, we discuss two of them in more detail, namely Bhabha and Møller scattering.

Written out in full we have[128]

$$S^{(2)} = -\frac{q^2}{2}\int\int d^4x_1\,d^4x_2\,T\left[\mathcal{N}\left[\bar\psi(x_1)\,A(x_1)\,\psi(x_1)\right]\mathcal{N}\left[\bar\psi(x_2)\,A(x_2)\,\psi(x_2)\right]\right].$$

$$(W.317)$$

On behalf of a more transparent notation, we use the abbreviation

$$\bar\psi(x_1)\,A(x_1)\,\psi(x_1) = \left(\bar\psi A\psi\right)_1 \qquad\qquad (W.318)$$

and can write

$$S^{(2)} = -\frac{q^2}{2}\int\int d^4x_1\,d^4x_2\,T\left[\mathcal{N}\left[(\bar\psi A\psi)_1\right]\mathcal{N}\left[(\bar\psi A\psi)_2\right]\right]. \qquad (W.319)$$

### W.11.1  Applying Wick's Theorem

We keep in mind that only the following contractions exist (see the last section, (W.285)):

$$\psi_\alpha(x)\,\bar\psi_\beta(y) = -\bar\psi_\beta(y)\,\psi_\alpha(x) = iS_{F\alpha\beta}(x - y) \qquad (W.320)$$
$$A^\mu(x)\,A^\nu(y) = iD_F^{\mu\nu}(x - y).$$

---

[127]Since $\Delta(x, y)$ depends on the difference $x - y$, it is sometimes written $\Delta(x - y)$.
[128]Note that $x_1$ and $x_2$ are dummy indices; interchanging them must leave the physics unchanged. The fact that either $x_1$ or $x_2$ can occur first is accounted for by time ordering.

All other contractions vanish as e.g. $\overline{\psi(x)A^\mu(y)} = 0$ or $\overline{\psi(x)\psi(y)} = 0$.

By means of Wick's theorem, we can replace the time ordering in (W.319) by the following sum of normal ordered strings:

$$\mathcal{T}\left[\mathcal{N}\left[(\bar{\psi}A\psi)_1\right]\mathcal{N}\left[(\bar{\psi}A\psi)_2\right]\right] = \mathcal{N}\left[(\bar{\psi}A\psi)_1\,(\bar{\psi}A\psi)_2\right] +$$

$$+\mathcal{N}\left[\overbrace{(\bar{\psi}A\psi)_1\,(\bar{\psi}A\psi)_2}\right] + \mathcal{N}\left[(\bar{\psi}A\psi)_1\,(\bar{\psi}A\psi)_2\right] + \mathcal{N}\left[(\bar{\psi}A\psi)_1\,(\bar{\psi}A\psi)_2\right] +$$

$$+\mathcal{N}\left[(\bar{\psi}A\psi)_1\,(\bar{\psi}A\psi)_2\right] + \mathcal{N}\left[(\bar{\psi}A\psi)_1\,(\bar{\psi}A\psi)_2\right] + \mathcal{N}\left[(\bar{\psi}A\psi)_1\,(\bar{\psi}A\psi)_2\right] +$$

$$+\mathcal{N}\left[(\bar{\psi}A\psi)_1\,(\bar{\psi}A\psi)_2\right].$$

$$\text{(W.321)}$$

As we see, there is one term without, three terms with one, three terms with two, and one term with three contractions.

Remember that a contraction, i.e., a propagator contains no destruction or creation operators. Hence we can move it out from normal ordered strings and from matrix elements:

$$\langle f|\mathcal{N}\left[\overbrace{(\bar{\psi}A\psi)_1\,(\bar{\psi}A\psi)_2}\right]|i\rangle = iD_F^{\mu\nu}(x_1 - x_2)\,\langle f|\mathcal{N}\left[(\bar{\psi}\gamma_\mu\psi)_1\,(\bar{\psi}\gamma_\nu\psi)_2\right]|i\rangle.$$

$$\text{(W.322)}$$

## W.11.2 Physical Interpretation

Without going into details for the present, we give now an overview of the physical meaning of the different terms in (W.321). In the following, electron/positron means electron *or* positron, and electron-positron stands for electron *and* positron.[129] The Feynman diagrams show typical processes.

### W.11.2.1 No Contraction

The term $\mathcal{N}\left[(\bar{\psi}A\psi)_1\,(\bar{\psi}A\psi)_2\right]$ represents two independent processes of first order, as we have discussed them above in section '$S$-matrix, first order'. Since all these processes are not real physical processes, we ignore the term.

---

[129] The considerations also apply to muons and taus and their antiparticles, of course.

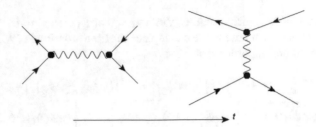

**Fig. W.7** Bhabha scattering type 1 (annihilation) and type 2 (direct scattering)

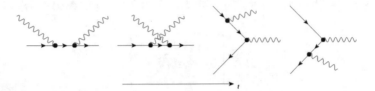

**Fig. W.8** Compton scattering and pair annihilation

### W.11.2.2   One Contraction

**Photon propagator** In this case, we have

$$S^{(2)} = -\frac{q^2}{2} \int \int d^4x_1 \, d^4x_2 \, \mathcal{N}\left[(\bar{\psi}A\psi)_1 (\bar{\psi}A\psi)_2\right]. \tag{W.323}$$

The contraction is given by the Feynman propagator $iD_F^{\mu\nu}(x_1 - x_2)$. Thus, the only creation or destruction operators are found in $\bar{\psi}$ and $\psi$. This means, that the only initial states that can be destroyed are electron/positron states, and the same holds true for the creation in the final states. This type of interaction is called four external lepton interaction, as e.g. scattering of electrons and positrons. As an example, we show Bhabha scattering in Fig. W.7, i.e., scattering of electron and positron. Below, we consider this case in more detail.

A remark concerning the naming. The in- and outgoing particles are called *external particles*, all other particles are called *internal particles*.

**Fermion propagator** There are two terms with a fermion propagator which can shown to be equal. So we have in this case

$$S^{(2)} = -q^2 \int \int d^4x_1 \, d^4x_2 \, \mathcal{N}\left[(\bar{\psi}A\psi)_1 (\bar{\psi}A\psi)_2\right]. \tag{W.324}$$

The contraction is given by the fermion propagator $iS_{F\alpha\beta}(x_1 - x_2)$. The states which can be destroyed or created comprise a photon and a electron/positron. Typical processes are Compton scattering or pair annihilation, see Fig. W.8.

**Fig. W.9** Electron and
positron loop

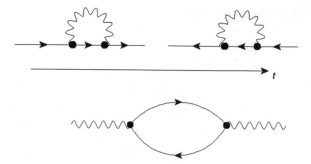

**Fig. W.10** Photon closed
loop

### W.11.2.3   Two Contractions

**One fermion and one photon propagator** There are two such terms in (W.321)
which again can be shown to be equal. So we have in this case

$$S^{(2)} = -q^2 \int\int d^4x_1\, d^4x_2\, \mathcal{N}\left[(\bar{\psi}\!\!\!\!\!A\psi)_1(\bar{\psi}\!\!\!\!\!A\psi)_2\right]. \qquad (W.325)$$

This represents destruction of an electron/positron, followed by the propagation
of both the electron/positron and the photon, and finally the creation of an elec-
tron/positron. These processes are called electron (or positron) closed loop. The
name stems from the pictorial representation, see Fig. W.9. Another name is self-
energy diagrams.

**Two fermion propagators** Here we have

$$S^{(2)} = -\frac{q^2}{2} \int\int d^4x_1\, d^4x_2\, \mathcal{N}\left[(\bar{\psi}\!\!\!\!\!A\psi)_1(\bar{\psi}\!\!\!\!\!A\psi)_2\right]. \qquad (W.326)$$

The two propagators are of the fermion propagator type, $iS_{F\alpha\beta}(x_1 - x_2)$. There is
a real incoming and outgoing photon; these two photons are connected by a virtual
electron-positron pair, see Fig. W.10. The process is called photon closed loop or
vacuum closed loop (since 'in between' the vacuum splits up into a negative and a
positive charged particle).

### W.11.2.4   Three Contractions

This term is given by

**Fig. W.11** Vacuum bubble

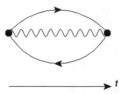

$$S^{(2)} = -\frac{q^2}{2} \int \int \mathrm{d}^4 x_1 \, \mathrm{d}^4 x_2 \, \mathcal{N} \left[ (\bar{\psi} \slashed{A} \psi)_1 (\bar{\psi} \slashed{A} \psi)_2 \right]. \qquad (\text{W.327})$$

There are only propagators, no annihilation and creation operators, i.e., only internal and no external lines. We can visualize the term as a process which starts and ends in the vacuum and splits up 'in between' in a virtual photon and a virtual electron-positron pair. The process is called vacuum bubble, see Fig. W.11.

## W.12   $S$-Matrix, 2. Order, 4 Lepton Scattering

As a worked out example, we consider in this section in some detail the case 'one photon propagator', i.e., four lepton scattering. We determine and discuss explicitly transition amplitudes for two physically different processes, namely Bhabha and Møller scattering. By these examples, we illustrate the one-to-one correspondence between transition amplitudes and Feynman diagrams. Finally, we consider the connection between transition amplitude and the differential scattering cross section.

We start with (W.323)

$$S^{(2)} = -\frac{q^2}{2} \int \int \mathrm{d}^4 x_1 \, \mathrm{d}^4 x_2 \, \mathcal{N} \left[ (\bar{\psi} \slashed{A} \psi)_1 (\bar{\psi} \slashed{A} \psi)_2 \right]. \qquad (\text{W.328})$$

The contraction is given by the Feynman propagator $i D_F^{\mu\nu} (x_1 - x_2)$. Thus, we can write

$$S^{(2)} = -\frac{q^2}{2} \int \int \mathrm{d}^4 x_1 \, \mathrm{d}^4 x_2 \, \mathcal{N} \left[ \left( \bar{\psi} \gamma_\mu \psi \right)_1 \left( \bar{\psi} \gamma_\nu \psi \right)_2 \right] i D_F^{\mu\nu} (x_1 - x_2). \qquad (\text{W.329})$$

As initial and final state we can choose electrons and positrons (of course, myons and tauons and their antiparticles would also be possible), whence the name *four external lepton interaction*. As a worked out example, we consider the case where initial and final states are an electron/positron pair:

$$|i\rangle = \left| e^- \left( \mathbf{p}_1 r_1 \right) \right\rangle \left| e^+ \left( \mathbf{p}_2 r_2 \right) \right\rangle \; ; \; |f\rangle = \left| e^- \left( \mathbf{p}_1' r_1' \right) \right\rangle \left| e^+ \left( \mathbf{p}_2' r_2' \right) \right\rangle. \qquad (\text{W.330})$$

This type of scattering (particle plus antiparticle) is called *Bhabha*[130] *scattering*. After this example, we transfer the results with the necessary modifications to describe the scattering of particle/particle pairs, e.g. two electrons, called *Møller*[131] *scattering*.

The purpose of the following considerations is twofold. First, we want to show step by step how to calculate transition amplitudes which is done for Bhabha scattering. Second, we will see that the Feynman diagrams (together with Feynman rules, see Table W.5) carry the same information as the mathematical formulation. This connection will be applied in case of Møller scattering, where we formulate the Feynman amplitude directly from the Feynman diagrams - and, hopefully, convince the reader that these diagrams are very functional and make life considerably easier.

## W.12.1 Bhabha Scattering, $e^+ e^- \to e^+ e^-$

We now start the calculation of $S^{(2)}$ for this process.[132] The decomposition of $\mathcal{N}\left[\left(\bar{\psi}\gamma_\mu\psi\right)_1 \left(\bar{\psi}\gamma_\nu\psi\right)_2\right]$ in (W.329) with respect to creation and deletion operators gives 16 terms. But we do not need to write them all down since only those will contribute which destroy incoming and create outgoing pairs of $e^-/e^+$.

Note that there are two possibilities in which in- and outgoing particles have the same individual momenta and spins. Either at e.g. $x_2$ both the $e^-$ and the $e^+$ are destroyed, a photon runs from $x_2$ to $x_1$, and at $x_1$ both the $e^-$ and the $e^+$ are created (plus the process with reversed roles of $x_2$ and $x_1$). In other words: we have first a pair annihilation producing a (virtual) photon, followed by pair production. We call this process Bhabha scattering type 1 or annihilation scattering; the transition amplitude is written as $S^{(2)}_{B1}$. Or the incoming $e^-$ is destroyed at e.g. $x_2$ and the outgoing $e^-$ is created at $x_2$; a photon starts at $x_2$ and ends in $x_1$ where the incoming $e^+$ is destroyed and again created (plus the process with reversed roles of $x_2$ and $x_1$). In other words: the two particles are directly scattered whereby the interaction is mediated by a (virtual) photon. We call this process Bhabha scattering type 2 or direct scattering; the transition amplitude is written as $S^{(2)}_{B2}$. The Feynman diagrams with the corresponding momenta are found in Fig. W.12. We repeat that the process with reserved role of $x_2$ and $x_1$ is also possible. Time ordering ensures the right chronological order of $x_1$ and $x_2$.

Note that we cannot distinguish the two processes by measurement. Indeed, we can only measure the incoming and outgoing particles, and hence we do not know which one of the two processes may have occurred. Consequently, quantum mechanics says

---

[130]Bhabha, Homi Jehangir; 1909–1966, Indian physicist.

[131]Møller, Christian; 1904–1980, Danish physicist.

[132]Note that Bhabha scattering is not only interesting from a theoretical point of view. Luminosity, i.e., the number of collisions per time and area, is a measure for the performance of a particle accelerator. In many colliders, luminosity is determined using Bhabha scattering at small angles.

**Fig. W.12** Bhabha scattering. Left annihilation (type 1), right direct scattering (type 2)

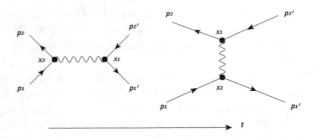

that we have to add the two amplitudes which means we have interference effects between the two possibilities.

In the discussion of the physical processes described by the first order $S$-matrix $S^{(1)}$ we have seen that one-vertex processes are not 'real'. However, they are the building blocks of real higher order processes. For instance, Bhabha scattering type 1 can be interpreted as composed of first order pair annihilation ($e^+ + e^- \rightarrow \gamma$), followed by first order pair production ($\gamma \rightarrow e^+ + e^-$). Incoming and outgoing particles are real whereas the photon is virtual.

### W.12.1.1   Bhabha Scattering Type 1

Warning: Be prepared for a lot of algebra! Calculating a transition amplitude step by step takes time and place. All the nicer then to work with Feynman diagrams.

For $x_2$ the only combination which destroys $\left|e^- (\mathbf{p}_1 r_1)\right\rangle$ and $\left|e^+ (\mathbf{p}_2 r_2)\right\rangle$ is given by $\left(\bar{\psi}^d \gamma_\nu \psi^d\right)_2$. The propagator creates a virtual photon at $x_2$ and propagates it to $x_1$ where it is destroyed. Then the outgoing pair of $e^-/e^+$ is created at $x_1$ by $\left(\bar{\psi}^c \gamma_\mu \psi^c\right)_1$.

The same consideration holds with reversed roles of $x_1$ and $x_2$. Thus, we have for $S_{B1}^{(2)} = \left\langle e^- \left(\mathbf{p}_1' r_1'\right) e^+ \left(\mathbf{p}_2' r_2'\right)\right| S^{(2)} \left|e^- (\mathbf{p}_1 r_1) e^+ (\mathbf{p}_2 r_2)\right\rangle$ the expression

$$S_{B1}^{(2)} = -\frac{q^2}{2} \left\langle e^- \left(\mathbf{p}_1' r_1'\right) e^+ \left(\mathbf{p}_2' r_2'\right)\right| \int\int d^4 x_1 \, d^4 x_2 \, iD_F^{\mu\nu}(x_1 - x_2) \cdot$$
$$\cdot \left\{ \mathcal{N}\left[\left(\bar{\psi}^c \gamma_\mu \psi^c\right)_1 \left(\bar{\psi}^d \gamma_\nu \psi^d\right)_2\right] + \mathcal{N}\left[\left(\bar{\psi}^d \gamma_\nu \psi^d\right)_1 \left(\bar{\psi}^c \gamma_\mu \psi^c\right)_2\right] \right\} \left|e^- (\mathbf{p}_1 r_1) e^+ (\mathbf{p}_2 r_2)\right\rangle.$$
$$\text{(W.331)}$$

**Normal ordering** The first term in the curled brackets is already normal ordered. In the second term, we switch $\psi^d$ first with $\bar{\psi}^c$ which gives a minus sign, and after that with $\psi^c$ which introduces a second minus sign; thus, we have no change altogether. The same holds for switching $\bar{\psi}^d$, and the result reads

$$S_{B1}^{(2)} = -\frac{q^2}{2} \left\langle e^- \left(\mathbf{p}_1' r_1'\right) e^+ \left(\mathbf{p}_2' r_2'\right)\right| \int\int d^4 x_1 \, d^4 x_2 \cdot$$
$$\cdot \left\{ \left(\bar{\psi}^c \gamma_\mu \psi^c\right)_1 \left(\bar{\psi}^d \gamma_\nu \psi^d\right)_2 + \left(\bar{\psi}^c \gamma_\mu \psi^c\right)_2 \left(\bar{\psi}^d \gamma_\nu \psi^d\right)_1 \right\} iD_F^{\mu\nu}(x_1 - x_2) \left|e^- (\mathbf{p}_1 r_1) e^+ (\mathbf{p}_2 r_2)\right\rangle.$$
$$\text{(W.332)}$$

For the second term, we switch the (dummy) integration variables[133] and arrive at

---

[133] Remember $D_F^{\mu\nu}(x_1 - x_2) = D_F^{\mu\nu}(x_2 - x_1)$.

$$S_{B1}^{(2)} = -q^2 \langle e^- \left(\mathbf{p}_1' r_1'\right) e^+ \left(\mathbf{p}_2' r_2'\right) | \int \int d^4x_1 \, d^4x_2 \cdot$$
$$\cdot \left(\bar{\psi}^c \gamma_\mu \psi^c\right)_1 i D_F^{\mu\nu} (x_1 - x_2) \left(\bar{\psi}^d \gamma_\nu \psi^d\right)_2 | e^- \left(\mathbf{p}_1 r_1\right) e^+ \left(\mathbf{p}_2 r_2\right) \rangle. \tag{W.333}$$

Here we have placed the propagator (which is just a $c$−number) between the field operators in order to display the physical process.

**Inserting the field operators, evaluation of the brackets** We now insert the field operators (discrete version) given by (see section 'Quantization of free fields, Dirac')[134]

$$\psi(x) = \sum_{\mathbf{p},r} \sqrt{\tfrac{m}{VE_\mathbf{p}}} \left(b_r(\mathbf{p}) u_r(\mathbf{p}) e^{-ipx} + d_r^\dagger(\mathbf{p}) w_r(\mathbf{p}) e^{ipx}\right)$$
$$\bar{\psi}(x) = \sum_{\mathbf{p},r} \sqrt{\tfrac{m}{VE_\mathbf{p}}} \left(b_r^\dagger(\mathbf{p}) \bar{u}_r(\mathbf{p}) e^{ipx} + d_r(\mathbf{p}) \bar{w}_r(\mathbf{p}) e^{-ipx}\right). \tag{W.334}$$

As we know from the discussion of the first order $S$-matrix (see section '$S$-matrix, first order'), applying the field operator $\left(\bar{\psi}^d \gamma_\nu \psi^d\right)_2$ to the initial state picks out the elements with the same quantum numbers,[135] i.e.,

$$\left(\bar{\psi}^d \gamma_\nu \psi^d\right)_2 | e^- \left(\mathbf{p}_1 r_1\right) e^+ \left(\mathbf{p}_2 r_2\right) \rangle = \sqrt{\tfrac{m}{VE_{\mathbf{p}_2}}} \bar{w}_{r_2}(\mathbf{p}_2) e^{-ip_2 x_2} \gamma_\nu \sqrt{\tfrac{m}{VE_{\mathbf{p}_1}}} u_{r_1}(\mathbf{p}_1) e^{-ip_1 x_2} |0\rangle. \tag{W.335}$$

In the same way, we have

$$\langle e^- \left(\mathbf{p}_1' r_1'\right) e^+ \left(\mathbf{p}_2' r_2'\right) | \left(\bar{\psi}^c \gamma_\mu \psi^c\right)_1 = \langle 0| \sqrt{\tfrac{m}{VE_{\mathbf{p}_1'}}} \bar{u}_{r_1'}(\mathbf{p}_1') e^{ip_1' x_1} \gamma_\mu \sqrt{\tfrac{m}{VE_{\mathbf{p}_2'}}} w_{r_2'}(\mathbf{p}_2') e^{ip_2' x_1}. \tag{W.336}$$

Combining the results yields

$$S_{B1}^{(2)} = -q^2 \langle 0| \int \int d^4x_1 \, d^4x_2 \sqrt{\tfrac{m}{VE_{\mathbf{p}_1'}}} \bar{u}_{r_1'}(\mathbf{p}_1') e^{ip_1' x_1} \gamma_\mu \sqrt{\tfrac{m}{VE_{\mathbf{p}_2'}}} \cdot$$
$$\cdot w_{r_2'}(\mathbf{p}_2') e^{ip_2' x_1} i D_F^{\mu\nu}(x_1 - x_2) \sqrt{\tfrac{m}{VE_{\mathbf{p}_2}}} \bar{w}_{r_2}(\mathbf{p}_2) e^{-ip_2 x_2} \gamma_\nu \sqrt{\tfrac{m}{VE_{\mathbf{p}_1}}} u_{r_1}(\mathbf{p}_1) e^{-ip_1 x_2} |0\rangle. \tag{W.337}$$

As we see, the term $\Theta$ between the bra-ket $\langle 0| \Theta |0\rangle$ contains neither creation nor destruction operators. Thus, we can write $\langle 0| \Theta |0\rangle = \Theta \langle 0| 0\rangle = \Theta$ due to $\langle 0| 0\rangle = 1$. It follows

$$S_{B1}^{(2)} = -q^2 \int \int d^4x_1 \, d^4x_2 \sqrt{\tfrac{m}{VE_{\mathbf{p}_1'}}} \sqrt{\tfrac{m}{VE_{\mathbf{p}_2'}}} \sqrt{\tfrac{m}{VE_{\mathbf{p}_2}}} \sqrt{\tfrac{m}{VE_{\mathbf{p}_1}}} \cdot$$
$$\cdot \bar{u}_{r_1'}(\mathbf{p}_1') \gamma_\mu w_{r_2'}(\mathbf{p}_2') \bar{w}_{r_2}(\mathbf{p}_2) \gamma_\nu u_{r_1}(\mathbf{p}_1) \cdot i D_F^{\mu\nu}(x_1 - x_2) e^{ip_1' x_1} e^{ip_2' x_1} e^{-ip_2 x_2} e^{-ip_1 x_2}. \tag{W.338}$$

**Calculating the $x$-integral** What remains is to calculate the $x_1/x_2$- integrals $X$

---

[134]For the continuous version replace the sum by an integral and $V$ by $(2\pi)^3$.

[135]Remember that $b_r^\dagger(\mathbf{p})$ / $b_r(\mathbf{p})$ creates /deletes an electron and $d_r^\dagger(\mathbf{p})$ / $d_r(\mathbf{p})$ creates /deletes a positron.

$$X = \int \int d^4x_1 \, d^4x_2 \, D_F^{\mu\nu} \, (x_1 - x_2) \, e^{iP_1'x_1} e^{-iPx_2} \;\; ; P = p_1 + p_2 \; ; \; P' = p_1' + p_2'.$$

(W.339)

The result reads (see exercises)

$$X = D_F^{\mu\nu} \, (P) \, (2\pi)^4 \, \delta^{(4)} \left(P - P'\right)$$

(W.340)

where $D_F^{\mu\nu} \, (P)$ is the Fourier transform of the propagator, i.e.,

$$D_F^{\mu\nu} \, (P) = -g^{\mu\nu} \frac{1}{P^2 + i\varepsilon}.$$

(W.341)

**Final result for $S_{B1}^{(2)}$** Thus, we have for the $S$-matrix the result

$$S_{B1}^{(2)} = \sqrt{\frac{m}{VE_{\mathbf{p}_1'}}} \sqrt{\frac{m}{VE_{\mathbf{p}_2'}}} \sqrt{\frac{m}{VE_{\mathbf{p}_2}}} \sqrt{\frac{m}{VE_{\mathbf{p}_1}}} \cdot (2\pi)^4 \, \delta^{(4)} \left(p_1 + p_2 - p_1' - p_2'\right) \cdot$$
$$\cdot \left\{ (-q^2) \, \bar{u}_{r_1'} \left(\mathbf{p}_1'\right) \gamma_\mu w_{r_2'} \left(\mathbf{p}_2'\right) i D_F^{\mu\nu} \left(p_1 + p_2\right) \bar{w}_{r_2} \left(\mathbf{p}_2\right) \gamma_\nu u_{r_1} \left(\mathbf{p}_1\right) \right\}.$$

(W.342)

The term in curly brackets is called *Feynman amplitude* and noted as $\mathcal{M}_{B1}^{(2)}$. Thus, we can write the $S$-matrix for Bhabha scattering type 1 as

$$S_{B1}^{(2)} = \left( \overset{\text{external fermions}}{\underset{\mathbf{p}}{\prod}} \sqrt{\frac{m}{VE_{\mathbf{p}}}} \right) (2\pi)^4 \, \delta^{(4)} \left(p_2 + p_1 - \left(p_2' + p_1'\right)\right) \mathcal{M}_{B1}^{(2)}$$

(W.343)

where

$$\mathcal{M}_{B1}^{(2)} = \left(-q^2\right) \bar{u}_{r_1'} \left(\mathbf{p}_1'\right) \gamma_\mu w_{r_2'} \left(\mathbf{p}_2'\right) i D_F^{\mu\nu} \left(p_1 + p_2\right) \bar{w}_{r_2} \left(\mathbf{p}_2\right) \gamma_\nu u_{r_1} \left(\mathbf{p}_1\right).$$

(W.344)

Let us concentrate on this Feynman amplitude. From right to left, i.e., in the direction of time, we can 'read' $\mathcal{M}_{B1}^{(2)}$ ) as follows: $u_{r_1} \left(\mathbf{p}_1\right)$ destroys an electron with quantum numbers $(r_1, \mathbf{p}_1)$, $\bar{w}_{r_2} \left(\mathbf{p}_2\right)$ destroys a positron $(r_2, \mathbf{p}_2)$. A virtual photon with $k = p_1 + p_2$ is the link to $w_{r_2'} \left(\mathbf{p}_2'\right)$, creating a positron with $(r_2', \mathbf{p}_2')$, and $\bar{u}_{r_1'} \left(\mathbf{p}_1'\right)$, creating an electron with $(r_1', \mathbf{p}_1')$. This order is precisely reflected in the corresponding Feynman diagram (W.12). Following the arrows, we have: At the left, we have an incoming electron $u_{r_1} \left(\mathbf{p}_1\right)$, then a vertex which contributes a factor $-iq\gamma_\nu$, followed by an incoming positron $\bar{w}_{r_2} \left(\mathbf{p}_2\right)$. A virtual photon with momentum $p_1 + p_2$ (i.e., $i D_F^{\mu\nu} \left(p_1 + p_2\right)$) is the connection to the outgoing channel at the right. Here we have a positron with $w_{r_2'} \left(\mathbf{p}_2'\right)$, a vertex with $-iq\gamma_\mu$ and an electron with $\bar{u}_{r_1'} \left(\mathbf{p}_1'\right)$.

In other words, we could have saved the last two pages of calculation, simply by reading off the information provided by the Feynman diagram. We see that the spinor factors ($\gamma$ matrices, fermionic propagators, 4-spinors) in (W.344) occur in exactly same order as following in Fig. W.12 the fermion lines in the direction of the arrows through the vertices. Indeed, between Feynman amplitude and Feynman diagram there is a one-to-one relation which evidently can save a lot of computing.

To test these statements with another (and last) example, we consider now Bhabha scattering type 2.

### W.12.1.2   Bhabha Scattering Type 2

Arguing along the same lines as for $S_{B1}^{(2)}$, given by (W.331), we can write for $S_{B2}^{(2)}$

$$S_{B2}^{(2)} = -\frac{q^2}{2} \langle e^- \left(\mathbf{p}_1' r_1'\right), e^+ \left(\mathbf{p}_2' r_2'\right)| \int \int d^4 x_1 \, d^4 x_2 \cdot$$
$$\cdot \left\{ \mathcal{N}\left[\left(\bar{\psi}^d \gamma_\mu \psi^c\right)_1 \left(\bar{\psi}^c \gamma_\nu \psi^d\right)_2\right] + \mathcal{N}\left[\left(\bar{\psi}^c \gamma_\mu \psi^d\right)_1 \left(\bar{\psi}^d \gamma_\nu \psi^c\right)_2\right]\right\} i D_F^{\mu\nu} \left(x_1 - x_2\right) |e^- \left(\mathbf{p}_1 r_1\right), e^+ \left(\mathbf{p}_2 r_2\right)\rangle.$$

$$\text{(W.345)}$$

**Normal ordering** Switching the integration variables $x_1$ and $x_2$ reveals that the two expressions in the curly brackets are identical. However, normal ordering these terms is a little bit more complicated as in the type 1 case. This is due to the mixed character of the brackets $\left(\bar{\psi}^d \gamma_\mu \psi^c\right)_1$ and $\left(\bar{\psi}^c \gamma_\nu \psi^d\right)_2$, where deletion and creation operator are found in one bracket. Thus, we can not simply interchange them but have to take into account the spinor indices. This means that we write[136]

$$S_{B2}^{(2)} = -q^2 \langle e^- \left(\mathbf{p}_1' r_1'\right), e^+ \left(\mathbf{p}_2' r_2'\right)| \int \int d^4 x_1 \, d^4 x_2 \cdot$$
$$\cdot \left\{ \mathcal{N}\left[\left(\bar{\psi}_\alpha^d \gamma_{\mu\alpha\beta} \psi_\beta^c\right)_1 \left(\bar{\psi}_\delta^c \gamma_{\nu\delta\eta} \psi_\eta^d\right)_2\right]\right\} i D_F^{\mu\nu} \left(x_1 - x_2\right) |e^- \left(\mathbf{p}_1 r_1\right), e^+ \left(\mathbf{p}_2 r_2\right)\rangle$$

$$\text{(W.346)}$$

where we use summation convention and $\gamma_{\mu\alpha\beta}$ is the element $(\alpha\beta)$ of $\gamma_\mu$. Now we can normal order:

$$\mathcal{N}\left[\left(\bar{\psi}_\alpha^d \gamma_{\mu\alpha\beta} \psi_\beta^c\right)_1 \left(\bar{\psi}_\delta^c \gamma_{\nu\delta\eta} \psi_\eta^d\right)_2\right] = \mathcal{N}\left[\left(\bar{\psi}_\delta^c\right)_2 \left(\bar{\psi}_\alpha^d \gamma_{\mu\alpha\beta} \psi_\beta^c\right)_1 \left(\gamma_{\nu\delta\eta} \psi_\eta^d\right)_2\right] =$$
$$= \mathcal{N}\left[-\left(\bar{\psi}_\delta^c\right)_2 \left(\gamma_{\mu\alpha\beta} \psi_\beta^c\right)_1 \left(\bar{\psi}_\alpha^d\right)_1 \left(\gamma_{\nu\delta\eta} \psi_\eta^d\right)_2\right].$$

$$\text{(W.347)}$$

The first ordering leaves the sign unchanged since the arguments differ ($x_1$ and $x_2$), the second ordering changes the sign, since the arguments are equal (both $x_1$). In this way we arrive at

$$S_{B2}^{(2)} = q^2 \langle e^- \left(\mathbf{p}_1' r_1'\right), e^+ \left(\mathbf{p}_2' r_2'\right)| \int \int d^4 x_1 \, d^4 x_2 \cdot$$
$$\cdot \left\{\left(\bar{\psi}_\delta^c\right)_2 \gamma_{\mu\alpha\beta} \left(\psi_\beta^c\right)_1 \left(\bar{\psi}_\alpha^d\right)_1 \gamma_{\nu\delta\eta} \left(\psi_\eta^d\right)_2\right\} i D_F^{\mu\nu} \left(x_1 - x_2\right) |e^- \left(\mathbf{p}_1 r_1\right), e^+ \left(\mathbf{p}_2 r_2\right)\rangle.$$

$$\text{(W.348)}$$

Note we have the same normal order as in type 1 scattering: the order from the right side moving leftward is $e^-$ deletion, $e^+$ deletion, $e^+$ creation, and $e^-$ creation, as it should be.

**Inserting the field operators, evaluation of the brackets** Inserting the field operators leads to

---

[136]We use the Einstein summation convention.

$$S_{B2}^{(2)} = q^2 \left\langle e^- \left(\mathbf{p}_1' r_1'\right), e^+ \left(\mathbf{p}_2' r_2'\right)\right| \int\int \mathrm{d}^4 x_1 \, \mathrm{d}^4 x_2 \cdot$$
$$\cdot \left\{ \left(\bar{\psi}_\delta^c\right)_2 \gamma_{\mu\alpha\beta} \left(\psi_\beta^c\right)_1 \left(\bar{\psi}_\alpha^d\right)_1 \gamma_{\nu\delta\eta} \left(\psi_\eta^d\right)_2 \right\} i D_F^{\mu\nu} (x_1 - x_2) \left| e^- (\mathbf{p}_1 r_1), e^+ (\mathbf{p}_2 r_2)\right\rangle.$$
$$\tag{W.349}$$

For a short auxiliary calculation, we neglect for a moment the integrals and the propagator,[137] thus defining an short-lived variable $K$ by

$$K = \left\langle e^- \left(\mathbf{p}_1' r_1'\right), e^+ \left(\mathbf{p}_2' r_2'\right)\right| \left\{ \left(\bar{\psi}_\delta^c\right)_2 \gamma_{\mu\alpha\beta} \left(\psi_\beta^c\right)_1 \left(\bar{\psi}_\alpha^d\right)_1 \gamma_{\nu\delta\eta} \left(\psi_\eta^d\right)_2 \right\} \left| e^- (\mathbf{p}_1 r_1), e^+ (\mathbf{p}_2 r_2)\right\rangle.$$
$$\tag{W.350}$$

Taking into account the above mentioned 'matching rules' (cf. section '$S$-matrix, first order'), the evaluation of this term gives

$$K = \langle 0| \left\{ \begin{array}{l} \sqrt{\dfrac{m}{VE_{\mathbf{p}_1'}}} \bar{u}_{r_1'} \left(\mathbf{p}_1'\right) e_2^{ip_1' x_2} \gamma_{\mu\alpha\beta} \sqrt{\dfrac{m}{VE_{\mathbf{p}_2'}}} w_{r_2'} \left(\mathbf{p}_2'\right) e^{ip_2' x_1} \cdot \\ \cdot \sqrt{\dfrac{m}{VE_{\mathbf{p}_2}}} \bar{w}_{r_2\alpha} \left(\mathbf{p}_2\right) e_1^{-ip_2 x_1} \gamma_{\nu\delta\eta} \sqrt{\dfrac{m}{VE_{\mathbf{p}_1}}} u_{r_1\eta} \left(\mathbf{p}_1\right) e^{-ip_1 x_2} \end{array} \right\} |0\rangle. \tag{W.351}$$

Again, the term inside the bra–ket $\langle 0| \, \Theta \, |0\rangle$ contains neither creation nor destruction operators whence due to $\langle 0 \, |0\rangle = 1$ follows $\langle 0| \, \Theta \, |0\rangle = \Theta \, \langle 0 \, |0\rangle = \Theta$.

Now we rearrange the indexed terms of this expression. Note that $\bar{u}_{r_{1\delta}'}$, $\gamma_{\mu\alpha\beta}$ etc. are single elements of the vectors and matrices, i.e., scalars which can be moved freely. We have[138]

$$\bar{u}_{r_{1\delta}'} \left(\mathbf{p}_1'\right) \gamma_{\mu\alpha\beta} w_{r_2'\beta} \left(\mathbf{p}_2'\right) \bar{w}_{r_2\alpha} \left(\mathbf{p}_2\right) \gamma_{\nu\delta\eta} u_{r_1\eta} \left(\mathbf{p}_1\right) =$$
$$= \bar{w}_{r_2\alpha} \left(\mathbf{p}_2\right) \gamma_{\mu\alpha\beta} w_{r_2'\beta} \left(\mathbf{p}_2'\right) \bar{u}_{r_{1\delta}'} \left(\mathbf{p}_1'\right) \gamma_{\nu\delta\eta} u_{r_1\eta} \left(\mathbf{p}_1\right) = \bar{w}_{r_2} \left(\mathbf{p}_2\right) \gamma_\mu w_{r_2'} \left(\mathbf{p}_2'\right) \cdot \bar{u}_{r_1'} \left(\mathbf{p}_1'\right) \gamma_\nu u_{r_1} \left(\mathbf{p}_1\right).$$
$$\tag{W.352}$$

Inserting these partial results into (W.349) yields

$$S_{B2}^{(2)} = q^2 \int\int \mathrm{d}^4 x_1 \, \mathrm{d}^4 x_2 \cdot \sqrt{\dfrac{m}{VE_{\mathbf{p}_1'}}} \sqrt{\dfrac{m}{VE_{\mathbf{p}_2'}}} \sqrt{\dfrac{m}{VE_{\mathbf{p}_2}}} \sqrt{\dfrac{m}{VE_{\mathbf{p}_1}}} \cdot$$
$$\cdot \left\{ \bar{w}_{r_2} \left(\mathbf{p}_2\right) \gamma_\mu w_{r_2'} \left(\mathbf{p}_2'\right) \cdot \bar{u}_{r_1'} \left(\mathbf{p}_1'\right) \gamma_\nu u_{r_1} \left(\mathbf{p}_1\right) \right\} e^{i(p_2' - p_2) x_1} e^{i(p_1' - p_1) x_2} i D_F^{\mu\nu} (x_1 - x_2).$$
$$\tag{W.353}$$

**Calculating the $x$-integral, final result for $S_{B2}^{(2)}$** Concerning the $x$-integration we can use the result just derived above, and obtain

$$\int\int \mathrm{d}^4 x_1 \, \mathrm{d}^4 x_2 \, e^{i(p_2' - p_2) x_1} e^{i(p_1' - p_1) x_2} i D_F^{\mu\nu} (x_1 - x_2) =$$
$$= i D_F^{\mu\nu} \left(p_2' - p_2\right) (2\pi)^4 \, \delta^{(4)} \left(p_1' - p_1 + \left(p_2' - p_2\right)\right). \tag{W.354}$$

Thus, we arrive finally at

$$S_{B2}^{(2)} = q^2 \sqrt{\dfrac{m}{VE_{\mathbf{p}_1'}}} \sqrt{\dfrac{m}{VE_{\mathbf{p}_2'}}} \sqrt{\dfrac{m}{VE_{\mathbf{p}_2}}} \sqrt{\dfrac{m}{VE_{\mathbf{p}_1}}} (2\pi)^4 \, \delta^{(4)} \left(p_1' + p_2' - (p_1 + p_2)\right) \cdot$$
$$\cdot \left\{ \bar{w}_{r_2} \left(\mathbf{p}_2\right) \gamma_\mu w_{r_2'} \left(\mathbf{p}_2'\right) i D_F^{\mu\nu} \left(p_2' - p_2\right) \bar{u}_{r_1'} \left(\mathbf{p}_1'\right) \gamma_\nu u_{r_1} \left(\mathbf{p}_1\right) \right\} \tag{W.355}$$

---

[137]Remember that the propagator does not contain any deletion or creation operators.
[138]Remind the rules of matrix multiplication, $xM y = \sum x_i M_{ij} y_j$.

which we write as[139]

$$S_{B2}^{(2)} = \left( \prod_{\mathbf{p}}^{\text{ext. ferm.}} \sqrt{\frac{m}{VE_{\mathbf{p}}}} \right) (2\pi)^4 \, \delta^{(4)} \left( p_2 + p_1 - \left( p_2' + p_1' \right) \right) \mathcal{M}_{B2}^{(2)} \qquad (\text{W.356})$$

with

$$\mathcal{M}_{B2}^{(2)} = q^2 \bar{w}_{r_2} \left( \mathbf{p}_2 \right) \gamma_\mu w_{r_2'} \left( \mathbf{p}_2' \right) i D_F^{\mu\nu} \left( p_2' - p_2 \right) \bar{u}_{r_1'} \left( \mathbf{p}_1' \right) \gamma_\nu u_{r_1} \left( \mathbf{p}_1 \right). \qquad (\text{W.357})$$

Again, the mathematical formulation (W.357) (from right to left) and the diagram (W.12) (in direction of the arrows) provide the same information: $u_{r_1} \left( \mathbf{p}_1 \right)$ describes an incoming electron with quantum numbers $(r_1, \mathbf{p}_1)$, the vertex contributes $-iq\gamma_\nu$, the term $\bar{u}_{r_1'} \left( \mathbf{p}_1' \right)$ stands for an outgoing electron with $(r_1', \mathbf{p}_1')$. A virtual photon with $k = p_2' - p_2$ is the link to $w_{r_2'} \left( \mathbf{p}_2' \right)$, creating a positron with $(r_2', \mathbf{p}_2')$, followed by the vertex $-iq\gamma_\mu$, and finally $\bar{w}_{r_2} \left( \mathbf{p}_2 \right)$, annihilating a positron $(r_2, \mathbf{p}_2)$.

### W.12.1.3   Total Bhabha Scattering

Now we can add up the two contributions for Bhaba scattering. Type 1 of (W.343) and (W.344), and type 2 of (W.356) and (W.357). The complete $S$-matrix element for the 2nd order ($n = 2$) Bhabha scattering is

$$S_{\text{Bhabha}}^{(2)} = S_{B1}^{(2)} + S_{B2}^{(2)} = \left( \prod_{\mathbf{p}}^{\text{ext. ferm.}} \sqrt{\frac{m}{VE_{\mathbf{p}}}} \right) (2\pi)^4 \, \delta^{(4)} \left( p_2 + p_1 - \left( p_2' + p_1' \right) \right) \mathcal{M}_{\text{Bhabha}}^{(2)}$$

$$(\text{W.358})$$

with $\mathcal{M}_{\text{Bhabha}}^{(2)} = \mathcal{M}_{B1}^{(2)} + \mathcal{M}_{B2}^{(2)}$ and

$$\mathcal{M}_{B1}^{(2)} = \left( -q^2 \right) \bar{u}_{r_1'} \left( \mathbf{p}_1' \right) \gamma_\mu w_{r_2'} \left( \mathbf{p}_2' \right) i D_F^{\mu\nu} \left( p_1 + p_2 \right) \bar{w}_{r_2} \left( \mathbf{p}_2 \right) \gamma_\nu u_{r_1} \left( \mathbf{p}_1 \right) \; ; \text{annihilation scattering}$$
$$\mathcal{M}_{B2}^{(2)} = \left( q^2 \right) \bar{w}_{r_2} \left( \mathbf{p}_2 \right) \gamma_\mu w_{r_2'} \left( \mathbf{p}_2' \right) i D_F^{\mu\nu} \left( p_2' - p_2 \right) \bar{u}_{r_1'} \left( \mathbf{p}_1' \right) \gamma_\nu u_{r_1} \left( \mathbf{p}_1 \right) \; ; \text{direct scattering.}$$
$$(\text{W.359})$$

We see that the real work in QED is in the calculation of the Feynman amplitude – the mass or normalization factors as well as the conservation of energy and momentum (i.e., the delta function) are easily written down. Thus, it is a great relief that this centerpiece of QED (or QFT) can be done by means of the easy to read Feynman diagrams.

By the way: In general, only the squared Feynman amplitude $|\mathcal{M}|^2$ occurs in applications. It follows that e.g. the sign of $q$ is irrelevant, but the relative sign between the subamplitudes matters, in this case $\mathcal{M}_{B1}^{(2)}$ and $\mathcal{M}_{B2}^{(2)}$.

---

[139]The formulation $\left( \prod_{\mathbf{p}}^{\text{ext. ferm.}} \sqrt{\frac{m}{VE_{\mathbf{p}}}} \right)$ instead of $\sqrt{\frac{m}{VE_{\mathbf{p}_1'}}} \sqrt{\frac{m}{VE_{\mathbf{p}_2'}}} \sqrt{\frac{m}{VE_{\mathbf{p}_2}}} \sqrt{\frac{m}{VE_{\mathbf{p}_1}}}$ allows also for muons, tauons and so on.

**Fig. W.13** Moller scattering, type 1 (exchange scattering) and type 2 (direct scattering)

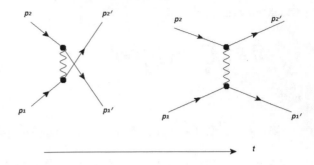

## W.12.2 Møller Scattering, $e^- e^- \rightarrow e^- e^-$

Møller scattering is defined as electron-electron scattering. We will not run through the complete calculation as for Bhabha scattering. Again, we have two types of scattering which are characterized by the following terms (Fig. W.13):

$$
\begin{aligned}
\text{exchange scattering, type 1} \quad & \left(\bar{\psi}_{1'}^c \gamma_\mu \psi_2^d\right)_1 \left(\bar{\psi}_{2'}^c \gamma_\nu \psi_1^d\right)_2 + \left(\bar{\psi}_{1'}^c \gamma_\mu \psi_2^d\right)_2 \left(\bar{\psi}_{2'}^c \gamma_\nu \psi_1^d\right)_1 \\
\text{direct scattering, type 2} \quad & \left(\bar{\psi}_{2'}^c \gamma_\mu \psi_2^d\right)_1 \left(\bar{\psi}_{1'}^c \gamma_\nu \psi_1^d\right)_2 + \left(\bar{\psi}_{2'}^c \gamma_\mu \psi_2^d\right)_2 \left(\bar{\psi}_{1'}^c \gamma_\nu \psi_1^d\right)_1
\end{aligned}
$$
$$\text{(W.360)}$$

For type 1, we have a virtual photon with $k = p_1 - p_2'$ (or equivalently $k = p_2 - p_1'$), for type 2 a photon with $k = p_1 - p_1'$ (or $k = p_2 - p_2'$). For better identification, the field operators in (W.360) are indexed in the same manner as the momenta.

This time, we do not perform the calculations as we have done for Bhabha scattering, but transform the diagrams directly into the transition amplitude. In a first step, we formulate the conservation of the 4-momentum and the mass and volume factors. This gives

$$
S_{\text{Møller}}^{(2)} = (2\pi)^4 \, \delta^{(4)} \left(p_2 + p_1 - (p_2' + p_1')\right) \left(\prod_{\mathbf{p}}^{\text{ext. ferm.}} \sqrt{\frac{m}{VE_{\mathbf{p}}}}\right) \mathcal{M}_{\text{Møller}}^{(2)} \quad \text{(W.361)}
$$

with $\mathcal{M}_{\text{Møller}}^{(2)} = \mathcal{M}_{M1}^{(2)} + \mathcal{M}_{M2}^{(2)}$ where $\mathcal{M}_{M1}^{(2)}$ and $\mathcal{M}_{M2}^{(2)}$ are the Feynman amplitudes for the two processes. As explained above, we have to add the amplitudes, not their squares, because we can not know which one of the two possibilities may have occurred.

The amplitudes themselves are read off from the diagrams. We repeat that for comparing the formulas with the diagrams, one has to read the formulas from right to left and the diagrams in direction of the arrows. For type 1, an incoming electron with $(\mathbf{p}_2, r_2)$ is switched into an outgoing electron with $(\mathbf{p}_1', r_1')$ and the other one with $(\mathbf{p}_1, r_1)$ into $(\mathbf{p}_2', r_2')$; for the virtual photon we have $k = p_1 - p_2'$. For type 2, there is an analogous formulation. Incorporating the vertices by $-iq\gamma_\mu$, we arrive at the Feynman amplitudes for Møller scattering:

$$\mathcal{M}_{M1}^{(2)} = (q^2)\,\bar{u}_{r_2'}\,(\mathbf{p}_2')\,\gamma_\mu u_{r_1}\,(\mathbf{p}_1)\,i D_F^{\mu\nu}\,(p_2 - p_1')\,\bar{u}_{r_1'}\,(\mathbf{p}_1')\,\gamma_\nu u_{r_2}\,(\mathbf{p}_2) \quad;\text{exchange scattering}$$
$$\mathcal{M}_{M2}^{(2)} = (-q^2)\,\bar{u}_{r_1'}\,(\mathbf{p}_1')\,\gamma_\mu u_{r_1}\,(\mathbf{p}_1)\,i D_F^{\mu\nu}\,(p_2 - p_2')\,\bar{u}_{r_2'}\,(\mathbf{p}_2')\,\gamma_\nu u_{r_2}\,(\mathbf{p}_2) \quad;\text{direct scattering.}$$

$$(W.362)$$

## W.12.3  Scattering Cross Section and Feynman Amplitude

In this way the Feynman amplitudes can be read directly from the corresponding diagrams. In principle, this holds for all processes in Figs. W.7, W.8, W.9, W.10 and W.11. However, for the diagrams with loops there is in addition a special feature which we will discuss only in the last section.

What remains is the explicit evaluation of the Feynman amplitudes as given e.g. for Bhabha and for Møller scattering by (W.359) and (W.362). To this end we have to insert and use the definitions and properties of the spinors $u_r\,(\mathbf{p})$ and $w_r\,(\mathbf{p})$ as given in (W.94)–(W.99) in section 'Quantization of free fields - Dirac'. The evaluation looks like an quite innocent task, but actually it is a very tedious and lengthy affair[140] which would consume too much space here. Thus, we omit it and refer to the literature. But we give now the results for Bhabha and Møller scattering in the form of the corresponding scattering cross sections.

### W.12.3.1  Scattering of Two Particles

In scattering experiments, the (differential) scattering cross section $\frac{d\sigma}{d\Omega}$ (cf. Vol. 2, Chap. 25) is central. We now look for the relation between $\frac{d\sigma}{d\Omega}$ and the Feynman amplitude $\mathcal{M}$.

We assume that there are two initial and two final particles. Then one can show, that in the center of mass system holds[141]

---

[140]The calculation is based to a large extent on trace techniques.

[141] A remark concerning the unit of the scattering cross section which should have units of area, i.e., length squared. In natural units ($\hbar = 1, c = 1$; see Vol. 1 Appendix B 'Units and Constants'), energy and mass have the same physical unit, since $E = mc^2 \rightarrow E = m$. The same holds for frequency $\omega$ and wave number $k$. Thus, $[E] = [m] = [\omega] = [k]$. Due to $\lambda = h/p \rightarrow \lambda = 1/p$ follows $[\lambda] = $ length $= 1/[E]$ or $[E] = 1/$length. Since $D_F^{\mu\nu} \sim 1/k^2$ we have $[\mathcal{M}] = [D_F^{\mu\nu}] = 1/[k]^2 = 1/[E]^2$. It follows from (W.364)

$$\left[\frac{d\sigma}{d\Omega}\right] = \left[\frac{1}{E}\right]^2 \left[\frac{|\mathbf{p}_1'|}{|\mathbf{p}_1|}\right][m]^4 \cdot \left[|\mathcal{M}|^2\right] = [E]^2\,[\mathcal{M}]^2 = [E]^2\,\frac{1}{[E]^4} = \frac{1}{[E]^2} = \text{length}^2 = \text{area}$$

$$(W.363)$$

as it should be.

$$\frac{d\sigma}{d\Omega} = \frac{1}{4} \frac{1}{(4\pi)^2 (E_1 + E_2)^2} \frac{|\mathbf{p}_1'|}{|\mathbf{p}_1|} \overset{\text{ext. fermions}}{\prod_i} (2m_i) \cdot |\mathcal{M}|^2 . \qquad (W.364)$$

The 4-momenta of the two particles before and after the scattering are given $p_i$ and $p_i'$. Note that in the center of mass system holds $\mathbf{p}_2 = -\mathbf{p}_1$. Equation (W.364) is valid for arbitrary fermions (electrons, muons, tauons and their antiparticles) in the incoming and outgoing channel, e.g. for the process $e^+ + e^- = \mu^+ + \mu^-$.

The essential point is the direct proportionality of $\frac{d\sigma}{d\Omega}$ and $|\mathcal{M}|^2$. In this respect, the Feynman amplitude resembles the scattering amplitude.[142] For the considered examples, Bhabha and Møller scattering, we have in each case two different processes with different subamplitudes and thus $\mathcal{M} = \mathcal{M}_1 + \mathcal{M}_2$. This means that we have a squared amplitude for each subprocess plus an additional interference term,

$$|\mathcal{M}|^2 = |\mathcal{M}_1|^2 + |\mathcal{M}_2|^2 + 2 \operatorname{Re} \mathcal{M}_1 \mathcal{M}_2^* . \qquad (W.365)$$

All following results are calculated assuming spin-averaging. This means that (1) that the spin orientations of the incoming particles are random and (2) the spin directions of the outgoing particles are not measured.[143] Thus, the cross section is a mean value of the cross sections (W.364) for all possible spin directions, i.e.,

$$|\mathcal{M}|^2 \rightarrow \frac{1}{4} \sum_{r_1, r_2, r_1', r_2'} |\mathcal{M}|^2 . \qquad (W.366)$$

As said above, the evaluation of the Feynman amplitudes would unfortunately consume too much place here. Thus, we report in the following just the results. The graphical representation is found in Fig. W.14.

### W.12.3.2  Møller Scattering

In case of Møller scattering, the general expression for the (unpolarized) scattering cross section reads[144]:

$$\left. \frac{d\sigma}{d\Omega} \right|_{\text{Møller}} = \frac{\alpha^2}{16E^2 \mathbf{p}^4} \cdot \left\{ \begin{array}{l} \frac{1}{\sin^4 \frac{\vartheta}{2}} \left[ m^4 + 4\mathbf{p}^2 m^2 \cos^2 \frac{\vartheta}{2} + 2\mathbf{p}^4 \left( 1 + \cos^4 \frac{\vartheta}{2} \right) \right] + \\ + \frac{1}{\cos^4 \frac{\vartheta}{2}} \left[ m^4 + 4\mathbf{p}^2 m^2 \sin^2 \frac{\vartheta}{2} + 2\mathbf{p}^4 \left( 1 + \sin^4 \frac{\vartheta}{2} \right) \right] + \\ + \frac{1}{\cos^2 \frac{\vartheta}{2} \sin^2 \frac{\vartheta}{2}} \left( 4\mathbf{p}^4 - m^4 \right) \end{array} \right\}$$

$$(W.367)$$

---

[142] Also decay rates for decay processes (e.g. $K^0 \rightarrow \pi^+ \pi^-$) are proportional to $|\mathcal{M}|^2$.

[143] Note that a particle beam with random spin orientations is often called 'unpolarized'. Thus, the term 'polarization' is used to include both fermion spin and photon polarizations.

[144] $\alpha$ is the fine structure constant, $\alpha = q^2/4\pi$.

where $\vartheta$ is the scattering angle. One by one, the summands are contributions of (1) direct scattering (type 2), (2) exchange scattering (type 1) and (3) of the interference between the two scattering processes.

One can rewrite the scattering cross section to give

$$\left.\frac{d\sigma}{d\Omega}\right|_{\text{Møller}} = \frac{\alpha^2 \left(2E^2 - m^2\right)^2}{4E^2 \left(E^2 - m^2\right)^2} \left[\frac{4}{\sin^4 \vartheta} - \frac{3}{\sin^2 \vartheta} + \frac{\left(E^2 - m^2\right)^2}{\left(2E^2 - m^2\right)^2}\left(1 + \frac{4}{\sin^2 \vartheta}\right)\right].$$
(W.368)

In the high relativistic limit $\mathbf{p}^2 \gg m^2$, we have

$$\left.\frac{d\sigma}{d\Omega}\right|_{\text{Møller, } \mathbf{p}^2 \gg m^2} = \frac{\alpha^2}{8E^2}\left\{\frac{1 + \cos^4 \frac{\vartheta}{2}}{\sin^4 \frac{\vartheta}{2}} + \frac{1 + \sin^4 \frac{\vartheta}{2}}{\cos^4 \frac{\vartheta}{2}} + \frac{2}{\cos^2 \frac{\vartheta}{2} \sin^2 \frac{\vartheta}{2}}\right\} = \frac{\alpha^2}{4E^2}\frac{\left(3 + \cos^2 \vartheta\right)^2}{\sin^4 \vartheta}$$
(W.369)

and in the nonrelativistic limit $\mathbf{p}^2 \ll m^2$ follows

$$\left.\frac{d\sigma}{d\Omega}\right|_{\text{Møller, } \mathbf{p}^2 \ll m^2} = \frac{\alpha^2 m^2}{16\mathbf{p}^4}\left[\frac{1}{\sin^4 \frac{\vartheta}{2}} + \frac{1}{\cos^4 \frac{\vartheta}{2}} - \frac{1}{\sin^2 \frac{\vartheta}{2} \cos^2 \frac{\vartheta}{2}}\right] = \frac{\alpha^2 m^2}{4\mathbf{p}^4}\left[\frac{1 + 3\cos^2 \vartheta}{\sin^4 \vartheta}\right].$$
(W.370)

Comparing this with the classical (i.e., nonrelativistic) Rutherford cross section (scattering of identical particles)

$$\left.\frac{d\sigma}{d\Omega}\right|_{\text{Rutherford}} = \frac{\alpha^2 m^2}{16\mathbf{p}^4}\left[\frac{1}{\sin^4 \frac{\vartheta}{2}} + \frac{1}{\cos^4 \frac{\vartheta}{2}}\right]$$
(W.371)

shows that QED gives an additional interference term. (See also Appendix O, Vol. 2, 'Scattering of identical particles').

### W.12.3.3 Bhabha Scattering

In case of Bhabha scattering, the general expression for the scattering cross section reads

$$\left.\frac{d\sigma}{d\Omega}\right|_{\text{Bhabha}} = \frac{\alpha^2}{16E^2 \mathbf{p}^4} \cdot \left\{\begin{array}{l} \frac{1}{\sin^4 \frac{\vartheta}{2}}\left[m^4 + 4\mathbf{p}^2 m^2 \cos^2 \frac{\vartheta}{2} + 2\mathbf{p}^4 \left(1 + \cos^4 \frac{\vartheta}{2}\right)\right] + \\ + \frac{\mathbf{p}^4}{E^4}\left[3m^4 + 4\mathbf{p}^2 m^2 + 2\mathbf{p}^4 \cos^2 \frac{\vartheta}{2}\right] - \\ - \frac{\mathbf{p}^2}{E^2 \sin^2 \frac{\vartheta}{2}}\left[3m^4 + 8\mathbf{p}^2 m^2 \cos^2 \frac{\vartheta}{2} + 4\mathbf{p}^4 \cos^4 \frac{\vartheta}{2}\right] \end{array}\right\}.$$
(W.372)

One by one, the summands are contributions of (1) direct scattering (type 2), (2) annihilation scattering (type 1) and (3) of the interference between the two scattering processes.

In the high relativistic limit with $\mathbf{p}^2 \gg m^2$, we have

**Fig. W.14** For better comparability, the cross sections are multiplied by $s = \frac{16\mathbf{p}^4}{\alpha^2 E^2}$

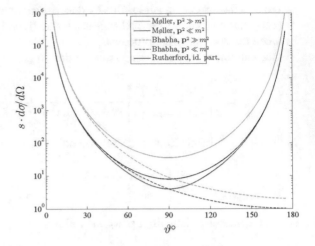

$$\frac{d\sigma}{d\Omega}\bigg|_{\text{Bhabha, } \mathbf{p}^2 \gg m^2} = \frac{\alpha^2}{8E^2}\left\{\frac{1+\cos^4\frac{\vartheta}{2}}{\sin^4\frac{\vartheta}{2}} + \cos^2\frac{\vartheta}{2} - \frac{2\cos^4\frac{\vartheta}{2}}{\sin^2\frac{\vartheta}{2}}\right\} = \frac{\alpha^2}{8E^2}\frac{\left(3+\cos^2\vartheta\right)^2}{2\left(1-\cos\vartheta\right)^2}.$$

(W.373)

In the nonrelativistic limit $\mathbf{p}^2 \ll m^2$ follows

$$\frac{d\sigma}{d\Omega}\bigg|_{\text{Bhabha, } \mathbf{p}^2 \ll m^2} = \frac{\alpha^2 m^2}{16\mathbf{p}^4}\frac{1}{\sin^4\frac{\vartheta}{2}}.$$

(W.374)

The contribution of the annihilation process and the interference is suppressed, since at low energies they are of order $O(\frac{\mathbf{p}^2}{m^2})$. As is seen, the cross section agrees with the classical Rutherford cross section (scattering of non-identical particles):

$$\frac{d\sigma}{d\Omega}\bigg|_{\text{Rutherford}} = \frac{\alpha^2 m^2}{16\mathbf{p}^4}\frac{1}{\sin^4\frac{\vartheta}{2}}$$

(W.375)

## W.12.4  Exercises and Solutions

1. Prove (W.340).
   Solution: We start with

$$X = \int\int d^4x_1\, d^4x_2\, D_F^{\mu\nu}\,(x_1 - x_2)\, e^{iP'_1 x_1}e^{-iPx_2}\;;\; P = p_1 + p_2\;;\; P' = p'_1 + p'_2.$$

(W.376)

Inserting the propagator ((W.286) in sec. 'contraction, propagator, Wick's theorem') brings

$$X = -\int\int d^4x_1\, d^4x_2\, g^{\mu\nu}\, \frac{1}{(2\pi)^4} \int d^4k\, \frac{1}{k^2+i\varepsilon} e^{-ik(x_1-x_2)} e^{iP_1'x_1} e^{-iPx_2} =$$

$$= -g^{\mu\nu}\, \frac{1}{(2\pi)^4} \int d^4x_1 \int d^4k\, \frac{1}{k^2+i\varepsilon} e^{-ikx_1} e^{iP_1'x_1} \int d^4x_2\, e^{ikx_2} e^{-iPx_2} =$$

$$= -g^{\mu\nu}\, \frac{1}{(2\pi)^4} \int d^4x_1 \int d^4k\, \frac{1}{k^2+i\varepsilon} e^{-ikx_1} e^{iP_1'x_1} (2\pi)^4\, \delta^{(4)}(k - P) = \quad (W.377)$$

$$= -g^{\mu\nu} \int d^4x_1\, \frac{1}{(p_1+p_2)^2+i\varepsilon} e^{-iPx_1} e^{iP_1'x_1} = -g^{\mu\nu}\, (2\pi)^4\, \frac{\delta^{(4)}(P-P')}{P^2+i\varepsilon} =$$

$$= -g^{\mu\nu}\, \frac{1}{P^2+i\varepsilon}\, (2\pi)^4\, \delta^{(4)}(P - P') = D_F^{\mu\nu}(P)\, (2\pi)^4\, \delta^{(4)}(P - P')$$

where $D_F^{\mu\nu}(p_1 + p_2)$ is the Fourier transform of the propagator, i.e.,

$$D_F^{\mu\nu}(P) = -g^{\mu\nu}\, \frac{1}{P^2 + i\varepsilon}. \qquad (W.378)$$

2. Formulate the Feynman amplitudes and the $S$-matrix for the processes in Fig. W.8.

   Solution: (1) Compton scattering. Labels are: incoming electron $p$, incoming photon $k$, outgoing electron $p'$, outgoing photon $k'$. It follows

$$\mathcal{M}_{C1}^{(2)} = -q^2 \varepsilon_{\mu,r'}(\mathbf{k}')\, \bar{u}_{s'}(\mathbf{p}')\, \gamma^\mu i S_F\, (q = p + k)\, \varepsilon_{\nu,r}(\mathbf{k})\, \gamma^\nu u_s(\mathbf{p}) \quad (W.379)$$

and

$$\mathcal{M}_{C2}^{(2)} = -q^2 \varepsilon_{\mu,r}(\mathbf{k})\, \bar{u}_{s'}(\mathbf{p}')\, \gamma^\mu i S_F\, (q = p - k')\, \varepsilon_{\nu,r'}(\mathbf{k}')\, \gamma^\nu u_s(\mathbf{p}) \quad (W.380)$$

and

$$S_C^{(2)} = \sqrt{\frac{m}{VE_{\mathbf{p}}}} \sqrt{\frac{m}{VE_{\mathbf{p}'}}} \sqrt{\frac{1}{2VE_{\mathbf{k}}}} \sqrt{\frac{1}{2VE_{\mathbf{k}'}}} (2\pi)^4\, \delta\,(p + k - p' - k')\, \left(\mathcal{M}_{C1}^{(2)} + \mathcal{M}_{C2}^{(2)}\right).$$

$$(W.381)$$

(2) Pair annihilation. Labels are: electron $p_1$, positron $p_2$. Photon at the electron-electron or positron-positron vertex $k_1$, photon at the electron-positron vertex $k_2$. It follows

$$\mathcal{M}_{PA1}^{(2)} = -q^2 \varepsilon_{\mu,r_2}(\mathbf{k}_2)\, \varepsilon_{\nu,r_1}(\mathbf{k}_1)\, \bar{w}_{r_2}(\mathbf{p}_2)\, \gamma^\mu i S_F\, (p_1 - k_1)\, \gamma^\nu u_{s_1}(\mathbf{p}_1) \quad (W.382)$$

and

$$\mathcal{M}_{PA2}^{(2)} = -q^2 \varepsilon_{\mu,r_1}(\mathbf{k}_1)\, \varepsilon_{\nu,r_2}(\mathbf{k}_2)\, \bar{w}_{r_2}(\mathbf{p}_2)\, \gamma^\mu i S_F\, (p_1 - k_2)\, \gamma^\nu u_{s_1}(\mathbf{p}_1) \quad (W.383)$$

and

$$S_{PA}^{(2)} = \sqrt{\frac{m}{VE_{\mathbf{p}_1}}} \sqrt{\frac{m}{VE_{\mathbf{p}_2}}} \sqrt{\frac{1}{2VE_{\mathbf{k}_1}}} \sqrt{\frac{1}{2VE_{\mathbf{k}_2}}} (2\pi)^4\, \delta\,(p_1 + p_2 - k_1 - k_2)\, \left(\mathcal{M}_{PA1}^{(2)} + \mathcal{M}_{PA2}^{(2)}\right).$$

$$(W.384)$$

3. Show

$$\frac{\alpha^2 m^2}{4\mathbf{p}^4}\frac{1+3\cos^2\vartheta}{\sin^4\vartheta} = \frac{\alpha^2 m^2}{16\mathbf{p}^4}\left[\frac{1}{\sin^4\frac{\vartheta}{2}} + \frac{1}{\cos^4\frac{\vartheta}{2}} - \frac{1}{\sin^2\frac{\vartheta}{2}\cos^2\frac{\vartheta}{2}}\right]. \quad \text{(W.385)}$$

Solution:

$$\begin{aligned}\text{r.h.s.} &= \frac{\alpha^2 m^2}{16\mathbf{p}^4}\left[\frac{1}{\sin^4\frac{\vartheta}{2}} + \frac{1}{\cos^4\frac{\vartheta}{2}} - \frac{1}{\sin^2\frac{\vartheta}{2}\cos^2\frac{\vartheta}{2}}\right] = \\ &= \frac{\alpha^2 m^2}{16\mathbf{p}^4}\frac{\cos^4\frac{\vartheta}{2}+\sin^4\frac{\vartheta}{2}-\sin^2\frac{\vartheta}{2}\cos^2\frac{\vartheta}{2}}{\sin^4\frac{\vartheta}{2}\cos^4\frac{\vartheta}{2}}.\end{aligned} \quad \text{(W.386)}$$

With $\sin\frac{\vartheta}{2} = \sqrt{\frac{1}{2}(1-\cos\vartheta)}$ and $\cos\frac{\vartheta}{2} = \sqrt{\frac{1}{2}(1+\cos\vartheta)}$ follows

$$\begin{aligned}\text{r.h.s.} &= \frac{\alpha^2 m^2}{16\mathbf{p}^4}\left[\frac{\left(\frac{1}{2}\right)^2(1+\cos\vartheta)^2+\left(\frac{1}{2}\right)^2(1-\cos\vartheta)^2-\left(\frac{1}{2}\right)^2(1-\cos^2\vartheta)}{\left(\frac{1}{2}\right)^4(1-\cos^2\vartheta)^2}\right] = \\ &= \frac{\alpha^2 m^2}{16\mathbf{p}^4}4\frac{1+3\cos^2\vartheta}{\sin^4\vartheta} = \text{l.h.s.}\end{aligned} \quad \text{(W.387)}$$

## W.13   High Precision and Infinities

As mentioned in the introduction to this chapter, we tried to set some stepping stones on the way to the realm of QFT, especially QED. Of course, we had to leave questions open, and there are some gaps. Nevertheless, we have accumulated a lot of material. As it turned out, the triad Feynman amplitudes, diagrams and rules is a centerpiece of QED.

In this last section, we want to complete the Feynman rules and show how (in principle) to formulate $S$-matrix elements of arbitrary order. After that, we discuss why QED is considered the most stringently proven theory in physics. And finally, we cursorily consider renormalization, i.e., how to handle divergent integrals which emerge in the calculation of so-called loop diagrams.

### W.13.1   Feynman Rules, Diagrams, Amplitudes for QED

We have seen in the previous sections that there is a one-to-one relation between Feynman diagrams and Feynman amplitudes.[145] The relation is given by the Feynman rules most of which we have already formulated. However, three are still lacking and are presented now. We will not derive them, since our main intention here is to

---

[145]Feynman diagrams are also used in other fields as e.g. quantum chromodynamics (QCD). In principle it is the same technique though there are new elements as e.g. gluons and ghosts and there is another coupling constant; instead of the fine structure constant $\alpha_{QED} \approx 1/137$ we have $\alpha_{QCD} \sim 1$.

complete the whole set of rules. However, below are found examples that at least illustrate how the new rules are used.

We can classify the diagrams by the order $n$ of the scattering matrix $S^{(n)}$ which equals the number of vertices. The rules formulated below are valid for all $n$.

Assume we have a Feynman diagram of order $n$. Then we can determine the Feynman amplitude of this diagram according to the following Feynman rules for QED.

1. For each vertex, there is a factor $-iq\gamma^{\mu}$ with $q = \sqrt{4\pi\alpha}$.
2. For each external line, there is one of the following spinors and polarization vectors:

    (a) for each incoming electron $u_r(\mathbf{p})$, for each outgoing electron $\bar{u}_r(\mathbf{p})$,
    (b) for each incoming positron $\bar{w}_r(\mathbf{p})$, for each outgoing positron $w_r(\mathbf{p})$,
    (c) for each incoming photon $\varepsilon_{\lambda\mu}(\mathbf{k})$, for each outgoing photon $\varepsilon_{\lambda\mu}(\mathbf{k})$.

3. For each internal

    (a) photon line (photon propagator), there is a factor $iD_F^{\mu\nu}(k) = i\frac{-g^{\mu\nu}}{k^2+i\varepsilon}$.
    (b) fermionic line (fermion propagator), there is a factor $iS_F(p) = i\frac{\not{p}+m}{p^2-m^2+i\varepsilon}$.

4. At each vertex, the 4-momenta are conserved.
5. The spinor factors ($\gamma$ matrices, fermionic propagators, 4-spinors) are ordered as follows: reading the formulas from right to left, they occur in the same sequence as following the fermion line in the direction of its arrows through the vertex. Note that order is crucial as it affects the spinor matrix multiplication.
6. For each interchange of neighboring fermionic operators which is required to bring the expression in appropriate normal order, one has to multiply the expression by a factor of $(-1)$. The word 'appropriate' means the following: if there are several subamplitudes (as in case of Bhabha scattering), each subamplitude must be in the same normal order.
7. For the interaction of a charged particle with a static external field, the vector potential is given by $A_\mu^e(\mathbf{k}) = \int d^3x\, e^{-i\mathbf{k}\mathbf{x}}A_\mu^e(\mathbf{x})$. In addition, the delta function $(2\pi)^4\,\delta^{(4)}\left(\sum p_f - \sum p_i\right)$ in $S_{fi}$ is replaced by $2\pi\delta\left(\sum E_f - \sum E_i\right)$.[146]

We have derived these rules in the preceding sections. As mentioned above, there are three more rules which have to do with loops and renormalization, i.e., with redefined values of mass and charge, see below. These rules are:

8. For each closed loop of fermion lines, one has to take the trace in spinor space of the resulting matrix and multiply by a factor of $(-1)$.
9. For each 4-momentum $p$ which is not fixed by 4-momentum conservation, one has to carry out the integration $\frac{1}{(2\pi)^4}\int d^4p$.[147]

---

[146]In other words: for external fields, the rules of the game are changed to some extent.

[147]Rule 8 and 9 are illustrated below by means of Bhabha scattering plus photon loop.

10. Renormalization of the mass leads to a mass counterterm diagram and, in the amplitudes, to an additional term equal to $i\delta m$.[148]

It is a very compact set - two handfuls of rules connect biuniquely the graphical and mathematical representation of all physical processes of QED.

Following these rules we can write down the exact analytical expression for the Feynman amplitude from the diagram.[149] In general, for a given order $n$ there are several subamplitudes $\mathcal{M}_i^{(n)}$ as e.g. in the case of Bhabha scattering where we found two different processes contributing to $\mathcal{M}_{\text{Bhabha}}^{(2)}$. Via $\mathcal{M}^{(n)} = \sum_i \mathcal{M}_i^{(n)}$ they contribute to the Feynman amplitude $\mathcal{M}^{(n)}$.

The next question is if we want to examine the scattering of order $n$ - then only $\mathcal{M} = \mathcal{M}^{(n)}$ is considered - or the entire process up to and including the order $N$ - then $\mathcal{M} = \sum_{n=1}^{N} \mathcal{M}^{(n)}$ is taken into account. Finally, the $S$-matrix element $\langle f | S | i \rangle = S_{fi}$ is given by (discrete version)

$$
S_{fi} = \delta_{fi} + \left[ (2\pi)^4 \, \delta^{(4)} \left( P_f - P_i \right) \cdot \overset{\text{ext. phot.}}{\prod} \sqrt{\frac{1}{2VE_\mathbf{k}}} \cdot \overset{\text{ext. ferm.}}{\prod} \sqrt{\frac{m}{VE_\mathbf{p}}} \right] \cdot \mathcal{M}. \quad \text{(W.388)}
$$

As mentioned above, we will not explicitly calculate the Feynman amplitudes by inserting and evaluating the spinors $u$ and $w$ and the polarization vectors $\varepsilon$. This simply would be too lengthy and can not be afforded within the scope of this brief introduction to QFT.

## W.13.2   Extraordinary Precision

The high-precision calculation of the $g$-factor (Landé factor) is one of the great success stories of QED. The Dirac equation predicts the magnetic moment of the electron $\boldsymbol{\mu}_e$ to be given by (see Appendix U, Vol. 1)

$$
\boldsymbol{\mu}_e = g \frac{q}{2m} \mathbf{s} \; ; \; g = 2. \quad \text{(W.389)}
$$

QED allows for an impressively accurate determination of the anomalous magnetic moment, i.e., the deviation from $g = 2$. In 1948, Julian Schwinger[150] showed that radiation and re-absorption of a single virtual photon (see e.g. Fig. W.15 for an electron in an external magnetic field) modifies the $g$-value from 2 to $2\left(1 + \frac{\alpha}{2\pi}\right) \approx 2,00332$. Incorporating effects of order $\alpha^3$ leads to $g = 2,002\,319\,304$. Meanwhile (April 2018), taking into account more and more higher order corrections (up to

---

[148]For $\delta m$ and a short explanation of the rule see below.

[149]Of course one can still calculate the amplitudes by the purely formal way, i.e., by inserting the field operators into $S^{(2)}$. But that is much more complicated and lengthy, as we have seen.

[150]Schwinger, Julian Seymour, 1918–1994, US-American physicist, nobel prize 1965.

**Fig. W.15** Vertex correction
(one loop contribution) of
the magnetic moment;
electron in an e.g. external
magnetic field

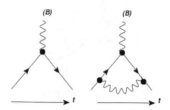

$\alpha^5$), the theoretical value is given by 2, 002 319 304 363 286(1528), whereby the experimental value is 2, 002 319 304 361 46(56).[151] The current theoretical limit is due to the fact that for higher orders $n$ there is a rapidly increasing number of integrals to evaluate (3th order 72 diagrams, 4th order 981, 5th order 12672).

As is seen, the theoretical and experimental values agree to more than 10 significant figures. This makes the magnetic moment of the electron the most precisely determined physical quantity and, hence, QED the most thoroughly tested theory.[152]

## W.13.3  Problematic Loops, Infinities, Renormalization

The extraordinary precision is considered proof by (almost) all physicists that QED is correct, although there are significant problems with infinities which we will present now. The origin are three diagrams containing loops which lead to divergent integrals. These infinities may be accounted for by a redefinition of mass and charge, called *renormalization*. This is a central issue of QED - in any textbook of QED, a substantial part is dedicated to this (quite technical) topic. Within the framework of our compressed considerations we can only cursorily touch on the topic.

### W.13.3.1  Three Problematic Loops

In QED, there are three types of problematic diagrams, illustrated in Fig. W.16. We have (1) 'bubble propagators': a real or virtual photon creates a fermion/antifermion pair (also called photon loop, photon self energy, vacuum polarization, closed fermion loop, (2) 'dressed fermions': a real or virtual fermion emits and reabsorbs a virtual photon (fermion loop, fermion self energy) and (3) vertex correction: a virtual photon connects fermions across a previous vertex (vertex loop correction, vertex modifi-

---

[151]From https://en.wikipedia.org/wiki/Anomalous_magnetic_dipole_moment. The National Institute of Standards and Technology gives for the experimental value 2, 002 319 304 361 82(52). Current status april 2018.

[152]The myon has a higher mass than the electron ($m_\mu \approx 200 m_e$). Its anomalous magnetic moment is therefore sensitive to interactions from physics beyond QED, even beyond the Standard Model. Thus, it may be used as a probe for new physics.

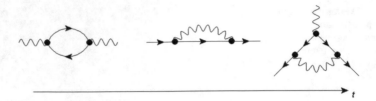

**Fig. W.16** Photon self energy, fermion self energy and vertex loop correction

cation). These three diagrams lead to divergent integrals.[153] The reason is that the momentum $p$ in the loops is not fixed which implies integration over $p$, see Feynman rule 9. However, as it turns out, these integrals are divergent. Such integrals are pretty unwelcome in any theory and usually imply the death of that theory, but not in QED.

### W.13.3.2 Renormalization, Basic Idea

In fact, these problems are healed by a redefinition of mass and charge, i.e., by renormalization. The redefinition is chosen in such a way that it compensates for the infinities and gives us finite answers. Following this idea leads to two kinds of charges and masses. There are the bare charge and bare mass $q_0$ and $m_0$. We cannot measure them since nature shows us only the 'ordinary' measurable quantities $q$ and $m$ (or $q_{\text{renorm}}$ and $m_{\text{renorm}}$), i.e., the renormalized quantities. They are so to speak real-world versions of the bare items, dressed up with corrections, i.e., contributions from one or several loops (Fig. W.17).

### W.13.3.3 Renormalization, Example

To illustrate the situation, we consider Bhabha scattering type 1 (annihilation scattering) plus a photon loop, see Fig. W.17. Following the Feynman rules, we can write down the transition amplitude for this scattering as (see rules 8 and 9):

$$\mathcal{M}_{B1}^{\text{photon loop}} = -\frac{q^4}{(2\pi)^4} \bar{u}_{r_1'}\left(\mathbf{p}_1'\right) \gamma^\rho w_{r_2}\left(\mathbf{p}_2'\right) D_{F\rho\eta}\left(k\right) \cdot$$
$$\cdot \left\{ Tr \int S_F\left(p\right) \gamma^\mu S_F\left(p - k\right) \gamma^\eta \mathrm{d}^4 p \right\} D_{F\mu\nu}\left(k\right) \bar{w}_{r_2}\left(\mathbf{p}_2\right) \gamma^\nu u_{r_1}\left(\mathbf{p}_1\right). \tag{W.390}$$

It is the integral which causes problems:

$$I^{\mu\eta} = \int S_F\left(p\right) \gamma^\mu S_F\left(p - k\right) \gamma^\eta \mathrm{d}^4 p = \int \frac{\not{p} + m}{p^2 - m^2 + i\varepsilon} \gamma^\mu \frac{\not{p} - \not{k} + m}{(p - k)^2 - m^2 + i\varepsilon} \gamma^\eta \mathrm{d}^4 p. \tag{W.391}$$

---

[153]Remember that we had to introduce normal ordering due to infinite vacuum energies. It seems that QED is indeed plagued by infinities.

**Fig. W.17**  Bhabha
scattering plus photon loop

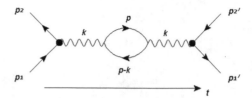

Since the evaluation is rather lengthy; we just state the result: this integral *diverges* with ln $p$ for high $p$.

Inspection of (W.390) shows that one can compensate for this infinity by e.g. (1) introducing a cut-off of the 4-momentum (the cut-off is often labeled $\Lambda$ or $\Pi$, the method is called 'regularization')[154] and (2) redefining the charge in such a way that it eliminates the problematic terms. The bottom line is that renormalizing the three loop integrals leads e.g. to a renormalized charge $q$ or $q_{\text{renorm}}$ which is related to the bare charge $q_0$ by[155]

$$q = q_0 \left[ 1 - \frac{\alpha}{3\pi} \ln \left( \frac{\Lambda}{m} \right)^2 + \mathcal{O}(\alpha^2) \right]. \qquad (W.392)$$

Note that the bare charge $q_0$ is greater than the renormalized, i.e., measurable charge $q_{\text{renorm}}$.[156]

Analogously, also the mass is renormalized which leads to changes in the Feynman amplitude and diagram. The reason is that the Lagrangian is formulated with the bare mass $m_0$. Substituting $m_0 = m - \delta m$ leads to an additional term $\delta m \bar{\psi}\psi$ (the mass counterterm of the Lagrangian; no photon involved, but sort of fermion self interaction) subsumed in the interaction part of the Lagrangian $\mathcal{L}^{\text{interaction}}$ and, hence, of the Hamiltonian $\mathcal{H}^{\text{interaction}} = -\mathcal{L}^{\text{interaction}}$. Thus, we have an extra term $\sim i\delta m$ in the amplitude and an extra diagram with an incoming and an outgoing fermion, usually represented by a fermionic line marked by a little centered cross '×'.

### W.13.3.4  Renormalization, Higher Orders and Question

Obviously, there are also loops in higher order terms. Figure W.18 shows as example the three (of infinite many ones) corrections for the photon propagator. One can show that renormalization can be performed in all orders and that the sum of higher order QED corrections converges.

---

[154]Remember that we have discussed a cut-off also in the context of infinite vacuum energies in the section 'Operator ordering'.

[155]For QED it is usually sufficient to ignore terms $\mathcal{O}(\alpha^2)$.

[156]As a consequence of renormalization, the value of the coupling 'constant' $\alpha$ (remind $q^2 = 4\pi\alpha$) depends on energy. As for numerical values: $\alpha$ increases with energy from $\alpha = 1/137$ at low energies to $\alpha = 1/128$ at 100 GeV. Therefore $\alpha$ is called *running coupling constant*.

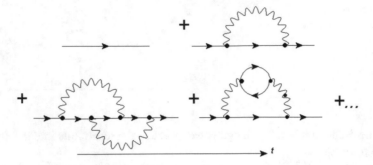

**Fig. W.18** Fermion propagator with higher order divergent corrections

### W.13.3.5 Renormalization, Vividly

We can get a vivid picture by looking at Fig. W.18. As is seen, the electron (or, more generally, fermion, real or virtual) constantly emits and reabsorbs photons. Hence, it is surrounded by a cloud of virtual photons (it is 'dressed') which carry a certain portion of energy and thus mass. Photons themselves (real and virtual ones) propagate while tearing fermion pairs from the vacuum. Thus, they are dressed by a cloud of fermion/antifermion pairs which modify the photon's amplitude for propagation between two points.

So we have fermions which emit and reabsorb photons which tear fermion pairs which emit and reabsorb photons which tear fermion pairs ... and so on, sort of a matryoshka doll.

We see that the vacuum is not a 'nothing', but an interacting medium. Dressing the bare photon in a cloud of fermion/antifermion pairs may be compared with a vacuum which interacts as a dielectric. In this context, one speaks of vacuum polarization or screening effect. We note that this effect can be and has been measured (affects the Lamb shift in the Hydrogen spectrum).

## W.13.4 Conclusion

For many, QED is just a good and high-precision theory; admittedly, there occur problems, but they are manageable. Others feel uncomfortable and are rather unhappy about this juggling with infinities So were, among others also the founding fathers Dirac and Feynman. Dirac criticized: "This so-called 'good theory' does involve neglecting infinities which appear in its equations, neglecting them in an arbitrary way. This is just not sensible mathematics. Sensible mathematics involves neglecting a quantity when it is small – not neglecting it just because it is infinitely great and you do not want it!" And Feynman pointed out: "It [renormalization] is still what I would call a dippy process! Having to resort to such hocus-pocus has prevented us from proving that the theory of quantum electrodynamics is mathematically self-consistent. ... I suspect that renormalization is not mathematically legitimate."

One can debate if the avoidance of divergences by renormalization is a more or less fruitful concept or a fundamental principle.

In any case, meanwhile there are new perspectives to renormalization. For instance, one of these approaches, instead of being concerned with good limiting behavior as the cut-off for great momenta, focuses on the behavior of the running constants, i.e., how e.g. charge or mass vary with increasing energy (i.e., decreasing distance).

Of course, in our short introduction we have to leave many questions open, for instance: Are there other theories besides QED which are renormalizable? Are there theories which are not renormalizable? What is the crucial property for a theory to be renormalizable?, to mention a few. Another interesting point would be the discussion of Haag's theorem which states that the interaction picture of a relativistic quantum field theory does not exist, thus shaking the foundations of QFT.

Basically, the issue expresses our ignorance of physics at very small distances. Is the electron really a mass point? Experimentally, this property could only be confirmed up to distances of about $10^{-18}$ m. But it is still an open question how the micro-world behaves at distances of say the Planck length $10^{-33}$ m. At this scale, there may be effects, still unknown to us, which, in principle, could repair the divergences.

# Appendix X
# Exercises and Solutions

## X.1 Exercises, Chap. 15

1. Given the potential step

$$V(x) = \begin{cases} 0 \\ V_0 > 0 \end{cases} \quad \text{for} \quad \begin{matrix} x > 0 \\ x \leq 0. \end{matrix} \tag{X.1}$$

The incident quantum object is described as a plane wave running from the right to the left with $E > V_0$. Determine the transmission and reflection coefficients.

2. Given a finite potential well of depth $V_0$ and width $L$; estimate the number of energy levels.

   Solution: We have the inequality (15.43):

$$V_0 - \frac{\hbar^2}{2m}\left(\frac{N}{2L}\pi\right)^2 < |E| < V_0 - \frac{\hbar^2}{2m}\left(\frac{N-1}{2L}\pi\right)^2. \tag{X.2}$$

   If $N_0$, but not $N_0+$ levels exist, this means that

$$V_0 - \frac{\hbar^2}{2m}\left(\frac{N_0-1}{2L}\pi\right)^2 > 0 \text{ and } V_0 - \frac{\hbar^2}{2m}\left(\frac{N_0}{2L}\pi\right)^2 < 0, \tag{X.3}$$

   i.e.

$$N_0 - 1 < \frac{2L}{\pi\hbar}\sqrt{2mV_0} < N_0. \tag{X.4}$$

   Solving for $N_0$, we find

$$\frac{2L}{\pi\hbar}\sqrt{2mV_0} < N_0 < 1 + \frac{2L}{\pi\hbar}\sqrt{2mV_0}. \tag{X.5}$$

© Springer Nature Switzerland AG 2018

J. Pade, *Quantum Mechanics for Pedestrians 2*, Undergraduate Lecture Notes in Physics, https://doi.org/10.1007/978-3-030-00467-5

3. Given a delta potential at $x = 0$; determine the spectrum (negative potential, $E < 0$) and the situation for scattering (positive potential, $E > 0$).
   Solution: The SEq reads

$$E\varphi = -\frac{\hbar^2}{2m}\varphi'' + V\delta(x)\varphi; \quad V = \frac{\hbar^2}{2m}\gamma. \tag{X.6}$$

We integrate this equation between $-\varepsilon$ and $\varepsilon$, where $\varepsilon$ is a small parameter which we subsequently allow to go to zero. First, it follows that

$$E\int_{-\varepsilon}^{\varepsilon}\varphi dx = -\frac{\hbar^2}{2m}\left[\varphi'(\varepsilon) - \varphi'(-\varepsilon)\right] + V\varphi(0). \tag{X.7}$$

We assume that $\varphi$ is continuous at the origin, but the derivative can have a discontinuity (as for the infinite potential wall). It follows for $\varepsilon \to 0$ that:

$$0 = -\frac{\hbar^2}{2m}\left[\varphi'(+0) - \varphi'(-0)\right] + V\varphi(0) \tag{X.8}$$

or

$$\varphi'(+0) - \varphi'(-0) = \frac{2m}{\hbar^2}V\varphi(0) = \gamma\varphi(0). \tag{X.9}$$

Concrete forms (bound or free) for the wavefunction have to be inserted into this equation.

(a) Bound case, i.e. $V < 0$: With

$$\varphi(x < 0) = Ae^{\kappa x}; \quad \varphi(x >) = Be^{-\kappa x}, \tag{X.10}$$

it follows that
$$A = B; \quad -\kappa B - \kappa A = \gamma A \tag{X.11}$$

and thus
$$\kappa = -\frac{1}{2}\gamma A. \tag{X.12}$$

That is, for $V < 0$ there is always exactly one bound state.

(b) Scattering, i.e. $V > 0$: The quantum object comes from the left. Then we have
$$\varphi(x < 0) = Ae^{ikx} + Be^{-ikx}; \quad \varphi(x > 0) = Ce^{ikx}. \tag{X.13}$$

This leads to
$$A + B = C; \quad ikC - (ikA - ikB) = \gamma C \tag{X.14}$$

**Fig. X.1** Scattering at the potential barrier

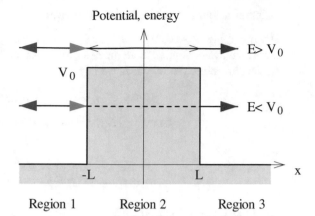

and it follows that

$$B = \frac{\gamma}{2ik - \gamma}A; \quad C = \frac{2ik}{2ik - \gamma}A. \tag{X.15}$$

This yields for the transmission and reflection coefficients:

$$T = \frac{4k^2}{4k^2 + \gamma^2}; \quad R = \frac{\gamma^2}{4k^2 + \gamma^2}. \tag{X.16}$$

4. Given the potential barrier

$$V(x) = \begin{cases} V_0 > 0 \\ 0 \end{cases} \text{ for } \begin{array}{l} -L < x < L \\ \text{otherwise}. \end{array} \tag{X.17}$$

The incident quantum object is described by a plane wave running from the left to the right. Determine the transmission and reflection coefficients.
Solution: The potential is outlined in Fig. X.1.
We treat the cases $E > V_0$ and $E < V_0$ together by setting

$$\gamma = \begin{cases} \kappa \\ ik' \end{cases} \text{ for } \begin{array}{l} E < V_0 \\ E > V_0 \end{array} \text{ with } \begin{array}{l} \kappa^2 = \frac{2m}{\hbar^2}(V_0 - E) \\ k'^2 = \frac{2m}{\hbar^2}(E - V_0). \end{array} \tag{X.18}$$

The solutions in the different regions are

$$\varphi_1 = Ae^{ikx} + Be^{-ikx}$$
$$\varphi_2 = Ce^{\gamma x} + De^{-\gamma x}$$
$$\varphi_3 = Fe^{ikx} + Ge^{-ikx} \tag{X.19}$$

with $k^2 = \frac{2m}{\hbar^2}E$.

**Determination of the Integration Constant**:

The incoming wave comes from the left with the amplitude $A$; thus, in region 3 there is no wave running from the right to the left and therefore we have $G = 0$. At the discontinuities $x = \pm L$, we have

$$Ae^{-ikL} + Be^{ikL} = Ce^{-\gamma L} + De^{\gamma L}$$
$$ikAe^{-ikL} - ikBe^{ikL} = \gamma Ce^{-\gamma L} - \gamma De^{\gamma L} \tag{X.20}$$

and

$$Ce^{\gamma L} + De^{-\gamma L} = Fe^{ikL}$$
$$\gamma Ce^{\gamma L} - \gamma De^{-\gamma L} = ikFe^{ikL} \tag{X.21}$$

From (X.20), it follows that

$$2ikAe^{-ikL} = (\gamma + ik)\, Ce^{-\gamma L} - (\gamma - ik)\, De^{\gamma L}, \tag{X.22}$$

and from (X.21) we have:

$$(\gamma - ik)\, Ce^{\gamma L} - (\gamma + ik)\, De^{-\gamma L} = 0. \tag{X.23}$$

This leads directly to

$$C = \frac{\gamma + ik}{\gamma - ik} De^{-2\gamma L}. \tag{X.24}$$

Insertion yields

$$2ikAe^{-ikL} = (\gamma + ik)\, \frac{(\gamma + ik)}{(\gamma - ik)} De^{-2\gamma L} e^{-\gamma L} - (\gamma - ik)\, De^{\gamma L}$$
$$\rightarrow 2ikAe^{-ikL} = \frac{e^{-2\gamma L}\, (\gamma + ik)^2 - e^{2\gamma L}\, (\gamma - ik)^2}{(\gamma - ik)} De^{-\gamma L}. \tag{X.25}$$

We introduce the shorthand notation

$$M = e^{-2\gamma L}\, (\gamma + ik)^2 - e^{2\gamma L}\, (\gamma - ik)^2. \tag{X.26}$$

We then have

$$2ikAe^{-ikL} = \frac{M}{(\gamma - ik)} De^{-\gamma L} \rightarrow$$
$$D = \frac{(\gamma - ik)}{M} 2ikAe^{-ikL} e^{\gamma L} = \frac{2ik\, (\gamma - ik)}{M} Ae^{-ikL} e^{\gamma L} \tag{X.27}$$

and from this with (X.24)

$$C = \frac{(\gamma + ik)}{(\gamma - ik)} De^{-2\gamma L} = \frac{(\gamma + ik)}{(\gamma - ik)} \frac{2ik(\gamma - ik)}{M} Ae^{-ikL}e^{\gamma L}e^{-2\gamma L}$$

$$= \frac{2ik(\gamma + ik)}{M} Ae^{-ikL}e^{-\gamma L}. \tag{X.28}$$

Due to

$$Be^{ikL} = Ce^{-\gamma L} + De^{\gamma L} - Ae^{-ikL}; \tag{X.29}$$

this leads for the constant $B$ to

$$B = \frac{2ik(\gamma + ik)e^{-2\gamma L} + 2ik(\gamma - ik)e^{2\gamma L} - M}{M} Ae^{-2ikL}$$

$$= \frac{[2ik - \gamma - ik](\gamma + ik)e^{-2\gamma L} + [2ik + \gamma - ik](\gamma - ik)e^{2\gamma L}}{M} Ae^{-2ikL}$$

$$= \frac{e^{2\gamma L} - e^{-2\gamma L}}{M} (\gamma^2 + k^2) Ae^{-2ikL}, \tag{X.30}$$

and due to

$$Fe^{ikL} = Ce^{\gamma L} + De^{-\gamma L}, \tag{X.31}$$

we find for the constant $F$:

$$F = \frac{2ik(\gamma + ik)}{M} Ae^{-2ikL}e^{-\gamma L}e^{\gamma L} + \frac{2ik(\gamma - ik)}{M} Ae^{-2ikL}e^{\gamma L}e^{-\gamma L}$$

$$= \frac{4ik\gamma}{M} Ae^{-2ikL}. \tag{X.32}$$

In sum

$$B = \frac{e^{2\gamma L} - e^{-2\gamma L}}{M} (\gamma^2 + k^2) Ae^{-2ikL}$$

$$F = \frac{4ik\gamma}{M} Ae^{-2ikL}, \tag{X.33}$$

where it holds that

$$\gamma = \begin{cases} \kappa \\ ik' \end{cases} \text{ for } \begin{array}{c} E < V_0 \\ E > V_0 \end{array} \tag{X.34}$$

$$M = \begin{cases} e^{-2\kappa L}(\kappa + ik)^2 - e^{2\kappa L}(\kappa - ik)^2 \\ e^{-2ik'L}(ik' + ik)^2 - e^{2ik'L}(ik' - ik)^2. \end{cases} \tag{X.35}$$

**Determination of $T$ and $R$:**

The partial waves of interest are

$$\varphi_{\text{ein}} = Ae^{ikx}; \ \varphi_{\text{refl}} = Be^{-ikx}; \ \varphi_{\text{trans}} = Fe^{ikx}. \tag{X.36}$$

We have

$$T = \frac{|F|^2}{|A|^2} = \frac{\left|\frac{4ik\gamma}{M} Ae^{-2ikL}\right|^2}{|A|^2} = \left|\frac{4k\gamma}{M}\right|^2 = \frac{16k^2\gamma\gamma^*}{MM^*} \tag{X.37}$$

and

$$R = \frac{|B|^2}{|A|^2} = \frac{\left|\frac{e^{2\gamma L}-e^{-2\gamma L}}{M}\left(\gamma^2 + k^2\right)Ae^{-2ikL}\right|^2}{|A|^2} \tag{X.38}$$

$$= \frac{\left(e^{2\gamma L} - e^{-2\gamma L}\right)\left(e^{2\gamma^* L} - e^{-2\gamma^* L}\right)\left(\gamma^2 + k^2\right)\left(\gamma^{*2} + k^2\right)}{MM^*}.$$

We confine the discussion to $T$. We have:

$$T = \frac{16k^2\gamma\gamma^*}{MM^*} = \begin{cases} \dfrac{16k^2\kappa^2}{MM^*} \\[2mm] \dfrac{16k^2k'^2}{MM^*} \end{cases} \text{ for } \begin{array}{c} E < V_0 \\ E > V_0 \end{array} \tag{X.39}$$

with

$$MM^* = \begin{cases} \left[e^{-2\kappa L}\left(\kappa + ik\right)^2 - e^{2\kappa L}\left(\kappa - ik\right)^2\right] \\ \left[e^{-2\kappa L}\left(\kappa - ik\right)^2 - e^{2\kappa L}\left(\kappa + ik\right)^2\right] \\ \left[-e^{-2ik'L}\left(k' + k\right)^2 + e^{2ik'L}\left(k' - k\right)^2\right] \\ \left[-e^{2ik'L}\left(k' + k\right)^2 + e^{-2ik'L}\left(k' - k\right)^2\right] \end{cases} \tag{X.40}$$

or

$$MM^* = \begin{cases} 2\cosh 4\kappa L \cdot (\kappa - ik)^2 (\kappa + ik)^2 - (\kappa + ik)^2 (\kappa + ik)^2 \\ \quad - (\kappa - ik)^2 (\kappa - ik)^2 \\ \left(k' + k\right)^4 + \left(k' - k\right)^4 - 2\cos 4ik'L \cdot \left(k' - k\right)^2 \left(k' + k\right)^2 \end{cases} \tag{X.41}$$

or

$$MM^* = \begin{cases} 2\left(\kappa^2 + k^2\right)^2 \cosh 4\kappa L - 2\left[\kappa^4 - 6\kappa^2 k^2 + k^4\right] \\ 2\left[k'^4 + 6k'^2 k^2 + k^4\right] - 2\left(k'^2 - k^2\right)^2 \cos 4k'L. \end{cases} \tag{X.42}$$

This gives

$$
T = \begin{cases} \dfrac{16k^2\kappa^2}{MM^*} = \dfrac{8k^2\kappa^2}{\left(\kappa^2 + k^2\right)^2 \cosh 4\kappa L - \left(\kappa^4 - 6\kappa^2 k^2 + k^4\right)} \\[4mm] \dfrac{16k^2 k'^2}{MM^*} = \dfrac{8k^2 k'^2}{\left(k'^4 + 6k'^2 k^2 + k^4\right) - \left(k'^2 - k^2\right)^2 \cos 4k'L} \end{cases} \text{for} \begin{array}{l} E < V_0 \\ E > V_0. \end{array}
$$

$$(X.43)$$

Finally, we insert $E$ and $V_0$; with

$$
k^2 = \frac{2m}{\hbar^2} E; \quad k'^2 = \frac{2m}{\hbar^2} (E - V_0); \quad \kappa^2 = \frac{2m}{\hbar^2} (V_0 - E) \qquad (X.44)
$$

we obtain

$$
T = \begin{cases} \dfrac{8E\,(E - V_0)}{8E\,(E - V_0) + V_0^2\,(1 - \cosh 4\kappa L)} \\[4mm] \dfrac{8E\,(E - V_0)}{8E\,(E - V_0) + V_0^2\,(1 - \cos 4k'L)} \end{cases} \text{for} \begin{array}{l} E < V_0 \\ E > V_0. \end{array}
$$

$$(X.45)$$

In order to write this more compactly, we introduce the abbreviations

$$
z = \frac{E}{V_0}; \quad \mu = \sqrt{\frac{2m}{\hbar^2} V_0 L^2}. \qquad (X.46)
$$

With

$$
k'L = \mu\sqrt{z - 1}; \quad \kappa L = \mu\sqrt{1 - z} \qquad (X.47)
$$

it follows finally that

$$
T = \begin{cases} \dfrac{8z\,(z - 1)}{8z\,(z - 1) + 1 - \cosh 4\mu\sqrt{1 - z}} \\[4mm] \dfrac{8z\,(z - 1)}{8z\,(z - 1) + 1 - \cos 4\mu\sqrt{z - 1}} \end{cases} \text{for} \begin{array}{l} E < V_0;\ 0 < z < 1 \\ E > V_0;\ z > 1 \end{array}
$$

$$(X.48)$$

or, with $\cos ix = \cosh x$ and in one formula:

$$
T = \frac{8z\,(z - 1)}{8z\,(z - 1) + 1 - \cosh 4\mu\sqrt{1 - z}} \quad \text{for } 0 < z = \frac{E}{V_0}. \qquad (X.49)
$$

The graphical representation of $T$ as a function of $z = \frac{E}{V_0}$ is found in Chap. 15.

**Fig. X.2** One-sided infinite
potential well

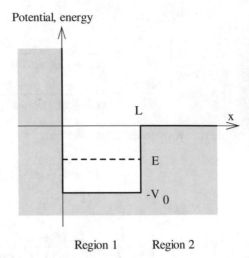

Potential, energy

L

x

E

-V₀

Region 1          Region 2

5. Given the one-sided infinite potential well

$$V(x) = \begin{cases} 0 & L < x \\ -V_0 & \text{for} \quad 0 < x \le L \\ \infty & x \le 0 \end{cases} \tag{X.50}$$

with $V_0 > 0$. For the energy, let $-V_0 < E < 0$. Sketch the potential. Determine
the stationary SEq in the different regions and deduce from them an *ansatz*
for the wavefunction. Adjust the wavefunctions at the discontinuities and show
that the allowed energy levels are defined by the equation $k \cot kL = -\kappa$ with
$k^2 = 2m(V_0 + E)/\hbar^2$ and $\kappa^2 = -2mE/\hbar^2$. Is there always (i.e. for all $V_0$) a
bound state?
Solution: The potential is outlined in Fig. X.2.
The stationary SEq are

$$E\varphi_1 = -\frac{\hbar^2}{2m}\varphi_1'' - V_0\varphi_1; \quad E\varphi_2 = -\frac{\hbar^2}{2m}\varphi_2''. \tag{X.51}$$

They have the solutions

$$\varphi_1 = Ae^{ikx} + Be^{-ikx}; \quad \varphi_2 = Ce^{\kappa x} + De^{-\kappa x}$$
$$k^2 = 2m(V_0 - |E|)/\hbar^2; \quad \kappa^2 = 2m|E|/\hbar^2. \tag{X.52}$$

For $x < 0$, $\varphi \equiv 0$.

Hence, the matching conditions at $x = 0$ are[157] $\varphi_1(0) = 0$. For $x > 0$, $C$ must vanish since otherwise the solution is not bounded. So we find

$$x = 0 : B = -A$$

$$x = L : \begin{array}{l} Ae^{ikL} - ABe^{-ikL} = 2iA \sin kL = De^{-\kappa L} \\ ikAe^{ikL} + ikAe^{-ikL} = 2ikA \cos kL = -\kappa De^{-\kappa L}. \end{array} \tag{X.53}$$

Division of the last two equations gives

$$k \cot kL = -\kappa \text{ or } \tan kL = -\frac{k}{\kappa}. \tag{X.54}$$

This is the quantization condition for the energy. The equation is not solvable in closed form, but with the following considerations we can obtain some more information.

The tangent is periodic in $\pi$; thus it holds that

$$\tan(kL + m\pi) = -\frac{k}{\kappa}; \ m = 0, \pm 1, \pm 2, \ldots \tag{X.55}$$

Since $k$, $\kappa$ and $L$ are positive, we can rewrite this equation as

$$kL = n\pi - \arctan\frac{k}{\kappa}; \ n = 1, 2, \ldots \tag{X.56}$$

We rewrite this again, as in Chap. 15, as an inequality, making use of $0 < \arctan x < \frac{\pi}{2}$ (due to $x > 0$). Then it follows initially that:

$$\frac{2n-1}{2}\pi < kL < n\pi. \tag{X.57}$$

Each of these three terms is positive; so we can square and insert the relation $k^2 = 2m(V_0 - |E|)/\hbar^2$. Resorting and solving for $|E|$, we find:

$$V_0 - \frac{1}{2m}\left(n\frac{\hbar\pi}{L}\right)^2 < |E| < V_0 - \frac{1}{2m}\left(\frac{2n-1}{2}\frac{\hbar\pi}{L}\right)^2. \tag{X.58}$$

In a finite potential well, there is always at least one bound state (see Chap. 15); does this apply here, too? We set $n = 1$ and obtain

$$V_0 - \frac{1}{2m}\left(\frac{\hbar\pi}{L}\right)^2 < |E| < V_0 - \frac{1}{2m}\left(\frac{\hbar\pi}{2L}\right)^2. \tag{X.59}$$

Hence, a bound state can exist only for

---

[157]Note: Due to the infinite jump of the potential, only the wavefunction can be matched at $x = 0$, but not the derivative.

$$V_0 > \frac{1}{2m} \left( \frac{\hbar\pi}{2L} \right)^2 . \tag{X.60}$$

Finally, we ask for the conditions that there are exactly $N$ energy levels. This is the case if the right side of (X.58) is positive for $n = N$, but negative for $n = N + 1$, i.e. if

$$V_0 - \frac{1}{2m} \left( \frac{2N-1}{2} \frac{\hbar\pi}{L} \right)^2 > 0; \quad V_0 - \frac{1}{2m} \left( \frac{2N+1}{2} \frac{\hbar\pi}{L} \right)^2 < 0. \tag{X.61}$$

It follows that

$$\left( N - \frac{1}{2} \right)^2 < \frac{2mL^2}{\hbar^2\pi^2} V_0 < \left( N + \frac{1}{2} \right)^2 . \tag{X.62}$$

6. Given the potential

$$V(x) = \begin{cases} \infty & x < 0 \\ V_0 > 0 & \text{for} \quad 0 \le x \le L . \\ 0 & L < x \end{cases} \tag{X.63}$$

An object described by a plane wave passes from the right towards the origin. Sketch the potential. Calculate the wavefunction for the case $E < V_0$. Which regions are classically allowed, which are not? Determine first the stationary SEq's in the different regions and solve them with an appropriate *ansatz*. Are all the mathematical solutions physically allowed? Determine the free constants using the continuity conditions at the discontinuities of the potential. Perform the calculations for the case $E > V_0$, also.

**Fig. X.3** The potential (X.63)

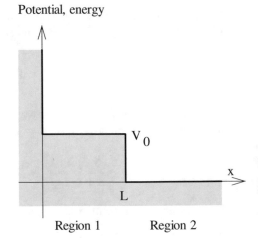

Solution: The potential is outlined in Fig. X.3. The region $x < 0$ as well as region 1 for $E < V_0$ are classically forbidden; region 2 as well as region 1 for $E > V_0$ are classically allowed. The SEq in regions 1 and 2 are

$$E\varphi_1 = -\frac{\hbar^2}{2m}\varphi_1'' + V_0\varphi_1; \quad E\varphi_2 = -\frac{\hbar^2}{2m}\varphi_2'', \tag{X.64}$$

and it follows that

$$\varphi_1'' = -\frac{2m}{\hbar^2}(E - V_0)\,\varphi_1 = \gamma^2\varphi_1; \quad \varphi_2'' = -\frac{2m}{\hbar^2}E\varphi_2 = -k^2\varphi_2 \tag{X.65}$$

with

$$\begin{array}{cc} \gamma^2 = -k'^2 \\ \gamma^2 = \kappa^2 \end{array} \text{ for } \begin{array}{c} E > V_0 \\ E < V_0 \end{array} \text{ or } \begin{array}{c} k'^2 = \frac{2m}{\hbar^2}(E - V_0) \\ \kappa^2 = \frac{2m}{\hbar^2}(V_0 - E). \end{array} \tag{X.66}$$

The solutions are

$$\varphi_1 = Ce^{\gamma x} + De^{-\gamma x}; \quad \varphi_2 = Fe^{ikx} + Ge^{-ikx}. \tag{X.67}$$

All four partial solutions are physically allowed (i.e. all four integration constants are nonzero). In particular, we have

$$\varphi_{\text{ein}} = Ge^{-ikx}; \quad \varphi_{\text{refl}} = Fe^{ikx}. \tag{X.68}$$

The matching conditions

$$\varphi_1(0) = 0; \quad \varphi_1(L) = \varphi_2(L); \quad \varphi_1'(L) = \varphi_2'(L) \tag{X.69}$$

yield

$$D = -C; \quad Ce^{\gamma L} + De^{-\gamma L} = Fe^{ikL} + Ge^{-ikL}$$
$$\gamma Ce^{\gamma L} - \gamma De^{-\gamma L} = ikFe^{ikL} - ikGe^{-ikL}. \tag{X.70}$$

Resolving gives

$$C = \frac{2k}{e^{\gamma L}(k + i\gamma) - e^{-\gamma L}(k - i\gamma)}Ge^{-ikL}$$
$$F = \frac{e^{\gamma L}(k - i\gamma) - e^{-\gamma L}(k + i\gamma)}{e^{\gamma L}(k + i\gamma) - e^{-\gamma L}(k - i\gamma)}Ge^{-2ikL}. \tag{X.71}$$

We calculate the reflection coefficient $R$ by using the probability current density:

$$j_{\text{ein}} = \frac{\hbar k}{m}|G|^2; \quad j_{\text{refl}} = \frac{\hbar k}{m}|F|^2; \quad R = \left|\frac{j_{\text{refl}}}{j_{\text{ein}}}\right| = \left|\frac{F}{G}\right|^2. \tag{X.72}$$

Since $R$ depends only on the term $F/G$, we consider in the following only $F$. For simplicity we rearrange:

$$F = \frac{k \sinh \gamma L - i\gamma \cosh \gamma L}{k \sinh \gamma L + i\gamma \cosh \gamma L} G e^{-2ikL} = -\frac{\cosh \gamma L + i\frac{k}{\gamma} \sinh \gamma}{\cosh \gamma L - i\frac{k}{\gamma} \sinh \gamma} G e^{-2ikL}.$$

(X.73)

Because of $\cosh(ix) = \cos x$ and $\sinh(ix) = \sin x$, both terms $\cosh \gamma L$ and $\frac{k}{\gamma} \sinh \gamma$ are real for $\gamma = \kappa$ and for $\gamma = ik'$. This means that

$$\cosh \gamma L + i\frac{k}{\gamma} \sinh \gamma = \sqrt{\cosh^2 \gamma L + \left(\frac{k}{\gamma}\right)^2 \sinh^2 \gamma} \cdot e^{i \arctan \frac{k}{\gamma} \tanh \gamma}, \quad (X.74)$$

and from this, it follows that

$$F = -e^{2i \arctan \frac{k}{\gamma} \tanh \gamma} e^{-2ikL} G.$$

(X.75)

Since $\frac{k}{\gamma} \tanh \gamma$ is real for $\gamma = \kappa$ and for $\gamma = ik'$, we see directly and without much arithmetic that for $E < V_0$ as well as for $E > V_0$, we have:

$$R = \left|\frac{F}{G}\right|^2 = 1$$

(X.76)

as indeed must hold.

7. Given a potential step embedded in an infinite potential well

$$V(x) = \begin{cases} 0 & 0 < x < L \\ V_0 > 0 & \text{for} \quad -L < x \leq 0. \\ \infty & x \geq |L| \end{cases}$$

(X.77)

Calculate the spectrum for $E > V_0$.

Solution: The potential is outlined in Fig. X.4. The wavefunctions are given by

$$\varphi_1 = A e^{ik_1 x} + B e^{-ik_1 x}; \quad \varphi_2 = C e^{ik_2 x} + D e^{-ik_2 x}$$

(X.78)

with

$$k_1^2 = \frac{2m}{\hbar^2}(E - V_0) = k_2^2 - \frac{2m}{\hbar^2}V_0; \quad k_2^2 = \frac{2m}{\hbar^2}E.$$

(X.79)

At $x = 0$, we find:

$$A + B = C + D$$
$$ik_1 A - ik_1 B = ik_2 C - ik_2 D$$

(X.80)

and at $x = \pm L$:

$$Ae^{-ik_1L} + Be^{ik_1L} = 0$$
$$Ce^{ik_2L} + De^{-ik_2L} = 0. \qquad (X.81)$$

These are four equations for four unknowns; in order that a solution exists, the determinant of the coefficients $D$ must vanish. This means that

$$D = \begin{vmatrix} 1 & 1 & -1 & -1 \\ k_1 & -k_1 & -k_2 & k_2 \\ e^{-ik_1L} & e^{ik_1L} & 0 & 0 \\ 0 & 0 & e^{ik_2L} & e^{-ik_2L} \end{vmatrix} = 0. \qquad (X.82)$$

The determinant is calculated to give (e.g. by expansion with respect to the third row):

$$D = 2i \left[ (k_2 + k_1) \sin (k_2L + k_1L) - (k_2 - k_1) \sin (k_2L - k_1L) \right]. \qquad (X.83)$$

Hence, the energy levels are determined by the equation

$$\sin (k_2L + k_1L) = \frac{k_2 - k_1}{k_2 + k_1} \sin (k_2L - k_1L). \qquad (X.84)$$

A closed solution does not exist; however, we can see directly from (X.84) that the spectrum is discrete (of course, this is due to the infinite potential walls).
In the following, we are concerned only with the approximate solution of (X.84).
First, we note that the ratio on the right side goes to zero for large $E$, since with (X.79) we have:

**Fig. X.4** The potential of (X.77)

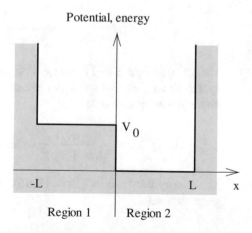

$$\frac{k_2 - k_1}{k_2 + k_1} = \frac{\sqrt{E} - \sqrt{E - V_0}}{\sqrt{E} + \sqrt{E - V_0}} = \frac{1 - \sqrt{1 - \frac{V_0}{E}}}{1 + \sqrt{1 - \frac{V_0}{E}}} \approx \frac{V_0}{4E}. \tag{X.85}$$

Hence, for sufficiently large $E$, the right side of (X.84) is approximately zero. Then it follows that

$$(k_2 + k_1) L = N\pi + \varepsilon_N, \tag{X.86}$$

where $\varepsilon_N$ is a small correction term and $N$ a sufficiently large natural number (because of $k_2 + k_1 \approx 2\sqrt{\frac{2m}{\hbar^2} E}$).
We rewrite (X.84) in the form

$$\sin\left((k_2 + k_1) L\right) = \frac{L^2 \left(k_2^2 - k_1^2\right)}{L^2 (k_2 + k_1)^2} \sin\left(\frac{L^2 \left(k_2^2 - k_1^2\right)}{L (k_2 + k_1)}\right). \tag{X.87}$$

With (X.79), it follows that:

$$L^2 \left(k_2^2 - k_1^2\right) = L^2 \frac{2m}{\hbar^2} V_0 = \mu^2 \tag{X.88}$$

and this gives

$$\sin\left((k_2 + k_1) L\right) = \frac{\mu^2}{L^2 (k_2 + k_1)^2} \sin\left(\frac{\mu^2}{L (k_2 + k_1)}\right); \tag{X.89}$$

and with (X.86)

$$\sin\left(N\pi + \varepsilon_N\right) = \frac{\mu^2}{(N\pi + \varepsilon_N)^2} \sin\left(\frac{\mu^2}{N\pi + \varepsilon_N}\right). \tag{X.90}$$

The left side equals $(-1)^N \sin \varepsilon_N$. For sufficiently large energies (which corresponds to large $N$ and small $\varepsilon_N$), we can use the approximation $\sin x \approx x$ and obtain the approximate result:

$$(-1)^N \varepsilon_N = \frac{\mu^4}{(N\pi + \varepsilon_N)^3} \quad \text{or} \quad \varepsilon_N = (-1)^N \frac{\mu^4}{(N\pi)^3} \sim \frac{1}{N^3}. \tag{X.91}$$

The energy levels follow from (X.86). We notice that due to (X.79), it holds generally that

$$k_2 + k_1 = \alpha \;\rightarrow\; E = \frac{V_0}{4}\left(\sqrt{\frac{\hbar^2}{2mV_0}}\,\alpha + \sqrt{\frac{2mV_0}{\hbar^2}}\frac{1}{\alpha}\right)^2$$

$$= \frac{V_0}{4}\left(\frac{\alpha}{\mu L} + \frac{\mu L}{\alpha}\right)^2. \tag{X.92}$$

With $\alpha = \frac{N\pi + \varepsilon_N}{L}$, it follows that

$$E_N = \frac{V_0}{4}\left(\frac{N\pi + \varepsilon_N}{\mu} + \frac{\mu}{N\pi + \varepsilon_N}\right)^2. \tag{X.93}$$

We expand the right side in terms of powers of $N$ and keep only the two largest terms. Due to $\varepsilon_N \sim N^{-3}$, we can neglect the correction term $\varepsilon_N$, and obtain approximately

$$E_N \approx \frac{V_0}{4}\frac{N^2\pi^2}{\mu^2}\left(1 + 2\frac{\mu^2}{N^2\pi^2}\right) = \frac{\hbar^2 N^2\pi^2}{8mL^2}\left(1 + \frac{4mV_0L^2}{N^2\pi^2\hbar^2}\right). \tag{X.94}$$

So we have a discrete energy spectrum which for sufficiently large $N$ (i.e. for sufficiently high energies) is by and large that of the infinite potential well. The existence of the potential step results essentially in a slight raising of all the energy levels.

8. (Resonances) Given a potential barrier in front of an infinite potential wall:

$$V(x) = \begin{cases} \infty & x < 0 \\ V_0 > 0 & \text{for} \quad a \leq x \leq b. \\ 0 & \text{otherwise}. \end{cases} \tag{X.95}$$

The incident quantum object has the energy $E < V_0$ and comes from the right. For which parameter values is the phase shift of the outgoing wave particularly large/does the phase change especially fast? What is the physical explanation? Solution: The potential is outlined in Fig. X.5.
The *ansatz* is (region 1: $0 \leq x \leq a$; region 2: $a \leq x \leq b$; region 3: $b \leq x$)

$$\begin{aligned} \psi_1 &= Ae^{ikx} + Be^{-ikx} \\ \psi_2 &= Ce^{\kappa x} + De^{-\kappa x} \\ \psi_3 &= Fe^{ikx} + Ge^{-ikx} \end{aligned} \tag{X.96}$$

with $k^2 = 2mE/\hbar^2$ and $\kappa^2 = 2m\left(V_0 - E\right)/\hbar^2$. The term $Ge^{-ikx}$ describes the incoming and $Fe^{ikx}$ the scattered (reflected) object.
At $x = 0$, we have $A = -B$. For the two other discontinuities, we find

$$Ae^{ika} - Ae^{-ika} = Ce^{\kappa a} + De^{-\kappa a}$$
$$ikAe^{ika} + ikAe^{-ika} = \kappa Ce^{\kappa a} - \kappa De^{-\kappa a} \tag{X.97}$$

and

$$Ce^{\kappa b} + De^{-\kappa b} = Fe^{ikb} + Ge^{-ikb}$$
$$\kappa Ce^{\kappa b} - \kappa De^{-\kappa b} = ikFe^{ikb} - ikGe^{-ikb}. \tag{X.98}$$

With the abbreviations

$$A' = 2iA; \quad C' = Ce^{\kappa a}; \quad D' = De^{-\kappa a}$$
$$F' = Fe^{ikb}; \quad G' = Ge^{-ikb}; \quad d = b - a, \tag{X.99}$$

it follows that

$$A' \sin ka = C' + D'$$
$$\frac{k}{\kappa} A' \cos ka = C' - D' \tag{X.100}$$

and

$$C'e^{\kappa d} + D'e^{-\kappa d} = F' + G'$$
$$\kappa C'e^{\kappa d} - \kappa D'e^{-\kappa d} = ikF' - ikG'. \tag{X.101}$$

This leads initially to

$$C' = A' \frac{\sin ka + \frac{k}{\kappa} \cos ka}{2}; \quad D' = A' \frac{\sin ka - \frac{k}{\kappa} \cos ka}{2} \tag{X.102}$$

and thus to

**Fig. X.5** The potential of (X.95)

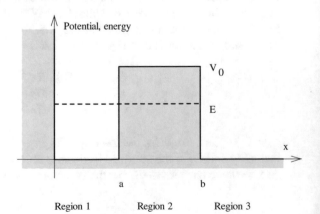

Potential, energy

$V_0$

E

x

a

b

Region 1          Region 2          Region 3

$$A' \frac{\sin ka + \frac{k}{\kappa}\cos ka}{2} e^{\kappa d} + A' \frac{\sin ka - \frac{k}{\kappa}\cos ka}{2} e^{-\kappa d} = F' + G'$$

$$\kappa A' \frac{\sin ka + \frac{k}{\kappa}\cos ka}{2} e^{\kappa d} - \kappa A' \frac{\sin ka - \frac{k}{\kappa}\cos ka}{2} e^{-\kappa d} = ikF' - ikG'.$$

$$(X.103)$$

It follows that

$$A' \frac{\sin ka + \frac{k}{\kappa}\cos ka}{2} e^{\kappa d} + A' \frac{\sin ka - \frac{k}{\kappa}\cos ka}{2} e^{-\kappa d} = F' + G'$$

$$A' \frac{\kappa}{ik} \frac{\sin ka + \frac{k}{\kappa}\cos ka}{2} e^{\kappa d} - A' \frac{\kappa}{ik} \frac{\sin ka - \frac{k}{\kappa}\cos ka}{2} e^{-\kappa d} = F' - G'$$

$$(X.104)$$

and therefore

$$2\frac{F'}{A'} = \frac{\sin ka + \frac{k}{\kappa}\cos ka}{2} e^{\kappa d} + \frac{\sin ka - \frac{k}{\kappa}\cos ka}{2} e^{-\kappa d}$$

$$+ \frac{\kappa}{ik}\frac{\sin ka + \frac{k}{\kappa}\cos ka}{2} e^{\kappa d} - \frac{\kappa}{ik}\frac{\sin ka - \frac{k}{\kappa}\cos ka}{2} e^{-\kappa d}$$

$$2\frac{G'}{A'} = \frac{\sin ka + \frac{k}{\kappa}\cos ka}{2} e^{\kappa d} + \frac{\sin ka - \frac{k}{\kappa}\cos ka}{2} e^{-\kappa d}$$

$$- \frac{\kappa}{ik}\frac{\sin ka + \frac{k}{\kappa}\cos ka}{2} e^{\kappa d} + \frac{\kappa}{ik}\frac{\sin ka - \frac{k}{\kappa}\cos ka}{2} e^{-\kappa d}$$

$$(X.105)$$

or

$$2\frac{F'}{A'} = \frac{\sin ka + \frac{k}{\kappa}\cos ka}{2}\left(1 + \frac{\kappa}{ik}\right)e^{\kappa d} + \frac{\sin ka - \frac{k}{\kappa}\cos ka}{2}\left(1 - \frac{\kappa}{ik}\right)e^{-\kappa d}$$

$$2\frac{G'}{A'} = \frac{\sin ka + \frac{k}{\kappa}\cos ka}{2}\left(1 - \frac{\kappa}{ik}\right)e^{\kappa d} + \frac{\sin ka - \frac{k}{\kappa}\cos ka}{2}\left(1 + \frac{\kappa}{ik}\right)e^{-\kappa d}$$

$$(X.106)$$

and finally

$$\frac{F'}{G'} = \frac{\frac{\sin ka + \frac{k}{\kappa}\cos ka}{2}\left(1 + \frac{\kappa}{ik}\right)e^{\kappa d} + \frac{\sin ka - \frac{k}{\kappa}\cos ka}{2}\left(1 - \frac{\kappa}{ik}\right)e^{-\kappa d}}{\frac{\sin ka + \frac{k}{\kappa}\cos ka}{2}\left(1 - \frac{\kappa}{ik}\right)e^{\kappa d} + \frac{\sin ka - \frac{k}{\kappa}\cos ka}{2}\left(1 + \frac{\kappa}{ik}\right)e^{-\kappa d}}. \quad (X.107)$$

We see directly that the right side has the form $\frac{z^*}{z}$ and therefore the absolute value 1 (as it must be; what comes in goes out again).
We rewrite the result:

$$\frac{F'}{G'} = \frac{(\kappa + k\cot ka)\left(1 + \frac{\kappa}{ik}\right)e^{\kappa d} + (\kappa - k\cot ka)\left(1 - \frac{\kappa}{ik}\right)e^{-\kappa d}}{(\kappa + k\cot ka)\left(1 - \frac{\kappa}{ik}\right)e^{\kappa d} + (\kappa - k\cot ka)\left(1 + \frac{\kappa}{ik}\right)e^{-\kappa d}} = e^{2i\theta}$$

$$(X.108)$$

with the phase

$$\theta = -\arctan \frac{\kappa}{k} \frac{(\kappa + k \cot ka) e^{\kappa d} - (\kappa - k \cot ka) e^{-\kappa d}}{(\kappa + k \cot ka) e^{\kappa d} + (\kappa - k \cot ka) e^{-\kappa d}}. \qquad \text{(X.109)}$$

We check first the case $d = 0$. It follows that

$$\theta = -\arctan \frac{\kappa}{k} \frac{2k \cot ka}{2\kappa} = -\arctan \cot ka$$

$$= -\arctan \tan \left( \frac{\pi}{2} + ka \right) = \frac{\pi}{2} + ka. \qquad \text{(X.110)}$$

This means that

$$\frac{F'}{G'} = -e^{2ika} \quad \text{or} \quad \frac{F}{G} = -e^{2ika-2ikb} = -e^{-2ikd} = -1 \qquad \text{(X.111)}$$

as expected.

Next we examine the case $\kappa d \gg 0$ (whereby we assume $\kappa > 0$). Then it follows from (X.109) due to $e^{-\kappa d} \approx 0$ that:

$$\theta \approx -\arctan \frac{\kappa}{k} \quad \text{for } \kappa + k \cot ka \neq 0$$

$$\to \frac{F'}{G'} = \frac{1 + \frac{\kappa}{ik}}{1 - \frac{\kappa}{ik}} \quad \text{or} \quad \frac{F}{G} = \frac{1 + \frac{\kappa}{ik}}{1 - \frac{\kappa}{ik}} e^{2ikb} \qquad \text{(X.112)}$$

$$\theta = \arctan \frac{\kappa}{k} \quad \text{for } \kappa + k \cot ka = 0$$

$$\to \frac{F'}{G'} = \frac{1 - \frac{\kappa}{ik}}{1 + \frac{\kappa}{ik}} \quad \text{or} \quad \frac{F}{G} = \frac{1 - \frac{\kappa}{ik}}{1 + \frac{\kappa}{ik}} e^{2ikb}. \qquad \text{(X.113)}$$

So we have a sudden change in the phase at those energies which are determined by $\kappa + k \cot ka = 0$. What is the physical reason?

The equation $\kappa + k \cot ka = 0$ gives the positions of the energy levels in the potential well of length $a$:

$$V(x) = \begin{cases} 0 & L < x \\ -V_0 & \text{for} \quad 0 < x \leq L \\ \infty & x \leq 0 \end{cases} \qquad \text{(X.114)}$$

with $V_0 > 0$. Now for the current problem, we do not have bound stable states, i.e. states of infinite lifetime, but nevertheless we find states which have a certain lifetime, called metastable states or resonances. Their energetic positions agree for sufficiently large $d$ approximately with the positions of the bound levels of the potential well (X.114).

Hence, the zeros of the phase (X.109) are crucial for the position of the resonances, i.e.

$$\kappa \sin ka + k \cos ka + (\kappa \sin ka - k \cos ka) e^{-2\kappa d} = 0. \qquad (X.115)$$

We rewrite this with the abbreviations $e^{-2\kappa d} = \varepsilon$ and $z = ka$. Due to $k^2 = 2mE/\hbar^2$ and $\kappa^2 = 2m(V_0 - E)/\hbar^2 = 2mV_0/\hbar^2 - k^2$, we arrive with $v^2 = 2ma^2 V_0/\hbar^2$ at

$$\sqrt{v^2 - z^2} \sin z + z \cos z + \left( \sqrt{v^2 - z^2} \sin z - z \cos z \right) \varepsilon = 0; \quad 0 \le z \le v. \qquad (X.116)$$

We insert $z = z_0 + \varepsilon z_1$ and compare equal powers of $\varepsilon$. In the zeroth approximation, the solutions are determined by

$$\sqrt{v^2 - z_0^2} \sin z_0 + z_0 \cos z_0 = 0; \quad 0 \le z_0 \le v. \qquad (X.117)$$

For the terms $\sim \varepsilon^1$, we find:

$$\left( \sqrt{v^2 - z_0^2} \cos z_0 + \cos z_0 - \frac{z_0 \sin z_0}{\sqrt{v^2 - z_0^2}} - z_0 \sin z_0 \right) z_1$$

$$+ \sqrt{v^2 - z_0^2} \sin z_0 - z_0 \cos z_0 = 0. \qquad (X.118)$$

We replace $\cos z_0$ with the help of (X.117); finally it follows that

$$z_1 = 2 z_0 \frac{v^2 - z_0^2}{v^2 \left( 1 + \sqrt{v^2 - z_0^2} \right)}. \qquad (X.119)$$

These corrections are always positive. This means that the resonances lie at somewhat higher energies than the stable energy levels of the potential (X.114).

9. In this chapter, a transcendental equation of the form

$$\tan kd = -\frac{k}{\kappa} = \; ; \; \kappa = \sqrt{\kappa_V^2 - k^2} \; ; \; k < \kappa_V \qquad (X.120)$$

occurs several times. Find an approximate solution for large $d$.
Solution: We have

$$\tan kd = -\frac{k}{\sqrt{\kappa_V^2 - k^2}} = -\tan \left( \arctan \frac{k}{\sqrt{\kappa_V^2 - k^2}} \right) = -\tan \left( \arcsin \frac{k}{\kappa_V} \right). \qquad (X.121)$$

The formulation with arcsin is simpler, because it contains no square root. It follows that:

$$kd + \arcsin \frac{k}{\kappa_V} = n\pi \; ; \; n = 1, 2, \ldots. \qquad (X.122)$$

**Fig. X.6** Double well
potential

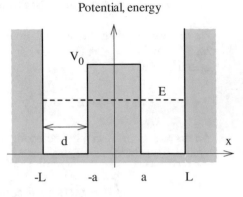

Region 1    Region 2    Region 3

Because of $0 \leq \arcsin x \leq \pi/2$ for $x \geq 0$, all solutions $k$ are confined to the
intervals $(n - 1/2)\,\pi < kd < n\pi$. Hence, for $d \to \infty$, we have $k \to 0$, and we
can use the power series expansion of the Arcus function for small arguments:

$$\arcsin x = x + \frac{x^3}{6} + O\left(x^5\right). \tag{X.123}$$

Neglecting terms of order 5 leads to

$$kd + \frac{k}{\kappa_V} + \frac{1}{6}\left(\frac{k}{\kappa_V}\right)^3 = n\pi \text{ or } k \cdot \frac{d\kappa_V + 1}{\kappa_V} = n\pi - \frac{1}{6}\left(\frac{k}{\kappa_V}\right)^3. \tag{X.124}$$

$\left(\frac{k}{\kappa_V}\right)^3$ is a very small term which we approximate using $k = n\pi\kappa_V/(d\kappa_V + 1)$.
Thus, we obtain

$$\frac{k}{\kappa_V} = \frac{n\pi}{d\kappa_V + 1}\left[1 - \frac{1}{6}\frac{(n\pi)^2}{(d\kappa_V + 1)^3}\right]. \tag{X.125}$$

10. Given the double well potential (see Fig. X.6):

$$\begin{array}{lll}
\text{region 1:} & -L \leq x \leq -a & V = 0 \\
\text{region 2:} & -a < x < a & V = V_0 > 0. \\
\text{region 3:} & a \leq x \leq L & V = 0
\end{array} \tag{X.126}$$

$V$ is infinite for $|x| > L$. We consider only energies $E$ for which $E < V_0$.

(a) Due to the symmetry of the problem $(H(x) = H(-x))$, there are symmetric
and antisymmetric eigenfunctions, sS and aS (cf. Chap. 21). Determine these
functions and their eigenvalue equations.
Solution: The *ansatz* for the wavefunction reads

$$\psi_1 = Ae^{ikx} + Be^{-ikx} \; ; \; \psi_2 = Ce^{\kappa x} + De^{-\kappa x} \; ; \; \psi_3 = Fe^{ikx} + Ge^{-ikx},$$
$$(X.127)$$

with $k, \kappa > 0, 0 < E < V_0$ and

$$k^2 = \frac{2m}{\hbar^2}E \; ; \; \kappa_V^2 = \frac{2m}{\hbar^2}V_0 \; ; \; \kappa^2 = \frac{2m}{\hbar^2}V_0 - \frac{2m}{\hbar^2}E = \kappa_V^2 - k^2 > 0.$$
$$(X.128)$$

The solutions sS and aS have to satisfy the following equation (the upper/lower sign denotes sS/aS)[158]:

$$\psi_1(x) = \pm\psi_3(-x) \; ; \; \psi_2(x) = \pm\psi_2(-x). \qquad (X.129)$$

This leads to (region 2 is classically forbidden):

$$\psi_1 = Ae^{ikx} + Be^{-ikx} \; ; \; \psi_2 = Ce^{\kappa x} \pm Ce^{-\kappa x} \; ; \; \psi_3 = \pm Be^{ikx} \pm Ae^{-ikx}.$$
$$(X.130)$$

The constants $B$ and $C$ are defined by means of the boundary conditions. At $x = -L$, we have

$$\psi_1(-L) = Ae^{-ikL} + Be^{ikL} = 0 \rightarrow B = -Ae^{-2ikL}, \qquad (X.131)$$

and at $x = -a$, we have

$$Ae^{-ika} + Be^{ika} = Ce^{-\kappa a} \pm Ce^{\kappa a} \; ; \; ikAe^{-ika} - ikBe^{ika} = \kappa Ce^{-\kappa a} \mp \kappa Ce^{\kappa a}.$$
$$(X.132)$$

In this way, we obtain for the wavefunction

$$
\text{sS:}
\begin{array}{l}
\psi_1 = A_s \sin k\,(x + L) \\
\psi_2 = C_s \cosh \kappa x \\
\psi_3 = -A_s \sin k\,(x - L) \\
C_s = \frac{\sin k(L-a)}{\cosh \kappa a} A_s
\end{array}
\quad ; \quad
\text{aS:}
\begin{array}{l}
\psi_1 = A_a \sin k\,(x + L) \\
\psi_2 = C_a \sinh \kappa x \\
\psi_3 = A_a \sin k\,(x - L) \\
C_a = -\frac{\sin k(L-a)}{\sinh \kappa a} A_a
\end{array}
\qquad (X.133)
$$

Therefore, the eigenvalue equations (obtained from (X.132) by division of the two equations) are:

$$
\begin{array}{l}
\text{sS:} \quad \tanh \kappa a \cdot \tan kd = -\frac{k}{\kappa} \\
\text{aS:} \quad \coth \kappa a \cdot \tan kd = -\frac{k}{\kappa}
\end{array}
\quad ; \; d = L - a, \qquad (X.134)
$$

which we write as

---

[158]It goes without saying that one can treat the problem without taking into account the symmetry properties right at the start. In this manner, the symmetry properties emerge by themselves in the course of the computation. However, the calculation is longer and more cumbersome—and the results are identical, of course.

$$\text{sS: } kd + \arctan\left(\tfrac{k}{\kappa}\coth\kappa a\right) = n\pi$$
$$\text{aS: } kd + \arctan\left(\tfrac{k}{\kappa}\tanh\kappa a\right) = n\pi \quad ; \ n = 1, 2, \ldots \quad \text{(X.135)}$$

(b) Show that there is no solution of the eigenvalue equations below a certain threshold value of $V_0$.

Solution: Due to $0 < \arctan x < \tfrac{\pi}{2}$ for $x > 0$, from (X.135), it follows immediately that

$$\left(n - \frac{1}{2}\right)\pi < kd < n\pi \ ; \ n = 1, 2, \ldots. \quad \text{(X.136)}$$

In particular, for $n = 1$, $\tfrac{\pi}{2} < kd < \pi$ holds. Thus, because of $0 < k < \kappa_V$, there is no solution of (X.135) for $d\kappa_V < \pi/2$, i.e. when the wells are too narrow and/or the potential $V_0$ is too low:

$$d^2 V_0 < \frac{\hbar^2\pi^2}{8m} \ \to \ \text{no solution.} \quad \text{(X.137)}$$

(c) Show that the ground state is symmetric.

Solution: Due to $0 < \tanh x < 1$ and $1 < \coth x < \infty$ for all $x > 0$, we can deduce from (X.135) the inequalities

$$\text{sS: } \left(n - \tfrac{1}{2}\right)\pi < kd < n\pi - \arctan\tfrac{k}{\kappa}$$
$$\text{aS: } \qquad n\pi - \arctan\tfrac{k}{\kappa} < kd < n\pi \quad ; \ n = 1, 2, \ldots. \quad \text{(X.138)}$$

We see immediately that the solutions for the symmetric cases are lower than those for the antisymmetric cases of the same order. In other words, the symmetric cases are energetically favorable and the ground state of the system is symmetric.

(d) Solve the eigenvalue equations approximately for the case of a 'thick' barrier, i.e. for very large $a$.

Solution: The limiting cases of the double well are (i) $\kappa a \to 0$ and (ii) $\kappa a \to \infty$, i.e. asymptotically (i) a single potential well of length $2L$ and (ii) two separated potential wells, each of length $d$. Of course, the position of the energy levels depends on the properties of the barrier. Schematically, this is shown in Fig. X.7.

We concentrate here on the case $\kappa a \gg 1$.

Due to $\tanh x = 1 - 2e^{-2x} + O\left(e^{-4x}\right)$ for $x \to \infty$, we can approximate (X.134) by (upper/lower sign for sS/aS):

$$kd + \arctan\left(\frac{k}{\kappa}\left(1 \pm 2e^{-2\kappa a}\right)\right) = n\pi \ ; \ n = 1, 2, \ldots. \quad \text{(X.139)}$$

For $\kappa a = \infty$, we have $kd + \arctan\tfrac{k}{\kappa} = n\pi$. Approximate solutions of this equation are given in (X.125); we denote them by $k_{\infty n}$ and write $\kappa_{\infty n} = $

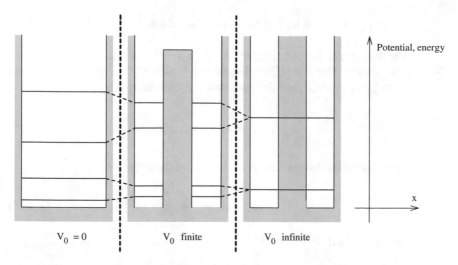

Potential, energy

x

$V_0 = 0$          $V_0$ finite          $V_0$ infinite

**Fig. X.7** Schematic representation of the energy levels in the double well for different barrier heights. Left for $V_0 = 0$, middle for $0 < V_0 < \infty$, right for $V_0 = \infty$. Not to scale

$\sqrt{\kappa_V^2 - k_{\infty n}^2}$. For an approximate solution of (X.139), we insert the *ansatz*

$$k = k_{\infty n} + \hat{k}_n \; ; \; \hat{k}_n = O\left(e^{-2a\kappa_{\infty n}}\right) \ll k_{\infty n} \qquad (X.140)$$

and retain only terms $O\left(\hat{k}_n^0\right)$ and $O\left(\hat{k}_n^1\right)$. The result reads (upper/lower sign for sS/aS):

$$\hat{k}_n = \mp k_{\infty n}\frac{\kappa_{\infty n}^2}{\kappa_V^2}\frac{2e^{-2\kappa_{\infty n}a}}{1 + d\kappa_{\infty n}} \text{ or } k = k_{\infty n}\left(1 \mp \frac{\kappa_{\infty n}^2}{\kappa_V^2}\frac{2e^{-2\kappa_{\infty n}a}}{1 + d\kappa_{\infty n}}\right).$$
$$(X.141)$$

This means that instead of the single energy level $E = \hbar^2 k_{\infty n}^2/2m$, we now have a doublet.[159]

As is seen, the determining factor is the exponential function $e^{-2\kappa_{\infty n}a}$, due to which the quantities may react very sensitively to small modifications of the potential. Hence, changing e.g. $a$ or $\kappa_V$ may have drastic effects.

(e) The initial state is assumed to be a linear combination of the symmetric and the antisymmetric states of the same order (for the sake of simplicity with equal amplitudes, $A_s = A_a = A$). Determine the time behavior of the wavefunction. Calculate the probabilities $P_i(t)$ of finding the object in region $i$.
Solution: The wavefunction for $t > 0$ is given by:

---

[159]Thus, in a double well, we observe a splitting of the energy levels into two terms. Correspondingly, in a triple well there is a splitting into three terms, and in an $n$-fold well into $n$ terms. For large $n$, this leads to the band structure of solids.

$$\Phi_3 (x, t) = -A \sin k_s (x - L) e^{i\omega_s t} + A \sin k_a (x - L) e^{i\omega_a t}, \qquad \text{(X.142)}$$

where $k_s$ and $k_a$ are solutions of (X.135); the coefficient may be written as $A = |A| e^{i\alpha}$. The probability density for locating the object in region 3 is given by $|\Phi_3 (x, t)|^2$, and the corresponding probability by

$$P_3 (t) = \int_a^L |\Phi_3 (x, t)|^2 \, dx. \qquad \text{(X.143)}$$

Carrying out the integration and introducing the abbreviation

$$\Delta \omega = \omega_a - \omega_s = \frac{\hbar}{2m} \left( k_a^2 - k_s^2 \right), \qquad \text{(X.144)}$$

we obtain finally

$$P_3 (t) \cdot \frac{2}{d} \frac{1}{|A|^2} = \left[ 1 - \frac{\sin (2k_s d)}{2k_s d} \right] + \left[ 1 - \frac{\sin (2k_a d)}{2k_a d} \right]$$
$$+ 2 \left[ \frac{\sin ((k_a + k_s)d)}{(k_a + k_s) d} - \frac{\sin ((k_a - k_s)d)}{(k_a - k_s) d} \right] \cdot \cos (\Delta \omega t).$$
$$\text{(X.145)}$$

As the calculation shows, $P_1 (t)$ has the same time-independent part as $P_3 (t)$, while the time-dependent part has the opposite sign from $P_3 (t)$. In region 2, the probability $P_2$ is time independent (but not the probability density):

$$P_2 = |C_s|^2 \left( \frac{\sinh 2\kappa_s a}{2\kappa_s} + a \right) + |C_a|^2 \left( \frac{\sinh 2\kappa_a a}{2\kappa_a} - a \right) > 0. \quad \text{(X.146)}$$

In this way, the total probability $P_1 (t) + P_2 + P_3 (t)$ is time independent, as it should be.[160]

We see that $P_3 (t)$ oscillates with a frequency $\Delta \omega / 2\pi$ about the time-independent part of $P_3 (t)$. Thus, a part of the position probability swings back and forth between regions 1 and 3. Such behavior is forbidden in classical mechanics, where for $E < V_0$, the two wells are strictly separated even for finite $V_0$. In quantum mechanics, the two regions 1 and 3 are 'coupled' due to the tunnel effect; this kind of barrier penetration occurs in many different physical situations without having a classical analogue.

(f) In the case of a thick barrier, it holds that $k_a - k_s \ll k_a + k_s$. Calculate up to and including quadratic terms in $k_a - k_s$ the quantities $R_{max}^{min} = \min (P_3) / \max (P_3)$ and $\Delta \omega$. Discuss your findings.

Solution: The extrema of $P_3 (t)$ are found at $\cos (\Delta \omega t) = \pm 1$. Inserting

---

[160]$|A|^2$ has to be chosen in such a way that the wavefunction is normalized, i.e. so that $P_1 (t) + P_2 + P_3 (t) = 1$ holds.

(X.141) into (X.145) and expanding in terms of powers of $\hat{k}_n = \frac{k_a - k_s}{2}$, we obtain up to and including terms $O\left(\hat{k}_n^2\right)$[161]:

$$R_{\max}^{\min} = \left(2\hat{k}_n d\right)^2 F\left(2k_{\infty n}d\right) \tag{X.147}$$

with

$$F\left(2k_{\infty n}d\right) = \frac{\frac{1}{6} - \frac{\sin 2k_{\infty n}d}{2k_{\infty n}d} \cdot \left(\frac{1}{(2k_{\infty n}d)^2} - \frac{1}{2}\right) + \frac{\cos 2k_{\infty n}d}{(2k_{\infty n}d)^2}}{2\left(1 - \frac{\sin 2k_{\infty n}d}{2k_{\infty n}d}\right)}. \tag{X.148}$$

Since $F\left(2k_{\infty n}d\right)$ is a bounded and well-behaved function,[162] the behavior of $R_{\max}^{\min}$ is essentially determined by the factor $\left(2\hat{k}_n d\right)^2$. Inserting $\hat{k}_n$ from (X.141), we obtain

$$R_{\max}^{\min} = \left(4dk_{\infty n} \frac{\kappa_{\infty n}^2}{\kappa_V^2} \frac{e^{-2\kappa_{\infty n}a}}{1 + d\kappa_{\infty n}}\right)^2 \cdot F\left(2k_{\infty n}d\right) \tag{X.149}$$

and

$$\Delta\omega = \frac{4\hbar}{m}k_{\infty n}^2 \frac{\kappa_{\infty n}^2}{\kappa_V^2} \frac{e^{-2\kappa_{\infty n}a}}{1 + d\kappa_{\infty n}}. \tag{X.150}$$

Thus, we have a periodic exchange of probabilities between regions 1 and 3 with the frequency $f = \Delta\omega/2\pi$. If $R_{\max}^{\min}$ is very small, $P_3$ becomes periodically 'practically' zero, although it never strictly vanishes, due to its definition (X.143). This situation resembles neutrino oscillations (cf. Chap. 8, Vol. 1, exchange of neutrino types) or beats in coupled pendulums (exchange of energy).[163] As mentioned above, the determining factor is the exponential function $e^{-2\kappa_{\infty n}a}$ due to which the quantities may react very sensitively to small changes in the potential.[164]

(g) In the ammonia molecule $NH_3$, the $N$ atom tunnels back and forth through the plane of the three $H$ atoms. This situation can be modelled by the double well potential with parameters $a = 0.2 \cdot 10^{-10}$ m, $d = 0.3 \cdot 10^{-10}$ m,

---

[161] Due to the choice $A_s = A_a = A$, there are no terms $O\left(\hat{k}_n^1\right)$.

[162] The function $F(x)$ oscillates with a period $x = 2\pi$; for $x > \pi$, we have $0.02 < F(x) < 0.14$ and $F(x) \to \frac{1}{12}$ for $x \to \infty$. The notation may be simplified somewhat by the use of spherical Bessel functions; see Appendix A, Vol. 2.

[163] Indeed, there may be a difference between the two examples mentioned and the double well, since in the latter, a substantial part of the probability may be contained in region 2.

[164] This sensitivity of the tunnel effect to the potential-barrier properties is also responsible for the enormous range of decay times observed for alpha decay.

$V_0 = 0.255$ eV and $m = 4 \cdot 10^{-27}$ kg (the reduced mass is $\frac{3m_H m_N}{3m_H + m_N}$).[165]
Compute numerical values for the ground-state levels, the frequency and
$R_{\max}^{\min}$. Discuss your findings.
Solution: Inserting the given data, we obtain

$$\kappa_V = 0.183 \cdot 10^{12} \text{ m}^{-1} \; ; k_{\infty 1} = 0.880 \cdot 10^{11} \text{ m}^{-1} \; ; \kappa_{\infty 1} = 0.160 \cdot 10^{12} \text{ m}^{-1}.$$

$$(X.151)$$

These data give $a\kappa_{\infty 1} = 3.2$, which means that we have a 'thick' barrier and
can apply (X.141). It follows (upper/lower signs for sS/aS) that:

$$k_{s,a} = 0.880 \cdot 10^{11} \left(1 \mp 0.433 \cdot 10^{-3}\right) \text{ m}^{-1}.$$

$$(X.152)$$

For the lowest (unsplit) energy level, we have[166]:

$$E_{\infty 1} = \frac{\hbar^2}{2m} k_{\infty 1}^2 = 0.591 \cdot 10^{-1} \text{ eV} \,\hat{=}\, 0.143 \cdot 10^{14} \text{ Hz}.$$

$$(X.153)$$

The energy splitting of the lowest level is given by

$$\Delta E_{a,s} = \frac{\hbar^2}{2m} \left(k_a^2 - k_s^2\right) = 0.102 \cdot 10^{-3} \text{ eV} \,\hat{=}\, 0.247 \cdot 10^{11} \text{ Hz} \,\hat{=}\, 0.824 \text{ cm}^{-1}.$$

$$(X.154)$$

Finally, the ratio $R_{\max}^{\min} = \min(P_3) / \max(P_3)$ has the value

$$R_{\max}^{\min} = 0.253 \cdot 10^{-6}.$$

$$(X.155)$$

For an intuitive picture, we discuss the findings in terms of coupled pendulums
(i.e. probability $\rightarrow$ energy).[167] The pendulums oscillate with the frequency
$0.143 \cdot 10^{14}$ Hz. The first pendulum pumps energy into the second pendulum
until all the energy is in the second pendulum and the first pendulum stops
(nearly stops, since we have $0.253 \cdot 10^{-6}$ and not exactly 0). Then the process
reverses, is repeated and so on. This continuous 'energy swapping' or beating
is a comparatively slow process—we find $\frac{0.143 \cdot 10^{14} \text{ Hz}}{0.247 \cdot 10^{11} \text{ Hz}} = 579$, i.e. a pendulum
oscillates several hundred times before the energy is swapped.

For ammonia, this means that the $N$ atom tunnels back and forth through the
$H_3$ plane with a frequency of $0.247 \cdot 10^{11}$ Hz (the 'real' value of this *inversion*

---

[165]This is a rather rough model, because $NH_3$ as a three-dimensional molecule has additional
degrees of freedom; furthermore, the potential would be better described by a coupling of two
harmonic potentials. Hence, one should not take the values for $a$ and $d$ too seriously - they represent
simply an order of magnitude. But it is possible to generate a double well potential like (X.126)
in the lab. Indeed, there are techniques to construct almost any arbitrary potentials using certain
semiconducting materials (keyword heterojunctions or heterostructures).

[166]For the conversion of the energy units, see e.g. Appendix A, Vol. 1.

[167]We can do this because the contribution of the classically-forbidden region 2 for the given data
is $P_2 \approx 0.02$, which we can neglect in this context.

*frequency* is $0.2387 \cdot 10^{11}$ Hz or 0.8 cm$^{-1}$).[168] The probability of finding the $N$ on one side of the $H_3$ plane varies periodically from nearly 1 to nearly 0 and back again during this process.[169]

11. For an illustration of the method of stationary phase, consider the (unnormalized) wavefunction

$$\psi(x,t) = \int_{-\infty}^{\infty} |A(k)| \, e^{i\varphi(k)} e^{i(kx - \omega t)} dk \qquad (X.156)$$

with

$$\omega = ck; \quad \varphi(k) = -x_0 k \qquad (X.157)$$

and

$$|A(k)| = \begin{cases} \kappa^2 - (k - K)^2 & \text{for} \quad 0 < K - \kappa \le k \le K + \kappa \\ 0 & \text{otherwise} \end{cases}. \qquad (X.158)$$

The constants $\kappa$, $K$ and $x_0$ are positive. Calculate explicitly $\psi(x,t)$ and discuss its properties. What is the physical significance of $x_0$?

Solution: We have to evaluate

$$\psi(x,t) = \int_{K-\kappa}^{K+\kappa} \left[ \kappa^2 - (k - K)^2 \right] e^{ik(x - ct - x_0)} dk. \qquad (X.159)$$

To simplify the notation, we introduce the abbreviation $\delta = x - ct - x_0$ and substitute $z = k - K$ in the integral. This leads to

$$\psi(x,t) = e^{iK\delta} \int_{-\kappa}^{\kappa} \left( \kappa^2 - z^2 \right) e^{iz\delta} dz. \qquad (X.160)$$

---

[168] We mention two applications: (i) This frequency is also used to identify ammonia in interstellar space (radio astronomy). (ii) Since $NH_3$ is polar, there is an oscillating dipole moment associated with the tunnelling through the $H_3$ plane, which fact is used in the ammonia *maser* (acronym for *Microwave Amplification by Stimulated Emission of Radiation*). The maser action in fact takes place between the two lowest levels considered above. Using external fields, one generates a population inversion with respect to these levels, followed by stimulated emission (with the frequency $0.247 \cdot 10^{11}$ Hz, i.e. in the microwave range). The extension of this concept to the realm of light (and, of course, to other materials) leads to the *laser* (*Light Amplification by Stimulated Emission of Radiation*).

[169] Structurally similar molecules demonstrate the sensitive dependence of tunnelling on the barrier potential in an exemplary manner. Thus, $NH_3$, $PH_3$ and $AsH_3$, with barrier heights of 0.25 eV, 0.75 eV and 1.39 eV, have inversion frequencies of $0.24 \cdot 10^{11}$ Hz, $0.14 \cdot 10^6$ Hz and $0.16 \cdot 10^{-7}$ Hz (i.e. 0.5 per year).

**Fig. X.8** The functions $h(y)$ (*black*), (X.162) and $h^2(y)$ (*red*)

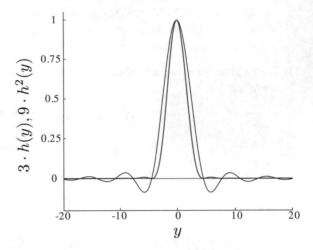

The integral may be evaluated by hand, by using software such as *Maple* or *mathematica*, or by an online integrator (e.g. http://integrals.wolfram.com/). The result can be brought into the form:

$$\psi(x,t) = 4e^{iK\delta}\kappa^3 \cdot \frac{\sin \kappa\delta - \kappa\delta \cos \kappa\delta}{(\kappa\delta)^3}. \tag{X.161}$$

As we see, $\psi(x,t)$ depends only on $\delta = x - ct - x_0$. Hence, the wavefunction moves along the $x$ axis without dispersion, i.e. without changing its shape in the course of time, $\psi(x,t) = \psi(x - t)$. The function

$$h(y) = \frac{\sin y - y \cos y}{y^3} \tag{X.162}$$

on the right-hand side of (X.161) determines the behavior of $\psi(x,t)$. It is shown in Fig. X.8.[170] The function has a maximum at $y = 0$ with $h(0) = \frac{1}{3}$, and a halfwidth of $\Delta y \approx 5$ (i.e. $h(\pm 2.498) \approx \frac{1}{6}$); the first zero lies at $y \approx 4.493$. With $\delta = x - ct - x_0$, this means that $\psi(x,t)$ has a pronounced maximum at $x - ct - x_0 = 0$. Thus, $x_0$ is the position of the maximum at the time $t=0$. The halfwidth of $\psi(x,t)$ is given by $\Delta x \approx \frac{5}{\kappa}$ and, correspondingly, the halfwidth of $|\psi(x,t)|^2$ is given by $\Delta x \approx \frac{3.6}{\kappa}$. In other words, the integral (X.156) yields essential contributions only in the neighborhood of the stationary phase $\frac{d}{dk}(kx - \omega t + \varphi(k)) = 0$.

---

[170]Note, by the way, that $h(y) = \frac{j_1(y)}{y}$, where $j_1(y)$ is a spherical Bessel function.

## X.2  Exercises, Chap. 16

1. For which $K, N, M$ are the spherical harmonics (in spherical coordinates)

$$f\left(\vartheta, \varphi\right) = \cos^K \vartheta \cdot \sin^M \vartheta \cdot e^{iN\varphi} \qquad (X.163)$$

eigenfunctions of $\mathbf{l}^2$?

2. Write out the spherical harmonics for $l = 1$ using Cartesian coordinates, $x, y, z$.
   Solution: With $x = r \sin \vartheta \cos \varphi$, $y = r \sin \vartheta \sin \varphi$, $z = r \cos \vartheta$, it follows that

$$Y_1^0\left(\vartheta, \varphi\right) = \sqrt{\frac{3}{4\pi}} \frac{z}{r}; \quad Y_1^{\pm 1}\left(\vartheta, \varphi\right) = \mp\sqrt{\frac{3}{8\pi}} \frac{x \pm iy}{r}. \qquad (X.164)$$

3. Show that:

$$\mathbf{l} \cdot \hat{\mathbf{r}} = \hat{\mathbf{r}} \cdot \mathbf{l} = 0 \qquad (X.165)$$

4. Show that the components of $\mathbf{l}$ are Hermitian.
   Solution: We have

$$l_x = \frac{\hbar}{i}\left(y\frac{\partial}{\partial z} - z\frac{\partial}{\partial y}\right) = y p_z - z p_y. \qquad (X.166)$$

On the right side, we have products of two commuting Hermitian operators. Hence, $l_x$ is Hermitian. Analogously for the two other components of $\mathbf{l}$.

5. Show that for the orbital angular momentum, it holds that

$$\left[l_x, l_y\right] = i\hbar l_z; \quad \left[l_y, l_z\right] = i\hbar l_x; \quad \left[l_z, l_x\right] = i\hbar l_y. \qquad (X.167)$$

Solution: We first consider $\left[l_x, l_y\right]$. We have:

$$
\begin{aligned}
\left[l_x, l_y\right] &= \left(\frac{\hbar}{i}\right)^2 \left(y\frac{\partial}{\partial z} - z\frac{\partial}{\partial y}\right)\left(z\frac{\partial}{\partial x} - x\frac{\partial}{\partial z}\right) - \left(\frac{\hbar}{i}\right)^2 \left(z\frac{\partial}{\partial x} - x\frac{\partial}{\partial z}\right)\left(y\frac{\partial}{\partial z} - z\frac{\partial}{\partial y}\right) \\
&= \left(\frac{\hbar}{i}\right)^2 \left(y\frac{\partial}{\partial z}z\frac{\partial}{\partial x} - y\frac{\partial}{\partial z}x\frac{\partial}{\partial z} - z\frac{\partial}{\partial y}z\frac{\partial}{\partial x} + z\frac{\partial}{\partial y}x\frac{\partial}{\partial z}\right) \\
&\quad - \left(\frac{\hbar}{i}\right)^2 \left(z\frac{\partial}{\partial x}y\frac{\partial}{\partial z} - z\frac{\partial}{\partial x}z\frac{\partial}{\partial y} - x\frac{\partial}{\partial z}y\frac{\partial}{\partial z} + x\frac{\partial}{\partial z}z\frac{\partial}{\partial y}\right) \\
&= \left(\frac{\hbar}{i}\right)^2 \left(y\frac{\partial}{\partial z}z\frac{\partial}{\partial x} + z\frac{\partial}{\partial y}x\frac{\partial}{\partial z} - z\frac{\partial}{\partial x}y\frac{\partial}{\partial z} - x\frac{\partial}{\partial z}z\frac{\partial}{\partial y}\right) \\
&= \left(\frac{\hbar}{i}\right)^2 \left(y\frac{\partial}{\partial z}\frac{\partial}{\partial x} + yz\frac{\partial}{\partial z}z\frac{\partial}{\partial x} + zx\frac{\partial}{\partial y}\frac{\partial}{\partial z} - zy\frac{\partial}{\partial x}\frac{\partial}{\partial z} - x\frac{\partial}{\partial z}z\frac{\partial}{\partial y} - xz\frac{\partial}{\partial z}z\frac{\partial}{\partial y}\right) \\
&= \left(\frac{\hbar}{i}\right)^2 \left(y\frac{\partial}{\partial x} - x\frac{\partial}{\partial y}\right) = -\frac{\hbar}{i}l_z = i\hbar l_z.
\end{aligned} \qquad (X.168)
$$

The other two relations follow by cyclic commutation.
An alternative derivation uses the commutators $[x, p_x] = i\hbar$, $[x, p_y] = [x, p_z] = 0$; correspondingly for $y$, $z$. We have:

$$[l_x, l_y] = [yp_z - zp_y, zp_x - xp_z]$$
$$= [yp_z, zp_x] - [yp_z, xp_z] - [zp_y, zp_x] + [zp_y, xp_z]. \quad (X.169)$$

The second and the third commutators vanish due to $[yp_z, xp_z] = yx$ $[p_z, p_z] = 0$ and $[zp_y, zp_x] = z^2[p_y, p_x] = 0$. Then it follows that:

$$[l_x, l_y] = y[p_z, zp_x] + x[zp_y, p_z] = y[p_z, zp_x] - x[p_z, zp_y]. \quad (X.170)$$

With $[A, BC] = B[A, C] + [A, B]C$ (see next exercise), we obtain

$$[l_x, l_y] = yz[p_z, p_x] + y[p_z, z]p_x - xz[p_z, p_y] - x[p_z, z]p_y. \quad (X.171)$$

The first and the third commutator vanish; due to $[p_z, z] = -i\hbar$, it follows that:

$$[l_x, l_y] = -i\hbar yp_x + i\hbar xp_y = i\hbar(xp_y - yp_x) = i\hbar l_z. \quad (X.172)$$

6. Show that $[A, BC] = B[A, C] + [A, B]C$ holds. Using this identity and the commutators $[l_x, l_y] = i\hbar l_z$ plus cyclic permutations, prove that $[l_x, \mathbf{l}^2] = 0$.
7. Show that:

$$[\mathbf{J}^2, J_\pm] = 0. \quad (X.173)$$

8. We have seen in the text that

$$J_\pm |j, m\rangle = c_{j,m}^\pm |j, m \pm 1\rangle. \quad (X.174)$$

Using

$$J_+ J_- |j, m\rangle = \hbar^2 [j(j+1) - m(m-1)]|j, m\rangle$$
$$J_- J_+ |j, m\rangle = \hbar^2 [j(j+1) - m(m+1)]|j, m\rangle, \quad (X.175)$$

show that for the coefficients $c_{j,m}^\pm$,

$$c_{j,m}^\pm = \hbar\sqrt{j(j+1) - m(m \pm 1)} \quad (X.176)$$

holds.
Solution: We consider the first equation of (X.175). It follows that

$$J_+ J_- |j, m\rangle = J_+ c_{j,m}^- |j, m-1\rangle = c_{j,m-1}^+ c_{j,m}^- |j, m\rangle \quad (X.177)$$

and thus,

$$c^+_{j,m-1}c^-_{j,m} = \hbar^2\left[j(j+1) - m(m-1)\right]. \qquad (X.178)$$

(The second equation yields the same result for the index $m + 1$ instead of $m$).
We use the *ansatz*

$$c^+_{j,m} = \hbar\sqrt{j(j+1) - d^+_{j,m}}; \quad c^-_{j,m} = \hbar\sqrt{j(j+1) - d^-_{j,m}} \qquad (X.179)$$

and obtain

$$\sqrt{j(j+1) - d^+_{j,m-1}}\sqrt{j(j+1) - d^-_{j,m}} = j(j+1) - m(m-1). \quad (X.180)$$

Squaring and multiplying yields

$$j(j+1)d^-_{j,m} + j(j+1)d^+_{j,m-1} - d^+_{j,m-1}d^-_{j,m}$$
$$= 2j(j+1)m(m-1) - m^2(m-1)^2. \qquad (X.181)$$

Comparing equal powers of $j$ leads to

$$d^+_{j,m-1} + d^-_{j,m} = 2m(m-1); \quad d^+_{j,m-1}d^-_{j,m} = m^2(m-1)^2 \qquad (X.182)$$

and this yields

$$d^+_{j,m-1} = d^-_{j,m} = m(m-1) \qquad (X.183)$$

or

$$c^{\pm}_{j,m} = \hbar\sqrt{j(j+1) - m(m\pm1)}. \qquad (X.184)$$

9. Given the Pauli matrices $\sigma_k$,

   (a) Show (once more) that

$$[\sigma_i, \sigma_j] = 2i\varepsilon_{ijk}\sigma_k; \quad \{\sigma_i, \sigma_j\} = 2\delta_{ij}; \quad \sigma_i^2 = 1; \quad \sigma_i\sigma_j = i\varepsilon_{ijk}\sigma_k; \quad (X.185)$$

   (b) Prove that

$$(\sigma A)(\sigma B) = AB + i\sigma(A \times B) \qquad (X.186)$$

   where $\sigma$ is the vector $\sigma = (\sigma_1, \sigma_2, \sigma_3)$ and $A, B$ are three-dimensional vectors;

   (c) Show that every $2 \times 2$ matrix can be expressed as a linear combination of the three Pauli matrices and the unit matrix.

10. Given the orbital angular momentum operator $\mathbf{l}$ and the spin operator $\mathbf{s}$, show that $[l_z, \mathbf{s} \cdot \mathbf{l}] \neq 0$; $[s_z, \mathbf{s} \cdot \mathbf{l}] \neq 0$; $[l_z + s_z, \mathbf{s} \cdot \mathbf{l}] = 0$.
   Solution: We have

$$[l_z, \mathbf{s} \cdot \mathbf{l}] = s_x [l_z, l_x] + s_y [l_z, l_y] = i\hbar \left(s_x l_y - s_y l_x\right)$$
$$[s_z, \mathbf{s} \cdot \mathbf{l}] = l_x [s_z, s_x] + l_y [s_z, s_y] = i\hbar \left(l_x s_y - l_y s_x\right)$$
(X.187)

and from this, $[l_z + s_z, \mathbf{s} \cdot \mathbf{l}] = 0$.

11. The ladder operators for a generalized angular momentum are given as $J_\pm = J_x \pm i J_y$.

(a) Show that $[J_z, J_+] = \hbar J_+, [J_z, J_-] = -\hbar J_-, [J_+, J_-] = 2\hbar J_z$, as well as $\mathbf{J}^2 = \frac{1}{2}(J_+ J_- + J_- J_+) + J_z^2$.

(b) Show that it follows from the last equation that:

$$J_+ J_- = \mathbf{J}^2 - J_z (J_z - \hbar); \quad J_- J_+ = \mathbf{J}^2 - J_z (J_z + \hbar) \qquad \text{(X.188)}$$

and hence

$$J_+ J_- |j, m\rangle = \hbar^2 [j(j+1) - m(m-1)] |j, m\rangle$$
$$J_- J_+ |j, m\rangle = \hbar^2 [j(j+1) - m(m+1)] |j, m\rangle .$$
(X.189)

(c) Show that from the last two equations, it follows that:

$$j(j+1) - m(m-1) = (j-m)(j+m+1) \geq 0$$
$$j(j+1) - m(m+1) = (j+m)(j-m+1) \geq 0$$
(X.190)

and hence

$$-j \leq m \leq j. \qquad \text{(X.191)}$$

12. What is the matrix representation of the orbital angular momentum for $l = 1$?
Solution: We start from

$$J_\pm |j, m\rangle = \hbar \sqrt{j(j+1) - m(m \pm 1)} |j, m \pm 1\rangle \qquad \text{(X.192)}$$

and obtain as a first step:

$$\langle 1, m' | L_\pm | 1, m\rangle = \hbar \sqrt{2 - m(m \pm 1)} \delta_{m', m \pm 1}$$
$$\langle 1, m' | L_z | 1, m\rangle = \hbar m \delta_{m', m}.$$
(X.193)

It follows that

$$\langle 1, 1 | L_+ | 1, 0\rangle = \hbar\sqrt{2}; \quad \langle 1, 0 | L_+ | 1, -1\rangle = \hbar\sqrt{2}$$
$$\langle 1, 0 | L_- | 1, 1\rangle = \hbar\sqrt{2}; \quad \langle 1, -1 | L_- | 1, 0\rangle = \hbar\sqrt{2}$$
$$\langle 1, 1 | L_z | 1, 1\rangle = \hbar; \quad \langle 1, -1 | L_z | 1, -1\rangle = -\hbar.$$
(X.194)

All other matrix elements vanish.

We expand (X.193) into

$$\sum_{m,m'} |1, m'\rangle\langle 1, m'| L_\pm |1, m\rangle \langle 1, m|$$

$$= \sum_{m,m'} \hbar\sqrt{2 - m\,(m \pm 1)}\delta_{m',m\pm 1} |1, m'\rangle\langle 1, m|$$

$$\sum_{m,m'} |1, m'\rangle\langle 1, m'| L_z |1, m\rangle \langle 1, m| = \sum_{m,m'} \hbar m \delta_{m',m} |1, m'\rangle\langle 1, m|.$$

$$(X.195)$$

Since $\{|1, m\rangle\}$ is a CONS, it follows that

$$L_\pm = \hbar \sum_m \sqrt{2 - m\,(m \pm 1)}\,|1, m \pm 1\rangle \langle 1, m|$$

$$L_z = \hbar \sum_m m\,|1, m\rangle \langle 1, m|,$$

$$(X.196)$$

or, explicitly,

$$L_+ = \hbar\sqrt{2}\,|1, 1\rangle \langle 1, 0| + \hbar\sqrt{2}\,|1, 0\rangle \langle 1, -1|$$

$$L_- = \hbar\sqrt{2}\,|1, 0\rangle \langle 1, 1| + \hbar\sqrt{2}\,|1, -1\rangle \langle 1, 0| = L_+^\dagger$$

$$L_z = \hbar\,|1, 1\rangle \langle 1, 1| - \hbar\,|1, -1\rangle \langle 1, -1|.$$

$$(X.197)$$

Evidently, the explicit matrix form of the angular momentum operators depends on the representation of the basis vectors. If we choose e.g.

$$|1, 1\rangle \cong \begin{pmatrix} 1 \\ 0 \\ 0 \end{pmatrix}; \quad |1, 0\rangle \cong \begin{pmatrix} 0 \\ 1 \\ 0 \end{pmatrix}; \quad |1, -1\rangle \cong \begin{pmatrix} 0 \\ 0 \\ 1 \end{pmatrix} \qquad (X.198)$$

then it follows that

$$L_+ \cong \hbar\sqrt{2} \begin{pmatrix} 0 & 1 & 0 \\ 0 & 0 & 1 \\ 0 & 0 & 0 \end{pmatrix}; \quad L_- \cong \hbar\sqrt{2} \begin{pmatrix} 0 & 0 & 0 \\ 1 & 0 & 0 \\ 0 & 1 & 0 \end{pmatrix}; \quad L_z \cong \hbar \begin{pmatrix} 1 & 0 & 0 \\ 0 & 0 & 0 \\ 0 & 0 & -1 \end{pmatrix}$$

$$(X.199)$$

and thus,

$$L_x \cong \frac{\hbar}{\sqrt{2}} \begin{pmatrix} 0 & 1 & 0 \\ 1 & 0 & 1 \\ 0 & 1 & 0 \end{pmatrix}; \quad L_y \cong \frac{\hbar}{\sqrt{2}} \begin{pmatrix} 0 & -i & 0 \\ i & 0 & -i \\ 0 & i & 0 \end{pmatrix}; \quad L_z \cong \hbar \begin{pmatrix} 1 & 0 & 0 \\ 0 & 0 & 0 \\ 0 & 0 & -1 \end{pmatrix}.$$

$$(X.200)$$

Another choice of the basis vectors is

$$|1, 1\rangle \cong \pm \frac{1}{\sqrt{2}} \begin{pmatrix} 1 \\ i \\ 0 \end{pmatrix} ; \quad |1, 0\rangle \cong \begin{pmatrix} 0 \\ 0 \\ 1 \end{pmatrix} ; \quad |1, -1\rangle \cong \mp \frac{1}{\sqrt{2}} \begin{pmatrix} 1 \\ -i \\ 0 \end{pmatrix} . \quad \text{(X.201)}$$

It leads to

$$L_+ \cong \pm \hbar \begin{pmatrix} 0 & 0 & 1 \\ 0 & 0 & i \\ -1 & -i & 0 \end{pmatrix} ; \quad L_- \cong \pm \hbar \begin{pmatrix} 0 & 0 & -1 \\ 0 & 0 & i \\ 1 & -i & 0 \end{pmatrix} ; \quad L_z \cong \hbar \begin{pmatrix} 0 & -i & 0 \\ i & 0 & 0 \\ 0 & 0 & 0 \end{pmatrix}$$
$$\text{(X.202)}$$

or

$$L_x \cong \pm \hbar \begin{pmatrix} 0 & 0 & 0 \\ 0 & 0 & i \\ 0 & -i & 0 \end{pmatrix} ; \quad L_y \cong \pm \hbar \begin{pmatrix} 0 & 0 & -i \\ 0 & 0 & 0 \\ i & 0 & 0 \end{pmatrix} ; \quad L_z \cong \hbar \begin{pmatrix} 0 & -i & 0 \\ i & 0 & 0 \\ 0 & 0 & 0 \end{pmatrix} .$$
$$\text{(X.203)}$$

13. Consider the orbital angular momentum $l = 1$. Express the operator $e^{-i\alpha L_z/\hbar}$ as sum over dyadic products (representation-free). Specify for the bases (X.198) and (X.201).

    Solution: We have initially

$$e^{-i\alpha L_z/\hbar} = \sum \frac{(-i\alpha)^n}{n!} \left( \frac{L_z}{\hbar} \right)^n . \quad \text{(X.204)}$$

With $L_z = \hbar [|1, 1\rangle \langle 1, 1| - |1, -1\rangle \langle 1, -1|]$ (see (X.197)), it follows that

$$\left( \frac{L_z}{\hbar} \right)^2 = [|1, 1\rangle \langle 1, 1| - |1, -1\rangle \langle 1, -1|]^2 = |1, 1\rangle \langle 1, 1| + |1, -1\rangle \langle 1, -1|$$
$$\left( \frac{L_z}{\hbar} \right)^3 = [|1, 1\rangle \langle 1, 1| - |1, -1\rangle \langle 1, -1|][|1, 1\rangle \langle 1, 1| + |1, -1\rangle \langle 1, -1|] = \frac{L_z}{\hbar} .$$
$$\text{(X.205)}$$

Due to

$$e^{-i\alpha L_z/\hbar} = 1 + \sum_{n=1} \frac{(-i\alpha)^{2n}}{(2n)!} \left( \frac{L_z}{\hbar} \right)^{2n} + \sum_{n=0} \frac{(-i\alpha)^{2n+1}}{(2n+1)!} \left( \frac{L_z}{\hbar} \right)^{2n+1} \quad \text{(X.206)}$$

we obtain

$$e^{-i\alpha L_z/\hbar} = 1 + \left( \frac{L_z}{\hbar} \right)^2 \sum_{n=1} (-1)^n \frac{\alpha^{2n}}{(2n)!} - i \left( \frac{L_z}{\hbar} \right) \sum_{n=0} (-1)^n \frac{\alpha^{2n+1}}{(2n+1)!} .$$
$$\text{(X.207)}$$

This yields

$$e^{-i\alpha L_z/\hbar} = 1 + \left(\frac{L_z}{\hbar}\right)^2 (\cos\alpha - 1) - i\left(\frac{L_z}{\hbar}\right)\sin\alpha$$

$$= |1,1\rangle\langle 1,1|\, e^{-i\alpha} + |1,0\rangle\langle 1,0| + |1,-1\rangle\langle 1,-1|\, e^{i\alpha}. \quad \text{(X.208)}$$

The choice of basis (X.198) yields the representation

$$e^{-i\alpha L_z/\hbar} \underset{\text{(W.162)}}{\cong} \begin{pmatrix} e^{-i\alpha} & 0 & 0 \\ 0 & 1 & 0 \\ 0 & 0 & e^{i\alpha} \end{pmatrix}, \quad \text{(X.209)}$$

and the choice of basis (X.201) yields the representation

$$e^{-i\alpha L_z/\hbar} \underset{\text{(W.165)}}{\cong} \begin{pmatrix} \cos\alpha & -\sin\alpha & 0 \\ \sin\alpha & \cos\alpha & 0 \\ 0 & 0 & 1 \end{pmatrix}. \quad \text{(X.210)}$$

14. Calculate the term

$$e^{-i\frac{\gamma \hat{a}L}{\hbar}} = e^{-i\gamma \hat{a}l} \quad \text{(X.211)}$$

for the orbital angular momentum $l = 1$ and the basis (16.73).[171] $\hat{a}$ is the rotation axis (a unit vector), $\gamma$ the rotation angle. For reasons of economy, use the 'simplified' angular momentum $\mathbf{l} = \mathbf{L}/\hbar$, i.e. the theoretical units system.

(a) Express the rotations about the $x$-, $y$- and $z$-axis as matrices.
    Solution: The matrix representation of the angular momentum is

$$l_x \cong \begin{pmatrix} 0 & 0 & 0 \\ 0 & 0 & i \\ 0 & -i & 0 \end{pmatrix}; \; l_y \cong \begin{pmatrix} 0 & 0 & -i \\ 0 & 0 & 0 \\ i & 0 & 0 \end{pmatrix}; \; l_z \cong \begin{pmatrix} 0 & -i & 0 \\ i & 0 & 0 \\ 0 & 0 & 0 \end{pmatrix}. \quad \text{(X.212)}$$

Due to

$$l_x^2 \cong \begin{pmatrix} 0 & 0 & 0 \\ 0 & 1 & 0 \\ 0 & 0 & 1 \end{pmatrix}; \; l_y^2 \cong \begin{pmatrix} 1 & 0 & 0 \\ 0 & 0 & 0 \\ 0 & 0 & 1 \end{pmatrix}; \; l_z^2 \cong \begin{pmatrix} 1 & 0 & 0 \\ 0 & 1 & 0 \\ 0 & 0 & 0 \end{pmatrix}, \quad \text{(X.213)}$$

we have the relations

$$l_x^3 = l_x; \; l_y^3 = l_y \; ; \; l_z^3 = l_z. \quad \text{(X.214)}$$

For $e^{-ibl_k}$, it follows due to $l_i^4 = l_i^2$ etc. that:

---

[171] Of course, all the calculations may also be performed representation-free.

$$e^{-ibl_k} = \sum_{n=0} \frac{(-ib)^n}{n!} l_i^n = 1 + l_k^2 \left[ \sum_{n=0} \frac{(-1)^n b^{2n}}{(2n)!} - 1 \right]$$

$$- i \sum_{n=0} \frac{(-1)^n b^{2n+1}}{(2n+1)!} l_k \qquad (X.215)$$

and from this:

$$e^{-ibl_k} = 1 + [\cos b - 1] \cdot l_k^2 - i \sin b \cdot l_k. \qquad (X.216)$$

(b) Express the rotations about an axis $\hat{a}$ with rotation angle $\gamma$ as matrices (the angles in spherical coordinates are $\theta$ and $\varphi$; see Fig. X.9).
Solution: With the representation (X.212), we obtain

$$\hat{a}\mathbf{l} = i \begin{pmatrix} 0 & -a_z & -a_y \\ a_z & 0 & a_x \\ a_y & -a_x & 0 \end{pmatrix}. \qquad (X.217)$$

Since $\hat{a}$ is an unit vector, we have $a_x^2 + a_y^2 + a_z^2 = 1$. For $(\hat{a}l)^2$, the calculation gives (check yourself):

$$(\hat{a}\mathbf{l})^2 = \begin{pmatrix} 1 - a_x^2 & -a_x a_y & a_x a_z \\ -a_x a_y & 1 - a_y^2 & a_y a_z \\ a_x a_z & a_y a_z & 1 - a_z^2 \end{pmatrix} \qquad (X.218)$$

and it follows that

$$(\hat{a}\mathbf{l})^3 = \hat{a}\mathbf{l} \qquad (X.219)$$

This yields the following relation, corresponding to (X.216):

$$e^{-i\gamma\hat{a}\mathbf{l}} = \sum_{n=0} \frac{(-i\gamma)^n}{n!} (\hat{a}\mathbf{l})^n = 1 + [\cos\gamma - 1] \cdot (\hat{a}\mathbf{l})^2 - i \sin\gamma \cdot (\hat{a}\mathbf{l}).$$

$$(X.220)$$

In matrix representation, we obtain

$$e^{-i\gamma\hat{a}\mathbf{l}} \cong 1 + [\cos\gamma - 1] \begin{pmatrix} 1 - a_x^2 & -a_x a_y & a_x a_z \\ -a_x a_y & 1 - a_y^2 & a_y a_z \\ a_x a_z & a_y a_z & 1 - a_z^2 \end{pmatrix}$$

$$+ \sin\gamma \begin{pmatrix} 0 & -a_z & -a_y \\ a_z & 0 & a_x \\ a_y & -a_x & 0 \end{pmatrix}. \qquad (X.221)$$

To cast this expression in a more familiar form, we rewrite:

**Fig. X.9** Rotation about an axis $\hat{a}$

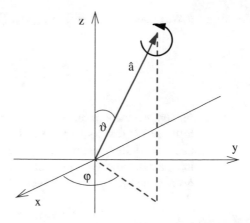

$$e^{-i\gamma\hat{a}\mathbf{l}} \cong \begin{pmatrix} \cos\gamma & -\sin\gamma & 0 \\ \sin\gamma & \cos\gamma & 0 \\ 0 & 0 & 1 \end{pmatrix} + (1-\cos\gamma)\begin{pmatrix} a_x^2 & a_x a_y & -a_x a_z \\ a_x a_y & a_y^2 & -a_y a_z \\ -a_x a_z & -a_y a_z & -\left(a_x^2 + a_y^2\right) \end{pmatrix}$$

$$+ \sin\gamma \begin{pmatrix} 0 & 1-a_z & -a_y \\ a_z-1 & 0 & a_x \\ a_y & -a_x & 0 \end{pmatrix}. \tag{X.222}$$

We recognize in the first matrix the well-known formulation for a two-dimensional rotation about the $z$-axis.

Finally, we can insert into (X.221) the components of the rotation axis explicitly, i.e. (see Fig. X.9):

$$a_x = \sin\theta\cos\varphi; \quad a_y = \sin\theta\sin\varphi; \quad a_z = \cos\theta. \tag{X.223}$$

We do not do this in the general case, as fairly extensive expressions[172] result; instead, we consider only two special cases.

First, we insert $\theta = 0$ or $a_z = 1$; then follows the familiar representation:

$$e^{-i\gamma\hat{a}\mathbf{l}} \underset{a_z=1}{\cong} \begin{pmatrix} \cos\gamma & -\sin\gamma & 0 \\ \sin\gamma & \cos\gamma & 0 \\ 0 & 0 & 1 \end{pmatrix}. \tag{X.224}$$

The next case is $\varphi = 0$ or $a_y = 0$; it follows that

$$e^{-i\gamma\hat{a}\mathbf{l}} \underset{a_y=0}{\cong} \begin{pmatrix} 1+(\cos\gamma-1)\cos^2\theta & -\sin\gamma\cos\theta & (\cos\gamma-1)\sin\theta\cos\theta \\ \sin\gamma\cos\theta & \cos\gamma & \sin\gamma\sin\theta \\ (\cos\gamma-1)\sin\theta\cos\theta & -\sin\gamma\sin\theta & 1+(\cos\gamma-1)\sin^2\theta \end{pmatrix}. \tag{X.225}$$

---

[172] It is remarkable how much more complicated it is to describe rotations in 3 than in 2 dimensions.

If we choose here $\theta = \frac{\pi}{2}$ or $a_x = 1$ (i.e. $a_y = a_z = 0$), we obtain

$$e^{-i\gamma\hat{a}\mathbf{l}} \underset{a_x=1}{\cong} \begin{pmatrix} 1 & 0 & 0 \\ 0 & \cos\gamma & \sin\gamma \\ 0 & -\sin\gamma & \cos\gamma \end{pmatrix}. \tag{X.226}$$

Finally, we want to remark that we can also describe the rotation $e^{-i\gamma\hat{a}\mathbf{l}}$ as follows: We first rotate the axis by the angle $-\varphi$ about the $z$-axis (then $\hat{a}$ lies around the $x$-axis), then by $-\theta$ about the $y$-axis ($\hat{a}$ now coincides with the $z$-axis). Now we can perform the rotation about the $z$-axis by $\gamma$ and then reverse the rotations by $\theta$ and $\varphi$. This yields

$$e^{-i\gamma\hat{a}\mathbf{l}} = e^{-i\varphi l_z} e^{-i\theta l_y} e^{-i\gamma l_z} e^{i\theta l_y} e^{i\varphi l_z}. \tag{X.227}$$

One can evaluate this expression with (X.216) and again obtains (X.221).

## X.3   Exercises, Chap. 17

1. Derive (17.14) from (17.11).
2. Show that

$$u_{E;l}(r) \underset{r\to\infty}{\sim} r^\alpha \text{ with } \alpha < -\frac{1}{2} \tag{X.228}$$

must hold.

Solution: In order that the wavefunction is square integrable, and due to $dV = r^2 dr \sin\vartheta \, d\vartheta \, d\varphi$, it must hold that

$$|\psi|^2 r^2 \sim R_{nl}^2 r^2 \sim r^b \text{ with } b < -1. \tag{X.229}$$

Because of $u_{E;l} = r R_{E;l}$, the proposition follows directly.

3. Hydrogen atom: the probability density of the electron in a volume element $d^3r = r^2 dr d\Omega$ around the point $(r, \vartheta, \varphi)$ is given by

$$d^3 w(r, \vartheta, \varphi) = |R_{nl}(r)|^2 \left|Y_l^m(\vartheta, \varphi)\right|^2 r^2 dr d\Omega = |u_{nl}(r)|^2 \left|Y_l^m(\vartheta, \varphi)\right|^2 dr d\Omega. \tag{X.230}$$

Find graphical representations, as illustrative as possible, of the probability densities for the various orbitals with $n = 1$ and $n = 2$.

## X.4   Exercises, Chap. 18

1. Show explicitly that the eigenvalues of $\hat{n}$ are positive.
   Solution: Since $\hat{n}$ is Hermitian, the eigenvalues are real. In addition, we have:

$$\langle \nu | \hat{n} | \nu \rangle = \nu \langle \nu | \nu \rangle = \nu$$
$$\langle \nu | \hat{n} | \nu \rangle = \langle \nu | a^\dagger a | \nu \rangle = \| a | \nu \rangle \|^2 \geq 0 \tag{X.231}$$

   and therefore $\nu \geq 0$.

2. Show that

$$a^l | \nu \rangle = \sqrt{(\nu - l)! \binom{\nu}{l}} \, | \nu - l \rangle \ \text{ and } \ a^{\dagger k} | \nu \rangle = \sqrt{\nu! \binom{\nu + k}{k}} \, | \nu + k \rangle .$$
$$\tag{X.232}$$

   Solution: We start from:

$$a | \nu \rangle = \sqrt{\nu} \, | \nu - 1 \rangle \, ; \ a^\dagger | \nu \rangle = \sqrt{\nu + 1} \, | \nu + 1 \rangle$$

   and thus

$$a^l | \nu \rangle = \sqrt{\nu (\nu - 1) \ldots (\nu - l + 1)} \, | \nu - l \rangle = \sqrt{(\nu - l)! \binom{\nu}{l}} \, | \nu - l \rangle$$
$$a^{\dagger k} | \nu \rangle = \sqrt{(\nu + 1) \ldots (\nu + k)} \, | \nu + k \rangle = \sqrt{\nu! \binom{\nu + k}{k}} \, | \nu + k \rangle .$$
$$\tag{X.233}$$

3. Determine $a^{\dagger k} a^l | \nu \rangle$ and $a^l a^{\dagger k} | \nu \rangle$.
   Solution:

$$a^{\dagger k} a^l | \nu \rangle = \sqrt{(\nu - l)! \binom{\nu}{l} (\nu - l)! \binom{\nu + k - l}{k}} \, | \nu + k - l \rangle$$
$$= \sqrt{\frac{\nu! (\nu + k - l)!}{l! k!}} \, | \nu + k - l \rangle$$
$$a^l a^{\dagger k} | \nu \rangle = \sqrt{\nu! \binom{\nu + k}{k} (\nu + k - l)! \binom{\nu + k}{l}} \, | \nu + k - l \rangle$$
$$= \sqrt{\frac{(\nu + k)! (\nu + k)!}{l! k!}} \, | \nu + k - l \rangle . \tag{X.234}$$

4. Show that the oscillator length $L$ yields essentially the position of the classical turning points.
   Solution: The classical turning points are determined by the equation

$$V = \frac{1}{2}m\omega^2 q_{\text{turning},n}^2 = E_n = \left(n + \frac{1}{2}\right)\hbar\omega. \qquad (\text{X}.235)$$

This leads to

$$q_{\text{turning},n} = \sqrt{(2n + 1)\frac{\hbar}{m\omega}} = \sqrt{2n + 1} \cdot L. \qquad (\text{X}.236)$$

For the ground state in particular, we have

$$q_{\text{turning},0} = L. \qquad (\text{X}.237)$$

5. Proofs by contradiction:

(a) Show by proof of contradiction: There is no largest eigenvalue $\nu_{\max}$.
Solution: If we assume that there is a greatest eigenvalue $\nu_{\max}$, then $a^\dagger |\nu_{\max}\rangle = 0$ must hold. From this it follows directly that:

$$0 = \langle \nu_{\max} | aa^\dagger |\nu_{\max}\rangle = \langle \nu_{\max} | 1 + a^\dagger a |\nu_{\max}\rangle = 1 + \nu_{\max}, \qquad (\text{X}.238)$$

or $\nu_{\max} = -1$. This is a contradiction since the eigenvalues $\nu$ are non-negative. Hence, there is no greatest eigenvalue.

(b) Show by proof of contradiction: The eigenvalues are integers.
Solution: If we assume that the eigenvalues are not integers, then there exists an eigenvector $|\mu\rangle$ with $\hat{n}|\mu\rangle = \mu|\mu\rangle = (m + \varepsilon)|\mu\rangle; m \in \mathbb{N}$ and $0 < \varepsilon < 1$. With the notation $a|\mu\rangle \equiv |a\mu\rangle$, we obtain

$$\langle a^{l+1}\mu | a^{l+1}\mu\rangle = \langle a^l \mu | a^\dagger a | a^l \mu\rangle = \langle a^l \mu | \hat{n} | a^l \mu\rangle. \qquad (\text{X}.239)$$

With (18.17), it follows that $\hat{n}a^l|\mu\rangle = a^l(m + \varepsilon - l)|\mu\rangle$, and therefore

$$\langle a^{l+1}\mu | a^{l+1}\mu\rangle = (m + \varepsilon - l)\langle a^l \mu | a^l \mu\rangle. \qquad (\text{X}.240)$$

If we start from $\langle a^l \mu | a^l \mu\rangle = \||a^l\mu\rangle\|^2 \neq 0$, it follows due to $0 < \varepsilon < 1$ that both sides of this equation are not zero for all $l$. For $l > m + \varepsilon$, the contradiction $\||a^{l+1}\mu\rangle\|^2 < 0$ would follow. Hence we have shown that the assumption $0 < \varepsilon < 1$ is incorrect.

(c) Show that to avoid negative eigenvalues, either (a) the smallest eigenvalue has to be zero or (b) there must be a state $|\nu_{\min}\rangle$ with $a|\nu_{\min}\rangle = 0$. Show that in case (b), $\nu_{\min} = 0$.

6. Show that $[q, p] = i\hbar$ ($q$ is the position, $p$ the momentum $\frac{\hbar}{i}\frac{d}{dq}$).
7. Given

$$a := \frac{1}{\sqrt{2\hbar}}\left\{\sqrt{m\omega}q + i\frac{p}{\sqrt{m\omega}}\right\}; \qquad (\text{X}.241)$$

(a) Derive

$$a^\dagger = \frac{1}{\sqrt{2\hbar}} \left\{ \sqrt{m\omega} q - i \frac{p}{\sqrt{m\omega}} \right\};$$

(X.242)

(b) Show that

$$\left[ a, a^\dagger \right] = 1$$

(X.243)

and

$$H = \hbar\omega \left\{ a^\dagger a + \frac{1}{2} \right\}.$$

(X.244)

(c) Given the eigenvalue problem

$$\hat{n} \, |\nu\rangle = \nu \, |\nu\rangle ; \quad \hat{n} = a^\dagger a,$$

(X.245)

show that

$$\| a \, |\nu\rangle \|^2 = \nu; \quad \| a^\dagger \, |\nu\rangle \|^2 = \nu + 1.$$

(X.246)

(d) Derive

$$\left[ \hat{n}, a \right] = -a; \quad \left[ \hat{n}, a^\dagger \right] = a^\dagger.$$

(X.247)

(e) Show that

$$\hat{n} a^l = a^l \left( \hat{n} - l \right); \quad \hat{n} a^{\dagger l} = a^{\dagger l} \left( \hat{n} + l \right); \quad l = 0, 1, 2, \ldots$$

(X.248)

(Proof by mathematical induction.)

(f) Prove

$$\hat{n} a \, |\nu\rangle = (\nu - 1) \, a \, |\nu\rangle ; \quad \hat{n} a^\dagger \, |\nu\rangle = (\nu + 1) \, a^\dagger \, |\nu\rangle .$$

(X.249)

(g) Derive

$$a \, |\nu\rangle = \sqrt{\nu} \, |\nu - 1\rangle ; \quad a^\dagger \, |\nu\rangle = \sqrt{\nu + 1} \, |\nu + 1\rangle .$$

(X.250)

(h) Show that

$$a^l \, |\nu\rangle = \sqrt{\nu (\nu - 1) \ldots (\nu - l + 1)} \, |\nu - l\rangle .$$

(X.251)

## X.5 Exercises, Chap. 19

1. Given

$$H = H^{(0)} + F(r) \mathbf{l} \cdot \mathbf{s} = \frac{\mathbf{p}^2}{2m} + V(r) + F(r) \mathbf{l} \cdot \mathbf{s}.$$

(X.252)

(a) Show that:

$$[H^{(0)}, l_z] = [H^{(0)}, s_z] = 0;$$ (X.253)

(b) Show that:

$$[H, l_z] \neq 0; \quad [H, s_z] \neq 0; \quad [H, j_z] = 0.$$ (X.254)

Hint: See the exercises for Chap. 16.

2. Expand the expression for the relativistic energy levels of the hydrogen atom:

$$E_{nj} = mc^2 \left\{ 1 + \alpha^2 \left[ n - j - \frac{1}{2} + \sqrt{\left(j + \frac{1}{2}\right)^2 - \alpha^2} \right]^{-2} \right\}^{-\frac{1}{2}} - mc^2$$
(X.255)

and compare with the approximation deduced in the text.
Solution: Series expansion with respect to powers of $x \ll 1$ with $x = \alpha^2$ and $y = j + \frac{1}{2}$ yields the expression

$$\left\{ 1 + x \left[ n - y + \sqrt{y^2 - x} \right]^{-2} \right\}^{-\frac{1}{2}} - 1$$

$$= -\frac{1}{2n^2} x + \left( -\frac{1}{2n^3 y} + \frac{3}{8n^4} \right) x^2 + \left( \frac{3}{4n^5 y} - \frac{n + 3y}{8n^4 y3} - \frac{5}{16n^6} \right) x^3 + \cdots$$
(X.256)

3. Given the Hamiltonian

$$H |\varphi\rangle = \left( H^{(0)} + W \right) |\varphi\rangle = \left( H^{(0)} + \varepsilon \hat{W} \right) |\varphi\rangle = E |\varphi\rangle,$$ (X.257)

where the states and the eigenvalues of $H^{(0)} |\varphi_n^{(0)}\rangle = E_n^{(0)} |\varphi_n^{(0)}\rangle$ are known (discrete, not degenerate). The initial state is $|\varphi_n^{(0)}\rangle$ and the corresponding energy is $E_n^{(0)}$. States and energies are expanded in terms of $\varepsilon$

$$|\varphi\rangle = |\varphi_n^{(0)}\rangle + \varepsilon |\varphi_n^{(1)}\rangle + \varepsilon^2 |\varphi_n^{(2)}\rangle + \cdots; \quad E = E_n^{(0)} + \varepsilon E_n^{(1)} + \varepsilon^2 E_n^{(2)} + \cdots$$
(X.258)

We can assume from the outset that the correction terms are orthogonal to the initial state, $\langle \varphi_n^{(0)} | \varphi_n^{(j)} \rangle = 0$ for $j \neq 0$. Calculate the corrections to the energy and the state to first order ($\sim \varepsilon^1$, repetition) and to second order ($\sim \varepsilon^2$).

4. We add a perturbation $\sim q^3$ to the Hamiltonian of the harmonic oscillator:

$$H = H^0 + W = -\frac{\hbar^2}{2m} \frac{d^2}{dq^2} + \frac{1}{2} m \omega^2 q^2 + \varepsilon q^3.$$ (X.259)

Calculate the correction term of the energy $E_n = \hbar \omega \left( n + \frac{1}{2} \right)$ to first order.

Solution: The correction term for the energy is $\langle\varphi_n^{(0)}|\,W\,|\varphi_n^{(0)}\rangle$. Since the eigenfunctions of the harmonic oscillator have well-defined parities (see Chap. 18), the integrand is point symmetrical and the integral disappears, $\langle\varphi_n^{(0)}|\,W\,|\varphi_n^{(0)}\rangle = 0$.

5. Finite nuclear size: For a hydrogen atom, we model the finite core size by the potential

$$V\,(r) = \begin{cases} -\frac{\gamma}{r} & \text{for } r \geq r_0 \\[2mm] \frac{\gamma}{2r_0}\left[\left(\frac{r}{r_0}\right)^2 - 3\right] & \text{for } r \leq r_0 \end{cases} \tag{X.260}$$

(Thus, we replace the point nucleus by a homogenously-charged sphere of radius $r_0$ with the charge density $\rho_0$.) Calculate the corrections to the energy in first order. Assume that the radial functions $R_{nl}\,(r)$ can be approximated for $r \leq r_0$ by $R_{nl}\,(0)$.

Solution: We have

$$W\,(r) = \begin{cases} \frac{\gamma}{2r_0}\left[\left(\frac{r}{r_0}\right)^2 - 3 + \frac{2r_0}{r}\right] & \text{for } r \leq r_0 \\[2mm] 0 & \text{otherwise.} \end{cases} \tag{X.261}$$

It follows that

$$\langle n, l, m|\,W\,|n, l', m'\rangle = \int d\Omega\, Y_l^{m*}\,(\vartheta, \varphi)\, Y_{l'}^{m'}\,(\vartheta, \varphi)$$

$$\times \int\limits_0^\infty r^2 dr R_{n,l}^*(r) R_{n',l'}(r) W(r). \tag{X.262}$$

This gives

$$\langle n, l, m|\,W\,|n, l', m'\rangle \approx \delta_{ll'}\delta_{mm'} \cdot |R_{n,l}(0)|^2 \int\limits_0^{r_0} r^2 dr\, W(r) \tag{X.263}$$

or

$$\langle n, l, m|\,W\,|n, l', m'\rangle \approx \delta_{ll'}\delta_{mm'} \cdot |R_{n,l}(0)|^2 \int\limits_0^{r_0} r^2 dr \frac{\gamma}{2r_0}\left[\left(\frac{r}{r_0}\right)^2 - 3 + \frac{2r_0}{r}\right]. \tag{X.264}$$

Evaluation of the integral yields the final result:

$$\langle n, l, m|\,W\,|n, l', m'\rangle \approx \delta_{ll'}\delta_{mm'} \cdot |R_{n,l}(0)|^2 \frac{\gamma r_0^2}{10}. \tag{X.265}$$

## X.6  Exercises, Chap. 20

1. Given two matrices $A$ and $B$ with

$$A = \begin{pmatrix} 1 & 3 \\ 2 & 1 \end{pmatrix}; \quad B = \begin{pmatrix} 1 & 0 \\ 2 & 1 \end{pmatrix}. \tag{X.266}$$

Determine $A \otimes B$.
Solution:

$$A \otimes B = \begin{pmatrix} 1B & 3B \\ 2B & 1B \end{pmatrix} = \begin{pmatrix} 1 & 0 & 3 & 0 \\ 2 & 1 & 6 & 3 \\ 2 & 0 & 1 & 0 \\ 4 & 2 & 2 & 1 \end{pmatrix}. \tag{X.267}$$

2. Represent the Bell states (20.14) as column vectors. Show in this representation that the Bell states are entangled and that they form a CONS.
Solution: With

$$|h\rangle \cong \begin{pmatrix} 1 \\ 0 \end{pmatrix}; \quad |v\rangle \cong \begin{pmatrix} 0 \\ 1 \end{pmatrix}, \tag{X.268}$$

it follows that

$$|hh\rangle \cong \begin{pmatrix} 1 \\ 0 \\ 0 \\ 0 \end{pmatrix}; \quad |hv\rangle \cong \begin{pmatrix} 0 \\ 1 \\ 0 \\ 0 \end{pmatrix}; \quad |vh\rangle \cong \begin{pmatrix} 0 \\ 0 \\ 1 \\ 0 \end{pmatrix}; \quad |vv\rangle \cong \begin{pmatrix} 0 \\ 0 \\ 0 \\ 1 \end{pmatrix} \tag{X.269}$$

and therefore

$$|\Psi^{\pm}\rangle \cong \frac{1}{\sqrt{2}} \begin{pmatrix} 0 \\ 1 \\ \pm 1 \\ 0 \end{pmatrix}; \quad |\Phi^{\pm}\rangle \cong \frac{1}{\sqrt{2}} \begin{pmatrix} 1 \\ 0 \\ 0 \\ \pm 1 \end{pmatrix}. \tag{X.270}$$

A column vector of the form $\begin{pmatrix} c_1 & c_2 & c_3 & c_4 \end{pmatrix}^T$ is factorizable, if it holds that $c_1 \cdot c_4 = c_2 \cdot c_3$; see (20.8). One can see directly that this condition is not satisfied for the Bell states.
In order to show that the Bell states are a CONS, we consider initially $|\Phi^{\pm}\rangle$. We have

$$\left|\Phi^+\right\rangle\left\langle\Phi^+\right| + \left|\Phi^-\right\rangle\left\langle\Phi^-\right|$$

$$= \frac{1}{2}\begin{pmatrix} 1 \\ 0 \\ 0 \\ 1 \end{pmatrix}\begin{pmatrix} 1\,0\,0\,1 \end{pmatrix} + \frac{1}{2}\begin{pmatrix} 1 \\ 0 \\ 0 \\ -1 \end{pmatrix}\begin{pmatrix} 1\,0\,0\,-1 \end{pmatrix}. \tag{X.271}$$

Multiplication yields

$$\left|\Phi^+\right\rangle\left\langle\Phi^+\right| + \left|\Phi^-\right\rangle\left\langle\Phi^-\right|$$

$$= \frac{1}{2}\begin{pmatrix} 1\,0\,0\,1 \\ 0\,0\,0\,0 \\ 0\,0\,0\,0 \\ 1\,0\,0\,1 \end{pmatrix} + \frac{1}{2}\begin{pmatrix} 1\ \,0\,0\,-1 \\ 0\ \,0\,0\ \,0 \\ 0\ \,0\,0\ \,0 \\ -1\,0\,0\ \,1 \end{pmatrix} = \begin{pmatrix} 1\,0\,0\,0 \\ 0\,0\,0\,0 \\ 0\,0\,0\,0 \\ 0\,0\,0\,1 \end{pmatrix}. \tag{X.272}$$

Consideration of $\left|\Psi^\pm\right\rangle$ yields the two missing diagonal elements.

3. Two photons are in the state

$$\left|\Psi\right\rangle = \frac{\left|hv\right\rangle - \left|vh\right\rangle}{\sqrt{2}}. \tag{X.273}$$

(a) Show explicitly that it is an entangled state.
(b) Photon 1 passes an analyzer for right-handed circular polarization (the corresponding state reads $\frac{\left|h\right\rangle + i\left|v\right\rangle}{\sqrt{2}}$). Show that through a measurement, the state $\left|\Psi\right\rangle$ is changed into a product state.

4. Show that the Bell states can be transformed into each other by applying the Pauli matrices to a subsystem.
Solution: We transform the system 1, i.e. we apply the operators $\sigma_i \otimes I$. Due to

$$\begin{aligned} \sigma_1\left|h\right\rangle &= \left|v\right\rangle\,;\; \sigma_1\left|v\right\rangle = \left|h\right\rangle \\ \sigma_2\left|h\right\rangle &= i\left|v\right\rangle\,;\; \sigma_2\left|v\right\rangle = -i\left|h\right\rangle \\ \sigma_3\left|h\right\rangle &= \left|h\right\rangle\,;\; \sigma_3\left|v\right\rangle = -\left|v\right\rangle\,, \end{aligned} \tag{X.274}$$

it follows that e.g.

$$\begin{aligned} (\sigma_1 \otimes I)\left|\Psi^\pm\right\rangle &= \frac{\sigma_1\left|h\right\rangle \otimes \left|v\right\rangle \pm \sigma_1\left|v\right\rangle \otimes \left|h\right\rangle}{\sqrt{2}} \\ &= \frac{\left|v\right\rangle \otimes \left|v\right\rangle \pm \left|h\right\rangle \otimes \left|h\right\rangle}{\sqrt{2}} = \pm\left|\Phi^\pm\right\rangle, \end{aligned} \tag{X.275}$$

and correspondingly for the other Pauli matrices or Bell states (as well as for the operators $I \otimes \sigma_i$).

5. Show that the Bell states are eigenvectors of products of the same Pauli matrices.
Solution: We consider as an example $(\sigma_2 \otimes \sigma_2)\left|\Psi^\pm\right\rangle$. It follows

$$(\sigma_2 \otimes \sigma_2) \left| \Psi^{\pm} \right\rangle = \frac{\sigma_2 \left| h \right\rangle \otimes \sigma_2 \left| v \right\rangle \pm \sigma_2 \left| v \right\rangle \otimes \sigma_2 \left| h \right\rangle}{\sqrt{2}} = \pm \left| \Psi^{\pm} \right\rangle; \quad (X.276)$$

and correspondingly for the other Pauli matrices or Bell states.

6. Transform the inequality (20.27)

$$\cos^2 (\alpha - \beta) \le \cos^2 (\alpha - \gamma) + \sin^2 (\beta - \gamma) \qquad (X.277)$$

for $\alpha = 0$ and $0 < \beta < \pi$ to give

$$\sin (\gamma - \beta) \cos \gamma \le 0. \qquad (X.278)$$

Solution: With $\alpha = 0$, we find

$$\cos^2 \beta \le \cos^2 \gamma + \sin^2 (\beta - \gamma). \qquad (X.279)$$

With the equation $\cos^2 y - \cos^2 x = \sin (x + y) \sin (x - y)$, we then obtain:

$$\cos^2 \beta - \cos^2 \gamma = \sin (\gamma + \beta) \sin (\gamma - \beta) \le \sin^2 (\gamma - \beta) \qquad (X.280)$$

or

$$\sin (\gamma - \beta) \left[ \sin (\gamma + \beta) - \sin (\gamma - \beta) \right] \le 0. \qquad (X.281)$$

Using the equation $\sin x - \sin y = 2 \cos \frac{x+y}{2} \sin \frac{x-y}{2}$, this gives

$$\sin (\gamma - \beta) \cos \gamma \sin \beta \le 0. \qquad (X.282)$$

For $0 < \beta < \pi$, we have $\sin \beta > 0$, and the inequality

$$\sin (\gamma - \beta) \cos \gamma \le 0 \qquad (X.283)$$

results.

7. Given the function

$$f (\gamma, \beta) = \sin (\gamma - \beta) \cos \gamma; \qquad (X.284)$$

determine the position of its zeros and the positions and values of its maxima with respect to $\gamma$.

Solution: Zeros exist (a) for $\cos \gamma = 0$, i.e. $\gamma = \frac{\pi}{2} + m_1 \pi$ and $\beta$ arbitrary, as well as for (b) $\sin (\gamma - \beta) = 0$, i.e. $\gamma = \beta + m_2 \pi$ with $m_1, m_2 \in \mathbb{Z}$. For the determination of the maxima, we use

$$f (\gamma, \beta) = \frac{\sin (2\gamma - \beta) - \sin \beta}{2}; \quad \frac{\partial f (\gamma, \beta)}{\partial \gamma} = \cos (2\gamma - \beta). \qquad (X.285)$$

Hence, we have extrema for $2\gamma - \beta = \frac{\pi}{2} + n\pi$ with $n \in \mathbb{Z}$, whereby there is a maximum/minimum for $n$ even/odd. Inserting this value of $\gamma$ gives the value for the maxima:

$$f_{\max}(\gamma, \beta) = \frac{1 - \sin \beta}{2}. \qquad (X.286)$$

We see that $f_{\max}(\gamma, \beta)$ is always positive for $\beta \neq \frac{\pi}{2}$.

8. A system of two photons is in one of the Bell states. The photon Q1 is incident on an analyzer for horizontal polarization, rotated by an angle $\alpha$. What is the probability that Q1 passes the analyzer?

Solution: Rotation by the angle $\alpha$ leads to the new horizontally- and vertically-polarized states:

$$|h_\alpha\rangle = \cos \alpha \, |h_0\rangle - \sin \alpha \, |v_0\rangle ; \quad |v_\alpha\rangle = \sin \alpha \, |h_0\rangle + \cos \alpha \, |v_0\rangle . \qquad (X.287)$$

First, we obtain Malus' law by evaluating $\langle h_0 \, | h_\alpha \rangle$:

$$|\langle h_0 \, | h_\alpha \rangle| = \cos^2 \alpha. \qquad (X.288)$$

The Bell states are

$$\left| \Psi^\pm \right\rangle = \frac{|hv\rangle \pm |vh\rangle}{\sqrt{2}}; \quad \left| \Phi^\pm \right\rangle = \frac{|hh\rangle \pm |vv\rangle}{\sqrt{2}}. \qquad (X.289)$$

Measurement of Q1 in the state $|h_{\alpha 1}\rangle$ and comparison with (X.287) yields (abbreviation: $c = \cos \alpha$, $s = \sin \alpha$):

$$\langle h_{\alpha 1} \, | \Psi^\pm \rangle = \frac{\langle h_{\alpha 1} \, | h_{01} v_{02} \rangle \pm \langle h_{\alpha 1} \, | v_{01} h_{02} \rangle}{\sqrt{2}} = \frac{c \, |v_{02}\rangle \pm s \, |h_{02}\rangle}{\sqrt{2}} = \frac{|v_{\mp \alpha 2}\rangle}{\sqrt{2}}$$

$$\langle h_{\alpha 1} \, | \Phi^\pm \rangle = \frac{\langle h_{\alpha 1} \, | h_{01} h_{02} \rangle \pm \langle h_{\alpha 1} \, | v_{01} v_{02} \rangle}{\sqrt{2}} = \frac{c \, |h_{02}\rangle \pm s \, |v_{02}\rangle}{\sqrt{2}} = \frac{|h_{\pm \alpha 2}\rangle}{\sqrt{2}}.$$

$$(X.290)$$

We thus have in each case

$$\left| \langle h_{\alpha 1} \, | \Psi^\pm \rangle \right|^2 = \left| \langle h_{\alpha 1} \, | \Phi^\pm \rangle \right|^2 = \frac{1}{2}. \qquad (X.291)$$

Hence, the probability that Q1 passes an arbitrarily adjusted analyzer is 50 % for all cases.

9. Two photons in the state $|h_0\rangle$ are rotated by the angles $\alpha$ and $\beta$ to give the states $|h_\alpha\rangle$ and $\left| h_\beta \right\rangle$. How does the projection operator referring to $\left| h_\alpha h_\beta \right\rangle$ act on the Bell states?

Solution: The state rotated by the angle $\alpha$ is

$$|h_\alpha\rangle = \cos\alpha\,|h_0\rangle - \sin\alpha\,|v_0\rangle\,. \qquad (X.292)$$

We start from the Bell states:

$$|\Psi^\pm\rangle = \frac{|hv\rangle \pm |vh\rangle}{\sqrt{2}}; \quad |\Phi^\pm\rangle = \frac{|hh\rangle \pm |vv\rangle}{\sqrt{2}}, \qquad (X.293)$$

which we measure in the state

$$\begin{aligned}
|h_\alpha h_\beta\rangle &= [\cos\alpha\,|h_0\rangle - \sin\alpha\,|v_0\rangle][\cos\beta\,|h_0\rangle - \sin\beta\,|v_0\rangle] \\
&= \cos\alpha\cos\beta\,|h_0 h_0\rangle - \cos\alpha\sin\beta\,|h_0 v_0\rangle - \sin\alpha\cos\beta\,|v_0 h_0\rangle \\
&\quad + \sin\alpha\sin\beta\,|v_0 v_0\rangle\,.
\end{aligned} \qquad (X.294)$$

We then have

$$\begin{aligned}
|h_\alpha h_\beta\rangle\langle h_\alpha h_\beta|\,\Psi^\pm\rangle &= |h_\alpha\beta\rangle\,\langle h_\alpha h_\beta|\,\frac{|h_0 v_0\rangle \pm |v_0 h_0\rangle}{\sqrt{2}} \\
&= -|h_\alpha h_\beta\rangle\,\frac{\cos\alpha\sin\beta \pm \sin\alpha\cos\beta}{\sqrt{2}} \\
&= -|h_\alpha h_\beta\rangle\,\frac{\sin(\beta \pm \alpha)}{\sqrt{2}}
\end{aligned} \qquad (X.295)$$

and

$$\begin{aligned}
|h_\alpha h_\beta\rangle\langle h_\alpha h_\beta|\,\Phi^\pm\rangle &= |h_\alpha h_\beta\rangle\langle h_\alpha h_\beta|\,\frac{|h_0 h_0\rangle \pm |v_0 v_0\rangle}{\sqrt{2}} \\
&= |h_\alpha h_\beta\rangle\,\frac{\cos\alpha\cos\beta \pm \sin\alpha\sin\beta}{\sqrt{2}} \\
&= |h_\alpha h_\beta\rangle\,\frac{\cos(\beta \mp \alpha)}{\sqrt{2}}\,.
\end{aligned} \qquad (X.296)$$

10. Given two quantum objects Q1 and Q2, with an $N$-dimensional CONS $\{|\varphi_i\rangle\}$ for Q1 and $\{|\psi_i\rangle\}$ for Q2 (due to this notation we can omit the index for the number of the quantum object). The initial state is

$$|\chi\rangle = \sum_{ij} c_{ij}\,|\varphi_i\rangle\,|\psi_j\rangle\,. \qquad (X.297)$$

What is the probability of measuring Q1 in some state $|\lambda\rangle$ (no matter which state)?
Solution:
Approach 1: The projection of the state $|\chi\rangle$ onto e.g. $|\lambda\rangle\,|\psi_n\rangle$ is

$$|\lambda\rangle \, |\psi_n\rangle \, \langle\lambda| \, \langle\psi_n| \, \chi\rangle = \sum_{ij} c_{ij} \, |\lambda\rangle \, \langle\lambda \, |\varphi_i\rangle \, |\psi_n\rangle \, \langle\psi_n \, |\psi_j\rangle$$

$$= \sum_{i} c_{in} \, \langle\lambda \, |\varphi_i\rangle \, |\lambda\rangle \, |\psi_n\rangle \, . \qquad \text{(X.298)}$$

The probability $w\,(\lambda,\,\psi_n)$ of measuring the state $|\lambda\rangle \, |\psi_n\rangle$ is thus given by

$$w\,(\lambda,\,\psi_n) = \left| \sum_{i} c_{in} \, \langle\lambda \, |\varphi_i\rangle \right|^2 . \qquad \text{(X.299)}$$

The probability $w\,(\lambda)$ of measuring Q1 in the state $|\lambda\rangle$ is the sum of the partial probabilities, i.e.

$$w\,(\lambda) = \sum_{n} W\,(\lambda,\,\psi_n) \, . \qquad \text{(X.300)}$$

It follows that

$$w\,(\lambda) = \sum_{n} \left| \sum_{i} c_{in} \, \langle\lambda \, |\varphi_i\rangle \right|^2$$

$$= \sum_{n} \sum_{ij} c_{in}^* c_{jn} \, \langle\varphi_i \, |\lambda\rangle \, \langle\lambda \, |\varphi_j\rangle$$

$$= \langle\lambda| \left( \sum_{ijn} c_{in}^* c_{jn} \, |\varphi_j\rangle \, \langle\varphi_i| \right) |\lambda\rangle \, . \qquad \text{(X.301)}$$

Approach 2: Alternatively, we can deduce $w\,(\lambda)$ by describing the measurement as

$$|\lambda\rangle \, \langle\lambda| \, \chi\rangle = \sum_{ij} c_{ij} \, |\lambda\rangle \, \langle\lambda \, |\varphi_i\rangle \, |\psi_j\rangle = \left( \sum_{ij} c_{ij} \, \langle\lambda \, |\varphi_i\rangle \, |\psi_j\rangle \right) |\lambda\rangle \, . \quad \text{(X.302)}$$

The probability of measuring $|\lambda\rangle$ is the squared value of the 'prefactor' of $|\lambda\rangle$, i.e.

$$w\,(\lambda) = \left| \sum_{ij} c_{ij} \, \langle\lambda \, |\varphi_i\rangle \, |\psi_j\rangle \right|^2 = \sum_{injm} c_{in}^* \, \langle\varphi_i \, |\lambda\rangle \, \langle\psi_n| \, c_{jm} \, \langle\lambda \, |\varphi_j\rangle \, |\psi_m\rangle$$

$$= \sum_{injm} c_{in}^* \, \langle\varphi_i \, |\lambda\rangle \, c_{jm} \, \langle\lambda \, |\varphi_j\rangle \, \delta_{nm} = \sum_{ijn} c_{in}^* c_{jn} \, \langle\varphi_i \, |\lambda\rangle \, \langle\lambda \, |\varphi_j\rangle \, ,$$

$$\text{(X.303)}$$

and we obtain the same result as in (X.301).

Special cases: for $c_{ij} = \delta_{ij}a_i$, we have:

$$|\chi\rangle = \sum_i a_i |\varphi_i\rangle |\psi_i\rangle ; \quad w(\lambda) = \langle\lambda| \left( \sum_i |a_i|^2 |\varphi_i\rangle \langle\varphi_i| \right) |\lambda\rangle ; \qquad \text{(X.304)}$$

and for $a_i = \frac{e^{i\alpha_i}}{\sqrt{N}}$, we obtain finally

$$|\chi\rangle = \frac{1}{\sqrt{N}} \sum_i |\varphi_i\rangle |\psi_i\rangle ; \quad w(\lambda) = \frac{1}{N}. \qquad \text{(X.305)}$$

11. Show that entangled states such as the Bell states cannot be 'disentangled' by a reversible transformation of the single-quantum-object basis; entanglement is preserved even in a different basis.

Solution: We begin with the transformation $T_i$

$$|h_i\rangle = a_i \left|h_i'\right\rangle + b_i \left|v_i'\right\rangle; \quad |v_i\rangle = c_i \left|h_i'\right\rangle + d_i \left|v_i'\right\rangle \qquad \text{(X.306)}$$

with $a_i d_i \neq b_i c_i$ (in order that $T_i^{-1}$ exists). The four Bell states

$$\left|\Psi^\pm\right\rangle = \frac{|hv\rangle \pm |vh\rangle}{\sqrt{2}}; \quad \left|\Phi^\pm\right\rangle = \frac{|hh\rangle \pm |vv\rangle}{\sqrt{2}} \qquad \text{(X.307)}$$

read in the new basis

$$\begin{aligned}
\sqrt{2}\left|\Psi^\pm\right\rangle &= [a_1c_2 \pm c_1a_2]\left|h'h'\right\rangle + [a_1d_2 \pm c_1b_2]\left|h'v'\right\rangle \\
&\quad + [b_1c_2 \pm d_1a_2]\left|v'h'\right\rangle + [b_1d_2 \pm d_1b_2]\left|v'v'\right\rangle \\
\sqrt{2}\left|\Phi^\pm\right\rangle &= [a_1a_2 \pm c_1c_2]\left|h'h'\right\rangle + [a_1b_2 \pm c_1d_2]\left|h'v'\right\rangle \\
&\quad + [b_1a_2 \pm d_1c_2]\left|v'h'\right\rangle + [b_1b_2 \pm d_1d_2]\left|v'v'\right\rangle.
\end{aligned} \qquad \text{(X.308)}$$

The states are factorizable in this basis, if it holds that $a_{h'h'} \cdot a_{v'v'} = a_{h'v'} \cdot a_{v'h'}$. This means that

$$\begin{aligned}
[a_1c_2 \pm c_1a_2][b_1d_2 \pm d_1b_2] &\overset{!}{=} [a_1d_2 \pm c_1b_2][b_1c_2 \pm d_1a_2] \\
[a_1a_2 \pm c_1c_2][b_1b_2 \pm d_1d_2] &\overset{!}{=} [a_1b_2 \pm c_1d_2][b_1a_2 \pm d_1c_2].
\end{aligned} \qquad \text{(X.309)}$$

Expanding the products and collecting terms leads to

$$\begin{aligned}
(a_1d_1 - c_1b_1)(c_2b_2 - d_2a_2) &\overset{!}{=} 0 \\
(a_1d_1 - c_1b_1)(a_2d_2 - b_2c_2) &\overset{!}{=} 0.
\end{aligned} \qquad \text{(X.310)}$$

This is a contradiction, since we have assumed an invertible ($a_i d_i \neq b_i c_i$) transformation. In other words, the entanglement is 'robust' under such transformations. We know that we can simplify linear combinations of states by a suitable change of basis. In contrast, one can *not* destroy the entanglement.

12. Determine the behavior of the Bell states under reversible transformations. Consider the case of rotations.

Solution: We begin as in exercise 11, with

$$|h_i\rangle = a_i \left|h_i'\right\rangle + b_i \left|v_i'\right\rangle \; ; \; |v_i\rangle = c_i \left|h_i'\right\rangle + d_i \left|v_i'\right\rangle \tag{X.311}$$

and $a_i d_i \neq b_i c_i$ (in order that $T_i^{-1}$ exists). We have the four basis states

$$\left|\Psi^{\pm}\right\rangle = \frac{|hv\rangle \pm |vh\rangle}{\sqrt{2}} \; ; \; \left|\Phi^{\pm}\right\rangle = \frac{|hh\rangle \pm |vv\rangle}{\sqrt{2}} \tag{X.312}$$

and want to link them with the transformed states

$$\left|\Psi^{\pm}\right\rangle' = \frac{|h'v'\rangle \pm |v'h'\rangle}{\sqrt{2}} \; ; \; \left|\Phi^{\pm}\right\rangle' = \frac{|h'h'\rangle \pm |v'v'\rangle}{\sqrt{2}}. \tag{X.313}$$

We can again deduce (X.308) and insert in it

$$
\begin{aligned}
\left|h'h'\right\rangle &= \frac{|\Phi^+\rangle' + |\Phi^-\rangle'}{\sqrt{2}} \; ; \; \left|v'v'\right\rangle = \frac{|\Phi^+\rangle' - |\Phi^-\rangle'}{\sqrt{2}} \\
\left|h'v'\right\rangle &= \frac{|\Psi^+\rangle' + |\Psi^-\rangle'}{\sqrt{2}} \; ; \; \left|v'h'\right\rangle = \frac{|\Psi^+\rangle' - |\Psi^-\rangle'}{\sqrt{2}}.
\end{aligned}
\tag{X.314}
$$

It follows that

$$
\begin{aligned}
2\left|\Psi^{\pm}\right\rangle = {} & [a_1 c_2 \pm c_1 a_2 + b_1 d_2 \pm d_1 b_2]\left|\Phi^+\right\rangle' \\
& + [a_1 c_2 \pm c_1 a_2 - b_1 d_2 \mp d_1 b_2]\left|\Phi^-\right\rangle' \\
& + [a_1 d_2 \pm c_1 b_2 + b_1 c_2 \pm d_1 a_2]\left|\Psi^+\right\rangle' \\
& + [a_1 d_2 \pm c_1 b_2 - b_1 c_2 \mp d_1 a_2]\left|\Psi^-\right\rangle'
\end{aligned}
\tag{X.315}
$$

$$
\begin{aligned}
2\left|\Phi^{\pm}\right\rangle = {} & [a_1 a_2 \pm c_1 c_2 + b_1 b_2 \pm d_1 d_2]\left|\Phi^+\right\rangle' \\
& + [a_1 a_2 \pm c_1 c_2 - b_1 b_2 \mp d_1 d_2]\left|\Phi^-\right\rangle' \\
& + [a_1 b_2 \pm c_1 d_2 + b_1 a_2 \pm d_1 c_2]\left|\Psi^+\right\rangle' \\
& + [a_1 b_2 \pm c_1 d_2 - b_1 a_2 \mp d_1 c_2]\left|\Psi^-\right\rangle'.
\end{aligned}
$$

We specialize to rotations of the single-object basis:

$$a_i = \cos \vartheta_i; \quad b_i = -\sin \vartheta_i; \quad c_i = -\sin \vartheta_i; \quad d_i = \cos \vartheta_i. \qquad \text{(X.316)}$$

Rearranging the trigonometric functions gives after some manipulations

$$
\begin{aligned}
\left|\Psi^+\right\rangle &= \cos(\vartheta_1 + \vartheta_2)\left|\Psi^+\right\rangle' + \sin(\vartheta_1 + \vartheta_2)\left|\Phi^-\right\rangle' \\
\left|\Psi^-\right\rangle &= \cos(\vartheta_1 - \vartheta_2)\left|\Psi^-\right\rangle' - \sin(\vartheta_1 - \vartheta_2)\left|\Phi^+\right\rangle' \\
\left|\Phi^+\right\rangle &= \cos(\vartheta_1 - \vartheta_2)\left|\Phi^+\right\rangle' + \sin(\vartheta_1 - \vartheta_2)\left|\Psi^-\right\rangle' \\
\left|\Phi^-\right\rangle &= \cos(\vartheta_1 + \vartheta_2)\left|\Phi^-\right\rangle' - \sin(\vartheta_1 + \vartheta_2)\left|\Psi^+\right\rangle'.
\end{aligned}
\qquad \text{(X.317)}
$$

In particular for $\vartheta_1 = \vartheta_2 = \vartheta$, it follows that

$$
\begin{aligned}
\left|\Psi^+\right\rangle &= \cos(2\vartheta)\left|\Psi^+\right\rangle' + \sin(2\vartheta)\left|\Phi^-\right\rangle' \\
\left|\Psi^-\right\rangle &= \left|\Psi^-\right\rangle' \\
\left|\Phi^+\right\rangle &= \left|\Phi^+\right\rangle' \\
\left|\Phi^-\right\rangle &= \cos(2\vartheta)\left|\Phi^-\right\rangle' - \sin(2\vartheta)\left|\Psi^+\right\rangle'.
\end{aligned}
\qquad \text{(X.318)}
$$

## X.7   Exercises, Chap. 21

1. Derive the commutation relation (21.26) for position and momentum.
   Solution: We have:

$$U^{-1}(a)\, XU(a) = X + a \quad \text{with } U = e^{-i\frac{Pa}{\hbar}}. \qquad \text{(X.319)}$$

Expanding in powers of $a$ yields

$$\left(1 + i\frac{Pa}{\hbar}\right) X \left(1 - i\frac{Pa}{\hbar}\right) + O(a^2) = X + a. \qquad \text{(X.320)}$$

It follows that

$$X + i\frac{Pa}{\hbar}X - Xi\frac{Pa}{\hbar} + O(a^2) = X + a \qquad \text{(X.321)}$$

or

$$i\frac{P}{\hbar}X - Xi\frac{P}{\hbar} + O(a) = 1 \qquad \text{(X.322)}$$

and thus for $a \to 0$

$$[X, P] = i\hbar. \qquad \text{(X.323)}$$

2. Consider the relation between symmetries and conserved quantities by means of the spatial translational invariance of an isolated system of two quantum objects whose interaction depends only on their distance $\mathbf{r}_1 - \mathbf{r}_2$.

Solution: After shifting by $\mathbf{a}$, the mean value of $H$ for the state $|\varphi\rangle$ must equal the mean value for the shifted state $|\varphi_\mathbf{a}\rangle = e^{-i\frac{\mathbf{pa}}{\hbar}}|\varphi\rangle$:

$$\langle \varphi_\mathbf{a}| H |\varphi_\mathbf{a}\rangle = \langle\varphi| e^{i\frac{\mathbf{pa}}{\hbar}} H e^{-i\frac{\mathbf{pa}}{\hbar}} |\varphi\rangle \overset{!}{=} \langle\varphi| H |\varphi\rangle. \qquad (X.324)$$

This equation must hold for all $\mathbf{a}$. Then it follows from the last equation that

$$\langle\varphi| 1 + i\frac{\mathbf{pa}}{\hbar}H - Hi\frac{\mathbf{pa}}{\hbar} |\varphi\rangle \overset{!}{=} \langle\varphi| H |\varphi\rangle \qquad (X.325)$$

and thus directly

$$\langle\varphi| \mathbf{pa}H - H\mathbf{pa} |\varphi\rangle \overset{!}{=} 0. \qquad (X.326)$$

Since this equation has to hold for all $|\varphi\rangle$, it follows that

$$\mathbf{pa}H - H\mathbf{pa} = \mathbf{a}\,(\mathbf{p}H - H\mathbf{p}) = \mathbf{a}\left[\mathbf{p},H\right] \overset{!}{=} 0. \qquad (X.327)$$

Since this equation has to hold for arbitrary $\mathbf{a}$, it follows finally as consequence of spatial translation invariance that

$$[H, \mathbf{p}] = 0. \qquad (X.328)$$

With this equation, it is also guaranteed that in the series expansion of the $e$-function in (X.324), all contributions of higher order in $|\mathbf{a}|$ vanish.

3. Let $B$ be a Hermitian operator and $U$ and $A$ a unitary and an antiunitary operator, resp. Show that:

$$e^{iUBU^{-1}} = Ue^{iB}U^{-1}; \quad e^{iABA^{-1}} = Ae^{-iB}A^{-1} \qquad (X.329)$$

Solution: Using the power series representation of the $e$-function, we have to show that

$$\left(iUBU^{-1}\right)^n = U\,(iB)^n\,U^{-1}; \quad \left(iABA^{-1}\right)^n = A\,(-iB)^n\,A^{-1}. \qquad (X.330)$$

This is done by mathematical induction. Evidently, the statement holds true for $n = 0$.

In the unitary case, we have

$$\left(iUBU^\dagger\right)^{n+1} = \left(iUBU^\dagger\right)^n iUBU^\dagger = U\,(iB)^n\,U^\dagger iUBU^\dagger$$
$$= U\,(iB)^n\,iBU^\dagger = U\,(iB)^{n+1}\,U^\dagger. \qquad (X.331)$$

In the antiunitary case, we find

$$\left(iABA^\dagger\right)^{n+1} = \left(iABA^\dagger\right)^n iABA^\dagger = A\left(-iB\right)^n A^\dagger iABA^\dagger =$$
$$= A\left(-iB\right)^n \left(-iA^\dagger\right) ABA^\dagger = A\left(-iB\right)^n \left(-iB\right) A^\dagger$$
$$= A\left(-iB\right)^{n+1} A^\dagger. \qquad (X.332)$$

4. Show with the help of the propagator $U$ that eigenvalues of $A$ are conserved, if $[H, A] = 0$.

   Solution: We proved the statement by means of the Ehrenfest theorem in Chap. 9, Vol. 1. Here we can argue as follows: $[H, A] = 0$ is equivalent to $[U(t - t_0), A] = 0$ ($H$ is independent of the time). It follows with $A|\varphi(t_0)\rangle = a|\varphi(t_0)\rangle$ that

$$A|\varphi(t)\rangle = AU|\varphi(t_0)\rangle = UA|\varphi(t_0)\rangle = Ua|\varphi(t_0)\rangle = a|\varphi(t)\rangle. \qquad (X.333)$$

   Hence, the eigenvalue $a$ is conserved in time—it is a good quantum number.

5. Consider the translation $\mathbf{r}' = \mathbf{r} + \mathbf{a}$ or $T(\mathbf{a})\mathbf{r} = \mathbf{r} + \mathbf{a}$. Show that it can be represented by the unitary transformation $U_{T(\mathbf{a})} = \lim_{n\to\infty}\left(1 - \frac{i}{\hbar}\frac{\mathbf{ap}}{n}\right)^n = e^{-\frac{i}{\hbar}\mathbf{ap}}$.

   Solution: For the wavefunctions, we have:

$$\psi(\mathbf{r}) \to \psi'(\mathbf{r}); \quad \psi'(\mathbf{r}) = U_{T(\mathbf{a})}\psi(\mathbf{r}), \qquad (X.334)$$

and it follows that

$$\psi'(\mathbf{r} + \mathbf{a}) = \psi'(T\mathbf{r}) = \psi(\mathbf{r})$$
$$\psi'(\mathbf{r}) = \psi(\mathbf{r} - \mathbf{a}) = U_{T(\mathbf{a})}\psi(\mathbf{r}). \qquad (X.335)$$

$U$ is unitary. In an infinitesimal transformation $\mathbf{r}' = \mathbf{r} + d\mathbf{a}$, it follows that

$$\psi'(\mathbf{r}) = \psi(\mathbf{r} - d\mathbf{a}) = \psi(\mathbf{r}) - d\mathbf{a}\nabla\psi(\mathbf{r}), \qquad (X.336)$$

and thus

$$\psi(\mathbf{r}) - d\mathbf{a}\nabla\psi(\mathbf{r}) = U_{T(d\mathbf{a})}\psi(\mathbf{r}). \qquad (X.337)$$

We write this in the form

$$U_{T(d\mathbf{a})} = 1 - d\mathbf{a}\nabla = 1 - \frac{i}{\hbar}d\mathbf{ap}. \qquad (X.338)$$

In view of the limiting process, we set $d\mathbf{a} = \frac{\mathbf{a}}{n}$ and obtain

$$U_{T(\mathbf{a})} = \lim_{n\to\infty}\left(1 - \frac{i}{\hbar}\frac{\mathbf{ap}}{n}\right)^n = e^{-\frac{i}{\hbar}\mathbf{ap}}. \qquad (X.339)$$

6. Determine the commutator of $P$ with an arbitrary function of $X$, without using $P = \frac{\hbar}{i}\frac{d}{dx}$ from the outset (this is to be derived). Use

$$U^{-1}(a)X^2U(a) = U^{-1}(a)XU(a)U^{-1}(a)XU(a) = (X+a)^2; \quad \text{(X.340)}$$

and analogously

$$U^{-1}(a)X^nU(a) = (X+a)^n \qquad \text{(X.341)}$$

as well as the power-series expansion of the function $f(X) = c_0 + c_1X + c_2X^2 + \cdots$.

Solution: We have

$$U^{-1}(a)XU(a) = X + a \qquad \text{(X.342)}$$

as well as

$$U^{-1}(a)X^nU(a) = U^{-1}(a)X^{n-1}U(a)U^{-1}(a)XU(a)$$
$$= U^{-1}(a)X^{n-1}U(a)(X+a). \qquad \text{(X.343)}$$

The proposition (X.341) follows by mathematical induction.
Hence, for deducing the commutator we can start from

$$e^{i\frac{Pa}{\hbar}}X^ne^{-i\frac{Pa}{\hbar}} = (X+a)^n. \qquad \text{(X.344)}$$

For sufficiently small $a$, it follows that

$$\left(1 + i\frac{Pa}{\hbar}\right)X^n\left(1 - i\frac{Pa}{\hbar}\right) + O(a^2) = X^n + a(n-1)X^{n-1} + O(a^2)$$
$$\text{(X.345)}$$

and thus

$$[P, X^n] = \frac{\hbar}{i}(n-1)X^{n-1}. \qquad \text{(X.346)}$$

For the commutator with the function $f(X)$, we find:

$$[P, f(X)] = \left[P, \sum c_nX^n\right] = \frac{\hbar}{i}\sum c_n(n-1)X^{n-1} = \frac{\hbar}{i}\frac{df(X)}{dX}. \quad \text{(X.347)}$$

We emphasize that at this point, $X$ and $P$ are still abstract operators. Analogously, one can derive that

$$[X, f(P)] = i\hbar\frac{df(P)}{dP} \qquad \text{(X.348)}$$

holds.
A note in passing: If we choose in particular $f(X) = e^{i\beta X}$, we obtain

$$e^{i\frac{Pa}{\hbar}} e^{i\beta X} e^{-i\frac{Pa}{\hbar}} = e^{i\beta X} e^{i\beta a}$$

i.e. the commutation relation in the *Weyl form*. It is mathematically clearly more well-behaved than the uncertainty principle $[X, P] = i\hbar$, because it contains with $e^{i\beta X}$ etc. only bounded operators (as opposed to $X$ and $P$, which are not bounded).

7. Show that a rotation through the angle $\varphi$ around the $z$ axis is represented by $e^{-i\alpha l_z}$.

   Solution: We assume a (sufficiently well-behaved) function $f(r, \vartheta, \varphi)$ (spherical coordinates). Taylor expansion yields

$$f(r, \vartheta, \varphi - \alpha) = \sum_n \frac{(-\alpha)^n}{n!} \frac{\partial^n}{\partial \varphi^n} f(r, \vartheta, \varphi). \tag{X.349}$$

With the definition of the $z$ component of the angular momentum:

$$l_z = \frac{\hbar}{i} \frac{\partial}{\partial \varphi}, \tag{X.350}$$

it follows that

$$f(r, \vartheta, \varphi - \alpha) = \sum_n \frac{(-\alpha)^n}{n!} \left(\frac{i}{\hbar}\right)^n l_z^n\, f(r, \vartheta, \varphi) = e^{-i\alpha l_z} f(r, \vartheta, \varphi). \tag{X.351}$$

The generalization to three-dimensional rotations and abstract angular momentum operators is analogous.

8. Using (21.32)

$$e^{-i\frac{\gamma \mathbf{j}\hat{\mathbf{a}}}{\hbar}} = e^{-i\frac{\varphi j_x}{\hbar}} e^{-i\frac{\gamma j_y}{\hbar}} e^{i\frac{\varphi j_x}{\hbar}}, \tag{X.352}$$

derive the commutation relation for the angular momentum.

Solution: We start with an infinitesimal $\gamma$. Then it follows that

$$1 - i\frac{\gamma \mathbf{j}\hat{\mathbf{a}}}{\hbar} = 1 - i e^{-i\frac{\varphi j_x}{\hbar}} \frac{\gamma j_y}{\hbar} e^{i\frac{\varphi j_x}{\hbar}} \text{ or } \mathbf{j}\hat{\mathbf{a}} = e^{-i\frac{\varphi j_x}{\hbar}} j_y e^{i\frac{\varphi j_x}{\hbar}}. \tag{X.353}$$

Because of $\hat{\mathbf{a}} = (0, \cos\varphi, \sin\varphi)$, we can write

$$\mathbf{j}\hat{\mathbf{a}} = \cos\varphi\, j_y + \sin\varphi\, j_z. \tag{X.354}$$

If $\varphi$ is infinitesimal, we obtain from the last two equations

$$j_y + \varphi\, j_z = \left(1 - i\frac{\varphi j_x}{\hbar}\right) j_y \left(1 + i\frac{\varphi j_x}{\hbar}\right) \tag{X.355}$$

or

$$j_y + \varphi \, j_z = j_y + \varphi \frac{i}{\hbar} j_y j_x - \varphi \frac{i}{\hbar} j_x j_y. \tag{X.356}$$

This leads directly to the commutation relation

$$i\hbar j_z = j_x j_y - j_y j_x = \left[ j_x, j_y \right]. \tag{X.357}$$

9. A scalar operator is defined as an operator whose mean value is invariant under a rotation. Derive the equation $[\mathbf{j}, S] = 0$.
   Solution: The rotation in $\mathcal{H}$ is $U(\mathcal{R})$, i.e. $|\varphi_\mathcal{R}\rangle = U(\mathcal{R}) |\varphi\rangle$. It must hold that

   $$\langle \varphi_\mathcal{R} | S | \varphi_\mathcal{R} \rangle = \langle \varphi | U^\dagger(\mathcal{R}) S U(\mathcal{R}) | \varphi \rangle = \langle \varphi | S | \varphi \rangle, \tag{X.358}$$

   and thus

   $$e^{i \frac{\gamma \mathbf{j} \hat{\mathbf{a}}}{\hbar}} S e^{-i \frac{\gamma \mathbf{j} \hat{\mathbf{a}}}{\hbar}} = S. \tag{X.359}$$

   Expansion for infinitesimal $\gamma$ gives immediately

   $$[\mathbf{j}, S] = 0. \tag{X.360}$$

10. A vector operator is an operator $\mathbf{V}$ whose mean value transforms like a vector $\mathbf{v}$ under a rotation through an angle $\gamma$ about an axis $\hat{\mathbf{a}}$, i.e. as

    $$\mathbf{v}' = \cos \gamma \cdot \mathbf{v} + \sin \gamma \cdot (\hat{\mathbf{a}} \times \mathbf{v}) + (1 - \cos \gamma) (\hat{\mathbf{a}} \cdot v) \cdot \hat{\mathbf{a}}. \tag{X.361}$$

    Derive $[j_i, V_k] = i\hbar \sum_l \varepsilon_{ikl} V_l$.
    Solution: The rotation in $\mathcal{H}$ is $U(\mathcal{R})$, i.e. $|\varphi_\mathcal{R}\rangle = U(\mathcal{R}) |\varphi\rangle$. For sufficiently small (infinitesimal) $\gamma$, we have

    $$\langle \varphi_\mathcal{R} | V_i | \varphi_\mathcal{R} \rangle = \langle \varphi | U^\dagger(\mathcal{R}) V_i U(\mathcal{R}) | \varphi \rangle = \langle \varphi | V_i' | \varphi \rangle, \tag{X.362}$$

    and accordingly for a rotation:

    $$e^{i \frac{\gamma \mathbf{j} \hat{\mathbf{a}}}{\hbar}} V_i e^{-i \frac{\gamma \mathbf{j} \hat{\mathbf{a}}}{\hbar}} = V_i'. \tag{X.363}$$

    Let $\hat{\mathbf{a}} = \hat{x}$ and $\gamma$ be infinitesimal. Then, due to

    $$\mathbf{v}' = \mathbf{v} + \gamma \cdot (\hat{\mathbf{a}} \times \mathbf{v}) + O(\gamma^2), \tag{X.364}$$

    for the transformed vector, it holds that:

    $$\mathbf{V}' = (V_x, V_y - \gamma V_z, V_z + \gamma V_y), \tag{X.365}$$

    and this gives for e.g. the $y$ component:

$$\left(1 + \frac{i}{\hbar}\gamma j_x\right) V_y \left(1 - \frac{i}{\hbar}\gamma j_x\right) = V_y - \gamma V_z, \qquad (X.366)$$

i.e. $i\left[j_x, V_y\right] = -\hbar V_z$. Analogously for the other components. In sum, it follows that

$$\left[j_x, V_x\right] = 0; \quad \left[j_x, V_y\right] = i\hbar V_z; \quad \left[j_x, V_z\right] = -i\hbar V_y; \quad \left[j_i, V_k\right] = i\hbar \sum_l \varepsilon_{ikl} V_l.$$
$$(X.367)$$

These relations hold e.g. if we insert the position or the momentum for **V**.

11. Formulate explicitly the unitary operator $e^{-i\frac{\gamma}{2}\sigma\hat{a}}$ for spin 1/2; $\sigma$ is the vector $\sigma = (\sigma_1, \sigma_1, \sigma_1)$ and $\hat{a}$ a 3-dimensional unit vector.

Solution: We have:

$$e^{-i\frac{\gamma}{2}\sigma\hat{a}} = \sum_n \frac{1}{n!}\left(-i\frac{\gamma}{2}\sigma\hat{a}\right)^n. \qquad (X.368)$$

With $(\sigma\mathbf{A})(\sigma\mathbf{B}) = \mathbf{A}\mathbf{B} + i\sigma(\mathbf{A}\times\mathbf{B})$ (already used in the exercises for Chap. 16), it follows that $(\sigma\mathbf{A})^2 = \mathbf{A}^2$, or $(\sigma\hat{a})^2 = 1$, and hence

$$e^{-i\frac{\gamma}{2}\sigma\hat{a}} = \sum_{n=0}^{\infty} \frac{1}{(2n)!}(-i)^{2n}\left(\frac{\gamma}{2}\right)^{2n}(\sigma\hat{a})^{2n}$$

$$+ \sum_{n=0}^{\infty} \frac{1}{(2n+1)!}(-i)^{2n+1}\left(\frac{\gamma}{2}\right)^{2n+1}(\sigma\hat{a})^{2n+1}$$

$$= \sum_{n=0}^{\infty} \frac{1}{(2n)!}(-i)^{2n}\left(\frac{\gamma}{2}\right)^{2n} + \sum_{n=0}^{\infty} \frac{1}{(2n+1)!}(-i)^{2n+1}\left(\frac{\gamma}{2}\right)^{2n+1}(\sigma\hat{a})$$

$$= \sum_{n=0}^{\infty} \frac{1}{(2n)!}(-1)^n\left(\frac{\gamma}{2}\right)^{2n} - i(\sigma\hat{a})\sum_{n=0}^{\infty} \frac{1}{(2n+1)!}(-1)^n\left(\frac{\gamma}{2}\right)^{2n+1}$$

$$= \cos\frac{\gamma}{2} - i\sigma\hat{a}\sin\frac{\gamma}{2}. \qquad (X.369)$$

## X.8  Exercises, Chap. 22

1. Write the density operator

$$\rho = \sum_n |\varphi_n\rangle\, p_n\, \langle\varphi_n| \qquad (X.370)$$

with normalized, but not necessarily orthogonal states $|\varphi_n\rangle$ when it is transformed unitarily.

Solution: The states transform according to

$$\left|\varphi_n'\right\rangle = U \left|\varphi_n\right\rangle \tag{X.371}$$

and we obtain

$$\rho' = U\rho U^{\dagger} = \sum_n U \left|\varphi_n\right\rangle p_n \left\langle\varphi_n\right| U^{\dagger} = \sum_n \left|\varphi_n'\right\rangle p_n \left\langle\varphi_n'\right|. \tag{X.372}$$

2. Show that $tr\,(AB) = tr\,(BA)$.
   Solution: With the CONS $\{|n\rangle\}$, we find that

$$tr\,(AB) = \sum_n \langle n|\,AB\,|n\rangle = \sum_{nm} \langle n|\,A\,|m\rangle \langle m|\,B\,|n\rangle = \sum_{nm} \langle m|\,B\,|n\rangle \langle n|\,A\,|m\rangle. \tag{X.373}$$

   The last step is possible since $\langle m|\,B\,|n\rangle$ and $\langle n|\,A\,|m\rangle$ are numbers. Thus we have

$$tr\,(AB) = \sum_{nm} \langle m|\,B\,|n\rangle \langle n|\,A\,|m\rangle = \sum_{nm} \langle m|\,BA\,|m\rangle = tr\,(BA). \tag{X.374}$$

3. Show that the trace is cyclically invariant, i.e.

$$tr\,(ABC) = tr\,(BCA) = tr\,(CAB). \tag{X.375}$$

   Solution: Due to $tr\,(AB) = tr\,(BA)$, we can write

$$\begin{aligned} tr\,(ABC) &= tr\,(A\,(BC)) = tr\,((BC)\,A) = tr\,(BCA) \\ tr\,(ABC) &= tr\,((AB)\,C) = tr\,(C\,(AB)) = tr\,(CAB). \end{aligned} \tag{X.376}$$

4. Show that the trace is invariant under unitary transformations.
   Solution: With the unitary matrix $U$, we obtain from the matrix $A$ the new matrix $A' = UAU^{-1}$. Using $tr(AB) = tr(BA)$, it follows that

$$tr\left(A'\right) = tr\left(UAU^{-1}\right) = tr\left(U^{-1}UA\right) = tr\,(A). \tag{X.377}$$

5. Show that the trace is independent of the basis. (This *must* apply, since a basis transformation is unitary.)
   Solution: Let two CONS, $\{|\varphi_n\rangle\}$ and $\{|\psi_n\rangle\}$ be given. Then it holds that

$$\begin{aligned} tr\,(A) &= \sum_m \langle\varphi_m|\,A\,|\varphi_m\rangle = \sum_{m,n} \langle\varphi_m|\,\psi_n\rangle \langle\psi_n|\,A\,|\varphi_m\rangle \\ &= \sum_{m,n} \langle\psi_n|\,A\,|\varphi_m\rangle \langle\varphi_m|\,\psi_n\rangle = \sum_n \langle\psi_n|\,A\,|\psi_n\rangle. \end{aligned} \tag{X.378}$$

6. Given a CONS $\{|n\rangle\}$ and a state $|\psi\rangle = \sum_n c_n |n\rangle$ with $\sum_n |c_n|^2 = 1$, show that the probability of finding the system in the state $m$ is given by $p_m = tr(\rho |m\rangle \langle m|) = tr(\rho P_m)$.

Solution: The probability sought is given by $|c_m|^2$. Due to $c_m = \langle m| \psi\rangle$, we have:

$$|c_m|^2 = \langle m| \psi\rangle \langle\psi| m\rangle = \langle m| \rho |m\rangle = \sum_n \langle m| n\rangle \langle n| \rho |m\rangle$$

$$= \sum_n \langle n| \rho |m\rangle \langle m| n\rangle = tr\left(\rho |m\rangle \langle m|\right). \tag{X.379}$$

7. Show that for the reduced density operator $\rho^{(1)}$, it holds in general that $tr([\rho^{(1)}]^2) \leq 1$; hence, we have a mixture if the strict inequality applies.

Solution: Being a Hermitian operator, $\rho^{(1)}$ is diagonalizable via a unitary transformation. Hence, there is a diagonal matrix $D$ and a unitary matrix $U$ such that:

$$\rho^{(1)} = UDU^\dagger. \tag{X.380}$$

About the matrix $D$, we know (since $\rho^{(1)}$ is positive) that its diagonal entries obey $0 \leq d_{nn} \leq 1$ and $tr(D) = 1$. We have (see the previous exercise):

$$tr(\rho^{(1)}) = tr(UDU^\dagger) = tr(D) = 1. \tag{X.381}$$

Moreover, it applies that

$$tr\left([\rho^{(1)}]^2\right) = tr\left(UD^2U^\dagger\right) = tr(D^2). \tag{X.382}$$

Due to $0 \leq d_{nn}^2 \leq d_{nn} \leq 1$, it follows that

$$tr\left([\rho^{(1)}]^2\right) = tr(D^2) \leq tr(D) = 1. \tag{X.383}$$

The equals sign can apply only if there is just one non-vanishing diagonal element (which, due to $tr(\rho) = 1$, must be 1).

8. Write the density operator in the position representation (cf. Chaps. 12 and 13, Vol. 1).

Solution: With

$$\rho = |\psi\rangle \langle\psi|, \tag{X.384}$$

it follows that

$$\langle x| \rho |x'\rangle = \rho\left(x, x'\right) = \langle x |\psi\rangle \langle\psi| x'\rangle = \psi(x)\psi^*(x'). \tag{X.385}$$

By applying $\rho$ to a state $|\varphi\rangle$, we obtain an state $|\chi\rangle$; in the abstract and position representations, we have:

$$|\chi\rangle = \rho |\varphi\rangle$$

$$\langle x\ |\chi\rangle = \chi\,(x) = \int\, \langle x|\, \rho\, |x'\rangle \langle x'\ |\varphi\rangle\, dx' = \int\, \rho\,(x, x')\, \varphi(x')dx'. \tag{X.386}$$

9. Show explicitly for

$$\rho = \begin{pmatrix} c_1 c_1^* & c_1 c_2^* \\ c_2 c_1^* & c_2 c_2^* \end{pmatrix} \tag{X.387}$$

that

$$\rho^2 = \rho \tag{X.388}$$

applies; using this matrix, show explicitly that $\rho^2 = \rho$. Here, it must hold that $|c_1|^2 + |c_2|^2 = 1$.

10. Show that the eigenvalues $\lambda_{1/2}$ of the matrix

$$\rho = \begin{pmatrix} c_1 c_1^* & c_1 c_2^* \\ c_2 c_1^* & c_2 c_2^* \end{pmatrix} \tag{X.389}$$

are 0 and 1.

Solution: The eigenvalues of the matrix are calculated by

$$\begin{vmatrix} |c_1|^2 - \lambda & c_1 c_2^* \\ c_2 c_1^* & |c_2|^2 - \lambda \end{vmatrix} = 0. \tag{X.390}$$

Expansion of the determinant gives

$$\left(|c_1|^2 - \lambda\right)\left(|c_2|^2 - \lambda\right) - c_1 c_2^* c_2 c_1^* = 0. \tag{X.391}$$

It follows that

$$\lambda^2 - \left(|c_1|^2 + |c_2|^2\right)\lambda + |c_1|^2\, |c_2|^2 - |c_1|^2\, |c_2|^2 = 0. \tag{X.392}$$

Due to $|c_1|^2 + |c_2|^2 = 1$, this yields

$$\lambda^2 - \lambda = 0 \tag{X.393}$$

and thus the proposition is demonstrated.

11. Given the density matrix for a statistical mixture in the form $\rho = p_h\, |h\rangle\, \langle h| + p_v\, |v\rangle\, \langle v|$ or

$$\rho = \begin{pmatrix} p_h & 0 \\ 0 & p_v \end{pmatrix}; \tag{X.394}$$

How does this read in the circularly-polarized basis?
Solution: We have:

$$|h\rangle = \frac{|r\rangle + |l\rangle}{\sqrt{2}}; \quad |v\rangle = \frac{|r\rangle - |l\rangle}{\sqrt{2}i}. \tag{X.395}$$

From this, it follows that:

$$\rho = p_h |h\rangle \langle h| + p_v |v\rangle \langle v| = p_h \frac{|r\rangle + |l\rangle}{\sqrt{2}} \frac{\langle r| + \langle l|}{\sqrt{2}} - p_v \frac{|r\rangle - |l\rangle}{\sqrt{2}i} \frac{\langle r| - \langle l|}{\sqrt{2}i}$$

$$= \frac{p_h + p_v}{2} |r\rangle \langle r| + \frac{p_h - p_v}{2} |r\rangle \langle l| + \frac{p_h - p_v}{2} |l\rangle \langle r| + \frac{p_h + p_v}{2} |l\rangle \langle l|, \tag{X.396}$$

and in matrix form:

$$\rho = \frac{1}{2} \begin{pmatrix} p_h + p_v & p_h - p_v \\ p_h - p_v & p_h + p_v \end{pmatrix}. \tag{X.397}$$

12. Given two quantum objects Q1 and Q2 with the respective $N$-dimensional CONS $\{|\varphi_i\rangle\}$ for Q1 and $\{|\psi_i\rangle\}$ for Q2 (by the choice of notation, we can omit the index for the number of the quantum object). The initial state is

$$|\chi\rangle = \sum_{ij} c_{ij} |\varphi_i\rangle |\psi_j\rangle. \tag{X.398}$$

Calculate the probability $w(\lambda)$ of measuring the quantum object 1 in a state $|\lambda\rangle$, and formulate it in terms of the reduced density operator $\rho^{(1)}$.
Solution: In an exercise for Chap. 20, we calculated the probability as

$$w(\lambda) = \sum_{ijn} c_{in}^* c_{jn} \langle \varphi_i |\lambda\rangle \langle \lambda |\varphi_j\rangle. \tag{X.399}$$

We now consider the density operator $\rho = |\chi\rangle \langle \chi|$; the reduced density operator $\rho^{(1)} = tr_2(\rho)$ is given by:

$$\rho^{(1)} = \sum_n \langle \psi_n |\chi\rangle \langle \chi |\psi_n\rangle$$

$$= \sum_n \langle \psi_n| \left( \sum_{ij} c_{ij} |\varphi_i\rangle |\psi_j\rangle \right) \left( \sum_{kl} c_{kl}^* \langle \varphi_k| \langle \psi_l| \right) |\psi_n\rangle$$

$$= \sum_{nijkl} \delta_{nj} c_{ij} |\varphi_i\rangle c_{kl}^* \langle \varphi_k| \delta_{nl}$$

$$= \sum_{ijk} c_{ij} |\varphi_i\rangle c_{kj}^* \langle \varphi_k|. \tag{X.400}$$

By comparison with (X.399), we see that

$$w(\lambda) = \langle \lambda | \rho^{(1)} | \lambda \rangle. \tag{X.401}$$

The probability of measuring the state $|\lambda\rangle$ is thus the expectation value of the reduced density operator, referred to this state.

13. Given the density operator $\rho = \sum_n p_n |\varphi_n\rangle \langle\varphi_n|$, where it holds that $i\hbar\partial_t |\varphi_n\rangle = H |\varphi_n\rangle$. Show that the time behavior of $\rho$ is described by the von-Neumann equation:

$$i\hbar\partial_t \rho = [H, \rho]. \tag{X.402}$$

Solution: It holds that

$$i\hbar\partial_t \rho = i\hbar\partial_t \sum_n p_n |\varphi_n\rangle \langle\varphi_n|$$

$$= H \sum_n p_n |\varphi_n\rangle \langle\varphi_n| - \sum_n p_n |\varphi_n\rangle \langle\varphi_n| H = [H, \rho]. \tag{X.403}$$

14. Using the example of a polarized photon, show explicitly that a given density matrix does not allow for a unique decomposition.

(a) First formulate the projection operators for the states $|h\rangle$, $|v\rangle$, $|r\rangle$ and $|l\rangle$.

Solution: With $|h\rangle = \begin{pmatrix} 1 \\ 0 \end{pmatrix}$, $|v\rangle = \begin{pmatrix} 0 \\ 1 \end{pmatrix}$, $|r\rangle = \frac{1}{\sqrt{2}}\begin{pmatrix} 1 \\ i \end{pmatrix}$ and $|l\rangle = \frac{1}{\sqrt{2}}\begin{pmatrix} 1 \\ -i \end{pmatrix}$, it follows that

$$P_h = \begin{pmatrix} 1 & 0 \\ 0 & 0 \end{pmatrix}; \; P_v = \begin{pmatrix} 0 & 0 \\ 0 & 1 \end{pmatrix} \tag{X.404}$$

and

$$P_r = \frac{1}{2}\begin{pmatrix} 1 & -i \\ i & 1 \end{pmatrix}; \; P_l = \frac{1}{2}\begin{pmatrix} 1 & i \\ -i & 1 \end{pmatrix}. \tag{X.405}$$

(b) Given the density matrix $\rho = \frac{1}{2}\begin{pmatrix} 1 & 0 \\ 0 & 1 \end{pmatrix}$; now formulate the decomposition of $\rho$ in terms of linearly- and circularly-polarized states.

Solution: It evidently applies that $\rho = \frac{1}{2}P_h + \frac{1}{2}P_v$ as well as $\rho = \frac{1}{2}P_r + \frac{1}{2}P_l$. Hence, from a given density matrix, one cannot uniquely determine the underlying states.

15. The spin state of an electron is represented (in the basis of eigenstates of the spin matrix $s_z = \frac{\hbar}{2}\sigma_z$) by the density matrix $\rho = \begin{pmatrix} a & 0 \\ 0 & b \end{pmatrix}$, with $a + b = 1$; $a \geq 0, b \geq 0$.

(a) What is the probability of obtaining $\pm\frac{\hbar}{2}$, if one measures $s_x$?

Solution: $s_x$ has the eigenvalues $\pm\frac{\hbar}{2}$ and the eigenvectors $|x_1\rangle = \frac{1}{\sqrt{2}}\begin{pmatrix} 1 \\ 1 \end{pmatrix}$

and $|x_2\rangle = \frac{1}{\sqrt{2}}\begin{pmatrix} 1 \\ -1 \end{pmatrix}$. Thereof follow the projectors

$$P_1 = \frac{1}{2}\begin{pmatrix} 1 \\ 1 \end{pmatrix}(1\ 1) = \frac{1}{2}\begin{pmatrix} 1 & 1 \\ 1 & 1 \end{pmatrix}$$

$$P_2 = \frac{1}{2}\begin{pmatrix} 1 \\ -1 \end{pmatrix}(1\ -1) = \frac{1}{2}\begin{pmatrix} 1 & -1 \\ -1 & 1 \end{pmatrix}, \tag{X.406}$$

and thus

$$tr\,(\rho P_1) = tr\left(\begin{pmatrix} a & 0 \\ 0 & b \end{pmatrix}\frac{1}{2}\begin{pmatrix} 1 & 1 \\ 1 & 1 \end{pmatrix}\right)$$

$$= \frac{1}{2}tr\begin{pmatrix} a & a \\ b & b \end{pmatrix} = \frac{a+b}{2} = \frac{1}{2}, \tag{X.407}$$

as well as

$$tr\,(\rho P_2) = tr\left(\begin{pmatrix} a & 0 \\ 0 & b \end{pmatrix}\frac{1}{2}\begin{pmatrix} 1 & -1 \\ -1 & 1 \end{pmatrix}\right)$$

$$= \frac{1}{2}tr\begin{pmatrix} a & -a \\ -b & b \end{pmatrix} = \frac{a+b}{2} = \frac{1}{2}. \tag{X.408}$$

Hence, the probability equals $\frac{1}{2}$ for both results.

(b) Calculate the expectation value of $s_x$ and compare it with the trace formalism.
Solution: From the previous part of the exercise, we know that we measure both the eigenvalues $\frac{\hbar}{2}$ and $-\frac{\hbar}{2}$ with a probability $\frac{1}{2}$. Accordingly, the mean value is zero. On the other hand, we obtain with $\langle A\rangle = tr\,(\rho A)$:

$$\langle s_x\rangle = tr\left(\begin{pmatrix} a & 0 \\ 0 & b \end{pmatrix}\frac{\hbar}{2}\begin{pmatrix} 0 & 1 \\ 1 & 0 \end{pmatrix}\right) = \frac{\hbar}{2}tr\begin{pmatrix} 0 & a \\ b & 0 \end{pmatrix} = 0; \tag{X.409}$$

i.e. agreement.

16. Given a system of two quantum objects; the basis states are in each case $|1\rangle$ and $|2\rangle$.

(a) How is the general total state $|\psi\rangle$ formulated?
Solution:

$$|\psi\rangle = c_{11}|1_1 1_2\rangle + c_{12}|1_1 2_2\rangle + c_{21}|2_1 1_2\rangle + c_{22}|2_1 2_2\rangle \quad \text{with} \quad \sum|c_{ij}|^2 = 1. \tag{X.410}$$

(b) Give explicitly the density matrix for this system.
   Solution: We have

$$\rho = \begin{pmatrix} c_{11}c_{11}^* & c_{11}c_{12}^* & c_{11}c_{21}^* & c_{11}c_{22}^* \\ c_{12}c_{11}^* & c_{12}c_{12}^* & c_{12}c_{21}^* & c_{12}c_{22}^* \\ c_{21}c_{11}^* & c_{21}c_{12}^* & c_{21}c_{21}^* & c_{21}c_{22}^* \\ c_{22}c_{11}^* & c_{22}c_{12}^* & c_{22}c_{21}^* & c_{22}c_{22}^*. \end{pmatrix} \qquad (X.411)$$

(c) Starting from this matrix, calculate the reduced density matrix $\rho^{(1)}$.
   Solution: The unity matrix in space 1 is $E_1$. It follows that

$$A_1 = E_1 \otimes |1_2\rangle \cong \begin{pmatrix} 1 & 0 \\ 0 & 1 \end{pmatrix}_1 \otimes \begin{pmatrix} 1 \\ 0 \end{pmatrix}_2 = \begin{pmatrix} 1 & 0 \\ 0 & 0 \\ 0 & 1 \\ 0 & 0 \end{pmatrix}$$

$$A_2 = E_1 \otimes |2_2\rangle \cong \begin{pmatrix} 1 & 0 \\ 0 & 1 \end{pmatrix}_1 \otimes \begin{pmatrix} 0 \\ 1 \end{pmatrix}_2 = \begin{pmatrix} 0 & 0 \\ 1 & 0 \\ 0 & 0 \\ 0 & 1 \end{pmatrix}. \qquad (X.412)$$

The reduced density matrix follows as

$$\rho^{(1)} = A_1^\dagger \rho A_1 + A_2^\dagger \rho A_2. \qquad (X.413)$$

Written out, this appears as:

$$\rho^{(1)} = \begin{pmatrix} 1 & 0 & 0 & 0 \\ 0 & 0 & 1 & 0 \end{pmatrix} \begin{pmatrix} c_{11}c_{11}^* & c_{11}c_{12}^* & c_{11}c_{21}^* & c_{11}c_{22}^* \\ c_{12}c_{11}^* & c_{12}c_{12}^* & c_{12}c_{21}^* & c_{12}c_{22}^* \\ c_{21}c_{11}^* & c_{21}c_{12}^* & c_{21}c_{21}^* & c_{21}c_{22}^* \\ c_{22}c_{11}^* & c_{22}c_{12}^* & c_{22}c_{21}^* & c_{22}c_{22}^* \end{pmatrix} \begin{pmatrix} 1 & 0 \\ 0 & 0 \\ 0 & 1 \\ 0 & 0 \end{pmatrix}$$

$$+ \begin{pmatrix} 0 & 1 & 0 & 0 \\ 0 & 0 & 0 & 1 \end{pmatrix} \begin{pmatrix} c_{11}c_{11}^* & c_{11}c_{12}^* & c_{11}c_{21}^* & c_{11}c_{22}^* \\ c_{12}c_{11}^* & c_{12}c_{12}^* & c_{12}c_{21}^* & c_{12}c_{22}^* \\ c_{21}c_{11}^* & c_{21}c_{12}^* & c_{21}c_{21}^* & c_{21}c_{22}^* \\ c_{22}c_{11}^* & c_{22}c_{12}^* & c_{22}c_{21}^* & c_{22}c_{22}^* \end{pmatrix} \begin{pmatrix} 0 & 0 \\ 1 & 0 \\ 0 & 0 \\ 0 & 1 \end{pmatrix}. \qquad (X.414)$$

After some calculations, we find

$$\rho^{(1)} = \begin{pmatrix} c_{11}c_{11}^* + c_{12}c_{12}^* & c_{11}c_{21}^* + c_{12}c_{22}^* \\ c_{21}c_{11}^* + c_{22}c_{12}^* & c_{21}c_{21}^* + c_{22}c_{22}^* \end{pmatrix}. \qquad (X.415)$$

(d) Show that $tr\left(\rho^{(1)}\right) = 1$ holds.

Solution: From this equation, we read off directly

$$tr\rho^{(1)} = |c_{11}|^2 + |c_{12}|^2 + |c_{21}|^2 + |c_{22}|^2 = 1. \qquad (X.416)$$

The last equals sign holds due to the normalization in (X.410).

(e) Show that $\rho^{(1)} = CC^\dagger$ with $C = \begin{pmatrix} c_{11} & c_{12} \\ c_{21} & c_{22} \end{pmatrix}$ holds.

Solution:

$$CC^\dagger = \begin{pmatrix} c_{11} & c_{12} \\ c_{21} & c_{22} \end{pmatrix} \begin{pmatrix} c_{11}^* & c_{21}^* \\ c_{12}^* & c_{22}^* \end{pmatrix} = \begin{pmatrix} c_{11}c_{11}^* + c_{12}c_{12}^* & c_{11}c_{21}^* + c_{12}c_{22}^* \\ c_{21}c_{11}^* + c_{22}c_{12}^* & c_{21}c_{21}^* + c_{22}c_{22}^* \end{pmatrix}. \qquad (X.417)$$

(f) Calculate $\rho^{(1)2}$.
Solution: We write, abbreviating

$$\rho^{(1)} = \begin{pmatrix} p & a \\ a^* & 1-p \end{pmatrix} \qquad (X.418)$$

and obtain

$$\rho^{(1)2} = \begin{pmatrix} p^2 + |a|^2 & a \\ a^* & (1-p)^2 + |a|^2 \end{pmatrix}. \qquad (X.419)$$

(g) Show that $tr\left(\rho^{(1)2}\right) = 1 - 2\,|\det C|^2$ is true.
Solution: We start from

$$\det \rho^{(1)} = p(1-p) - |a|^2 = \det\left(CC^\dagger\right) = \det C \cdot \det C^\dagger = |\det C|^2. \qquad (X.420)$$

From (X.419), we can read off directly that

$$tr\left(\rho^{(1)2}\right) = 1 - 2p + 2p^2 + 2\,|a|^2. \qquad (X.421)$$

With $|a|^2 = p(1-p) - \det \rho^{(1)}$, it follows that

$$tr\left(\rho^{(1)2}\right) = 1 - 2\det\rho^{(1)} = 1 - 2\,|\det C|^2. \qquad (X.422)$$

17. $\{|\varphi_i\rangle\,, i = 1, \ldots, N\}$ are normalized, but not necessarily orthogonal states. Show that the density matrix $\rho = \frac{1}{N}\sum_{i=1}^{N} |\varphi_i\rangle\langle\varphi_i|$ describes a pure state, iff these $N$ states are equal up to a phase.

(a) Let $|\varphi_n\rangle = e^{i\delta_n} |\varphi\rangle$. Show that $\rho^2 = \rho$.
Solution: It follows that $\rho = \frac{1}{N}\sum_{i=1}^{N} |\varphi\rangle\langle\varphi| = |\varphi\rangle\langle\varphi|$ and $\rho^2 = \rho$.

(b) Let $\rho^2 = \rho$; show that the $N$ states $|\varphi_i\rangle$ are equal up to a phase.
Solution: We introduce a CONS $\{|m\rangle\,, m = 1, \ldots, N\}$. Then it follows that:

$$\rho = \frac{1}{N} \sum_n |\varphi_n\rangle \langle \varphi_n| = \sum_m |m\rangle \, p_m \, \langle m| \, , \qquad (X.423)$$

and the probabilities are given by

$$p_m = \langle m| \, \rho \, |m\rangle = \frac{1}{N} \sum_n \langle m \, |\varphi_n\rangle \langle \varphi_n| \, m\rangle = \frac{1}{N} \sum_n |\langle m \, |\varphi_n\rangle|^2 \, . \quad (X.424)$$

Due to $\rho^2 = \rho$, all $p_m$ have to vanish except one of them, i.e. $p_m = \delta_{mM}$. This means that

$$p_m = \frac{1}{N} \sum_n |\langle m \, |\varphi_n\rangle|^2 = \delta_{mM} \, . \qquad (X.425)$$

All the terms in the sum are greater than or equal to zero. Thus, it follows that

$$\begin{array}{l} m \neq M \; :\rightarrow \; |\langle m \, |\varphi_n\rangle|^2 = 0 \; \forall n \\ m = M \; :\rightarrow \; \frac{1}{N} \sum_n |\langle M \, |\varphi_n\rangle|^2 = 1 \end{array} \, . \qquad (X.426)$$

Because of $|\langle m \, |\varphi_n\rangle|^2 = 0$ for all $m \neq M$, it follows that all the states $|\varphi_n\rangle$ have to be proportional to $|M\rangle$, i.e. $|\varphi_n\rangle = c_n |M\rangle$. Since the $|\varphi_n\rangle$ as well as $|M\rangle$ are normalized, it follows that $|c_n|^2 = 1$ or $c_n = e^{i\alpha_n} |M\rangle$. Thus, we have shown that all states are equal, apart from a phase. The second equation in (X.426) is also satisfied, of course, since

$$m = M \; :\rightarrow \; \frac{1}{N} \sum_n |\langle M \, |\varphi_n\rangle|^2 = \frac{1}{N} \sum_n |e^{i\alpha_n}|^2 = 1. \qquad (X.427)$$

## X.9   Exercises, Chap. 23

1. Two identical quantum objects are in the states $|\alpha_1\rangle$ and $|\alpha_2\rangle$. Show that the total state must be symmetrical or antisymmetrical,

$$|\psi_\pm\rangle = \frac{|1 : \alpha_1, 2 : \alpha_2\rangle \pm |1 : \alpha_2, 2 : \alpha_1\rangle}{\sqrt{2}}. \qquad (X.428)$$

Solution: There are initially two equivalent possibilities to describe the product state, namely $|1 : \alpha_1, 2 : \alpha_2\rangle$ and $|1 : \alpha_2, 2 : \alpha_1\rangle$. Since these descriptions are indistinguishable (and we cannot exclude one of them), the total state must be a linear combination, i.e.

$$|\Psi\rangle = a \, |1 : \alpha_1, 2 : \alpha_2\rangle + b \, |1 : \alpha_2, 2 : \alpha_1\rangle \, . \qquad (X.429)$$

As usual, the state is normalized

$$1 = \langle \Psi \mid \Psi \rangle = |a|^2 + |b|^2. \tag{X.430}$$

Since the product states in (X.429) are equivalent, it follows that

$$a = \frac{e^{i\alpha}}{\sqrt{2}}; \ b = \frac{e^{i\beta}}{\sqrt{2}} \tag{X.431}$$

with yet undetermined phases $\alpha$ and $\beta$.

Apart from (X.429), there is a further equivalent representation of the total state, namely (commutation of the coefficients $a$ and $b$)

$$|\Phi\rangle = b \, |1 : \alpha_1, 2 : \alpha_2\rangle + a \, |1 : \alpha_2, 2 : \alpha_1\rangle. \tag{X.432}$$

Since the states $|\Psi\rangle$ and $|\Phi\rangle$ describe the same facts, they may differ at most by a phase. So it must hold that

$$|\Phi\rangle = e^{i\delta} \, |\Psi\rangle. \tag{X.433}$$

Insertion of (X.429) and (X.432) and comparison leads to

$$e^{i\alpha} = e^{i\delta} e^{i\beta}; \ e^{i\beta} = e^{i\delta} e^{i\alpha}. \tag{X.434}$$

This yields

$$e^{i\delta} = e^{i(\alpha - \beta)}; \ e^{i\delta} = e^{i(\beta - \alpha)}, \tag{X.435}$$

and, due to $e^{i\delta} = e^{-i\delta}$ or $e^{2i\delta} = 1$, we obtain

$$e^{i\delta} = e^{i(\beta - \alpha)} = \pm 1. \tag{X.436}$$

Thus, it follows that

$$
\begin{aligned}
|\Psi\rangle &= e^{i\alpha} \frac{|1 : \alpha_1, 2 : \alpha_2\rangle + e^{i(\beta - \alpha)} |1 : \alpha_2, 2 : \alpha_1\rangle}{\sqrt{2}} \\
&= e^{i\alpha} \frac{|1 : \alpha_1, 2 : \alpha_2\rangle \pm |1 : \alpha_2, 2 : \alpha_1\rangle}{\sqrt{2}}.
\end{aligned}
\tag{X.437}
$$

Since the global phase does not play any physical role, we can write finally

$$|\psi_\pm\rangle = \frac{|1 : \alpha_1, 2 : \alpha_2\rangle \pm |1 : \alpha_2, 2 : \alpha_1\rangle}{\sqrt{2}}. \tag{X.438}$$

Hence, there are only these two possibilities, bosons ($+$) and fermions ($-$).

2. Two identical particles are in the states $|a\rangle$ and $|b\rangle$. What is the correct expression for the total state $|\psi\rangle$?

Solution:

$$|\psi\rangle = \frac{|1:a,2:b\rangle \pm |1:b,2:a\rangle}{\sqrt{2}} \quad \text{or, written compactly,} \quad \frac{|ab\rangle \pm |ba\rangle}{\sqrt{2}}.$$
(X.439)

3. Let $|\varphi\rangle = |1:\alpha_1, 2:\alpha_2, 3:\alpha_3\rangle$. Determine $P_{12}P_{23}|\varphi\rangle$ and $P_{23}P_{12}|\varphi\rangle$. Under what conditions do $P_{12}$ and $P_{23}$ commute?
Solution:

$$P_{12}P_{23}|\varphi\rangle = P_{12}|1:\alpha_1, 2:\alpha_3, 3:\alpha_2\rangle = |1:\alpha_3, 2:\alpha_1, 3:\alpha_2\rangle$$
$$P_{23}P_{12}|\varphi\rangle = P_{23}|1:\alpha_2, 2:\alpha_1, 3:\alpha_3\rangle = |1:\alpha_2, 2:\alpha_3, 3:\alpha_1\rangle.$$
(X.440)

If these two states are to be equal (i.e. they commute), in general it must hold that $\alpha_1 = \alpha_2 = \alpha_3$.

4. Write down explicitly the normalized states $|1:\alpha_1, 2:\alpha_2, \ldots, N:\alpha_N\rangle_{norm}^{(\pm)}$ for 2 and 3 particles.

5. Given 3 identical particles; to save paperwork, we denote the product states simply by $|1,2,3\rangle$ instead of $|1:\alpha_1, 2:\alpha_2, 3:\alpha_3\rangle$; $|1:\alpha_2, 2:\alpha_1, 3:\alpha_3\rangle$ is then $|2,1,3\rangle$ etc.

   (a) Write down all 6 product states.
   (b) Show explicitly that for the total (anti)symmetrical state, $P_{12}|\psi\rangle^{\pm} = \eta_{12}|\psi\rangle^{\pm}$. Determine $\eta_{12}$.
   (c) Given the state $|\varphi\rangle = |1,2,3\rangle - |1,3,2\rangle + |2,1,3\rangle - |2,3,1\rangle + |3,1,2\rangle - |3,2,1\rangle$, show explicitly that $P_{12}|\varphi\rangle$ cannot be written as $c|\varphi\rangle$.

6. Show explicitly that $P_{ni}P_{mj}P_{nm}P_{ni}P_{mj} = P_{ij}$.
Solution: We perform each step explicitly by rearranging the state $|\ldots, i:\alpha_i, \ldots, j:\alpha_j, \ldots, m:\alpha_m, \ldots, n:\alpha_n, \ldots\rangle$ correspondingly. This is a bit clumsy and lengthy, but it may help in understanding the problem:

$$P_{mj}|\ldots, i:\alpha_i, \ldots, j:\alpha_j, \ldots, m:\alpha_m, \ldots, n:\alpha_n, \ldots\rangle$$
$$= |\ldots, i:\alpha_i, \ldots, j:\alpha_m, \ldots, m:\alpha_j, \ldots, n:\alpha_n, \ldots\rangle$$
$$P_{ni}|\ldots, i:\alpha_i, \ldots, j:\alpha_m, \ldots, m:\alpha_j, \ldots, n:\alpha_n, \ldots\rangle$$
$$= |\ldots, i:\alpha_n, \ldots, j:\alpha_m, \ldots, m:\alpha_j, \ldots, n:\alpha_i, \ldots\rangle$$
$$P_{nm}|\ldots, i:\alpha_n, \ldots, j:\alpha_m, \ldots, m:\alpha_j, \ldots, n:\alpha_i, \ldots\rangle$$
$$= |\ldots, i:\alpha_n, \ldots, j:\alpha_m, \ldots, m:\alpha_i, \ldots, n:\alpha_j, \ldots\rangle$$
$$P_{mj}|\ldots, i:\alpha_n, \ldots, j:\alpha_m, \ldots, m:\alpha_i, \ldots, n:\alpha_j, \ldots\rangle$$
$$= |\ldots, i:\alpha_n, \ldots, j:\alpha_i, \ldots, m:\alpha_m, \ldots, n:\alpha_j, \ldots\rangle$$
$$P_{ni}|\ldots, i:\alpha_n, \ldots, j:\alpha_i, \ldots, m:\alpha_m, \ldots, n:\alpha_j, \ldots\rangle$$
$$= |\ldots, i:\alpha_j, \ldots, j:\alpha_i, \ldots, m:\alpha_m, \ldots, n:\alpha_n, \ldots\rangle.$$
(X.441)

We can arrange this in a somewhat clearer manner by using the convention that the quantum numbers for the particles $i, j, m, n$ are always written at the positions 1, 2, 3, 4, i.e.

$|\ldots, i : \alpha_u, \ldots, j : \alpha_v, \ldots, m : \alpha_w, \ldots, n : \alpha_x, \ldots\rangle \equiv |\alpha_u, \alpha_v, \alpha_w, \alpha_x\rangle$ as an example. Then the calculation reads

$$
\begin{array}{c}
\text{Place } i \ \ j \ \ n \ \ m \\
P_{mj} \left|\alpha_i, \alpha_j, \alpha_m, \alpha_n\right\rangle = \left|\alpha_i, \alpha_m, \alpha_j, \alpha_n\right\rangle \\
P_{ni} \left|\alpha_i, \alpha_m, \alpha_j, \alpha_n\right\rangle = \left|\alpha_n, \alpha_m, \alpha_j, \alpha_i\right\rangle \\
P_{nm} \left|\alpha_n, \alpha_m, \alpha_j, \alpha_i\right\rangle = \left|\alpha_n, \alpha_m, \alpha_i, \alpha_j\right\rangle \\
P_{mj} \left|\alpha_n, \alpha_m, \alpha_i, \alpha_j\right\rangle = \left|\alpha_n, \alpha_i, \alpha_m, \alpha_j\right\rangle \\
P_{ni} \left|\alpha_n, \alpha_i, \alpha_m, \alpha_j\right\rangle = \left|\alpha_j, \alpha_i, \alpha_m, \alpha_n\right\rangle .
\end{array}
\tag{X.442}
$$

7. Show that

$$
\begin{aligned}
E_{100;nlm}^{(1)} &= \frac{e^2}{4\pi\varepsilon_0} \int d^3 r_1 d^3 r_2 \frac{\left|\psi_{100}(\mathbf{r}_1)\,\psi_{nlm}(\mathbf{r}_2) \pm \psi_{nlm}(\mathbf{r}_1)\,\psi_{100}(\mathbf{r}_2)\right|^2}{2\,|\mathbf{r}_1 - \mathbf{r}_2|} \\
&= C_{nl} \pm A_{nl}.
\end{aligned}
\tag{X.443}
$$

Solution: We have

$$
\begin{aligned}
&\int d^3 r_1 d^3 r_2 \frac{\left|\psi_{100}(\mathbf{r}_1)\,\psi_{nlm}(\mathbf{r}_2) \pm \psi_{nlm}(\mathbf{r}_1)\,\psi_{100}(\mathbf{r}_2)\right|^2}{2\,|\mathbf{r}_1 - \mathbf{r}_2|} \\
&= \int d^3 r_1 d^3 r_2 \frac{\left[\psi_{100}(\mathbf{r}_1)\,\psi_{nlm}(\mathbf{r}_2) \pm \psi_{nlm}(\mathbf{r}_1)\,\psi_{100}(\mathbf{r}_2)\right]\left[\psi_{100}^*(\mathbf{r}_1)\,\psi_{nlm}^*(\mathbf{r}_2) \pm \psi_{nlm}^*(\mathbf{r}_1)\,\psi_{100}^*(\mathbf{r}_2)\right]}{2\,|\mathbf{r}_1 - \mathbf{r}_2|} \\
&= \int d^3 r_1 d^3 r_2 \frac{\left|\psi_{100}(\mathbf{r}_1)\,\psi_{nlm}(\mathbf{r}_2)\right|^2 + \left|\psi_{nlm}(\mathbf{r}_1)\,\psi_{100}(\mathbf{r}_2)\right|^2}{2\,|\mathbf{r}_1 - \mathbf{r}_2|} \\
&\quad \pm \int d^3 r_1 d^3 r_2 \frac{\psi_{100}(\mathbf{r}_1)\,\psi_{nlm}(\mathbf{r}_2)\,\psi_{nlm}^*(\mathbf{r}_1)\,\psi_{100}^*(\mathbf{r}_2) + \psi_{nlm}(\mathbf{r}_1)\,\psi_{100}(\mathbf{r}_2)\,\psi_{100}^*(\mathbf{r}_1)\,\psi_{nlm}^*(\mathbf{r}_2)}{2\,|\mathbf{r}_1 - \mathbf{r}_2|} \\
&= \int d^3 r_1 d^3 r_2 \frac{\left|\psi_{100}(\mathbf{r}_1)\,\psi_{nlm}(\mathbf{r}_2)\right|^2}{|\mathbf{r}_1 - \mathbf{r}_2|} \pm \int d^3 r_1 d^3 r_2 \frac{\psi_{100}(\mathbf{r}_1)\,\psi_{nlm}(\mathbf{r}_2)\,\psi_{nlm}^*(\mathbf{r}_1)\,\psi_{100}^*(\mathbf{r}_2)}{|\mathbf{r}_1 - \mathbf{r}_2|}.
\end{aligned}
\tag{X.444}
$$

8. Prove (23.71), i.e.

$$
w_n = \sum_m \left|{}^{(-)}\langle\varphi_n\psi_m | \Phi\Psi\rangle^{(-)}\right|^2 = \left|\langle\varphi_n | \Phi\rangle\right|^2 .
\tag{X.445}
$$

Solution: We begin with the relation

$$
\begin{aligned}
&4 \sum_m \left|{}^{(-)}\langle\varphi_n\psi_m | \Phi\Psi\rangle^{(-)}\right|^2 \\
&= \sum_m \left|\langle\varphi_n^{(1)} | \Phi^{(1)}\rangle\langle\psi_m^{(2)} | \Psi^{(2)}\rangle + \langle\varphi_n^{(2)} | \Phi^{(2)}\rangle\langle\psi_m^{(1)} | \Psi^{(1)}\rangle\right|^2 ,
\end{aligned}
\tag{X.446}
$$

where we have used (23.68). Written out, this is

$$4 \sum_m \left| {}^{(-)} \langle \varphi_n \psi_m | \Phi \Psi \rangle^{(-)} \right|^2$$

$$= \sum_m \left\{ \begin{array}{l} [\langle 1:\Phi| 1:\varphi_n \rangle \langle 2:\Psi| 2:\psi_m \rangle + \langle 2:\Phi| 2:\varphi_n \rangle \langle 1:\Psi| 1:\psi_m \rangle] \\ \cdot [\langle 1:\varphi_n| 1:\Phi \rangle \langle 2:\psi_m| 2:\Psi \rangle + \langle 2:\varphi_n| 2:\Phi \rangle \langle 1:\psi_m| 1:\Psi \rangle] \end{array} \right\} \cdot$$

(X.447)

We expand and obtain

$$4 \sum_m \left| {}^{(-)} \langle \varphi_n \psi_m | \Phi \Psi \rangle^{(-)} \right|^2$$

$$= \sum_m \left\{ \begin{array}{l} \langle 1:\Phi| 1:\varphi_n \rangle \langle 2:\Psi| 2:\psi_m \rangle \langle 1:\varphi_n| 1:\Phi \rangle \langle 2:\psi_m| 2:\Psi \rangle \\ + \langle 1:\Phi| 1:\varphi_n \rangle \langle 2:\Psi| 2:\psi_m \rangle \langle 2:\varphi_n| 2:\Phi \rangle \langle 1:\psi_m| 1:\Psi \rangle \\ + \langle 2:\Phi| 2:\varphi_n \rangle \langle 1:\Psi| 1:\psi_m \rangle \langle 1:\varphi_n| 1:\Phi \rangle \langle 2:\psi_m| 2:\Psi \rangle \\ + \langle 2:\Phi| 2:\varphi_n \rangle \langle 1:\Psi| 1:\psi_m \rangle \langle 2:\varphi_n| 2:\Phi \rangle \langle 1:\psi_m| 1:\Psi \rangle \end{array} \right\} \cdot$$

(X.448)

We extract from the sum all the terms which are independent of the summation index:

$$4 \sum_m \left| {}^{(-)} \langle \varphi_n \psi_m | \Phi \Psi \rangle^{(-)} \right|^2$$

$$= \left\{ \begin{array}{l} |\langle 1:\varphi_n| 1:\Phi \rangle|^2 \sum_m \langle 2:\Psi| 2:\psi_m \rangle \langle 2:\psi_m| 2:\Psi \rangle \\ + \langle 1:\Phi| 1:\varphi_n \rangle \langle 2:\varphi_n| 2:\Phi \rangle \sum_m \langle 2:\Psi| 2:\psi_m \rangle \langle 1:\psi_m| 1:\Psi \rangle \\ + \langle 1:\varphi_n| 1:\Phi \rangle \langle 2:\Phi| 2:\varphi_n \rangle \sum_m \langle 1:\Psi| 1:\psi_m \rangle \langle 2:\psi_m| 2:\Psi \rangle \\ + |\langle 2:\varphi_n| 2:\Phi \rangle|^2 \sum_m \langle 1:\Psi| 1:\psi_m \rangle \langle 1:\psi_m| 1:\Psi \rangle \end{array} \right\}$$

(X.449)

Since $\{|\varphi_n \rangle\}$ and $\{|\psi_m \rangle\}$ are CONS, the terms

$$\begin{array}{l} \sum_m \langle 2:\Psi| 2:\psi_m \rangle \langle 2:\psi_m| 2:\Psi \rangle = \langle 2:\Psi| 2:\Psi \rangle = 1 \\ \sum_m \langle 1:\Psi| 1:\psi_m \rangle \langle 1:\psi_m| 1:\Psi \rangle = \langle 1:\Psi| 1:\Psi \rangle = 1 \end{array}$$

(X.450)

can be calculated immediately. Due to

$$\langle 1:\psi_m| 1:\Psi \rangle = \langle 2:\psi_m| 2:\Psi \rangle,$$

(X.451)

completeness holds also for the two remaining terms, and from this follows the desired result.

## X.10   Exercises, Chap. 24

1. Given the density matrix

$$\rho = \begin{pmatrix} |c_1|^2 & c_1 c_2^* e^{i\omega t} \\ c_1^* c_2 e^{-i\omega t} & |c_2|^2 \end{pmatrix}. \tag{X.452}$$

Calculate $\frac{1}{T} \int\limits_0^T \rho \, dt$.

Solution: The diagonal elements are clear. For the off-diagonal elements, we have e.g.

$$\frac{1}{T} \int\limits_0^T e^{i\omega t} \, dt = \frac{e^{i\omega T} - 1}{i\omega T} = e^{i\omega T/2} \frac{e^{i\omega T/2} - e^{-i\omega T/2}}{i\omega T} = e^{i\omega T/2} \frac{\sin \omega T/2}{\omega T/2}$$

$$\tag{X.453}$$

and correspondingly for $e^{-i\omega t}$. The off-diagonal terms tend to zero for $T \to \infty$, and we obtain

$$\frac{1}{T} \int\limits_0^T \rho \, dt \underset{T \to \infty}{=} \begin{pmatrix} |c_1|^2 & 0 \\ 0 & |c_2|^2 \end{pmatrix}. \tag{X.454}$$

2. Consider the reduced density matrix $\rho_{S,\text{red}} = CC^\dagger$ of (24.16), where $C$ is given as an $M \times N$ matrix:

$$C = (c_{mn}) = \begin{pmatrix} c_{11} & c_{12} & \cdots & c_{1N} \\ c_{21} & c_{22} & \cdots & c_{2N} \\ \vdots & \vdots & \ddots & \vdots \\ c_{M1} & c_{M2} & \cdots & c_{MN} \end{pmatrix}. \tag{X.455}$$

Hence, the system has $M$ states, the environment $N$. Estimate the order of magnitude of the elements of $\rho_{S,\text{red}}$.

Solution: With the $N$-dimensional row vectors

$$\mathbf{c}_1 = \begin{pmatrix} c_{11} & c_{12} & c_{13} & \cdots & c_{1N} \end{pmatrix}, \tag{X.456}$$

we can write

$$\rho_{S,\text{red}} = CC^\dagger = \begin{pmatrix} \mathbf{c}_1 \mathbf{c}_1^\dagger & \mathbf{c}_1 \mathbf{c}_2^\dagger & \cdots & \mathbf{c}_1 \mathbf{c}_M^\dagger \\ \mathbf{c}_2 \mathbf{c}_1^\dagger & \mathbf{c}_2 \mathbf{c}_2^\dagger & \cdots & \mathbf{c}_2 \mathbf{c}_M^\dagger \\ \vdots & \vdots & \ddots & \vdots \\ \mathbf{c}_M \mathbf{c}_1^\dagger & \mathbf{c}_M \mathbf{c}_2^\dagger & \cdots & \mathbf{c}_M \mathbf{c}_M^\dagger \end{pmatrix}. \tag{X.457}$$

Due to $tr\left(\rho_{S,\text{red}}\right) = \sum_{k=1}^M \mathbf{c}_k \mathbf{c}_k^\dagger = 1$, the average value of the diagonal elements is $1/M$. Since each diagonal element consists of $N$ positive summands, we can assume that $\left|c_{jk}\right| = O\left(\frac{1}{\sqrt{M \cdot N}}\right)$. In contrast to the diagonal terms, the off-diagonal elements do not consist of only positive summands; the real and imaginary parts can assume positive and negative values and will (normal distribution provided) cancel for sufficiently large $N$ on average, or add up to zero; and this with the usual statistical error $\sim 1/\sqrt{N}$. Thus we obtain in summary:

$$(CC^\dagger)_{ij} = O\left(\frac{1}{M}\right)\left(\delta_{ij} + O\left(\frac{1}{\sqrt{N}}\right)\right). \tag{X.458}$$

In order to illuminate the argument from another side, we consider the elements $\mathbf{c}_1 \mathbf{c}_k^\dagger$. Without loss of generality, we position the coordinate system in such a way that $\mathbf{c}_1 = (c_{11}\ 0\ 0\ \ldots\ 0)$. Then according to the above, we have $c_{11} = O\left(\frac{1}{\sqrt{M}}\right)$ or $\mathbf{c}_1 \mathbf{c}_1^\dagger = O\left(\frac{1}{M}\right)$. For $k \neq 1$, it holds that $\mathbf{c}_k = (c_{k1}\ c_{k2}\ \ldots\ c_{kN})$, whereby the individual components have the average value $\frac{1}{\sqrt{M \cdot N}}$. From this, an estimate follows for the scalar product $\mathbf{c}_1 \mathbf{c}_k^\dagger$:

$$\mathbf{c}_1 \mathbf{c}_k^\dagger = c_{11} \cdot c_{k1} = O\left(\frac{1}{\sqrt{M}}\right) \cdot O\left(\frac{1}{\sqrt{M \cdot N}}\right) = O\left(\frac{1}{M\sqrt{N}}\right) \text{ for } k \neq 1 \tag{X.459}$$

i.e. again the result (X.458).

3. Calculate explicitly the eigenvalues of the density matrix

$$\rho = \begin{pmatrix} |c_1|^2 & c_1 c_2^* \\ c_1^* c_2 & |c_2|^2 \end{pmatrix} \tag{X.460}$$

with $|c_1| + |c_2|^2 = 1$.

Solution: The eigenvalues are the solutions of the equation

$$\left[|c_1|^2 - \lambda\right]\left[|c_2|^2 - \lambda\right] - |c_1|^2 |c_2|^2 = 0. \tag{X.461}$$

It follows that

$$\lambda^2 - \lambda\left[|c_1| + |c_2|^2\right] = 0 \text{ or } \lambda^2 - \lambda = 0 \tag{X.462}$$

with the solutions $\lambda = 0, 1$.

4. We consider two quantum objects with $\mathcal{H} = \mathcal{H}_1 \otimes \mathcal{H}_1$. The CONS $\{|0\rangle, |1\rangle\}$ is a basis of $\mathcal{H}_1$.

   (a) Show that the states

$$|\pm\rangle = \frac{|0\rangle \pm |1\rangle}{\sqrt{2}} \tag{X.463}$$

   are also a CONS in $\mathcal{H}_1$.
   Solution: We have

$$\langle\pm|\pm\rangle = \frac{\langle 0|\pm\langle 1|}{\sqrt{2}}\frac{|0\rangle\pm|1\rangle}{\sqrt{2}} = 1; \quad \langle\pm|\mp\rangle = \frac{\langle 0|\pm\langle 1|}{\sqrt{2}}\frac{|0\rangle\mp|1\rangle}{\sqrt{2}} = 0$$

$$\text{(X.464)}$$

and

$$|+\rangle\langle+| + |-\rangle\langle-| = \frac{|0\rangle+|1\rangle}{\sqrt{2}}\frac{\langle 0|+\langle 1|}{\sqrt{2}} + \frac{|0\rangle-|1\rangle}{\sqrt{2}}\frac{\langle 0|-\langle 1|}{\sqrt{2}}$$

$$= \frac{2|0\rangle\langle 0| + 2|1\rangle\langle 1|}{2} = 1. \qquad \text{(X.465)}$$

(b) Write down the states

$$|\psi^{\pm}\rangle = \frac{|01\rangle\pm|10\rangle}{\sqrt{2}} \qquad \text{(X.466)}$$

in the basis $\{|+\rangle, |-\rangle\}$.
Solution: With

$$|0\rangle = \frac{|+\rangle+|-\rangle}{\sqrt{2}}; \quad |1\rangle = \frac{|+\rangle-|-\rangle}{\sqrt{2}}, \qquad \text{(X.467)}$$

we obtain

$$|\psi^{\pm}\rangle = \frac{|01\rangle\pm|10\rangle}{\sqrt{2}}$$

$$= \frac{|++\rangle-|+-\rangle+|-+\rangle-|--\rangle}{2\sqrt{2}} \pm \frac{|++\rangle+|+-\rangle-|-+\rangle-|--\rangle}{2\sqrt{2}},$$

$$\text{(X.468)}$$

and thus

$$|\psi^{+}\rangle = \frac{|++\rangle-|--\rangle}{\sqrt{2}}; \quad |\psi^{-}\rangle = \frac{|-+\rangle-|+-\rangle}{\sqrt{2}}. \qquad \text{(X.469)}$$

(c) As assumed in the text, the effect of the environment is to add to each basis state a corresponding random phase. How are the new states $|\psi^{\pm}\rangle$ formulated?
Solution: For $\{|0\rangle, |1\rangle\}$, it holds that $|0\rangle \to e^{i\varphi_0}|0\rangle$ and $|1\rangle \to e^{i\varphi_1}|1\rangle$. It follows that

$$|\psi^{\pm}\rangle \to e^{i(\varphi_0+\varphi_1)}\frac{|01\rangle\pm|10\rangle}{\sqrt{2}}. \qquad \text{(X.470)}$$

For $\{|+\rangle, |-\rangle\}$, we have $|+\rangle \to e^{i\varphi_+}|+\rangle$ and $|-\rangle \to e^{i\varphi_-}|-\rangle$. It follows that

$$|\psi^{+}\rangle \to e^{2i\varphi_+}\frac{|++\rangle-e^{2i(\varphi_--\varphi_+)}|--\rangle}{\sqrt{2}}$$

$$|\psi^{-}\rangle \to e^{i(\varphi_-+\varphi_+)}\frac{|-+\rangle-|+-\rangle}{\sqrt{2}}. \qquad \text{(X.471)}$$

Only the state $|\psi^+\rangle$ in the basis $\{|+\rangle, |-\rangle\}$ is not decoherence-free under these assumptions.

5. Show that

$$\sum_{i=(m,n)} A_i^\dagger(t) A_i(t) = 1; \quad A_{i=(m,n)}(t) = \sqrt{p_n} \langle m| \hat{U}(t) |n\rangle. \qquad (X.472)$$

See (24.34).
Solution: We have

$$\sum_{i=(m,n)} A_i^\dagger(t) A_i(t) = \sum_{m,n} A_{mn}^\dagger(t) A_{mn}(t)$$

$$= \sum_{m,n} p_n \langle n| \hat{U}^\dagger(t) |m\rangle \langle m| \hat{U}(t) |n\rangle. \qquad (X.473)$$

Since the environment states are a CONS, we have $\sum_m |m\rangle \langle m| = 1$. This yields

$$\sum_{i=(m,n)} A_i^\dagger(t) A_i(t) = \sum_n p_n \langle n| \hat{U}^\dagger(t) \hat{U}(t) |n\rangle = \sum_n p_n = 1. \qquad (X.474)$$

6. Two quantum objects each have a two-dimensional Hilbert space with the orthonormal basis vectors $|0\rangle$ and $|1\rangle$. They are in the ground state:

$$|\psi\rangle = c_0 |0\rangle |0\rangle + c_1 |1\rangle |1\rangle. \qquad (X.475)$$

We now perform a change of basis via

$$|0\rangle = a_{11} |+\rangle + a_{12} |-\rangle; \quad |1\rangle = a_{21} |+\rangle + a_{22} |-\rangle, \qquad (X.476)$$

where $|+\rangle$ and $|-\rangle$ are also an orthonormal basis. Under which conditions does $|\psi\rangle = d_+ |+\rangle |+\rangle + d_- |-\rangle |-\rangle$ hold?
Solution: We first note that a change of basis is a unitary transformation:

$$U = \begin{pmatrix} a_{11} & a_{12} \\ a_{21} & a_{22} \end{pmatrix}; \quad U^\dagger = U^{-1} \qquad (X.477)$$

which among other things means

$$\det U = e^{i\delta}; \quad \delta \in \mathbb{R} \text{ and } a_{22} = \det U \cdot a_{11}^*; \quad a_{21} = -\det U \cdot a_{12}^*, \qquad (X.478)$$

where we assume in the following $a_{11} \neq 0$ and $a_{12} \neq 0$ in order to exclude trivial cases.
We now insert the basis transformations (X.476) into (X.475) and obtain initially

$$|\psi\rangle = c_0 \left( a_{11}^2 |+\rangle |+\rangle + a_{11}a_{12} |+\rangle |-\rangle + a_{12}a_{11} |-\rangle |+\rangle + a_{12}^2 |-\rangle |-\rangle \right)$$
$$+ c_1 \left( a_{21}^2 |+\rangle |+\rangle + a_{21}a_{22} |+\rangle |-\rangle + a_{22}a_{21} |-\rangle |+\rangle + a_{22}^2 |-\rangle |-\rangle \right).$$
$$\text{(X.479)}$$

According to our premises, the coefficients of $|+\rangle |-\rangle$ and $|-\rangle |+\rangle$ vanish; thus it follows that:

$$c_0 a_{11} a_{12} + c_1 a_{21} a_{22} \overset{!}{=} 0. \tag{X.480}$$

With (X.478), this leads to

$$c_0 a_{11} a_{12} - c_1 e^{2i\delta} \cdot a_{12}^* a_{11}^* \overset{!}{=} 0, \tag{X.481}$$

and this means apparently (due to $a_{11}a_{12} \neq 0$) that:

$$|c_0| \overset{!}{=} |c_1|. \tag{X.482}$$

Hence, the question is answered. But we still want to determine the coefficients of $|+\rangle |+\rangle$ and $|-\rangle |-\rangle$. We have

$$|\psi\rangle = \left( c_0 a_{11}^2 + c_1 a_{21}^2 \right) |+\rangle |+\rangle + \left( c_0 a_{12}^2 + c_1 a_{22}^2 \right) |-\rangle |-\rangle. \tag{X.483}$$

Because of $a_{21}a_{22} \neq 0$, it follows from (X.480) that:

$$c_1 = -c_0 \frac{a_{11} a_{12}}{a_{21} a_{22}} \tag{X.484}$$

and therefore

$$|\psi\rangle = c_0 \left( a_{11}a_{22} - a_{12}a_{21} \right) \left[ \frac{a_{11}}{a_{22}} |+\rangle |+\rangle - \frac{a_{12}}{a_{21}} |-\rangle |-\rangle \right]$$
$$= c_0 \left[ \frac{a_{11}}{a_{11}^*} |+\rangle |+\rangle ' + \frac{a_{12}}{a_{12}^*} |-\rangle |-\rangle \right]; \tag{X.485}$$

$$d_+ = c_0 \frac{a_{11}}{a_{11}^*}; \quad d_- = c_0 \frac{a_{12}}{a_{12}^*}, \quad |d_+| = |d_-|. \tag{X.486}$$

## X.11   Exercises, Chap. 25

1. Show that:

$$|\mathbf{r} - \mathbf{r}'| \underset{r \to \infty}{\to} r - \hat{\mathbf{r}} \cdot \mathbf{r}'. \tag{X.487}$$

Solution:

$$|\mathbf{r} - \mathbf{r}'| = \sqrt{(\mathbf{r} - \mathbf{r}')^2} = \sqrt{r^2 - 2\mathbf{r}\mathbf{r}' + r'^2} = r\sqrt{1 - 2\frac{\mathbf{r}\mathbf{r}'}{r^2} + \frac{r'^2}{r^2}}$$

$$|\mathbf{r} - \mathbf{r}'| \underset{r\to\infty}{\to} r\left(1 - \frac{\mathbf{r}\mathbf{r}'}{r^2} + O\left(\frac{1}{r^2}\right)\right) \underset{r\to\infty}{\to} r - \hat{\mathbf{r}} \cdot \mathbf{r}'. \tag{X.488}$$

2. Prove that:

$$\frac{e^{ik|\mathbf{r}-\mathbf{r}'|}}{|\mathbf{r} - \mathbf{r}'|} \underset{r\to\infty}{\to} \frac{e^{ikr}}{r} e^{-i\mathbf{k}' \cdot \mathbf{r}'}. \tag{X.489}$$

Solution: From Exercise 1, it follows that

$$\frac{e^{ik|\mathbf{r}-\mathbf{r}'|}}{|\mathbf{r} - \mathbf{r}'|} \underset{r\to\infty}{\to} \frac{e^{ik\left(1 - \frac{\mathbf{r}\mathbf{r}'}{r^2} + O\left(\frac{1}{r^2}\right)\right)}}{r\left(1 - \frac{\mathbf{r}\mathbf{r}'}{r^2} + O\left(\frac{1}{r^2}\right)\right)}$$

$$= \frac{e^{ikr} e^{-ik\frac{\mathbf{r}\mathbf{r}'}{r}} e^{ikO\left(\frac{1}{r}\right)}}{r}\left(1 + O\left(\frac{1}{r}\right)\right) \underset{r\to\infty}{\to} \frac{e^{ikr}}{r} e^{-ik\hat{\mathbf{r}}\mathbf{r}'} = \frac{e^{ikr}}{r} e^{-i\mathbf{k}'\mathbf{r}'}. \tag{X.490}$$

3. Calculate explicitly the asymptotic form of the current density for the scattered wave.
   Solution: With

$$\varphi_{\text{scatt}}(\mathbf{r}) \underset{r\to\infty}{\to} f(\vartheta, \varphi) \frac{e^{ikr}}{r} \tag{X.491}$$

and

$$\mathbf{j} = \frac{\hbar}{2mi}\left(\psi^* \nabla\psi - \psi\nabla\psi^*\right), \tag{X.492}$$

it follows asymptotically that:

$$\mathbf{j}_{\text{scatt}} \underset{r\to\infty}{\to} \frac{\hbar}{2mi}\left(f^*(\vartheta, \varphi)\frac{e^{-ikr}}{r}\nabla f(\vartheta, \varphi)\frac{e^{ikr}}{r} - f(\vartheta, \varphi)\frac{e^{ikr}}{r}\nabla f^*(\vartheta, \varphi)\frac{e^{-ikr}}{r}\right). \tag{X.493}$$

For the evaluation we use the representation of the gradient in spherical coordinates (see Appendix D, Vol. 1)

$$\nabla = \frac{\partial}{\partial r}\mathbf{e}_r + \frac{1}{r}\frac{\partial}{\partial\vartheta}\mathbf{e}_\vartheta + \frac{1}{r\sin\vartheta}\frac{\partial}{\partial\varphi}\mathbf{e}_\varphi. \tag{X.494}$$

First, we have:

$$\nabla f(\vartheta, \varphi) \frac{e^{ikr}}{r} = f(\vartheta, \varphi) \frac{\partial}{\partial r} \frac{e^{ikr}}{r} \mathbf{e}_r + \frac{e^{ikr}}{r^2} \frac{\partial}{\partial \vartheta} f(\vartheta, \varphi) \mathbf{e}_\vartheta$$

$$= f(\vartheta, \varphi) \left(ik - \frac{1}{r}\right) \frac{e^{ikr}}{r} \mathbf{e}_r + \frac{e^{ikr}}{r^2} \frac{\partial f(\vartheta, \varphi)}{\partial \vartheta} \mathbf{e}_\vartheta + \frac{e^{ikr}}{r^2 \sin \vartheta} \frac{\partial f(\vartheta, \varphi)}{\partial \varphi} \mathbf{e}_\varphi.$$

$$\text{(X.495)}$$

This leads to

$$f^*(\vartheta, \varphi) \frac{e^{-ikr}}{r} \nabla f(\vartheta, \varphi) \frac{e^{ikr}}{r}$$

$$= |f(\vartheta, \varphi)|^2 \left(ik - \frac{1}{r}\right) \frac{\mathbf{e}_r}{r^2} + O\left(\frac{1}{r^3}\right) \mathbf{e}_\vartheta + O\left(\frac{1}{r^3}\right) \mathbf{e}_\varphi \to |f(\vartheta, \varphi)|^2 ik \frac{\mathbf{e}_r}{r^2},$$

$$\text{(X.496)}$$

where in the last step we have assumed a sufficiently large $r$. Then we obtain

$$\mathbf{j}_{\text{scatt}} \underset{r \to \infty}{\to} \frac{\hbar k}{m} |f(\vartheta, \varphi)|^2 \frac{\mathbf{e}_r}{r^2}. \tag{X.497}$$

4. Determine the general relation between scattering amplitude and scattering phases.

Solution: We have for the wavefunction the expression:

$$\varphi(\mathbf{r}) = \sum_{l=0}^{\infty} \frac{u_l(r)}{r} P_l(\cos \vartheta). \tag{X.498}$$

We know that on the other hand that the following expression must hold asymptotically:

$$\varphi_{\text{asy}}(\mathbf{r}) = e^{ikz} + f(\vartheta) \frac{e^{ikr}}{r} = \sum_{l=0}^{\infty} \left[ (2l+1) i^l j_l(kr) + f_l(\vartheta) \frac{e^{ikr}}{r} \right] P_l(\cos \vartheta). \tag{X.499}$$

Due to the linear independence of the Legendre polynomials, we arrive at

$$\frac{u_l(r)}{r} \to (2l+1) i^l j_l(kr) + f_l(\vartheta) \frac{e^{ikr}}{r}. \tag{X.500}$$

We insert the asymptotic expressions:

$$j_l(kr) \underset{r \to \infty}{\sim} \frac{\sin\left(kr - \frac{l\pi}{2}\right)}{kr}; \quad u_l(r) \underset{r \to \infty}{\sim} c_l \sin\left(kr - \frac{l\pi}{2} + \delta_l\right), \tag{X.501}$$

and obtain

$$\frac{c_l \sin \left(kr - \frac{l\pi}{2} + \delta_l\right)}{r} = (2l + 1) i^l \frac{\sin \left(kr - \frac{l\pi}{2}\right)}{kr} + f_l (\vartheta) \frac{e^{ikr}}{r}. \quad \text{(X.502)}$$

We now expand the sine in terms of plane waves

$$c_l \left[ e^{i\left(kr - \frac{l\pi}{2} + \delta_l\right)} - e^{-i\left(kr - \frac{l\pi}{2} + \delta_l\right)} \right]$$

$$= \frac{(2l + 1)}{k} i^l \left[ e^{i\left(kr - \frac{l\pi}{2}\right)} - e^{-i\left(kr - \frac{l\pi}{2}\right)} \right] + 2i f_l (\vartheta) e^{ikr}. \quad \text{(X.503)}$$

Due to the linear independence of the in- and outgoing partial waves, the equations

$$c_l e^{i\left(kr - \frac{l\pi}{2} + \delta_l\right)} = \frac{(2l + 1)}{k} i^l e^{i\left(kr - \frac{l\pi}{2}\right)} + 2i f_l (\vartheta) e^{ikr}$$

$$c_l e^{-i\left(kr - \frac{l\pi}{2} + \delta_l\right)} = \frac{(2l + 1)}{k} i^l e^{-i\left(kr - \frac{l\pi}{2}\right)} \quad \text{(X.504)}$$

must hold. The second equation yields

$$c_l = \frac{(2l + 1)}{k} i^l e^{i\delta_l}, \quad \text{(X.505)}$$

and it follows that:

$$f_l (\vartheta) = \frac{(2l + 1)}{2ik} \left(e^{2i\delta_l} - 1\right) = \frac{(2l + 1)}{k} e^{i\delta_l} \sin \delta_l. \quad \text{(X.506)}$$

Hence, the connection between scattering amplitude and scattering phases is

$$f (\vartheta) = \sum_{l=0}^{\infty} f_l (\vartheta) P_l (\cos \vartheta) = \frac{1}{k} \sum_{l=0}^{\infty} (2l + 1) e^{i\delta_l} \sin \delta_l P_l (\cos \vartheta). \quad \text{(X.507)}$$

5. Determine the radial equations for a general potential $V (\mathbf{r})$.
   Solution: We start from the SEq in the form

$$\left(\nabla^2 + k^2 - v_{\text{eff}} (\mathbf{r})\right) \psi (\mathbf{r}) = 0. \quad \text{(X.508)}$$

The multipole expansions of the potential and the wavefunction read (we use here the abbreviation $\hat{\mathbf{r}} = \vartheta, \varphi$)

$$v_{\text{eff}} (\mathbf{r}) = \sum_{lm} w_{lm} (r) Y_l^m (\hat{\mathbf{r}}) ; \quad \psi (\mathbf{r}) = \sum_{lm} \frac{u_{lm} (r)}{r} Y_l^m (\hat{\mathbf{r}}). \quad \text{(X.509)}$$

Insertion in the SEq yields

$$\sum_{lm} \left( \frac{d^2}{dr^2} - \frac{l(l+1)}{r^2} + k^2 \right) u_{lm}(r) Y_l^m(\hat{\mathbf{r}}) - \sum_{l_1 m_1} w_{l_1 m_1}(r) Y_{l_1}^{m_1}(\hat{\mathbf{r}})$$

$$\times \sum_{l_2 m_2} u_{l_2 m_2}(r) Y_{l_2}^{m_2}(\hat{\mathbf{r}}) = 0. \tag{X.510}$$

We transform the last expression by means of (cf. Chap. 16 and Appendix B, Vol. 2):

$$Y_{l_1}^{m_1}(\hat{\mathbf{r}}) Y_{l_2}^{m_2}(\hat{\mathbf{r}}) = \sum_{L=|l_1 - l_2|}^{l_1 + l_2} \sqrt{\frac{(2l_1 + 1)(2l_2 + 1)}{4\pi(2L + 1)}} \langle l_1 l_2 00 | L0 \rangle \langle l_1 l_2 m_1 m_2 | L\, m_1 + m_2 \rangle Y_L^M(\hat{\mathbf{r}}), \tag{X.511}$$

and obtain initially

$$0 = \sum_{lm} \left( \frac{d^2}{dr^2} - \frac{l(l+1)}{r^2} + k^2 \right) u_{lm}(r) Y_l^m(\hat{\mathbf{r}})$$

$$- \sum_{l_1 m_1 l_2 m_2 L} \sqrt{\frac{(2l_1 + 1)(2l_2 + 1)}{4\pi(2L + 1)}} \langle l_1 l_2 00 | L0 \rangle \langle l_1 l_2 m_1 m_2 | L\, m_1 + m_2 \rangle w_{l_1 m_1}(r) u_{l_2 m_2}(r) Y_L^M(\hat{\mathbf{r}}). \tag{X.512}$$

Due to the orthogonality of the spherical harmonics $\left( \int \left[ Y_l^m(\hat{r}) \right]^* Y_{l'}^{m'}(\hat{r}) \, d\Omega = \delta_{ll'} \delta_{mm'} \right)$, we obtain from this:

$$0 = \left( \frac{d^2}{dr^2} - \frac{l(l+1)}{r^2} + k^2 \right) u_{lm}(r)$$

$$- \sum_{l_1 m_1} \sum_{l_2 = |l - l_1|}^{l + l_1} \sqrt{\frac{(2l_1 + 1)(2l_2 + 1)}{4\pi(2l + 1)}} \langle l_1 l_2 00 | l0 \rangle \langle l_1 l_2 m_1\, m - m_1 | l\, m \rangle w_{l_1 m_1}(r) u_{l_2 m - m_1}(r). \tag{X.513}$$

We see that in the second term, there are radial functions coupled for different angular momentum indices.

6. The Yukawa potential (also called the screened Coulomb potential) has the form

$$V(r) = V_0 \frac{e^{-r/a}}{r}; \quad a > 0. \tag{X.514}$$

The range of the potential is of order $a$. Determine the scattering amplitude for the potential in the Born approximation. The Coulomb potential follows for $a \to \infty$ (infinite range of the Coulomb potential). Calculate also in this case the scattering cross section (*Rutherford scattering cross section*).

Solution: The scattering amplitude in the Born approximation is given by

$$f_{\text{Born}}(\vartheta, \varphi) = -\frac{2m}{q\hbar^2} \int_0^\infty dr\, rV(r) \sin qr$$

$$= -\frac{2mV_0}{q\hbar^2} \int_0^\infty dr\, e^{-r/a} \sin qr = -\frac{2mV_0}{\hbar^2} \frac{a^2}{1 + a^2 q^2}. \quad (\text{X.515})$$

With $q = 2k \sin \frac{\vartheta}{2}$, it follows that:

$$f_{\text{Born}}(\vartheta, \varphi) = -\frac{2mV_0}{\hbar^2} \frac{a^2}{1 + 4a^2 k^2 \sin^2 \frac{\vartheta}{2}} \quad (\text{X.516})$$

or

$$\frac{d\sigma}{d\Omega}\bigg|_{\text{Born}} = \left(\frac{2mV_0}{\hbar^2}\right)^2 \left(\frac{a^2}{1 + 4a^2 k^2 \sin^2 \frac{\vartheta}{2}}\right)^2. \quad (\text{X.517})$$

The Coulomb potential follows in the limit $a \to \infty$ (infinite range of the Coulomb potential), and the scattering cross section in this case is given by:

$$\frac{d\sigma}{d\Omega}\bigg|_{\text{Born}} = \left(\frac{2mV_0}{\hbar^2}\right)^2 \frac{1}{16k^4 \sin^4 \frac{\vartheta}{2}}^2. \quad (\text{X.518})$$

With

$$E = \frac{\hbar^2 k^2}{2m}, \quad V_0 = q^2, \quad (\text{X.519})$$

we find the usual form of the Rutherford scattering cross section:

$$\frac{d\sigma}{d\Omega}\bigg|_{\text{Born}} = \left(\frac{2mq^2}{\hbar^2}\right)^2 \left(\frac{\hbar^2}{2mE}\right)^2 \frac{1}{16 \sin^4 \frac{\vartheta}{2}} = q^4 \frac{1}{16E^2 \sin^4 \frac{\vartheta}{2}}. \quad (\text{X.520})$$

7. In this exercise, we address the transformation between the abstract representation and the position representation. We recall that this topic is discussed in more detail in Chap. 12, Vol. 1.

   (a) Transform the equation

   $$|\psi\rangle = |\psi_0\rangle + Gv\,|\psi\rangle \quad (\text{X.521})$$

   into the position representation.
   Solution: We multiply by $\langle \mathbf{r}|$ from the left and insert the 1 (i.e. $\int d^3 r'\, |\mathbf{r}'\rangle \langle \mathbf{r}'|$) twice on the right-hand side. This gives

$$\langle \mathbf{r} \, | \psi \rangle = \langle \mathbf{r} \, | \psi_0 \rangle + \int d^3 r' \int d^3 r'' \, \langle \mathbf{r} | \, G \, | \mathbf{r}' \rangle \langle \mathbf{r}' | \, v \, | \mathbf{r}'' \rangle \langle \mathbf{r}'' | \, \psi \rangle . \quad \text{(X.522)}$$

The potential operator $v$ is local:

$$\langle \mathbf{r}' | \, v \, | \mathbf{r}'' \rangle = \langle \mathbf{r}' | \, v \, | \mathbf{r}'' \rangle \delta \left( \mathbf{r}' - \mathbf{r}'' \right) = v \left( \mathbf{r}' \right) \delta \left( \mathbf{r}' - \mathbf{r}'' \right) , \quad \text{(X.523)}$$

and with

$$\langle \mathbf{r} \, | \psi \rangle = \psi \left( \mathbf{r} \right) ; \quad \langle \mathbf{r} | \, G \, | \mathbf{r}' \rangle = -\frac{1}{4\pi} \frac{e^{ik|\mathbf{r} - \mathbf{r}'|}}{|\mathbf{r} - \mathbf{r}'|} \quad \text{(X.524)}$$

we obtain

$$\psi \left( \mathbf{r} \right) = \psi_0 \left( \mathbf{r} \right) - \frac{1}{4\pi} \int d^3 r' \frac{e^{ik|\mathbf{r} - \mathbf{r}'|}}{|\mathbf{r} - \mathbf{r}'|} v \left( \mathbf{r}' \right) \psi \left( \mathbf{r}' \right) . \quad \text{(X.525)}$$

(b) Write the right-hand side of the following equation:

$$f_{\text{Born}} \left( \vartheta, \varphi \right) = -\frac{1}{4\pi} \langle \mathbf{k}' | \, v \, | \mathbf{k} \rangle \quad \text{(X.526)}$$

explicitly in the position representation.
Solution: On the right-hand side, we insert the 1 (i.e. $\int d^3 r \, | \mathbf{r} \rangle \langle \mathbf{r} |$) twice:

$$f_{\text{Born}} \left( \vartheta, \varphi \right) = -\frac{1}{4\pi} \int d^3 r' \int d^3 r'' \langle \mathbf{k}' | \, \mathbf{r}' \rangle \langle \mathbf{r}' | \, v \, | \mathbf{r}'' \rangle \langle \mathbf{r}'' | \, \mathbf{k} \rangle . \quad \text{(X.527)}$$

We then have

$$\langle \mathbf{r} \, | \mathbf{k} \rangle = e^{i\mathbf{k}\mathbf{r}}; \quad \langle \mathbf{k} | \, \mathbf{r} \rangle = \langle \mathbf{r} \, | \mathbf{k} \rangle^\dagger = \langle \mathbf{r} \, | \mathbf{k} \rangle^* = e^{-i\mathbf{k}\mathbf{r}} \quad \text{(X.528)}$$

(recall that here, we omit the normalization factor $(2\pi)^{-3/2}$, in contrast to Chap. 12, Vol. 1). Since the potential operator $v$ is local, it follows that:

$$f_{\text{Born}} \left( \vartheta, \varphi \right) = -\frac{1}{4\pi} \int d^3 r' \, e^{-i\mathbf{k}'\mathbf{r}'} v \left( \mathbf{r}' \right) e^{i\mathbf{k}\mathbf{r}'} . \quad \text{(X.529)}$$

With $v \left( \mathbf{r} \right) = \frac{2m}{\hbar^2} V \left( \mathbf{r} \right)$ and $\mathbf{q} := \mathbf{k} - \mathbf{k}'$, we arrive at

$$f_{\text{Born}} \left( \vartheta, \varphi \right) = -\frac{m}{2\pi\hbar^2} \int d^3 r' \, V \left( \mathbf{r}' \right) e^{i\mathbf{q}\mathbf{r}'} . \quad \text{(X.530)}$$

## X.12 Exercises, Chap. 26

1. Above, it was proposed that you yourself try to find the prime factorization of
268898680104636581 and 170699960169639253. Did you find it?
If not, here is the solution: The result is $268898680104636581 = 998653 \cdot 998681 \cdot 269617$ and $170699960169639253 = 413158511 \cdot 413158523$.

2. Pauli matrices and qubits

    (a) How do the Pauli matrices act on the qubit states $|0\rangle$ and $|1\rangle$?
    Solution: In the following we do not distinguish between $\cong$ and $=$. We have:

$$\sigma_x |0\rangle = \begin{pmatrix} 0 & 1 \\ 1 & 0 \end{pmatrix} \begin{pmatrix} 1 \\ 0 \end{pmatrix} = \begin{pmatrix} 0 \\ 1 \end{pmatrix} = |1\rangle$$

$$\sigma_x |1\rangle = \begin{pmatrix} 0 & 1 \\ 1 & 0 \end{pmatrix} \begin{pmatrix} 0 \\ 1 \end{pmatrix} = \begin{pmatrix} 1 \\ 0 \end{pmatrix} = |0\rangle ; \tag{X.531}$$

and analogously:

$$\sigma_y |0\rangle = i |1\rangle ; \ \sigma_y |1\rangle = -i |0\rangle$$
$$\sigma_z |0\rangle = |0\rangle ; \ \sigma_z |1\rangle = -|1\rangle . \tag{X.532}$$

    (b) How do the Pauli matrices act on the qubit state $|\varphi\rangle = c |0\rangle + d |1\rangle$?
    Solution: With the results of Part (a), we obtain:

$$\sigma_x |\varphi\rangle = d |0\rangle + c |1\rangle$$
$$\sigma_y |\varphi\rangle = -id |0\rangle + ic |1\rangle \tag{X.533}$$
$$\sigma_z |\varphi\rangle = c |0\rangle - d |1\rangle .$$

3. Calculate the full expression containing $N$ terms:

$$|z\rangle = \frac{|0\rangle - |1\rangle}{\sqrt{2}} \otimes \frac{|0\rangle - |1\rangle}{\sqrt{2}} \otimes \cdots \otimes \frac{|0\rangle - |1\rangle}{\sqrt{2}}. \tag{X.534}$$

Solution: We consider first as an example the case $N = 3$. Expanding yields:

$$|z\rangle = \frac{|000\rangle - |001\rangle - |010\rangle + |011\rangle - |100\rangle + |101\rangle + |110\rangle - |111\rangle}{2\sqrt{2}}. \tag{X.535}$$

We see that in the binary representation, the sign of $|a_1 a_2 a_3\rangle$ is given by $(-1)^{a_1 + a_2 + a_3}$. Generalizing, we can conclude that for an arbitrary natural number $k$ (in any representation, whereby we confine ourselves here to decimal numbers), the sign $(-1)^{t_k}$ can be calculated as follows: We represent $k$ as a binary number. If the number of 1's is odd, then $t_k = 1$; if it is even, we have $t_k = 0$. (For this reason, the numbers with $t_k = 1$ and $t_k = 0$ are sometimes called *od*ious numbers and *ev*il numbers.) For a general state of the form (X.534), we have in the decimal representation:

$$|z\rangle = \frac{|0\rangle - |1\rangle}{\sqrt{2}} \otimes \frac{|0\rangle - |1\rangle}{\sqrt{2}} \otimes \cdots \otimes \frac{|0\rangle - |1\rangle}{\sqrt{2}} = \frac{1}{2^{N/2}} \sum_{k=0}^{2^N-1} (-1)^{t_k} |k\rangle . \quad \text{(X.536)}$$

Remark: The series of the $t_k$ begins with

$$0110100110010110100101100110101001\ldots . \quad \text{(X.537)}$$

It is called the *Thue–Morse series* and occurs in different areas (number theory, combinatorics, fractals, computer-generated music, etc.).[173]

4. Show that:

$$|q\rangle \to H|q\rangle = \frac{|1-q\rangle + (-1)^q |q\rangle}{\sqrt{2}}; \quad q \in \{0, 1\} \quad \text{(X.538)}$$

where $H$ is the Hadamard matrix.
Solution: With

$$|0\rangle \cong \begin{pmatrix} 1 \\ 0 \end{pmatrix}; \quad |1\rangle \cong \begin{pmatrix} 0 \\ 1 \end{pmatrix} \quad \text{(X.539)}$$

it follows that:

$$\begin{aligned} H|0\rangle &\cong \frac{1}{\sqrt{2}} \begin{pmatrix} 1 & 1 \\ 1 & -1 \end{pmatrix} \begin{pmatrix} 1 \\ 0 \end{pmatrix} = \frac{1}{\sqrt{2}} \begin{pmatrix} 1 \\ 1 \end{pmatrix} = \frac{|0\rangle + |1\rangle}{\sqrt{2}} \\ H|1\rangle &\cong \frac{1}{\sqrt{2}} \begin{pmatrix} 1 & 1 \\ 1 & -1 \end{pmatrix} \begin{pmatrix} 0 \\ 1 \end{pmatrix} = \frac{1}{\sqrt{2}} \begin{pmatrix} 1 \\ -1 \end{pmatrix} = \frac{|0\rangle - |1\rangle}{\sqrt{2}}. \end{aligned} \quad \text{(X.540)}$$

One can combine the last two equations in various ways; one possibility is

$$H|q\rangle = \frac{|1-q\rangle + (-1)^q |q\rangle}{\sqrt{2}}; \quad q \in \{0, 1\}. \quad \text{(X.541)}$$

5. Calculate explicitly

$$\Phi_\varphi H \Phi_\vartheta H \quad \text{(X.542)}$$

where $H$ is the Hadamard transformation and $\Phi$ the phase shifter.
Solution: We first use the matrix representation:

$$H \cong \frac{1}{\sqrt{2}} \begin{pmatrix} 1 & 1 \\ 1 & -1 \end{pmatrix}; \quad \Phi_\varphi \cong \begin{pmatrix} 1 & 0 \\ 0 & e^{i\varphi} \end{pmatrix}. \quad \text{(X.543)}$$

---

[173] 'Construction principle': Starting with 0, we replace in every step a 0 by 01 and a 1 by 10. Thus we obtain: $0 \to 01 \to 0110 \to 01101001 \to \cdots$.

This gives

$$\Phi_\varphi H \Phi_\vartheta H = \frac{1}{2}\begin{pmatrix} 1 & 0 \\ 0 & e^{i\varphi} \end{pmatrix}\begin{pmatrix} 1 & 1 \\ 1 & -1 \end{pmatrix}\begin{pmatrix} 1 & 0 \\ 0 & e^{i\vartheta} \end{pmatrix}\begin{pmatrix} 1 & 1 \\ 1 & -1 \end{pmatrix}$$

$$= \frac{1}{2}\begin{pmatrix} 1 & 1 \\ e^{i\varphi} & -e^{i\varphi} \end{pmatrix}\begin{pmatrix} 1 & 1 \\ e^{i\vartheta} & -e^{i\vartheta} \end{pmatrix} = \frac{1}{2}\begin{pmatrix} 1+e^{i\vartheta} & 1-e^{i\vartheta} \\ e^{i\varphi}\left(1-e^{i\vartheta}\right) & e^{i\varphi}\left(1+e^{i\vartheta}\right) \end{pmatrix}.$$

$$\text{(X.544)}$$

By extracting the factor $e^{i\vartheta/2}e^{i\varphi/2}$, we can write (X.544) as

$$\Phi_\varphi H \Phi_\vartheta H = e^{i\vartheta/2}\begin{pmatrix} \cos\frac{\vartheta}{2} & -i\sin\frac{\vartheta}{2} \\ -ie^{i\varphi}\sin\frac{\vartheta}{2} & e^{i\varphi}\cos\frac{\vartheta}{2}. \end{pmatrix} \qquad \text{(X.545)}$$

Another possibility is offered by the relation

$$|q\rangle \to H|q\rangle = \frac{|1-q\rangle + (-1)^q|q\rangle}{\sqrt{2}}; \quad |q\rangle \to \Phi_\varphi|q\rangle = e^{iq\varphi}|q\rangle; \quad q \in \{0,1\}.$$

$$\text{(X.546)}$$

From it, we find a representation which is completely equivalent to (X.544):

$$\Phi_\varphi H \Phi_\vartheta H |q\rangle = \Phi_\varphi H \Phi_\vartheta \frac{|1-q\rangle + (-1)^q|q\rangle}{\sqrt{2}} = \Phi_\varphi H \frac{e^{i(1-q)\vartheta}|1-q\rangle + (-1)^q e^{iq\vartheta}|q\rangle}{\sqrt{2}}$$

$$= \Phi_\varphi \frac{\left[e^{i(1-q)\vartheta} + e^{iq\vartheta}\right]|q\rangle + (-1)^{1-q}\left[e^{i(1-q)\vartheta} - e^{iq\vartheta}\right]|1-q\rangle}{2}$$

$$= \frac{e^{iq\varphi}\left[e^{i(1-q)\vartheta} + e^{iq\vartheta}\right]|q\rangle + (-1)^{1-q}e^{i(1-q)\varphi}\left[e^{i(1-q)\vartheta} - e^{iq\vartheta}\right]|1-q\rangle}{2}$$

$$= \frac{e^{iq\varphi}\left[1 + e^{i\vartheta}\right]|q\rangle + e^{i(1-q)\varphi}\left[1 - e^{i\vartheta}\right]|1-q\rangle}{2}$$

$$= e^{i\vartheta/2}\left(e^{iq\varphi}\cos\frac{\vartheta}{2}|q\rangle - ie^{i(1-q)\varphi}\sin\frac{\vartheta}{2}|1-q\rangle\right). \qquad \text{(X.547)}$$

6. Kickback and Grover's algorithm: Given that

$$f(k) = \delta_{k\kappa}; \quad k = 0, 1, \ldots, d-1; \quad d = 2^n; \quad 0 \le \kappa \le d-1. \qquad \text{(X.548)}$$

The effect of the kickback may be written as:

$$|k\rangle \to (-1)^{f(k)}|k\rangle \text{ or } U_\kappa|k\rangle = (-1)^{f(k)}|k\rangle, \qquad \text{(X.549)}$$

where $\{|k\rangle\}$ is a CONS. Show that:

$$U_\kappa = 1 - 2\,|\kappa\rangle\,\langle\kappa|\,. \qquad (\text{X.550})$$

Solution: It holds that

$$U_\kappa\,|k\rangle\,\langle k| = (-1)^{f(k)}\,|k\rangle\,\langle k|\,. \qquad (\text{X.551})$$

We sum over the states, use $\sum|k\rangle\,\langle k| = 1$, and obtain:

$$U_\kappa = \sum_{k=0}^{d-1}(-1)^{f(k)}\,|k\rangle\,\langle k| = \sum_{k=0,\neq\kappa}^{d-1}|k\rangle\,\langle k| - |\kappa\rangle\,\langle\kappa|$$

$$= \sum_{k=0}^{d-1}|k\rangle\,\langle k| - 2\,|\kappa\rangle\,\langle\kappa| = 1 - 2\,|\kappa\rangle\,\langle\kappa|\,. \qquad (\text{X.552})$$

7. Given the normalized states $|x\rangle$ and $|y\rangle$, with $\langle x\,|y\rangle = 0$; show that the operator
   $U = 2\,|x\rangle\,\langle x| - 1$ describes a reflection at $|x\rangle$ and $-U$ a reflection at $|y\rangle$.
   Solution: We can represent an arbitrary state as $|z\rangle = a\,|x\rangle + b\,|y\rangle$. Then we
   have:

$$U\,|z\rangle = 2\,|x\rangle\,\langle x\,|z\rangle - |z\rangle = 2a\,|x\rangle - a\,|x\rangle - b\,|y\rangle = a\,|x\rangle - b\,|y\rangle$$
$$-U\,|z\rangle = -a\,|x\rangle + b\,|y\rangle\,. \qquad (\text{X.553})$$

Hence, the operator $U$ leaves the prefactor of $|x\rangle$ unchanged and modifies that
of $|y\rangle$; accordingly, it is a reflection on $|x\rangle$. Analogously, one infers that $-U$
describes a reflection on $|y\rangle$; cf. Fig. X.10.

8. Given the normalized state

$$|\psi\rangle = \sum_{n=1}^{N}c_n\,|\varphi_n\rangle \quad\text{with}\quad \langle\varphi_n\,|\varphi_m\rangle = \delta_{nm}. \qquad (\text{X.554})$$

The probability of measuring the state $|\varphi_k\rangle$ is thus given by $|c_k|^2$. We selectively
amplify the amplitude $c_m \neq 0$ by the following unitary transformation $U$:

$$U : c_n \to \alpha c_n \text{ for } n \neq m;\ c_m \to \beta c_m \text{ for } n = m \qquad (\text{X.555})$$

with suitably chosen $\alpha, \beta$.

(a) How are $\alpha$ and $\beta$ connected?
    Solution: We have

$$|\psi'\rangle = U\,|\psi\rangle = \sum_{n=1}^{N}[\alpha + (\beta - \alpha)\,\delta_{nm}]\,c_n\,|\varphi_n\rangle\,, \qquad (\text{X.556})$$

and, due to the orthonormality of $\{|\varphi_n\rangle\}$, it follows that

**Fig. X.10** The action of $-U = 1 - 2 |x\rangle \langle x|$ on a general state

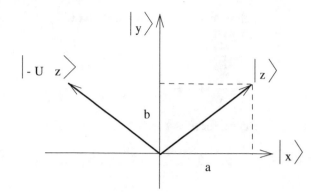

$$\langle \psi' | \psi' \rangle = 1 = \sum_{n=1}^{N} |\alpha + (\beta - \alpha) \delta_{nm}|^2 |c_n|^2 = |\alpha|^2 + \left[ |\beta|^2 - |\alpha|^2 \right] |c_m|^2.$$

$$(X.557)$$

This yields

$$1 = |\alpha|^2 + \left[ |\beta|^2 - |\alpha|^2 \right] |c_m|^2 \quad \text{or} \quad |\beta|^2 = \frac{1 - |\alpha|^2}{|c_m|^2} + |\alpha|^2; \quad (X.558)$$

and for the probability, it follows

$$|c_n|^2 \rightarrow |\alpha|^2 |c_n|^2 \quad \text{for } n \neq m;$$
$$|c_m|^2 \rightarrow |\beta|^2 |c_m|^2 = 1 - |\alpha|^2 + |\alpha|^2 |c_m|^2 \quad \text{for } n = m. \quad (X.559)$$

(b) How do the measurement probabilities behave under a $k$-fold iteration of $U$?

Solution: Clearly, a multiple application of the transformation gives

$$|c_n|^2 \rightarrow |\alpha|^2 |c_n|^2 \rightarrow |\alpha|^4 |c_n|^2 \rightarrow \cdots \quad (X.560)$$

and therefore

$$|c_m|^2 \rightarrow 1 - |\alpha|^2 + |\alpha|^2 |c_m|^2 \rightarrow 1 - |\alpha|^4 + |\alpha|^4 |c_m|^2 \rightarrow \cdots \quad (X.561)$$

With $k$ iterations, we thus have:

$$|c_n|^2 \rightarrow |\alpha|^{2k} |c_n|^2; \quad |c_m|^2 \rightarrow 1 - |\alpha|^{2k} + |\alpha|^{2k} |c_m|^2 \rightarrow \cdots \quad (X.562)$$

(c) Specialize to the case of an initially uniform distribution $c_n = \frac{1}{\sqrt{N}}$ and $\alpha = \frac{1}{4}$. How often does one have to iterate in order to measure the state $m$ with a probability of $w > 1 - 10^{-6}$ (assuming $N \gg 1$)?

Solution: With $k$ iterations, we have

$$
\frac{1}{N} \to \frac{1}{4^{2k}} \frac{1}{N} \quad \text{for } n \ne m;
$$

$$
\frac{1}{N} \to 1 - \frac{1}{4^{2k}} + \frac{1}{4^{2k}} \frac{1}{N} = 1 - \frac{1}{4^{2k}} \frac{N-1}{N} \quad \text{for } n = m. \tag{X.563}
$$

Hence, for $w > 1 - 10^{-6}$, it must hold that

$$
1 - \frac{1}{4^{2k}} \frac{N-1}{N} > 1 - 10^{-6} \text{ or } 10^{-6} > \frac{1}{4^{2k}} \frac{N-1}{N} \approx \frac{1}{4^{2k}}, \tag{X.564}
$$

and, due to $4^{2k} > 10^6$, it follows that

$$
k > \frac{3}{\log 4} = 4.98 \dots \approx 5. \tag{X.565}
$$

## X.13  Exercises, Chap. 27

1. A system is in the polarization state $|r\rangle$. Using $w_P = tr(\rho P)$, calculate the probability of measuring the system in the state $|h\rangle$.
   Solution: The density operator is $\rho = |r\rangle \langle r|$ and the projection operator $P = |h\rangle \langle h|$. It follows that:

$$
w_P = \langle |h\rangle \langle h| \rangle = Sp(\rho |h\rangle \langle h|) = Sp(|r\rangle \langle r |h\rangle \langle h|)
$$

$$
= \frac{1}{\sqrt{2}} Sp(|r\rangle \langle h|) = \frac{1}{\sqrt{2}} [\langle h |r\rangle \langle h |h\rangle + \langle v |r\rangle \langle h |v\rangle] = \frac{1}{2}. \tag{X.566}
$$

2. A mixture is described by $\rho = \sum p_n |\varphi_n\rangle \langle \varphi_n|$, where $\{|\varphi_n\rangle\}$ is a CONS. Using $w_P = tr(\rho P)$, calculate the probability of measuring the system in the state $|\varphi_N\rangle$.
   Solution: The projection operator is $P = |\varphi_N\rangle \langle \varphi_N|$. It follows that:

$$
w_P = \langle P\rangle = Sp(\rho P) = Sp(\rho |\varphi_N\rangle \langle \varphi_N|)
$$

$$
= \sum_n p_n Sp(|\varphi_n\rangle \langle \varphi_n| \varphi_N\rangle \langle \varphi_N|) = p_N Sp(|\varphi_N\rangle \langle \varphi_N|) = p_N. \tag{X.567}
$$

3. The value function $V_{|\psi\rangle}$ is defined by $V_{|\psi\rangle}(F(A)) = F(V_{|\psi\rangle}(A))$.

   (a) Prove for $[A, B] = 0$ the sum rule $V_{|\psi\rangle}(A + B) = V_{|\psi\rangle}(A) + V_{|\psi\rangle}(B)$.
       Solution: Due to $[A, B] = 0$, there exists an operator $C$ such that $A = F(C)$ and $B = G(C)$. From this, it follows that $A + B = (F + G)(C)$, and thus

$$V_{|\psi\rangle}(A + B) = V_{|\psi\rangle}((F + G)(C)) = (F + G) V_{|\psi\rangle}(C)$$
$$= F V_{|\psi\rangle}(C) + G V_{|\psi\rangle}(C) = V_{|\psi\rangle}(F(C)) + V_{|\psi\rangle}(G(C))$$
$$= V_{|\psi\rangle}(A) + V_{|\psi\rangle}(B). \tag{X.568}$$

(b) Prove for $[A, B] = 0$ the product rule $V_{|\psi\rangle}(A \cdot B) = V_{|\psi\rangle}(A) \cdot V_{|\psi\rangle}(B)$.
Solution: Due to the definition of the value function, we have:

$$V_{|\psi\rangle}(A^2) = V_{|\psi\rangle}^2(A) \text{ or } V_{|\psi\rangle}(A^n) = V_{|\psi\rangle}^n(A). \tag{X.569}$$

We can again assume $A = F(C)$ and $B = G(C)$. Expanding the functions in power series, we get

$$V_{|\psi\rangle}(A \cdot B) = V_{|\psi\rangle}(F(C) \cdot G(C))$$
$$= V_{|\psi\rangle}\left(\sum_n F_n C^n \cdot \sum_m G_m C^m\right) = V_{|\psi\rangle}\left(\sum_{n,m} F_n G_m C^{n+m}\right)$$
$$= \sum_{n,m} F_n G_m V_{|\psi\rangle}\left(C^{n+m}\right) = \sum_{n,m} F_n G_m V_{|\psi\rangle}^{n+m}(C)$$
$$= \sum_{n,m} F_n V_{|\psi\rangle}^n(C) G_m V_{|\psi\rangle}^m(C) = \sum_n F_n V_{|\psi\rangle}^n(C) \cdot \sum_m G_m V_{|\psi\rangle}^m(C)$$
$$= \sum_n F_n V_{|\psi\rangle}(C^n) \cdot \sum_m G_m V_{|\psi\rangle}(C^m)$$
$$= V_{|\psi\rangle}\left(\sum_n F_n C^n\right) \cdot V_{|\psi\rangle}^m\left(\sum_m G_m C^m\right)$$
$$= V_{|\psi\rangle}(A) \cdot V_{|\psi\rangle}^m(B). \tag{X.570}$$

(c) Show that $V_{|\psi\rangle}(1) = 1$.
Solution: Let 1 be the unit operator. Due to the product rule, it holds that $V_{|\psi\rangle}(B) = V_{|\psi\rangle}(1 \cdot B) = V_{|\psi\rangle}(1) \cdot V_{|\psi\rangle}(B)$, and it follows that $V_{|\psi\rangle}(1) = 1$, where it is supposed that there is at least one quantity $B$ for which $V_{|\psi\rangle}(B) \neq 0$.

4. Given the polarization operators $P_L, P_{L'}$ and $P_C$ (or the corresponding Pauli matrices, see (27.11)):

(a) Determine (once more) their eigenvalues and eigenvectors.
Solution: Due to $P_A^2 = 1$, the eigenvalues are given by $\lambda = \pm 1$. For e.g. $P_{L'}$, it holds that

$$\begin{pmatrix} 0 & 1 \\ 1 & 0 \end{pmatrix}\begin{pmatrix} a \\ b \end{pmatrix} = \pm \begin{pmatrix} a \\ b \end{pmatrix}, \tag{X.571}$$

and this gives the normalized eigenvectors:

$$P_{L'} : |h'\rangle = \frac{1}{\sqrt{2}} \begin{pmatrix} 1 \\ 1 \end{pmatrix} \text{ for } \lambda_{L'} = 1; \quad |v'\rangle = \frac{1}{\sqrt{2}} \begin{pmatrix} -1 \\ 1 \end{pmatrix} \text{ for } \lambda_{L'} = -1. \quad (X.572)$$

Remark: As always, the vectors are defined up to a phase; we choose it in such a way that $|h'\rangle$ and $|v'\rangle$ arise from $|h\rangle$ and $|v\rangle$ by an active rotation. Analogously, it follows that

$$P_C : |r\rangle = \frac{1}{\sqrt{2}} \begin{pmatrix} 1 \\ i \end{pmatrix} \text{ for } \lambda_C = 1; \quad |l\rangle = \frac{1}{\sqrt{2}} \begin{pmatrix} 1 \\ -i \end{pmatrix} \text{ for } \lambda_Z = -1$$

$$P_L : |h\rangle = \begin{pmatrix} 1 \\ 0 \end{pmatrix} \text{ for } \lambda_L = 1; \quad |v\rangle = \begin{pmatrix} 0 \\ 1 \end{pmatrix} \text{ for } \lambda_L = -1. \quad (X.573)$$

(b) Express the eigenvectors of $P_C$ and $P_{L'}$ in terms of those of $P_L$.
Solution: We have

$$|h'\rangle = \frac{|h\rangle + |v\rangle}{\sqrt{2}}; \quad |v'\rangle = \frac{-|h\rangle + |v\rangle}{\sqrt{2}}$$

$$|r\rangle = \frac{|h\rangle + i\,|v\rangle}{\sqrt{2}}; \quad |l\rangle = \frac{|h\rangle - i\,|v\rangle}{\sqrt{2}}. \quad (X.574)$$

The inversion reads:

$$|h\rangle = \frac{|h'\rangle - |v'\rangle}{\sqrt{2}}; \quad |v\rangle = \frac{|h'\rangle + |v'\rangle}{\sqrt{2}}$$

$$|h\rangle = \frac{|r\rangle + |l\rangle}{\sqrt{2}}; \quad |v\rangle = \frac{|r\rangle - |l\rangle}{\sqrt{2}i}. \quad (X.575)$$

5. Given the GHZ state

$$|\psi\rangle_\pm = \frac{|h, h, h\rangle \pm |v, v, v\rangle}{\sqrt{2}} \quad (X.576)$$

corresponding to an $LLL$ measurement; rewrite this for a $CCL'$ measurement (plus $CL'C$ and $L'CC$) (27.12) and for an $L'L'L'$ measurement (27.14).

(a) Solution for $CCL'$: With the results from the previous exercise, we have

$$\sqrt{2}|\psi\rangle_\pm = \left( \frac{|r\rangle + |l\rangle}{\sqrt{2}} \right)_1 \left( \frac{|r\rangle + |l\rangle}{\sqrt{2}} \right)_2 \left( \frac{|h'\rangle - |v'\rangle}{\sqrt{2}} \right)_3$$

$$\pm \left( \frac{|r\rangle - |r\rangle}{\sqrt{2}i} \right)_1 \left( \frac{|r\rangle - |l\rangle}{\sqrt{2}i} \right)_2 \left( \frac{|h'\rangle + |v'\rangle}{\sqrt{2}} \right)_3. \quad (X.577)$$

Expanding the first two ratios yields:

$$\sqrt{2}\,|\psi\rangle_{\pm} = \frac{|r,r\rangle + |r,l\rangle + |l,r\rangle + |l,l\rangle}{2}\left(\frac{|h'\rangle - |v'\rangle}{\sqrt{2}}\right)_3$$

$$\mp \frac{|r,r\rangle - |r,l\rangle - |l,r\rangle + |l,l\rangle}{2}\left(\frac{|h'\rangle + |v'\rangle}{\sqrt{2}}\right)_3. \qquad \text{(X.578)}$$

From this, it follows that

$$|\psi\rangle_{+} = \frac{|r,l,h'\rangle + |l,r,h'\rangle - |r,r,v'\rangle - |l,l,v'\rangle}{2}$$

$$|\psi\rangle_{-} = \frac{|r,r,h'\rangle + |l,l,h'\rangle - |r,l,v'\rangle - |l,r,v'\rangle}{2}. \qquad \text{(X.579)}$$

The results for $CL'C$ and $L'CC$ follow by cyclic permutation. It is clear that if we have two readings, we can predict the third with certainty; the combination $|r,r,?\rangle$ can be only $|r,r,v'\rangle$ for $|\psi\rangle_{+}$, $|l,?,h'\rangle$ only $|l,r,h'\rangle$, etc.

(b) Solution for $L'L'L'$: We have initially:

$$\sqrt{2}\,|\psi\rangle_{\pm} = \left(\frac{|h'\rangle - |v'\rangle}{\sqrt{2}}\right)_1\left(\frac{|h'\rangle - |v'\rangle}{\sqrt{2}}\right)_2\left(\frac{|h'\rangle - |v'\rangle}{\sqrt{2}}\right)_3$$

$$\pm \left(\frac{|h'\rangle + |v'\rangle}{\sqrt{2}}\right)_1\left(\frac{|h'\rangle + |v'\rangle}{\sqrt{2}}\right)_2\left(\frac{|h'\rangle + |v'\rangle}{\sqrt{2}}\right)_3. \qquad \text{(X.580)}$$

We expand again the first two ratios:

$$\sqrt{2}\,|\psi\rangle_{\pm} = \frac{|h',h'\rangle - |v',h'\rangle - |h',v'\rangle + |v',v'\rangle}{2}\left(\frac{|h'\rangle - |v'\rangle}{\sqrt{2}}\right)_3$$

$$\pm \frac{|h',h'\rangle + |v',h'\rangle + |h',v'\rangle + |v',v'\rangle}{2}\left(\frac{|h'\rangle + |v'\rangle}{\sqrt{2}}\right)_3,$$

$$\text{(X.581)}$$

and obtain

$$|\psi\rangle_{+} = \frac{|h',h',h'\rangle + |v',v',h'\rangle + |v',h',v'\rangle + |h',v',v'\rangle}{2}$$

$$|\psi\rangle_{-} = -\frac{|v',h',h'\rangle + |h',v',h'\rangle + |h',h',v'\rangle + |v',v',v'\rangle}{2}. \qquad \text{(X.582)}$$

Two readings again with certainty determine the third one. All in all we have an odd number of states $|h'\rangle$ for $|\psi\rangle_{+}$, and an even number for $|\psi\rangle_{-}$.

6. The following combinations of the polarization operators (27.10) are given:

$$Q_1 = P_{1L'}P_{2C}P_{3C}; \quad Q_2 = P_{1C}P_{2L'}P_{3C}$$
$$Q_3 = P_{1C}P_{2C}P_{3L'}; \quad Q = P_{1L'}P_{2L'}P_{3L'}. \tag{X.583}$$

The numerical index denotes the space in which the particular polarization operator acts. We use in the following the fact that operators from different spaces commute, e.g. $P_{1L'}P_{2C} = P_{2C}P_{1L'}$. In addition, we have $P_{nL'}P_{nC} = -P_{nC}P_{nL'}$ as well as $P_{nC}^2 = P_{nL'}^2 = 1$.

(a) Show that the three operators $Q_i$ have the eigenvalues $\pm 1$.
   Solution: Evidently, we have

$$Q_i^2 = 1, \tag{X.584}$$

   and the proposition follows immediately from this.

(b) Show that the three operators $Q_i$ commute pairwise.
   Solution: We have e.g.

$$Q_1 Q_2 = P_{1L'}P_{2C}P_{3C}P_{1C}P_{2L'}P_{3C} = P_{1L'}P_{2C}P_{1C}P_{2L'}$$
$$= -P_{1C}P_{2C}P_{1L'}P_{2L'} = P_{1C}P_{2L'}P_{1L'}P_{2C}$$
$$= P_{1C}P_{2L'}P_{3C}P_{1L'}P_{2C}P_{3C} = Q_2 Q_1. \tag{X.585}$$

(c) Show that the states

$$|\psi\rangle_\pm = \frac{|h, h, h\rangle \pm |v, v, v\rangle}{\sqrt{2}} \tag{X.586}$$

   are common eigenstates of the three operators $Q_i$ with the eigenvalues $\mp 1$, as well as eigenstates of the operator $Q$ with the eigenvalues $\pm 1$.
   Solution: With

$$P_{L'}|h\rangle = |v\rangle ; \quad P_{L'}|v\rangle = |h\rangle$$
$$P_C |h\rangle = i|v\rangle ; \quad P_C |v\rangle = -i|h\rangle , \tag{X.587}$$

   it follows that e.g.

$$Q_1 |\psi\rangle_\pm = P_{1L'}P_{2C}P_{3C} \frac{|h, h, h\rangle \pm |v, v, v\rangle}{\sqrt{2}}$$
$$= \frac{i^2 |v, v, v\rangle \pm (-i)^2 |h, h, h\rangle}{\sqrt{2}} = \mp |\psi\rangle_\pm , \tag{X.588}$$

   and analogously for $Q_2$ and $Q_3$. For $Q$, we have:

$$Q |\psi\rangle_\pm = P_{1L'}P_{2L'}P_{3L'} \frac{|h, h, h\rangle \pm |v, v, v\rangle}{\sqrt{2}}$$
$$= \frac{|v, v, v\rangle \pm |h, h, h\rangle}{\sqrt{2}} = \pm |\psi\rangle_\pm . \tag{X.589}$$

# Further Reading

1. J. Audretsch (ed.), *Verschränkte Welt-Faszination der Quanten (Entangled World-Fascination of Quantum, in German)* (Wiley-VCH, Weinheim, 2002)
2. J. Audretsch, *Entangled Systems* (Wiley-VCH, Weinheim, 2007)
3. J.-L. Basdevant, J. Dalibard, *Quantum Mechanics* (Springer, Berlin, 2002)
4. D.R. Bes, *Quantum Mechanics* (Springer, Berlin, 2004)
5. B.H. Bransden, C.J. Joachain, *Quantum Mechanics* (Pearson Education Limited, Harlow, 2000)
6. C. Cohen-Tannoudji, B. Diu, F. Laloë, *Quantum Mechanics*, vols. 1 & 2 (Hermann Paris/Wiley, New York, 1977)
7. F. Embacher, Homepage with much material about quantum theory (also for school), University Vienna (2012), http://homepage.univie.ac.at/franz.embacher/
8. R.P. Feynman, R.B. Leighton, M. Sand, *Quantum Mechanics. The Feynman Lectures on Physics*, vol. 3 (Addison-Wesley Reading, Massachusetts, 1965)
9. T. Fließbach, *Quantenmechanik (Quantum Mechanics, in German)*, 3rd edn. (Spektrum Akademischer Verlag, Heidelberg, 2000)
10. K. Gottfried, T.-M. Yan, *Quantum Mechanics: Fundamentals* (Springer, New York, 2006)
11. K.T. Hecht, *Quantum Mechanics* (Springer, New York, 2000)
12. C.J. Isham, *Quantum Theory-Mathematical and Structural Foundations* (Imperial College Press, London, 2008)
13. T. Lancaster, S.J. Blundell, *Quantum Field Theory for the Gifted Amateur* (Oxford University Press, Oxford, 2014)
14. R.D. Klauber, *Student Friendly Quantum Field Theory*, 2nd edn. (Sandrove Press, Fairfield, 2015)
15. M. Le Bellac, *Quantum Physics* (Cambridge University Press, Cambridge, 2006)
16. H. Lüth, *Quantenphysik in der Nanowelt (Quantum Physics in the Nanoworld, in German)* (Springer, Berlin, 2009)
17. E. Merzbacher, *Quantum Mechanics*, 3rd edn. (Wiley, New York, 1998)
18. A. Messiah, *Quantum Mechanics*, vols. 1 & 2 (North-Holland Publishing Company, Amsterdam, 1964)
19. Münchener Internetprojekt zur Lehrerfortbildung in Quantenmechanik (Munich internet project for teacher training in quantum mechanics, in German) (2012), http://homepages. physik.uni-muenchen.de/milq/
20. G. Münster, *Quantentheorie (Quantum Theory, in German)* (Walter de Gruyter, Berlin, 2006)
21. W. Nolting, *Grundkurs Theoretische Physik 5, Quantenmechanik, Teil 1: Grundlagen und Quantenmechanik Teil 2: Methoden und Anwendungen (Quantum Mechanics, Part 1: Basics*

© Springer Nature Switzerland AG 2018

J. Pade, *Quantum Mechanics for Pedestrians 2*, Undergraduate Lecture
Notes in Physics, https://doi.org/10.1007/978-3-030-00467-5

*and Part 2: Methods and Applications, in German*) (Verlag Zimmermann-Neufang, Ulmen, 1992)

22. A. Peres, *Quantum Theory-Concepts and Methods* (Kluwer Academic Publishers, Doordrecht, 1995)

23. A.I.M. Rae, *Quantum Mechanics*, 5th edn. (Taylor and Francis, New York, 2008)

24. H. Rollnik, *Quantentheorie 1 & 2 (Quantum Theory 1 & 2, in German)*, 2nd edn. (Springer, Berlin, 2003)

25. H. Schulz, *Physik mit Bleistift (Physics with a Pencil, in German)*, 4th edn. (Verlag Harri Deutsch, Frankfurt am Main, 2001)

26. F. Schwabl, *Quantum Mechanics*, 3rd edn. (Springer, Berlin, 2002)

27. N. Zettili, *Quantum Mechanics, Concepts and Applications*, 2nd edn. (Wiley, New York, 2009)

# Index of Volume 1

© Springer Nature Switzerland AG 2018
J. Pade, *Quantum Mechanics for Pedestrians 2*, Undergraduate Lecture
Notes in Physics, https://doi.org/10.1007/978-3-030-00467-5

# Index of Volume 2

© Springer Nature Switzerland AG 2018
J. Pade, *Quantum Mechanics for Pedestrians 2*, Undergraduate Lecture
Notes in Physics, https://doi.org/10.1007/978-3-030-00467-5

Printed in the United States
by Baker & Taylor Publisher Services